ENCYCLOPEDIA OF MATHEMATICS AND ITS APPLICATIONS

Volume 70

Orthonormal Systems and Banach Space Geometry

ENCYCLOPEDIA OF MATHEMATICS AND ITS APPLICATIONS

4 W. Miller, Jr. *Symmetry and separation of variables*
6 H. Minc *Permanents*
11 W. B. Jones and W. J. Thron *Continued fractions*
12 N. F. G. Martin and J. W. England *Mathematical theory of entropy*
18 H. O. Fattorini *The Cauchy problem*
19 G. G. Lorentz, K. Jetter, and S. D. Riemenschneider *Birkhoff interpolation*
21 W. T. Tutte *Graph theory*
22 J. R. Bastida *Field extensions and Galois theory*
23 J. R. Cannon *The one-dimensional heat equation*
25 A. Salomaa *Computation and automata*
26 N. White (ed.) *Theory of matroids*
27 N. H. Bingham, C. M. Goldie, and J. L. Teugels *Regular variation*
28 P. P. Petrushev and V. A Popov *Rational approximation of real functions*
29 N. White (ed.) *Combinatorial geometrics*
30 M. Pohst and H. Zassenhaus *Algorithmic algebraic number theory*
31 J. Aczel and J. Dhombres *Functional equations containing several variables*
32 M. Kuczma, B. Chozewski, and R. Ger *Iterative functional equations*
33 R. V. Ambartzumian *Factorization calculus and geometric probability*
34 G. Gripenberg, S.-O. Londen, and O. Staffans *Volterra integral and functional equations*
35 G. Gasper and M. Rahman *Basic hypergeometric series*
36 E. Torgersen *Comparison of statistical experiments*
37 A. Neumaier *Interval methods for systems of equations*
38 N. Korneichuk *Exact constants in approximation theory*
39 R. A. Brualdi and H. J. Ryser *Combinatorial matrix theory*
40 N. White (ed.) *Matroid applications*
41 S. Sakai *Operator algebras in dynamical systems*
42 W. Hodges *Model theory*
43 H. Stahl and V. Totik *General orthogonal polynomials*
44 R. Schneider *Convex bodies*
45 G. Da Prato and J. Zabczyk *Stochastic equations in infinite dimensions*
46 A. Bjorner, M. Las Vergnas, B. Sturmfels, N. White, and G. Ziegler *Oriented matroids*
47 E. A. Edgar and L. Sucheston *Stopping times and directed processes*
48 C. Sims *Computation with finitely presented groups*
49 T. Palmer *Banach algebras and the general theory of *-algebras*
50 F. Borceux *Handbook of categorical algebra I*
51 F. Borceux *Handbook of categorical algebra II*
52 F. Borceux *Handbook of categorical algebra III*
54 A Katok and B. Hassleblatt *Introduction to the modern theory of dynamical systems*
55 V. N. Sachkov *Combinatorial methods in discrete mathematics*
56 V. N. Sachkov *Probabilistic methods in discrete mathematics*
57 P. M. Cohn *Skew Fields*
58 Richard J. Gardner *Geometric tomography*
59 George A. Baker, Jr., and Peter Graves-Morris *Padé approximants*
60 Jan Krajicek *Bounded arithmetic, propositional logic, and complex theory*
61 H. Gromer *Geometric applications of Fourier series and spherical harmonics*
62 H. O. Fattorini *infinite dimensional optimization and control theory*
63 A. C. Thompson *Minkowski geometry*
64 R. B. Bapat and T. E. S. Raghavan *Nonnegative matrices and applications*
65 K. Engel *Sperner theory*
66 D. Cvetkovic, P. Rowlinson and S. Simic *Eigenspaces of graphs*
67 F. Bergeron, G. Labelle and P. Leroux *Combinatorial species and tree-like structures*

ENCYCLOPEDIA OF MATHEMATICS AND ITS APPLICATIONS

Orthonormal Systems and Banach Space Geometry

ALBRECHT PIETSCH & JÖRG WENZEL

University of Jena

CAMBRIDGE
UNIVERSITY PRESS

PUBLISHED BY THE PRESS SYNDICATE OF THE UNIVERSITY OF CAMBRIDGE
The Pitt Building, Trumpington Street, Cambridge CB2 1RP, United Kingdom

CAMBRIDGE UNIVERSITY PRESS
The Edinburgh Building, Cambridge CB2 2RU, United Kingdom
40 West 20th Street, New York, NY 10011-4211, USA
10 Stamford Road, Oakleigh, Melbourne 3166, Australia

First published 1998

Printed in the United Kingdom at the University Press, Cambridge

A catalogue record of this book is available from the British Library

ISBN 0 521 62462 2 hardback

Contents

Preface

This book is based on the pioneering work of (in chronological order) R. C. James, S. Kwapień, B. Maurey, G. Pisier, D. L. Burkholder and J. Bourgain.

We have done our best to unify and simplify the material. All participants of the Jenaer Seminar *'Operatorenideale'* contributed their ideas and their patience. Above all, we are indebted to A. Hinrichs who made several significant improvements. From S. Geiss we learnt many results and techniques related to the theory of martingales. Particular gratitude goes to H. Jarchow (Zürich) for various helpful remarks.

For many years, our research on this subject was supported by the Deutsche Forschungsgemeinschaft, contracts Ko 962/3-1 and Pi 322/1-1.

Finally, we thank Cambridge University Press for their excellent collaboration in publishing this book.

Jena, October 1997

Albrecht Pietsch
Jörg Wenzel

Preface

Which
Banach spaces
can be distinguished
with the help of
orthonormal systems?

Which
orthonormal systems
can be distinguished
with the help of
Banach spaces?

Introduction

The main goal of functional analysis is to provide powerful tools for a unified treatment of differential and integral equations, integral transforms, expansions and approximations of functions, and various other topics. A basic idea consists in extending classical results about real or complex functions to operators acting between topological linear spaces. Another important goal is the classification of objects, like spaces and operators. Luckily, these goals, the practical and the theoretical, are closely related to each other.

A significant trend in Banach space theory is the search for numerical parameters that can be used to quantify special properties. Certainly, everybody would agree that Hilbert spaces are the most beautiful among all Banach spaces. Thus it is important to decide whether a given Banach space admits an equivalent norm induced by an inner product. If so, this space is called *Hilbertian.* If such renormings do not exist, then we may ask for a measure of non-Hilbertness. To what extent is the sequence space l_4 closer to l_2 than l_{1892}?

We illustrate our point of view by asking whether Bessel's inequality also holds for functions $\boldsymbol{f}$ with values in a Banach space X. Do we have

$$\left(\sum_{k=1}^{n}\left\|\frac{1}{2\pi}\int_{-\pi}^{+\pi}\boldsymbol{f}(t)\exp(-ikt)\,dt\right\|^2\right)^{1/2}\leq\left(\frac{1}{2\pi}\int_{-\pi}^{+\pi}\|\boldsymbol{f}(t)\|^2\,dt\right)^{1/2}\ ?\qquad\text{(b)}$$

As observed by S. Bochner in 1933, this is not so in general. Later, it became clear that the validity of (b), even if it is only true for some fixed $n\geq 2$, characterizes Hilbert spaces isometrically; see [AMI, p. 51]. An isomorphic analogue of this criterion was established by S. Kwapień in 1972. He showed that X is Hilbertian if and only if there exists a constant $c\geq 1$ such that

$$\left(\sum_{k=1}^{n}\left\|\frac{1}{2\pi}\int_{-\pi}^{+\pi}\boldsymbol{f}(t)\exp(-ikt)\,dt\right\|^2\right)^{1/2}\leq c\left(\frac{1}{2\pi}\int_{-\pi}^{+\pi}\|\boldsymbol{f}(t)\|^2\,dt\right)^{1/2}\qquad\text{(B)}$$

for all square integrable X-valued functions $\boldsymbol{f}$ and $n=1,2,\ldots$. Now it is only a minor step to fix n and ask for the least constant $c \geq 1$ such that (B) holds in a given Banach space X. Denote, for the moment, this quantity by $\varphi_n(X)$. Then we have $\varphi_n(X) \leq \sqrt{n}$ and

$$\varphi_n(l_r) \asymp n^{|1/r-1/2|} \quad \text{for } 1 \leq r \leq \infty;$$

see p. 23 for the definition of $\asymp$. This observation suggests the following: With every exponent $0 \leq \lambda \leq 1/2$ we associate the class F_λ consisting of all Banach spaces X such that $\varphi_n(X) \prec n^\lambda$. Then $l_r \in \mathsf{F}_\lambda \setminus \mathsf{F}_{\lambda-\varepsilon}$ for $\lambda = |1/r - 1/2| \geq \varepsilon > 0$. Thus F_λ strictly increases with λ, and we have obtained a useful classification of Banach spaces. Remember that, by Kwapień's criterion, F_0 is the class of all Hilbertian Banach spaces.

Of course, we may wonder what happens when the trigonometric system is replaced by any other orthonormal system, complete or not. Important examples are Haar and Walsh functions, on the one hand, and Rademacher functions and Gaussian random variables, on the other hand.

We also mention that there are many different ways to obtain quantities similar to $\varphi_n(X)$. For instance, given two orthonormal systems $\mathcal{A}_n = (a_1, \dots, a_n)$ and $\mathcal{B}_n = (b_1, \dots, b_n)$ in Hilbert spaces $L_2(M, \mu)$ and $L_2(N, \nu)$, respectively, we can look for the least constant $c \geq 1$ such that

$$\left(\int_N \left\| \sum_{k=1}^n x_k b_k(t) \right\|^2 d\nu(t) \right)^{1/2} \leq c \left(\int_M \left\| \sum_{k=1}^n x_k a_k(s) \right\|^2 d\mu(s) \right)^{1/2},$$

where $x_1, \dots, x_n$ range over a Banach space X. An obvious modification allows us to extend this definition even to (bounded linear) operators acting from a Banach space X into a Banach space Y.

The asymptotic behaviour of the sequence $(\varphi_n(X))$ is invariant under isomorphisms. However, there are non-isomorphic Banach spaces X and Y such that $\varphi_n(X) \asymp \varphi_n(Y)$. For example, $\varphi_n(l_r \oplus l_2) \asymp \varphi_n(l_r)$. Thus we may ask which differences between X and Y are realized by φ_n. Roughly speaking, $\varphi_n(X)$ is determined by the *'worst'* n-dimensional subspace of X, where badness means large deviation from l_2^n. More generally, we may say that $\varphi_n(X)$ only depends on the collection of all n-dimensional subspaces, but neither on their position inside X nor on how often a specific subspace occurs.

In this book, we present a theory of orthonormal expansions with vector-valued coefficients and describe its interplay with Banach space geometry. Many results were obtained by straightforward extension of those concerned with Rademacher functions and Gaussian random variables. However, we hope that our general view yields more insight even into such well-known concepts as type and cotype of Banach spaces, B-convexity, superreflexivity, the vector-valued Fourier transform, the vector-valued Hilbert transform and the unconditionality property for martingale differences (UMD).

It is our hope that this treatise will be read not only by an esoteric group of specialists, but also by some graduate students interested in functional analysis. We have included many unsolved problems which show that there remains something to do for the future. Large parts of the presentation should be understandable with a basic knowledge in Banach space theory together with an elementary background in real analysis, probability and algebra. Exceptions prove the rule!

The proofs in this treatise require techniques from the fields just mentioned. Besides classical inequalities, we use various properties of special functions. Clearly, harmonic analysis serves as the basic pattern. It will turn out that orthonormal systems consisting of characters on compact Abelian groups possess many advantages because of the underlying algebraic structure. Another important feature is the use of probabilistic concepts, like random variables and martingales. In the theory of superreflexivity we employ Ramsey's theorem from combinatorics. Ultraproducts will prove to be an indispensable tool. Further key-words are: interpolation, extrapolation and averaging. Last but not least, we present many tricks and non-straightforward ideas. Of course, lengthy manipulations cannot be avoided. However, we have done our best to make things as easy as possible, and we hope the final result provides a colourful picture.

Basically, we have adopted standard notation and terminology from Banach space theory. It may nevertheless happen that experts well-acquainted with some special results are shocked by the symbols $\boldsymbol{\varrho}(T|\mathcal{B}_n, \mathcal{A}_n)$ and $\boldsymbol{\delta}(T|\mathcal{B}_n, \mathcal{A}_n)$ or even $\boldsymbol{\varrho}_u^{(v)}(T|\mathcal{B}_n, \mathcal{A}_n)$ and $\boldsymbol{\delta}_u^{(v)}(T|\mathcal{B}_n, \mathcal{A}_n)$. Hopefully, this displeasure will gradually be replaced by the understanding that our *lengthy notation* is indeed quite economical and suggestive. Of course, it seems better at first glance to denote the Rademacher cotype q constant computed with n vectors simply by $C_q(X, n)$ or $C_{q,n}(X)$, as done in [DIE*a, p. 290], [MIL*, p. 51] and [TOM, p. 188]. However, there occur similar quantities related to Gaussian random variables, various trigonometric functions, etc. Thus in the traditional way, we would run out of letters very quickly. To help the patient reader, a fairly complete list of symbols is included, pp. 514–522.

0

Preliminaries

This chapter provides some elementary facts from the theory of Banach spaces and the basic terminology. For more information, we recommend the following books:

Beauzamy .... *Introduction to Banach spaces and their geometry* [BEA 2],
Day .... *Normed linear spaces* [DAY],
Dunford/Schwartz .... *Linear operators, vol. I* [DUN*1],
Lindenstrauss/Tzafriri ... *Classical Banach spaces, vols. I and II* [LIN*1, LIN*2].

0.1 Banach spaces and operators

0.1.1 Throughout this book, X, Y and Z denote **Banach spaces** over $\mathbb{K}$ (synonym of the real field $\mathbb{R}$ or the complex field $\mathbb{C}$). Whenever it is necessary to indicate that x is an element of the Banach space X, then we denote its norm by $\|x|X\|$.

The **closed unit ball** of X is defined by $U_X := \{x \in X : \|x\| \leq 1\}$.

CONVENTION. Unless otherwise stated, all Banach spaces under consideration are assumed to be different from $\{o\}$, where o denotes the zero element.

0.1.2 We write L for the **class of all Banach spaces.**

0.1.3 The **dual Banach space** X' consists of all (bounded linear) **functionals** $x' : X \to \mathbb{K}$. The value of x' at $x \in X$ is denoted by $\langle x, x' \rangle$, and we let

$$\|x'\| := \sup\{|\langle x, x' \rangle| : x \in U_X\}.$$

Moreover, U_X° stands for the closed unit ball of X'.

When dealing with duals of higher order, besides $\langle x, x' \rangle$ we use the symbols $\langle x'', x' \rangle$ and $\langle x'', x''' \rangle$. That is, $x \in X$ and $x'' \in X''$ are placed left, while $x' \in X'$ and $x''' \in X'''$ are placed right.

0.1.4 Throughout this book, T denotes a (bounded linear) **operator** from X into Y. The **null space** and the **range** of T are defined by

$$N(T) := \{ x \in X : Tx = o \} \quad \text{and} \quad M(T) := \{ Tx \in Y : x \in X \},$$

respectively. The **operator norm** is given by

$$\|T\| := \sup\{ \|Tx\| : x \in U_X \}.$$

Whenever it is advisable to indicate that the operator T acts from X into Y, then we use the more precise notation $\|T : X \to Y\|$. We denote the identity map of X by I_X. If $\|T\| \le 1$, then T is called a **contraction**.

The Banach space of all operators from X into Y is denoted by $\mathfrak{L}(X, Y)$. To simplify matters, we write $\mathfrak{L}(X)$ instead of $\mathfrak{L}(X, X)$.

0.1.5 Let $\mathfrak{L}$ denote the **class of all operators** acting between arbitrary Banach spaces. This means that

$$\mathfrak{L} = \bigcup_{X,Y} \mathfrak{L}(X, Y),$$

where X and Y range over L.

0.1.6 For $T \in \mathfrak{L}(X, Y)$, the **dual operator** $T' \in \mathfrak{L}(Y', X')$ is defined by

$$\langle x, T'y' \rangle = \langle Tx, y' \rangle \quad \text{for } x \in X \text{ and } y' \in Y'.$$

0.1.7 For fixed $x \in X$, the rule

$$K_X x : x' \longrightarrow \langle x, x' \rangle$$

defines a functional on X'. In this way, we obtain the **natural embedding** K_X from X into X'', which is a linear isometry. Note that the diagram

$$\begin{array}{ccc} X & \xrightarrow{\quad T \quad} & Y \\ {\scriptstyle K_X}\big\downarrow & & \big\downarrow{\scriptstyle K_Y} \\ X'' & \xrightarrow[\quad T'' \quad]{} & Y'' \end{array}$$

commutes for every operator $T \in \mathfrak{L}(X, Y)$.

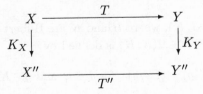

0.1.8 For a proof of the following classical result, we refer the reader to [DEF*, p. 73] and [PIE 2, p. 383].

HELLY'S LEMMA. *Let $x'' \in X''$. Then, given $x'_1, \ldots, x'_n \in X'$ and $\varepsilon > 0$, there exists $x \in X$ such that*

$$\|x\| \le (1+\varepsilon)\|x''\| \quad \text{and} \quad \langle x, x'_k\rangle = \langle x'', x'_k\rangle \quad \text{for } k = 1, \ldots, n.$$

0.1.9 An operator $J \in \mathfrak{L}(X, Y)$ is an **injection** if there exists a constant $c > 0$ such that

$$\|Jx\| \ge c\,\|x\| \quad \text{for all } x \in X.$$

A **metric injection** is defined by the property that $\|Jx\| = \|x\|$.

An operator $Q \in \mathfrak{L}(X, Y)$ is a **surjection** if $Q(X) = Y$. By definition, a **metric surjection** $Q \in \mathfrak{L}(X, Y)$ maps the open unit ball of X onto the open unit ball of Y; see [PIE 2, pp. 26–28].

By a **subspace** M of a Banach space X we always mean a closed linear subset. The canonical (metric) injection from M into X is denoted by J_M^X. If N is a subspace of X, then Q_N^X stands for the canonical (metric) surjection from X onto the **quotient space** X/N.

An operator $P \in \mathfrak{L}(X)$ is called a **projection** if $P^2 = P$. Subspaces M of X that can be obtained as the range of a projection are said to be **complemented**.

0.1.10 A real or complex **Hilbert space** with the **inner product** $(\cdot, \cdot)$ will always be denoted by H or K.

0.1.11 With every element $y \in H$ we associate the functional

$$\overline{y} : x \to (x, y).$$

By the Riesz representation theorem, the map $C_H : y \to \overline{y}$ is a conjugate-linear isometry from H onto H'.

0.1.12 Let $T \in \mathfrak{L}(H, K)$, where H and K are Hilbert spaces. Then the **adjoint operator** $T^* \in \mathfrak{L}(K, H)$ is defined by

$$(x, T^*y) = (Tx, y) \quad \text{for } x \in H \text{ and } y \in K.$$

This means that $T^* = C_H^{-1} T' C_K$.

0.2 Finite dimensional spaces and operators

0.2.1 The dimension of a **finite dimensional** linear space M is denoted by $\dim(M)$. Given elements $x_1, \ldots, x_n$ in any linear space X, then $\dim[x_1, \ldots, x_n]$ stands for the dimension of $\mathrm{span}(x_1, \ldots, x_n)$, the linear span.

A subspace N of a linear space X is said to be **finite codimensional** if $\mathrm{cod}(N) := \dim(X/N)$ is finite.

0.2.2 For a Banach space X the collection of all subspaces M with $\dim(M) \le n$ is denoted by $\mathrm{DIM}_{\le n}(X)$. Analogously, $\mathrm{COD}_{\le n}(X)$ stands for the collection of all subspaces N with $\mathrm{cod}(N) \le n$. We write

$$\mathrm{DIM}(X) := \bigcup_{n=0}^{\infty} \mathrm{DIM}_{\le n}(X) \quad \text{and} \quad \mathrm{COD}(X) := \bigcup_{n=0}^{\infty} \mathrm{COD}_{\le n}(X).$$

0.2.3 The **Banach–Mazur distance** of n-dimensional Banach spaces X and Y is defined by

$$d(X,Y) := \inf\Big\{ \|T\|\,\|T^{-1}\| : T \in \mathfrak{L}(X,Y), \text{ bijection}\Big\}.$$

We have a multiplicative triangle inequality $d(X,Z) \le d(X,Y)\, d(Y,Z)$. Moreover, X and Y are isometric if and only if $d(X,Y) = 1$.

Whenever there exist $T \in \mathfrak{L}(X,Y)$ and $0 < c < 1$ such that

$$\Big| \|Tx\| - \|x\| \Big| \le c\|x\| \quad \text{for} \quad x \in X,$$

then $\|Tx\| \le (1+c)\|x\|$ and $(1-c)\|x\| \le \|Tx\|$. Hence $d(X,Y) \le \frac{1+c}{1-c}$.

0.2.4 Without proof, we state an extremely important result; see [joh], [PIE 2, p. 385] and [TOM, p. 54]. As usual, l_2^n denotes the n-dimensional Hilbert space; see 0.3.2.

JOHN'S THEOREM. $d(X, l_2^n) \le \sqrt{n}$ *whenever* $\dim(X) = n$.

0.2.5 An operator $T \in \mathfrak{L}(X,Y)$ has **finite rank** if its range

$$M(T) := \{Tx : x \in X\}$$

is finite dimensional. Then we write $\mathrm{rank}(T) = \dim(M(T))$. The set of all finite rank operators from X into Y is denoted by $\mathfrak{F}(X,Y)$.

0.3 Classical sequence spaces

0.3.1 Given any set $\mathbb{I}$, by an **$\mathbb{I}$-tuple** we mean a family of objects indexed by $i \in \mathbb{I}$. The letter $\mathbb{F}$ always stands for a finite index set, and $|\mathbb{F}|$ denotes its cardinality.

0.3.2 Let $1 \le r < \infty$, and consider any $\mathbb{I}$-tuple of Banach spaces X_i. Then $[l_r(\mathbb{I}), X_i]$ consists of all $\mathbb{I}$-tuples (x_i) with $x_i \in X_i$ for which

$$\|(x_i)|l_r(\mathbb{I})\| := \Big(\sum_{\mathbb{I}} \|x_i\|^r\Big)^{1/r}$$

is finite. In the limiting case $r = \infty$, the $\mathbb{I}$-tuples (x_i) are assumed to be bounded, and we let

$$\|(x_i)|l_\infty(\mathbb{I})\| := \sup_{\mathbb{I}} \|x_i\|.$$

To simplify matters, we write $[l_r, X_i]$ and $[l_r^n, X_i]$ when the index set $\mathbb{I}$ is $\{1, 2, \dots\}$ and $\{1, \dots, n\}$, respectively. In the scalar-valued case, the usual symbols $l_r(\mathbb{I})$, l_r and l_r^n will be used. The Banach space $[l_r(\mathbb{I}), X_i]$ is called the **$l_r(\mathbb{I})$-sum** of (X_i). If $X_i = X$ for all $i \in \mathbb{I}$, we refer to $[l_r(\mathbb{I}), X]$ as the **$l_r(\mathbb{I})$-multiple** of X. In this case, the underlying norm will sometimes be denoted by the more precise symbol $\|(x_i)|[l_r(\mathbb{I}), X]\|$.

0.3.3 The **natural injection** J_k^X from X_k into $X := [l_r(\mathbb{I}), X_i]$ takes $x \in X_k$ into the $\mathbb{I}$-tuple (x_i) with $x_k = x$ and $x_i = o$ for $i \neq k$. The **natural surjection** Q_k^X from X onto X_k is defined by $Q_k^X(x_i) := x_k$.

0.3.4 Let (T_i) be any $\mathbb{I}$-tuple of operators $T_i \in \mathfrak{L}(X_i, Y_i)$. Then the rule

$$[l_r(\mathbb{I}), T_i] : (x_i) \longrightarrow (T_i x_i)$$

yields a **diagonal operator** from $[l_r(\mathbb{I}), X_i]$ into $[l_r(\mathbb{I}), Y_i]$ provided that

$$\|[l_r(\mathbb{I}), T_i]\| = \sup_{\mathbb{I}} \|T_i\|$$

is finite. With the natural injections and surjections introduced above, we have

$$T_j = Q_j^Y [l_r(\mathbb{I}), T_i] J_j^X \quad \text{whenever } j \in \mathbb{I}.$$

The operators

$$\begin{aligned} B_\alpha &: (x_k) \longrightarrow ((1 + \log k)^{-\alpha} x_k), \\ C_\alpha &: (x_k) \longrightarrow (k^{-\alpha} x_k), \\ D_\alpha &: (x_k) \longrightarrow (2^{-k\alpha} x_k), \end{aligned}$$

defined for $(x_k) \in [l_2, l_\infty^{2^k}]$ and $\alpha \ge 0$ will play an important role as examples; see 1.2.12.

0.4 Classical function spaces

0.4.1 Throughout this book, (M,μ) and (N,ν) are assumed to be σ-finite measure spaces. To simplify notation, we suppress the underlying σ-algebras of measurable subsets. Scalar-valued functions will be denoted by $f,g,\dots$, while $\boldsymbol{f},\boldsymbol{g},\dots$ stand for vector-valued functions. With every scalar-valued function f on $M\times N$ we associate the function $\boldsymbol{f}$ which assigns to any fixed $s\in M$ the function $t\to f(s,t)$.

As usual, Banach function spaces are constituted by equivalence classes of functions which coincide almost everywhere.

0.4.2 If $f_1,\dots,f_n$ are scalar-valued functions on M and $x_1,\dots,x_n\in X$, then we let

$$\sum_{k=1}^{n} f_k\otimes x_k \,:\, s\longrightarrow \sum_{k=1}^{n} f_k(s)\,x_k.$$

0.4.3 A function $\boldsymbol{f}: M\to X$ is called **simple** if it can be written in the form

$$\boldsymbol{f}=\sum_{k=1}^{n}\chi_k\otimes x_k,$$

where $\chi_1,\dots,\chi_n$ are characteristic functions of measurable subsets $A_1,\dots,A_n$. We denote by $S_0(M,\mu)\otimes X$ the collection of such simple functions for which $\mu(A_1),\dots,\mu(A_n)$ are finite.

0.4.4 A function $\boldsymbol{f}: M\to X$ that coincides almost everywhere with the pointwise limit of a sequence of simple X-valued functions is said to be **measurable**.

0.4.5 For $1\le r<\infty$, we denote by $[L_r(M,\mu),X]$ the Banach space of all measurable functions $\boldsymbol{f}: M\to X$ for which

$$\|\boldsymbol{f}|L_r\| := \Big(\int_M \|\boldsymbol{f}(s)\|^r\,d\mu(s)\Big)^{1/r}$$

is finite. In the limiting case $r=\infty$, we let $[L_\infty(M,\mu),X]$ denote the Banach space of all essentially bounded measurable functions $\boldsymbol{f}: M\to X$ equipped with the norm

$$\|\boldsymbol{f}|L_\infty\| := \operatorname*{ess-sup}_M \|\boldsymbol{f}(s)\|.$$

When X is the scalar field $\mathbb{K}$, we simply write $L_r(M,\mu)$ instead of $[L_r(M,\mu),\mathbb{K}]$.

We denote by $L_r(M,\mu)\otimes X$ the collection of all X-valued functions

$$\sum_{k=1}^{n} f_k \otimes x_k$$

with $f_1,\dots,f_n \in L_r(M,\mu)$ and $x_1,\dots,x_n \in X$. Note that $L_r(M,\mu)\otimes X$ is a dense linear subset of $[L_r(M,\mu),X]$.

REMARK. For further information about Banach spaces $[L_r(M,\mu),X]$, the reader is referred to [DUN*1, pp. 119 and 146].

CONVENTIONS. From a pedantic point of view, the norm on $[L_r,X]$ should be denoted by $\|\cdot|[L_r,X]\|$. But, for simplicity, we will mostly use the shorter symbol $\|\cdot|L_r\|$. However, if a vector-valued function $\boldsymbol{f}\in[L_u(M,\mu),L_v(N,\nu)]$ with $1\le u,v<\infty$ is regarded as a scalar-valued function f on $M\times N$, then we will write

$$\|\boldsymbol{f}|[L_u,L_v]\| := \left[\int_M \left(\int_N |f(s,t)|^v \, d\nu(t)\right)^{u/v} d\mu(s)\right]^{1/u}.$$

When Banach function spaces $L_r(M,\mu)$ are used as examples, we simply write L_r and assume tacitly that they are infinite dimensional.

0.4.6 For $1<r<\infty$, the **dual exponent** r' is defined by $\frac{1}{r}+\frac{1}{r'}=1$. In the cases $r=1$ and $r=\infty$, we let $r'=\infty$ and $r'=1$, respectively.

The famous **Hölder inequality** says that

$$\left|\int_M f(s)g(s)\,d\mu(s)\right| \le \left(\int_M |f(s)|^r \, d\mu(s)\right)^{1/r} \left(\int_M |g(s)|^{r'} \, d\mu(s)\right)^{1/r'}$$

for $f\in L_r(M,\mu)$ and $g\in L_{r'}(M,\mu)$. If g is not the zero function, then equality holds if and only if there exist constants $c\ge 0$ and $\tau\in\mathbb{R}$ such that

$$|f(s)|^r = c\,|g(s)|^{r'} \quad \text{and} \quad f(s)g(s)e^{i\tau}\ge 0$$

for almost all $s\in M$ and $g(s)\ne 0$; see [ZYG, vol. I, p. 18].

0.4.7 Given $\boldsymbol{f}\in[L_r(M,\mu),X]$ and $\boldsymbol{g}\in[L_{r'}(M,\mu),X']$, we write

$$\langle \boldsymbol{f},\boldsymbol{g}\rangle := \int_M \Big\langle \boldsymbol{f}(s),\boldsymbol{g}(s)\Big\rangle \, d\mu(s).$$

If $\boldsymbol{g}\in[L_{r'}(M,\mu),X']$ is fixed, then the rule

$$\boldsymbol{f} \longrightarrow \langle \boldsymbol{f},\boldsymbol{g}\rangle$$

defines a functional on $[L_r(M,\mu),X]$, which is also denoted by $\boldsymbol{g}$. In the case of scalar-valued functions, this map yields an isometric isomorphism between $L_{r'}(M,\mu)$ and $L_r(M,\mu)'$.

REMARK. In general, we only get an isometric embedding, which is onto if and only if X has the Radon–Nikodým property with respect to the measure μ; see [DIN, p. 232] and [DIE*b, p. 98].

0.4.8 If H denotes a Hilbert space, then $[L_2(M,\mu),H]$ is also a Hilbert space with respect to the inner product

$$(\boldsymbol{f},\boldsymbol{g}) := \int\limits_M \big(\boldsymbol{f}(s),\boldsymbol{g}(s)\big)\,d\mu(s) \quad \text{for } \boldsymbol{f},\boldsymbol{g}\in[L_2(M,\mu),H],$$

where $\big(\boldsymbol{f}(s),\boldsymbol{g}(s)\big)$ means the inner product in H. Note that

$$\|\boldsymbol{f}\| = \|\boldsymbol{f}|L_2\| = \Big(\int\limits_M \|\boldsymbol{f}(s)\|^2 d\mu(s)\Big)^{1/2}.$$

In the scalar-valued case, the inner product on $L_2(M,\mu)$ is defined by

$$(f,g) := \int\limits_M f(s)\overline{g(s)}\,d\mu(s).$$

0.4.9 If $W\in\mathfrak{L}(L_u(M,\mu),L_v(N,\nu))$ and $T\in\mathfrak{L}(X,Y)$, then

$$W\otimes T : \sum_{k=1}^n f_k\otimes x_k \longrightarrow \sum_{k=1}^n Wf_k\otimes Tx_k$$

yields a well-defined map from $L_u(M,\mu)\otimes X$ into $L_v(N,\nu)\otimes Y$. Note that, in general, $W\otimes T$ does not extend continuously to an operator from $[L_u(M,\mu),X]$ into $[L_v(N,\nu),Y]$. This is the case, if and only if there exists a constant $c\geq\|W\|\,\|T\|$ such that

$$\Big\|\sum_{k=1}^n Wf_k\otimes Tx_k\Big|L_v\Big\| \leq c\,\Big\|\sum_{k=1}^n f_k\otimes x_k\Big|L_u\Big\| \qquad (*)$$

for $f_1,\dots,f_n\in L_u(M,\mu)$ and $x_1,\dots,x_n\in X$. Then we say that T is **compatible** with W, and the resulting extension will be denoted by $[W,T]$. To simplify matters, we write $[W,X]$ instead of $[W,I_X]$, where I_X is the identity map of a Banach space X.

If condition $(*)$ is satisfied for all operators T, then W is called **totally compatible**. This is, in particular, the case when W has finite rank. Indeed, using a representation

$$Wf = \sum_{h=1}^m \langle f,a_h\rangle\, b_h$$

with $a_1, \ldots, a_m \in L_{u'}(M, \mu)$ and $b_1, \ldots, b_m \in L_v(N, \nu)$, we have

$$[W,T] : \boldsymbol{f} \longrightarrow \sum_{h=1}^{m} b_h \otimes T \langle \boldsymbol{f}, a_h \rangle,$$

where

$$\langle \boldsymbol{f}, a_h \rangle := \int_M \boldsymbol{f}(s) a_h(s) \, d\mu(s).$$

Positive operators are also totally compatible. This fact can be verified as follows. First of all, a density argument tells us that it is enough to check condition $(*)$ for all functions of the form

$$\boldsymbol{f} = \sum_{k=1}^{n} \chi_k \otimes x_k,$$

where $\chi_1, \ldots, \chi_n$ are characteristic functions of pairwise disjoint subsets. In this case, however, we conclude from $|W\chi_k| = W\chi_k$ that

$$\begin{aligned}
\|[W,T]\boldsymbol{f} | L_v\| &= \left(\int_N \Big\| \sum_{k=1}^{n} W\chi_k(t) T x_k \Big\|^v d\nu(t) \right)^{1/v} \\
&\leq \left(\int_N \Big[\sum_{k=1}^{n} |W\chi_k(t)| \, \|Tx_k\| \Big]^v d\nu(t) \right)^{1/v} \\
&\leq \|T\| \left(\int_N \Big[W\Big(\sum_{k=1}^{n} \chi_k \|x_k\| \Big)(t) \Big]^v d\nu(t) \right)^{1/v} \\
&\leq \|W\| \, \|T\| \left(\int_M \Big[\sum_{k=1}^{n} \chi_k(s) \|x_k\| \Big]^u d\mu(s) \right)^{1/u} \\
&= \|W\| \, \|T\| \left(\int_M \Big\| \sum_{k=1}^{n} \chi_k(s) x_k \Big\|^u d\mu(s) \right)^{1/u} = \|W\| \, \|T\| \, \|\boldsymbol{f} | L_u\|.
\end{aligned}$$

Hence $\|[W,T]\| \leq \|W\| \, \|T\|$. Of course, we even have equality.

In particular, every $T \in \mathfrak{L}(X,Y)$ is compatible with the identity map of any function space $L_r(M, \mu)$. The extension operator from $[L_r(M, \mu), X]$ into $[L_r(M, \mu), Y]$ acts as follows:

$$[L_r, T] : \boldsymbol{f}(s) \longrightarrow T\boldsymbol{f}(s)$$

0.4.10 Let $0 < u \le v < \infty$, and assume that f is a $\mu \times \nu$-measurable scalar-valued function on $M \times N$. Then applying

$$\Big\| \int_M \boldsymbol{F}(s)\, d\mu(s) \Big\| \le \int_M \|\boldsymbol{F}(s)\|\, d\mu(s)$$

to the $L_{v/u}$-valued function $\boldsymbol{F} : s \to |f(s,\cdot)|^u$, it follows that

$$\left[\int_N \Big(\int_M |f(s,t)|^u d\mu(s) \Big)^{v/u} d\nu(t) \right]^{1/v} \le \left[\int_M \Big(\int_N |f(s,t)|^v d\nu(t) \Big)^{u/v} d\mu(s) \right]^{1/u} \qquad \text{(J)}$$

whenever the right-hand integral is finite. We refer to (J) as **Jessen's inequality**; see [jes].

0.5 Lorentz spaces

The theory of Lorentz spaces is presented in the following monographs: [BENN*, pp. 216–226], [BER*, pp. 6–13 and 113], [STEI*, pp. 188–205] and [TRIE, pp. 131–134].

0.5.1 Let (M, μ) be a σ-finite measure space, and write $\mathbb{R}_+ := (0, \infty)$. Then, given any measurable function $\boldsymbol{f} : M \to X$, the **distribution function** $\mu_{\boldsymbol{f}}$ is defined by

$$\mu_{\boldsymbol{f}}(u) := \mu\Big\{ s \in M : \|\boldsymbol{f}(s)\| > u \Big\} \quad \text{for } u \in \mathbb{R}_+.$$

Throughout, we assume that $\mu_{\boldsymbol{f}}(u)$ is finite. In this case, $\lim_{u\to\infty} \mu_{\boldsymbol{f}}(u) = 0$. Letting

$$f^*(t) := \inf \Big\{ u \in \mathbb{R}_+ : \mu_{\boldsymbol{f}}(u) \ge t \Big\} \quad \text{for } t \in \mathbb{R}_+$$

yields the **non-increasing rearrangement** f^*. This terminology is justified by the fact that $\|\boldsymbol{f}\|$ and f^* are identically distributed. More precisely, we have

$$\mu\Big\{ s \in M : \|\boldsymbol{f}(s)\| > u \Big\} = \lambda\Big\{ t \in \mathbb{R}_+ : f^*(t) > u \Big\} \quad \text{for } u \in \mathbb{R}_+,$$

where λ denotes the Lebesgue measure on $\mathbb{R}_+$. Hence

$$\int_M \|\boldsymbol{f}(s)\|^r\, d\mu(s) = \int_0^\infty f^*(t)^r\, dt.$$

0.5.2 For $1<r<\infty$ and $1\leq w\leq\infty$, we denote by $[L_{r,w}(M,\mu),X]$ the quasi-Banach space of all measurable functions $\boldsymbol{f}:M\to X$ for which the **Lorentz quasi-norm**

$$\|\boldsymbol{f}|L_{r,w}\|^* := \left(\int_0^\infty \left[t^{1/r}\boldsymbol{f}^*(t)\right]^w \frac{dt}{t}\right)^{1/w} = \left(r\int_0^\infty \left[u\,\mu_{\boldsymbol{f}}(u)^{1/r}\right]^w \frac{du}{u}\right)^{1/w}$$

is finite. The right-hand equation follows from the formula of integration by parts, since the functions $\mu_{\boldsymbol{f}}$ and $\boldsymbol{f}^*$ are, roughly speaking, inverse to each other. In the limiting case $w=\infty$, the usual modification yields

$$\|\boldsymbol{f}|L_{r,\infty}\|^* := \sup_{t>0} t^{1/r}\boldsymbol{f}^*(t) = \sup_{u>0} u\,\mu_{\boldsymbol{f}}(u)^{1/r}.$$

When X is the scalar field $\mathbb{K}$, we simply write $L_{r,w}(M,\mu)$ instead of $[L_{r,w}(M,\mu),\mathbb{K}]$.

In general, the expression $\|\cdot|L_{r,w}\|^*$ is a quasi-norm only. However, replacing $\boldsymbol{f}^*$ by

$$\boldsymbol{f}^{**}(t) := \frac{1}{t}\int_0^t \boldsymbol{f}^*(\tau)\,d\tau,$$

we obtain a norm $\|\cdot|L_{r,w}\|^{**}$. Thus the **Lorentz space** $[L_{r,w}(M,\mu),X]$ is even a Banach space.

In what follows, besides the classical spaces $[L_r(M,\mu),X]$ with $w=r$, we mainly need $[L_{r,1}(M,\mu),X]$ and $[L_{r,\infty}(M,\mu),X]$. Then

$$[L_{r,1}(M,\mu),X]\subseteq[L_r(M,\mu),X]\subseteq[L_{r,\infty}(M,\mu),X].$$

0.5.3 Now we give some more information about the case when M is a finite set equipped with the counting measure.

Fix any finite set $\mathbb{F}$, and write $n:=|\mathbb{F}|$. Given any $\mathbb{F}$-tuple (x_h) in a Banach space X, there exists a bijection $\pi:\{1,\dots,n\}\to\mathbb{F}$ such that $\|x_{\pi(1)}\|\geq\ldots\geq\|x_{\pi(n)}\|$. Then the **non-increasing rearrangement** is defined by $x_k^*=x_{\pi(k)}$.

For $1<r<\infty$, we let

$$\|(x_k)|l_{r,1}(\mathbb{F})\| := \sum_{k=1}^n k^{-1/r'}\|x_k^*\|$$

and

$$\|(x_k)|l_{r,\infty}(\mathbb{F})\| := \max_{1\leq k\leq n}\left\{k^{1/r}\|x_k^*\|\right\}.$$

Since

$$\|(x_k)|l_{r,1}(\mathbb{F})\| = \max_{\pi}\left\{\sum_{k=1}^n k^{-1/r'}\|x_{\pi(k)}\|\right\},$$

the expression $\|\cdot|l_{r,1}(\mathbb{F})\|$ is even a norm. However, $\|\cdot|l_{r,\infty}(\mathbb{F})\|$ is only a quasi-norm. Hence we refer to these objects as **Lorentz norms** and **Lorentz quasi-norms**, respectively.

In the special case when $\mathbb{F}=\{1,\dots,n\}$, we write $\|\cdot|l_{r,1}^n\|$ and $\|\cdot|l_{r,\infty}^n\|$ instead of $\|\cdot|l_{r,1}(\mathbb{F})\|$ and $\|\cdot|l_{r,\infty}(\mathbb{F})\|$.

0.5.4 LEMMA. $\|\cdot|l_{r,\infty}^n\| \le \|\cdot|l_r^n\| \le \|\cdot|l_{r,1}^n\|$.

PROOF. The left-hand inequality follows from the fact that

$$h\,\|x_h^*\|^r \le \sum_{k=1}^{h} \|x_k^*\|^r \le \|(x_k)|l_r^n\|^r \quad \text{for } h=1,\dots,n.$$

Since

$$\|(x_k)|l_{r,w}^n\| := \left(\sum_{k=1}^{n}\left[k^{1/r}\|x_k^*\|\right]^w \frac{1}{k}\right)^{1/w}$$

is logarithmically convex as a function of w, we have

$$\|(x_k)|l_{r,r}^n\| \le \|(x_k)|l_{r,1}^n\|^{1-\theta}\|(x_k)|l_{r,\infty}^n\|^{\theta} \le \|(x_k)|l_{r,1}^n\|^{1-\theta}\|(x_k)|l_{r,r}^n\|^{\theta},$$

where θ is defined by $1/r=(1-\theta)/1+\theta/\infty$. This proves the right-hand inequality.

0.5.5 LEMMA.
$\|\cdot|l_{r,1}^n\| \le (1+\log n)^{1/r'}\|\cdot|l_r^n\|$ *and* $\|\cdot|l_r^n\| \le (1+\log n)^{1/r}\|\cdot|l_{r,\infty}^n\|$.

PROOF. As stated in 0.10.2, we have

$$\sum_{k=1}^{n}\frac{1}{k} \le 1+\log n.$$

Hence

$$\begin{aligned}\|(x_k)|l_{r,1}^n\| &= \sum_{k=1}^{n} k^{-1/r'}\|x_k^*\| \le \left(\sum_{k=1}^{n}\frac{1}{k}\right)^{1/r'}\left(\sum_{k=1}^{n}\|x_k^*\|^r\right)^{1/r}\\ &\le (1+\log n)^{1/r'}\|(x_k)|l_r^n\|\end{aligned}$$

and

$$\|(x_k)|l_r^n\| = \left(\sum_{k=1}^{n}\frac{1}{k}[k^{1/r}\|x_k^*\|]^r\right)^{1/r} \le (1+\log n)^{1/r}\|(x_k)|l_{r,\infty}^n\|.$$

0.5.6 LEMMA. *There exists a constant $A(r,s)\ge 1$ such that*

$$\|\cdot|l_{r,1}^n\| \le A(r,s)\,n^{1/r-1/s}\|\cdot|l_s^n\| \quad \textit{if} \quad 1<r<s\le\infty$$

and

$$\|\cdot|l_{r,1}^n\| \le A(r,s)\|\cdot|l_s^n\| \quad \textit{if} \quad 1<s<r<\infty.$$

PROOF. It follows from 0.5.4, 0.10.2 and

$$\|(x_k)|l_{r,1}^n\| = \sum_{k=1}^{n} k^{1/r-1/s-1} k^{1/s} \|x_k^*\| \le \sum_{k=1}^{n} k^{1/r-1/s-1} \|(x_k)|l_{s,\infty}^n\|$$

that the required inequalities hold with $A(r,s) := \frac{2rs}{|r-s|}$.

0.5.7 For $1<r<s<\infty$, we define the **Lorentz quasi-norm**

$$\|(x_k)|l_{r,1}^n\|_s := \min\left\{ \sum_{h=1}^{m} |\mathbb{F}_h|^{1/r-1/s} \left(\sum_{k\in\mathbb{F}_h} \|x_k\|^s \right)^{1/s} \right\},$$

the minimum being taken over all partitions of $\{1,\dots,n\}$ into pairwise disjoint subsets $\mathbb{F}_h$ with $h=1,\dots,m$ and $m\le n$. In the limiting case $s=\infty$, the usual modification is required.

REMARK. Note that $\|\cdot|l_{r,1}^n\|_s$ is a quasi-norm, but not a norm.

0.5.8 LEMMA. *For $1<r<s\le\infty$, there exists a constant $A(r,s)\ge 1$ such that*

$$\tfrac{1}{2}\|\cdot|l_{r,1}^n\|_s \le \|\cdot|l_{r,1}^n\| \le A(r,s)\,\|\cdot|l_{r,1}^n\|_s.$$

PROOF. It follows from 0.5.6 that

$$\|(x_k)|l_{r,1}(\mathbb{F})\| \le A(r,s)|\mathbb{F}|^{1/r-1/s} \left(\sum_{k\in\mathbb{F}} \|x_k\|^s \right)^{1/s}$$

for all finite sets $\mathbb{F}$. Consider any partition of $\{1,\dots,n\}$ into pairwise disjoint subsets F_h with $h=1,\dots,m$ and $m\le n$. Since $\|\cdot|l_{r,1}^n\|$ is a norm, we obtain

$$\begin{aligned}\|(x_k)|l_{r,1}^n\| &\le \sum_{h=1}^{m} \|(x_k)|l_{r,1}(\mathbb{F}_h)\| \\ &\le A(r,s) \sum_{h=1}^{m} |\mathbb{F}_h|^{1/r-1/s} \left(\sum_{k\in\mathbb{F}_h} \|x_k\|^s \right)^{1/s}.\end{aligned}$$

This proves the right-hand inequality. Assume that $\|x_1\|\ge\ldots\ge\|x_n\|$, and consider the partition given by

$$\mathbb{F}_h := \{\, k\in\mathbb{N} : 2^h \le k < 2^{h+1},\ k\le n \,\} \quad \text{with} \quad 1\le 2^h\le n.$$

In view of 0.10.3, the left-hand inequality now follows from

$$\begin{aligned}\|(x_k)|l_{r,1}^n\| &= \sum_{k=1}^{n} k^{-1/r'} \|x_k\| \ge \|x_1\| + \sum_{2\le 2^h\le n} 2^{h-1} 2^{-h/r'} \|x_{2^h}\| \\ &\ge \frac{1}{2} \sum_{1\le 2^h\le n} 2^{h/r} \|x_{2^h}\|\end{aligned}$$

and

$$\sum_{1\le 2^h\le n} |\mathbb{F}_h|^{1/r-1/s}\Big(\sum_{k\in\mathbb{F}_h}\|x_k\|^s\Big)^{1/s} \le \sum_{1\le 2^h\le n} |\mathbb{F}_h|^{1/r}\|x_{2^h}\| \le \sum_{1\le 2^h\le n} 2^{h/r}\|x_{2^h}\|.$$

0.5.9 For $1<s<r<\infty$, we define the **Lorentz norm**

$$\|(x_k)|l^n_{r,\infty}\|_s := \max\left\{|\mathbb{F}|^{1/r-1/s}\Big(\sum_{k\in\mathbb{F}}\|x_k\|^s\Big)^{1/s}\right\},$$

the maximum being taken over all subsets $\mathbb{F}$ of $\{1,\dots,n\}$.

0.5.10 We now establish a counterpart of 0.5.8.

LEMMA. *For $1<s<r<\infty$, there exists a constant $A(r,s)\ge 1$ such that*

$$\|\cdot|l^n_{r,\infty}\| \le \|\cdot|l^n_{r,\infty}\|_s \le A(r,s)\,\|\cdot|l^n_{r,\infty}\|.$$

PROOF. Assume that $\|x_1\|\ge\dots\ge\|x_n\|$, and let $\mathbb{F}_m:=\{1,\dots,m\}$ for $m=1,\dots,n$. Then

$$m^{1/s}\|x_m\| \le \Big(\sum_{k\in\mathbb{F}_m}\|x_k\|^s\Big)^{1/s}.$$

Hence

$$m^{1/r}\|x_m\| \le |\mathbb{F}_m|^{1/r-1/s}\Big(\sum_{k\in\mathbb{F}_m}\|x_k\|^s\Big)^{1/s} \le \|(x_k)|l^n_{r,\infty}\|_s$$

which proves the left-hand inequality. On the other hand, by 0.10.2, we have

$$\Big(\sum_{k\in\mathbb{F}}\|x_k\|^s\Big)^{1/s} \le \Big(\sum_{k=1}^{|\mathbb{F}|}\|x_k\|^s\Big)^{1/s} \le \Big(\sum_{k=1}^{|\mathbb{F}|}k^{-s/r}\Big)^{1/s}\|(x_k)|l^n_{r,\infty}\| \le \Big(\frac{1}{1-s/r}\Big)^{1/s}|\mathbb{F}|^{1/s-1/r}\|(x_k)|l^n_{r,\infty}\|$$

for every subset $\mathbb{F}$ of $\{1,\dots,n\}$. Thus the right-hand inequality holds with the constant $A(r,s):=\big(\frac{r}{r-s}\big)^{1/s}$.

0.6 Interpolation methods

Concerning interpolation theory the reader is referred to the monographs [BENN*], [BER*], [BUT*a], [KRE*] and [TRIE]. In what follows, we only fix some terminology and notation.

0.6.1 By an **interpolation couple** we mean a pair of Banach spaces X_0 and X_1 which are continuously embedded into another Banach space. Then $X_0 \cap X_1$ and $X_0 + X_1$ become Banach spaces under the norms

$$\|x|X_0 \cap X_1\| := \max\{\, \|x|X_0\|,\ \|x|X_1\| \,\}$$

and

$$\|x|X_0+X_1\| := \inf\{\, \|x_0|X_0\|+\|x_1|X_1\| \, : \, x_0 \in X_0,\ x_1 \in X_1,\ x = x_0+x_1 \,\},$$

respectively.

0.6.2 The **complex interpolation space** $X_\theta := [X_0, X_1]_\theta$ depends on **one** parameter $0 < \theta < 1$. This method has the advantage that it produces the classical Banach spaces with the original norms. That is,

$$[L_{r_0}(M,\mu), L_{r_1}(M,\mu)]_\theta = L_r(M,\mu)$$

whenever $1/r = (1-\theta)/r_0 + \theta/r_1$.

0.6.3 The **real interpolation space** $X_{\theta,w} := (X_0, X_1)_{\theta,w}$ depends on **two** parameters $0 < \theta < 1$ and $1 \le w \le \infty$. This method also yields the Lorentz spaces, since

$$\big(L_{r_0,w_0}(M,\mu), L_{r_1,w_1}(M,\mu)\big)_{\theta,w} = L_{r,w}(M,\mu)$$

whenever $1/r = (1-\theta)/r_0 + \theta/r_1$ and $1 \le w,\ w_0,\ w_1 \le \infty$. In this case, however, we only get equivalent norms.

0.6.4 Assume that the operator $T \in \mathfrak{L}(X_0 + X_1, Y_0 + Y_1)$ transforms X_0 into Y_0 and X_1 into Y_1. Define

$$T_0 : X_0 \xrightarrow{T} Y_0 \quad \text{and} \quad T_1 : X_1 \xrightarrow{T} Y_1.$$

Then T induces operators

$$\underset{\text{(complex method)}}{T_\theta : X_\theta \xrightarrow{T} Y_\theta} \quad \text{and} \quad \underset{\text{(real method)}}{T_{\theta,w} : X_{\theta,w} \xrightarrow{T} Y_{\theta,w}}$$

such that

$$\|T_\theta\| \le \|T_0\|^{1-\theta}\, \|T_1\|^\theta \quad \text{and} \quad \|T_{\theta,w}\| \le \|T_0\|^{1-\theta}\, \|T_1\|^\theta.$$

0.7 Summation operators

0.7.1 We define the infinite **summation operator** $\Sigma \in \mathfrak{L}(l_1, l_\infty)$ and its transpose $\Sigma^t \in \mathfrak{L}(l_1, l_\infty)$ by

$$\Sigma : (\xi_k) \longrightarrow \Big(\sum_{k=1}^{h} \xi_k\Big) \quad \text{and} \quad \Sigma^t : (\xi_k) \longrightarrow \Big(\sum_{k=h}^{\infty} \xi_k\Big).$$

The representing matrices have the form

$$\Sigma := \begin{pmatrix} 1 & 0 & 0 & 0 & \dots \\ 1 & 1 & 0 & 0 & \dots \\ 1 & 1 & 1 & 0 & \dots \\ 1 & 1 & 1 & 1 & \dots \\ \vdots & \vdots & \vdots & \vdots & \ddots \end{pmatrix} \quad \text{and} \quad \Sigma^t := \begin{pmatrix} 1 & 1 & 1 & 1 & \dots \\ 0 & 1 & 1 & 1 & \dots \\ 0 & 0 & 1 & 1 & \dots \\ 0 & 0 & 0 & 1 & \dots \\ \vdots & \vdots & \vdots & \vdots & \ddots \end{pmatrix}.$$

Note that

$$\|\Sigma : l_1 \to l_\infty\| = 1 \quad \text{and} \quad \|\Sigma^t : l_1 \to l_\infty\| = 1.$$

Viewing Σ^t as an operator from l_1 into c_0, we get $\Sigma = (\Sigma^t)'$.

0.7.2 LEMMA. *There exists $U \in \mathfrak{L}(l_1)$ such that*

$$\Sigma^t = \Sigma U \qquad \textit{and} \qquad \Sigma = U^t \Sigma^t,$$
$$\|U : l_1 \to l_1\| = 2 \qquad \textit{and} \qquad \|U^t : l_\infty \to l_\infty\| = 2.$$

PROOF. Use the operator induced by the matrix

$$U := \begin{pmatrix} 1 & 1 & 1 & 1 \dots \\ -1 & 0 & 0 & 0 \dots \\ 0 & -1 & 0 & 0 \dots \\ 0 & 0 & -1 & 0 \dots \\ \vdots & \vdots & \vdots & \vdots\ddots \end{pmatrix}$$

0.7.3 The finite **summation operator** $\Sigma_n \in \mathfrak{L}(l_1^n, l_\infty^n)$ and its transpose $\Sigma_n^t \in \mathfrak{L}(l_1^n, l_\infty^n)$ are defined in the same way. Letting

$$J_n : (\xi_1, \dots, \xi_n) \longrightarrow (\xi_1, \dots, \xi_n, 0, \dots)$$

and

$$Q_n : (\xi_1, \dots, \xi_n, \xi_{n+1}, \dots) \longrightarrow (\xi_1, \dots, \xi_n),$$

we have

$$\Sigma_n = Q_n \Sigma J_n \quad \text{and} \quad \Sigma_n^t = Q_n \Sigma^t J_n.$$

Moreover, Σ_n and Σ_n^t are dual to each other. That is, $\Sigma_n^t = \Sigma_n'$.

0.7.4 LEMMA. *There exists $R_n \in \mathfrak{L}(l_1^n)$ such that*

$$\Sigma_n^t = R_n \Sigma_n R_n \quad \textit{and} \quad \Sigma_n = R_n \Sigma_n^t R_n,$$
$$\|R_n : l_1^n \to l_1^n\| = 1 \quad \textit{and} \quad \|R_n : l_\infty^n \to l_\infty^n\| = 1.$$

PROOF. Use the rearrangement $R_n(\xi_k) := (\xi_{n-k+1})$, and note that $R_n^t = R_n$.

0.7.5 We consider the map $\mathcal{M}_n$ which assigns to every formal power series

$$A(\zeta) = \sum_{k=0}^{\infty} \alpha_k \zeta^k$$

the subdiagonal (n, n)-matrix

$$\Sigma_n^A := (\sigma_{hk}^A), \quad \text{where} \quad \sigma_{hk}^A := \begin{cases} \alpha_{h-k} & \text{if} \quad h \geq k, \\ 0 & \text{if} \quad h < k. \end{cases}$$

Note that $\mathcal{M}_n$ is a ring homomorphism. Since the binomial series

$$\frac{1}{(1-\zeta)^\lambda} = \sum_{k=0}^{\infty} \sigma_k(\lambda) \zeta^k \quad \text{with} \quad \lambda \in \mathbb{R} \quad \text{and} \quad \sigma_k(\lambda) := (-1)^k \binom{-\lambda}{k}$$

constitute a 1-parameter group, the same statement holds for the associated (n, n)-matrices Σ_n^λ. This means that

$$\Sigma_n^\lambda \Sigma_n^\mu = \Sigma_n^{\lambda+\mu} \quad \text{for all } \lambda, \mu \in \mathbb{R}.$$

In the case $\lambda = 1$, the matrix Σ_n^1 yields the summation operator Σ_n.

0.8 Finite representability and ultrapowers

0.8.1 We say that a Banach space Y is **finitely representable** in a Banach space X if for every finite dimensional subspace F of Y and $\varepsilon > 0$ there exists a finite dimensional subspace E of X such that

$$\dim(E) = \dim(F) \quad \text{and} \quad d(E, F) \leq 1 + \varepsilon.$$

Note that this relation is transitive.

0.8.2 For every Banach space X and every index set $\mathbb{I}$, we denote by $[l_\infty(\mathbb{I}), X]$ the Banach space of all bounded $\mathbb{I}$-tuples (x_i) in X equipped with the norm

$$\|(x_i)|l_\infty(\mathbb{I})\| := \sup_{\mathbb{I}} \|x_i\|.$$

Fix any ultrafilter $\mathcal{U}$ on $\mathbb{I}$. Let

$$N^{\mathcal{U}} := \Big\{ (x_i^\circ) \in [l_\infty(\mathbb{I}), X] : \mathcal{U}\text{-}\lim_i \|x_i^\circ\| = 0 \Big\},$$

form the quotient space $X^{\mathcal{U}} := [l_\infty(\mathbb{I}), X]/N^{\mathcal{U}}$, and define

$$\|(x_i)^{\mathcal{U}}\| := \inf\Big\{ \|(x_i - x_i^\circ)|l_\infty(\mathbb{I})\| : (x_i^\circ) \in N^{\mathcal{U}} \Big\} = \mathcal{U}\text{-}\lim_i \|x_i\|,$$

where $(x_i)^{\mathcal{U}}$ stands for the equivalence class generated by (x_i). The Banach space $X^{\mathcal{U}}$ obtained in this way is said to be an **ultrapower** of X.

Note that every element $\boldsymbol{x} \in X^{\mathcal{U}}$ admits a representation $\boldsymbol{x} = (x_i)^{\mathcal{U}}$ such that $\|x_i\| = \|\boldsymbol{x}\|$ for all $i \in \mathbb{I}$.

REMARK. Ultrafilters of the form $\mathcal{U} = \{ U \subseteq \mathbb{I} : \{i_0\} \in U \}$ with fixed $i_0 \in \mathbb{I}$ are called trivial. In this case, $X^{\mathcal{U}}$ and X can be identified.

0.8.3 The two concepts just introduced are closely connected. Proofs of this fact are given in [DIE*a, pp. 176–177] and [GUE, pp. 72–78]; see also [BEA 2, pp. 220–225].

PROPOSITION.

(1) *X can be embedded isometrically into every ultrapower $X^{\mathcal{U}}$.*

(2) *Every ultrapower $X^{\mathcal{U}}$ is finitely representable in X.*

(3) *Y is finitely representable in X if and only if there exists an isometric embedding from Y into some ultrapower $X^{\mathcal{U}}$.*

0.8.4 Let (T_i) be a bounded $\mathbb{I}$-tuple of operators from X into Y. Then

$$(T_i)^{\mathcal{U}} : (x_i)^{\mathcal{U}} \longrightarrow (T_i x_i)^{\mathcal{U}}$$

is an operator from $X^{\mathcal{U}}$ into $Y^{\mathcal{U}}$ such that $\|(T_i)^{\mathcal{U}}\| = \mathcal{U}\text{-}\lim_i \|T_i\|$. In the case when $T_i = T$, we call $T^{\mathcal{U}}$ an **ultrapower** of T.

0.9 Extreme points

0.9.1 If A is a set in a (real or complex) linear space, then $\operatorname{conv}(A)$ denotes its **convex hull**. In the special case when A is a subset of $\mathbb{R}^n$, Carathéodory's theorem says that

$$\operatorname{conv}(A) = \Big\{ \sum_{k=1}^{n+1} \lambda_k x_k : \sum_{k=1}^{n+1} \lambda_k = 1,\ \lambda_k \geq 0,\ x_k \in A \Big\}.$$

Because of this fundamental fact, compactness carries over from A to $\operatorname{conv}(A)$; see [ROC, p. 158].

0.9.2 Let A be a convex set in a linear space X. We say that $x \in A$ is an **extreme point** of A if there is no representation $x = \frac{x_+ + x_-}{2}$ with $x_\pm \in A$ and $x_+ \neq x_-$.

0.9.3 A classical result of H. Minkowski asserts that every point x in a compact convex subset A of $\mathbb{R}^n$ is a convex combination of extreme points $x_1, \dots, x_N$:

$$x = \sum_{k=1}^{N} \lambda_k x_k \quad \text{with} \sum_{k=1}^{N} \lambda_k = 1 \text{ and } \lambda_k > 0.$$

Hence, given any convex continuous function $f : A \to \mathbb{R}$, we have

$$f(x) \le \sum_{k=1}^{N} \lambda_k f(x_k) \le \max_{1 \le k \le N} f(x_k).$$

Thus f attains its maximum at some extreme point; see [ROC, p. 344]. This conclusion also holds in the complex case, since $\mathbb{C}^n$ can be regarded as $\mathbb{R}^{2n}$.

0.9.4 Next, we present the most important examples:
The extreme points of U_∞^n, the closed unit ball of l_∞^n, have the form

$$x = (\xi_k) \quad \text{with } |\xi_k| = 1.$$

The extreme points of U_1^n, the closed unit ball of l_1^n, have the form

$$x = \xi u_k^{(n)} \quad \text{with } |\xi| = 1$$

and a unit vector $u_k^{(n)} = (0, \dots, 0, 1, 0, \dots, 0)$.
See also 2.1.6.

0.9.5 We now treat another example which is quite useful for later applications.

EXTREME POINT LEMMA. *The coordinates x_k of all extreme points in the closed ball*

$$\Big\{ (x_k) \in [l_r^n, l_1^N] : \sum_{k=1}^{n} \|x_k\|^r \le \varrho^r \Big\}$$

are multiples of unit vectors.

PROOF. Given any extreme point (x_k), we fix $h = 1, \dots, n$ and let

$$\varrho_h^r := \varrho^r - \sum_{\substack{k=1 \\ k \ne h}}^{n} \|x_k\|^r = \|x_h\|^r.$$

Then the coordinate x_h is an extreme point of $\{x \in l_1^N : \|x\| \le \varrho_h\}$. Thus it must have the form $x_h = \xi_h u_i^{(N)}$.

0.10 Various tools

0.10.1 Let (α_n) and (β_n) be sequences of positive numbers. We write $\alpha_n \prec \beta_n$ if there exists a constant $c > 0$ such that

$$\alpha_n \le c\beta_n \quad \text{for } n = 1, 2, \ldots .$$

Furthermore, $\alpha_n \asymp \beta_n$ means that $\alpha_n \prec \beta_n$ and $\beta_n \prec \alpha_n$.

The relation $\alpha_n \prec \beta_n$ can also be expressed by the **Landau symbol** O. That is, $\alpha_n = O(\beta_n)$. Moreover, $\alpha_n = o(\beta_n)$ says that $\lim_{n\to\infty} \alpha_n/\beta_n = 0$.

0.10.2 For $n = 1, 2, \ldots$, we have

$$C + \log n < \sum_{k=1}^{n} \frac{1}{k} \le 1 + \log n,$$

where $C = 0.5772\ldots$ denotes the Euler–Mascheroni constant. If $0 < \lambda \le 1$, then

$$n^\lambda \le \sum_{k=1}^{n} k^{\lambda-1} \le \frac{1}{\lambda} n^\lambda \quad \text{and} \quad \sum_{k=1}^{\infty} k^{-\lambda-1} < \frac{2}{\lambda}.$$

0.10.3 Next, we provide the basic inequalities from the proof of Cauchy's condensation test.

LEMMA. *Let* $\xi_1 \ge \xi_2 \ge \ldots \ge \xi_{2^n} \ge 0$. *Then*

$$\xi_1 + \sum_{h=1}^{n} 2^{h-1}\xi_{2^h} \le \sum_{k=1}^{2^n} \xi_k \le \sum_{h=0}^{n-1} 2^h \xi_{2^h} + \xi_{2^n}.$$

PROOF. We have

$$\begin{aligned}\sum_{k=1}^{2^n} \xi_k &= \xi_1 + (\xi_2 + \xi_3) + \ldots + (\xi_{2^{n-1}} + \ldots + \xi_{2^n-1}) + \xi_{2^n} \\ &\le \xi_1 + \quad 2\xi_2 \quad + \ldots + \quad 2^{n-1}\xi_{2^{n-1}} \quad + \xi_{2^n}\end{aligned}$$

and

$$\begin{aligned}\sum_{k=1}^{2^n} \xi_k &= \xi_1 + \xi_2 + (\xi_3 + \xi_4) + \ldots + (\xi_{2^{n-1}+1} + \ldots + \xi_{2^n}) \\ &\ge \xi_1 + \xi_2 + \quad 2\xi_4 \quad + \ldots + \quad 2^{n-1}\xi_{2^n} \quad .\end{aligned}$$

0.10.4 We add some exercises in calculus:

$$\left|\frac{e^t - 1}{e^t + 1}\right| \le \min\{1, \tfrac{1}{2}|t|\} \quad \text{for} \quad t \in \mathbb{R},$$

$$0 < \frac{1}{\sin t} - \frac{1}{t} \le 1 - \frac{2}{\pi} \quad \text{and} \quad 0 < \frac{1}{t} - \cot t \le \frac{2}{\pi} \quad \text{for } 0 < t \le \pi/2.$$

0.10.5 Stirling's formula asserts that

$$\Gamma(t) = \sqrt{2\pi}\, t^{t-1/2} \exp(-t + \tfrac{\theta}{12t}),$$

where $0 < \theta < 1$ depends on $t > 0$.

0.10.6 A non-decreasing function $F : \mathbb{N} \to \mathbb{R}_+$ is said to be **very slowly growing** if there exists a constant $c > 1$ such that

$$F(2^n) \le c\, F(n) \quad \text{for} \quad n = 1, 2, \ldots .$$

This terminology, used in 4.3.16 and 4.5.16, will be justified by the following considerations.

The *iterated exponential functions* are inductively defined by

$$P_0(m) := m \quad \text{and} \quad P_{k+1}(m) := 2^{P_k(m)}.$$

In other terms, we let

$$P_k(m) := 2^{2^{\cdot^{\cdot^{\cdot^{2^m}}}}},$$

where the right-hand tower is built from k digits 2. For example,

$$P_0(1) = 1,\ P_1(1) = 2,\ P_2(1) = 4,\ P_3(1) = 16,\ P_4(1) = 65536,$$
$$P_5(1) = \text{out of memory}.$$

Clearly, for large k, the functions $P_k(m)$ are rapidly increasing as $m \to \infty$. Hence the *iterated logarithms*

$$L_k(n) := \min\{\, m : n \le P_k(m) \,\}$$

grow slowly. Induction on k yields $F(P_k(m)) \le c^k F(m)$. In particular, $F(P_k(1)) \le c^k F(1)$. For $k = 1, 2, \ldots$ and $n = 1, 2, \ldots$, let $m := L_k(n)$ and choose h such that $P_h(1) \le m < P_{h+1}(1)$. Obviously, $n \to \infty$ implies $m \to \infty$ and $h \to \infty$. Since $n \le P_k(m)$, we get

$$\frac{F(n)}{L_k(n)} \le \frac{F(P_k(m))}{m} \le \frac{c^k\, F(m)}{m} \le \frac{c^k\, F(P_{h+1}(1))}{P_h(1)} \le \frac{c^{h+k+1}}{P_h(1)}\, F(1).$$

This proves that

$$\lim_{n \to \infty} \frac{F(n)}{L_k(n)} = 0.$$

1

Ideal norms and operator ideals

In this chapter, we collect some specific results from the theory of operator ideals. For more information, the following books are recommended:

Defant/Floret	*Tensor norms and operator ideals* [DEF*],
Diestel/Uhl	*Vector measures* [DIE*b],
König	*Eigenvalue distribution of compact operators* [KOE],
Pietsch.................	*Operator ideals* [PIE 2],
Pietsch.................	*Eigenvalues and s-numbers* [PIE 3],
Pisier	*Factorization of linear operators and geometry of Banach spaces* [PIS 1],
Tomczak-Jaegermann...	*Banach–Mazur distances and finite dimensional operator ideals* [TOM].

1.1 Ideal norms

1.1.1 An **ideal norm** $\boldsymbol{\alpha}$ is a function that assigns to every operator T between arbitrary Banach spaces a non-negative number $\boldsymbol{\alpha}(T)$ such that the following conditions are satisfied:

(1) $\boldsymbol{\alpha}(S+T) \le \boldsymbol{\alpha}(S) + \boldsymbol{\alpha}(T)$ for $S,T \in \mathfrak{L}(X,Y)$.

(2) $\boldsymbol{\alpha}(BTA) \le \|B\|\,\boldsymbol{\alpha}(T)\,\|A\|$ for $T \in \mathfrak{L}(X,Y)$, $A \in \mathfrak{L}(X_0,X)$ and $B \in \mathfrak{L}(Y,Y_0)$.

(3) $\boldsymbol{\alpha}(T)=0$ implies that T is a zero operator.

Note that

$$\boldsymbol{\alpha}(\lambda T) = |\lambda|\boldsymbol{\alpha}(T) \quad \text{for } T \in \mathfrak{L}(X,Y) \text{ and } \lambda \in \mathbb{K}.$$

Whenever it is necessary to indicate that the operator T acts from X into Y, the more precise notation $\boldsymbol{\alpha}(T : X \to Y)$ will be used. To simplify matters, $\boldsymbol{\alpha}(X)$ stands for $\boldsymbol{\alpha}(I_X)$, where I_X is the identity map of a Banach space X.

1.1.2 Let $\boldsymbol{\alpha}$ be an ideal norm and let $c > 0$. We write $\boldsymbol{\alpha} \leq c$ if

$$\boldsymbol{\alpha}(X) \leq c \quad \text{for all } X \in \mathsf{L}.$$

Then it follows from $\boldsymbol{\alpha}(T) \leq \|T\|\, \boldsymbol{\alpha}(X)$ that

$$\boldsymbol{\alpha}(T) \leq c\,\|T\| \quad \text{for all } T \in \mathfrak{L}.$$

Analogously, we write $c \leq \boldsymbol{\alpha}$ if

$$c \leq \boldsymbol{\alpha}(X) \quad \text{for all } X \in \mathsf{L}.$$

In view of the next proposition, this is already true when $c \leq \boldsymbol{\alpha}(I_{\mathbb{K}})$. In this case, we have

$$c\,\|T\| \leq \boldsymbol{\alpha}(T) \quad \text{for all } T \in \mathfrak{L}.$$

In particular, $\boldsymbol{\alpha} = 1$ means that $\boldsymbol{\alpha}$ coincides with the operator norm.

1.1.3 The following inequality is an easy consequence of the Hahn–Banach theorem.

PROPOSITION. *Let $\boldsymbol{\alpha}$ be any ideal norm. Then $\boldsymbol{\alpha}(\mathbb{K}) \leq \boldsymbol{\alpha}$.*

PROOF. Given $T \in \mathfrak{L}(X, Y)$ and $\varepsilon > 0$, we choose $x \in X$ and $y' \in Y'$ such that

$$\tfrac{1}{1+\varepsilon}\|T\| \leq |\,\langle Tx, y'\rangle\,|, \qquad \|x\| \leq 1 \quad \text{and} \quad \|y'\| \leq 1.$$

Define $A \in \mathfrak{L}(\mathbb{K}, X)$ and $B \in \mathfrak{L}(Y, \mathbb{K})$ by

$$A\xi := \xi x \quad \text{for } \xi \in \mathbb{K} \qquad \text{and} \qquad By := \langle y, y'\rangle \quad \text{for } y \in Y.$$

Then $\|A\| \leq 1$ and $\|B\| \leq 1$. Looking at the diagram

$$\begin{array}{ccc} X & \xrightarrow{\quad T \quad} & Y \\ A\big\uparrow & & \big\downarrow B \\ \mathbb{K} & \xrightarrow[\langle Tx, y'\rangle\, I_{\mathbb{K}}]{} & \mathbb{K}, \end{array}$$

we conclude that

$$\tfrac{1}{1+\varepsilon}\|T\|\boldsymbol{\alpha}(\mathbb{K}) \leq |\langle Tx, y'\rangle|\, \boldsymbol{\alpha}(I_{\mathbb{K}}) \leq \|B\|\boldsymbol{\alpha}(T)\|A\| \leq \boldsymbol{\alpha}(T).$$

Letting $\varepsilon \to 0$ yields $\boldsymbol{\alpha}(\mathbb{K})\|T\| \leq \boldsymbol{\alpha}(T)$.

1.1.4 Define

$$\boldsymbol{\alpha}(1) := \inf\Big\{\, \boldsymbol{\alpha}(X) : X \in \mathsf{L} \Big\} \quad \text{and} \quad \boldsymbol{\alpha}(\infty) := \sup\Big\{\, \boldsymbol{\alpha}(X) : X \in \mathsf{L} \Big\}.$$

The motivation for the choice of the symbols $\boldsymbol{\alpha}(1)$ and $\boldsymbol{\alpha}(\infty)$ will become clear in 9.3.2.

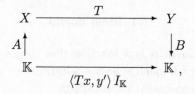

1.1.5 Thanks to 1.1.3, we have $\boldsymbol{\alpha}(\mathbb{K}) \le \boldsymbol{\alpha}(X)$ for all Banach spaces X. So $\boldsymbol{\alpha}(1) = \boldsymbol{\alpha}(\mathbb{K})$. Next, we show that the supremum in the definition of $\boldsymbol{\alpha}(\infty)$ is attained as well.

PROPOSITION. *For every ideal norm $\boldsymbol{\alpha}$, there exists a Banach space X_0 such that*

$$\boldsymbol{\alpha}(X) \le \boldsymbol{\alpha}(X_0) \quad \textit{whenever} \quad X \in \mathsf{L}.$$

PROOF. Choose a sequence of Banach spaces $X_n \in \mathsf{L}$ with

$$\lim_{n\to\infty} \boldsymbol{\alpha}(X_n) = \boldsymbol{\alpha}(\infty).$$

Let $X_0 := [l_2, X_n]$; see 0.3.2. Then it follows from $I_{X_n} = Q_n^{X_0} I_{X_0} J_n^{X_0}$ that

$$\boldsymbol{\alpha}(X_n) \le \boldsymbol{\alpha}(X_0) \quad \text{for} \quad n = 1, 2, \ldots .$$

1.1.6 Combining the above observations yields the following fact.

THEOREM. *Every ideal norm $\boldsymbol{\alpha}$ is equivalent to the operator norm. More precisely, we have*

$$\boldsymbol{\alpha}(1) \le \boldsymbol{\alpha} \le \boldsymbol{\alpha}(\infty).$$

1.1.7 Given ideal norms $\boldsymbol{\alpha}$, $\boldsymbol{\beta}$ and $\boldsymbol{\gamma}$, then $\boldsymbol{\alpha} \le \boldsymbol{\beta} \circ \boldsymbol{\gamma}$ means that

$$\boldsymbol{\alpha}(ST) \le \boldsymbol{\beta}(S)\,\boldsymbol{\gamma}(T) \quad \text{for } T \in \mathfrak{L}(X, Y) \text{ and } S \in \mathfrak{L}(Y, Z),$$

where X, Y and Z are arbitrary Banach spaces.

1.1.8 Let $(\boldsymbol{\alpha}_n)$ and $(\boldsymbol{\beta}_n)$ be sequences of ideal norms. We write

$$\boldsymbol{\alpha}_n \prec \boldsymbol{\beta}_n$$

if there exists a constant $c > 0$ such that

$$\boldsymbol{\alpha}_n(T) \le c\,\boldsymbol{\beta}_n(T) \quad \text{for all } T \in \mathfrak{L} \text{ and } n = 1, 2, \ldots .$$

Furthermore,

$$\boldsymbol{\alpha}_n \asymp \boldsymbol{\beta}_n$$

means that $\boldsymbol{\alpha}_n \prec \boldsymbol{\beta}_n$ and $\boldsymbol{\beta}_n \prec \boldsymbol{\alpha}_n$. In this case, $(\boldsymbol{\alpha}_n)$ and $(\boldsymbol{\beta}_n)$ are said to be **uniformly equivalent**. This property is extremely important for the following considerations.

1.1.9 With every ideal norm $\boldsymbol{\alpha}$ we associate the **dual ideal norm** $\boldsymbol{\alpha}'$ defined by

$$\boldsymbol{\alpha}'(T) := \boldsymbol{\alpha}(T').$$

An ideal norm $\boldsymbol{\alpha}$ is called **symmetric** if $\boldsymbol{\alpha}' = \boldsymbol{\alpha}$.

1.1.10 An ideal norm $\boldsymbol{\alpha}$ is said to be **injective** if

$$\boldsymbol{\alpha}(JT) = \boldsymbol{\alpha}(T)$$

for $T \in \mathfrak{L}(X,Y)$ and any metric injection $J \in \mathfrak{L}(Y,Y_0)$; see 0.1.9.

An ideal norm $\boldsymbol{\alpha}$ is said to be **surjective** if

$$\boldsymbol{\alpha}(TQ) = \boldsymbol{\alpha}(T)$$

for $T \in \mathfrak{L}(X,Y)$ and any metric surjection $Q \in \mathfrak{L}(X_0,X)$; see 0.1.9.

1.1.11 PROPOSITION. *If an ideal norm $\boldsymbol{\alpha}$ is injective (surjective), then $\boldsymbol{\alpha}'$ is surjective (injective).*

PROOF. Since the dual Q' of every metric surjection $Q \in \mathfrak{L}(X_0,X)$ is a metric injection, injectivity of $\boldsymbol{\alpha}$ implies that

$$\boldsymbol{\alpha}'(TQ) = \boldsymbol{\alpha}(Q'T') = \boldsymbol{\alpha}(T') = \boldsymbol{\alpha}'(T).$$

This proves the first part of the proposition. The second part can be obtained analogously.

1.1.12 An ideal norm $\boldsymbol{\alpha}$ is said to be **l_2-stable** if

$$\boldsymbol{\alpha}\big([l_2(\mathbb{I}),T_i]\big) = \sup_{\mathbb{I}} \boldsymbol{\alpha}(T_i)$$

for every $\mathbb{I}$-tuple of operators $T_i \in \mathfrak{L}(X_i,Y_i)$ such that $\sup_{\mathbb{I}} \|T_i\| < \infty$. The definition of the diagonal operator $[l_2(\mathbb{I}),T_i]$ was given in 0.3.4.

1.1.13 PROPOSITION. *If the ideal norm $\boldsymbol{\alpha}$ is l_2-stable, then so is the dual ideal norm $\boldsymbol{\alpha}'$.*

PROOF. Identifying $[l_2(\mathbb{I}),X_i]'$ with $[l_2(\mathbb{I}),X_i']$, we have

$$\boldsymbol{\alpha}'([l_2(\mathbb{I}),T_i]) = \boldsymbol{\alpha}([l_2(\mathbb{I}),T_i]') = \boldsymbol{\alpha}([l_2(\mathbb{I}),T_i']) = \sup_{\mathbb{I}} \boldsymbol{\alpha}(T_i') = \sup_{\mathbb{I}} \boldsymbol{\alpha}'(T_i).$$

1.2 Operator ideals

1.2.1 A subclass $\mathfrak{A}$ of $\mathfrak{L}$ is called an **operator ideal** if the components

$$\mathfrak{A}(X,Y) := \mathfrak{A} \cap \mathfrak{L}(X,Y)$$

satisfy the following conditions:

(1) $S+T \in \mathfrak{A}(X,Y)$ for $S,T \in \mathfrak{A}(X,Y)$.

(2) $BTA \in \mathfrak{A}(X_0,Y_0)$ for $T \in \mathfrak{A}(X,Y)$, $A \in \mathfrak{L}(X_0,X)$ and $B \in \mathfrak{L}(Y,Y_0)$.

Note that

$$\lambda T \in \mathfrak{A}(X,Y) \quad \text{for } T \in \mathfrak{A}(X,Y) \text{ and } \lambda \in \mathbb{K}.$$

1.2.2 An **ideal norm** $\|\cdot|\mathfrak{A}\|$ on an operator ideal $\mathfrak{A}$ is a function that assigns to every operator $T \in \mathfrak{A}$ a non-negative number $\|T|\mathfrak{A}\|$ such that the following conditions are satisfied:

(1) $\|S+T|\mathfrak{A}\| \leq \|S|\mathfrak{A}\| + \|T|\mathfrak{A}\|$ for $S,T \in \mathfrak{A}(X,Y)$.

(2) $\|BTA|\mathfrak{A}\| \leq \|B\|\|T|\mathfrak{A}\|\|A\|$ for $T \in \mathfrak{A}(X,Y)$, $A \in \mathfrak{L}(X_0,X)$ and $B \in \mathfrak{L}(Y,Y_0)$.

(3) $\|T|\mathfrak{A}\| = 0$ implies $T = O$.

Note that

$$\|\lambda T|\mathfrak{A}\| = |\lambda|\|T|\mathfrak{A}\| \quad \text{for } T \in \mathfrak{A}(X,Y) \text{ and } \lambda \in \mathbb{K}.$$

Whenever it is necessary to indicate that the operator T acts from X into Y, then we use the more precise notation $\|T : X \to Y|\mathfrak{A}\|$.

Remark. In 1.1.1 we have already introduced the special concept of an ideal norm defined for *all* operators. Although there is no danger of confusion, different symbols will be used. Namely,

$\boldsymbol{\alpha}, \boldsymbol{\beta}, \ldots$: ideal norms on $\mathfrak{L}$,
$\|\cdot|\mathfrak{A}\|, \|\cdot|\mathfrak{B}\|, \ldots$: ideal norms on $\mathfrak{A}$, $\mathfrak{B}, \ldots$.

To unify the above concepts, any ideal norm given on $\mathfrak{A}$ may be defined on all of $\mathfrak{L}$ by assigning to every operator outside $\mathfrak{A}$ the value $+\infty$. From this point of view and in analogy with measure theory, one could distinguish between **finite** and **extended** ideal norms. But we will not do so.

1.2.3 A **Banach operator ideal** is an operator ideal $\mathfrak{A}$ equipped with an ideal norm $\|\cdot|\mathfrak{A}\|$ such that all components $\mathfrak{A}(X,Y)$ are complete.

1.2.4 The **product** $\mathfrak{B} \circ \mathfrak{A}$ of operator ideals $\mathfrak{A}$ and $\mathfrak{B}$ consists of all operators $T \in \mathfrak{L}(X,Z)$ that can be decomposed in the form $T = BA$ with $A \in \mathfrak{A}(X,Y)$ and $B \in \mathfrak{B}(Y,Z)$, where Y is a suitable Banach space.

1.2.5 Let $\mathfrak{A}$ be a Banach operator ideal. Then

$$\mathfrak{A}' := \left\{ T \in \mathfrak{L} : T' \in \mathfrak{A} \right\}$$

becomes a Banach operator ideal with respect to the ideal norm $\|T|\mathfrak{A}'\| := \|T'|\mathfrak{A}\|$. We refer to $\mathfrak{A}'$ as the **dual Banach operator ideal**.

A Banach operator ideal $\mathfrak{A}$ is said to be **symmetric** if $\mathfrak{A}$ and $\mathfrak{A}'$ coincide. Then the ideal norms $\|\cdot|\mathfrak{A}\|$ and $\|\cdot|\mathfrak{A}'\|$ are equivalent.

1.2.6 A Banach operator ideal $\mathfrak{A}$ is called **injective** if, for every operator $T \in \mathfrak{L}(X,Y)$ and any injection $J \in \mathfrak{L}(Y,Y_0)$, it follows from

$$JT \in \mathfrak{A}(X,Y_0) \quad \text{that} \quad T \in \mathfrak{A}(X,Y).$$

Roughly speaking, this means that the decision whether or not an operator $T \in \mathfrak{L}(X,Y)$ belongs to $\mathfrak{A}$ is independent of the target space Y.

A Banach operator ideal $\mathfrak{A}$ is called **surjective** if, for every operator $T \in \mathfrak{L}(X,Y)$ and any surjection $Q \in \mathfrak{L}(X_0,X)$, it follows from

$$TQ \in \mathfrak{A}(X_0,Y) \quad \text{that} \quad T \in \mathfrak{A}(X,Y).$$

In this case, the decision whether or not an operator $T \in \mathfrak{L}(X,Y)$ belongs to $\mathfrak{A}$ is independent of the source space X.

1.2.7 We have the following analogue of 1.1.11; see [PIE 2, p. 112].

PROPOSITION. *If a Banach operator ideal $\mathfrak{A}$ is injective (surjective), then $\mathfrak{A}'$ is surjective (injective).*

1.2.8 Given any sequence of ideal norms $\boldsymbol{\alpha}_1, \boldsymbol{\alpha}_2, \ldots$, we let

$$\mathfrak{L}[\boldsymbol{\alpha}_n] := \Big\{ T \in \mathfrak{L} : \boldsymbol{\alpha}_n(T) = O(1) \Big\}$$

and

$$\mathfrak{L}_0[\boldsymbol{\alpha}_n] := \Big\{ T \in \mathfrak{L} : \boldsymbol{\alpha}_n(T) = o(1) \Big\}.$$

Define

$$\|T|\mathfrak{L}[\boldsymbol{\alpha}_n]\| := \sup_n \boldsymbol{\alpha}_n(T) \quad \text{for } T \in \mathfrak{L}[\boldsymbol{\alpha}_n].$$

REMARK. Of course, it may happen that $\mathfrak{L}[\boldsymbol{\alpha}_n]$ or $\mathfrak{L}_0[\boldsymbol{\alpha}_n]$ consists of all operators or of zero operators only. These trivial ideals are denoted by $\mathfrak{L}$ and $\mathfrak{O}$, respectively. In order to avoid such situations, the asymptotic behaviour of $(\boldsymbol{\alpha}_n)$ must be appropriate. This can be achieved by passing from $(\boldsymbol{\alpha}_n)$ to $(\lambda_n^{-1}\boldsymbol{\alpha}_n)$, where (λ_n) is a sequence of weights such that

$$\boldsymbol{\alpha}_n(1) := \min\Big\{\boldsymbol{\alpha}(X) : X \in \mathsf{L}\Big\} \le \lambda_n \le \max\Big\{\boldsymbol{\alpha}(X) : X \in \mathsf{L}\Big\} =: \boldsymbol{\alpha}_n(\infty).$$

Of special significance are the limiting cases

$$\Big\{T \in \mathfrak{L} : \boldsymbol{\alpha}_n(T) = O(\boldsymbol{\alpha}_n(1))\Big\}$$

and

$$\Big\{T \in \mathfrak{L} : \boldsymbol{\alpha}_n(T) = o(\boldsymbol{\alpha}_n(\infty))\Big\},$$

while the ideals

$$\Big\{T \in \mathfrak{L} : \boldsymbol{\alpha}_n(T) = o(\boldsymbol{\alpha}_n(1))\Big\}$$

and

$$\Big\{T \in \mathfrak{L} : \boldsymbol{\alpha}_n(T) = O(\boldsymbol{\alpha}_n(\infty))\Big\}$$

are trivial, since they coincide with $\mathfrak{O}$ and $\mathfrak{L}$, respectively.

1.2.9 The next result can be checked by straightforward arguments.

THEOREM. *$\mathfrak{L}[\boldsymbol{\alpha}_n]$ and $\mathfrak{L}_0[\boldsymbol{\alpha}_n]$ are Banach operator ideals.*

REMARK. $\mathfrak{L}_0[\boldsymbol{\alpha}_n]$ is equipped with the ideal norm induced by $\mathfrak{L}[\boldsymbol{\alpha}_n]$.

1.2.10 An operator ideal $\mathfrak{A}$ is said to be **closed** if all components $\mathfrak{A}(X,Y)$ are closed in $\mathfrak{L}(X,Y)$. In this case, $\mathfrak{A}$ is a Banach operator ideal with respect to the usual operator norm.

1.2.11 PROPOSITION. *Assume that $\boldsymbol{\alpha}_n \le 1$. Then the operator ideal $\mathfrak{L}_0[\boldsymbol{\alpha}_n]$ is closed.*

PROOF. Let T be an operator in the closed hull of $\mathfrak{L}_0[\boldsymbol{\alpha}_n](X,Y)$. Then for every $\varepsilon > 0$ there exists $T_0 \in \mathfrak{L}_0[\boldsymbol{\alpha}_n](X,Y)$ with $\|T - T_0\| \le \varepsilon$. Next, we choose n_0 such that

$$\boldsymbol{\alpha}_n(T_0) \le \varepsilon \quad \text{for } n \ge n_0.$$

Consequently,

$$\boldsymbol{\alpha}_n(T) \le \boldsymbol{\alpha}_n(T - T_0) + \boldsymbol{\alpha}_n(T_0) \le \|T - T_0\| + \boldsymbol{\alpha}_n(T_0) \le 2\,\varepsilon$$

for $n \ge n_0$.

1.2.12 In order to show that ideals are different, we may use **diagonal operators** acting on the Banach space $[l_2, l_\infty^{2^k}]$. For any sequence $t = (\tau_k)$ with $\tau_1 \ge \tau_2 \ge \ldots \ge 0$, define

$$D_t : (x_k) \longrightarrow (\tau_k x_k),$$

where $(x_k) \in l_\infty^{2^k}$. Let $\alpha \ge 0$. The most important examples are

$$\begin{aligned} B_\alpha &: (x_k) \longrightarrow ((1+\log k)^{-\alpha} x_k), \\ C_\alpha &: (x_k) \longrightarrow (k^{-\alpha} x_k), \\ D_\alpha &: (x_k) \longrightarrow (2^{-k\alpha} x_k). \end{aligned}$$

We now consider a sequence of l_2-stable ideal norms $\boldsymbol{\alpha}_n$ for which

$$\boldsymbol{\alpha}_n(l_\infty^N) \asymp \min\{f(n), F(N)\}.$$

The functions $f : \mathbb{N} \to \mathbb{R}_+$ and $F : \mathbb{N} \to \mathbb{R}_+$ are subject to the following conditions. We assume that $(\tau_k F(2^k))$ is non-decreasing. Moreover, $\varphi(n) := \min\{k : F(2^k) \ge f(n)\}$ must satisfy the relation $F(2^{\varphi(n)}) \asymp f(n)$. Then

$$\boldsymbol{\alpha}_n(D_t) \asymp \tau_{\varphi(n)} f(n).$$

Indeed, we have

$$\begin{aligned} \boldsymbol{\alpha}_n(D_t) &= \sup_k \left\{ \tau_k \boldsymbol{\alpha}_n(l_\infty^{2^k}) \right\} \asymp \sup_k \left\{ \tau_k \min\{f(n), F(2^k)\} \right\} \\ &= \max\left\{ \sup_{k<\varphi(n)} \{\tau_k F(2^k)\},\ \sup_{k \ge \varphi(n)} \{\tau_k f(n)\} \right\}. \end{aligned}$$

Hence

$$\tau_{\varphi(n)} f(n) \prec \boldsymbol{\alpha}_n(D_t) \prec \max\left\{\tau_{\varphi(n)} F(2^{\varphi(n)}), \tau_{\varphi(n)} f(n)\right\} \prec \tau_{\varphi(n)} f(n).$$

The following results, which will be quoted as $(\triangle)$, are quite useful:

$$\begin{aligned}
&\boldsymbol{\alpha}_n(B_\alpha) \asymp n^{1/2}(1+\log n)^{-\alpha} && \text{if } \boldsymbol{\alpha}_n(l_\infty^N) \asymp \min\{\sqrt{n}, \sqrt{1+\log N}\},\\
&\boldsymbol{\alpha}_n(C_\alpha) \asymp n^{1/2-\alpha} && \text{if } \boldsymbol{\alpha}_n(l_\infty^N) \asymp \min\{\sqrt{n}, \sqrt{1+\log N}\},\\
&\boldsymbol{\alpha}_n(C_\alpha) \asymp n^{1/2}(1+\log n)^{-\alpha} && \text{if } \boldsymbol{\alpha}_n(l_\infty^N) \asymp \min\{\sqrt{n}, \sqrt{N}\},\\
&\boldsymbol{\alpha}_n(C_\alpha) \asymp (1+\log n)^{1/2-\alpha} && \text{if } \boldsymbol{\alpha}_n(l_\infty^N) \asymp \min\{\sqrt{1+\log n}, \sqrt{1+\log N}\},\\
&\boldsymbol{\alpha}_n(D_\alpha) \asymp n^{1/2-\alpha} && \text{if } \boldsymbol{\alpha}_n(l_\infty^N) \asymp \min\{\sqrt{n}, \sqrt{N}\}.
\end{aligned}$$

If the exponent $1/2 - \alpha$ occurs, then it is assumed to be non-negative. That is, $0 \le \alpha \le 1/2$.

1.3 Classes of Banach spaces

1.3.1 With every operator ideal $\mathfrak{A}$ we associate the class of all Banach spaces X such that the identity map I_X belongs to that ideal:

$$\mathsf{A} := \left\{ X \in \mathsf{L} : I_X \in \mathfrak{A} \right\}.$$

Throughout this book, we fix the following principle of notation:

Bold Gothic capitals (possibly decorated with sub- or superscripts) stand for operator ideals, and the associated classes of Banach spaces are denoted by the corresponding bold sans serif capitals:

$$\begin{array}{cccccccc}
\mathfrak{A}, & \mathfrak{B}, & \mathfrak{L}[\boldsymbol{\alpha}_n], & \mathfrak{RT}_p, & \mathfrak{RT}_p^{weak}, & \mathfrak{GR}, & \mathfrak{UMD}, & \dots \\
\mathsf{A}, & \mathsf{B}, & \mathsf{L}[\boldsymbol{\alpha}_n], & \mathsf{RT}_p, & \mathsf{RT}_p^{weak}, & \mathsf{GR}, & \mathsf{UMD}, & \dots .
\end{array}$$

The same terminology will be used when we refer to operators $T \in \mathfrak{A}$ and to Banach spaces $X \in \mathsf{A}$, respectively. For example, $T \in \mathfrak{RT}_p$ and $X \in \mathsf{RT}_p$ are said to have Rademacher type p. Moreover, $T \in \mathfrak{UMD}$ is called a UMD-operator, while $X \in \mathsf{UMD}$ is called a UMD-space.

For a SUMMARY of results we refer to p. 509.

REMARK. It happens quite often that the classes A and B coincide while the underlying operator ideals $\mathfrak{A}$ and $\mathfrak{B}$ are different. This means that the life of operator theorists is much more complicated than that of those who only love spaces.

1.3.2 A class of Banach spaces, denoted by A, is said to be **stable under the formation of l_2-multiples**, if $[l_2(\mathbb{I}), X] \in \mathsf{A}$ whenever $X \in \mathsf{A}$.

REMARK. In general, this property does not mean that $[l_2(\mathbb{I}), X_i] \in \mathsf{A}$ if $X_i \in \mathsf{A}$ for all $i \in \mathbb{I}$. Such an implication only holds under an extra uniformity condition.

1.3.3 The following result is obvious, but nevertheless important.

THEOREM. $\mathsf{L}[\boldsymbol{\alpha}_n]$ *and* $\mathsf{L}_0[\boldsymbol{\alpha}_n]$ *are stable under the formation of finite direct sums and when passing to complemented subspaces.*

(1) *If the ideal norms $\boldsymbol{\alpha}_n$ are injective, then $\mathsf{L}[\boldsymbol{\alpha}_n]$ and $\mathsf{L}_0[\boldsymbol{\alpha}_n]$ are stable when passing to subspaces.*

(2) *If the ideal norms $\boldsymbol{\alpha}_n$ are surjective, then $\mathsf{L}[\boldsymbol{\alpha}_n]$ and $\mathsf{L}_0[\boldsymbol{\alpha}_n]$ are stable when passing to quotient spaces.*

(3) *If the ideal norms $\boldsymbol{\alpha}_n$ are symmetric, then $\mathsf{L}[\boldsymbol{\alpha}_n]$ and $\mathsf{L}_0[\boldsymbol{\alpha}_n]$ are stable when passing to duals, and conversely.*

(4) *If the ideal norms $\boldsymbol{\alpha}_n$ are l_2-stable, then $\mathsf{L}[\boldsymbol{\alpha}_n]$ and $\mathsf{L}_0[\boldsymbol{\alpha}_n]$ are stable under the formation of l_2-multiples.*

1.3.4 A sequence of ideal norms $\boldsymbol{\alpha}_n$ is called **submultiplicative** if

$$\boldsymbol{\alpha}_{mn} \le \boldsymbol{\alpha}_m \circ \boldsymbol{\alpha}_n \quad \text{for } m, n = 1, 2, \ldots .$$

1.3.5 The use of submultiplicativity is based on the following lemma which will be applied in 4.4.1, 4.6.1, 5.6.27, 6.3.10, and 7.7.8.

LEMMA. *Let $(\boldsymbol{\alpha}_n)$ be a submultiplicative sequence of ideal norms. Assume that there exists a uniformly equivalent sequence of ideal norms which is non-decreasing. If*

$$\boldsymbol{\alpha}_{n_0}(X) < n_0^{\lambda}$$

for a fixed $\lambda > 0$ and **some** *$n_0 > 1$, then there exist $\varepsilon > 0$ and $c > 0$ such that*

$$\boldsymbol{\alpha}_n(X) \le c\, n^{\lambda - \varepsilon} \quad \textit{for } \textbf{all } n = 1, 2, \ldots .$$

PROOF. Let $(\boldsymbol{\beta}_n)$ be a non-decreasing sequence of ideal norms with

$$\boldsymbol{\alpha}_n \le \alpha \boldsymbol{\beta}_n \quad \text{and} \quad \boldsymbol{\beta}_n \le \beta \boldsymbol{\alpha}_n,$$

where $\alpha > 0$ and $\beta > 0$. For any Banach space X, we have

$$0 < \boldsymbol{\beta}_1(X) \le \boldsymbol{\beta}_{n^k}(X) \le \beta \boldsymbol{\alpha}_{n^k}(X) \le \beta \boldsymbol{\alpha}_n(X)^k.$$

This implies that

$$\boldsymbol{\alpha}_n(X) \ge 1 \quad \text{for } n = 1, 2, \ldots .$$

By assumption, $\boldsymbol{\alpha}_{n_0}(X) < n_0^\lambda$. Thus $\varepsilon > 0$ can be defined by

$$\boldsymbol{\alpha}_{n_0}(X) = n_0^{\lambda-\varepsilon}.$$

Note that $\lambda - \varepsilon \geq 0$, since $\boldsymbol{\alpha}_{n_0}(X) \geq 1$. Given $n = 1, 2, \ldots$, we choose k such that $n_0^{k-1} \leq n < n_0^k$. Then

$$\begin{aligned}\boldsymbol{\alpha}_n(X) &\leq \alpha\boldsymbol{\beta}_n(X) \leq \alpha\boldsymbol{\beta}_{n_0^k}(X) \leq \alpha\beta\boldsymbol{\alpha}_{n_0^k}(X)\\ &\leq \alpha\beta\boldsymbol{\alpha}_{n_0}(X)^k = \alpha\beta n_0^{\lambda-\varepsilon} n_0^{(k-1)(\lambda-\varepsilon)} \leq \alpha\beta n_0^{\lambda-\varepsilon} n^{\lambda-\varepsilon}.\end{aligned}$$

This is the desired inequality with $c := \alpha\beta n_0^{\lambda-\varepsilon}$.

REMARK. We conclude from $\boldsymbol{\alpha}_{n_0}(X) = n_0^{\lambda-\varepsilon}$ that

$$\varepsilon = \lambda\left[1 - \frac{\log \boldsymbol{\alpha}_{n_0}(X)}{\log n_0^\lambda}\right].$$

Thus, in order to get ε as large as possible, we should try to choose n_0 such that

$$\frac{\log \boldsymbol{\alpha}_{n_0}(X)}{\log n_0^\lambda}$$

becomes as small as possible.

1.3.6 Finally, we provide an auxiliary result which can be proved by a simple, but nice, trick.

LEMMA. *Let $\boldsymbol{\alpha}$ be an ideal norm with the property that*

$$\boldsymbol{\alpha}(l_q^n) \leq f(q)\, n^{\mu/q+\nu} \quad \text{for } n = 1, 2, \ldots$$

where $f : [q_0, \infty) \to (0, \infty)$ is a given function and $2 \leq q_0 \leq q < \infty$. Then

$$\boldsymbol{\alpha}(l_\infty^n) \prec f(1 + \log n)\, n^\nu.$$

PROOF. We have

$$\boldsymbol{\alpha}(l_\infty^n) \leq \|Id : l_\infty^n \to l_q^n\|\, \boldsymbol{\alpha}(l_q^n)\, \|Id : l_q^n \to l_\infty^n\| \leq n^{1/q} f(q)\, n^{\mu/q+\nu}.$$

Letting $q := 1 + \log n \geq q_0$ completes the proof.

2

Ideal norms associated with matrices

This is the last chapter of preliminary nature. We introduce Parseval and Kwapień ideal norms generated by finite matrices. These quantities are used to characterize 2-summing operators and operators factored by a Hilbert space. As an interesting example, we deal with finite Hilbert matrices. For more information, the following books are recommended:

Diestel/Jarchow/Tonge. *Absolutely summing operators* [DIE*a],
Jameson .... *Summing and nuclear norms in Banach space theory* [JAM],
Pietsch................. *Operator ideals* [PIE 2],
Pisier .... *Factorization of linear operators and geometry of Banach spaces* [PIS 1],
Tomczak-Jaegermann... *Banach–Mazur distances and finite dimensional operator ideals* [TOM].

2.1 Matrices

2.1.1 To indicate that (σ_{hk}) is an (m,n)-matrix, we use the notation S_{mn}. The same symbol stands for the associated operator acting from l_2^n into l_2^m. Note that the dual operator S'_{mn} is induced by the **transpose** $S^t_{mn} = (\sigma^t_{hk})$ with $\sigma^t_{hk} := \sigma_{kh}$. The **adjoint matrix** has the form $S^*_{mn} = (\sigma^*_{hk})$, where $\sigma^*_{hk} = \overline{\sigma_{kh}}$. The **rank** of S_{mn} is denoted by $\mathrm{rank}(S_{mn})$. In the case $m = n$, we simply write S_n. In particular, I_n denotes the **unit matrix** of order n.

REMARK. In the present situation, there is no difference between S'_{mn} and S^t_{mn}. However, when the associated operators act between Banach spaces, some care is required; see 0.7.1.

2.1.2 The **operator norm** of S_{mn} is defined by

$$\|S_{mn}\| := \|(\sigma_{hk}) : l_2^n \to l_2^m\|.$$

Furthermore, let

$$|||S_{mn}||| := \|(|\sigma_{hk}|) : l_2^n \to l_2^m\|.$$

The **Hilbert–Schmidt norm** of S_{mn} is given by

$$\|S_{mn}|\mathfrak{S}\| := \left(\sum_{h=1}^{m}\sum_{k=1}^{n}|\sigma_{hk}|^2\right)^{1/2}.$$

2.1.3 We now compare the above matrix norms with each other.

LEMMA. $\|S_{mn}\| \le |||S_{mn}||| \le \|S_{mn}|\mathfrak{S}\| \le \min(\sqrt{m}, \sqrt{n})\|S_{mn}\|$.

PROOF. In view of the Cauchy–Schwarz inequality,

$$\left(\sum_{h=1}^{m}\left|\sum_{k=1}^{n}\sigma_{hk}\xi_k\right|^2\right)^{1/2} \le \left(\sum_{h=1}^{m}\sum_{k=1}^{n}|\sigma_{hk}|^2\sum_{k=1}^{n}|\xi_k|^2\right)^{1/2} \quad \text{for } (\xi_k) \in l_2^n.$$

Thus $\|S_{mn}\| \le \|S_{mn}|\mathfrak{S}\|$. Since the Hilbert–Schmidt norm depends only on $(|\sigma_{hk}|)$, we even have $|||S_{mn}||| \le \|S_{mn}|\mathfrak{S}\|$. Moreover, it follows from

$$\sum_{k=1}^{n}|\sigma_{hk}|^2 \le \|S_{mn}\|^2 \quad \text{for } h = 1, \ldots, m$$

that

$$\|S_{mn}|\mathfrak{S}\|^2 = \sum_{h=1}^{m}\sum_{k=1}^{n}|\sigma_{hk}|^2 \le m\|S_{mn}\|^2.$$

In the same way, we get $\|S_{mn}|\mathfrak{S}\|^2 \le n\|S_{mn}\|^2$. This proves the right-hand inequality.

2.1.4 Next, we establish a special case of the famous **Schur test**.

LEMMA.

$$|||S_{mn}||| \le \left(\max_h \sum_{k=1}^{n}|\sigma_{hk}|\right)^{1/2}\left(\max_k \sum_{h=1}^{m}|\sigma_{hk}|\right)^{1/2}.$$

PROOF. Let $(\xi_k) \in l_2^n$. Then, by the Cauchy–Schwarz inequality,

$$\begin{aligned}\|S_{mn}(\xi_k)|l_2^m\|^2 &= \sum_{h=1}^{m}\left|\sum_{k=1}^{n}\sigma_{hk}\xi_k\right|^2 \le \sum_{h=1}^{m}\left(\sum_{k=1}^{n}|\sigma_{hk}|^{1/2}|\sigma_{hk}|^{1/2}|\xi_k|\right)^2\\ &\le \sum_{h=1}^{m}\left(\sum_{k=1}^{n}|\sigma_{hk}|\right)\left(\sum_{k=1}^{n}|\sigma_{hk}||\xi_k|^2\right)\\ &\le \left(\max_h \sum_{k=1}^{n}|\sigma_{hk}|\right)\sum_{k=1}^{n}\left(\sum_{h=1}^{m}|\sigma_{hk}|\right)|\xi_k|^2\\ &\le \left(\max_h \sum_{k=1}^{n}|\sigma_{hk}|\right)\left(\max_k \sum_{h=1}^{m}|\sigma_{hk}|\right)\|(\xi_k)|l_2^n\|^2.\end{aligned}$$

This proves that

$$\|S_{mn}\| \le \left(\max_h \sum_{k=1}^{n} |\sigma_{hk}|\right)^{1/2} \left(\max_k \sum_{h=1}^{m} |\sigma_{hk}|\right)^{1/2}.$$

The assertion follows, since the right-hand side depends only on $(|\sigma_{hk}|)$.

2.1.5 The set of all **contracting** (m,n)**-matrices**, $\|S_{mn}\| \le 1$, will be denoted by $\mathbb{L}^{mn}$. If $m=n$, then we simply write $\mathbb{L}^n$.

2.1.6 When $m \ge n$, a matrix S_{mn} is said to be **isometric** if

$$\|S_{mn}(\xi_k)|l_2^m\| = \|(\xi_k)|l_2^n\| \quad \text{for } (\xi_k) \in l_2^n.$$

This means that $S_{mn}^* S_{mn} = I_n$. If $m=n$, then we speak of **orthogonal matrices** (real case) and **unitary matrices** (complex case).

The set of all isometric (m,n)-matrices will be denoted by $\mathbb{U}^{mn}$. In the case $m=n$, we use the symbol $\mathbb{U}^n$. Note that $\mathbb{U}^{mn}$ coincides with the set of extreme points of $\mathbb{L}^{mn}$; see [kadi] or [HAL, p. 107]. Therefore $\mathbb{L}^{mn}$ is the convex hull of $\mathbb{U}^{mn}$.

2.1.7 Let S_{mn} be any (m,n)-matrix. For $x_1,\dots,x_n \in X$, we write

$$\|(x_k)|S_{mn}\| := \left(\sum_{h=1}^{m}\left\|\sum_{k=1}^{n}\sigma_{hk}x_k\right\|^2\right)^{1/2}.$$

This expression yields a semi-norm on the n-fold Cartesian power of X. There is no reason to use the more precise symbol $\|(x_k)|[S_{mn}, X]\|$.

2.1.8 Next, we prove an auxiliary result.

LEMMA. *For $x_1,\dots,x_n \in X$ and $x_1^\circ,\dots,x_m^\circ \in X$, the following are equivalent:*

(1) $$\sum_{h=1}^{m} |\langle x_h^\circ, x'\rangle|^2 \le \sum_{k=1}^{n} |\langle x_k, x'\rangle|^2 \quad \textit{for } x' \in X'.$$

(2) *There exists a matrix $S_{mn} = (\sigma_{hk}) \in \mathbb{L}^{mn}$ such that*

$$x_h^\circ = \sum_{k=1}^{n} \sigma_{hk}x_k \quad \textit{for } h = 1,\dots,m.$$

In this case, it can be arranged that

$$\operatorname{rank}(S_{mn}) = \dim[x_1^\circ,\dots,x_m^\circ] \le \dim[x_1,\dots,x_n].$$

PROOF. Let

$$M := \left\{ \, (\langle x_k, x'\rangle) \in l_2^n \; : \; x' \in X' \right\},$$

and denote the embedding map from M into l_2^n by A_n. Then $A_n A_n^*$ is the orthogonal projection from l_2^n onto M. It follows from **(1)** that

$$B_m : (\langle x_k, x' \rangle) \longrightarrow (\langle x_h^\circ, x' \rangle)$$

defines an operator from M into l_2^m with $\|B_m\| \le 1$, and the representing matrix S_{mn} of $B_m A_n^* \in \mathfrak{L}(l_2^n, l_2^m)$ has the required properties. Obviously,

$$\text{rank}(S_{mn}) = \dim[x_1^\circ, \ldots, x_m^\circ] \le \dim[x_1, \ldots, x_n].$$

Conversely, **(2)** implies that

$$\|(\langle x_h^\circ, x' \rangle)|l_2^m\| \le \|S_{mn}\| \, \|(\langle x_k, x' \rangle)|l_2^n\|$$

and

$$\dim[x_1^\circ, \ldots, x_m^\circ] \le \text{rank}(S_{mn}).$$

2.1.9 Finally, we state a special case of the preceding result.

LEMMA. *Let $x_1, \ldots, x_n \in X$ and $m := \dim[x_1, \ldots, x_n]$. Then there exists a matrix $P_n = (\pi_{hk}) \in \mathbb{L}^n$ such that* $\text{rank}(P_n) = m$ *and*

$$x_h = \sum_{k=1}^{n} \pi_{hk} x_k \quad \textit{for} \quad h = 1, \ldots, n.$$

More precisely, we have $P_n = A_{nm} A_{nm}^*$ with $A_{nm} = (\alpha_{hi}) \in \mathbb{U}^{nm}$.

PROOF. Choose any orthonormal basis of

$$M := \Big\{ \big(\langle x_k, x' \rangle \big) \in l_2^n : x' \in X' \Big\},$$

and let $A_{nm} = (\alpha_{hi})$ be the representing matrix of the embedding map from M into l_2^n.

2.2 Parseval ideal norms and 2-summing operators

2.2.1 For $x_1, \ldots, x_n \in X$, we let

$$\|(x_k)|w_2^n\| := \sup\left\{ \left(\sum_{k=1}^{n} |\langle x_k, x' \rangle|^2 \right)^{1/2} : x' \in U_X^\circ \right\},$$

where w stands for *weak*. Occasionally we will use the more precise symbol $\|(x_k)|[w_2^n, X]\|$. If the operator $A_n \in \mathfrak{L}(l_2^n, X)$ is defined by

$$A_n(\xi_k) := \sum_{k=1}^{n} \xi_k x_k \quad \text{for } (\xi_k) \in l_2^n,$$

then

$$A_n' x' = (\langle x_k, x' \rangle) \quad \text{for } x' \in X'.$$

Hence

$$\|(x_k)|w_2^n\| = \|A_n' : X' \to l_2^n\| = \|A_n : l_2^n \to X\|.$$

2.2.2 Let $S_{mn}=(\sigma_{hk})$ be any non-zero (m,n)-matrix. For $T\in\mathfrak{L}(X,Y)$, we denote by $\boldsymbol{\pi}(T|S_{mn})$ the least constant $c\geq 0$ such that

$$\left(\sum_{h=1}^{m}\left\|\sum_{k=1}^{n}\sigma_{hk}Tx_k\right\|^2\right)^{1/2}\leq c\sup\left\{\left(\sum_{k=1}^{n}|\langle x_k,x'\rangle|^2\right)^{1/2}: x'\in U_X^\circ\right\}$$

whenever $x_1,\ldots,x_n\in X$. More concisely, the preceding inequality reads as follows:

$$\|(Tx_k)|S_{mn}\|\leq c\,\|(x_k)|w_2^n\|.$$

We refer to

$$\boldsymbol{\pi}(S_{mn}): T\longrightarrow\boldsymbol{\pi}(T|S_{mn})$$

as a **Parseval ideal norm.**

2.2.3 PROPOSITION. $\|S_{mn}\|\leq\boldsymbol{\pi}(S_{mn})\leq\|S_{mn}|\mathfrak{S}\|$.

PROOF. Given $x_1,\ldots,x_n\in X$, we choose $y'_1,\ldots,y'_m\in Y'$ such that

$$\left\langle\sum_{k=1}^{n}\sigma_{hk}Tx_k,y'_h\right\rangle=\left\|\sum_{k=1}^{n}\sigma_{hk}Tx_k\right\|^2=\|y'_h\|^2\quad\text{for } h=1,\ldots,m.$$

Then

$$\|(Tx_k)|S_{mn}\|=\|(y'_h)|l_2^m\|.$$

Hence

$$\begin{aligned}\|(Tx_k)|S_{mn}\|^2&=\sum_{h=1}^{m}\left\|\sum_{k=1}^{n}\sigma_{hk}Tx_k\right\|^2=\sum_{h=1}^{m}\sum_{k=1}^{n}\sigma_{hk}\langle Tx_k,y'_h\rangle\\&\leq\left(\sum_{h=1}^{m}\sum_{k=1}^{n}|\sigma_{hk}|^2\right)^{1/2}\left(\sum_{h=1}^{m}\sum_{k=1}^{n}|\langle x_k,T'y'_h\rangle|^2\right)^{1/2}\\&\leq\|S_{mn}|\mathfrak{S}\|\,\|(T'y'_h)|l_2^m\|\,\|(x_k)|w_2^n\|\\&\leq\|S_{mn}|\mathfrak{S}\|\,\|T'\|\,\|(y'_h)|l_2^m\|\,\|(x_k)|w_2^n\|\\&=\|S_{mn}|\mathfrak{S}\|\,\|T\|\,\|(Tx_k)|S_{mn}\|\,\|(x_k)|w_2^n\|.\end{aligned}$$

This implies the upper estimate. The lower estimate follows from 1.1.3.

2.2.4 The next result shows that the above inequality is sharp.

PROPOSITION. $\boldsymbol{\pi}(H|S_{mn})=\|S_{mn}|\mathfrak{S}\|$ *for every Hilbert space* H *with* $\dim(H)\geq n$.

PROOF. Let $(x_1,\ldots,x_n)$ be orthonormal. Then $\|(x_k)|w_2^n\|=1$ and

$$\|(x_k)|S_{mn}\|=\left(\sum_{h=1}^{m}\left\|\sum_{k=1}^{n}\sigma_{hk}x_k\right\|^2\right)^{1/2}=\left(\sum_{h=1}^{m}\sum_{k=1}^{n}|\sigma_{hk}|^2\right)^{1/2}=\|S_{mn}|\mathfrak{S}\|.$$

Hence

$$\|S_{mn}|\mathfrak{S}\| \leq \boldsymbol{\pi}(H|S_{mn}).$$

The reverse inequality follows from the previous proposition.

2.2.5 We proceed with a trivial observation.

PROPOSITION. *The ideal norm $\boldsymbol{\pi}(S_{mn})$ is injective.*

REMARK. For $n \geq 2$, Parseval ideal norms are neither surjective nor symmetric.

2.2.6 An operator $T \in \mathfrak{L}(X,Y)$ is called **2-summing** if there exists a constant $c \geq 0$ such that

$$\Big(\sum_{k=1}^{n} \|Tx_k\|^2\Big)^{1/2} \leq c \sup\left\{\Big(\sum_{k=1}^{n} |\langle x_k, x'\rangle|^2\Big)^{1/2} : x' \in U_X^\circ\right\}$$

for $x_1, \ldots, x_n \in X$ and $n = 1, 2, \ldots$. More concisely, the preceding inequality reads as follows:

$$\|(Tx_k)|l_2^n\| \leq c\, \|(x_k)|w_2^n\|.$$

We let

$$\|T|\mathfrak{P}_2\| := \inf c,$$

where the infimum is taken over all constants $c \geq 0$ with the above property.

The class of 2-summing operators is a Banach operator ideal, denoted by $\mathfrak{P}_2$. Using the notation introduced in 1.2.8 and the Parseval ideal norm associated with the unit matrix I_n, we have

$$\mathfrak{P}_2 = \mathfrak{L}[\boldsymbol{\pi}(I_n)] \quad \text{and} \quad \|T|\mathfrak{P}_2\| = \sup_n \boldsymbol{\pi}(T|I_n).$$

REMARK. For more information about 2-summing operators, we refer to [DIE*a], [PIE 2] and [PIE 3].

2.2.7 Let $T \in \mathfrak{L}(H,K)$, where H and K are Hilbert spaces. Assume that $\dim(H) = n$. Then we have

$$\|T|\mathfrak{P}_2\| = \Big(\sum_{k=1}^{n} \|Tu_k\|^2\Big)^{1/2}$$

for every orthonormal system $(u_1, \ldots, u_n)$ in H; see [PIE 3, p. 55]. The same formula does not hold if K is replaced by an arbitrary Banach space. However, N. Tomczak-Jaegermann [tom 2] proved an 'ersatz' which is extremely useful.

LEMMA. *Let $T \in \mathfrak{L}(H, X)$ and $\dim(H) = n$. Then there exists an orthonormal system $(u_1, \ldots, u_n)$ in H such that*

$$\|T|\mathfrak{P}_2\| \le \left(2\sum_{k=1}^{n} \|Tu_k\|^2\right)^{1/2}.$$

PROOF. By homogeneity, we may assume that $\|T|\mathfrak{P}_2\| = 1$. A basic theorem about 2-summing operators yields a factorization

$$T : H \xrightarrow{A} K \xrightarrow{B} X$$

with $\|A|\mathfrak{P}_2\| = 1$ and $\|B\| = 1$; see [PIE 3, p. 58].

By induction, we construct an orthonormal system $(u_1, \ldots, u_n)$ in H such that

$$\|Tu_k\| = \|B_k\|\|Au_k\|$$

for $k = 1, \ldots, n$, where B_k denotes the restriction of B to

$$M_k := \Big\{ Au \in K : (u, u_i) = 0 \text{ for } i = 1, \ldots, k-1 \Big\}.$$

Indeed, if $u_1, \ldots, u_{k-1}$ are already found, then there exists $v_k \in M_k$ with

$$\|Bv_k\| = \|B_k\| \quad \text{and} \quad \|v_k\| = 1.$$

Choosing $u_k \in H$ and $\alpha_k \ge 0$ such that

$$Au_k = \alpha_k v_k, \quad \|u_k\| = 1 \quad \text{and} \quad (u_k, u_i) = 0 \quad \text{for } i = 1, \ldots, k-1,$$

we have

$$Tu_k = \alpha_k Bv_k \quad \text{and} \quad \|Au_k\| = \alpha_k.$$

Hence

$$\|Tu_k\| = \alpha_k \|Bv_k\| = \|Au_k\|\,\|B_k\|.$$

Write

$$\sigma_k := \left(\sum_{i=k}^{n} \alpha_i^2\right)^{1/2} \quad \text{and} \quad \beta_k := \|B_k\|.$$

Obviously,

$$1 = \|A|\mathfrak{P}_2\|^2 = \sum_{k=1}^{n} \|Au_k\|^2 = \sum_{k=1}^{n} \alpha_k^2.$$

We now define on H orthogonal projections P_k and Q_k by

$$P_k u := \sum_{i=1}^{k-1} (u, u_i)\, u_i \quad \text{and} \quad Q_k u := \sum_{i=k}^{n} (u, u_i)\, u_i.$$

Note that $AQ_k u \in M_k$ for $u \in H$. Thus, if $0 < \lambda < 1$, then

$$\begin{aligned} 1 = \|T|\mathfrak{P}_2\| &\le \|BAP_k + (1-\lambda)BAQ_k|\mathfrak{P}_2\| + \|\lambda BAQ_k|\mathfrak{P}_2\| \\ &\le \|B\|\,\|AP_k + (1-\lambda)AQ_k|\mathfrak{P}_2\| + \lambda\|B_k\|\,\|AQ_k|\mathfrak{P}_2\|. \end{aligned}$$

Since

$$\begin{aligned}\|AP_k + (1-\lambda)AQ_k|\mathfrak{P}_2\|^2 &= \sum_{i=1}^{k-1}\|Au_i\|^2 + (1-\lambda)^2\sum_{i=k}^{n}\|Au_i\|^2 \\ &= 1 + (\lambda^2 - 2\lambda)\sigma_k^2\end{aligned}$$

and

$$\|AQ_k|\mathfrak{P}_2\|^2 = \sigma_k^2,$$

we obtain

$$1 \le [1 + (\lambda^2 - 2\lambda)\sigma_k^2]^{1/2} + \lambda\beta_k\sigma_k.$$

It follows from $0 < \lambda < 1$, $0 \le \beta_k \le \|B\| = 1$ and $0 \le \sigma_k \le 1$ that

$$0 \le 1 - \lambda\beta_k\sigma_k \le [1 + (\lambda^2 - 2\lambda)\sigma_k^2]^{1/2}.$$

Hence

$$\lambda^2\sigma_k^2(\beta_k^2 - 1) \le 2\lambda\sigma_k(\beta_k - \sigma_k).$$

If $\sigma_k > 0$, then

$$\lambda\sigma_k(\beta_k^2 - 1) \le 2(\beta_k - \sigma_k).$$

Letting $\lambda \to 0$ yields

$$\sigma_k \le \beta_k.$$

Of course, the same result holds if $\sigma_k = 0$. Put

$$\alpha^2 := \sum_{h \ge k} \alpha_h^2\alpha_k^2.$$

Then

$$2\alpha^2 = \sum_{h=1}^{n}\sum_{k=1}^{n}\alpha_h^2\alpha_k^2 + \sum_{i=1}^{n}\alpha_i^4 > \sum_{h=1}^{n}\alpha_h^2 \cdot \sum_{k=1}^{n}\alpha_k^2 = 1.$$

Hence

$$\begin{aligned}\sum_{k=1}^{n}\|Tu_k\|^2 &= \sum_{k=1}^{n}\|Au_k\|^2\|B_k\|^2 = \sum_{k=1}^{n}\alpha_k^2\beta_k^2 \\ &\ge \sum_{k=1}^{n}\alpha_k^2\sigma_k^2 = \sum_{h\ge k}\alpha_h^2\alpha_k^2 = \alpha^2 > \frac{1}{2}\,.\end{aligned}$$

2.2.8 We are now able to establish the main result of this section.

THEOREM. *Let $T \in \mathfrak{F}(X,Y)$. Then*

$$\|T|\mathfrak{P}_2\| \le \sqrt{2}\,\boldsymbol{\pi}(T|\mathfrak{I}_N) \quad \textit{whenever} \quad \operatorname{rank}(T) = N.$$

PROOF. Given $x_1, \ldots, x_n \in X$, we define $A_n \in \mathfrak{L}(l_2^n, X)$ by

$$A_n(\xi_k) := \sum_{k=1}^{n}\xi_k x_k \quad \text{for } (\xi_k) \in l_2^n.$$

Let $m := \dim[Tx_1, \ldots, Tx_n] \le N$. Then 2.1.9 provides us with a matrix $A_{nm} = (\alpha_{hi}) \in \mathbb{U}^{nm}$ such that

$$Tx_h = \sum_{i=1}^{m}\sum_{k=1}^{n} \alpha_{hi}\alpha_{ik}^* Tx_k \quad \text{for } h = 1, \ldots, n.$$

So we get the factorization

$$TA_n : l_2^n \xrightarrow{A_{nm}^*} l_2^m \xrightarrow{A_{nm}} l_2^n \xrightarrow{A_n} X \xrightarrow{T} Y.$$

Applying the preceding lemma to the operator $TA_nA_{nm} \in \mathfrak{L}(l_2^m, Y)$, we finally arrive at

$$\begin{aligned}\|(Tx_k)|l_2^n\| &= \|(TA_nA_{nm}A_{nm}^* u_k^{(n)})|l_2^n\| \\ &\le \|TA_nA_{nm}|\mathfrak{P}_2\|\,\|(A_{nm}^* u_k^{(n)})|w_2^n\| \le \sqrt{2}\,\boldsymbol{\pi}(TA_nA_{nm}|I_m) \\ &\le \sqrt{2}\,\boldsymbol{\pi}(T|I_N)\|A_n : l_2^n \to X\| = \sqrt{2}\,\boldsymbol{\pi}(T|I_N)\|(x_k)|w_2^n\|.\end{aligned}$$

2.2.9 Let

$$\boldsymbol{\pi}_n(T) := \sup\Big\{\, \boldsymbol{\pi}(T|S_{mn}) : S_{mn} \in \mathbb{L}^{mn} \text{ with } m \ge n \Big\}.$$

For obvious reasons, we refer to

$$\boldsymbol{\pi}_n : T \longrightarrow \boldsymbol{\pi}_n(T)$$

as a **universal Parseval ideal norm**.

An extreme point argument shows that

$$\boldsymbol{\pi}_n(T) = \sup\Big\{\, \boldsymbol{\pi}(T|A_{mn}) : A_{mn} \in \mathbb{U}^{mn} \text{ with } m \ge n \Big\},$$

where $\mathbb{U}^{mn}$ denotes the set of all isometric (m,n)-matrices; see 2.1.6.

2.2.10 PROPOSITION. *The sequence $(\boldsymbol{\pi}_n)$ is non-decreasing.*

2.2.11 PROPOSITION. $\boldsymbol{\pi}_n(T) \le \|T|\mathfrak{P}_2\|$ *for $T \in \mathfrak{P}_2(X,Y)$.*

PROOF. If $S_{mn} \in \mathbb{L}^{mn}$ and $x_1, \ldots, x_n \in X$, then

$$\begin{aligned}\|(Tx_k)|S_{mn}\| = \left\|\left(\sum_{k=1}^{n} \sigma_{hk}Tx_k\right)\Big|l_2^m\right\| &\le \|T|\mathfrak{P}_2\|\left\|\left(\sum_{k=1}^{n} \sigma_{hk}x_k\right)\Big|w_2^m\right\| \\ &\le \|T|\mathfrak{P}_2\|\,\|(x_k)|w_2^n\|.\end{aligned}$$

Hence $\boldsymbol{\pi}(T|S_{mn}) \le \|T|\mathfrak{P}_2\|$.

2.2.12 Replacing $\boldsymbol{\pi}(I_n)$ by $\boldsymbol{\pi}_n$, we get rid of the factor $\sqrt{2}$ in 2.2.8.

PROPOSITION. *Let $T \in \mathfrak{F}(X,Y)$. Then*

$$\|T|\mathfrak{P}_2\| = \boldsymbol{\pi}_N(T) \quad \textit{whenever} \quad \operatorname{rank}(T) = N.$$

PROOF. Given $x_1,\dots,x_n \in X$, we have $m := \dim[Tx_1,\dots,Tx_n] \le N$. Then 2.1.9 yields a matrix $A_{nm} = (\alpha_{hi}) \in \mathbb{U}^{nm}$ such that

$$Tx_h = \sum_{i=1}^{m}\sum_{k=1}^{n} \alpha_{hi}\alpha_{ik}^* Tx_k \quad \text{for } h=1,\dots,n.$$

Let

$$x_i^\circ := \sum_{k=1}^{n} \alpha_{ik}^* x_k.$$

Since

$$Tx_h = \sum_{i=1}^{m} \alpha_{hi} Tx_i^\circ \quad \text{for } h=1,\dots,n,$$

we get

$$\begin{aligned}\|(Tx_h)|l_2^n\| &= \|(Tx_i^\circ)|A_{nm}\| \le \boldsymbol{\pi}(T|A_{nm})\, \|(x_i^\circ)|w_2^m\| \\ &\le \boldsymbol{\pi}_m(T)\, \|(x_k)|w_2^n\| \le \boldsymbol{\pi}_N(T)\, \|(x_k)|w_2^n\|.\end{aligned}$$

This proves that $\|T|\mathfrak{P}_2\| \le \boldsymbol{\pi}_N(T)$. The reverse inequality follows from the previous proposition.

2.2.13 We denote by $\mathrm{DIM}_{\le n}(X)$ the set of all subspaces M of X such that $\dim(M) \le n$. Moreover, J_M^X is the embedding map from M into X.

THEOREM. $\boldsymbol{\pi}_n(T) = \sup\Big\{ \|TJ_M^X|\mathfrak{P}_2\| : M \in \mathrm{DIM}_{\le n}(X) \Big\}$.

PROOF. If $M \in \mathrm{DIM}_{\le n}(X)$, then $\mathrm{rank}(TJ_M^X) \le n$. Hence, in view of the previous proposition,

$$\|TJ_M^X|\mathfrak{P}_2\| = \boldsymbol{\pi}_n(TJ_M^X) \le \boldsymbol{\pi}_n(T).$$

So the right-hand supremum is less than or equal to $\boldsymbol{\pi}_n(T)$.

Let $S_{mn} \in \mathbb{L}^{mn}$. Given $x_1,\dots,x_n \in X$, we denote the linear span by M. Again by the previous proposition,

$$\|(Tx_k)|S_{mn}\| \le \boldsymbol{\pi}(TJ_M^X|S_{mn})\, \|(x_k)|w_2^n\| \le \|TJ_M^X|\mathfrak{P}_2\|\, \|(x_k)|w_2^n\|.$$

Hence

$$\boldsymbol{\pi}(T|S_{mn}) \le \sup\Big\{ \|TJ_M^X|\mathfrak{P}_2\| : M \in \mathrm{DIM}_{\le n}(X) \Big\}.$$

2.2.14 We now combine 2.2.8 and 2.2.13.

THEOREM. $\boldsymbol{\pi}(I_n) \le \boldsymbol{\pi}_n \le \sqrt{2}\,\boldsymbol{\pi}(I_n)$.

2.2.15 Let h^2 be the real 2-dimensional Banach space with the norm

$$\|(\xi,\eta)|h^2\| = \max\{\, |\xi|, \tfrac{1}{2}|\xi+\eta\sqrt{3}|, \tfrac{1}{2}|\xi-\eta\sqrt{3}|\,\}.$$

Then U_{h^2} is a regular hexagon. We consider the identity map from l_2^2 onto h^2. The following example shows that the ideal norms $\boldsymbol{\pi}_n$ and $\boldsymbol{\pi}(I_n)$ are different.

EXAMPLE.

$$\pi(Id : l_2^2 \to h^2|I_2) = \tfrac{1+\sqrt{3}}{2} \quad \textit{and} \quad \pi_2(Id : l_2^2 \to h^2) = \sqrt{2}.$$

PROOF. An extreme point argument yields that

$$\pi(Id : l_2^2 \to h^2|I_2) = \sup_t \Big(\|(\cos t, \sin t)|h^2\|^2 + \|(\sin t, -\cos t)|h^2\|^2 \Big)^{1/2}.$$

This supremum is attained for $t = \frac{\pi}{4}$. On the other hand, taking the matrix

$$S_{3,2} = \sqrt{\tfrac{2}{3}} \begin{pmatrix} \frac{1}{2} & +\sqrt{\frac{3}{4}} \\ 1 & 0 \\ \frac{1}{2} & -\sqrt{\frac{3}{4}} \end{pmatrix}$$

and the unit vectors $(1,0)$ and $(0,1)$, we obtain $\sqrt{2}$ as a lower bound of $\pi_2(Id : l_2^2 \to h^2)$, which completes the proof.

The situation is illustrated by the following figure:

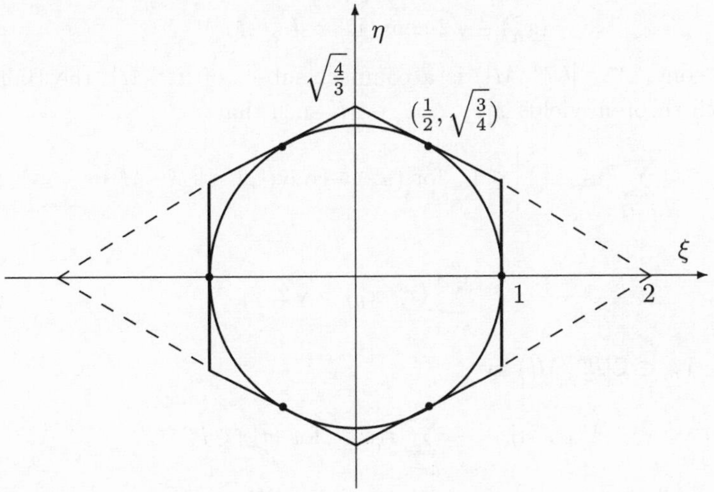

Note that the contact points of the hexagon and the inscribed unit circle have coordinates $(\pm 1, 0)$ and $(\pm\frac{1}{2}, \pm\sqrt{\frac{3}{4}})$.

REMARK. Accidentally, $\pi_2(Id : l_2^2 \to h^2) = \|Id : l_2^2 \to h^2|\mathfrak{P}_2\|$. So we take the opportunity to warn the reader that the universal Parseval ideal norm π_2 should not be mistaken for the 2-summing norm $\|\cdot|\mathfrak{P}_2\|$.

2.2.16 We conclude this section with an auxiliary result which will be used in 3.10.8, 3.10.9 and 9.3.13. Let

$$[U_2^n, X] := \left\{ (x_k) : \sum_{k=1}^{n} \|x_k\|^2 \leq 1 \right\}$$

be the closed unit ball of $[l_2^n, X]$. We denote by $\mathbb{L}^{mn} \circ [U_2^n, X]$ the set of all m-tuples (x_h°) in X which can be written in the form

$$x_h^\circ = \sum_{k=1}^{n} \sigma_{hk} x_k \quad \text{for } h = 1, \ldots, m,$$

where $S_{mn} = (\sigma_{hk}) \in \mathbb{L}^{mn}$ and $(x_k) \in [U_2^n, X]$.

LEMMA. $\mathbb{L}^{mn} \circ [U_2^n, X] \subseteq \sqrt{2}\ \mathrm{conv}(\mathbb{U}^m \circ [U_2^m, X])$.

PROOF. Let $(x_h^\circ) \in \mathbb{L}^{mn} \circ [U_2^n, X]$. Then there are $S_{mn} = (\sigma_{hk}) \in \mathbb{L}^{mn}$ and $(x_k) \in [U_2^n, X]$ such that

$$x_h^\circ = \sum_{k=1}^{n} \sigma_{hk} x_k \quad \text{for } h = 1, \ldots, m.$$

Denote the linear span of $x_1, \ldots, x_n$ by M, and assume that

$$(x_h^\circ) \notin \sqrt{2}\ \mathrm{conv}(\mathbb{L}^m \circ [U_2^m, M]).$$

Since $\mathrm{conv}(\mathbb{L}^m \circ [U_2^m, M])$ is a compact subset of $[l_2^m, M]$, the Hahn–Banach theorem yields $x_1', \ldots, x_m' \in M'$ such that

$$\left| \sum_{h=1}^{m} \langle u_h^\circ, x_h' \rangle \right| \leq 1 \quad \text{for } (u_h^\circ) \in \mathrm{conv}(\mathbb{L}^m \circ [U_2^m, M]) \tag{1}$$

and

$$\sum_{h=1}^{m} \langle x_h^\circ, x_h' \rangle > \sqrt{2}. \tag{2}$$

Define $V_m \in \mathfrak{L}(l_2^m, M')$ by

$$V_m : (\tau_h) \longrightarrow \sum_{h=1}^{m} \tau_h x_h' \quad \text{for } (\tau_h) \in l_2^m.$$

Given $t_1 = (\tau_{h1}), \ldots, t_m = (\tau_{hm}) \in l_2^m$, we let $T_m := (\tau_{hi}) \in \mathfrak{L}(l_2^m)$. Then $\|T_m\| = \|(t_i)|w_2^m\|$. Choose $u_1, \ldots, u_m \in M$ such that

$$\langle u_i, V_m t_i \rangle = \|V_m t_i\|^2 = \|u_i\|^2 \quad \text{for } i = 1, \ldots, m.$$

Setting

$$u_h^\circ := (\|T_m\|\, \|(u_i)|l_2^m\|)^{-1} \sum_{i=1}^{m} \tau_{hi} u_i \quad \text{for } h = 1, \ldots, m,$$

we get $(u_h^\circ) \in \mathbb{L}^m \circ [U_2^m, M]$. Consequently, by (1),

$$\|(u_i)|l_2^m\|^2 = \sum_{i=1}^m \|u_i\|^2 = \sum_{i=1}^m \langle u_i, V_m t_i\rangle = \sum_{i=1}^m \sum_{h=1}^m \tau_{hi} \langle u_i, x_h'\rangle$$
$$= \|T_m\| \, \|(u_i)|l_2^m\| \sum_{h=1}^m \langle u_h^\circ, x_h'\rangle \le \|(u_i)|l_2^m\| \, \|(t_i)|w_2^m\|.$$

This implies that

$$\|(V_m t_i)|l_2^m\| = \|(u_i)|l_2^m\| \le \|(t_i)|w_2^m\|.$$

Hence $\boldsymbol{\pi}(V_m|I_m) \le 1$. So it follows from 2.2.8 that $\|V_m|\mathfrak{P}_2\| \le \sqrt{2}$. Since $V_m u_h^{(m)} = x_h'$, we finally obtain

$$\sum_{h=1}^m \langle x_h^\circ, x_h'\rangle = \sum_{h=1}^m \sum_{k=1}^n \sigma_{hk} \langle x_k, V_m u_h^{(m)}\rangle = \sum_{k=1}^n \Big\langle x_k, V_m \Big[\sum_{h=1}^m \sigma_{hk} u_h^{(m)}\Big]\Big\rangle$$
$$\le \|(x_k)|l_2^n\| \, \Big\|\Big(V_m \Big[\sum_{h=1}^m \sigma_{hk} u_h^{(m)}\Big]\Big)\Big| l_2^n\Big\|$$
$$\le \|V_m|\mathfrak{P}_2\| \, \Big\|\Big(\sum_{h=1}^m \sigma_{hk} u_h^{(m)}\Big)\Big| w_2^m\Big\| = \|V_m|\mathfrak{P}_2\| \, \|S_{mn}^t\| \le \sqrt{2},$$

which contradicts (2). Therefore $(x_h^\circ) \in \sqrt{2}\,\mathrm{conv}(\mathbb{L}^m \circ [U_2^m, M])$. The assertion now follows from the fact that $\mathbb{L}^m$ is the convex hull of $\mathbb{U}^m$.

2.3 Kwapień ideal norms and Hilbertian operators

2.3.1 Let $S_{mn} = (\sigma_{hk})$ be any non-zero (m,n)-matrix. For $T \in \mathfrak{L}(X, Y)$, we denote by $\boldsymbol{\kappa}(T|S_{mn})$ the least constant $c \ge 0$ such that

$$\left(\sum_{h=1}^m \Big\|\sum_{k=1}^n \sigma_{hk} T x_k\Big\|^2\right)^{1/2} \le c \left(\sum_{k=1}^n \|x_k\|^2\right)^{1/2}$$

whenever $x_1, \ldots, x_n \in X$. More concisely, the preceding inequality reads as follows:

$$\|(Tx_k)|S_{mn}\| \le c\, \|(x_k)|l_2^n\|.$$

We refer to

$$\boldsymbol{\kappa}(S_{mn}) : T \longrightarrow \boldsymbol{\kappa}(T|S_{mn})$$

as a **Kwapień ideal norm.**

REMARK. Note that $\boldsymbol{\kappa}(T|S_{mn})$ is just the norm of the operator $[S_{mn}, T]$ from $[l_2^n, X]$ into $[l_2^m, Y]$ defined by

$$[S_{mn}, T] : (x_k) \longrightarrow \left(\sum_{k=1}^n \sigma_{hk} T x_k\right).$$

Consequently, $\boldsymbol{\kappa}(S_{mn})$ is indeed an ideal norm.

2.3.2 PROPOSITION.

$$\|S_{mn}\| \le \boldsymbol{\kappa}(S_{mn}) \le |||S_{mn}||| \le \min(\sqrt{m}, \sqrt{n})\|S_{mn}\|.$$

PROOF. Since $\boldsymbol{\kappa}(I_{\mathbb{K}}|S_{mn}) = \|S_{mn}\|$, we deduce from 1.1.3 that

$$\|S_{mn}\|\|T\| \le \boldsymbol{\kappa}(T|S_{mn}).$$

The other inequalities follow from

$$\sum_{h=1}^{m}\Big\|\sum_{k=1}^{n}\sigma_{hk}Tx_k\Big\|^2 \le \sum_{h=1}^{m}\Big(\sum_{k=1}^{n}|\sigma_{hk}|\|Tx_k\|\Big)^2$$

$$\le |||S_{mn}|||^2\sum_{k=1}^{n}\|Tx_k\|^2 \le |||S_{mn}|||^2\|T\|^2\sum_{k=1}^{n}\|x_k\|^2$$

and 2.1.3.

2.3.3 The next example shows that not only the lower estimate, but also the upper estimate of the preceding proposition is sharp.

EXAMPLE. $\boldsymbol{\kappa}(l_1^n|S_{mn}) = |||S_{mn}|||$.

PROOF. Let $u_1^{(n)}, \dots, u_n^{(n)}$ be the unit vectors in l_1^n, and fix $(\xi_k) \in l_2^n$. Substituting $x_k := \xi_k u_k^{(n)}$ in the defining inequality of $\boldsymbol{\kappa}(l_1^n|S_{mn})$, we get

$$\left[\sum_{h=1}^{m}\Big(\sum_{k=1}^{n}|\sigma_{hk}||\xi_k|\Big)^2\right]^{1/2} = \left(\sum_{h=1}^{m}\Big\|\sum_{k=1}^{n}\sigma_{hk}x_k\Big\|^2\right)^{1/2}$$

$$\le \boldsymbol{\kappa}(l_1^n|S_{mn})\Big(\sum_{k=1}^{n}|\xi_k|^2\Big)^{1/2}.$$

So $|||S_{mn}||| \le \boldsymbol{\kappa}(l_1^n|S_{mn})$. The reverse inequality follows from 2.3.2.

2.3.4 PROPOSITION. $\boldsymbol{\kappa}(H|S_{mn}) = \|S_{mn}\|$ *for all Hilbert spaces H.*

PROOF. Given elements $x_1, \dots, x_n \in H$, we choose an orthonormal system $(u_1, \dots, u_r)$ such that all x_k can be written in the form

$$x_k = \sum_{i=1}^{r}\xi_{ki}u_i.$$

Let

$$\eta_{hi} := \sum_{k=1}^{n}\sigma_{hk}\xi_{ki}$$

for $h = 1, \dots, m$ and $i = 1, \dots, r$. Then

$$\sum_{h=1}^{m}\Big\|\sum_{k=1}^{n}\sigma_{hk}x_k\Big\|^2 = \sum_{h=1}^{m}\sum_{p=1}^{n}\sum_{q=1}^{n}\sigma_{hp}\overline{\sigma_{hq}}(x_p, x_q)$$

$$= \sum_{h=1}^{m}\sum_{i=1}^{r}\Big(\sum_{p=1}^{n}\sigma_{hp}\xi_{pi}\Big)\Big(\sum_{q=1}^{n}\overline{\sigma_{hq}\xi_{qi}}\Big) = \sum_{i=1}^{r}\sum_{h=1}^{m}|\eta_{hi}|^2$$

$$\le \|S_{mn}\|^2 \sum_{i=1}^{r}\sum_{k=1}^{n}|\xi_{ki}|^2 = \|S_{mn}\|^2 \sum_{k=1}^{n}\|x_k\|^2.$$

Hence $\boldsymbol{\kappa}(H|S_{mn}) \le \|S_{mn}\|$. The reverse inequality follows from

$$\|S_{mn}\| = \boldsymbol{\kappa}(I_{\mathbb{K}}|S_{mn}) \le \boldsymbol{\kappa}(H|S_{mn}).$$

2.3.5 The next result can be checked by a standard application of the Hahn–Banach theorem.

PROPOSITION. *Let $T\in\mathfrak{L}(X,Y)$. Then $\boldsymbol{\kappa}(T|S_{mn})$ is the least constant $c\ge 0$ such that*

$$\Big|\sum_{h=1}^{m}\sum_{k=1}^{n}\sigma_{hk}\,\langle Tx_k, y_h'\rangle\Big| \le c\Big(\sum_{k=1}^{n}\|x_k\|^2\Big)^{1/2}\Big(\sum_{h=1}^{m}\|y_h'\|^2\Big)^{1/2}$$

whenever $x_1,\dots,x_n\in X$ and $y_1',\dots,y_m'\in Y'$.

2.3.6 We now establish a consequence of the preceding observation and Helly's lemma 0.1.8.

PROPOSITION. $\boldsymbol{\kappa}'(S_{mn}) = \boldsymbol{\kappa}(S_{mn}^t)$.

PROOF. Let $x_1'',\dots,x_m''\in X''$ and $y_1',\dots,y_n'\in Y'$. Then, for $\varepsilon>0$, there are $x_1,\dots,x_m\in X$ such that

$$\|x_h\| \le (1+\varepsilon)\|x_h''\| \quad\text{and}\quad \Big\langle x_h, \sum_{k=1}^{n}\sigma_{hk}T'y_k'\Big\rangle = \Big\langle x_h'', \sum_{k=1}^{n}\sigma_{hk}T'y_k'\Big\rangle$$

for $h=1,\dots,m$. Hence it follows from

$$\Big|\sum_{h=1}^{m}\sum_{k=1}^{n}\sigma_{hk}\,\langle x_h'', T'y_k'\rangle\Big| = \Big|\sum_{h=1}^{m}\sum_{k=1}^{n}\sigma_{hk}\,\langle Tx_h, y_k'\rangle\Big|$$

$$\le \boldsymbol{\kappa}(T|S_{mn}^t)\Big(\sum_{h=1}^{m}\|x_h\|^2\Big)^{1/2}\Big(\sum_{k=1}^{n}\|y_k'\|^2\Big)^{1/2}$$

$$\le (1+\varepsilon)\,\boldsymbol{\kappa}(T|S_{mn}^t)\Big(\sum_{h=1}^{m}\|x_h''\|^2\Big)^{1/2}\Big(\sum_{k=1}^{n}\|y_k'\|^2\Big)^{1/2}$$

that $\boldsymbol{\kappa}(T'|S_{mn}) \le (1+\varepsilon)\,\boldsymbol{\kappa}(T|S_{mn}^t)$. Letting $\varepsilon\to 0$ yields

$$\boldsymbol{\kappa}(T'|S_{mn}) \le \boldsymbol{\kappa}(T|S_{mn}^t).$$

So we are done, since the inequality $\boldsymbol{\kappa}(T|S_{mn}^t) \le \boldsymbol{\kappa}(T'|S_{mn})$ is obvious.

2.3.7 PROPOSITION. *The ideal norm $\boldsymbol{\kappa}(S_{mn})$ is injective and surjective.*

PROOF. Injectivity is trivial, while surjectivity follows from 1.1.11 and the previous result.

2.3.8 PROPOSITION. *The ideal norm $\boldsymbol{\kappa}(S_{mn})$ is l_2-stable.*

PROOF. For $x_1 = (x_i^{(1)}), \dots, x_n = (x_i^{(n)}) \in [l_2(\mathbb{I}), X_i]$ and $T := [l_2(\mathbb{I}), T_i]$, we have

$$\sum_{h=1}^{m} \Big\| \sum_{k=1}^{n} \sigma_{hk} T x_k \Big| l_2(\mathbb{I}) \Big\|^2 = \sum_{h=1}^{m} \sum_{\mathbb{I}} \Big\| \sum_{k=1}^{n} \sigma_{hk} T_i x_i^{(k)} \Big\|^2$$
$$\leq \sum_{\mathbb{I}} \boldsymbol{\kappa}(T_i|S_{mn})^2 \sum_{k=1}^{n} \|x_i^{(k)}\|^2 \leq \sup_{\mathbb{I}} \boldsymbol{\kappa}(T_i|S_{mn})^2 \sum_{k=1}^{n} \|x_k|l_2(\mathbb{I})\|^2.$$

This proves that

$$\boldsymbol{\kappa}(T|S_{mn}) \leq \sup_{\mathbb{I}} \boldsymbol{\kappa}(T_i|S_{mn}).$$

The reverse estimate follows from $T_i = Q_i^Y T J_i^X$, since $\|Q_i^Y\| \leq 1$ and $\|J_i^X\| \leq 1$; see 0.3.4.

2.3.9 Kwapień ideal norms behave very well with respect to complex interpolation; see Sections 0.6 and 9.1.

PROPOSITION. *Let (X_0, X_1) and (Y_0, Y_1) be interpolation couples. Assume that $T \in \mathfrak{L}(X_0 + X_1, Y_0 + Y_1)$ transforms X_0 into Y_0 and X_1 into Y_1. If $0 < \theta < 1$, then*

$$\boldsymbol{\kappa}(T_\theta|S_{mn}) \leq \boldsymbol{\kappa}(T_0|S_{mn})^{1-\theta} \boldsymbol{\kappa}(T_1|S_{mn})^{\theta}.$$

PROOF. The conclusion follows from 2.3.1 (Remark) and the isometry

$$\Big[l_2^n, [X_0, X_1]_\theta\Big] = \Big[[l_2^n, X_0], [l_2^n, X_1]\Big]_\theta ;$$

see [BER*, pp. 107 and 123].

REMARK. The same result holds if the complex interpolation spaces $[X_0, X_1]_\theta$ and $[Y_0, Y_1]_\theta$ are replaced by their real counterparts $(X_0, X_1)_{\theta,2}$ and $(Y_0, Y_1)_{\theta,2}$; see [BER*, p. 130].

2.3.10 Next, we compare Kwapień ideal norms associated with different non-zero (m, n)-matrices A_{mn} and B_{mn}.

PROPOSITION.

$$\boldsymbol{\kappa}(B_{mn}) \leq \left(1 + \frac{|||A_{mn} - B_{mn}|||}{\|A_{mn}\|}\right) \boldsymbol{\kappa}(A_{mn}).$$

PROOF. By 2.3.2, we have

$$\boldsymbol{\kappa}(T|A_{mn}-B_{mn}) \leq |||A_{mn}-B_{mn}|||\, \|T\| \leq \frac{|||A_{mn}-B_{mn}|||}{\|A_{mn}\|}\boldsymbol{\kappa}(T|A_{mn}).$$

Therefore the assertion follows from

$$\boldsymbol{\kappa}(T|B_{mn}) \leq \boldsymbol{\kappa}(T|A_{mn}) + \boldsymbol{\kappa}(T|A_{mn}-B_{mn}).$$

2.3.11 PROPOSITION.

$$\boldsymbol{\kappa}(A_{mn}) \leq \boldsymbol{\kappa}\begin{pmatrix} A_{mn} B_{mq} \\ C_{pn}\; D_{pq}\end{pmatrix} \leq \sqrt{\boldsymbol{\kappa}(A_{mn})^2 + \boldsymbol{\kappa}(B_{mq})^2 + \boldsymbol{\kappa}(C_{pn})^2 + \boldsymbol{\kappa}(D_{pq})^2}.$$

PROOF. Let $A_{mn}=(\alpha_{hk})$, $B_{mq}=(\beta_{hj})$, $C_{pn}=(\gamma_{ik})$ and $D_{pq}=(\delta_{ij})$. Then, for $x_1 \ldots, x_n \in X$ and $y_1, \ldots, y_q \in X$, we have

$$\left(\sum_{h=1}^{m}\left\|\sum_{k=1}^{n}\alpha_{hk}Tx_k + \sum_{j=1}^{q}\beta_{hj}Ty_j\right\|^2\right)^{1/2} \leq$$

$$\leq \left(\sum_{h=1}^{m}\left[\left\|\sum_{k=1}^{n}\alpha_{hk}Tx_k\right\| + \left\|\sum_{j=1}^{q}\beta_{hj}Ty_j\right\|\right]^2\right)^{1/2}$$

$$\leq \left(\sum_{h=1}^{m}\left\|\sum_{k=1}^{n}\alpha_{hk}Tx_k\right\|^2\right)^{1/2} + \left(\sum_{h=1}^{m}\left\|\sum_{j=1}^{q}\beta_{hj}Ty_j\right\|^2\right)^{1/2}$$

$$\leq \boldsymbol{\kappa}(T|A_{mn})\|(x_k)|l_2^n\| + \boldsymbol{\kappa}(T|B_{mq})\|(y_j)|l_2^q\|$$

$$\leq \Big(\boldsymbol{\kappa}(T|A_{mn})^2 + \boldsymbol{\kappa}(T|B_{mq})^2\Big)^{1/2}\Big(\|(x_k)|l_2^n\|^2 + \|(y_j)|l_2^q\|^2\Big)^{1/2}.$$

Hence

$$\sum_{h=1}^{m}\left\|\sum_{k=1}^{n}\alpha_{hk}Tx_k + \sum_{j=1}^{q}\beta_{hj}Ty_j\right\|^2 \leq$$

$$\leq \Big(\boldsymbol{\kappa}(T|A_{mn})^2 + \boldsymbol{\kappa}(T|B_{mq})^2\Big)\Big(\|(x_k)|l_2^n\|^2 + \|(y_j)|l_2^q\|^2\Big).$$

Analogously, we obtain

$$\sum_{i=1}^{p}\left\|\sum_{k=1}^{n}\gamma_{ik}Tx_k + \sum_{j=1}^{q}\delta_{ij}Ty_j\right\|^2 \leq$$

$$\leq \Big(\boldsymbol{\kappa}(T|C_{pn})^2 + \boldsymbol{\kappa}(T|D_{pq})^2\Big)\Big(\|(x_k)|l_2^n\|^2 + \|(y_j)|l_2^q\|^2\Big).$$

Adding these inequalities yields the right-hand part. The left-hand part is obvious.

2.3.12 An operator $T \in \mathfrak{L}(X,Y)$ is called **Hilbertian** if there exists a factorization $T = BA$ with $A \in \mathfrak{L}(X,H)$ and $B \in \mathfrak{L}(H,Y)$, where H is a Hilbert space. We let

$$\|T|\mathfrak{H}\| := \inf \|B\|\,\|A\|,$$

the infimum being taken over all factorizations described above. In this way, we get a Banach operator ideal, denoted by $\mathfrak{H}$.

The same terminology will be used when we consider Banach spaces belonging to the associated class H. Obviously, a Banach space is **Hilbertian** if and only if it can be renormed into a Hilbert space.

2.3.13 PROPOSITION. *Let $\boldsymbol{\alpha}$ be any ideal norm such that $\boldsymbol{\alpha}(H) \le c$ for all Hilbert spaces H. Then*

$$\boldsymbol{\alpha}(T) \le c\,\|T|\mathfrak{H}\| \quad \textit{whenever } T \in \mathfrak{H}(X,Y).$$

PROOF. Given $\varepsilon>0$, we choose $A \in \mathfrak{L}(X,H)$ and $B \in \mathfrak{L}(H,Y)$ such that

$$T = BA \quad \text{and} \quad \|B\|\,\|A\| \le (1+\varepsilon)\,\|T|\mathfrak{H}\|.$$

Then

$$\boldsymbol{\alpha}(T) \le \|B\|\boldsymbol{\alpha}(H)\|A\| \le (1+\varepsilon)\,c\,\|T|\mathfrak{H}\|.$$

Letting $\varepsilon \to 0$ completes the proof.

REMARK. In particular, by John's theorem 0.2.4, we get $\alpha(X) \le c\sqrt{n}$ for all n-dimensional Banach spaces X.

2.3.14 Next, we establish one of the most famous theorems in Banach space theory. Our approach is adopted from G. Pisier. The basic idea consists in using a special version of the Hahn–Banach theorem; see [bon], [JAM, p. 54] and [pis 20, pp. 254–256].

THEOREM. *Let X_0 be any subspace of X, and denote the embedding map from X_0 into X by J. Let $T_0 \in \mathfrak{L}(X_0, Y)$ and $c \ge 0$. Then the following are equivalent:*

(1) *There exists an extension $T \in \mathfrak{H}(X,Y)$ such that*

$$T_0 = TJ \quad \textit{and} \quad \|T|\mathfrak{H}\| \le c.$$

(2) *Let $S_{mn} = (\sigma_{hk}) \in \mathbb{L}^{mn}$, $x_1, \dots, x_n \in X$, and $m,n = 1,2,\dots$. If*

$$\sum_{k=1}^{n} \sigma_{hk} x_k \in X_0 \quad \textit{for } h = 1,\dots,m,$$

then

$$\left(\sum_{h=1}^{m} \Big\| T_0\Big(\sum_{k=1}^{n} \sigma_{hk} x_k\Big)\Big\|^2\right)^{1/2} \le c\left(\sum_{k=1}^{n} \|x_k\|^2\right)^{1/2}.$$

PROOF. **(1)⇒(2)**: Combine the preceding proposition with 2.3.4.

(2)⇒(1): We denote by F the collection of all functions $f : X' \to \mathbb{R}$ for which there exist $x_1, \ldots, x_n \in X$ such that

$$|f(x')| \leq \sum_{k=1}^{n} |\langle x_k, x' \rangle|^2 \quad \text{whenever } x' \in X'. \tag{P}$$

Let

$$p(f) := c^2 \inf \sum_{k=1}^{n} \|x_k\|^2,$$

where the infimum is taken over all choices of elements $x_1, \ldots, x_n \in X$ with property (P). Furthermore, we denote by F_+ the cone of all non-negative functions $f_+ \in F$. Let

$$q(f_+) := \sup \sum_{h=1}^{m} \|T_0 x_h^\circ\|^2,$$

where the supremum ranges over all $x_1^\circ, \ldots, x_m^\circ \in X_0$ such that

$$f_+(x') \geq \sum_{h=1}^{m} |\langle x_h^\circ, x' \rangle|^2 \quad \text{whenever } x' \in X'. \tag{Q}$$

Assume that (P) and (Q) hold for some $f_+ \in F_+$. Then

$$\sum_{h=1}^{m} |\langle x_h^\circ, x' \rangle|^2 \leq \sum_{k=1}^{n} |\langle x_k, x' \rangle|^2 \quad \text{whenever } x' \in X'.$$

Thus, by 2.1.8 and **(2)**, we have

$$\sum_{h=1}^{m} \|T_0 x_h^\circ\|^2 \leq c^2 \sum_{k=1}^{n} \|x_k\|^2.$$

This implies that

$$q(f_+) \leq p(f_+).$$

Obviously, p is sublinear on F and q is superlinear on F_+. If $f \in F$ and $f_+ \in F_+$, then it follows from

$$q(f_+) \leq p(f_+) \leq p(f + f_+) + p(-f)$$

that

$$-p(-f) \leq p(f + f_+) - q(f_+).$$

Hence the right-hand infimum in the following definition is finite:

$$r(f) := \inf \left\{ p(f + f_+) - q(f_+) : f_+ \in F_+ \right\} \quad \text{for } f \in F.$$

Since r is sublinear on F, the analytic version of the Hahn–Banach theorem provides us with a linear functional $\mu : F \to \mathbb{R}$ such that

$$\mu(f) \leq r(f) \quad \text{for } f \in F.$$

Therefore

$$\mu(f) \le p(f) \quad \text{for } f \in F. \tag{P$_0$}$$

We conclude from

$$-\mu(f_+) = \mu(-f_+) \le r(-f_+) \le p(-f_+ + f_+) - q(f_+)$$

that

$$q(f_+) \le \mu(f_+) \quad \text{for } f_+ \in F_+. \tag{Q$_0$}$$

This shows, in particular, that μ is non-negative.

Let

$$H_0 := \left\{ f : X' \to \mathbb{K} : |f|^2 \in F \right\}.$$

Since

$$2|f\overline{g}| \le |f|^2 + |g|^2 \quad \text{for } f, g \in H_0,$$

we can define

$$(f, g) := \mu(f\overline{g}).$$

In the complex case, the canonical extension of μ to the complexification $F + iF$ is required. Obviously, (f, g) yields a semi-inner product on H_0. Factoring by the null space $N := \{ f \in H_0 : \mu(|f|^2) = 0 \}$ and passing to the completion, we obtain a Hilbert space H with norm

$$|||\hat{f}||| = \mu(|f|^2)^{1/2},$$

where $\hat{f}$ denotes the equivalence class determined by $f \in H_0$.

Given $x \in X$, we let

$$f_x(x') := \langle x, x' \rangle \quad \text{for } x' \in X'.$$

Then $\hat{f}_x \in H$ and, by (P$_0$),

$$|||\hat{f}_x|||^2 = \mu(|f_x|^2) \le p(|f_x|^2) \le c^2 \|x\|^2.$$

Hence $A : x \to \hat{f}_x$ yields an operator $A \in \mathfrak{L}(X, H)$ with $\|A\| \le c$. On the other hand, for $x \in X_0$, it follows from (Q$_0$) that

$$\|T_0 x\|^2 \le q(|f_x|^2) \le \mu(|f_x|^2) = |||\hat{f}_x|||^2.$$

Consequently, $B_0 : \hat{f}_x \to T_0 x$ is a well-defined map from the linear space

$$M_0 := \left\{ \hat{f}_x \in H : x \in X_0 \right\}$$

into Y which admits a unique continuous extension $B := \overline{B_0}$ on the closed hull $M := \overline{M_0}$. Note that $\|B\| \le 1$. Finally, we obtain the required operator by letting

$$T := BQA,$$

where Q is the orthogonal projection from H onto M.

2.3.15 Let

$$\boldsymbol{\kappa}_n(T) := \sup\Big\{\boldsymbol{\kappa}(T|S_n) : S_n \in \mathbb{L}^n\Big\}.$$

For obvious reasons, we refer to

$$\boldsymbol{\kappa}_n : T \to \boldsymbol{\kappa}_n(T)$$

as a **universal Kwapień ideal norm.**

An extreme point argument shows that

$$\boldsymbol{\kappa}_n(T) = \sup\Big\{\boldsymbol{\kappa}(T|S_n) : S_n \in \mathbb{U}^n\Big\},$$

where $\mathbb{U}^n$ denotes the collection of all orthogonal (real case) or unitary (complex case) (n, n)-matrices; see 2.1.6.

2.3.16 It is unknown whether a counterpart of 2.2.14 holds in the setting of Kwapień ideal norms.

PROBLEM. *Do there exist 'universal' matrices $M_n \in \mathbb{L}^n$ such that $\boldsymbol{\kappa}_n \prec \boldsymbol{\kappa}(M_n)$?*

REMARK. Reasoning as in 3.11.9, it turns out that the stronger condition $\boldsymbol{\kappa}_n = \boldsymbol{\kappa}(M_n)$ cannot be fulfilled.

2.3.17 We proceed with an elementary observation.

PROPOSITION. *The sequence $(\boldsymbol{\kappa}_n)$ is non-decreasing.*

2.3.18 Here are some more trivial properties.

PROPOSITION. *The ideal norm $\boldsymbol{\kappa}_n$ is injective, surjective, symmetric and l_2-stable.*

PROOF. Use 2.3.6, 2.3.7 and 2.3.8.

2.3.19 Universal Kwapień ideal norms behave very well with respect to complex interpolation.

PROPOSITION. *Let (X_0, X_1) and (Y_0, Y_1) be interpolation couples, and assume that $T \in \mathfrak{L}(X_0 + X_1, Y_0 + Y_1)$ transforms X_0 into Y_0 and X_1 into Y_1. If $0 < \theta < 1$, then*

$$\boldsymbol{\kappa}_n(T_\theta) \leq \boldsymbol{\kappa}_n(T_0)^{1-\theta}\boldsymbol{\kappa}_n(T_1)^{\theta}.$$

PROOF. In view of

$$\boldsymbol{\kappa}_n(T) = \sup\Big\{\boldsymbol{\kappa}(T|S_n) : S_n \in \mathbb{L}^n\Big\},$$

the assertion follows from 2.3.9.

2.3.20 When the underlying Banach spaces are complex, we get different universal Kwapień ideal norms $\boldsymbol{\kappa}_n^{\mathbb{R}}$ and $\boldsymbol{\kappa}_n^{\mathbb{C}}$ by taking the suprema over all orthogonal and all unitary matrices, respectively.

PROPOSITION. $\boldsymbol{\kappa}_n^{\mathbb{R}} \le \boldsymbol{\kappa}_n^{\mathbb{C}} \le \boldsymbol{\kappa}_{2n}^{\mathbb{R}} \le 2\,\boldsymbol{\kappa}_n^{\mathbb{R}}$.

PROOF. The inequality $\boldsymbol{\kappa}_n^{\mathbb{R}} \le \boldsymbol{\kappa}_n^{\mathbb{C}}$ is trivial. Decompose $C_n = (\gamma_{hk}) \in \mathbb{U}^n$ into real and imaginary parts: $C_n = A_n + iB_n$ with $A_n = (\alpha_{hk})$ and $B_n = (\beta_{hk})$. Then the real $(2n, 2n)$-matrix

$$V_{2n} := \begin{pmatrix} A_n & B_n \\ -B_n & A_n \end{pmatrix}$$

is orthogonal. Given $(x_1, \ldots, x_n)$, we substitute the $2n$-tuple

$$(x_1, \ldots, x_n; ix_1, \ldots, ix_n)$$

in the defining inequality of $\boldsymbol{\kappa}(T|V_{2n})$. This yields

$$\sum_{h=1}^{n}\Big\|\sum_{k=1}^{n}\alpha_{hk}Tx_k+\sum_{k=1}^{n}i\beta_{hk}Tx_k\Big\|^2+\sum_{h=1}^{n}\Big\|\sum_{k=1}^{n}-\beta_{hk}Tx_k+\sum_{k=1}^{n}i\alpha_{hk}Tx_k\Big\|^2 \le$$

$$\le \boldsymbol{\kappa}(T|V_{2n})^2\Big(\sum_{k=1}^{n}\|x_k\|^2+\sum_{k=1}^{n}\|ix_k\|^2\Big).$$

So $\boldsymbol{\kappa}_n^{\mathbb{C}}(T) \le \boldsymbol{\kappa}_{2n}^{\mathbb{R}}(T)$. The inequality $\boldsymbol{\kappa}_{2n}^{\mathbb{R}} \le 2\,\boldsymbol{\kappa}_n^{\mathbb{R}}$ follows from 2.3.11.

2.3.21 We call $n = 1, 2, \ldots$ a **Hadamard number** if there exists an (n, n)-matrix $A_n = (\alpha_{hk})$ such that

$$A_n A_n^t = nI_n \quad \text{and} \quad \alpha_{hk} = \pm 1.$$

Then $n^{-1/2}A_n$ is orthogonal.

Hadamard numbers greater than 2 are necessarily multiples of 4. However, it is a long-standing open problem whether this property also suffices. In any case, all powers of 2 are Hadamard. The required matrices can be obtained by induction:

$$A_{2^{n+1}} := \begin{pmatrix} +A_{2^n} & +A_{2^n} \\ +A_{2^n} & -A_{2^n} \end{pmatrix} \quad \text{and} \quad A_1 := (1);$$

see 6.1.5. For more information, we refer the reader to [WAL*].

2.3.22 PROPOSITION.

$$\boldsymbol{\kappa}_n(l_r^n) \asymp \boldsymbol{\kappa}_n(L_r) \asymp n^{|1/r-1/2|} \qquad \textit{(real case)},$$

$$\boldsymbol{\kappa}_n(l_r^n) = \boldsymbol{\kappa}_n(L_r) = n^{|1/r-1/2|} \qquad \textit{(complex case)}.$$

PROOF. We know from 2.3.2 that $\boldsymbol{\kappa}_n(X) \leq \sqrt{n}$ for any Banach space X. By interpolation, this fact combined with $\boldsymbol{\kappa}_n(L_2) = 1$ yields

$$\boldsymbol{\kappa}_n(L_r) \leq \boldsymbol{\kappa}_n(L_1)^{1-\theta}\, \boldsymbol{\kappa}_n(L_2)^{\theta}$$

whenever $1 \leq r \leq 2$ and $1/r = (1-\theta)/1 + \theta/2$. Hence

$$\boldsymbol{\kappa}_n(l_r^n) \leq \boldsymbol{\kappa}_n(L_r) \leq n^{1/r-1/2}.$$

In the complex case, we use the (n,n)-matrix $E_n^\circ = n^{-1/2}\left(\exp(\frac{2\pi i}{n} hk)\right)$, which is unitary. Then, substituting the unit vectors $u_1^{(n)}, \dots, u_n^{(n)}$ in the defining inequality of $\boldsymbol{\kappa}(l_r^n | E_n^\circ)$, we obtain

$$n^{1/r-1/2} \leq \boldsymbol{\kappa}(l_r^n | E_n^\circ) \leq \boldsymbol{\kappa}_n(l_r^n).$$

In the real case, the same reasoning works for Hadamard numbers, in particular, for $n = 2^k$. If n is arbitrary, we pick k such that $2^k \leq n < 2^{k+1}$. Then

$$\tfrac{1}{\sqrt{2}} n^{1/r-1/2} \leq 2^{k(1/r-1/2)} = \boldsymbol{\kappa}_{2^k}(l_r^{2^k}) \leq \boldsymbol{\kappa}_n(l_r^n).$$

This completes the proof for $1 \leq r \leq 2$. The rest can be settled by duality; see 2.3.18.

2.3.23 We now establish a fundamental criterion that goes back to S. Kwapień; see [kwa 2, p. 586].

THEOREM. *The Banach ideal* $\mathfrak{H}$ *consists of all operators* T *for which* $\sup\limits_n \boldsymbol{\kappa}_n(T)$ *is finite. If so, then this quantity equals* $\|T|\mathfrak{H}\|$.

PROOF. The assertion follows from 2.3.14 by letting $X_0 = X$.

2.3.24 Modifying the condition in the preceding theorem, we are led to the following definition:
Let $1 < p < 2$. An operator T has **weak Kwapień type** p if

$$\|T|\mathfrak{KT}_p^{weak}\| := \sup_n n^{1/2-1/p} \boldsymbol{\kappa}_n(T)$$

is finite. These operators form the Banach ideal

$$\mathfrak{KT}_p^{weak} := \mathfrak{L}[n^{1/2-1/p}\, \boldsymbol{\kappa}_n(T)].$$

REMARKS. The attribute *weak* is added in order to be consistent with the theories of weak Rademacher type, weak Fourier type, weak Haar type, etc. Unfortunately, we have no idea how to define a concept of *ordinary* Kwapień type p. Of course, we could use the condition that

$$\|(Tx_k)|S_n\| \leq c\, \|(x_k)|l_p^n\|$$

for $x_1, \dots, x_n \in X$, all $S_n \in \mathbb{L}^n$ and $n = 1, 2, \dots$; where $c \geq 0$ is a constant; see [pis*2, p. 513]. However, in view of 4.3.6, 5.6.6, 6.3.5 and 7.5.7, it seems that this is not the right approach. Geometric characterizations of Banach spaces belonging to KT_p^{weak} will be given in 3.10.11.

2.3.25 We proceed with an immediate consequence of 2.3.18.

PROPOSITION. *The operator ideal $\mathfrak{KT}_p^{weak}$ is injective, surjective, symmetric and l_2-stable.*

REMARK. In the setting of spaces, the above result implies that the class KT_p^{weak} is stable when passing to subspaces, quotients and duals. We also have stability under the formation of finite direct sums and l_2-multiples.

2.3.26 The next observation follows from 2.3.22.

EXAMPLE.

$$L_p,\, L_{p'} \in \mathsf{KT}_p^{weak} \setminus \mathsf{KT}_{p_0}^{weak} \quad \textit{if} \quad 1 < p < p_0 < 2.$$

2.3.27 An operator T has **Kwapień subtype** if

$$\boldsymbol{\kappa}_n(T) = o(\sqrt{n}).$$

These operators form the closed ideal

$$\mathfrak{KT} := \mathfrak{L}_0[n^{-1/2}\boldsymbol{\kappa}_n].$$

2.3.28 We now state a supplement to 2.3.25.

PROPOSITION. *The operator ideal $\mathfrak{KT}$ is injective, surjective, symmetric and l_2-stable.*

2.3.29 The following problem, which has been open for a long time, was raised by G. Pisier [pis 8].

PROBLEM. *Is it true that* $\mathsf{KT} = \bigcup_{1<p<2} \mathsf{KT}_p^{weak}$ *?*

2.4 Ideal norms associated with Hilbert matrices

CONVENTION. The dash in the symbol $\sum_{k=1}^{n}{}'$ indicates that the sum ranges over all k except $k = h$, where h is a given index.

2.4.1 Define

$$\chi_m := \begin{cases} \frac{1}{m} & \text{if } m \neq 0, \\ 0 & \text{if } m = 0. \end{cases}$$

Then we refer to $H_n := \frac{1}{\pi}(\chi_{h-k})$ and $K_n := \frac{1}{\pi}(\chi_{h+k-1})$ as the n-th **Hilbert matrices**. In order to get $H_1 \neq O$, we let $H_1 := (\frac{1}{\pi})$.

2.4.2 First of all, we recall a classical result.

LEMMA.

$\frac{1}{\pi} \le \|H_n\| \le 1, \quad \frac{1}{\pi} \le |||H_n||| \le 1+\log n \quad$ *and* $\quad \frac{1}{\pi} \le \|K_n\| = |||K_n||| \le 1.$

PROOF. Thanks to Schur's test 2.1.4, the second inequality is implied by

$$\frac{1}{\pi}\sum_{k=1}^{n}{}' \frac{1}{|h-k|} < \frac{2}{\pi}\sum_{k=1}^{n}\frac{1}{k} < 1+\log n.$$

The first and the third inequality can be found in [HAR*, p. 212].

2.4.3 The next result follows from 2.3.2 and 2.4.2.

PROPOSITION. $\boldsymbol{\kappa}(H_n) \le 1+\log n$ *and* $\boldsymbol{\kappa}(K_n) \le 1.$

REMARK. We see from the previous proposition that the ideal norms $\boldsymbol{\kappa}(K_n)$ deserve no further interest, since they are uniformly equivalent to the operator norm. Thus, from now on, we concentrate on the study of $\boldsymbol{\kappa}(H_n)$.

2.4.4 The following example shows that the estimate of $\boldsymbol{\kappa}(H_n)$ given above is asymptotically best possible. For the definition of the finite summation operator Σ_n we refer to 0.7.3.

EXAMPLE. $\boldsymbol{\kappa}(\Sigma_n : l_1^n \to l_\infty^n | H_n) \asymp 1+\log n.$

PROOF. Let $u_1^{(n)}, \dots, u_n^{(n)}$ denote the unit vectors in l_1^n, and write

$$c_n := \pi\, \boldsymbol{\kappa}(\Sigma_n : l_1^n \to l_\infty^n | H_n).$$

Then we have

$$\sum_{h=1}^{n}\Big\| \sum_{k=1}^{n}{}' \frac{1}{h-k}\Sigma_n u_k^{(n)} \Big| l_\infty^n \Big\|^2 \le c_n^2 \sum_{k=1}^{n} \|u_k^{(n)}|l_1^n\|^2.$$

Since

$$\Sigma_n u_k^{(n)} = \sum_{l=k}^{n} u_l^{(n)},$$

it follows that

$$\sum_{h=1}^{n} \max_{1\le l\le n}\Big| \sum_{k=1}^{l}{}' \frac{1}{h-k}\Big|^2 = \sum_{h=1}^{n}\Big\| \sum_{l=1}^{n}\Big(\sum_{k=1}^{l}{}' \frac{1}{h-k}\Big) u_l^{(n)} \Big| l_\infty^n \Big\|^2 \le n\, c_n^2.$$

Next, we conclude from

$$\max_{1\le l\le n}\Big| \sum_{k=1}^{l}{}' \frac{1}{h-k}\Big| \ge \sum_{k=1}^{h}{}' \frac{1}{h-k} = \sum_{m=1}^{h-1}\frac{1}{m}$$

that

$$n^{1/2}\Big(\sum_{h=1}^{n}\max_{1\le l\le n}\Big|{\sum_{k=1}^{l}}'\frac{1}{h-k}\Big|^2\Big)^{1/2}\ge\sum_{h=1}^{n}\max_{1\le l\le n}\Big|{\sum_{k=1}^{l}}'\frac{1}{h-k}\Big|$$
$$\ge\sum_{m=1}^{n-1}\frac{n-m}{m}=n\sum_{m=1}^{n-1}\frac{1}{m}-n+1\succ n(1+\log n).$$

So $1+\log n\prec c_n$. The reverse estimate is obvious, by 2.4.3.

2.4.5 PROPOSITION. *The sequence $(\boldsymbol{\kappa}(H_n))$ is non-decreasing. Moreover, we have*

$$\boldsymbol{\kappa}(H_{2n})\le 5\,\boldsymbol{\kappa}(H_n).$$

PROOF. Let $\widetilde{K}_n=\frac{1}{\pi}(\widetilde{\kappa}_{hk})$ be the (n,n)-matrix obtained from K_n by reversing the order of the columns. That is,

$$\widetilde{\kappa}_{hk}:=\frac{1}{h+n-k}.$$

Since

$$H_{2n}=\begin{pmatrix}H_n & -\widetilde{K}_n^t\\ \widetilde{K}_n & H_n\end{pmatrix}\quad\text{and}\quad\boldsymbol{\kappa}(\widetilde{K}_n)=\boldsymbol{\kappa}(\widetilde{K}_n^t)=\boldsymbol{\kappa}(K_n)\le 1\le\pi\,\boldsymbol{\kappa}(H_n)$$

it follows from 2.3.11 that

$$\boldsymbol{\kappa}(H_{2n})\le\sqrt{2+2\pi^2}\,\boldsymbol{\kappa}(H_n)\le 5\,\boldsymbol{\kappa}(H_n).$$

2.4.6 PROPOSITION. *The ideal norm $\boldsymbol{\kappa}(H_n)$ is symmetric.*

PROOF. The assertion follows from 2.3.6 and $H_n^t=-H_n$.

2.4.7 We refer to T as an **HM-operator** if

$$\|T|\mathfrak{HM}\|:=\sup_n\boldsymbol{\kappa}(T|H_n)$$

is finite. These operators form the Banach ideal

$$\mathfrak{HM}:=\mathfrak{L}[\boldsymbol{\kappa}(H_n)].$$

REMARK. It will turn out in 5.4.19 that $\mathfrak{HM}$ is nothing but the class of operators compatible with all different kinds of Hilbert transforms.

2.4.8 Next, we state a corollary of 2.3.7, 2.3.8 and 2.4.6.

PROPOSITION. *The operator ideal $\mathfrak{HM}$ is injective, surjective, symmetric and l_2-stable.*

REMARK. In the setting of spaces, the above result implies that the class HM is stable when passing to subspaces, quotients and duals. We also have stability under the formation of finite direct sums and l_2-multiples.

2.4.9 Quasi-HM-operators are defined by the property

$$\boldsymbol{\kappa}(T|H_n) = o(1+\log n).$$

They form the closed ideal

$$\mathfrak{QHM} := \mathfrak{L}_0[(1+\log n)^{-1}\,\boldsymbol{\kappa}(H_n)].$$

REMARK. At present, we have no interesting results about this ideal apart from the fact that it consists of super weakly compact operators; see 8.5.24.

2.4.10 EXAMPLE.

$$\boldsymbol{\kappa}(l_1^n|H_n) = \boldsymbol{\kappa}(l_\infty^n|H_n) \asymp 1+\log n,$$
$$\boldsymbol{\kappa}(L_1|H_n) = \boldsymbol{\kappa}(L_\infty|H_n) \asymp 1+\log n.$$

PROOF. By 2.4.4, we conclude from

$$\boldsymbol{\kappa}(\Sigma_n : l_1^n \to l_\infty^n|H_n) \le \boldsymbol{\kappa}(l_1^n|H_n)\|\Sigma_n : l_1^n \to l_\infty^n\|$$

that

$$1+\log n \prec \boldsymbol{\kappa}(l_1^n|H_n) \le \boldsymbol{\kappa}(L_1|H_n).$$

The upper estimate follows from 2.4.3, and L_∞ can be treated by duality.

2.4.11 We now state a deep result which is closely related to the famous M. Riesz theorem. In our presentation it is obtained as a corollary of 5.4.19 and 5.4.20.

EXAMPLE. $\boldsymbol{\kappa}(L_r|H_n) \asymp 1$ *if* $1 < r < \infty$.

2.4.12 The preceding observations can be summarized as follows.

EXAMPLE. $L_r \in \mathsf{HM}$ *if* $1 < r < \infty$, *and* $L_1, L_\infty \notin \mathsf{QHM}$.

2.4.13 Next, we provide an auxiliary result which will be used in the subsequent proof. The operator $\Sigma_n^{1/2}$ was defined in 0.7.5.

EXAMPLE.

$$\|\Sigma_n^{1/2} : l_1^n \to l_2^n\| \le \sqrt{1+\log n} \quad \textit{and} \quad \|\Sigma_n^{1/2} : l_2^n \to l_\infty^n\| \le \sqrt{1+\log n}.$$

PROOF. Note that

$$\sigma_0(\tfrac{1}{2}) = 1 \quad \text{and} \quad \sigma_k(\tfrac{1}{2}) = (-1)^k\binom{-\frac{1}{2}}{k} = \tfrac{1\cdot 3\cdot 5\cdot\ldots\cdot(2k-1)}{2^k\cdot 1\cdot 2\cdot 3\cdot\ldots\cdot k} \quad \text{for } k = 1, 2, \ldots.$$

Hence

$$\sigma_k(\tfrac{1}{2}) = \tfrac{2k-1}{2k}\sigma_{k-1}(\tfrac{1}{2}).$$

Now it follows by induction that

$$|\sigma_k(\tfrac{1}{2})| \le \frac{1}{\sqrt{k+1}} \quad \text{for } k = 0, 1, 2, \ldots.$$

Therefore

$$\|\Sigma_n^{1/2} : l_1^n \to l_2^n\| = \max_k \|\Sigma_n^{1/2}(u_k^{(n)})|l_2^n\| = \left(\sum_{k=0}^{n-1} |\sigma_k(\tfrac{1}{2})|^2\right)^{1/2}$$
$$\leq \left(\sum_{k=1}^{n} \frac{1}{k}\right)^{1/2} \leq \sqrt{1+\log n}.$$

By duality, the same can be shown for $\|\Sigma_n^{1/2} : l_2^n \to l_\infty^n\|$.

2.4.14 EXAMPLE. $\|\Sigma_n : l_1^n \to l_\infty^n | \mathfrak{H}\| \asymp 1+\log n.$

PROOF. We consider the factorization

$$\Sigma_n : l_1^n \xrightarrow{\ \Sigma_n^{1/2}\ } l_2^n \xrightarrow{\ \Sigma_n^{1/2}\ } l_\infty^n.$$

Then 2.4.13 implies that

$$\|\Sigma_n : l_1^n \to l_\infty^n|\mathfrak{H}\| \leq \|\Sigma_n^{1/2} : l_1^n \to l_2^n\|\,\|\Sigma_n^{1/2} : l_2^n \to l_\infty^n\| \leq 1+\log n.$$

Since $\boldsymbol{\kappa}(T|H_n) \leq \|T|\mathfrak{H}\|$ for all T, we obtain from 2.4.4 and 2.3.13 that

$$1+\log n \asymp \boldsymbol{\kappa}(\Sigma_n : l_1^n \to l_\infty^n|H_n) \leq \|\Sigma_n : l_1^n \to l_\infty^n|\mathfrak{H}\|.$$

2.4.15 Finally, we show that the scale $\mathsf{L}[(1+\log n)^{-\theta}\,\boldsymbol{\kappa}(H_n)]$ with $0 \leq \theta \leq 1$ is strictly increasing.

EXAMPLE. *For* $0 < \theta < 1$, *there exists a Banach space* X_θ *such that*

$$\boldsymbol{\kappa}(X_\theta|H_n) \asymp (1+\log n)^\theta.$$

PROOF. Using the operators $\Sigma_n^{1/2}$ and $\Sigma_n^{-1/2}$, we define on $\mathbb{K}^n$ the norms

$$\|(\xi_k)|\Sigma_n^{1/2}(l_1^n)\| := \|\Sigma_n^{-1/2}(\xi_k)|l_1^n\|$$

and

$$\|(\xi_k)|\Sigma_n^{-1/2}(l_\infty^n)\| := \|\Sigma_n^{1/2}(\xi_k)|l_\infty^n\|.$$

This transfer yields the Banach spaces $\Sigma_n^{1/2}(l_1^n)$ and $\Sigma_n^{-1/2}(l_\infty^n)$, and we get the diagram

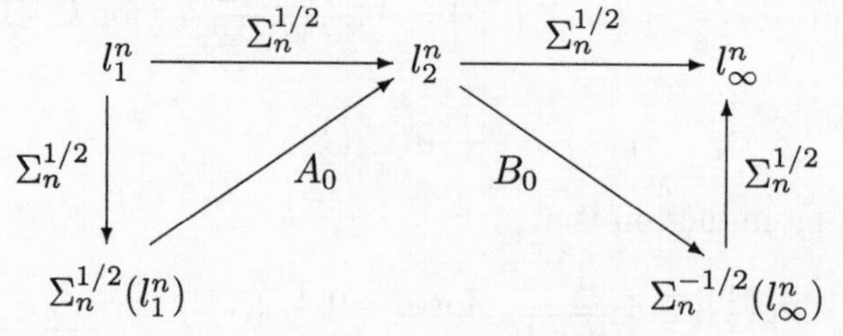

,

in which A_0 and B_0 denote the identity maps from $\Sigma_n^{1/2}(l_1^n)$ into l_2^n and from l_2^n into $\Sigma_n^{-1/2}(l_\infty^n)$, respectively. By 2.4.13, it follows that

$$\|(\xi_k)|l_2^n\| \le \|\Sigma_n^{1/2} : l_1^n \to l_2^n\| \|\Sigma_n^{-1/2}(\xi_k)|l_1^n\| \le \sqrt{1+\log n}\, \|(\xi_k)|\Sigma_n^{1/2}(l_1^n)\|$$

and

$$\begin{aligned}\|(\xi_k)|\Sigma_n^{-1/2}(l_\infty^n)\| &= \|\Sigma_n^{1/2}(\xi_k)|l_\infty^n\| \le \|\Sigma_n^{1/2} : l_2^n \to l_\infty^n\| \|(\xi_k)|l_2^n\| \\ &\le \sqrt{1+\log n}\, \|(\xi_k)|l_2^n\|.\end{aligned}$$

Thus we have

$$\|A_0 : \Sigma_n^{1/2}(l_1^n) \to l_2^n\| \le \sqrt{1+\log n}$$

and

$$\|B_0 : l_2^n \to \Sigma_n^{-1/2}(l_\infty^n)\| \le \sqrt{1+\log n}.$$

Moreover, $\|\Sigma_n : l_1^n \to l_\infty^n\| = 1$ implies that

$$\begin{aligned}\|(\xi_k)|\Sigma_n^{-1/2}(l_\infty^n)\| &= \|\Sigma_n^{1/2}(\xi_k)|l_\infty^n\| \le \|\Sigma_n : l_1^n \to l_\infty^n\|\, \|\Sigma_n^{-1/2}(\xi_k)|l_1^n\| \\ &\le \|(\xi_k)|\Sigma_n^{1/2}(l_1^n)\|.\end{aligned}$$

Hence the norm of the identity map from $\Sigma_n^{1/2}(l_1^n)$ into $\Sigma_n^{-1/2}(l_\infty^n)$ is less than or equal to 1. Therefore we can choose an intermediate Banach space X_1^n such that

$$\|A_1 : \Sigma_n^{1/2}(l_1^n) \to X_1^n\| \le 1 \quad \text{and} \quad \|B_1 : X_1^n \to \Sigma_n^{-1/2}(l_\infty^n)\| \le 1,$$

where A_1 and B_1 denote identity maps.

Writing X_0^n instead of l_2^n, the preceding considerations are summarized in the diagrams

$$\Sigma_n^{1/2}(l_1^n) \xrightarrow{A_0} X_0^n \xrightarrow{B_0} \Sigma_n^{-1/2}(l_\infty^n)$$

and

$$\Sigma_n^{1/2}(l_1^n) \xrightarrow{A_1} X_1^n \xrightarrow{B_1} \Sigma_n^{-1/2}(l_\infty^n).$$

Complex interpolation yields a Banach space $X_\theta^n := [X_0^n, X_1^n]_\theta$, and we have

$$\Sigma_n^{1/2}(l_1^n) \xrightarrow{A_\theta} X_\theta^n \xrightarrow{B_\theta} \Sigma_n^{-1/2}(l_\infty^n)$$

with

$$\big\|A_\theta : \Sigma_n^{1/2}(l_1^n) \to X_\theta^n\big\| \le (1+\log n)^{(1-\theta)/2}$$

and

$$\big\|B_\theta : X_\theta^n \to \Sigma_n^{-1/2}(l_\infty^n)\big\| \le (1+\log n)^{(1-\theta)/2}.$$

By 2.3.4, 2.4.2 and 2.4.3,

$$\boldsymbol{\kappa}(X_0^n|H_m) \le 1 \quad \text{and} \quad \boldsymbol{\kappa}(X_1^n|H_m) \le 1+\log m.$$

It now follows from 2.3.9 that

$$\boldsymbol{\kappa}(X_\theta^n|H_m) \le c_\theta(1+\log m)^\theta.$$

for $m,n=1,2,\ldots$. On the other hand, by 2.4.4 and the preceding estimates,

$$\begin{aligned} c_0(1+\log n) &\le \boldsymbol{\kappa}(\Sigma_n : l_1^n \to l_\infty^n | H_n) \\ &= \boldsymbol{\kappa}\big(B_\theta A_\theta : \Sigma_n^{1/2}(l_1^n) \to \Sigma_n^{-1/2}(l_\infty^n) \big| H_n\big) \\ &\le \big\| A_\theta : \Sigma_n^{1/2}(l_1^n) \to X_\theta^n \big\| \, \boldsymbol{\kappa}(X_\theta^n | H_n) \, \big\| B_\theta : X_\theta^n \to \Sigma_n^{-1/2}(l_\infty^n) \big\| \\ &\le (1+\log n)^{1-\theta} \boldsymbol{\kappa}(X_\theta^n | H_n) \end{aligned}$$

yields

$$c_0(1+\log n)^\theta \le \boldsymbol{\kappa}(X_\theta^n | H_n).$$

Finally, let $X_\theta := [l_2, X_\theta^m]$. Then we know from 2.3.8 that

$$\boldsymbol{\kappa}(X_\theta | H_n) = \sup_m \boldsymbol{\kappa}(X_\theta^m | H_n).$$

Hence

$$c_0(1+\log n)^\theta \le \boldsymbol{\kappa}(X_\theta^n | H_n) \le \boldsymbol{\kappa}(X_\theta | H_n) \le c_\theta (1+\log n)^\theta.$$

This proves that X_θ has the desired property.

REMARK. By a proper choice of X_1^n, we can even arrange that X_θ has Rademacher type 2 and Rademacher cotype q, where $q>2$ is prescribed; see [pis*1].

3

Ideal norms associated with orthonormal systems

Having finished the preliminaries, we now come to the proper object of this book. Our first concern is to collect some elementary facts about ideal norms generated from arbitrary orthonormal systems. These are the Riemann ideal norms $\varrho(\mathcal{B}_n, \mathcal{A}_n)$ and the Dirichlet ideal norms $\delta(\mathcal{B}_n, \mathcal{A}_n)$. Of particular interest are type and cotype ideal norms. The main purpose of such a general theory is to provide a unified language. In this way, we are able to elaborate parallel results for Rademacher functions, Gaussian random variables, trigonometric functions, Walsh functions and Haar functions.

It will turn out that orthonormal systems consisting of characters on a compact Abelian group can be handled much more easily by using the underlying algebraic structure. Other useful properties of orthonormal systems are unimodularity and unconditionality. We also describe a useful tensor product technique.

Local considerations in Banach spaces stress a special point of view in the theory of orthonormal systems. Typically, we deal with inequalities for systems of n functions and look at the asymptotic behaviour of the constants involved as $n \to \infty$. Classical examples are Khintchine inequalities.

For more information about orthogonal systems of functions, we refer to the following books:

Hewitt/Ross.......... *Abstract harmonic analysis, vols. I and II* [HEW*1, HEW*2],
Kaczmarz/Steinhaus... *Theorie der Orthogonalreihen* [KAC*],
Kashin/Saakyan....... *Orthogonal series* [KAS*],
Olevskii............... *Fourier series with respect to general orthonormal systems* [OLE],
Rudin.................. *Fourier analysis on groups* [RUD].

3.1 Orthonormal systems

3.1.1 Throughout, $\mathcal{A}_n = (a_1, \ldots, a_n)$ is an **orthonormal system** in a Hilbert space $L_2(M, \mu)$.

In most cases, $\mathcal{A}_n$ will be the n-th section of an orthonormal sequence $\mathcal{A} = (a_1, a_2, \ldots)$. More generally, given an orthonormal system $\mathcal{A} = (a_i)$ and a finite subset $\mathbb{F}$ of the index set $\mathbb{I}$, we denote by $\mathcal{A}(\mathbb{F})$ the subsystem consisting of all a_i with $i \in \mathbb{F}$. We call $\overline{\mathcal{A}}_n = (\overline{a_1}, \ldots, \overline{a_n})$ the **conjugate orthonormal system**.

3.1.2 For $\boldsymbol{f} \in [L_2(M, \mu), X]$, the i-th **Fourier coefficient** with respect to an orthonormal system $\mathcal{A} = (a_i)$ is defined by

$$\langle \boldsymbol{f}, \overline{a_i} \rangle = (\boldsymbol{f}, a_i) := \int_M \boldsymbol{f}(s) \overline{a_i(s)} \, d\mu(s).$$

The set of all $i \in \mathbb{I}$ with $\langle \boldsymbol{f}, \overline{a_i} \rangle \neq 0$ is said to be the **$\mathcal{A}$-spectrum** of $\boldsymbol{f}$.

3.1.3 Assume that the orthonormal system $\mathcal{A}_n = (a_1, \ldots, a_n)$ belongs to $L_2(M, \mu) \cap L_r(M, \mu)$ with $1 \leq r < \infty$.

For $x_1, \ldots, x_n \in X$, we consider the r-th **average**

$$\|(x_k)|\mathcal{A}_n\|_r := \Big\| \sum_{k=1}^n a_k \otimes x_k \Big| L_r \Big\| = \left(\int_M \Big\| \sum_{k=1}^n x_k a_k(s) \Big\|^r d\mu(s) \right)^{1/r}.$$

In the limiting case $r = \infty$, the usual modification is required. Whenever it is necessary to indicate that $x_1, \ldots, x_n$ are elements of the Banach space X, we use the more precise notation $\|(x_k)|[\mathcal{A}_n, X]\|_r$. The expression $\|(x_k)|\mathcal{A}_n\|_r$ yields a norm on the n-fold Cartesian power of X.

This book mainly deals with the quadratic case. Therefore, to simplify notation, we write $\|(x_k)|\mathcal{A}_n\|$ instead of $\|(x_k)|\mathcal{A}_n\|_2$.

REMARK. Occasionally, we will need the **Lorentz average**

$$\|(x_k)|\mathcal{A}_n\|_{r,w} := \Big\| \sum_{k=1}^n a_k \otimes x_k \Big| L_{r,w} \Big\|.$$

3.1.4 If H is a Hilbert space, then so is $[L_2(M, \mu), H]$; see 0.4.8. Using this fact, easy manipulations yield **Parseval's equality**

$$\int_M \Big\| \sum_{k=1}^n x_k a_k(s) \Big\|^2 d\mu(s) = \sum_{k=1}^n \|x_k\|^2 \quad \text{for } x_1, \ldots, x_n \in H.$$

Thus, in contrast to the situation in arbitrary Banach spaces, the average $\|(x_k)|\mathcal{A}_n\|$ does not depend on the orthonormal system $\mathcal{A}_n$.

Furthermore, given $\boldsymbol{f} \in [L_2(M,\mu), H]$, we have

$$\Big(\boldsymbol{f} - \sum_{k=1}^{n} a_k\,(\boldsymbol{f}, a_k), \sum_{k=1}^{n} a_k\,(\boldsymbol{f}, a_k)\Big) = 0.$$

By Pythagoras's theorem, this orthogonality implies that

$$\Big\|\boldsymbol{f} - \sum_{k=1}^{n} a_k\,(\boldsymbol{f}, a_k)\Big\|^2 + \Big\|\sum_{k=1}^{n} a_k\,(\boldsymbol{f}, a_k)\Big\|^2 = \|\boldsymbol{f}\|^2.$$

In analogy with the scalar-valued case, we now get **Bessel's inequality**

$$\sum_{k=1}^{n} \|(\boldsymbol{f}, a_k)\|^2 \le \int_M \|\boldsymbol{f}(s)\|^2\, d\mu(s).$$

3.1.5 Recall from 2.2.1 that

$$\|(x_k)|w_2^n\| := \sup\Big\{\Big(\sum_{k=1}^{n} |\langle x_k, x'\rangle|^2\Big)^{1/2} : x' \in U_X^\circ\Big\}.$$

With this notation, we have the following elementary result.

LEMMA. $\|(x_k)|w_2^n\| \le \|(x_k)|\mathcal{A}_n\|$.

Moreover,

$\|x_h\| \le \|(x_k)|\mathcal{A}_n\|$ *for* $h = 1, \dots, n$ *and* $\Big\|\sum_{k=1}^{n} x_k\Big\| \le \sqrt{n}\,\|(x_k)|\mathcal{A}_n\|$.

PROOF. By Parseval's equality,

$$\sum_{k=1}^{n} |\langle x_k, x'\rangle|^2 = \int_M \Big|\sum_{k=1}^{n} \langle x_k, x'\rangle\, a_k(s)\Big|^2 d\mu(s) \le \int_M \Big\|\sum_{k=1}^{n} x_k a_k(s)\Big\|^2 d\mu(s)$$

whenever $x' \in U_X^\circ$, which proves the first inequality. Now it follows that

$$\|x_h\| = \sup\{|\langle x_h, x'\rangle| : x' \in U_X^\circ\} \le \|(x_k)|w_2^n\| \le \|(x_k)|\mathcal{A}_n\|$$

and

$$\Big\|\sum_{k=1}^{n} x_k\Big\| \le \sup\Big\{\sum_{k=1}^{n} |\langle x_k, x'\rangle| : x' \in U_X^\circ\Big\}$$
$$\le \sqrt{n}\,\|(x_k)|w_2^n\| \le \sqrt{n}\,\|(x_k)|\mathcal{A}_n\|.$$

3.1.6 Orthonormal systems $\mathcal{A}_n = (a_1, \dots, a_n)$ and $\mathcal{B}_n = (b_1, \dots, b_n)$ in the Hilbert spaces $L_2(M,\mu)$ and $L_2(N,\nu)$, respectively, are said to have **identical distributions** if

$$\mu\Big\{\, s \in M : (a_k(s)) \in \Omega \,\Big\} = \nu\Big\{\, t \in N : (b_k(t)) \in \Omega \,\Big\} \qquad (*)$$

for all Borel sets $\Omega \subseteq \mathbb{K}^n$.

REMARK. Note that the following result remains true if condition $(*)$ is satisfied only for all absolutely convex (Borel) sets. In Section 3.8, we will consider the intermediate case when the Borel sets Ω are **symmetric**. That is, $(\xi_k) \in \Omega$ and $|\lambda| = 1$ imply $(\lambda\xi_k) \in \Omega$.

3.1.7 PROPOSITION. *If $\mathcal{A}_n = (a_1, \dots, a_n)$ and $\mathcal{B}_n = (b_1, \dots, b_n)$ have identical distributions, then*

$$\|(x_k)|\mathcal{A}_n\|_r = \|(x_k)|\mathcal{B}_n\|_r$$

for $x_1, \dots, x_n \in X$, all Banach spaces X and $1 \le r \le \infty$.

PROOF. Define the (absolutely convex) Borel sets

$$\Omega_u := \left\{ (\xi_k) \in \mathbb{K}^n : \left\| \sum_{k=1}^{n} \xi_k x_k \right\| \le u \right\} \quad \text{with } u > 0.$$

Then it follows from

$$\mu\Big\{ s \in M : (a_k(s)) \in \Omega_u \Big\} = \nu\Big\{ t \in N : (b_k(t)) \in \Omega_u \Big\}$$

that the functions

$$f(s) := \left\| \sum_{k=1}^{n} a_k(s) x_k \right\| \quad \text{and} \quad g(t) := \left\| \sum_{k=1}^{n} b_k(t) x_k \right\|$$

are identically distributed. This observation completes the proof.

REMARK. The above proposition means that, for our purpose, any orthonormal system $\mathcal{A}_n$ can be replaced by the orthonormal system $(p_1^{(n)}, \dots, p_n^{(n)})$ consisting of the coordinate functionals (projections)

$$p_k^{(n)}(s) := \sigma_k \quad \text{for } s = (\sigma_1, \dots, \sigma_n),$$

where the underlying set $\mathbb{K}^n$ is equipped with the measure

$$\mu \circ \mathcal{A}_n^{-1}(\Omega) := \mu\Big\{ s \in M : (a_1(s), \dots, a_n(s)) \in \Omega \Big\}.$$

In other words, it would be enough to consider only the special systems $(p_1^{(n)}, \dots, p_n^{(n)})$ in a Hilbert space $L_2(\mathbb{K}^n, \mu^n)$ provided that the measure μ^n satisfies the orthonormality relations

$$\int\limits_{\mathbb{K}^n} \sigma_h \overline{\sigma_k} \, d\mu^n(s) = \begin{cases} 1 & \text{if } h = k, \\ 0 & \text{if } h \ne k. \end{cases}$$

However, in most cases it is better to work with orthonormal systems that are given in their original form.

3.1.8 Let $\mathcal{A}_n = (a_1, \dots, a_n)$ be an orthonormal system in a Hilbert space $L_2(M, \mu)$. Write

$$a(s) := \left(\frac{1}{n} \sum_{k=1}^{n} |a_k(s)|^2 \right)^{1/2},$$

and define $\mu_\circ$ on

$$M_\circ := \left\{ s \in M : a(s) > 0 \right\}$$

by $d\mu_\circ := a^2 \, d\mu$. Then $\mathcal{N}\mathcal{A}_n := (a_1/a, \dots, a_n/a)$ is an orthonormal system in $L_2(M_\circ, \mu_\circ)$ such that

$$\|(x_k)|\mathcal{N}\mathcal{A}_n\| = \|(x_k)|\mathcal{A}_n\|$$

for $x_1, \dots, x_n \in X$ and any Banach space X. This means that, when we deal with quadratic averages, $\mathcal{A}_n$ can be replaced by $\mathcal{N}\mathcal{A}_n$ with the advantage that $\mu_\circ$ is a probability measure.

3.1.9 Let

$$\mathcal{I}_n := (u_1^{(n)}, \dots, u_n^{(n)})$$

denote the **unit vector system** in l_2^n,

$$u_k^{(n)} := \underbrace{(0, \dots, 0, \overset{k}{1}, 0, \dots, 0)}_{n}.$$

Then we have

$$\|(x_k)|\mathcal{I}_n\| = \|(x_k)|l_2^n\| = \left(\sum_{k=1}^{n} \|x_k\|^2 \right)^{1/2},$$

and the above construction yields the orthonormal system

$$\mathcal{N}\mathcal{I}_n := (\sqrt{n}\, u_1^{(n)}, \dots, \sqrt{n}\, u_n^{(n)})$$

in the Hilbert space L_2^n equipped with the inner product

$$(x, y) := \frac{1}{n} \sum_{k=1}^{n} \xi_k \overline{\eta_k}$$

for $x = (\xi_k)$ and $y = (\eta_k)$. Of course, $\mathcal{I}_n$ can be identified with the n-th section of the canonical basis $\mathcal{I} := (u_1, u_2, \dots)$ in l_2,

$$u_k := (0, \dots, 0, \overset{k}{1}, 0, \dots).$$

3.2 Khintchine constants

CONVENTION. Throughout this book, when we use Khintchine constants $K'_p(\mathcal{A}_n)$ and $K_q(\mathcal{A}_n)$ for some orthonormal system $\mathcal{A}_n = (a_1, \dots, a_n)$ in a Hilbert space $L_2(M, \mu)$ it will be tacitly assumed that $\mu(M) = 1$. Furthermore, the reader should observe that an upper Khintchine inequality only makes sense if $a_1, \dots, a_n \in L_q(M, \mu)$ with $2 \le q \le \infty$.

3.2.1 For $2 \le q < \infty$, there exists a positive constant K such that the **upper Khintchine inequality**

$$\left(\int\limits_M \left| \sum_{k=1}^n \xi_k a_k(s) \right|^q d\mu(s) \right)^{1/q} \le K \left(\sum_{k=1}^n |\xi_k|^2 \right)^{1/2} \tag{K}$$

holds for all scalars $\xi_1, \dots, \xi_n \in \mathbb{K}$. More concisely, the preceding inequality reads as follows:

$$\|(\xi_k)|\mathcal{A}_n\|_q \le K \, \|(\xi_k)|l_2^n\|.$$

In the limiting case $q = \infty$, the usual modification is required.

The quantity

$$K_q(\mathcal{A}_n) := \inf K$$

is said to be the **upper Khintchine constant** of $\mathcal{A}_n$ with respect to the exponent q. Parseval's equality implies that $K_2(\mathcal{A}_n) = 1$.

3.2.2 For $1 \le p \le 2$, there exists a positive constant K' such that the **lower Khintchine inequality**

$$\left(\sum_{k=1}^n |\xi_k|^2 \right)^{1/2} \le K' \left(\int\limits_M \left| \sum_{k=1}^n \xi_k a_k(s) \right|^p d\mu(s) \right)^{1/p} \tag{K$'$}$$

holds for all scalars $\xi_1, \dots, \xi_n \in \mathbb{K}$. More concisely, the preceding inequality reads as follows:

$$\|(\xi_k)|l_2^n\| \le K' \, \|(\xi_k)|\mathcal{A}_n\|_p.$$

The quantity

$$K'_p(\mathcal{A}_n) := \inf K'$$

is said to be the **lower Khintchine constant** of $\mathcal{A}_n$ with respect to the exponent p. Parseval's equality implies that $K'_2(\mathcal{A}_n) = 1$.

3.2.3 Let

$$M_r(\mathcal{A}_n) := \left(\int\limits_M \left| \frac{1}{\sqrt{n}} \sum_{k=1}^n a_k(s) \right|^r d\mu(s) \right)^{1/r}$$

and

$$N_r(\mathcal{A}_n) := \Big(\int\limits_M \Big(\frac{1}{n} \sum_{k=1}^{n} |a_k(s)|^r \Big)^{2/r} d\mu(s) \Big)^{1/2}.$$

In the limiting case $r=\infty$, the usual modifications are required. Clearly,

$$M_2(\mathcal{A}_n) = N_2(\mathcal{A}_n) = 1.$$

We have

$$M_q(\mathcal{A}_n) \le K_q(\mathcal{A}_n) \quad \text{if } 2 \le q \le \infty$$

and

$$M_p(\mathcal{A}_n)^{-1} \le K'_p(\mathcal{A}_n) \quad \text{if } 1 \le p \le 2.$$

The significance of the quantity $N_r(\mathcal{A}_n)$ is due to the formula

$$\|(u_k^{(n)})|[\mathcal{A}_n, l_r^n]\| = n^{1/r} N_r(\mathcal{A}_n),$$

where $u_1^{(n)}, \dots, u_n^{(n)}$ are the unit vectors in l_r^n.

3.2.4 For a complex orthonormal system $\mathcal{A}_n$, the scalar fields of the Banach spaces under consideration are assumed to be complex as well. Note that

$$K_q(\overline{\mathcal{A}}_n) = K_q(\mathcal{A}_n) \quad \text{and} \quad K'_p(\overline{\mathcal{A}}_n) = K'_p(\mathcal{A}_n).$$

Fortunately, if $\mathcal{A}_n$ is real, the Khintchine constants do not increase when passing from real to complex scalars. This observation follows from Jessen's inequality 0.4.10. Indeed, letting $\zeta_k = \xi_k + i\eta_k$ with $\xi_k, \eta_k \in \mathbb{R}$, we have

$$\Big(\int\limits_M \Big| \sum_{k=1}^{n} \zeta_k a_k(s) \Big|^q d\mu(s) \Big)^{1/q} =$$

$$= \Big[\int\limits_M \Big(\Big| \sum_{k=1}^{n} \xi_k a_k(s) \Big|^2 + \Big| \sum_{k=1}^{n} \eta_k a_k(s) \Big|^2 \Big)^{q/2} d\mu(s) \Big]^{1/q}$$

$$\le \Big[\Big(\int\limits_M \Big| \sum_{k=1}^{n} \xi_k a_k(s) \Big|^q d\mu(s) \Big)^{2/q} + \Big(\int\limits_M \Big| \sum_{k=1}^{n} \eta_k a_k(s) \Big|^q d\mu(s) \Big)^{2/q} \Big]^{1/2}$$

$$\le K_q(\mathcal{A}_n) \Big(\sum_{k=1}^{n} |\xi_k|^2 + \sum_{k=1}^{n} |\eta_k|^2 \Big)^{1/2} = K_q(\mathcal{A}_n) \Big(\sum_{k=1}^{n} |\zeta_k|^2 \Big)^{1/2}.$$

The case of the lower Khintchine constant can be treated similarly.

3.2.5 LEMMA. $K'_p(\mathcal{A}_n) \le K_{p'}(\mathcal{A}_n)$.

PROOF. Let $1 \le p \le 2$, and define $A_n \in \mathfrak{L}(l_2^n, L_{p'}(M,\mu))$ by

$$A_n x := \sum_{k=1}^n \xi_k a_k \quad \text{for } x = (\xi_k) \in l_2^n.$$

Then the dual operator $A'_n \in \mathfrak{L}(L_p(M,\mu), l_2^n)$ is given by

$$A'_n f = (\langle f, a_k \rangle) \quad \text{for } f \in L_p(M,\mu).$$

The assertion now follows from

$$K'_p(\mathcal{A}_n) \le \|A'_n : L_p \to l_2^n\| = \|A_n : l_2^n \to L_{p'}\| = K_{p'}(\mathcal{A}_n).$$

In the case $p = 1$ an obvious modification is required.

REMARK. In general, we have $K'_p(\mathcal{A}_n) < K_{p'}(\mathcal{A}_n)$. However, there exist orthonormal systems for which equality holds. In particular, this happens if $(a_1, \ldots, a_n)$ is a basis of $L_2(M,\mu)$.

3.2.6 The orthonormal system $\mathcal{NJ}_n := (\sqrt{n}\, u_1^{(n)}, \ldots, \sqrt{n}\, u_n^{(n)})$, defined in 3.1.9, will be treated as an example. It follows from

$$\|(\xi_k)|\mathcal{NJ}_n\|_r = n^{1/2-1/r} \|(\xi_k)|l_r^n\|$$

that

$$K'_p(\mathcal{NJ}_n) = K_{p'}(\mathcal{NJ}_n) = n^{1/p-1/2} \quad \text{if } 1 \le p \le 2.$$

3.3 Riemann ideal norms

3.3.1 Assume that $1 \le u, v < \infty$. Let

$$\mathcal{A}_n = (a_1, \ldots, a_n) \quad \text{and} \quad \mathcal{B}_n = (b_1, \ldots, b_n)$$

be orthonormal systems in $L_2(M,\mu) \cap L_u(M,\mu)$ and $L_2(N,\nu) \cap L_v(N,\nu)$, respectively.

For $T \in \mathfrak{L}(X,Y)$, we denote by $\varrho_u^{(v)}(T|\mathcal{B}_n, \mathcal{A}_n)$ the least constant $c \ge 0$ such that

$$\left(\int_N \left\| \sum_{k=1}^n T x_k b_k(t) \right\|^v d\nu(t) \right)^{1/v} \le c \left(\int_M \left\| \sum_{k=1}^n x_k a_k(s) \right\|^u d\mu(s) \right)^{1/u} \qquad \text{(R)}$$

whenever $x_1, \ldots, x_n \in X$. More concisely, the preceding inequality reads as follows:

$$\|(Tx_k)|\mathcal{B}_n\|_v \le c\, \|(x_k)|\mathcal{A}_n\|_u.$$

In the limiting cases $u = \infty$ or $v = \infty$, the usual modifications are required. We refer to

$$\varrho_u^{(v)}(\mathcal{B}_n, \mathcal{A}_n) : T \longrightarrow \varrho_u^{(v)}(T|\mathcal{B}_n, \mathcal{A}_n)$$

as a **Riemann ideal norm.**

This book mainly deals with the quadratic case. Therefore, to simplify notation, we write $\varrho(\mathcal{B}_n, \mathcal{A}_n)$ instead of $\varrho_2^{(2)}(\mathcal{B}_n, \mathcal{A}_n)$.

3.3.2 PROPOSITION. $1 \le \varrho(\mathcal{B}_n, \mathcal{A}_n) \le \sqrt{n}$.

PROOF. Substituting $x_1 = x$ and $x_2 = \ldots = x_n = o$ in the defining inequality of $\varrho(T|\mathcal{B}_n, \mathcal{A}_n)$ gives the lower estimate. In order to check the upper estimate, we fix $t \in N$ and choose $y' \in U_Y^\circ$ such that

$$\Big\|\sum_{k=1}^n Tx_k b_k(t)\Big\| = \Big\langle \sum_{k=1}^n Tx_k b_k(t), y' \Big\rangle = \sum_{k=1}^n \langle Tx_k, y'\rangle\, b_k(t).$$

Then it follows that

$$\begin{aligned}\Big\|\sum_{k=1}^n Tx_k b_k(t)\Big\| &\le \Big(\sum_{k=1}^n |\langle Tx_k, y'\rangle|^2\Big)^{1/2} \Big(\sum_{k=1}^n |b_k(t)|^2\Big)^{1/2} \\ &\le \|T\|\, \|(x_k)|w_2^n\| \Big(\sum_{k=1}^n |b_k(t)|^2\Big)^{1/2}.\end{aligned}$$

Integrating over t and applying Lemma 3.1.5 yields

$$\begin{aligned}\|(Tx_k)|\mathcal{B}_n\|^2 &= \int_N \Big\|\sum_{k=1}^n Tx_k b_k(t)\Big\|^2 d\nu(t) \\ &\le \|T\|^2 \|(x_k)|w_2^n\|^2 \Big(\sum_{k=1}^n \int_N |b_k(t)|^2\, d\nu(t)\Big) \le n\|T\|^2 \|(x_k)|\mathcal{A}_n\|^2.\end{aligned}$$

3.3.3 PROPOSITION. *The ideal norm $\varrho(\mathcal{B}_n, \mathcal{A}_n)$ is injective.*

REMARK. The ideal norm $\varrho(\mathcal{I}_n, \mathcal{R}_n)$ fails to be surjective; see 4.2.5.

3.3.4 PROPOSITION. *The ideal norm $\varrho(\mathcal{B}_n, \mathcal{A}_n)$ is l_2-stable.*

PROOF. Let $x_1 = (x_i^{(1)}), \ldots, x_n = (x_i^{(n)}) \in [l_2(\mathbb{I}), X_i]$ and $T := [l_2(\mathbb{I}), T_i]$. The assertion now follows from

$$\|(x_k)|\mathcal{A}_n\|^2 = \sum_{\mathbb{I}} \|(x_i^{(k)})|\mathcal{A}_n\|^2, \quad \|(Tx_k)|\mathcal{B}_n\|^2 = \sum_{\mathbb{I}} \|(T_i x_i^{(k)})|\mathcal{B}_n\|^2,$$

and

$$\|(T_i x_i^{(k)})|\mathcal{B}_n\| \le \varrho(T_i|\mathcal{B}_n, \mathcal{A}_n)\, \|(x_i^{(k)})|\mathcal{A}_n\|.$$

3.3.5 The following **triangle inequality** is obvious. The symbol $\circ$ was defined in 1.1.7.

PROPOSITION. *Let $\mathcal{A}_n$, $\mathcal{B}_n$ and $\mathcal{C}_n$ be orthonormal systems. Then*

$$\varrho(\mathcal{C}_n, \mathcal{A}_n) \le \varrho(\mathcal{C}_n, \mathcal{B}_n) \circ \varrho(\mathcal{B}_n, \mathcal{A}_n).$$

3.3.6 We now compare Riemann ideal norms of different order. To this end, let

$$\mathcal{A}_n = (a_1, \ldots, a_n), \quad \mathcal{A}_{n+1} = (a_1, \ldots, a_n, a_{n+1})$$

and

$$\mathcal{B}_n = (b_1, \ldots, b_n), \quad \mathcal{B}_{n+1} = (b_1, \ldots, b_n, b_{n+1})$$

be orthonormal systems in the Hilbert spaces $L_2(M, \mu)$ and $L_2(N, \nu)$, respectively.

PROPOSITION. $\varrho(\mathcal{B}_n, \mathcal{A}_n) \le \varrho(\mathcal{B}_{n+1}, \mathcal{A}_{n+1}) \le 3\, \varrho(\mathcal{B}_n, \mathcal{A}_n)$.

PROOF. The left-hand inequality is obvious. In view of

$$\|x_{n+1}\| \le \|(x_k)|\mathcal{A}_{n+1}\| \qquad \text{(by 3.1.5)},$$

$$\|(x_k)|\mathcal{A}_n\| \le \|(x_k)|\mathcal{A}_{n+1}\| + \|x_{n+1}\| \qquad \text{(trivial)},$$

and

$$\|T\| \le \varrho(T|\mathcal{B}_n, \mathcal{A}_n) \qquad \text{(by 3.3.2)},$$

we have

$$\begin{aligned} \|(Tx_k)|\mathcal{B}_{n+1}\| &\le \|(Tx_k)|\mathcal{B}_n\| + \|Tx_{k+1}\| \\ &\le \varrho(T|\mathcal{B}_n, \mathcal{A}_n)\, \|(x_k)|\mathcal{A}_n\| + \|T\|\, \|x_{n+1}\| \\ &\le 3\, \varrho(T|\mathcal{B}_n, \mathcal{A}_n)\, \|(x_k)|\mathcal{A}_{n+1}\|. \end{aligned}$$

This proves the right-hand inequality.

3.3.7 EXAMPLE. *Let $1 \le p \le 2$ and $\mu(M) = 1$. Then*

$$\varrho(L_p|\mathcal{B}_n, \mathcal{A}_n) \le K'_p(\mathcal{A}_n).$$

PROOF. Given $f_1, \ldots, f_n \in L_p(\Omega, \omega)$, we have

$$\|(f_k)|[\mathcal{B}_n, L_p]\| = \left[\int\limits_N \left(\int\limits_\Omega \Big| \sum_{k=1}^n f_k(\xi) b_k(t) \Big|^p d\omega(\xi) \right)^{2/p} d\nu(t) \right]^{1/2} \le$$

(by definition)

$$\le \left[\int\limits_\Omega \left(\int\limits_N \Big| \sum_{k=1}^n f_k(\xi) b_k(t) \Big|^2 d\nu(t) \right)^{p/2} d\omega(\xi) \right]^{1/p}$$

(by Jessen's inequality)

$$= \left[\int_\Omega \left(\sum_{k=1}^n |f_k(\xi)|^2 \right)^{p/2} d\omega(\xi) \right]^{1/p}$$

(by Parseval's equality)

$$\leq K'_p(\mathcal{A}_n) \left[\int_\Omega \left(\int_M \left| \sum_{k=1}^n f_k(\xi) a_k(s) \right|^p d\mu(s) \right)^{p/p} d\omega(\xi) \right]^{1/p}$$

(by Khintchine's inequality)

$$= K'_p(\mathcal{A}_n) \left[\int_M \left(\int_\Omega \left| \sum_{k=1}^n f_k(\xi) a_k(s) \right|^p d\omega(\xi) \right)^{p/p} d\mu(s) \right]^{1/p}$$

(by Fubini's theorem)

$$\leq K'_p(\mathcal{A}_n) \left[\int_M \left(\int_\Omega \left| \sum_{k=1}^n f_k(\xi) a_k(s) \right|^p d\omega(\xi) \right)^{2/p} d\mu(s) \right]^{1/2}$$

(by $\| \cdot | L_p(M,\mu) \| \leq \| \cdot | L_2(M,\mu) \|$)

$$= K'_p(\mathcal{A}_n) \, \| (f_k) | [\mathcal{A}_n, L_p] \|.$$

(by definition)

REMARK. From 3.4.10 and 3.4.15 we would obtain

$$\varrho(L_p | \mathcal{B}_n, \mathcal{A}_n) \leq K_{p'}(\mathcal{A}_n).$$

This is a weaker result, since $K'_p(\mathcal{A}_n) \leq K_{p'}(\mathcal{A}_n)$.

3.3.8 EXAMPLE. *Let $2 \leq q \leq \infty$ and $\nu(N) = 1$. Then*

$$\varrho(L_q | \mathcal{B}_n, \mathcal{A}_n) \leq K_q(\mathcal{B}_n).$$

PROOF. Assume that q is finite. Given $f_1, \ldots, f_n \in L_q(\Omega, \omega)$, we have

$$\| (f_k) | [\mathcal{B}_n, L_q] \| = \left[\int_N \left(\int_\Omega \left| \sum_{k=1}^n f_k(\xi) b_k(t) \right|^q d\omega(\xi) \right)^{2/q} d\nu(t) \right]^{1/2} \leq$$

(by definition)

$$\leq \left[\int_N \left(\int_\Omega \left| \sum_{k=1}^n f_k(\xi) b_k(t) \right|^q d\omega(\xi) \right)^{q/q} d\nu(t) \right]^{1/q}$$

(by $\| \cdot | L_2(N,\nu) \| \leq \| \cdot | L_q(N,\nu) \|$)

$$= \left[\int_\Omega \left(\int_N \left| \sum_{k=1}^n f_k(\xi) b_k(t) \right|^q d\nu(t) \right)^{q/q} d\omega(\xi) \right]^{1/q}$$

(by Fubini's theorem)

$$\leq K_q(\mathcal{B}_n)\left[\int\limits_{\Omega}\Big(\sum_{k=1}^{n}|f_k(\xi)|^2\Big)^{q/2}d\omega(\xi)\right]^{1/q}$$

(by Khintchine's inequality)

$$= K_q(\mathcal{B}_n)\left[\int\limits_{\Omega}\Big(\int\limits_{M}\Big|\sum_{k=1}^{n}f_k(\xi)a_k(s)\Big|^2 d\mu(s)\Big)^{q/2}d\omega(\xi)\right]^{1/q}$$

(by Parseval's equality)

$$\leq K_q(\mathcal{B}_n)\left[\int\limits_{M}\Big(\int\limits_{\Omega}\Big|\sum_{k=1}^{n}f_k(\xi)a_k(s)\Big|^q d\omega(\xi)\Big)^{2/q}d\mu(s)\right]^{1/2}$$

(by Jessen's inequality)

$$= K_q(\mathcal{B}_n)\,\|(f_k)|[\mathcal{A}_n, L_q]\|.$$

(by definition)

The case $q=\infty$ can be treated in the same way.

REMARK. This result could also be derived from 3.4.10 and 3.4.16.

It is amusing to compare the order of comments in the two preceding proofs: $\downarrow\uparrow$.

3.4 Dirichlet ideal norms

3.4.1 Assume that $1\leq u,v<\infty$. Let

$$\mathcal{A}_n=(a_1,\ldots,a_n) \quad\text{and}\quad \mathcal{B}_n=(b_1,\ldots,b_n)$$

be orthonormal systems in $L_2(M,\mu)\cap L_{u'}(M,\mu)$ and $L_2(N,\nu)\cap L_v(N,\nu)$, respectively.

For $T\in\mathfrak{L}(X,Y)$, we denote by $\boldsymbol{\delta}_u^{(v)}(T|\mathcal{B}_n,\mathcal{A}_n)$ the least constant $c\geq 0$ such that

$$\Big(\int\limits_{N}\Big\|\sum_{k=1}^{n}T\,\langle \boldsymbol{f},\overline{a_k}\rangle\, b_k(t)\Big\|^v d\nu(t)\Big)^{1/v}\leq c\Big(\int\limits_{M}\|\boldsymbol{f}(s)\|^u\, d\mu(s)\Big)^{1/u} \qquad \text{(D)}$$

whenever $\boldsymbol{f}\in[L_u(M,\mu),X]$. More concisely, the preceding inequality reads as follows:

$$\|(T\,\langle \boldsymbol{f},\overline{a_k}\rangle)\,|\mathcal{B}_n\|_v\leq c\,\|\boldsymbol{f}|L_u\|.$$

In the limiting cases $u=\infty$ or $v=\infty$, the usual modifications are required. We refer to

$$\boldsymbol{\delta}_u^{(v)}(\mathcal{B}_n,\mathcal{A}_n)\,:\,T\longrightarrow\boldsymbol{\delta}_u^{(v)}(T|\mathcal{B}_n,\mathcal{A}_n)$$

as a **Dirichlet ideal norm**.

This book mainly deals with the quadratic case. Therefore, to simplify notation, we write $\boldsymbol{\delta}(\mathcal{B}_n, \mathcal{A}_n)$ instead of $\boldsymbol{\delta}_2^{(2)}(\mathcal{B}_n, \mathcal{A}_n)$.

3.4.2 Define $A_n \in \mathfrak{L}(l_2^n, L_2(M,\mu))$ by

$$A_n x := \sum_{k=1}^{n} \xi_k a_k \quad \text{for} \quad x = (\xi_k) \in l_2^n.$$

Then the adjoint operator $A_n^* \in \mathfrak{L}(L_2(M,\mu), l_2^n)$ is given by

$$A_n^* f := (\langle f, \overline{a_k}\rangle) \quad \text{for } f \in L_2(M,\nu).$$

Let $B_n \in \mathfrak{L}(l_2^n, L_2(N,\nu))$ be defined analogously.

We refer to $B_n A_n^* \in \mathfrak{L}(L_2(M,\nu), L_2(N,\nu))$ as the **Dirichlet operator** associated with the orthonormal systems $\mathcal{A}_n$ and $\mathcal{B}_n$. Note that $B_n A_n^*$ is the integral operator

$$B_n A_n^* : f(s) \longrightarrow g(t) := \int\limits_M f(s) B_n A_n^*(s,t)\, d\mu(s)$$

induced by the **Dirichlet kernel**

$$B_n A_n^*(s,t) := \sum_{k=1}^{n} \overline{a_k}(s) b_k(t).$$

For $T \in \mathfrak{L}(X,Y)$ and $\boldsymbol{f} \in [L_2(M,\mu), X]$, we have

$$[B_n A_n^*, T] : \boldsymbol{f}(s) \longrightarrow \boldsymbol{g}(t) := \int\limits_M T\boldsymbol{f}(s) B_n A_n^*(s,t)\, d\mu(s).$$

The definition of the Dirichlet ideal norm can now be restated as follows:

$$\boldsymbol{\delta}(T|\mathcal{B}_n, \mathcal{A}_n) := \Big\| [B_n A_n^*, T] : [L_2(M,\mu), X] \longrightarrow [L_2(N,\nu), Y] \Big\|.$$

REMARK. We refer to $A_n A_n^*$ as the **Dirichlet projection** associated with the orthonormal system $\mathcal{A}_n$.

3.4.3 Next, we formulate an elementary fact which illustrates the significance of the Dirichlet ideal norms.

PROPOSITION. *Let $\mathcal{A} = (a_1, a_2, \ldots)$ be an orthonormal basis in the Hilbert space $L_2(M,\mu)$. Then, for every Banach space X, the following are equivalent:*

(1) *All X-valued functions $\boldsymbol{f} \in [L_2(M,\mu), X]$ admit expansions*

$$\boldsymbol{f} = \sum_{k=1}^{\infty} (\boldsymbol{f}, a_k) a_k$$

convergent with respect to $\|\cdot|L_2\|$.

(2) $\boldsymbol{\delta}(X|\mathcal{A}_n, \mathcal{A}_n) = O(1)$.

PROOF. Since $L_2(M,\mu)\otimes X$ is dense in $[L_2(M,\mu),X]$, the assertion can be deduced from the principle of uniform boundedness.

3.4.4 PROPOSITION. $1 \le \boldsymbol{\delta}(\mathcal{B}_n,\mathcal{A}_n) \le \sqrt{n}$.

PROOF. The classical Parseval equality states that

$$\int\limits_M \Big| \sum_{k=1}^n \overline{a_k}(s) b_k(t) \Big|^2 d\mu(s) = \sum_{k=1}^n |b_k(t)|^2$$

which implies

$$\int\limits_N\int\limits_M \Big| \sum_{k=1}^n \overline{a_k}(s) b_k(t) \Big|^2 d\mu(s)\, d\nu(t) = n.$$

Hence, for $T \in \mathfrak{L}(X,Y)$ and $\boldsymbol{f} \in [L_2(M,\mu),X]$, we obtain

$$\begin{aligned}
\int\limits_N \Big\| \sum_{k=1}^n T\,\langle \boldsymbol{f}, \overline{a_k}\rangle\, b_k(t) \Big\|^2 d\nu(t) \le \\
&\le \|T\|^2 \int\limits_N \Big\| \int\limits_M \boldsymbol{f}(s) \sum_{k=1}^n \overline{a_k}(s) b_k(t)\, d\mu(s) \Big\|^2 d\nu(t) \\
&\le \|T\|^2 \int\limits_M \|\boldsymbol{f}(s)\|^2 d\mu(s) \int\limits_N\int\limits_M \Big| \sum_{k=1}^n \overline{a_k}(s) b_k(t) \Big|^2 d\mu(s)\, d\nu(t) \\
&= n\,\|T\|^2 \int\limits_M \|\boldsymbol{f}(s)\|^2 d\mu(s).
\end{aligned}$$

This proves that $\boldsymbol{\delta}(T|\mathcal{B}_n,\mathcal{A}_n) \le \sqrt{n}\,\|T\|$. By 1.1.3, the lower estimate $\|T\| \le \boldsymbol{\delta}(T|\mathcal{B}_n,\mathcal{A}_n)$ follows from $\boldsymbol{\delta}(I_{\mathbb{K}}|\mathcal{B}_n,\mathcal{A}_n) = 1$.

3.4.5 PROPOSITION. *The ideal norm $\boldsymbol{\delta}(\mathcal{B}_n,\mathcal{A}_n)$ is injective and surjective.*

PROOF. The injectivity of $\boldsymbol{\delta}(\mathcal{B}_n,\mathcal{A}_n)$ is obvious, and the surjectivity can be deduced from 1.1.11 and the next proposition.

3.4.6 The Dirichlet ideal norms behave very nicely with respect to duality.

PROPOSITION. $\boldsymbol{\delta}'(\mathcal{A}_n,\mathcal{B}_n) = \boldsymbol{\delta}(\overline{\mathcal{B}}_n,\overline{\mathcal{A}}_n)$.

PROOF. For $T \in \mathfrak{L}(X,Y)$ and $\boldsymbol{g} \in [L_2(N,\nu),Y']$, we let

$$c := \bigg(\int\limits_M \Big\| \sum_{k=1}^n T'\,\langle \boldsymbol{g}, \overline{b_k}\rangle\, a_k(s) \Big\|^2 d\mu(s) \bigg)^{1/2} = \|(T'\,\langle \boldsymbol{g}, \overline{b_k}\rangle)|\mathcal{A}_n\|.$$

Given $\varepsilon > 0$, by [DIN, p. 232] or [DUN*2, p. 1173], there exists a function $\boldsymbol{f} \in [L_2(M,\mu), X]$ such that

$$c^2 = \int\limits_M \Big\langle \boldsymbol{f}(s), \sum_{k=1}^n T' \langle \boldsymbol{g}, \overline{b_k} \rangle \, a_k(s) \Big\rangle \, d\mu(s)$$

and

$$\|\boldsymbol{f}|L_2\| \le (1+\varepsilon)c.$$

We now obtain

$$\begin{aligned} c^2 &= \int\limits_M\int\limits_N \sum_{k=1}^n \langle T\boldsymbol{f}(s), \boldsymbol{g}(t)\rangle \, a_k(s)\overline{b_k(t)} \, d\mu(s)d\nu(t) \\ &= \int\limits_N \Big\langle \sum_{k=1}^n T\langle \boldsymbol{f}, a_k\rangle \, \overline{b_k}(t), \boldsymbol{g}(t) \Big\rangle \, d\nu(t) \\ &\le \Big(\int\limits_N \Big\| \sum_{k=1}^n T\langle \boldsymbol{f}, a_k\rangle \, \overline{b_k}(t) \Big\|^2 d\nu(t) \Big)^{1/2} \Big(\int\limits_N \|\boldsymbol{g}(t)\|^2 d\nu(t) \Big)^{1/2} \\ &\le (1+\varepsilon)\, c\, \boldsymbol{\delta}(T|\overline{\mathcal{B}}_n, \overline{\mathcal{A}}_n) \, \|\boldsymbol{g}|L_2\|. \end{aligned}$$

Letting $\varepsilon \to 0$ and dividing by c yields

$$\| \big(T' \langle \boldsymbol{g}, \overline{b_k}\rangle\big) | \mathcal{A}_n \| \le \boldsymbol{\delta}(T|\overline{\mathcal{B}}_n, \overline{\mathcal{A}}_n) \|\boldsymbol{g}|L_2\|.$$

This proves that

$$\boldsymbol{\delta}(T'|\mathcal{A}_n, \mathcal{B}_n) \le \boldsymbol{\delta}(T|\overline{\mathcal{B}}_n, \overline{\mathcal{A}}_n).$$

We know from 0.1.7 that $T''K_X = K_Y T$, where K_X and K_Y denote the canonical embeddings from X into X'' and from Y into Y'', respectively. Using the injectivity of $\boldsymbol{\delta}(\mathcal{B}_n, \mathcal{A}_n)$, we finally conclude that

$$\begin{aligned} \boldsymbol{\delta}(T|\overline{\mathcal{B}}_n, \overline{\mathcal{A}}_n) &= \boldsymbol{\delta}(K_Y T|\overline{\mathcal{B}}_n, \overline{\mathcal{A}}_n) = \boldsymbol{\delta}(T''K_X|\overline{\mathcal{B}}_n, \overline{\mathcal{A}}_n) \\ &\le \boldsymbol{\delta}(T''|\overline{\mathcal{B}}_n, \overline{\mathcal{A}}_n) \le \boldsymbol{\delta}(T'|\mathcal{A}_n, \mathcal{B}_n) \le \boldsymbol{\delta}(T|\overline{\mathcal{B}}_n, \overline{\mathcal{A}}_n). \end{aligned}$$

3.4.7 PROPOSITION. *The ideal norm $\boldsymbol{\delta}(\mathcal{B}_n, \mathcal{A}_n)$ is l_2-stable.*

PROOF. Let $X := [l_2(\mathbb{I}), X_i]$ and $T := [l_2(\mathbb{I}), T_i]$. Since $\boldsymbol{f} \in [L_2(M,\mu), X]$ can be viewed as an $\mathbb{I}$-tuple of functions $\boldsymbol{f}_i \in [L_2(M,\mu), X_i]$, the assertion follows from

$$\|\boldsymbol{f}|L_2\|^2 = \sum_{\mathbb{I}} \|\boldsymbol{f}_i|L_2\|^2,$$

$$\|(\langle T\boldsymbol{f}, \overline{a_k}\rangle)|\mathcal{B}_n\|^2 = \sum_{\mathbb{I}} \|(\langle T_i \boldsymbol{f}_i, \overline{a_k}\rangle)|\mathcal{B}_n\|^2,$$

and

$$\|(\langle T_i \boldsymbol{f}_i, \overline{a_k}\rangle)|\mathcal{B}_n\| \le \boldsymbol{\delta}(T_i|\mathcal{B}_n, \mathcal{A}_n) \, \|\boldsymbol{f}_i|L_2\|.$$

3.4.8 We now establish counterparts of the **triangle inequality** stated in 3.3.5.

PROPOSITION. *Let $\mathcal{A}_n$, $\mathcal{B}_n$ and $\mathcal{C}_n$ be orthonormal systems. Then*

$$\boldsymbol{\delta}(\mathcal{C}_n, \mathcal{A}_n) \leq \boldsymbol{\varrho}(\mathcal{C}_n, \mathcal{B}_n) \circ \boldsymbol{\delta}(\mathcal{B}_n, \mathcal{A}_n)$$

and

$$\boldsymbol{\delta}(\mathcal{C}_n, \mathcal{A}_n) \leq \boldsymbol{\delta}(\mathcal{C}_n, \mathcal{B}_n) \circ \boldsymbol{\varrho}'(\overline{\mathcal{A}}_n, \overline{\mathcal{B}}_n).$$

PROOF. The inequalities

$$\boldsymbol{\delta}(\mathcal{C}_n, \mathcal{A}_n) \leq \boldsymbol{\varrho}(\mathcal{C}_n, \mathcal{B}_n) \circ \boldsymbol{\delta}(\mathcal{B}_n, \mathcal{A}_n)$$

and

$$\boldsymbol{\delta}(\overline{\mathcal{A}}_n, \overline{\mathcal{C}}_n) \leq \boldsymbol{\varrho}(\overline{\mathcal{A}}_n, \overline{\mathcal{B}}_n) \circ \boldsymbol{\delta}(\overline{\mathcal{B}}_n, \overline{\mathcal{C}}_n)$$

are obvious. By duality, the last one passes into

$$\boldsymbol{\delta}(\mathcal{C}_n, \mathcal{A}_n) \leq \boldsymbol{\delta}(\mathcal{C}_n, \mathcal{B}_n) \circ \boldsymbol{\varrho}'(\overline{\mathcal{A}}_n, \overline{\mathcal{B}}_n).$$

3.4.9 Dirichlet ideal norms behave very well with respect to (complex) interpolation; see Sections 0.6 and 9.1.

PROPOSITION. *Let (X_0, X_1) and (Y_0, Y_1) be interpolation couples. Assume that $T \in \mathfrak{L}(X_0 + X_1, Y_0 + Y_1)$ transforms X_0 into Y_0 and X_1 into Y_1. If $0 < \theta < 1$, then*

$$\boldsymbol{\delta}(T_\theta | \mathcal{B}_n, \mathcal{A}_n) \leq \boldsymbol{\delta}(T_0 | \mathcal{B}_n, \mathcal{A}_n)^{1-\theta} \boldsymbol{\delta}(T_1 | \mathcal{B}_n, \mathcal{A}_n)^{\theta}.$$

PROOF. We recall the fact, stated in 3.4.2, that $\boldsymbol{\delta}(T|\mathcal{B}_n, \mathcal{A}_n)$ coincides with the norm of the operator

$$[B_n A_n^*, T] : [L_2(M, \mu), X] \longrightarrow [L_2(N, \nu), Y].$$

The conclusion now follows from the isometry

$$\Big[L_2(M, \mu), [X_0, X_1]_\theta\Big] = \Big[[L_2(M, \mu), X_0], [L_2(M, \mu), X_1]\Big]_\theta;$$

see 9.1.3.

REMARK. The same result holds if the complex interpolation spaces $[X_0, X_1]_\theta$ and $[Y_0, Y_1]_\theta$ are replaced by their real counterparts $(X_0, X_1)_{\theta,2}$ and $(Y_0, Y_1)_{\theta,2}$; see 9.1.6 (Remark).

3.4.10 Next, we compare the Dirichlet ideal norm with the Riemann ideal norm.

PROPOSITION. $\boldsymbol{\varrho}(\mathcal{B}_n, \mathcal{A}_n) \leq \boldsymbol{\delta}(\mathcal{B}_n, \mathcal{A}_n)$.

PROOF. Substituting the function

$$\boldsymbol{f} := \sum_{k=1}^{n} a_k \otimes x_k \quad \text{with} \quad x_1, \ldots, x_n \in X$$

in condition (D) from 3.4.1 yields (R) from 3.3.1.

3.4.11 PROPOSITION. *$\varrho(H|\mathcal{B}_n,\mathcal{A}_n) = \delta(H|\mathcal{B}_n,\mathcal{A}_n) = 1$ for all Hilbert spaces H.*

PROOF. This follows from the fact that, in Hilbert spaces, Parseval's equality and Bessel's inequality hold with constants 1; see 3.1.4.

3.4.12 PROPOSITION. *If $\delta(X|\mathcal{A}_n,\mathcal{A}_n)=1$ for all Banach spaces X, then $\varrho(\mathcal{B}_n,\mathcal{A}_n)=\delta(\mathcal{B}_n,\mathcal{A}_n)$.*

PROOF. We have

$$\delta(T|\mathcal{B}_n,\mathcal{A}_n) \le \varrho(T|\mathcal{B}_n,\mathcal{A}_n) \circ \delta(X|\mathcal{A}_n,\mathcal{A}_n).$$

REMARK. The condition $\delta(X|\mathcal{A}_n,\mathcal{A}_n)=1$ means that

$$[A_n A_n^*, X] : \boldsymbol{f} \longrightarrow \sum_{k=1}^{n} a_k \otimes \langle \boldsymbol{f}, \overline{a_k}\rangle$$

is a projection of norm 1 in $[L_2(M,\mu),X]$. This is, in particular, the case when $a_1,\dots,a_n$ form a basis of $L_2(M,\mu)$.

3.4.13 We now compare Dirichlet ideal norms of different order.

PROPOSITION. *Let*

$$\mathcal{A}_{n+m} = (\overbrace{a_1,\dots,a_n}^{\mathcal{A}_n},\overbrace{a_{n+1},\dots,a_{n+m}}^{\mathcal{A}_m})$$

and

$$\mathcal{B}_{n+m} = (\overbrace{b_1,\dots,b_n}^{\mathcal{B}_n},\overbrace{b_{n+1},\dots,b_{n+m}}^{\mathcal{B}_m}).$$

Then

$$\delta(\mathcal{B}_{n+m},\mathcal{A}_{n+m}) \le \delta(\mathcal{B}_n,\mathcal{A}_n) + \delta(\mathcal{B}_m,\mathcal{A}_m),$$
$$\delta(\mathcal{B}_n,\mathcal{A}_n) \le \delta(\mathcal{B}_{n+m},\mathcal{A}_{n+m}) + \delta(\mathcal{B}_m,\mathcal{A}_m).$$

In particular,

$$\delta(\mathcal{B}_{n+1},\mathcal{A}_{n+1}) \le 2\,\delta(\mathcal{B}_n,\mathcal{A}_n) \quad \textit{and} \quad \delta(\mathcal{B}_n,\mathcal{A}_n) \le 2\,\delta(\mathcal{B}_{n+1},\mathcal{A}_{n+1}).$$

PROOF. For $\boldsymbol{f} \in [L_2(M,\mu),X]$, we have

$$\begin{aligned}\|(T\langle \boldsymbol{f},\overline{a_k}\rangle)|\mathcal{B}_{n+m}\| &\le \|(T\langle \boldsymbol{f},\overline{a_k}\rangle)|\mathcal{B}_n\| + \|(T\langle \boldsymbol{f},\overline{a_k}\rangle)|\mathcal{B}_m\| \\ &\le [\delta(T|\mathcal{B}_n,\mathcal{A}_n) + \delta(T|\mathcal{B}_m,\mathcal{A}_m)]\,\|\boldsymbol{f}|L_2\|.\end{aligned}$$

This proves the first inequality. The second one can be checked analogously. Since $\|T\| \le \delta(T|\mathcal{B}_n,\mathcal{A}_n)$, the rest is obtained by letting $m=1$.

3.4.14 In the following, we consider the operators

$$A_n^* \in \mathfrak{L}(L_r(M,\mu), l_2^n) \quad \text{and} \quad B_n \in \mathfrak{L}(l_2^n, L_s(N,\nu))$$

defined by

$$A_n^* f := (\langle f, \overline{a_k} \rangle) \quad \text{for } f \in L_r(M, \mu)$$

and

$$B_n x := \sum_{k=1}^{n} \xi_k b_k \quad \text{for } x = (\xi_k) \in l_2^n.$$

LEMMA.

$$\| B_n A_n^* : L_p \to L_2 \| \le K_{p'}(\mathcal{A}_n) \quad \textit{if } 1 \le p \le 2 \textit{ and } \mu(M) = 1$$

and

$$\| B_n A_n^* : L_2 \to L_q \| \le K_q(\mathcal{B}_n) \quad \textit{if } 2 \le q \le \infty \textit{ and } \nu(N) = 1.$$

Moreover, whenever $1 \le p \le 2 \le q \le \infty$ *and* $\mu(M) = \nu(N) = 1$, *then*

$$\| B_n A_n^* : L_p \to L_q \| \le K_{p'}(\mathcal{A}_n)\, K_q(\mathcal{B}_n).$$

PROOF. We have

$$\| B_n A_n^* : L_p \to L_q \| \le \| A_n^* : L_p \to l_2^n \| \, \| B_n : l_2^n \to L_q \| = K_{p'}(\mathcal{A}_n)\, K_q(\mathcal{B}_n).$$

REMARK. In the limiting cases $p = 2$ and $q = 2$ the assumptions $\mu(M) = 1$ and $\nu(N) = 1$, respectively, are not needed.

3.4.15 EXAMPLE. *Let* $1 \le p \le 2$ *and* $\mu(M) = 1$. *Then*

$$\boldsymbol{\delta}(L_p | \mathcal{B}_n, \mathcal{A}_n) \le K_{p'}(\mathcal{A}_n).$$

PROOF. Given $\boldsymbol{f} \in [L_2(M,\mu), L_p(\Omega,\omega)]$, we let $\boldsymbol{g} := [B_n A_n^*, L_p]\boldsymbol{f}$. Note that the vector-valued functions $\boldsymbol{f}$ and $\boldsymbol{g}$ can be viewed as scalar-valued functions f and g on $M \times \Omega$ and $N \times \Omega$, respectively. Lemma 3.4.14 tells us that

$$\Big(\int_N |g(t,\xi)|^2 \, d\nu(t) \Big)^{1/2} \le K_{p'}(\mathcal{A}_n) \Big(\int_M |f(s,\xi)|^p \, d\mu(s) \Big)^{1/p} \qquad (*)$$

for fixed $\xi \in \Omega$. We now obtain

$$\| \boldsymbol{g} | [L_2, L_p] \| = \Big[\int_N \Big(\int_\Omega \big| g(t,\xi) \big|^p d\omega(\xi) \Big)^{2/p} d\nu(t) \Big]^{1/2} \le$$

(by definition)

$$\le \Big[\int_\Omega \Big(\int_N \big| g(t,\xi) \big|^2 d\nu(t) \Big)^{p/2} d\omega(\xi) \Big]^{1/p}$$

(by Jessen's inequality)

$$\leq K_{p'}(\mathcal{A}_n)\left[\int_{\Omega}\left(\int_{M}\left|f(s,\xi)\right|^{p}d\mu(s)\right)^{p/p}d\omega(\xi)\right]^{1/p}$$

(by (*))

$$= K_{p'}(\mathcal{A}_n)\left[\int_{M}\left(\int_{\Omega}|f(s,\xi)|^{p}d\omega(\xi)\right)^{p/p}d\mu(s)\right]^{1/p}$$

(by Fubini's theorem)

$$\leq K_{p'}(\mathcal{A}_n)\left[\int_{M}\left(\int_{\Omega}\left|f(s,\xi)\right|^{p}d\omega(\xi)\right)^{2/p}d\mu(s)\right]^{1/2}$$

(by $\|\cdot|L_p(M,\mu)\|\leq\|\cdot|L_2(M,\mu)\|$)

$$= K_{p'}(\mathcal{A}_n)\,\|\boldsymbol{f}|[L_2,L_p]\|.$$

(by definition)

3.4.16 By 3.4.6, the next result is dual to the previous one. Hence the proof could be omitted. However, we waste paper in favour of symmetry.

EXAMPLE. *Let* $2\leq q\leq\infty$ *and* $\nu(N)=1$. *Then*

$$\boldsymbol{\delta}(L_q|\mathcal{B}_n,\mathcal{A}_n)\leq K_q(\mathcal{B}_n).$$

PROOF. Assume that q is finite. Given $\boldsymbol{f}\in[L_2(M,\mu),L_q(\Omega,\omega)]$, we let $\boldsymbol{g}:=[B_nA_n^*,L_p]\boldsymbol{f}$. Lemma 3.4.14 tells us that

$$\left(\int_{N}|g(t,\xi)|^{q}\,d\nu(t)\right)^{1/q}\leq K_q(\mathcal{B}_n)\left(\int_{M}|f(s,\xi)|^{2}\,d\mu(s)\right)^{1/2} \qquad (*)$$

for fixed $\xi\in\Omega$. We now obtain

$$\|\boldsymbol{g}|[L_2,L_q]\| = \left[\int_{N}\left(\int_{\Omega}\left|g(t,\xi)\right|^{q}d\omega(\xi)\right)^{2/q}d\nu(t)\right]^{1/2}$$

(by definition)

$$\leq\left[\int_{N}\left(\int_{\Omega}\left|g(t,\xi)\right|^{q}d\omega(\xi)\right)^{q/q}d\nu(t)\right]^{1/q}$$

(by $\|\cdot|L_2(N,\nu)\|\leq\|\cdot|L_q(N,\nu)\|$)

$$=\left[\int_{\Omega}\left(\int_{N}\left|g(t,\xi)\right|^{q}d\nu(t)\right)^{q/q}d\omega(\xi)\right]^{1/q}$$

(by Fubini's theorem)

$$\leq K_q(\mathcal{B}_n)\left[\int_\Omega\left(\int_M \left|f(s,\xi)\right|^2 d\mu(s)\right)^{q/2} d\omega(\xi)\right]^{1/q}$$

(by (*))

$$\leq K_q(\mathcal{B}_n)\left[\int_M\left(\int_\Omega \left|f(s,\xi)\right|^q d\omega(\xi)\right)^{2/q} d\mu(s)\right]^{1/2}$$

(by Jessen's inequality)

$$= K_q(\mathcal{B}_n)\,\|\boldsymbol{f}|[L_2, L_q]\|.$$

(by definition)

The case $q = \infty$ can be treated in the same way.

3.4.17 The following estimate is very coarse. Nevertheless, it yields sharp results in some important cases.

EXAMPLE. *If* $1 \leq r \leq \infty$ *then*

$$\boldsymbol{\varrho}(L_r|\mathcal{B}_n, \mathcal{A}_n) \leq \boldsymbol{\delta}(L_r|\mathcal{B}_n, \mathcal{A}_n) \leq n^{|1/r-1/2|}.$$

PROOF. First of all, we recall from 3.4.10 that

$$\boldsymbol{\varrho}(L_r|\mathcal{B}_n, \mathcal{A}_n) \leq \boldsymbol{\delta}(L_r|\mathcal{B}_n, \mathcal{A}_n).$$

For $1 \leq r \leq 2$, the right-hand inequality follows from

$$\boldsymbol{\delta}(L_1|\mathcal{B}_n, \mathcal{A}_n) \leq \sqrt{n} \quad \text{and} \quad \boldsymbol{\delta}(L_2|\mathcal{B}_n, \mathcal{A}_n) = 1,$$

by interpolation; see 3.4.9. In view of 3.4.6, the case $2 \leq r \leq \infty$ can be treated by duality.

3.4.18 Finally, we contrast the properties of Riemann ideal norms with those of Dirichlet ideal norms.

property	$\boldsymbol{\varrho}(\mathcal{B}_n, \mathcal{A}_n)$	$\boldsymbol{\delta}(\mathcal{B}_n, \mathcal{A}_n)$
monotonicity	3.3.6	fails
triangle inequality	3.3.5	3.4.8
duality	fails	3.4.6
injectivity	3.3.3	3.4.5
surjectivity	fails	3.4.5
l_2-stability	3.3.4	3.4.7
interpolation	fails	3.4.9
tensor products	3.6.3	fails

The last line is stated in anticipation of Section 3.6. See also 3.9.8.

3.5 Orthonormal systems with special properties

3.5.1 An orthonormal system $\mathcal{A}_n = (a_1, \ldots, a_n)$ in a Hilbert space $L_2(M, \mu)$ is called **unimodular** if

$$\mu(M) = 1 \quad \text{and} \quad |a_1(s)| = \ldots = |a_n(s)| = 1 \quad \text{for } s \in M.$$

The most important examples are given by systems of characters on compact Abelian groups; see Section 3.8.

When it is necessary to distinguish between the real and the complex case, then we will use the more specific attributes $\mathbb{R}$-unimodular and $\mathbb{C}$-unimodular; see 3.6.2.

3.5.2 If $\mathcal{A}_n$ is unimodular, then

$$\sup_{s\in M} \Big| \sum_{k=1}^{n} \xi_k a_k(s) \Big| \le \sum_{k=1}^{n} |\xi_k|.$$

By complex interpolation, this fact combined with Parseval's equality

$$\Big(\int_M \Big| \sum_{k=1}^{n} \xi_k a_k(s) \Big|^2 d\mu(s) \Big)^{1/2} = \Big(\sum_{k=1}^{n} |\xi_k|^2 \Big)^{1/2}$$

yields the **Hausdorff–Young inequality**

$$\Big(\int_M \Big| \sum_{k=1}^{n} \xi_k a_k(s) \Big|^{p'} d\mu(s) \Big)^{1/p'} \le \Big(\sum_{k=1}^{n} |\xi_k|^p \Big)^{1/p}$$

which holds for $\xi_1, \ldots, \xi_n \in \mathbb{K}$ and $1 \le p \le 2$. Consequently, by 3.2.5 and Hölder's inequality,

$$K'_p(\mathcal{A}_n) \le K_{p'}(\mathcal{A}_n) \le n^{1/p-1/2}.$$

3.5.3 An orthonormal system $\mathcal{A}_n = (a_1, \ldots, a_n)$ is $\mathbb{R}$-**unconditional** if all orthonormal systems $(\varepsilon_1 a_1, \ldots, \varepsilon_n a_n)$ with $\varepsilon_k = \pm 1$ have identical distributions. Replacing ε_k by $e^{i\tau_k}$ with $\tau_k \in \mathbb{R}$, we get the concept of $\mathbb{C}$-**unconditionality**.

REMARK. The most important examples, $\mathcal{R}_n$, $\mathcal{P}_n(\pm 1)$ and $\mathcal{P}_n(e^{i\tau})$, will be introduced in Section 4.1.

3.5.4 The following important observation is due to J.-P. Kahane; see [KAH, p. 20].

PRINCIPLE OF CONTRACTION. *Let $\alpha_1, \ldots, \alpha_n \in \mathbb{K}$ be scalars such that $|\alpha_k| \le 1$. If $\mathcal{A}_n = (a_1, \ldots, a_n)$ is $\mathbb{K}$-unconditional, then*

$$\|(\alpha_k x_k)|\mathcal{A}_n\| \le \|(x_k)|\mathcal{A}_n\|.$$

In the mixed case when $\mathcal{A}_n = (a_1, \dots, a_n)$ is $\mathbb{R}$-unconditional, but the scalars $(\alpha_1, \dots, \alpha_n)$ are still complex, we have

$$\|(\alpha_k x_k)|\mathcal{A}_n\| \le \tfrac{\pi}{2}\, \|(x_k)|\mathcal{A}_n\|.$$

PROOF. The first part of the assertion follows from the fact that, in the real and in the complex case, the extreme points $z = (\zeta_1, \dots, \zeta_n)$ of the compact unit ball in l_∞^n are characterized by the condition $|\zeta_k| = 1$.

The second part could be obtained by splitting into real and imaginary components. But this would yield the factor 2. The sharper estimate is a consequence of Lemma 4.11.5.

3.5.5 PROPOSITION. *Let*

$$\mathcal{A}_{n+m} = (\overbrace{a_1, \dots, a_n}^{\mathcal{A}_n}, a_{n+1} \dots, a_{n+m})$$

and

$$\mathcal{B}_{n+m} = (\overbrace{b_1, \dots, b_n}^{\mathcal{B}_n}, b_{n+1} \dots, b_{n+m}).$$

If $\mathcal{B}_{n+m}$ is $\mathbb{R}$-unconditional, then

$$\boldsymbol{\delta}(\mathcal{B}_n, \mathcal{A}_n) \le \boldsymbol{\delta}(\mathcal{B}_{n+m}, \mathcal{A}_{n+m}).$$

PROOF. Let $\alpha_k = 1$ if $1 \le k \le n$ and $\alpha_k = 0$ if $n+1 \le k \le n+m$. Then

$$\|(T\langle \boldsymbol{f}, \overline{a_k}\rangle)|\mathcal{B}_n\| = \|(\alpha_k T\langle \boldsymbol{f}, \overline{a_k}\rangle)|\mathcal{B}_{n+m}\| \le \boldsymbol{\delta}(T|\mathcal{B}_{n+m}, \mathcal{A}_{n+m})\|\boldsymbol{f}|L_2\|.$$

3.6 Tensor products of orthonormal systems

3.6.1 In the following, $\mathcal{F}_n = (f_1, \dots, f_n)$ is a normalized system in some Hilbert space $L_2(\Omega, \omega)$ which need not be orthogonal. For any orthonormal system $\mathcal{A}_n = (a_1, \dots, a_n)$ in a Hilbert space $L_2(M, \mu)$, the **tensor product** $\mathcal{A}_n \otimes \mathcal{F}_n$ is the orthonormal system in $L_2(M \times \Omega, \mu \times \omega)$ formed by the functions $a_k \otimes f_k : (s, \xi) \to a_k(s) f_k(\xi)$ with $k = 1, \dots, n$. Then

$$\|(x_k)|\mathcal{A}_n \otimes \mathcal{F}_n\|_r = \left(\iint\limits_{M\,\Omega} \left\| \sum_{k=1}^n x_k a_k(s) f_k(\xi) \right\|^r d\mu(s)\, d\omega(\xi) \right)^{1/r}.$$

3.6.2 PROPOSITION. *Assume that $\mathcal{A}_n$ is $\mathbb{K}$-unimodular and that $\mathcal{B}_n$ is $\mathbb{K}$-unconditional. Then*

$$\|(x_k)|\mathcal{A}_n \otimes \mathcal{B}_n\| = \|(x_k)|\mathcal{B}_n\|.$$

In the mixed case when $\mathcal{A}_n$ is $\mathbb{C}$-unimodular and $\mathcal{B}_n$ is $\mathbb{R}$-unconditional, we have

$$\tfrac{2}{\pi}\, \|(x_k)|\mathcal{B}_n\| \le \|(x_k)|\mathcal{A}_n \otimes \mathcal{B}_n\| \le \tfrac{\pi}{2}\, \|(x_k)|\mathcal{B}_n\|.$$

PROOF. We only check the first part of the last inequality, which is an easy consequence of the contraction principle. Indeed, 3.5.4 tells us that

$$\int_N \Big\| \sum_{k=1}^n x_k \overline{a_k(s)} b_k(t) \Big\|^2 d\nu(t) \le \Big(\frac{\pi}{2}\Big)^2 \int_N \Big\| \sum_{k=1}^n x_k b_k(t) \Big\|^2 d\nu(t)$$

for fixed $s \in M$. Substituting $x_k a_k(s)$ for x_k yields

$$\int_N \Big\| \sum_{k=1}^n x_k b_k(t) \Big\|^2 d\nu(t) \le \Big(\frac{\pi}{2}\Big)^2 \int_N \Big\| \sum_{k=1}^n x_k a_k(s) b_k(t) \Big\|^2 d\nu(t).$$

The desired inequality now follows by integration over $s \in M$.

3.6.3 We now provide a tool which will be useful to formalize various proofs. Recall that $\mathcal{F}_n = (f_1, \dots, f_n)$ denotes a normalized system in a Hilbert space $L_2(\Omega, \omega)$.

PROPOSITION. $\varrho(\mathcal{B}_n \otimes \mathcal{F}_n, \mathcal{A}_n \otimes \mathcal{F}_n) \le \varrho(\mathcal{B}_n, \mathcal{A}_n)$.

PROOF. Substituting $(x_k f_k(\xi))$ with $\xi \in \Omega$ in the defining inequality of $\varrho(T|\mathcal{B}_n, \mathcal{A}_n)$, we obtain

$$\int_N \Big\| \sum_{k=1}^n T x_k b_k(t) f_k(\xi) \Big\|^2 d\nu(t) \le$$

$$\le \varrho(T|\mathcal{B}_n, \mathcal{A}_n)^2 \int_M \Big\| \sum_{k=1}^n x_k a_k(s) f_k(\xi) \Big\|^2 d\mu(s).$$

Integration over ξ yields

$$\|(Tx_k)|\mathcal{B}_n \otimes \mathcal{F}_n\| \le \varrho(T|\mathcal{B}_n, \mathcal{A}_n) \, \|(x_k)|\mathcal{A}_n \otimes \mathcal{F}_n\|.$$

3.6.4 Let $\mathcal{A}_m = (a_1, \dots, a_m)$ and $\mathcal{B}_n = (b_1, \dots, b_n)$ be orthonormal systems in the Hilbert spaces $L_2(M, \mu)$ and $L_2(N, \nu)$, respectively. Then the **full tensor product**

$$\mathcal{A}_m \overset{\text{full}}{\otimes} \mathcal{B}_n := \begin{pmatrix} a_1 \otimes b_1 & \cdots & a_1 \otimes b_n \\ \vdots & \cdots & \vdots \\ a_m \otimes b_1 & \cdots & a_m \otimes b_n \end{pmatrix}$$

is an orthonormal system in $L_2(M \times N, \mu \times \nu)$.

3.6.5 PROPOSITION. $\varrho(\mathcal{A}_m \overset{\text{full}}{\otimes} \mathcal{B}_n, \mathcal{I}_{mn}) \le \varrho(\mathcal{A}_m, \mathcal{I}_m) \circ \varrho(\mathcal{B}_n, \mathcal{I}_n)$.

PROOF. By integration over $t \in N$, it follows from

$$\int_M \Big\| \sum_{h=1}^m S\Big[\sum_{k=1}^n Tx_{hk} b_k(t)\Big] a_h(s) \Big\|^2 d\mu(s) \le$$
$$\le \varrho(S|\mathcal{A}_m, \mathcal{I}_m)^2 \sum_{h=1}^m \Big\| \sum_{k=1}^n Tx_{hk} b_k(t) \Big\|^2$$

that

$$\|(STx_{hk})|\mathcal{A}_m \overset{\text{full}}{\otimes} \mathcal{B}_n\|^2 = \int_M \int_N \Big\| \sum_{h=1}^m \sum_{k=1}^n STx_{hk} a_h(s) b_k(t) \Big\|^2 d\mu(s) d\nu(t)$$
$$\le \varrho(S|\mathcal{A}_m, \mathcal{I}_m)^2 \sum_{h=1}^m \int_N \Big\| \sum_{k=1}^n Tx_{hk} b_k(t) \Big\|^2 d\nu(t)$$
$$\le \varrho(S|\mathcal{A}_m, \mathcal{I}_m)^2 \, \varrho(T|\mathcal{B}_n, \mathcal{I}_n)^2 \sum_{h=1}^m \sum_{k=1}^n \|x_{hk}\|^2.$$

3.6.6 PROPOSITION.

$$K_p'(\mathcal{A}_m \overset{\text{full}}{\otimes} \mathcal{B}_n) \le K_p'(\mathcal{A}_m)\, K_p'(\mathcal{B}_n) \qquad \text{if } 1 \le p \le 2,$$
$$K_q(\mathcal{A}_m \overset{\text{full}}{\otimes} \mathcal{B}_n) \le K_q(\mathcal{A}_m)\, K_q(\mathcal{B}_n) \qquad \text{if } 2 \le q < \infty.$$

PROOF. By integration over $t \in N$ and Jessen's inequality 0.4.10, we infer from

$$\int_M \Big| \sum_{h=1}^m \Big[\sum_{k=1}^n \xi_{hk} b_k(t)\Big] a_h(s) \Big|^q d\mu(s) \le K_q(\mathcal{A}_m)^q \Big[\sum_{h=1}^m \Big| \sum_{k=1}^n \xi_{hk} b_k(t) \Big|^2 \Big]^{q/2}$$

that

$$\Big(\int_M \int_N \Big| \sum_{h=1}^m \sum_{k=1}^n \xi_{hk} a_h(s) b_k(t) \Big|^q d\mu(s) d\nu(t) \Big)^{1/q} \le$$
$$\le K_q(\mathcal{A}_m) \Big(\int_N \Big[\sum_{h=1}^m \Big| \sum_{k=1}^n \xi_{hk} b_k(t) \Big|^2 \Big]^{q/2} d\nu(t) \Big)^{1/q}$$
$$\le K_q(\mathcal{A}_m) \Big(\sum_{h=1}^m \Big[\int_N \Big| \sum_{k=1}^n \xi_{hk} b_k(t) \Big|^q d\nu(t) \Big]^{2/q} \Big)^{1/2}$$
$$\le K_q(\mathcal{A}_m) K_q(\mathcal{B}_n) \Big(\sum_{h=1}^m \sum_{k=1}^n |\xi_{hk}|^2 \Big)^{1/2}.$$

This proves the second inequality. The first one can be checked similarly.

3.7 Type and cotype ideal norms

3.7.1 Assume that $1 \le u, v < \infty$. Let $\mathcal{A}_n = (a_1, \dots, a_n)$ be any orthonormal system in $L_2(M, \mu) \cap L_v(M, \mu)$.

For $T \in \mathfrak{L}(X, Y)$, we denote by $\varrho_u^{(v)}(T|\mathcal{A}_n, \mathfrak{I}_n)$ the least constant $c \ge 0$ such that

$$\left(\int\limits_M \left\| \sum_{k=1}^n T x_k a_k(s) \right\|^v d\mu(s) \right)^{1/v} \le c \left(\sum_{k=1}^n \|x_k\|^u \right)^{1/u} \tag{T}$$

whenever $x_1, \dots, x_n \in X$. More concisely, the preceding inequality reads as follows:

$$\|(Tx_k)|\mathcal{A}_n\|_v \le c\,\|(x_k)|l_u^n\|.$$

In the limiting cases $u = \infty$ and $v = \infty$, the usual modifications are required. We refer to

$$\varrho_u^{(v)}(\mathcal{A}_n, \mathfrak{I}_n) : T \longrightarrow \varrho_u^{(v)}(T|\mathcal{A}_n, \mathfrak{I}_n)$$

as a **type ideal norm**. In the quadratic case, we will write $\varrho(\mathcal{A}_n, \mathfrak{I}_n)$ instead of $\varrho_2^{(2)}(\mathcal{A}_n, \mathfrak{I}_n)$.

3.7.2 The above definition has the following counterpart:

Assume that $1 \le u, v < \infty$. Let $\mathcal{A}_n = (a_1, \dots, a_n)$ be any orthonormal system in $L_2(M, \mu) \cap L_u(M, \mu)$.

For $T \in \mathfrak{L}(X, Y)$, we denote by $\varrho_u^{(v)}(T|\mathfrak{I}_n, \mathcal{A}_n)$ the least constant $c \ge 0$ such that

$$\left(\sum_{k=1}^n \|Tx_k\|^v \right)^{1/v} \le c \left(\int\limits_M \left\| \sum_{k=1}^n x_k a_k(s) \right\|^u d\mu(s) \right)^{1/u} \tag{C}$$

whenever $x_1, \dots, x_n \in X$. More concisely, the preceding inequality reads as follows:

$$\|(Tx_k)|l_v^n\| \le c\,\|(x_k)|\mathcal{A}_n\|_u.$$

In the limiting cases $u = \infty$ and $v = \infty$, the usual modifications are required. We refer to

$$\varrho_u^{(v)}(\mathfrak{I}_n, \mathcal{A}_n) : T \longrightarrow \varrho_u^{(v)}(T|\mathfrak{I}_n, \mathcal{A}_n)$$

as a **cotype ideal norm**. In the quadratic case, we will write $\varrho(\mathfrak{I}_n, \mathcal{A}_n)$ instead of $\varrho_2^{(2)}(\mathfrak{I}_n, \mathcal{A}_n)$.

3.7.3 In view of the following observation, it is more or less superfluous to consider the corresponding Dirichlet ideal norms.

PROPOSITION.

$$\boldsymbol{\delta}(\mathcal{A}_n, \mathcal{I}_n) = \boldsymbol{\varrho}(\mathcal{A}_n, \mathcal{I}_n) \quad \textit{and} \quad \boldsymbol{\delta}(\mathcal{I}_n, \mathcal{A}_n) = \boldsymbol{\varrho}'(\overline{\mathcal{A}}_n, \mathcal{I}_n).$$

PROOF. Since $\boldsymbol{\delta}(X|\mathcal{I}_n, \mathcal{I}_n) = 1$ for all Banach spaces X, the left-hand equation follows from 3.4.12. Moreover, by 3.4.6 and $\overline{\mathcal{I}}_n = \mathcal{I}_n$, we have

$$\boldsymbol{\delta}(\mathcal{I}_n, \mathcal{A}_n) = \boldsymbol{\delta}'(\overline{\mathcal{A}}_n, \overline{\mathcal{I}}_n) = \boldsymbol{\varrho}'(\overline{\mathcal{A}}_n, \mathcal{I}_n).$$

3.7.4 In what follows, we denote by $\mathcal{A}_n$ the n-th section of an orthonormal sequence

$$\mathcal{A} = (\overbrace{a_1, \ldots, a_n}^{\mathcal{A}_n}, a_{n+1}, \ldots).$$

PROPOSITION. *The sequences* $(\boldsymbol{\varrho}(\mathcal{A}_n, \mathcal{I}_n))$ *and* $(\boldsymbol{\varrho}(\mathcal{I}_n, \mathcal{A}_n))$ *are non-decreasing.*

3.7.5 Next, we compare type and cotype ideal norms.

PROPOSITION.

$$\boldsymbol{\varrho}(\mathcal{I}_n, \mathcal{A}_n) \leq \boldsymbol{\varrho}'(\overline{\mathcal{A}}_n, \mathcal{I}_n) \leq \boldsymbol{\varrho}(\mathcal{I}_n, \mathcal{A}_n) \circ \boldsymbol{\delta}(\mathcal{A}_n, \mathcal{A}_n).$$

PROOF. By 3.4.10 and 3.4.8,

$$\boldsymbol{\varrho}(\mathcal{I}_n, \mathcal{A}_n) \leq \boldsymbol{\delta}(\mathcal{I}_n, \mathcal{A}_n) \leq \boldsymbol{\varrho}(\mathcal{I}_n, \mathcal{A}_n) \circ \boldsymbol{\delta}(\mathcal{A}_n, \mathcal{A}_n).$$

On the other hand, we know from 3.7.3 that $\boldsymbol{\delta}(\mathcal{I}_n, \mathcal{A}_n) = \boldsymbol{\varrho}'(\overline{\mathcal{A}}_n, \mathcal{I}_n)$.

3.7.6 We now state a special case of 3.3.3 and 3.4.5.

PROPOSITION. *The type ideal norm* $\boldsymbol{\varrho}(\mathcal{A}_n, \mathcal{I}_n)$ *is injective and surjective, while the cotype ideal norm* $\boldsymbol{\varrho}(\mathcal{I}_n, \mathcal{A}_n)$ *is, in general, injective only.*

PROOF. Use $\boldsymbol{\varrho}(\mathcal{A}_n, \mathcal{I}_n) = \boldsymbol{\delta}(\mathcal{A}_n, \mathcal{I}_n)$.

REMARK. Note that $\boldsymbol{\varrho}(\mathcal{I}_n, \mathcal{R}_n)$ fails to be surjective; see 4.2.5.

3.7.7 LEMMA. $\|(x_k)|\mathcal{A}_n \otimes \mathcal{I}_n\| = \|(x_k)|\mathcal{I}_n\|$.

PROOF.

$$\|(x_k)|\mathcal{A}_n \otimes \mathcal{I}_n\|^2 = \sum_{k=1}^{n} \int_M \|x_k a_k(s)\|^2 \, d\mu(s) = \sum_{k=1}^{n} \|x_k\|^2 = \|(x_k)|\mathcal{I}_n\|^2.$$

3.7.8 PROPOSITION.

$$\boldsymbol{\varrho}(\mathcal{A}_n \otimes \mathcal{B}_n, \mathcal{I}_n) \leq \boldsymbol{\varrho}(\mathcal{A}_n, \mathcal{I}_n) \quad \textit{and} \quad \boldsymbol{\varrho}(\mathcal{I}_n, \mathcal{A}_n \otimes \mathcal{B}_n) \leq \boldsymbol{\varrho}(\mathcal{I}_n, \mathcal{A}_n).$$

PROOF. In view of the preceding lemma,

$$\boldsymbol{\varrho}(\mathcal{A}_n \otimes \mathcal{B}_n, \mathcal{I}_n) = \boldsymbol{\varrho}(\mathcal{A}_n \otimes \mathcal{B}_n, \mathcal{I}_n \otimes \mathcal{B}_n),$$

and 3.6.3 tells us that

$$\varrho(\mathcal{A}_n \otimes \mathcal{B}_n, \mathcal{I}_n \otimes \mathcal{B}_n) \le \varrho(\mathcal{A}_n, \mathcal{I}_n).$$

Combining these observations yields the left-hand inequality. The right-hand one can be checked analogously.

3.7.9 PROPOSITION. *Assume that $\mathcal{A}_n$ is $\mathbb{K}$-unimodular and that $\mathcal{B}_n$ is $\mathbb{K}$-unconditional. Then*

$$\varrho(\mathcal{B}_n, \mathcal{I}_n) \le \varrho(\mathcal{A}_n, \mathcal{I}_n) \quad \text{and} \quad \varrho(\mathcal{I}_n, \mathcal{B}_n) \le \varrho(\mathcal{I}_n, \mathcal{A}_n).$$

PROOF. By 3.6.2 and 3.7.8, we have

$$\varrho(\mathcal{B}_n, \mathcal{I}_n) = \varrho(\mathcal{A}_n \otimes \mathcal{B}_n, \mathcal{I}_n) \le \varrho(\mathcal{A}_n, \mathcal{I}_n)$$

and

$$\varrho(\mathcal{I}_n, \mathcal{B}_n) = \varrho(\mathcal{I}_n, \mathcal{A}_n \otimes \mathcal{B}_n) \le \varrho(\mathcal{I}_n, \mathcal{A}_n).$$

3.7.10 Next, we state a special case of 3.4.17.

EXAMPLE. *For any orthonormal system $\mathcal{A}_n$,*

$$\varrho(l_p^n|\mathcal{A}_n, \mathcal{I}_n) \le \varrho(L_p|\mathcal{A}_n, \mathcal{I}_n) \le n^{1/p-1/2} \quad \text{if } 1 \le p \le 2,$$
$$\varrho(l_q^n|\mathcal{I}_n, \mathcal{A}_n) \le \varrho(L_q|\mathcal{I}_n, \mathcal{A}_n) \le n^{1/2-1/q} \quad \text{if } 2 \le q \le \infty.$$

3.7.11 For a large class of orthonormal systems, the above inequalities are even equalities.

EXAMPLE. *Let $\mathcal{A}_n$ be unimodular. Then*

$$\varrho(l_p^n|\mathcal{A}_n, \mathcal{I}_n) = \varrho(L_p|\mathcal{A}_n, \mathcal{I}_n) = n^{1/p-1/2} \quad \text{if } 1 \le p \le 2,$$
$$\varrho(l_q^n|\mathcal{I}_n, \mathcal{A}_n) = \varrho(L_q|\mathcal{I}_n, \mathcal{A}_n) = n^{1/2-1/q} \quad \text{if } 2 \le q \le \infty.$$

PROOF. The lower estimate

$$n^{1/2-1/q} \le \varrho(l_q^n|\mathcal{I}_n, \mathcal{A}_n) \le \varrho(l_{q'}^n|\mathcal{A}_n, \mathcal{I}_n)$$

follows from

$$\|(u_k^{(n)})|[\mathcal{I}_n, l_q^n]\| \le \varrho(l_q^n|\mathcal{I}_n, \mathcal{A}_n)\, \|(u_k^{(n)})|[\mathcal{A}_n, l_q^n]\|,$$
$$\|(u_k^{(n)})|[\mathcal{I}_n, l_q^n]\| = n^{1/2} \quad \text{and} \quad \|(u_k^{(n)})|[\mathcal{A}_n, l_q^n]\| = n^{1/q}.$$

3.7.12 We now use ultraproduct techniques to establish a result which will be crucial in 4.4.5 and 4.11.20.

PROPOSITION. *Let $\mathcal{A}_n = (a_1, \dots, a_n)$ be an orthonormal system in $L_2(M, \mu)$. The following are equivalent:*

(1) *$|a_1(s)| = \ldots = |a_n(s)|$ for almost all $s \in M$.*
(2) *$\varrho(l_1^n|\mathcal{A}_n, \mathcal{I}_n) = \sqrt{n}$.*
(3) *There is a Banach space X such that $\varrho(X|\mathcal{A}_n, \mathcal{I}_n) = \sqrt{n}$.*

In particular, if $\varrho(X|\mathcal{A}_n,\mathcal{I}_n)=\sqrt{n}$, then we can find elements $\boldsymbol{x}_1,\dots,\boldsymbol{x}_n$ in some ultrapower $X^{\mathcal{U}}$ such that $\|\boldsymbol{x}_1\|=\dots=\|\boldsymbol{x}_n\|=1$ and

$$\Big\|\frac{1}{n}\sum_{k=1}^{n}\boldsymbol{x}_k a_k(s)\Big\| = |a_1(s)| = \dots = |a_n(s)| \quad \textit{for almost all } s\in M.$$

PROOF. **(1)⇒(2)**: The lower estimate $\sqrt{n}\le\varrho(l_1^n|\mathcal{A}_n,\mathcal{I}_n)$ follows from

$$\|(u_k^{(n)})|[\mathcal{A}_n,l_1^n]\| \le \varrho(l_1^n|\mathcal{A}_n,\mathcal{I}_n)\,\|(u_k^{(n)})|[\mathcal{I}_n,l_1^n]\|,$$

$$\|(u_k^{(n)})|[\mathcal{A}_n,l_1^n]\| = \left[\int_M\Big(\sum_{k=1}^{n}|a_k(s)|\Big)^2 d\mu(s)\right]^{1/2} = n$$

and

$$\|(u_k^{(n)})|[\mathcal{I}_n,l_1^n]\| = \sqrt{n}.$$

(2)⇒(3) is trivial.

(3)⇒(1): For $\delta>0$, there exist $x_1^{(\delta)},\dots,x_n^{(\delta)}\in X$ such that

$$\left(\int_M\Big\|\sum_{k=1}^{n}x_k^{(\delta)}a_k(s)\Big\|^2 d\mu(s)\right)^{1/2} \ge \frac{1}{1+\delta}\sqrt{n}\Big(\sum_{k=1}^{n}\|x_k^{(\delta)}\|^2\Big)^{1/2}.$$

By scaling, it can be arranged that

$$\frac{n^2}{(1+\delta)^2}\le\int_M\Big\|\sum_{k=1}^{n}x_k^{(\delta)}a_k(s)\Big\|^2 d\mu(s) \quad\text{and}\quad \sum_{k=1}^{n}\|x_k^{(\delta)}\|^2=n.$$

Fix an ultrafilter $\mathcal{U}$ on $\mathbb{R}_+$ which contains all intervals $(0,\delta)$ with $\delta>0$. Passing to the equivalence classes $\boldsymbol{x}_k=(x_k^{(\delta)})^{\mathcal{U}}$ in the ultrapower $X^{\mathcal{U}}$ yields

$$n^2\le\int_M\Big\|\sum_{k=1}^{n}\boldsymbol{x}_k a_k(s)\Big\|^2 d\mu(s) \quad\text{and}\quad \sum_{k=1}^{n}\|\boldsymbol{x}_k\|^2=n.$$

From

$$\Big\|\sum_{k=1}^{n}\boldsymbol{x}_k a_k(s)\Big\| \le \sum_{k=1}^{n}\|\boldsymbol{x}_k\|\,|a_k(s)| \le \Big(\sum_{k=1}^{n}\|\boldsymbol{x}_k\|^2\Big)^{1/2}\Big(\sum_{k=1}^{n}|a_k(s)|^2\Big)^{1/2} \qquad (*)$$

we obtain

$$n^2\le\int_M\Big\|\sum_{k=1}^{n}\boldsymbol{x}_k a_k(s)\Big\|^2 d\mu(s) \le \int_M\Big(\sum_{k=1}^{n}\|\boldsymbol{x}_k\|\,|a_k(s)|\Big)^2 d\mu(s)$$

$$\le\sum_{k=1}^{n}\|\boldsymbol{x}_k\|^2\int_M\sum_{k=1}^{n}|a_k(s)|^2 d\mu(s)=n^2.$$

Thus, for almost all $s \in M$, equality holds in $(*)$. Hence, by 0.4.6, there exists $a(s) \geq 0$ such that $|a_k(s)| = \|\boldsymbol{x}_k\|\, a(s)$. We now conclude from

$$1 = \int\limits_M |a_k(s)|^2\, d\mu(s) = \|\boldsymbol{x}_k\|^2 \int\limits_M |a(s)|^2\, d\mu(s)$$

that $\|\boldsymbol{x}_1\| = \ldots = \|\boldsymbol{x}_n\|$ which in turn yields $\|\boldsymbol{x}_k\| = 1$, $|a_k(s)| = a(s)$ and

$$\Big\| \sum_{k=1}^n \boldsymbol{x}_k a_k(s) \Big\|^2 = n \sum_{k=1}^n |a_k(s)|^2 \quad \text{for almost all } s \in M.$$

REMARK. Condition **(1)** means that the modified orthonormal system $\mathcal{N}\mathcal{A}_n$ is unimodular; see 3.1.8.

3.7.13 We proceed with a dual version of the previous result.

PROPOSITION. *Let $\mathcal{A}_n = (a_1, \ldots, a_n)$ be an orthonormal system in $L_2(M, \mu)$. The following are equivalent:*

(1) $|a_1(s)| = \ldots = |a_n(s)|$ *for almost all* $s \in M$.

(2) $\varrho(l_\infty^n | \mathcal{I}_n, \mathcal{A}_n) = \sqrt{n}$.

(3) *There is a Banach space X such that* $\varrho(X | \mathcal{I}_n, \mathcal{A}_n) = \sqrt{n}$.

In particular, if $\varrho(X | \mathcal{I}_n, \mathcal{A}_n) = \sqrt{n}$, then we can find elements $\boldsymbol{x}_1, \ldots, \boldsymbol{x}_n$ in some ultrapower $X^{\mathcal{U}}$ such that $\|\boldsymbol{x}_1\| = \ldots = \|\boldsymbol{x}_n\| = 1$ and

$$\Big\| \sum_{k=1}^n \boldsymbol{x}_k a_k(s) \Big\| = |a_1(s)| = \ldots = |a_n(s)| \quad \textit{for almost all } s \in M.$$

PROOF. **(1)⇒(2)**: The lower estimate $\sqrt{n} \leq \varrho(l_\infty^n | \mathcal{I}_n, \mathcal{A}_n)$ follows from

$$\|(u_k^{(n)}) | [\mathcal{I}_n, l_\infty^n]\| \leq \varrho(l_\infty^n | \mathcal{I}_n, \mathcal{A}_n)\, \|(u_k^{(n)}) | [\mathcal{A}_n, l_\infty^n]\|,$$

$$\|(u_k^{(n)}) | [\mathcal{I}_n, l_\infty^n]\| = \sqrt{n}$$

and

$$\|(u_k^{(n)}) | [\mathcal{A}_n, l_\infty^n]\| = \Big(\int\limits_M \sup_{1 \leq k \leq n} |a_k(s)|^2 d\mu(s) \Big)^{1/2} = 1.$$

(2)⇒(3) is trivial.

(3)⇒(1): For $\delta > 0$, there exist $x_1^{(\delta)}, \ldots, x_n^{(\delta)} \in X$ such that

$$\Big(\sum_{k=1}^n \|x_k^{(\delta)}\|^2 \Big)^{1/2} \geq \frac{1}{1+\delta} \sqrt{n} \Big(\int\limits_M \Big\| \sum_{k=1}^n x_k^{(\delta)} a_k(s) \Big\|^2 d\mu(s) \Big)^{1/2}.$$

By scaling, it can be arranged that

$$\sum_{k=1}^n \|x_k^{(\delta)}\|^2 = n \quad \text{and} \quad \int\limits_M \Big\| \sum_{k=1}^n x_k^{(\delta)} a_k(s) \Big\|^2 d\mu(s) \leq (1+\delta)^2.$$

Fix an ultrafilter $\mathcal{U}$ on $\mathbb{R}_+$ which contains all intervals $(0,\delta)$ with $\delta>0$. Passing to the equivalence classes $\boldsymbol{x}_k=(x_k^{(\delta)})^{\mathcal{U}}$ in the ultrapower $X^{\mathcal{U}}$ yields

$$\sum_{k=1}^n \|\boldsymbol{x}_k\|^2 = n \quad \text{and} \quad \int\limits_M \Big\|\sum_{k=1}^n \boldsymbol{x}_k a_k(s)\Big\|^2 d\mu(s) \le 1.$$

We conclude from

$$\boldsymbol{x}_k = \int\limits_M \Big(\sum_{h=1}^n \boldsymbol{x}_h a_h(s)\Big)\overline{a_k(s)}\, d\mu(s)$$

that

$$\begin{aligned} \|\boldsymbol{x}_k\| &\le \int\limits_M \Big\|\sum_{h=1}^n \boldsymbol{x}_h a_h(s)\Big\| |a_k(s)|\, d\mu(s) \\ &\le \Big(\int\limits_M \Big\|\sum_{h=1}^n \boldsymbol{x}_h a_h(s)\Big\|^2 d\mu(s)\Big)^{1/2} \le 1. \end{aligned} \tag{1}$$

Hence

$$1 = \Big(\frac{1}{n}\sum_{k=1}^n \|\boldsymbol{x}_k\|^2\Big)^{1/2} \le \Big(\int\limits_M \Big\|\sum_{h=1}^n \boldsymbol{x}_h a_h(s)\Big\|^2 d\mu(s)\Big)^{1/2} \le 1. \tag{2}$$

So all inequalities in (1) and (2) are even equalities. In particular, we have

$$\|\boldsymbol{x}_1\| = \ldots = \|\boldsymbol{x}_n\| = \int\limits_M \Big\|\sum_{h=1}^n \boldsymbol{x}_h a_h(s)\Big\|^2 d\mu(s) = 1. \tag{3}$$

Moreover, by 0.4.6, there exist positive constants $c_1,\ldots,c_n$ such that

$$|a_k(s)| = c_k \Big\|\sum_{h=1}^n \boldsymbol{x}_h a_h(s)\Big\| \quad \text{for almost all } s\in M \text{ and } k=1,\ldots,n.$$

However, it follows from $\|a_k|L_2\|=1$ and (3) that $c_k=1$, which proves **(1)**.

3.7.14 Recall that Σ_n denotes the finite summation operator; see 0.7.3.

PROPOSITION.

$$\begin{aligned} &\boldsymbol{\varrho}(\Sigma_n : l_1^n \to l_\infty^n|\mathcal{A}_n,\mathfrak{I}_n) = \\ &= \sup\Big\{\Big(\int\limits_M \max_{1\le h\le n}\Big|\sum_{\pi(k)\le h}\xi_k a_k(s)\Big|^2 d\mu(s)\Big)^{1/2} : \pi,\ \|(\xi_k)|l_2^n\|\le 1\Big\}, \end{aligned}$$

where π ranges over all permutations of $\{1,\ldots,n\}$.

PROOF. Choose $x_1,\ldots,x_n\in l_1^n$ with

$$\|(\Sigma_n x_k)|[\mathcal{A}_n, l_\infty^n]\| = \boldsymbol{\varrho}(\Sigma_n : l_1^n\to l_\infty^n|\mathcal{A}_n,\mathfrak{I}_n) \quad\text{and}\quad \|(x_k)|[l_2^n, l_1^n]\| = 1.$$

By the extreme point lemma 0.9.5, we may assume that $x_1, \dots, x_n$ are multiples of unit vectors. Thus there exist a map π of $\{1, \dots, n\}$ into itself and coefficients $\xi_1, \dots, \xi_n \in \mathbb{K}$ such that $x_k = \xi_k u^{(n)}_{\pi(k)}$. Note that $\|(\xi_k)|l_2^n\| = \|(x_k)|[l_2^n, l_1^n]\| = 1$. The desired formula now follows from

$$\Big\|\sum_{k=1}^n \Sigma_n x_k a_k(s)\Big|l_\infty^n\Big\| = \Big\|\sum_{k=1}^n \xi_k \sum_{\pi(k)\le h} u_h^{(n)} a_k(s)\Big|l_\infty^n\Big\|$$

$$= \Big\|\sum_{h=1}^n \Big(\sum_{\pi(k)\le h} \xi_k a_k(s)\Big) u_h^{(n)}\Big|l_\infty^n\Big\|$$

$$= \max_{1\le h\le n}\Big|\sum_{\pi(k)\le h} \xi_k a_k(s)\Big|.$$

Finally, we observe that it is enough to take the supremum over all permutations π of $\{1, \dots, n\}$.

REMARK. We deduce from 2.3.13, 3.4.11, and 2.4.14 that

$$\varrho(\Sigma_n : l_1^n \to l_\infty^n | \mathcal{A}_n, \mathfrak{I}_n) \le \|\Sigma_n : l_1^n \to l_\infty^n | \mathfrak{H}\| \asymp 1+\log n.$$

This yields the Rademacher–Menshov theorem:

$$\sup\left\{\Big(\int_M \max_{1\le h\le n}\Big|\sum_{k=1}^h \xi_k a_k(s)\Big|^2 d\mu(s)\Big)^{1/2} : \|(\xi_k)|l_2^n\| \le 1\right\} \prec 1+\log n.$$

Classical proofs can be found in [KAC*, p. 164] and [KAS*, p. 251]. There are orthonormal systems $\mathcal{A}_n = (a_1, \dots, a_n)$ for which the above estimate is sharp; see [KAS*, p. 255] and [OLE, p. 42].

3.7.15 For every sequence $t = (\tau_k)$ with $\tau_1 \ge \tau_2 \ge \ldots \ge 0$, the associated **diagonal operator** is defined by $D_t : (\xi_k) \to (\tau_k \xi_k)$.

EXAMPLE. *Let $\mathcal{A}_n$ be any unimodular orthonormal system. Then*

$$\varrho_p^{(v)}(D_t : l_1 \to l_1 | \mathcal{A}_n, \mathfrak{I}_n) = \Big(\sum_{k=1}^n \tau_k^{p'}\Big)^{1/p'} \qquad \text{if } 1 < p \le 2,\ 1 \le v \le p',$$

and

$$\varrho_u^{(q)}(D_t : l_\infty \to l_\infty | \mathfrak{I}_n, \mathcal{A}_n) = \Big(\sum_{k=1}^n \tau_k^{q}\Big)^{1/q} \qquad \text{if } 2 \le q < \infty,\ q' \le u \le \infty.$$

PROOF. We have

$$\|(D_t u_k)|[\mathfrak{I}_n, l_\infty]\|_q = \Big(\sum_{k=1}^n \|D_t u_k\|^q\Big)^{1/q} = \Big(\sum_{k=1}^n \tau_k^q\Big)^{1/q}$$

and

$$\|(u_k)|[\mathcal{A}_n, l_\infty]\|_u = \Big(\int\limits_M \Big\| \sum_{k=1}^n u_k a_k(s) \Big\|^u d\mu(s) \Big)^{1/u} = 1.$$

Hence

$$\Big(\sum_{k=1}^n \tau_k^q \Big)^{1/q} \le \boldsymbol{\varrho}_u^{(q)}(D_t : l_\infty \to l_\infty | \mathfrak{I}_n, \mathcal{A}_n). \qquad (1)$$

Fix any natural number N, and let $D_t^{(N)}$ denote the restriction of D_t to l_1^N, the N-th section of l_1. Choose $x_1, \ldots, x_n \in l_1^N$ such that

$$\|(D_t^{(N)} x_k)|[\mathcal{A}_n, l_1^N]\|_v = \boldsymbol{\varrho}_p^{(v)}(D_t^{(N)} : l_1^N \to l_1^N | \mathcal{A}_n, \mathfrak{I}_n)$$

and

$$\|(x_k)|[l_p^n, l_1^N]\| = 1.$$

By the extreme point lemma 0.9.5, we may assume that $x_1, \ldots, x_n$ are multiples of unit vectors. Thus there exist a map π from $\{1, \ldots, n\}$ into $\{1, \ldots, N\}$ and coefficients $\xi_1, \ldots, \xi_n \in \mathbb{K}$ such that $x_k = \xi_k u_{\pi(k)}^{(N)}$. Note that $\|(\xi_k)|l_p^n\| = \|(x_k)|[l_p^n, l_1^N]\| = 1$. In view of the Hausdorff–Young inequality 3.5.2, we now obtain

$$\boldsymbol{\varrho}_p^{(v)}(D_t^{(N)} : l_1^N \to l_1^N | \mathcal{A}_n, \mathfrak{I}_n) = \Big(\int\limits_M \Big\| \sum_{k=1}^n D_t^{(N)} x_k a_k(s) \Big\|^v d\mu(s) \Big)^{1/v}$$

$$= \Big(\int\limits_M \Big\| \sum_{i=1}^N \tau_i \Big[\sum_{\pi(k)=i} \xi_k a_k(s) \Big] u_i^{(N)} \Big\|^v d\mu(s) \Big)^{1/v}$$

$$= \Big(\int\limits_M \Big[\sum_{i=1}^N \tau_i \Big| \sum_{\pi(k)=i} \xi_k a_k(s) \Big| \Big]^v d\mu(s) \Big)^{1/v}$$

$$\le \sum_{i=1}^N \tau_i \Big(\int\limits_M \Big| \sum_{\pi(k)=i} \xi_k a_k(s) \Big|^v d\mu(s) \Big)^{1/v}$$

$$\le \sum_{i=1}^N \tau_i \Big(\int\limits_M \Big| \sum_{\pi(k)=i} \xi_k a_k(s) \Big|^{p'} d\mu(s) \Big)^{1/p'} \le \sum_{i=1}^N \tau_i \Big(\sum_{\pi(k)=i} |\xi_k|^p \Big)^{1/p}.$$

Since at most n sums $\sum\limits_{\pi(k)=i} |\xi_k|^p$ do not equal 0, we conclude from $\tau_1 \ge \tau_2 \ge \ldots \ge 0$ that

$$\boldsymbol{\varrho}_p^{(v)}(D_t^{(N)} : l_1^N \to l_1^N | \mathcal{A}_n, \mathfrak{I}_n) \le \Big(\sum_{k=1}^n |\tau_k|^{p'} \Big)^{1/p'} \Big(\sum_{i=1}^N \sum_{\pi(k)=i} |\xi_k|^p \Big)^{1/p}$$

$$\le \Big(\sum_{k=1}^n |\tau_k|^{p'} \Big)^{1/p'}.$$

Finally,

$$\varrho_p^{(v)}(D_t : l_1 \to l_1|\mathcal{A}_n, \mathfrak{I}_n) = \lim_{N\to\infty} \varrho_p^{(v)}(D_t^{(N)} : l_1^N \to l_1^N|\mathcal{A}_n, \mathfrak{I}_n)$$

implies that

$$\varrho_p^{(v)}(D_t : l_1 \to l_1|\mathcal{A}_n, \mathfrak{I}_n) \le \Big(\sum_{k=1}^n |\tau_k|^{p'}\Big)^{1/p'}. \tag{2}$$

If $p = q'$ and $v = u'$, then a generalization of 3.7.5 says that

$$\varrho_u^{(q)}(D_t : l_\infty \to l_\infty|\mathfrak{I}_n, \mathcal{A}_n) \le \varrho_p^{(v)}(D_t : l_1 \to l_1|\overline{\mathcal{A}}_n, \mathfrak{I}_n). \tag{3}$$

Combining (1), (2) and (3) completes the proof.

3.7.16 EXAMPLE. *Let $\mathcal{A}_n$ be any unimodular orthonormal system. Then*

$$\varrho_p^{(p')}(L_r|\mathcal{A}_n, \mathfrak{I}_n) = 1 \quad \text{if } p \le r \le p',$$
$$\varrho_{q'}^{(q)}(L_r|\mathfrak{I}_n, \mathcal{A}_n) = 1 \quad \text{if } q' \le r \le q.$$

PROOF. Fix $f_1, \dots, f_n \in L_r(\Omega, \omega)$ and $\xi \in \Omega$. By the Hausdorff–Young inequality 3.5.2, we have

$$\Big(\int_M \Big|\sum_{k=1}^n f_k(\xi)a_k(s)\Big|^{p'} d\mu(s)\Big)^{1/p'} \le \Big(\sum_{k=1}^n |f_k(\xi)|^p\Big)^{1/p}.$$

Integration over $\xi \in \Omega$ yields

$$\Big[\int_\Omega\Big(\int_M\Big|\sum_{k=1}^n f_k(\xi)a_k(s)\Big|^{p'}d\mu(s)\Big)^{r/p'} d\omega(\xi)\Big]^{1/r} \le \Big[\int_\Omega\Big(\sum_{k=1}^n|f_k(\xi)|^p\Big)^{r/p} d\omega(\xi)\Big]^{1/r}.$$

Since $p \le r$ and $r \le p'$, Jessen's inequality can be applied to both sides of the above formula. Hence

$$\Big[\int_M\Big(\int_\Omega\Big|\sum_{k=1}^n f_k(\xi)a_k(s)\Big|^{r}d\omega(\xi)\Big)^{p'/r} d\mu(s)\Big]^{1/p'} \le \Big[\sum_{k=1}^n\Big(\int_\Omega|f_k(\xi)|^r d\omega(\xi)\Big)^{p/r}\Big]^{1/p}.$$

More concisely,

$$\Big(\int_M \|\sum_{k=1}^n f_k a_k(s)|L_r\|^{p'} d\mu(s)\Big)^{1/p'} \le \Big(\sum_{k=1}^n \|f_k|L_r\|^p\Big)^{1/p}$$

which implies that $\varrho_p^{(p')}(L_r|\mathcal{A}_n, \mathfrak{I}_n) \le 1$. The reverse estimate is obvious. This proves the assertion for type ideal norms. Letting $p := q'$, the cotype case can be treated by duality,

$$\varrho_{q'}^{(q)}(L_r|\mathfrak{I}_n, \mathcal{A}_n) \le \varrho_p^{(p')}(L_{r'}|\overline{\mathcal{A}}_n, \mathfrak{I}_n).$$

3.8 Characters on compact Abelian groups

3.8.1 This section is devoted to **compact Abelian groups**, denoted by $\mathbb{G}$ or $\mathbb{H}$. In the abstract setting, the algebraic operation is written as addition, and o stands for the zero element. However, in concrete examples we use the traditional notation.

3.8.2 On every compact Abelian group $\mathbb{G}$ there exists a normalized (unique) **Haar measure** μ which, by definition, is translation invariant:

$$\int_{\mathbb{G}} f(s-t)\,d\mu(s) = \int_{\mathbb{G}} f(s)\,d\mu(s) \quad \text{whenever } f \in C(\mathbb{G}) \text{ and } t \in \mathbb{G}.$$

Here $C(\mathbb{G})$ is the Banach space of all continuous functions on $\mathbb{G}$. For $1 \le r < \infty$, we let $L_r(\mathbb{G})$ denote the Banach space of r-integrable functions, and the inner product on the Hilbert space $L_2(\mathbb{G})$ is given by

$$(f,g) := \int_{\mathbb{G}} f(s)\overline{g(s)}\,d\mu(s).$$

3.8.3 The most important examples of compact Abelian groups are the **circle group**

$$\mathbb{T} := \Big\{ \zeta \in \mathbb{C} : |\zeta| = 1 \Big\}$$

and the **cyclic groups**

$$\mathbb{E}_n := \Big\{ \exp(\tfrac{2\pi i}{n} h) \ : \ h = 1, \dots, n \Big\} = \Big\{ \zeta \in \mathbb{C} \ : \ \zeta^n = 1 \Big\}$$

formed by the n-th roots of unity and their direct products.

3.8.4 By a **character** we mean a continuous map $a : \mathbb{G} \to \mathbb{T}$ such that

$$a(s+t) = a(s)\,a(t) \quad \text{for } s,\ t \in \mathbb{G}.$$

Note that $a(o) = 1$.

3.8.5 The collection of all characters on $\mathbb{G}$ becomes a discrete Abelian group, denoted by $\mathbb{G}'$. In this case, we mostly write the algebraic operation as multiplication:

$$ab(s) := a(s)b(s) \quad \text{for } s \in \mathbb{G}.$$

It is fundamental that the characters form an orthonormal system (basis) in $L_2(\mathbb{G})$. In particular, by [RUD, p. 10], we have

$$\int_{\mathbb{G}} a(s)\,d\mu(s) = 0 \quad \text{whenever } a \neq 1.$$

CONVENTION. Throughout this book, if $\mathcal{A}_n = (a_1, \ldots, a_n)$ is a system of characters, then we tacitly assume that its elements are pairwise different, $a_h \neq a_k$ for $h \neq k$.

3.8.6 The characters of $\mathbb{T}$ are given by

$$e_k : \zeta \longrightarrow \zeta^k.$$

Thus $\mathbb{T}'$ is isomorphic to the (additive) group $\mathbb{Z} := \{0, \pm 1, \pm 2, \ldots\}$. Writing ζ in the form $\zeta = \exp(it)$, the character e_k can be identified with the function $e_k(t) = \exp(ikt)$.

The characters of $\mathbb{E}_n$ have the same form with the understanding that $e_h = e_k$ if $h \equiv k \pmod{n}$. Thus $\mathbb{E}_n$ and $\mathbb{E}'_n$ are isomorphic.

3.8.7 The structure theorem tells us that every **finitely generated Abelian group** $\mathbb{G}$ is the direct product of cyclic groups; see [HEW*1, p. 451], [HUP, p. 80] and [LAN, pp. 47–48]. This means that $\mathbb{G}$ can be written in the form

$$\mathbb{Z}^N \times \mathbb{E}_{N_1} \times \ldots \times \mathbb{E}_{N_j},$$

where $N_1, \ldots, N_j$ are powers of primes. In particular, every finite Abelian group $\mathbb{G}_n$ is isomorphic with its character group $\mathbb{G}'_n$, and we have $|\mathbb{G}_n| = |\mathbb{G}'_n|$.

3.8.8 Taking arbitrary enumerations

$$\mathbb{G}_n = \{s_1, \ldots, s_n\} \quad \text{and} \quad \mathbb{G}'_n = \{a_1, \ldots, a_n\},$$

we obtain the unitary **character matrix**

$$\Gamma_n = \left(\gamma_{hk}^{(n)}\right) \quad \text{with} \quad \gamma_{hk}^{(n)} := n^{-1/2} a_k(s_h) \quad \text{for } h, k = 1, \ldots, n.$$

In the particular case of the cyclic group $\mathbb{E}_n$, the character matrix is given by

$$E_n^\circ := \left(\varepsilon_{hk}^{(n)}\right) \quad \text{with} \quad \varepsilon_{hk}^{(n)} := n^{-1/2} \exp\left(\tfrac{2\pi i}{n} hk\right) \quad \text{for } h, k = 1, \ldots, n.$$

Thus we have symmetry. Thanks to the structure theorem, there exists an isomorphism $s_i \leftrightarrow a_i$ which yields symmetry in general, $\gamma_{hk}^{(n)} = \gamma_{kh}^{(n)}$.

3.8.9 Let

$$\mathbb{G}^n := \left\{ s = (\sigma_1, \ldots, \sigma_n) : \sigma_k \in \mathbb{G} \quad \text{for } k = 1, \ldots, n \right\}$$

be the n-fold direct power equipped with the product measure μ^n. The coordinate functionals (projections) are defined by

$$p_k^{(n)}(s) := \sigma_k \quad \text{for } k = 1, \ldots, n.$$

Of particular interest are the n-th **Cantor group**

$$\mathbb{E}_2^n := \Big\{\, e = (\varepsilon_1, \ldots, \varepsilon_n) \,:\, \varepsilon_k = \pm 1 \Big\}$$

with $|\mathbb{E}_2^n| = 2^n$ and its complex counterpart

$$\mathbb{T}^n := \Big\{\, z = (\zeta_1, \ldots, \zeta_n) \,:\, |\zeta_k| = 1 \Big\}.$$

In these cases, we refer to $(p_1^{(n)}, \ldots, p_n^{(n)})$ as the **Bernoulli system** $\mathcal{P}_n(\pm 1)$ and the **Steinhaus system** $\mathcal{P}_n(e^{i\tau})$, respectively. Another definition of these orthonormal systems will be given in Section 4.1, where we take a probabilistic point of view.

When the real and the complex case are treated simultaneously, we use $\mathbb{E}^n$ as a synonym of $\mathbb{E}_2^n$ and $\mathbb{T}^n$. Moreover, $\mathcal{P}_n$ stands for both, the Bernoulli system as well as the Steinhaus system, and λ^n denotes the underlying Haar measure.

REMARK. Since this treatise mainly deals with finite orthonormal systems, the (infinite) **Cantor group**

$$\mathbb{E}_2^\infty := \Big\{\, e = (\varepsilon_1, \varepsilon_2, \ldots) \,:\, \varepsilon_1 = \pm 1,\ \varepsilon_2 = \pm 1, \ldots \,\Big\}$$

is less important for our considerations.

3.8.10 The following results demonstrate the great advantages which can be achieved by exploiting the group structure. We begin with a trivial observation.

PROPOSITION. *If $\mathcal{A}_n = (a_1, \ldots, a_n)$ is any system of characters on a compact Abelian group $\mathbb{G}$, then*

$$\|(x_k)|\overline{\mathcal{A}}_n\|_r = \|(x_k)|\mathcal{A}_n\|_r \quad \textit{and} \quad \|(x_k)|\mathcal{A}_n \otimes \mathcal{A}_n\|_r = \|(x_k)|\mathcal{A}_n\|_r.$$

PROOF. Note that $\overline{a(s)} = a(-s)$ and $a_k(s)a_k(t) = a_k(s+t)$ for $s, t \in \mathbb{G}$. The desired relations now follow from the invariance properties of the Haar measure.

3.8.11 For the definition of $M_1(\mathcal{A}_n)$, the reader is referred to 3.2.3.

PROPOSITION. *Let $\mathcal{A}_n = (a_1, \ldots, a_n)$ be any system of characters on a compact Abelian group $\mathbb{G}$. Then*

$$\boldsymbol{\delta}(\mathcal{A}_n, \mathcal{A}_n) \le \int\limits_{\mathbb{G}} \Big| \sum_{k=1}^n a_k(s) \Big| \, d\mu(s) = M_1(\mathcal{A}_n)\sqrt{n}.$$

PROOF. First of all, we note that the Dirichlet kernel, defined in 3.4.2, has the form

$$A_n A_n^*(s,t) = \sum_{k=1}^{n} \overline{a_k}(s) a_k(t) = \sum_{k=1}^{n} a_k(t-s).$$

Hence, for $T \in \mathfrak{L}(X,Y)$, the expression $\boldsymbol{\delta}(T|\mathcal{A}_n, \mathcal{A}_n)$ is nothing but the norm of the convolution operator

$$\boldsymbol{f}(s) \longrightarrow \int_{\mathbb{G}} T\boldsymbol{f}(s) \sum_{k=1}^{n} a_k(t-s)\, d\mu(s)$$

acting from $[L_2(\mathbb{G}), X]$ into $[L_2(\mathbb{G}), Y]$. This implies that

$$\boldsymbol{\delta}(T|\mathcal{A}_n, \mathcal{A}_n) \le \|T\| \int_{\mathbb{G}} \Big| \sum_{k=1}^{n} a_k(s) \Big|\, d\mu(s),$$

by Young's inequality; see 4.11.8 or [BENN*, p. 199].

3.8.12 The next example shows that the preceding inequality is sharp.

EXAMPLE. *Let $\mathcal{A}_n = (a_1, \dots, a_n)$ be any system of characters on a compact Abelian group $\mathbb{G}$. Then*

$$\boldsymbol{\delta}(L_1(\mathbb{G})|\mathcal{A}_n, \mathcal{A}_n) = \int_{\mathbb{G}} \Big| \sum_{k=1}^{n} a_k(s) \Big|\, d\mu(s) = M_1(\mathcal{A}_n)\sqrt{n}.$$

PROOF. If U is a zero neighbourhood, then we may choose a non-negative (measurable or even continuous) function h_U vanishing outside U such that

$$\int_{\mathbb{G}} h_U(v)\, d\mu(v) = 1.$$

Write $f_U(s,u) := h_U(s-u)$ for $s, u \in \mathbb{G}$, and view this scalar-valued function as a vector-valued function $\boldsymbol{f}_U \in [L_2(\mathbb{G}), L_1(\mathbb{G})]$. Obviously,

$$\|\boldsymbol{f}_U|[L_2, L_1]\| = \left[\int_{\mathbb{G}} \left(\int_{\mathbb{G}} h_U(s-u)\, d\mu(u) \right)^2 d\mu(s) \right]^{1/2} = 1.$$

On the other hand, if $\boldsymbol{g}_U$ denotes the image of $\boldsymbol{f}_U$ under the Dirichlet projection, then the substitution $v = s-u$ yields

$$\begin{aligned} g_U(t,u) &= \int_{\mathbb{G}} h_U(s-u) \sum_{k=1}^{n} a_k(t-s)\, d\mu(s) \\ &= \sum_{k=1}^{n} \int_{\mathbb{G}} h_U(v) \overline{a_k(v)}\, d\mu(v)\, a_k(t-u). \end{aligned}$$

Therefore, it follows from

$$\lim_U \int_{\mathbb{G}} h_U(v)\overline{a_k(v)}\, d\mu(v) = 1$$

and

$$\int_{\mathbb{G}} \Big| \sum_{k=1}^{n} a_k(t-u) \Big| \, d\mu(u) = \int_{\mathbb{G}} \Big| \sum_{k=1}^{n} a_k(-u) \Big| \, d\mu(u)$$

that

$$\lim_U \| \boldsymbol{g}_U | [L_2, L_1] \| = \int_{\mathbb{G}} \Big| \sum_{k=1}^{n} a_k(u) \Big| \, d\mu(u) = M_1(\mathcal{A}_n)\, \sqrt{n}.$$

In view of

$$\| \boldsymbol{g}_U | [L_2, L_1] \| \le \boldsymbol{\delta}(L_1(\mathbb{G}) | \mathcal{A}_n, \mathcal{A}_n) \, \| \boldsymbol{f}_U | [L_2, L_1] \|,$$

this proves the required lower estimate.

REMARK. Note that

$$L(\mathcal{A}_n) := \int_{\mathbb{G}} \Big| \sum_{k=1}^{n} a_k(s) \Big| \, d\mu(s)$$

is called the **Lebesgue constant** of $\mathcal{A}_n$.

3.8.13 Type and cotype ideal norms may be quite different. However, there is a remarkable case in which they coincide.

PROPOSITION. *Let $\mathcal{A}_n$ be the system formed by* **all** *characters of any finite Abelian group $\mathbb{G}_n$. Then*

$$\boldsymbol{\varrho}(\mathcal{A}_n, \mathcal{I}_n) = \boldsymbol{\delta}(\mathcal{A}_n, \mathcal{I}_n) = \boldsymbol{\varrho}(\mathcal{I}_n, \mathcal{A}_n) = \boldsymbol{\delta}(\mathcal{I}_n, \mathcal{A}_n).$$

PROOF. Choose enumerations

$$\mathbb{G}_n = \{s_1, \ldots, s_n\} \quad \text{and} \quad \mathbb{G}_n' = \{a_1, \ldots, a_n\}$$

such that $a_k(s_h) = a_h(s_k)$. Given $x_1, \ldots, x_n$, we define

$$x_h^\circ := \frac{1}{\sqrt{n}} \sum_{k=1}^{n} x_k a_k(s_h) \quad \text{for } h = 1, \ldots, n.$$

Since $\big(\frac{1}{\sqrt{n}} a_k(s_h)\big)$ is unitary, we may conclude from $\overline{a_k(s_h)} = a_h(-s_k)$ that

$$x_k = \frac{1}{\sqrt{n}} \sum_{h=1}^{n} x_h^\circ \, \overline{a_k(s_h)} = \frac{1}{\sqrt{n}} \sum_{h=1}^{n} x_h^\circ \, a_h(-s_k).$$

Hence

$$\| (Tx_k) | \mathcal{A}_n \| = \Big(\frac{1}{n} \sum_{h=1}^{n} \Big\| \sum_{k=1}^{n} T x_k a_k(s_h) \Big\|^2 \Big)^{1/2} = \| (Tx_h^\circ) | l_2^n \|$$

and

$$\|(x_k)|l_2^n\| = \left(\frac{1}{n}\sum_{k=1}^{n}\left\|\sum_{h=1}^{n} x_h^\circ \overline{a_k(s_h)}\right\|^2\right)^{1/2} = \|(x_h^\circ)|\mathcal{A}_n\|.$$

Thus the inequalities

$$\|(Tx_k)|\mathcal{A}_n\| \le c\,\|(x_k)|l_2^n\| \quad\text{and}\quad \|(Tx_h^\circ)|l_2^n\| \le c\,\|(x_h^\circ)|\mathcal{A}_n\|$$

are equivalent, which proves that

$$\boldsymbol{\varrho}(\mathcal{A}_n, \mathfrak{I}_n) = \boldsymbol{\varrho}(\mathfrak{I}_n, \mathcal{A}_n).$$

The remaining equalities follow from 3.4.12.

3.8.14 For many orthonormal systems, type and cotype ideal norms fail to be symmetric. Nevertheless, there is the following exception.

PROPOSITION. *Let $\mathcal{A}_n$ be the system formed by* **all** *characters of any finite Abelian group $\mathbb{G}_n$. Then $\boldsymbol{\varrho}(\mathcal{A}_n, \mathfrak{I}_n)$ is symmetric.*

PROOF. By 3.4.6, 3.8.13 and 3.8.10, we have

$$\boldsymbol{\varrho}'(\mathcal{A}_n, \mathfrak{I}_n) = \boldsymbol{\delta}'(\mathcal{A}_n, \mathfrak{I}_n) = \boldsymbol{\delta}(\mathfrak{I}_n, \overline{\mathcal{A}}_n) = \boldsymbol{\varrho}(\overline{\mathcal{A}}_n, \mathfrak{I}_n) = \boldsymbol{\varrho}(\mathcal{A}_n, \mathfrak{I}_n).$$

3.8.15 Next, we show that the inequalities proved in 3.3.7 and 3.3.8 are sharp.

EXAMPLE. *Let $\mathcal{A}_n = (a_1, \dots, a_n)$ be any system of characters on a compact Abelian group $\mathbb{G}$. Then*

$$\begin{aligned} \boldsymbol{\varrho}(L_p(\mathbb{G})|\mathfrak{I}_n, \mathcal{A}_n) &= K_p'(\mathcal{A}_n) \quad \textit{if } 1 \le p \le 2, \\ \boldsymbol{\varrho}(L_q(\mathbb{G})|\mathcal{A}_n, \mathfrak{I}_n) &= K_q(\mathcal{A}_n) \quad \textit{if } 2 \le q \le \infty. \end{aligned}$$

PROOF. For $1 \le r < \infty$ and $\xi_1, \dots, \xi_n \in \mathbb{C}$, we have

$$\begin{aligned} \|(\xi_k a_k)|[\mathcal{A}_n, L_r(\mathbb{G})]\| &= \left[\int_{\mathbb{G}} \left\| \sum_{k=1}^{n} \xi_k a_k\, a_k(t) \middle| L_r(\mathbb{G})\right\|^2 d\mu(t)\right]^{1/2} \\ &= \left[\int_{\mathbb{G}} \left(\int_{\mathbb{G}} \left|\sum_{k=1}^{n} \xi_k a_k(s)\, a_k(t)\right|^r d\mu(s)\right)^{2/r} d\mu(t)\right]^{1/2}. \end{aligned}$$

Since $a_k(s)\, a_k(t) = a_k(s+t)$, the invariance of μ yields

$$\|(\xi_k a_k)|[\mathcal{A}_n, L_r(\mathbb{G})]\| = \left(\int_{\mathbb{G}} \left|\sum_{k=1}^{n} \xi_k a_k(s)\right|^r d\mu(s)\right)^{1/r}.$$

With the usual modification, this formula also holds for $r = \infty$. On the other hand,

$$\|(\xi_k a_k)|[\mathfrak{I}_n, L_r(\mathbb{G})]\| = \Big(\sum_{k=1}^n |\xi_k|^2\Big)^{1/2}.$$

These observations imply the Khintchine inequalities

$$\Big(\sum_{k=1}^n |\xi_k|^2\Big)^{1/2} \le \boldsymbol{\varrho}(L_p(\mathbb{G})|\mathfrak{I}_n, \mathcal{A}_n)\Big(\int\limits_{\mathbb{G}} \Big|\sum_{k=1}^n \xi_k a_k(s)\Big|^p d\mu(s)\Big)^{1/p}$$

and

$$\Big(\int\limits_{\mathbb{G}} \Big|\sum_{k=1}^n \xi_k a_k(s)\Big|^q d\mu(s)\Big)^{1/q} \le \boldsymbol{\varrho}(L_q(\mathbb{G})|\mathcal{A}_n, \mathfrak{I}_n)\Big(\sum_{k=1}^n |\xi_k|^2\Big)^{1/2}.$$

Hence

$$K_p'(\mathcal{A}_n) \le \boldsymbol{\varrho}(L_p(\mathbb{G})|\mathfrak{I}_n, \mathcal{A}_n) \quad \text{and} \quad K_q(\mathcal{A}_n) \le \boldsymbol{\varrho}(L_q(\mathbb{G})|\mathcal{A}_n, \mathfrak{I}_n).$$

3.8.16 The preceding example extends to arbitrary infinite dimensional function spaces.

EXAMPLE. *Let $\mathcal{A}_n = (a_1, \dots, a_n)$ be any system of characters on a compact Abelian group $\mathbb{G}$. Then*

$$\begin{aligned} \boldsymbol{\varrho}(L_p|\mathfrak{I}_n, \mathcal{A}_n) &= K_p'(\mathcal{A}_n) \quad \text{if } 1 \le p \le 2, \\ \boldsymbol{\varrho}(L_q|\mathcal{A}_n, \mathfrak{I}_n) &= K_q(\mathcal{A}_n) \quad \text{if } 2 \le q \le \infty. \end{aligned}$$

PROOF. Fix $\varepsilon > 0$. By 3.8.15, an approximation argument yields simple functions $f_1^\circ, \dots, f_n^\circ \in L_p(\mathbb{G})$ such that

$$\Big(\sum_{k=1}^n \|f_k^\circ|L_p\|^2\Big)^{1/2} \ge \tfrac{1}{1+\varepsilon}\, K_p'(\mathcal{A}_n)\bigg(\int\limits_{\mathbb{G}} \Big\|\sum_{k=1}^n f_k^\circ\, a_k(s)\Big|L_p\Big\|^2 d\mu(s)\bigg)^{1/2}.$$

Clearly, these functions belong to an N-dimensional subspace spanned by characteristic functions of pairwise disjoint subsets. This subspace is isometric to l_p^N. Therefore

$$\boldsymbol{\varrho}(l_p^N|\mathfrak{I}_n, \mathcal{A}_n) \ge \tfrac{1}{1+\varepsilon}\, K_p'(\mathcal{A}_n).$$

Since every infinite dimensional space L_p contains isometric copies of l_p^N for all N, we obtain

$$\boldsymbol{\varrho}(L_p|\mathfrak{I}_n, \mathcal{A}_n) \ge \tfrac{1}{1+\varepsilon}\, K_p'(\mathcal{A}_n).$$

Letting $\varepsilon \to 0$ shows that

$$\boldsymbol{\varrho}(L_p|\mathfrak{I}_n, \mathcal{A}_n) \ge K_p'(\mathcal{A}_n).$$

The reverse estimate follows from 3.3.7. This proves the first part of the assertion. The second part can be checked analogously.

3.8.17 Let $\mathcal{A}_n = (a_1, \ldots, a_n)$ be a system of characters on a compact Abelian group $\mathbb{G}$. Then the associated operator $A_n \in \mathfrak{L}(l_1^n, C(\mathbb{G}))$ is given by

$$A_n x := \sum_{k=1}^{n} \xi_k a_k \quad \text{for } x = (\xi_k) \in l_1^n.$$

Obviously,

$$\|A_n : l_1^n \to C(\mathbb{G})\| = 1 \quad \text{and} \quad \varrho(A_n : l_1^n \to C(\mathbb{G}) | \mathcal{A}_n, \mathcal{I}_n) = \sqrt{n}.$$

Furthermore, for every permutation π of $\{1, \ldots, n\}$ and $a \in \mathbb{G}'$, we define the rearranged system and the shifted system

$$\pi \circ \mathcal{A}_n := (a_{\pi(1)}, \ldots, a_{\pi(n)}) \quad \text{and} \quad a \circ \mathcal{A}_n = (aa_1, \ldots, aa_n),$$

respectively.

3.8.18 The remaining results of this section are taken from [hin 3].

PROPOSITION. *Let $\mathcal{A}_n = (a_1, \ldots, a_n)$ and $\mathcal{B}_n = (b_1, \ldots, b_n)$ be systems of characters on compact Abelian groups $\mathbb{G}$ and $\mathbb{H}$, respectively. Then the following are equivalent:*

(1) $\varrho(\mathcal{B}_n, \mathcal{I}_n) \leq \varrho(\mathcal{A}_n, \mathcal{I}_n)$.

(2) $\varrho(B_n : l_1^n \to C(\mathbb{H}) | \mathcal{A}_n, \mathcal{I}_n) = \sqrt{n}$.

(3) *There exist a map $\varphi : \mathbb{G} \to \mathbb{H}$ and a permutation π of $\{1, \ldots, n\}$ such that*

$$\frac{a_k(s)}{a_1(s)} = \frac{b_{\pi(k)}(\varphi(s))}{b_{\pi(1)}(\varphi(s))} \quad \text{for } s \in \mathbb{G} \text{ and } k = 1, \ldots, n.$$

(4) *There exists a permutation π of $\{1, \ldots, n\}$ which guarantees that $(a_1 \otimes b_{\pi(1)}, \ldots, a_n \otimes b_{\pi(n)})$ and $(b_{\pi(1)}, \ldots, b_{\pi(n)})$ have identical distributions with respect to symmetric Borel sets.*

PROOF. **(1)⇒(2)**: This implication follows from

$$\begin{aligned} \sqrt{n} &= \varrho(B_n : l_1^n \to C(\mathbb{H}) | \mathcal{B}_n, \mathcal{I}_n) \\ &\leq \varrho(B_n : l_1^n \to C(\mathbb{H}) | \mathcal{A}_n, \mathcal{I}_n) \\ &\leq \|B_n : l_1^n \to C(\mathbb{H})\| \, \varrho(l_1^n | \mathcal{A}_n, \mathcal{I}_n) \leq \sqrt{n}. \end{aligned}$$

(2)⇒(3): First of all, choose $x_1, \ldots, x_n \in l_1^n$ with

$$\|(B_n x_k) | [\mathcal{A}_n, C(\mathbb{H})]\| = n \quad \text{and} \quad \|(x_k) | [l_2^n, l_1^n]\| = \sqrt{n}.$$

Applying the extreme point argument provided in 0.9.5, we may assume that $x_1, \ldots, x_n$ are multiples of unit vectors. Thus there exist a map π from $\{1, \ldots, n\}$ into itself and $\xi_1, \ldots, \xi_n \in \mathbb{K}$ such that $x_k = \xi_k u_{\pi(k)}^{(n)}$. Let

$M := \text{span}\{x_1, \dots, x_n\}$, and write $m := \dim(M)$. Since M is isometric to l_1^m, we have $\varrho(M|\mathcal{A}_n, \mathcal{I}_n) \le \sqrt{m}$. It follows from

$$\begin{aligned} n &= \|(B_n x_k)|[\mathcal{A}_n, C(\mathbb{H})]\| \\ &\le \|B_n : M \to C(\mathbb{H})\|\, \varrho(M|\mathcal{A}_n, \mathcal{I}_n)\, \|(x_k)|[l_2^n, l_1^n]\| \le \sqrt{m}\sqrt{n} \end{aligned}$$

that $m = n$. Hence π is even a permutation. Since

$$\sum_{k=1}^n \xi_k a_k \otimes b_{\pi(k)}$$

is uniformly continuous on $\mathbb{G} \times \mathbb{H}$, we see that

$$f(s) := \sup_{t \in \mathbb{H}} \left| \sum_{k=1}^n \xi_k a_k(s) b_{\pi(k)}(t) \right|$$

defines a continuous function on $\mathbb{G}$. Now

$$\int_{\mathbb{G}} f(s)^2 \, d\mu(s) = n^2 \quad \text{and} \quad f(s) \le \sum_{k=1}^n |\xi_k| \le \sqrt{n} \left(\sum_{k=1}^n |\xi_k|^2 \right)^{1/2} \le n$$

imply that the non-negative function $n^2 - f^2$ has mean 0. Consequently, $f(s) = n$ for all $s \in \mathbb{G}$. By continuity and compactness, there exists a map $\psi : \mathbb{G} \to \mathbb{H}$ such that

$$f(s) = \left| \sum_{k=1}^n \xi_k a_k(s) b_{\pi(k)}(\psi(s)) \right| = n.$$

Next, we deduce from

$$n \le \sum_{k=1}^n |\xi_k| \le \sqrt{n} \left(\sum_{k=1}^n |\xi_k|^2 \right)^{1/2} \le n$$

that $|\xi_1| = \dots = |\xi_n| = 1$. Hence

$$\xi_1 a_1(s) b_{\pi(1)}(\psi(s)) = \dots = \xi_n a_n(s) b_{\pi(n)}(\psi(s)),$$

and we obtain from

$$\frac{\xi_k a_k(o) b_{\pi(k)}(\psi(o))}{\xi_1 a_1(o) b_{\pi(1)}(\psi(o))} = 1 = \frac{\xi_k a_k(s) b_{\pi(k)}(\psi(s))}{\xi_1 a_1(s) b_{\pi(1)}(\psi(s))}$$

that

$$\frac{a_k(s)}{a_1(s)} = \frac{b_{\pi(k)}(\psi(o) - \psi(s))}{b_{\pi(1)}(\psi(o) - \psi(s))} \quad \text{for } s \in \mathbb{G} \text{ and } k = 1, \dots, n.$$

Define $\varphi : \mathbb{G} \to \mathbb{H}$ by $\varphi(s) := \psi(o) - \psi(s)$. Then

$$\frac{a_k(s)}{a_1(s)} = \frac{b_{\pi(k)}(\varphi(s))}{b_{\pi(1)}(\varphi(s))} \quad \text{for } s \in \mathbb{G} \text{ and } k = 1, \dots, n.$$

(3)⇒(4): Multiplying the equality

$$a_k(s)b_{\pi(1)}(\varphi(s)) = a_1(s)b_{\pi(k)}(\varphi(s))$$

by $b_{\pi(k)}(t)$, we obtain

$$a_k(s)b_{\pi(1)}(\varphi(s))b_{\pi(k)}(t) = a_1(s)b_{\pi(k)}(\varphi(s)+t).$$

Hence the invariance of the Haar measure ν implies that

$$\begin{aligned}\nu\Big\{t\in\mathbb{H} : \Big(a_k(s)b_{\pi(k)}(t)\Big)\in\Omega\Big\} &= \\ &= \nu\Big\{t\in\mathbb{H} : b_{\pi(1)}(\varphi(s))\Big(a_k(s)b_{\pi(k)}(t)\Big)\in\Omega\Big\} \\ &= \nu\Big\{t\in\mathbb{H} : a_1(s)\Big(b_{\pi(k)}(\varphi(s)+t)\Big)\in\Omega\Big\} \\ &= \nu\Big\{t\in\mathbb{H} : \Big(b_{\pi(k)}(t)\Big)\in\Omega\Big\}\end{aligned}$$

for $s\in\mathbb{G}$ and any symmetric Borel set Ω in $\mathbb{C}^n$. Thus integration over $s\in\mathbb{G}$ yields

$$\mu\otimes\nu\Big\{(s,t)\in\mathbb{G}\times\mathbb{H} : \Big(a_k(s)b_{\pi(k)}(t)\Big)\in\Omega\Big\} =$$
$$= \int_{\mathbb{G}} \nu\Big\{t\in\mathbb{H} : \Big(b_{\pi(k)}(t)\Big)\in\Omega\Big\}\,d\mu(s) = \nu\Big\{t\in\mathbb{H} : \Big(b_{\pi(k)}(t)\Big)\in\Omega\Big\}.$$

(4)⇒(1): By assumption, $\|(x_k)|\pi\circ\mathcal{B}_n\| = \|(x_k)|\mathcal{A}_n\otimes(\pi\circ\mathcal{B}_n)\|$. Hence, in view of 3.7.8,

$$\varrho(\mathcal{B}_n,\mathcal{I}_n) = \varrho(\pi\circ\mathcal{B}_n,\mathcal{I}_n) = \varrho(\mathcal{A}_n\otimes(\pi\circ\mathcal{B}_n),\mathcal{I}_n) \le \varrho(\mathcal{A}_n,\mathcal{I}_n).$$

3.8.19 PROPOSITION. *Let $\mathcal{A}_n = (a_1,\ldots,a_n)$ and $\mathcal{B}_n = (b_1,\ldots,b_n)$ be systems of characters on compact Abelian groups $\mathbb{G}$ and $\mathbb{H}$, respectively. Then the following are equivalent:*

(1) $\varrho(\mathcal{B}_n,\mathcal{A}_n) = 1$.

(2) *There exists a map $\varphi:\mathbb{G}\to\mathbb{H}$ such that*

$$\frac{a_k(s)}{a_1(s)} = \frac{b_k(\varphi(s))}{b_1(\varphi(s))} \quad \textit{for } s\in\mathbb{G} \textit{ and } k=1,\ldots,n.$$

(3) *The systems $(a_1\otimes b_1,\ldots,a_n\otimes b_n)$ and $(b_1,\ldots,b_n)$ have identical distributions with respect to symmetric Borel sets.*

PROOF. **(1)⇒(2)**: It follows from $\varrho(C(\mathbb{H})|\mathcal{B}_n,\mathcal{A}_n) = 1$ and

$$\begin{aligned}\|(b_k)|[\mathcal{B}_n, C(\mathbb{H})]\| &= \Big(\int_{\mathbb{H}} \sup_{v\in\mathbb{H}}\Big|\sum_{k=1}^n b_k(v)b_k(t)\Big|^2 d\nu(t)\Big)^{1/2} \\ &= \Big(\int_{\mathbb{H}} \sup_{v\in\mathbb{H}}\Big|\sum_{k=1}^n b_k(v)\Big|^2 d\nu(t)\Big)^{1/2} = n\end{aligned}$$

that

$$\|(b_k)|[\mathcal{A}_n, C(\mathbb{H})]\| = \Big(\int\limits_{\mathbb{H}} \sup_{t\in\mathbb{H}} \Big| \sum_{k=1}^{n} a_k(s) b_k(t) \Big|^2 d\mu(s) \Big)^{1/2} = n.$$

From now on, we may proceed as in the previous proof.

3.8.20 Let $\mathcal{A}_n = (a_1, \dots, a_n)$ be a system of characters on a compact Abelian group $\mathbb{G}$. We say that $\mathcal{A}_n$ **separates points** if for $s, t \in \mathbb{G}$ with $s \neq t$ there exists k_0 such that $a_{k_0}(s) \neq a_{k_0}(t)$. In this case, the character group $\mathbb{G}'$ is generated by $a_1, \dots, a_n$. Therefore $\mathbb{G}'$ is isomorphic to the direct product of a finite number of cyclic groups, and by duality, $\mathbb{G}$ can be identified with a group of the form

$$\mathbb{T}^N \times \mathbb{E}_{N_1} \times \dots \times \mathbb{E}_{N_j}, \qquad (*)$$

where $N_1, \dots, N_j$ are powers of primes; see 3.8.7.

3.8.21 If the underlying systems of characters separate points, then statement **(3)** in Proposition 3.8.18 can be improved significantly.

PROPOSITION. *Let $\mathcal{A}_n = (a_1, \dots, a_n)$ and $\mathcal{B}_n = (b_1, \dots, b_n)$ be systems of characters on compact Abelian groups $\mathbb{G}$ and $\mathbb{H}$, respectively. Assume that $\mathcal{B}_n$ separates points. Then every map $\varphi : \mathbb{G} \to \mathbb{H}$ with the property that*

$$a_k = b_k \circ \varphi \quad \textit{for } k = 1, \dots, n$$

is a continuous homomorphism. Moreover, if $\mathcal{A}_n$ also separates points, then φ is even injective.

PROOF. First of all, it follows from

$$b_k(\varphi(u+v)) = a_k(u+v) = a_k(u)a_k(v) = b_k(\varphi(u))b_k(\varphi(v)) = b_k(\varphi(u)+\varphi(v))$$

that $\varphi(u+v) = \varphi(u) + \varphi(v)$ for $u, v \in \mathbb{G}$. Next, we define

$$U_\varepsilon := \Big\{ s \in \mathbb{G} : \max_{1\le k\le n} |a_k(s) - 1| < \varepsilon \Big\},$$

$$V_\varepsilon := \Big\{ t \in \mathbb{H} : \max_{1\le k\le n} |b_k(t) - 1| < \varepsilon \Big\}.$$

Then $\varphi(U_\varepsilon) \subseteq V_\varepsilon$. Moreover, taking $\{V_\varepsilon : \varepsilon > 0\}$ as a base of zero neighbourhoods in $\mathbb{H}$, we obtain a Hausdorff topology that is weaker than the original one. Hence both topologies coincide; see [KEL, p. 141]. This proves the continuity we are looking for.

3.8.22 Using an elementary process, we may quite often arrange that the underlying systems of characters separate points.

If $\mathcal{A}_n = (a_1, \ldots, a_n)$ is arbitrary, then we factor $\mathbb{G}$ by the closed subgroup

$$\mathbb{N}(\mathcal{A}_n) := \Big\{ s \in \mathbb{G} : a_h(s) = a_k(s) \quad \text{for } h, k = 1, \ldots, n \Big\}.$$

Denote this quotient group by $\mathbb{G}/\mathcal{A}_n$, and write $\widetilde{s} := s + \mathbb{N}(\mathcal{A}_n)$ for its elements. Of course, the characters

$$\widetilde{a}_k(\widetilde{s}) := a_1^{-1}(s) a_k(s) \quad \text{for } \widetilde{s} = s + \mathbb{N}(\mathcal{A}_n) \in \mathbb{G}/\mathcal{A}_n$$

are well-defined. We call $\widetilde{\mathcal{A}}_n = (\widetilde{a}_1, \ldots, \widetilde{a}_n)$ the **reduced system** associated with the original system $\mathcal{A}_n = (a_1, \ldots, a_n)$.

3.8.23 Owing to this construction, the next observations are obvious.

LEMMA. *Let $\widetilde{\mathcal{A}}_n = (\widetilde{a}_1, \ldots, \widetilde{a}_n)$ denote the reduced system associated with $\mathcal{A}_n = (a_1, \ldots, a_n)$. Then the following holds:*

(1) *$\mathcal{A}_n = (a_1, \ldots, a_n)$ and $\widetilde{\mathcal{A}}_n = (\widetilde{a}_1, \ldots, \widetilde{a}_n)$ have identical distributions with respect to symmetric Borel sets.*

(2) *$\widetilde{\mathcal{A}}_n = (\widetilde{a}_1, \ldots, \widetilde{a}_n)$ separates points.*

(3) *$\widetilde{a}_1 = 1$.*

PROOF. We only note that

$$\begin{aligned} &\mu\Big\{ s \in \mathbb{G} : \big(a_1(s), \ldots, a_n(s)\big) \in \Omega \Big\} = \\ &\qquad = \mu\Big\{ s \in \mathbb{G} : a_1(s)^{-1}\big(a_1(s), \ldots, a_n(s)\big) \in \Omega \Big\} \\ &\qquad = \widetilde{\mu}\Big\{ \widetilde{s} \in \mathbb{G} : \big(\widetilde{a}_1(\widetilde{s}), \ldots, \widetilde{a}_n(\widetilde{s})\big) \in \Omega \Big\}, \end{aligned}$$

where Ω is a symmetric Borel set in $\mathbb{C}^n$ and $\widetilde{\mu}$ stands for the Haar measure on $\mathbb{G}/\mathcal{A}_n$.

3.8.24 The next result says that a system of characters is essentially specified by its quadratic averages. In other terms, systems of characters can be distinguished with the help of Banach spaces.

PROPOSITION. *Let $\mathcal{A}_n = (a_1, \ldots, a_n)$ and $\mathcal{B}_n = (b_1, \ldots, b_n)$ be systems of characters on compact Abelian groups $\mathbb{G}$ and $\mathbb{H}$, respectively. Then*

$$\|(x_k)|\mathcal{A}_n\| = \|(x_k)|\mathcal{B}_n\| \quad \textit{for } x_1, \ldots, x_n \in X \textit{ and all Banach spaces } X$$

if and only if the systems $\widetilde{\mathcal{A}}_n = (\widetilde{a}_1, \ldots, \widetilde{a}_n)$ and $\widetilde{\mathcal{B}}_n = (\widetilde{b}_1, \ldots, \widetilde{b}_n)$ are isomorphic, which means that there exists an isomorphism $\widetilde{\varphi}$ from $\mathbb{G}/\mathcal{A}_n$ onto $\mathbb{H}/\mathcal{B}_n$ such that

$$\widetilde{a}_k = \widetilde{b}_k \circ \widetilde{\varphi} \quad \textit{for } k = 1, \ldots, n.$$

PROOF. Apply 3.8.19 in two directions and use 3.8.21.

3.8.25 PROPOSITION. *Let $\mathcal{A}_n = (a_1, \ldots, a_n)$ and $\mathcal{B}_n = (b_1, \ldots, b_n)$ be systems of characters on compact Abelian groups $\mathbb{G}$ and $\mathbb{H}$, respectively. Then*

$$\varrho(\mathcal{A}_n, \mathcal{I}_n) = \varrho(\mathcal{B}_n, \mathcal{I}_n)$$

if and only if there exist an isomorphism $\widetilde{\varphi}$ from $\mathbb{G}/\mathcal{A}_n$ onto $\mathbb{H}/\mathcal{B}_n$ and a permutation π of $\{1, \ldots, n\}$ such that

$$\widetilde{a}_k = \widetilde{b}_{\pi(k)} \circ \widetilde{\varphi} \quad \textit{for } k = 1, \ldots, n.$$

PROOF. In order to show the non-trivial implication, we pass to the reduced systems $\widetilde{\mathcal{A}}_n = (\widetilde{a}_1, \ldots, \widetilde{a}_n)$ and $\widetilde{\mathcal{B}}_n = (\widetilde{b}_1, \ldots, \widetilde{b}_n)$. Then, in view of 3.8.18 and 3.8.21, there exist an injective continuous homomorphism $\widetilde{\varphi} : \mathbb{G}/\mathcal{A}_n \to \mathbb{H}/\mathcal{B}_n$ and a permutation π of $\{1, \ldots, n\}$ such that

$$\widetilde{a}_k = \widetilde{b}_{\pi(k)} \circ \widetilde{\varphi} \quad \text{for } k = 1, \ldots, n.$$

However, since the assumption is symmetric, we can also find an injective continuous homomorphism $\widetilde{\psi} : \mathbb{H}/\mathcal{B}_n \to \mathbb{G}/\mathcal{A}_n$. Hence $\widetilde{\psi} \circ \widetilde{\varphi}$ must be an automorphism of $\mathbb{G}/\mathcal{A}_n$; see [HUP, p. 80].

3.8.26 We now illustrate the above considerations in a specific case.

EXAMPLE. *Let $\mathcal{A}_n = (a_1, \ldots, a_n)$ and $\mathcal{B}_n = (b_1, \ldots, b_n)$ be systems of characters on $\mathbb{T}$. Then the following are equivalent:*

(1) $\varrho(\mathcal{B}_n, \mathcal{I}_n) = \varrho(\mathcal{A}_n, \mathcal{I}_n)$.
(2) $\varrho(\mathcal{B}_n, \mathcal{I}_n) \le \varrho(\mathcal{A}_n, \mathcal{I}_n)$.
(3) *Viewing $\mathcal{A}_n = (\zeta^{a_1}, \ldots, \zeta^{a_n})$ and $\mathcal{B}_n = (\zeta^{b_1}, \ldots, \zeta^{b_n})$ as subsets of $\mathbb{Z}$, there exist a permutation π of $\{1, \ldots, n\}$ and integers $p \neq 0$, $q \neq 0$ and r such that*

$$pa_k + qb_{\pi(k)} = r.$$

PROOF. Since the other implications are trivial, we only show **(2)**$\Rightarrow$**(3)**. First of all, note that the quotient $\mathbb{T}/\mathcal{A}_n$ can be identified with $\mathbb{T}$. Moreover, the natural surjection has the form $\sigma(\zeta) = \zeta^a$, where a denotes the greatest common divisor of $a_1 - a_1, \ldots, a_n - a_1$. Then the reduced system has the form

$$\widetilde{\mathcal{A}}_n = \Big(\frac{a_1 - a_1}{a}, \ldots, \frac{a_n - a_1}{a} \Big).$$

Replacing a by b, the corresponding formulas hold for the systems $\mathcal{B}_n$ and $\widetilde{\mathcal{B}}_n$. In view of **(3)** in 3.8.18, there exist a map $\varphi : \mathbb{T} \to \mathbb{T}$ and a permutation π of $\{1, \ldots, n\}$ such that

$$\zeta^{(a_k - a_1)/a} = \varphi(\zeta)^{(b_{\pi(k)} - b_{\pi(1)})/b} \quad \text{for } \zeta \in \mathbb{T} \text{ and } k = 1, \ldots, n.$$

Next, 3.8.21 tells us that φ must be a continuous homomorphism. Thus, in our special context, it is a character. Consequently, $\varphi(\zeta) = \zeta^m$. Finally, the injectivity, which also follows from 3.8.21, implies that $\varphi(\zeta) = \zeta^{\pm 1}$. Hence we have

$$b(a_k - a_1) = \pm a(b_{\pi(k)} - b_{\pi(1)}) \quad \text{for } k = 1, \ldots, n.$$

3.9 Discrete orthonormal systems

3.9.1 An orthonormal system $\mathcal{A}_n = (a_1, \ldots, a_n)$ in a Hilbert space $L_2(M, \mu)$ is called **discrete** if it consists of simple functions only.

3.9.2 To begin with, we show that discrete orthonormal systems are closely related to isometric matrices. Quite often, this observation makes life easier.

PROPOSITION. *If $\mathcal{A}_n = (a_1, \ldots, a_n)$ is discrete, then there exists an isometric matrix $A_{mn} = (\alpha_{hk})$ such that*

$$\|(x_k)|\mathcal{A}_n\| = \|(x_k)|A_{mn}\|.$$

Conversely, if $A_{mn} = (\alpha_{hk}) \in \mathbb{U}^{mn}$, then the vectors

$$a_k := (\alpha_{1k}, \ldots, \alpha_{mk}) \quad \textit{with } k = 1, \ldots, n$$

form an orthonormal system in l_2^m.

PROOF. Write

$$a_i = \sum_{h=1}^{m} \alpha_{hi}\, \mu(A_h)^{-1/2} \chi_h \quad \text{and} \quad a_j = \sum_{k=1}^{m} \alpha_{kj}\, \mu(A_k)^{-1/2} \chi_k$$

where $\chi_1, \ldots, \chi_m$ are characteristic functions of pairwise disjoint and measurable non-zero subsets $A_1, \ldots, A_m$ of M. We conclude from

$$\begin{aligned}(a_i, a_j) &= \sum_{h=1}^{m}\sum_{k=1}^{m} \alpha_{hi}\, \overline{\alpha_{kj}}\, \mu(A_h)^{-1/2} \mu(A_k)^{-1/2} (\chi_h, \chi_k) \\ &= \sum_{h=1}^{m} \alpha_{hi}\, \overline{\alpha_{hj}} = \sum_{h=1}^{m} \alpha_{jh}^* \, \alpha_{hi}\end{aligned}$$

that $A_{mn}^* A_{mn} = I_n$. Hence $A_{mn} \in \mathbb{U}^{mn}$. The formula

$$\|(x_k)|\mathcal{A}_n\| = \|(x_k)|A_{mn}\|$$

is obvious.

3.9.3 We now establish an elementary inequality which shows how the norm $\| \cdot |\mathcal{A}_n\|$ behaves under perturbations.

LEMMA. *Let $\mathcal{A}_n = (a_1, \ldots, a_n)$ and $\mathcal{B}_n = (b_1, \ldots, b_n)$ be orthonormal systems in a Hilbert space $L_2(M, \mu)$. Then*

$$\|(x_k)|\mathcal{B}_n\| \leq \left[1 + \left(n \sum_{k=1}^{n} \|b_k - a_k|L_2\|^2\right)^{1/2}\right] \|(x_k)|\mathcal{A}_n\|.$$

PROOF. We have

$$\|(x_k)|\mathcal{B}_n\| \leq \|(x_k)|\mathcal{A}_n\| + \left(\int_M \left\|\sum_{k=1}^{n} x_k\big(b_k(s) - a_k(s)\big)\right\|^2 d\mu(s)\right)^{1/2}.$$

Using the Cauchy–Schwarz inequality and 3.1.5, it follows that

$$\begin{aligned}\left(\int_M \left\|\sum_{k=1}^{n} x_k\big(b_k(s) - a_k(s)\big)\right\|^2 d\mu(s)\right)^{1/2} &\leq \\ &\leq \left(\sum_{k=1}^{n} \|x_k\|^2\right)^{1/2} \left(\sum_{k=1}^{n} \|b_k - a_k|L_2\|^2\right)^{1/2} \\ &\leq \sqrt{n}\, \|(x_k)|\mathcal{A}_n\| \left(\sum_{k=1}^{n} \|b_k - a_k|L_2\|^2\right)^{1/2}.\end{aligned}$$

This observation completes the proof.

3.9.4 The collection of discrete orthonormal systems will be denoted by $\boldsymbol{\Delta}_n$, while $\boldsymbol{\Omega}_n$ stands for the collection of all orthonormal systems consisting of n functions. Roughly speaking, the following considerations show that $\boldsymbol{\Delta}_n$ is dense in $\boldsymbol{\Omega}_n$.

LEMMA. *For every orthonormal system $\mathcal{A}_n = (a_1, \ldots, a_n)$ and $\varepsilon > 0$ there exists a discrete orthonormal system $\mathcal{A}_n^\circ = (a_1^\circ, \ldots, a_n^\circ)$ such that*

$$\left(\sum_{k=1}^{n} \|a_k - a_k^\circ|L_2\|^2\right)^{1/2} \leq \varepsilon.$$

PROOF. If $(f_1, \ldots, f_n)$ is a linearly independent system in $L_2(M, \mu)$, then the Gram–Schmidt process yields an orthonormal system $(g_1, \ldots, g_n)$. We have

$$g_k = (D_{k-1}D_k)^{-1/2} \det \begin{pmatrix} (f_1, f_1) & \cdots & (f_1, f_{k-1}) & f_1 \\ \vdots & \vdots & \vdots & \vdots \\ (f_k, f_1) & \cdots & (f_k, f_{k-1}) & f_k \end{pmatrix},$$

where $D_0 = 1$ and

$$D_k = \det \begin{pmatrix} (f_1, f_1) & \cdots & (f_1, f_k) \\ \vdots & \vdots & \vdots \\ (f_k, f_1) & \cdots & (f_k, f_k) \end{pmatrix} \quad \text{for } k = 1, \ldots, n.$$

This construction is stable under sufficiently small perturbations. Thus the desired orthonormal system $\mathcal{A}_n^\circ = (a_1^\circ, \ldots, a_n^\circ)$ can be found as follows. First of all, $a_1, \ldots, a_n$ are approximated by simple functions and, secondly, we apply the Gram–Schmidt process.

3.9.5 LEMMA. *For an arbitrary orthonormal system $\mathcal{A}_n$ and $\varepsilon > 0$, there exists a discrete orthonormal system $\mathcal{A}_n^\circ$ such that*

$$\varrho(\mathcal{A}_n, \mathcal{A}_n^\circ) \le 1 + \varepsilon \quad \textit{and} \quad \varrho(\mathcal{A}_n^\circ, \mathcal{A}_n) \le 1 + \varepsilon.$$

PROOF. Applying Lemma 3.9.4, we find a discrete orthonormal system $\mathcal{A}_n^\circ = (a_1^\circ, \ldots, a_n^\circ)$ such that

$$\Big(n \sum_{k=1}^{n} \|a_k - a_k^\circ | L_2\|^2 \Big)^{1/2} \le \varepsilon,$$

which has the required property, by 3.9.3.

3.9.6 PROPOSITION. *Let $\mathcal{A}_n$ and $\mathcal{B}_n$ be orthonormal systems. Then, for $\varepsilon > 0$, there exist discrete orthonormal systems $\mathcal{A}_n^\circ$ and $\mathcal{B}_n^\circ$ such that*

$$\varrho(\mathcal{B}_n, \mathcal{A}_n) \le (1+\varepsilon)\, \varrho(\mathcal{B}_n^\circ, \mathcal{A}_n^\circ) \quad \textit{and} \quad \varrho(\mathcal{B}_n^\circ, \mathcal{A}_n^\circ) \le (1+\varepsilon)\, \varrho(\mathcal{B}_n, \mathcal{A}_n),$$
$$\boldsymbol{\delta}(\mathcal{B}_n, \mathcal{A}_n) \le (1+\varepsilon)\, \boldsymbol{\delta}(\mathcal{B}_n^\circ, \mathcal{A}_n^\circ) \quad \textit{and} \quad \boldsymbol{\delta}(\mathcal{B}_n^\circ, \mathcal{A}_n^\circ) \le (1+\varepsilon)\, \boldsymbol{\delta}(\mathcal{B}_n, \mathcal{A}_n).$$

PROOF. Thanks to 3.3.5 and 3.4.8, the assertion follows from the above lemma.

3.9.7 For any orthonormal system $\mathcal{A}_n = (a_1, \ldots, a_n)$, the quadratic average $\|(x_k)|\mathcal{A}_n\|$ is a norm on the n-fold Cartesian power of X. The Banach space so obtained will be denoted by $[\mathcal{A}_n, X]$. It follows from

$$\frac{1}{\sqrt{n}} \|(x_k)|l_2^n\| \le \|(x_k)|\mathcal{A}_n\| \le \sqrt{n}\, \|(x_k)|l_2^n\|$$

that $[\mathcal{A}_n, X]$ and $[l_2^n, X]$ are isomorphic. So, as a linear space, $[\mathcal{A}_n, X]''$ can be identified with the n-fold Cartesian power of X''. But what about its norm?

LEMMA. *$[\mathcal{A}_n, X]''$ and $[\mathcal{A}_n, X'']$ coincide isometrically.*

PROOF. First we consider the case when $\mathcal{A}_n$ is discrete. Then there exists an isometric matrix $A_{mn} = (\alpha_{hk})$ as described in 3.9.2. This means that

$$[A_{mn}, X] : (x_k) \longrightarrow \left(\sum_{k=1}^{n} \alpha_{hk} x_k \right)$$

is a metric injection from $[\mathcal{A}_n, X]$ into $[l_2^m, X]$. Regarding the elements of $[l_2^m, X]'$ and $[\mathcal{A}_n, X]''$ as n-tuples (x'_h) and (x''_k), respectively, we get

$$[A_{mn}, X]' : (x'_h) \longrightarrow \left(\sum_{h=1}^{m} \alpha_{hk} x'_h \right)$$

and

$$[A_{mn}, X]'' : (x''_k) \longrightarrow \left(\sum_{k=1}^{n} \alpha_{hk} x''_k \right).$$

Since $[A_{mn}, X]''$ is a metric injection, we are done.

The general case can be treated by approximation. Thanks to 3.9.4, for $\varepsilon > 0$ there exists a discrete orthonormal system $\mathcal{A}_n^\circ$ such that

$$\left(n \sum_{k=1}^{n} \|a_k - a_k^\circ | L_2\|^2 \right)^{1/2} \leq \varepsilon.$$

Now it follows from

$$\|Id : [\mathcal{A}_n, X] \to [\mathcal{A}_n^\circ, X]\| \leq 1+\varepsilon, \quad \|Id : [\mathcal{A}_n^\circ, X''] \to [\mathcal{A}_n, X'']\| \leq 1+\varepsilon,$$

and

$$\|Id : [\mathcal{A}_n, X]'' \to [\mathcal{A}_n, X'']\| \leq \begin{bmatrix} \|Id : [\mathcal{A}_n, X]'' \to [\mathcal{A}_n^\circ, X]''\| \\ \times \\ \|Id : [\mathcal{A}_n^\circ, X]'' \to [\mathcal{A}_n^\circ, X'']\| \\ \times \\ \|Id : [\mathcal{A}_n^\circ, X''] \to [\mathcal{A}_n, X'']\| \end{bmatrix}$$

that

$$\|Id : [\mathcal{A}_n, X]'' \to [\mathcal{A}_n, X'']\| \leq (1+\varepsilon)^2.$$

In the same way we obtain

$$\|Id : [\mathcal{A}_n, X''] \to [\mathcal{A}_n, X]''\| \leq (1+\varepsilon)^2.$$

Thus

$$1 \leq \|Id : [\mathcal{A}_n, X]'' \to [\mathcal{A}_n, X'']\| \, \|Id : [\mathcal{A}_n, X''] \to [\mathcal{A}_n, X]''\| \leq (1+\varepsilon)^4.$$

Letting $\varepsilon \to 0$ proves that the identity map from $[\mathcal{A}_n, X]''$ onto $[\mathcal{A}_n, X'']$ is an isometry.

3.9.8 Riemann ideal norms may fail to be symmetric. However, they are *reflexive.*

PROPOSITION. $\varrho''(\mathcal{B}_n,\mathcal{A}_n)=\varrho(\mathcal{B}_n,\mathcal{A}_n)$.

PROOF. Observe that $\varrho(T|\mathcal{B}_n,\mathcal{A}_n)$ is nothing but the norm of the diagonal operator $\boldsymbol{T}=(T,\ldots,T)$ acting from $[\mathcal{A}_n,X]$ into $[\mathcal{B}_n,Y]$. By the previous lemma, the bidual operator $\boldsymbol{T}''=(T'',\ldots,T'')$ can be viewed as a map from $[\mathcal{A}_n,X'']$ into $[\mathcal{B}_n,Y'']$. The assertion is now a consequence of $\|\boldsymbol{T}''\|=\|\boldsymbol{T}\|$.

REMARK. By 3.4.6, we also have $\boldsymbol{\delta}''(\mathcal{B}_n,\mathcal{A}_n)=\boldsymbol{\delta}'(\overline{\mathcal{B}}_n,\overline{\mathcal{A}}_n)=\boldsymbol{\delta}(\mathcal{B}_n,\mathcal{A}_n)$.

3.10 Some universal ideal norms

3.10.1 Recall from 3.9.4 that $\boldsymbol{\Omega}_n$ stands for the collection of all orthonormal systems consisting of n functions. We let

$$\begin{aligned}
\boldsymbol{\tau}_n(T) &:= \sup\Big\{\,\varrho(T|\mathcal{A}_n,\mathcal{I}_n) : \mathcal{A}_n\in\boldsymbol{\Omega}_n\Big\},\\
\boldsymbol{\omega}_n(T) &:= \sup\Big\{\,\varrho(T|\mathcal{I}_n,\mathcal{A}_n) : \mathcal{A}_n\in\boldsymbol{\Omega}_n\Big\},\\
\boldsymbol{\varrho}_n(T) &:= \sup\Big\{\,\varrho(T|\mathcal{B}_n,\mathcal{A}_n) : \mathcal{A}_n,\mathcal{B}_n\in\boldsymbol{\Omega}_n\Big\},\\
\boldsymbol{\delta}_n(T) &:= \sup\Big\{\,\boldsymbol{\delta}(T|\mathcal{B}_n,\mathcal{A}_n) : \mathcal{A}_n,\mathcal{B}_n\in\boldsymbol{\Omega}_n\Big\}.
\end{aligned}$$

These definitions yield

$\boldsymbol{\tau}_n : T\longrightarrow\boldsymbol{\tau}_n(T)$, **universal type ideal norm,**
$\boldsymbol{\omega}_n : T\longrightarrow\boldsymbol{\omega}_n(T)$, **universal cotype ideal norm,**
$\boldsymbol{\varrho}_n : T\longrightarrow\boldsymbol{\varrho}_n(T)$, **universal Riemann ideal norm,**
$\boldsymbol{\delta}_n : T\longrightarrow\boldsymbol{\delta}_n(T)$, **universal Dirichlet ideal norm.**

REMARK. We have shown in 3.9.6 that $\boldsymbol{\Delta}_n$ is a dense subset of $\boldsymbol{\Omega}_n$. Thus it suffices to take the above suprema over all discrete orthonormal systems.

3.10.2 We denote by $\mathrm{DIM}_{\leq n}(X)$ the set of all subspaces M of X such that $\dim(M)\leq n$. Moreover, J_M^X is the embedding map from M into X.

THEOREM. $\boldsymbol{\varrho}_n(T)=\sup\Big\{\,\|TJ_M^X|\mathfrak{H}\| : M\in\mathrm{DIM}_{\leq n}(X)\Big\}$.

PROOF. First of all, we conclude from 2.3.13 and 3.4.11 that

$$\begin{aligned}
\varrho(T|\mathcal{B}_n,\mathcal{A}_n) &= \sup\Big\{\,\varrho(TJ_M^X|\mathcal{B}_n,\mathcal{A}_n) : M\in\mathrm{DIM}_{\leq n}(X)\Big\}\\
&\leq \sup\Big\{\,\|TJ_M^X|\mathfrak{H}\| : M\in\mathrm{DIM}_{\leq n}(X)\Big\}
\end{aligned}$$

which yields the $\leq$ part. Thanks to 2.3.23, we know that

$$\|TJ_M^X|\mathfrak{H}\| = \sup\Big\{ \boldsymbol{\kappa}(TJ_M^X|S_m) : S_m \in \mathbb{U}^m \text{ with } m = 1, 2, \dots \Big\}.$$

Fix $M \in \text{DIM}_{\leq n}(X)$. If $x_1, \dots, x_m \in M$, then 2.1.9 provides us with a matrix $A_{mn} = (\alpha_{ki}) \in \mathbb{U}^{mn}$ such that

$$x_k = \sum_{h=1}^{m} \sum_{i=1}^{n} \alpha_{ki} \alpha_{ih}^* x_h \quad \text{for } k = 1, \dots, m.$$

As described in 3.9.2, we associate with A_{mn} and $S_m A_{mn}$ discrete orthonormal systems $\mathcal{A}_n$ and $\mathcal{B}_n$. Letting

$$x_i^\circ := \sum_{h=1}^{m} \alpha_{ih}^* x_h \quad \text{for } i = 1, \dots, n,$$

it follows from

$$\bigg(\sum_{h=1}^{m} \Big\| \sum_{i=1}^{n} \sum_{k=1}^{m} \sigma_{hk} \alpha_{ki} T x_i^\circ \Big\|^2 \bigg)^{1/2} \leq \boldsymbol{\varrho}(T|\mathcal{B}_n, \mathcal{A}_n) \bigg(\sum_{k=1}^{m} \Big\| \sum_{i=1}^{n} \alpha_{ki} x_i^\circ \Big\|^2 \bigg)^{1/2}$$

that

$$\bigg(\sum_{h=1}^{m} \Big\| \sum_{k=1}^{m} \sigma_{hk} T x_k \Big\|^2 \bigg)^{1/2} \leq \boldsymbol{\varrho}(T|\mathcal{B}_n, \mathcal{A}_n) \bigg(\sum_{k=1}^{m} \|x_k\|^2 \bigg)^{1/2}.$$

Hence

$$\boldsymbol{\kappa}(TJ_M^X|S_m) \leq \sup\Big\{ \boldsymbol{\varrho}(T|\mathcal{B}_n, \mathcal{A}_n) : \mathcal{A}_n, \mathcal{B}_n \in \boldsymbol{\Delta}_n \Big\} = \boldsymbol{\varrho}_n(T).$$

3.10.3 The next observation shows that the ideal norms $\boldsymbol{\delta}_n$ and $\boldsymbol{\kappa}_n$ are built quite similarly; see 2.3.15.

PROPOSITION.

$$\boldsymbol{\delta}_n(T) = \sup\Big\{ \boldsymbol{\kappa}(T|S_m) : S_m \in \mathbb{L}^m \text{ } \textit{with} \text{ } \text{rank}(S_m) \leq n \text{ } \textit{and} \text{ } m = 1, 2, \dots \Big\}.$$

PROOF. Recall that

$$\boldsymbol{\delta}_n(T) = \sup\Big\{ \boldsymbol{\delta}(T|\mathcal{B}_n, \mathcal{A}_n) : \mathcal{A}_n, \mathcal{B}_n \in \boldsymbol{\Delta}_n \Big\}.$$

Let $A_{mn} = (\alpha_{hi})$ and $B_{mn} = (\beta_{hi})$ be isometric matrices associated with $\mathcal{A}_n$ and $\mathcal{B}_n$; see 3.9.2. It easily turns out that

$$\boldsymbol{\delta}(T|\mathcal{B}_n, \mathcal{A}_n) = \boldsymbol{\kappa}(T|B_{mn} A_{mn}^*).$$

Since every matrix $S_m \in \mathbb{L}^m$ with $\text{rank}(S_m) \leq n$ belongs to $\mathbb{L}^{mn} \circ \mathbb{L}^{nm}$, an extreme point argument shows that the quantities

$$\sup\Big\{ \boldsymbol{\kappa}(T|B_{mn} A_{mn}^*) : A_{mn}, B_{mn} \in \mathbb{U}^{mn} \quad \text{with } m \geq n \Big\}$$

and

$$\sup\Big\{ \boldsymbol{\kappa}(T|S_m) : S_m \in \mathbb{L}^m \text{ with } \text{rank}(S_m) \leq n \text{ and } m = 1, 2, \dots \Big\}$$

are equal.

3.10.4 The following characterization of the universal Dirichlet ideal norm goes back to G. Pisier, who introduced the fixed index n in Kwapień's Hilbert space criterion and in Maurey's extension theorem; see [pis 8] and [pis 10, p. 278].

THEOREM. *For* $T \in \mathfrak{L}(X,Y)$, $n = 1, 2, \ldots,$ *and* $c \geq 0$, *the following are equivalent:*

(1) $$\boldsymbol{\delta}_n(T) \leq c.$$

(2) *Let* $S_m = (\sigma_{hk}) \in \mathbb{L}^m$ *such that* $\operatorname{rank}(S_m) \leq n$, $x_1, \ldots, x_m \in X$, *and* $m = 1, 2, \ldots$. *Then*

$$\Big(\sum_{h=1}^{m}\Big\|\sum_{k=1}^{m}\sigma_{hk}Tx_k\Big\|^2\Big)^{1/2} \leq c\Big(\sum_{k=1}^{m}\|x_k\|^2\Big)^{1/2}.$$

(3) *Given* $M \in \mathrm{DIM}_{\leq n}(X)$, *there exists* $T_M \in \mathfrak{H}(X,Y)$ *such that*

$$\|T_M|\mathfrak{H}\| \leq c \quad \text{and} \quad T_M x = Tx \quad \text{for } x \in M.$$

(4) $$\|TA|\mathfrak{P}_2\| \leq c\,\|A'|\mathfrak{P}_2\| \quad \text{for } A \in \mathfrak{L}(l_2^n, X).$$

PROOF. The equivalence **(1)** $\Longleftrightarrow$ **(2)** follows from 3.10.3 and the definition of the Kwapień ideal norm $\boldsymbol{\kappa}(S_m)$.

(2)⇒(3): Fix $M \in \mathrm{DIM}_{\leq n}(X)$ and let $T_0 := TJ_M^X$. If $x_1, \ldots, x_m \in X$ and $S_m = (\sigma_{hk}) \in \mathbb{L}^m$ are such that

$$x_h^\circ := \sum_{k=1}^{m}\sigma_{hk}x_k \in M \quad \text{for} \quad h = 1, \ldots, m,$$

then, by 2.1.9, we can find $P_m = (\pi_{ih}) \in \mathbb{L}^m$ with $\operatorname{rank}(P_m) \leq n$ and

$$x_i^\circ = \sum_{h=1}^{m}\pi_{ih}x_h^\circ \quad \text{for } i = 1, \ldots, m. \tag{$*$}$$

Since $\operatorname{rank}(P_m S_m) \leq n$, it follows that

$$\Big(\sum_{i=1}^{m}\Big\|T_0\Big[\sum_{k=1}^{m}\Big(\sum_{h=1}^{m}\pi_{ih}\,\sigma_{hk}\Big)x_k\Big]\Big\|^2\Big)^{1/2} \leq c\Big(\sum_{k=1}^{m}\|x_k\|^2\Big)^{1/2}.$$

Thus $(*)$ yields

$$\Big(\sum_{h=1}^{m}\Big\|T_0\Big[\sum_{k=1}^{m}\sigma_{hk}x_k\Big]\Big\|^2\Big)^{1/2} \leq c\Big(\sum_{k=1}^{m}\|x_k\|^2\Big)^{1/2}.$$

Hence, by 2.3.14, there exists the required extension T_M of T_0. This proves that c has the property we are looking for.

(3)⇒(4): Let $M := A(l_2^n)$, and choose the operator $T_M \in \mathfrak{H}(X,Y)$ as described in **(3)**. Then for $\varepsilon > 0$ there is a factorization $T_M = VU$ through a Hilbert space H such that $\|V\|\,\|U\| \le (1+\varepsilon)\,c$. Since UA acts between Hilbert spaces, we have

$$\|UA|\mathfrak{P}_2\| = \|A'U'|\mathfrak{P}_2\|;$$

see [PIE 3, pp. 55–57]. Hence

$$\|TA|\mathfrak{P}_2\| \le \|V\|\,\|UA|\mathfrak{P}_2\| = \|V\|\,\|A'U'|\mathfrak{P}_2\| \le (1+\varepsilon)\,c\,\|A'|\mathfrak{P}_2\|.$$

Letting $\varepsilon \to 0$ yields **(4)**.

(4)⇒(2): Given $x_1, \dots, x_m \in X$, we define $A \in \mathfrak{L}(l_2^m, X)$ by

$$A(\xi_k) := \sum_{k=1}^{m} \xi_k x_k.$$

Then

$$\|A'|\mathfrak{P}_2\| \le \left(\sum_{k=1}^{m} \|x_k\|^2 \right)^{1/2}.$$

Since $\operatorname{rank}(S_m) \le n$, there exists a factorization $S_m = V_{mn}U_{nm}$ with $V_{mn} \in \mathbb{L}^{mn}$ and $U_{nm} \in \mathbb{L}^{nm}$. Hence $AU'_{nm} \in \mathfrak{L}(l_2^n, X)$, and it follows from **(4)** that

$$\|TAS'_m|\mathfrak{P}_2\| \le \|TAU'_{nm}|\mathfrak{P}_2\| \le c\,\|U_{nm}A'|\mathfrak{P}_2\| \le c\,\|A'|\mathfrak{P}_2\|.$$

Consequently, we have

$$\left(\sum_{h=1}^{m} \|TAS'_m\, u_h^{(m)}\|^2 \right)^{1/2} \le c \left(\sum_{k=1}^{m} \|x_k\|^2 \right)^{1/2}.$$

This proves **(2)**, in view of

$$TAS'_m\, u_h^{(m)} = \sum_{k=1}^{m} \sigma_{hk} T x_k \quad \text{for } h = 1, \dots, m.$$

3.10.5 For later application, we dualize property **(4)** in the preceding theorem.

PROPOSITION. *For* $T \in \mathfrak{L}(X,Y)$, $n = 1, 2, \dots,$ *and* $c \ge 0$, *the following are equivalent:*

(4) $\quad \|TA|\mathfrak{P}_2\| \le c\,\|A'|\mathfrak{P}_2\|$ *for* $A \in \mathfrak{L}(l_2^n, X)$.

(4′) $\quad \|T'B'|\mathfrak{P}_2\| \le c\,\|B|\mathfrak{P}_2\|$ *for* $B \in \mathfrak{L}(Y, l_2^n)$.

PROOF. Replacing T by T', property **(4)** passes into

($4'_0$) $\quad \|T'A|\mathfrak{P}_2\| \le c\,\|A'|\mathfrak{P}_2\|$ for $A \in \mathfrak{L}(l_2^n, Y')$.

However, all operators $A \in \mathfrak{L}(l_2^n, Y')$ can be written in the form $A = B'$ with $B \in \mathfrak{L}(Y, l_2^n)$, and we have

$$\|B''|\mathfrak{P}_2\| = \|B|\mathfrak{P}_2\|;$$

see [DIE*a, p. 50] or [PIE 2, p. 229]. Thus $(\mathbf{4}_0')$ is equivalent to $(\mathbf{4}')$. Finally, we observe that, by 3.4.6, property $(\mathbf{1})$ in 3.10.4 holds for T if and only if it holds for T'.

3.10.6 PROPOSITION.

$$\boldsymbol{\tau}_n(T) = \sup\Big\{ \boldsymbol{\kappa}(T|S_{mn}) \,:\, S_{mn} \in \mathbb{L}^{mn} \quad \textit{with } m = 1, 2, \dots \Big\}.$$

PROOF. By definition,

$$\boldsymbol{\tau}_n(T) := \sup\Big\{ \boldsymbol{\varrho}(T|\mathcal{A}_n, \mathcal{I}_n) \,:\, \mathcal{A}_n \in \boldsymbol{\Omega}_n \Big\}.$$

We conclude from 3.9.6 that it is enough to take the right-hand supremum over all discrete orthonormal systems $\mathcal{A}_n$. Let $A_{mn} \in \mathbb{U}^{mn}$ be the associated isometric (m, n)-matrix; see 3.9.2. Then

$$\boldsymbol{\varrho}(T|\mathcal{A}_n, \mathcal{I}_n) = \boldsymbol{\kappa}(T|A_{mn}).$$

An extreme point argument shows that the quantities

$$\sup\Big\{ \boldsymbol{\kappa}(T|A_{mn}) \,:\, A_{mn} \in \mathbb{U}^{mn} \quad \text{with } m \geq n \Big\}$$

and

$$\sup\Big\{ \boldsymbol{\kappa}(T|S_{mn}) \,:\, S_{mn} \in \mathbb{L}^{mn} \quad \text{with } m = 1, 2, \dots \Big\}$$

are equal.

3.10.7 We now describe the order relations between the ideal norms under consideration.

PROPOSITION. $\quad \boldsymbol{\kappa}_n \leq \genfrac{}{}{0pt}{}{\boldsymbol{\tau}_n}{\boldsymbol{\omega}_n} \leq \boldsymbol{\varrho}_n \leq \boldsymbol{\delta}_n.$

PROOF. For $A_n = (\alpha_{hk}) \in \mathbb{U}^n$, the inequalities

$$\left(\sum_{h=1}^{n} \Big\| \sum_{k=1}^{n} \alpha_{hk} T x_k \Big\|^2 \right)^{1/2} \leq c \left(\sum_{k=1}^{n} \|x_k\|^2 \right)^{1/2}$$

and

$$\left(\sum_{h=1}^{n} \|T x_h^\circ\|^2 \right)^{1/2} \leq c \left(\sum_{k=1}^{m} \Big\| \sum_{h=1}^{n} \alpha_{kh}^* T x_k \Big\|^2 \right)^{1/2}$$

are equivalent via

$$x_h^\circ = \sum_{k=1}^{n} \alpha_{hk} x_k \quad \text{and} \quad x_k = \sum_{k=1}^{n} \alpha_{kh}^* x_h^\circ.$$

Hence $\boldsymbol{\kappa}(\mathcal{A}_n)$ is a cotype ideal norm, which proves that $\boldsymbol{\kappa}_n \le \boldsymbol{\omega}_n$. From

$$\boldsymbol{\varrho}(\mathcal{I}_n, \mathcal{A}_n) \le \boldsymbol{\varrho}'(\overline{\mathcal{A}}_n, \mathcal{I}_n) \qquad \text{(by 3.7.5)}$$

we deduce that $\boldsymbol{\omega}_n \le \boldsymbol{\tau}'_n$. Hence, thanks to 2.3.18 and 3.9.8,

$$\boldsymbol{\kappa}_n = \boldsymbol{\kappa}'_n \le \boldsymbol{\omega}'_n \le \boldsymbol{\tau}''_n = \boldsymbol{\tau}_n.$$

More simply, the inequality $\boldsymbol{\kappa}_n \le \boldsymbol{\tau}_n$ follows from 3.10.6. Next, a quick look at the definitions of $\boldsymbol{\tau}_n$, $\boldsymbol{\omega}_n$ and $\boldsymbol{\varrho}_n$ shows that $\boldsymbol{\tau}_n \le \boldsymbol{\varrho}_n$ and $\boldsymbol{\omega}_n \le \boldsymbol{\varrho}_n$. Finally, $\boldsymbol{\varrho}_n \le \boldsymbol{\delta}_n$ is a consequence of 3.4.10. It can also be deduced from 3.10.2 and 3.10.4.

3.10.8 PROPOSITION. $\quad \boldsymbol{\tau}_n \le \sqrt{2}\,\boldsymbol{\kappa}_n.$

PROOF. In a first step we show that

$$\|(Tx_k^\circ)|l_2^n\| \le \boldsymbol{\kappa}_n(T) \quad \text{whenever } (x_k^\circ) \in \operatorname{conv}(\mathbb{L}^n \circ [U_2^n, X]). \qquad (*)$$

Clearly, it is enough to check this claim for all $(u_k^\circ) \in \mathbb{L}^n \circ [U_2^n, X]$. Then there exist $L_n = (\lambda_{ki}) \in \mathbb{L}^n$ and $(u_i) \in [U_2^n, X]$ such that

$$u_k^\circ = \sum_{i=1}^{n} \lambda_{ki} u_i \quad \text{for } k = 1, \ldots, n,$$

and we have indeed

$$\|(Tu_k^\circ)|l_2^n\| = \|(Tu_i)|L_n\| \le \boldsymbol{\kappa}_n(T).$$

Fix $S_{mn} = (\sigma_{hk}) \in \mathbb{L}^{mn}$. Given $(x_h) \in [U_2^m, X]$, we let

$$x_k^\circ := \sum_{h=1}^{m} \sigma_{hk} x_h \quad \text{for } k = 1, \ldots, n.$$

Then, in view of 2.2.16,

$$(x_k^\circ) \in \mathbb{L}^{nm} \circ [U_2^m, X] \subseteq \sqrt{2}\operatorname{conv}(\mathbb{L}^n \circ [U_2^n, X]).$$

Now $(*)$ yields

$$\|(Tx_h)|S_{mn}^t\| = \|(Tx_k^\circ)|l_2^n\| \le \sqrt{2}\,\boldsymbol{\kappa}_n(T)$$

and, therefore,

$$\boldsymbol{\kappa}(T|S_{mn}^t) \le \sqrt{2}\,\boldsymbol{\kappa}_n(T).$$

Next, replacing T by T', we conclude that

$$\boldsymbol{\kappa}(T|S_{mn}) = \boldsymbol{\kappa}(T'|S_{mn}^t) \le \sqrt{2}\,\boldsymbol{\kappa}_n(T') = \sqrt{2}\,\boldsymbol{\kappa}_n(T),$$

by 2.3.6 and 2.3.18. Finally, the assertion follows from

$$\boldsymbol{\tau}_n(T) = \sup\Big\{\, \boldsymbol{\kappa}(T|S_{mn}) : S_{mn} \in \mathbb{L}^{mn} \text{ with } m = 1, 2, \ldots \Big\} \quad \text{(by 3.10.6).}$$

3.10.9 The following proof is in some sense dual to the previous one.

PROPOSITION. $\delta_n \le \sqrt{2}\,\tau_n$.

PROOF. Let $(x_k) \in [U_2^m, X]$ and $S_m = (\sigma_{hk}) \in \mathbb{L}^m$ such that $\operatorname{rank}(S_m) \le n$. Choose some factorization $S_m = B_{mn} A_{nm}$ with $A_{nm} = (\alpha_{ik}) \in \mathbb{L}^{nm}$ and $B_{mn} = (\beta_{hi}) \in \mathbb{L}^{mn}$.

In order to show that

$$\|(Tx_i^\circ)|B_{mn}\| \le \tau_n(T) \quad \text{for } (x_i^\circ) \in \operatorname{conv}(\mathbb{L}^n \circ [U_2^n, X]), \qquad (*)$$

we need check this claim only for all $(u_i^\circ) \in \mathbb{L}^n \circ [U_2^n, X]$. Then there exist $L_n = (\lambda_{ij}) \in \mathbb{L}^n$ and $(u_j) \in [U_2^n, X]$ such that

$$u_i^\circ = \sum_{j=1}^{n} \lambda_{ij} u_j \quad \text{for } i = 1, \ldots, n,$$

and, by 3.10.6, we have indeed

$$\|(Tu_i^\circ)|B_{mn}\| = \|(Tu_j)|B_{mn} L_n\| \le \tau_n(T).$$

Let

$$x_i^\circ := \sum_{k=1}^{m} \alpha_{ik} x_k \quad \text{for } i = 1, \ldots, n.$$

Then, in view of 2.2.16,

$$(x_i^\circ) \in \mathbb{L}^{nm} \circ [U_2^m, X] \subseteq \sqrt{2}\operatorname{conv}(\mathbb{L}^n \circ [U_2^n, X]).$$

Now $(*)$ yields

$$\|(Tx_k)|S_m\| = \|(Tx_i^\circ)|B_{mn}\| \le \sqrt{2}\,\tau_n(T).$$

and, therefore,

$$\kappa(T|S_m) \le \sqrt{2}\,\tau_n(T).$$

Finally, the assertion follows from 3.10.3.

3.10.10 We now summarize the preceding considerations.

THEOREM. *The sequences of the ideal norms κ_n, τ_n, ω_n, ϱ_n and δ_n are uniformly equivalent.*

REMARK. As a consequence, the ideal norm κ_n in Definition 2.3.24 can be replaced by τ_n, ω_n, ϱ_n or δ_n. In particular, we have

$$\mathfrak{KT}_p^{weak} = \Big\{ T \in \mathfrak{L} : \tau_n(T) = O(n^{1/p-1/2}) \Big\}.$$

This motivates the terminology *weak Kwapień type p*. If $1/p + 1/q = 1$, then $\mathfrak{KT}_p^{weak}$ coincides with

$$\mathfrak{KC}_q^{weak} = \Big\{ T \in \mathfrak{L} : \omega_n(T) = O(n^{1/2-1/q}) \Big\}.$$

Thus, with the same justification, one could refer to $T \in \mathfrak{KC}_q^{weak}$ as an operator of *weak Kwapień cotype q*.

3.10.11 In view of the previous observation, we are now able to establish an important criterion which is mainly due to G. Pisier; see [pis 8] and [pis 10, p. 278].

THEOREM. *Let $1 < p < 2$. An infinite dimensional Banach space X belongs to KT_p^{weak} if and only if one of the following equivalent conditions is satisfied:*

(1) *There exists a constant $c \geq 1$ such that $d(M, l_2^n) \leq c\, n^{1/p-1/2}$ for all n-dimensional subspaces M of X and $n = 1, 2, \ldots$.*

(2) *There exists a constant $c \geq 1$ such that all finite dimensional subspaces M of X are the range of a projection $P \in \mathfrak{L}(X)$ with $\|P|\mathfrak{H}\| \leq c \dim(M)^{1/p-1/2}$.*

PROOF. Condition **(1)** is a reformulation of 3.10.2 in the setting of spaces, and condition **(2)** follows from 3.10.4. Indeed, its sufficiency is clear. In order to see its necessity, we let $n := \dim(M)$ and choose $c > n^{1/2-1/p}\boldsymbol{\delta}_n(X)$ for $n = 1, 2, \ldots$. Then there exists $T_M \in \mathfrak{H}(X)$ such that

$$\|T_M|\mathfrak{H}\| < c\, n^{1/p-1/2} \quad \text{and} \quad T_M x = x \quad \text{for } x \in M.$$

Take a factorization $T_M = BA$ through a Hilbert space H with

$$\|A : X \to H\| \, \|B : H \to X\| < c\, n^{1/p-1/2},$$

and let $P_0 \in \mathfrak{L}(H)$ denote the orthogonal projection onto $A(M)$. Then $P := BP_0A$ has the required properties.

REMARKS. For completeness, we add another equivalent condition; see [koe*b, p. 111] and [pie 4, p. 67]:

(3) *There exists a constant $c \geq 1$ such that all finite dimensional subspaces M of X are the range of a projection $P \in \mathfrak{L}(X)$ with $\|P\| \leq c \dim(M)^{1/p-1/2}$.*

In the limiting case $p = 2$, conditions **(1)** and **(2)** characterize Hilbertian Banach spaces; see [joi], [lin*c] and [kad*a]. We stress that **(2)** is closely related to the complemented subspace theorem: *A Banach space is Hilbertian if and only if all of its subspaces are complemented.*

3.10.12 Finally, we formulate an analogue of Problem 2.3.16.

PROBLEM. *Do there exist 'universal' orthonormal systems $\mathcal{A}_n$ and $\mathcal{B}_n$ such that*

$$\boldsymbol{\tau}_n \prec \boldsymbol{\varrho}(\mathcal{B}_n, \mathcal{I}_n), \;\; \boldsymbol{\omega}_n \prec \boldsymbol{\varrho}(\mathcal{I}_n, \mathcal{A}_n), \;\; \boldsymbol{\varrho}_n \prec \boldsymbol{\varrho}(\mathcal{B}_n, \mathcal{A}_n) \;\; \textit{or} \;\; \boldsymbol{\delta}_n \prec \boldsymbol{\delta}(\mathcal{B}_n, \mathcal{A}_n)\,?$$

REMARK. The stronger conditions $\boldsymbol{\tau}_n = \boldsymbol{\varrho}(\mathcal{B}_n, \mathcal{I}_n)$ and $\boldsymbol{\omega}_n = \boldsymbol{\varrho}(\mathcal{I}_n, \mathcal{A}_n)$ are impossible.

3.11 Parseval ideal norms

3.11.1 Let $\mathcal{A}_n = (a_1, \dots, a_n)$ be an orthonormal system in the Hilbert space $L_2(M, \mu)$.

For $T \in \mathfrak{L}(X, Y)$, we denote by $\boldsymbol{\pi}(T|\mathcal{A}_n)$ the least constant $c \geq 0$ such that

$$\left(\int_M \Big\| \sum_{k=1}^n T x_k a_k(s) \Big\|^2 d\mu(s) \right)^{1/2} \leq c \sup \left\{ \left(\sum_{k=1}^n |\langle x_k, x' \rangle|^2 \right)^{1/2} : x' \in U_X^\circ \right\} \quad \text{(P)}$$

whenever $x_1, \dots, x_n \in X$. More concisely, the preceding inequality reads as follows:

$$\|(Tx_k)|\mathcal{A}_n\| \leq c\, \|(x_k)|w_2^n\|.$$

We refer to

$$\boldsymbol{\pi}(\mathcal{A}_n) : T \longrightarrow \boldsymbol{\pi}(T|\mathcal{A}_n)$$

as a **Parseval ideal norm**; see 2.2.2.

3.11.2 First of all, we state a trivial consequence of 2.2.1.

LEMMA.

$$\boldsymbol{\pi}(T|\mathcal{A}_n) = \sup \Big\{ \|(T A_n u_k^{(n)})|\mathcal{A}_n\| : \|A_n : l_2^n \to X\| \leq 1 \Big\}.$$

In the particular case when $T \in \mathfrak{L}(l_2^n, X)$, we have

$$\boldsymbol{\pi}(T|\mathcal{A}_n) = \sup \Big\{ \|(T A_n u_k^{(n)})|\mathcal{A}_n\| : A_n \in \mathbb{U}^n \Big\}.$$

3.11.3 Next, some elementary properties are listed.

PROPOSITION. $1 \leq \boldsymbol{\pi}(\mathcal{A}_n) \leq \sqrt{n}$.

3.11.4 As in 2.2.4, we can show that the above inequality is sharp.

PROPOSITION. $\boldsymbol{\pi}(H|\mathcal{A}_n) = \sqrt{n}$ *for all Hilbert spaces H with* $\dim(H) \geq n$.

3.11.5 We also have a **triangle inequality**; see 3.3.5 and 3.4.8.

PROPOSITION. $\boldsymbol{\pi}(\mathcal{B}_n) \leq \boldsymbol{\varrho}(\mathcal{B}_n, \mathcal{A}_n) \circ \boldsymbol{\pi}(\mathcal{A}_n)$.

3.11.6 **PROPOSITION.** *The ideal norm $\boldsymbol{\pi}(\mathcal{A}_n)$ is injective.*

REMARK. For $n \geq 2$, Parseval ideal norms are neither surjective nor symmetric.

3.11.7 For any matrix $A_n = (\alpha_{hk}) \in \mathbb{U}^n$, every orthonormal system $\mathcal{A}_n = (a_1, \dots, a_n)$ is transformed into the orthonormal system

$$A_n \circ \mathcal{A}_n := \left(\sum_{k=1}^n \alpha_{1k}\, a_k, \dots, \sum_{k=1}^n \alpha_{nk}\, a_k \right).$$

Parseval ideal norms are invariant under such transformations.

PROPOSITION. $\pi(A_n \circ \mathcal{A}_n) = \pi(\mathcal{A}_n)$.

3.11.8 The identity map from l_2^n onto l_∞^n can successfully be used to distinguish Parseval ideal norms.

PROPOSITION. *For every orthonormal system* $\mathcal{A}_n$ *the properties*

$$\pi(\mathcal{A}_n) = \pi(\mathcal{I}_n) \quad \textit{and} \quad \pi(Id : l_2^n \to l_\infty^n | \mathcal{A}_n) = \sqrt{n}$$

are equivalent.

PROOF (A. Hinrichs/J. Seigner). The implication $\Rightarrow$ follows from

$$\pi(Id : l_2^n \to l_\infty^n | \mathcal{I}_n) = \sqrt{n}.$$

We now assume that

$$\pi(Id : l_2^n \to l_\infty^n | \mathcal{A}_n) = \sqrt{n}.$$

Then 3.11.2 yields a matrix $A_n = (\alpha_{hk}) \in \mathbb{U}^n$ such that

$$\|(A_n u_k^{(n)}) | [\mathcal{A}_n, l_\infty^n]\| = \sqrt{n}.$$

Letting

$$b_h := \sum_{k=1}^{n} \alpha_{hk}\, a_k,$$

this means that

$$\int_M \max_{1 \le h \le n} |b_h(s)|^2 d\mu(s) = n = \int_M \sum_{h=1}^{n} |b_h(s)|^2 d\mu(s).$$

Thus the supports $B_h := \{ s \in M : b_h(s) \neq 0 \}$ are almost disjoint:

$$\mu(B_h \cap B_k) = 0 \quad \text{if } h \neq k.$$

Consequently, $\|(x_k) | \mathcal{B}_n\| = \|(x_k) | l_2^n\|$. Finally, in view of 3.11.7,

$$\pi(\mathcal{A}_n) = \pi(A_n \circ \mathcal{A}_n) = \pi(\mathcal{B}_n) = \pi(\mathcal{I}_n).$$

3.11.9 We know from 2.2.14 that $\pi_n \le \sqrt{2}\, \pi(\mathcal{I}_n)$. Thus the sequence $(\pi(\mathcal{I}_n))$ is almost maximal. But this is the best result we may hope for.

PROPOSITION. *There is no orthonormal system* $\mathcal{A}_n$ *with* $\pi_n = \pi(\mathcal{A}_n)$.

PROOF. Assume the contrary. Then

$$\pi(Id : l_2^n \to l_\infty^n | \mathcal{A}_n) = \pi_n(Id : l_2^n \to l_\infty^n) = \sqrt{n}.$$

So it follows from 3.11.8 that $\pi_n = \pi(\mathcal{A}_n) = \pi(\mathcal{I}_n)$. However, as shown in 2.2.15, this is impossible.

3.11.10 Next, we give a characterization of universal Parseval ideal norms, which were defined in 2.2.9.

PROPOSITION. $\boldsymbol{\pi}_n(T) = \sup\Big\{\boldsymbol{\pi}(T|\mathcal{A}_n) : \mathcal{A}_n \in \boldsymbol{\Omega}_n\Big\}.$

PROOF. The assertion follows from 3.9.2 and

$$\boldsymbol{\pi}_n(T) := \sup\Big\{\boldsymbol{\pi}(T|A_{mn}) : A_{mn} \in \mathbb{U}^{mn} \text{ with } m \geq n\Big\},$$

since the discrete orthonormal systems form a dense subset of $\boldsymbol{\Omega}_n$.

3.11.11 EXAMPLE. $\boldsymbol{\pi}(l_q^N|\mathcal{A}_n) \leq K_q(\mathcal{A}_n)\, N^{1/q}$ *if* $2 \leq q < \infty$.

PROOF. For $x_1 = (\xi_h^{(1)}), \ldots, x_n = (\xi_h^{(n)}) \in l_q^N$, we have

$$\Big(\sum_{k=1}^n |\xi_h^{(k)}|^2\Big)^{1/2} = \Big(\sum_{k=1}^n |\langle x_k, u_h^{(N)}\rangle|^2\Big)^{1/2} \leq \|(x_k)|w_2^n\| \quad \text{for } h = 1, \ldots, N. \quad (*)$$

Hence

$$\|(x_k)|[\mathcal{A}_n, l_q^N]\| = \left[\int_M \Big(\sum_{h=1}^N \Big|\sum_{k=1}^n \xi_h^{(k)} a_k(s)\Big|^q\Big)^{2/q} d\mu(s)\right]^{1/2}$$

(by definition)

$$\leq \left[\int_M \Big(\sum_{h=1}^N \Big|\sum_{k=1}^n \xi_h^{(k)} a_k(s)\Big|^q\Big)^{q/q} d\mu(s)\right]^{1/q}$$

(by $\|\cdot|L_2(M,\mu)\| \leq \|\cdot|L_q(M,\mu)\|$)

$$= \left[\sum_{h=1}^N \Big(\int_M \Big|\sum_{k=1}^n \xi_h^{(k)} a_k(s)\Big|^q d\mu(s)\Big)^{q/q}\right]^{1/q}$$

(by Fubini's theorem)

$$\leq K_q(\mathcal{A}_n)\left[\sum_{h=1}^N \Big(\sum_{k=1}^n |\xi_h^{(k)}|^2\Big)^{q/2}\right]^{1/q}$$

(by Khintchine's inequality)

$$\leq K_q(\mathcal{A}_n)\, N^{1/q}\, \|(x_k)|[w_2^n, l_q^N]\|.$$

(by (*))

REMARK. Of course, $\boldsymbol{\pi}(l_p^N|\mathcal{A}_n) \leq \sqrt{N}$ for $1 \leq p \leq 2$.

4

Rademacher and Gauss ideal norms

The classical examples of ideal norms,

$$\boldsymbol{\varrho}(\mathcal{R}_n, \mathcal{I}_n),\ \boldsymbol{\varrho}(\mathcal{I}_n, \mathcal{R}_n) \text{ and } \boldsymbol{\delta}(\mathcal{R}_n, \mathcal{R}_n),$$

are obtained from the Rademacher system. In the complex case, it is favourable to work with the Steinhaus system. Useful counterparts of the Rademacher and Steinhaus systems are the (real and complex) Gauss systems. There is a close connection with systems of characters on compact Abelian groups having small Sidon constants.

Among others, the following classes of Banach spaces will be treated:

RT_p	:	Banach spaces of Rademacher type p,
GT_p	:	Banach spaces of Gauss type p,
RT	:	Banach spaces of Rademacher subtype,
GT	:	Banach spaces of Gauss subtype,
B	:	B-convex Banach spaces,
RC_q	:	Banach spaces of Rademacher cotype q,
GC_q	:	Banach spaces of Gauss cotype q,
RC	:	Banach spaces of Rademacher subcotype,
GC	:	Banach spaces of Gauss subcotype,
MP	:	MP-convex Banach spaces.

We are going to prove that

$$\mathsf{RT}_p = \mathsf{GT}_p, \quad \mathsf{RC}_q = \mathsf{GC}_q, \qquad \mathsf{RT} = \mathsf{GT}, \quad \mathsf{RC} = \mathsf{GC},$$

$$\bigcup_{1<p\leq 2} \mathsf{RT}_p = \mathsf{RT} = \mathsf{B} \subset \mathsf{MP} = \mathsf{RC} = \bigcup_{2\leq q<\infty} \mathsf{RC}_q,$$

$$\mathsf{RT}_2 \cap \mathsf{RC}_2 = \mathsf{H}.$$

The duality between Rademacher type and cotype is described by

$$(\mathsf{RT}_p)' = \mathsf{RC}_{p'} \cap \mathsf{B} \quad \text{and} \quad (\mathsf{RC}_q \cap B)' = \mathsf{RT}_{q'}.$$

A further highlight of this chapter is the relation

$$\bigcup_{2\le q<\infty} \mathfrak{RC}_q = \bigcup_{2\le q<\infty} \mathfrak{GC}_q \subset \mathfrak{GR} \subset \mathfrak{QGR} = \mathfrak{GC} = \mathfrak{RC},$$

which summarizes results stated in 4.8.20, 4.10.2 and 4.10.11. We also show how the celebrated Dvoretzky theorem can be formulated in terms of ideal norms.

The theory of Rademacher type and cotype is presented in various monographs:

Diestel/Jarchow/Tonge... *Absolutely summing operators* [DIE*a],
Lindenstrauss/Tzafriri ... *Classical Banach spaces, vol. II* [LIN*2],
Kadets/Kadets........... *Rearrangement of series in Banach spaces* [KAD*],
Milman/Schechtman ..... *Asymptotic theory of finite dimensional normed spaces* [MIL*],
Tomczak-Jaegermann .... *Banach–Mazur distances and finite dimensional operator ideals* [TOM].

For more information about Rademacher functions, Gaussian random variables and Sidon sets, we refer to the following books:

Hewitt/Ross .... *Abstract harmonic analysis* [HEW*1, HEW*2],
Ledoux/Talagrand... *Probability in Banach spaces* [LED*],
López/Ross .... *Sidon sets* [LOP*].

4.1 Rademacher functions

4.1.1 For $k = 1, 2, \dots$ and $j = 1, \dots, 2^k$, let

$$\Delta_k^{(j)} := \big[\tfrac{j-1}{2^k}, \tfrac{j}{2^k}\big).$$

These sets are called **dyadic intervals**.

We define the k-th **Rademacher function** by

$$r_k(t) := (-1)^{j+1} \quad \text{for } t \in \Delta_k^{(j)}.$$

The **Rademacher system**

$$\mathcal{R}_n := (r_1, \dots, r_n)$$

is $\mathbb{R}$-unconditional, $\mathbb{R}$-unimodular and orthonormal in $L_2[0,1)$.

4.1.2 The Khintchine constants $K_q(\mathcal{R}_n)$ and $K'_p(\mathcal{R}_n)$ of the finite Rademacher systems are unknown. However, the limits

$$K_q(\mathcal{R}) := \lim_{n\to\infty} K_q(\mathcal{R}_n) \quad \text{and} \quad K'_p(\mathcal{R}) := \lim_{n\to\infty} K'_p(\mathcal{R}_n)$$

have been computed by U. Haagerup; see [haag]:

$$K_q(\mathcal{R}) = \sqrt{2}\left[\frac{\Gamma(\frac{q+1}{2})}{\Gamma(\frac{1}{2})}\right]^{1/q} \quad \text{if } 2 \le q < \infty$$

and

$$K'_p(\mathcal{R})^{-1} = \begin{cases} \sqrt{2}\left[\frac{\Gamma(\frac{p+1}{2})}{\Gamma(\frac{1}{2})}\right]^{1/p} & \text{if } p_0 \le p \le 2, \\ 2^{1/2-1/p} & \text{if } 1 \le p \le p_0, \end{cases}$$

where $p_0 = 1.8474\ldots$ is the unique solution of $\Gamma(\frac{p+1}{2}) = \Gamma(\frac{3}{2})$ in $(0,2)$. It follows that

$$K_q(\mathcal{R}) \le \sqrt{q-1} \quad \text{and} \quad 1 \le K'_p(\mathcal{R}) \le \sqrt{2}.$$

The proofs of the above statements are quite elaborate. Fortunately, the following weaker results are sufficient for our purpose:

$$1 \le K_q(\mathcal{R}) \le \sqrt{q};$$

see [KAC*], [LIN*2, p. 66] and [PIE 1, p. 43]. A very elegant proof of $K'_1(\mathcal{R}) = \sqrt{2}$ is due to [lat*].

REMARK. Note that

$$M_r^{\mathbb{R}} = \sqrt{2}\left[\frac{\Gamma(\frac{r+1}{2})}{\Gamma(\frac{1}{2})}\right]^{1/r}$$

is the r-th absolute moment of the real Gaussian measure; see 4.7.2.

4.1.3 Next, we describe a general process to construct orthonormal systems that behave in the same way as the Rademacher system; see also 3.8.9.

Let μ be a probability measure on a set M, and fix any measurable scalar-valued function a on M such that

$$\int_M a(\sigma)\, d\mu(\sigma) = 0 \quad \text{and} \quad \int_M |a(\sigma)|^2\, d\mu(\sigma) = 1.$$

Take the product measure μ^n on the n-fold Cartesian power

$$M^n := \Big\{ s = (\sigma_1, \ldots, \sigma_n) \,:\, \sigma_k \in M \quad \text{for} \quad k = 1, \ldots, n \Big\},$$

and define the coordinate functionals (projections)

$$p_k^{(n)}(s) := \sigma_k \quad \text{for } k = 1, \ldots, n.$$

Then

$$\mathcal{P}_n(a) := (a \circ p_1^{(n)}, \ldots, a \circ p_n^{(n)})$$

is an orthonormal system in $L_2(M^n, \mu^n)$. From the probabilistic point of view the **random variables** $a \circ p_1^{(n)}, \dots, a \circ p_n^{(n)}$ are independent and identically distributed.

4.1.4 LEMMA. $\|(x_k)|\mathcal{P}_n(a)\| \le \|(x_k)|\mathcal{P}_{n+1}(a)\|$.

PROOF. Writing $s = (\sigma_1, \dots, \sigma_n)$, we have

$$\|(x_k)|\mathcal{P}_n(a)\| = \left(\int\limits_{M^n} \Big\| \sum_{k=1}^{n} x_k\, a(\sigma_k) \Big\|^2 d\mu^n(s) \right)^{1/2}$$

$$= \left(\int\limits_{M^n} \Big\| \sum_{k=1}^{n+1} x_k \int\limits_{M} a(\sigma_k)\, d\mu(\sigma_{n+1}) \Big\|^2 d\mu^n(s) \right)^{1/2}$$

$$\le \left(\int\limits_{M^n} \int\limits_{M} \Big\| \sum_{k=1}^{n+1} x_k\, a(\sigma_k) \Big\|^2 d\mu^n(s)\, d\mu(\sigma_{n+1}) \right)^{1/2} = \|(x_k)|\mathcal{P}_{n+1}(a)\|.$$

4.1.5 The simplest example of the preceding construction is the 2-point set $\{+1, -1\}$ equipped with the measure $\mu\{\pm 1\} = \frac{1}{2}$ together with the identity map which takes ± 1 into ± 1. In this case, we get the **Bernoulli system** $\mathcal{P}_n(\pm 1) := (p_1^{(n)}, \dots, p_n^{(n)})$ defined on the n-th Cantor group

$$\mathbb{E}_2^n := \Big\{ e = (\varepsilon_1, \dots, \varepsilon_n) \,:\, \varepsilon_k = \pm 1 \Big\},$$

which is $\mathbb{R}$-unconditional and $\mathbb{R}$-unimodular. Note that

$$\|(x_k)|\mathcal{P}_n(\pm 1)\| = \left(\frac{1}{2^n} \sum_{\mathbb{E}_2^n} \Big\| \sum_{k=1}^{n} \varepsilon_k x_k \Big\|^2 \right)^{1/2}.$$

Since $\mathcal{R}_n$ and $\mathcal{P}_n(\pm 1)$ are identically distributed, we have

$$\|(x_k)|\mathcal{R}_n\| = \|(x_k)|\mathcal{P}_n(\pm 1)\|.$$

4.1.6 A complex counterpart of the Bernoulli system is obtained as follows:

Taking the function $e^{i\tau}$ on the interval $[-\pi, +\pi)$ equipped with the normalized Lebesgue measure, we get the **Steinhaus system**

$$\mathcal{P}_n(e^{i\tau}) := (e^{i\tau_1}, \dots, e^{i\tau_n})$$

which is $\mathbb{C}$-unconditional and $\mathbb{C}$-unimodular. Note that

$$\|(x_k)|\mathcal{P}_n(e^{i\tau})\| = \left(\frac{1}{(2\pi)^n} \int\limits_{-\pi}^{+\pi} \cdots \int\limits_{-\pi}^{+\pi} \Big\| \sum_{k=1}^{n} \exp(i\tau_k) x_k \Big\|^2 d\tau_1 \dots d\tau_n \right)^{1/2}.$$

4.1.7 Substituting $e^{i\tau_k}\xi_k$ in the Khintchine inequalities for $\mathcal{P}_n(\pm 1)$, integration over $\tau_1, \dots, \tau_n$ yields the corresponding inequalities for the Steinhaus system $\mathcal{P}_n(e^{i\tau})$. Hence

$$K_q(\mathcal{P}_n(e^{i\tau})) \le K_q(\mathcal{P}_n(\pm 1)) \quad \text{and} \quad K'_p(\mathcal{P}_n(e^{i\tau})) \le K'_p(\mathcal{P}_n(\pm 1)).$$

4.1.8 Write

$$K_q(\mathcal{P}(e^{i\tau})) := \lim_{n\to\infty} K_q(\mathcal{P}_n(e^{i\tau})) \quad \text{and} \quad K'_p(\mathcal{P}(e^{i\tau})) := \lim_{n\to\infty} K'_p(\mathcal{P}_n(e^{i\tau})).$$

We conjecture that

$$K_q(\mathcal{P}(e^{i\tau})) = \Gamma(\tfrac{q+2}{2})^{1/q} \quad \text{if } 2 \le q < \infty$$

and

$$K'_p(\mathcal{P}(e^{i\tau}))^{-1} = \Gamma(\tfrac{p+2}{2})^{1/p} \quad \text{if } 1 \le p \le 2;$$

see [saw].

4.1.9 CONVENTION. In what follows, whenever the real case and the complex case are treated simultaneously, the symbol $\mathcal{P}_n$ is used as a synonym of $\mathcal{P}_n(\pm 1)$ and $\mathcal{P}_n(e^{i\tau})$.

4.1.10 Next, we state the famous **Khintchine–Kahane inequality**; see [LED*, p. 100] and [LIN*2, p. 74].

PROPOSITION. *Let $0 < u < v < \infty$. Then there exists a constant $K(u,v) \ge 1$ such that*

$$\left(\int_0^1 \Big\| \sum_{k=1}^n x_k r_k(t) \Big\|^v dt \right)^{1/v} \le K(u,v) \left(\int_0^1 \Big\| \sum_{k=1}^n x_k r_k(t) \Big\|^u dt \right)^{1/u}$$

for $x_1, \dots, x_n \in X$ and any Banach space X.

REMARK. We may take

$$K(u,v) = \begin{cases} \sqrt{2} & \text{if } 1 \le u < v \le 2, \\ \sqrt{v-1} & \text{if } 2 \le u < v < \infty; \end{cases}$$

see [lat*] and [pis 6]. The reverse inequality

$$\left(\int_0^1 \Big\| \sum_{k=1}^n x_k r_k(t) \Big\|^u dt \right)^{1/u} \le \left(\int_0^1 \Big\| \sum_{k=1}^n x_k r_k(t) \Big\|^v dt \right)^{1/v}$$

is obvious.

4.2 Rademacher type and cotype ideal norms

Repetitio est mater studiorum!

4.2.1 For $T \in \mathfrak{L}(X,Y)$, we denote by $\varrho(T|\mathcal{R}_n, \mathcal{I}_n)$ the least constant $c \geq 0$ such that

$$\left(\int_0^1 \left\| \sum_{k=1}^n Tx_k r_k(t) \right\|^2 dt \right)^{1/2} \leq c \left(\sum_{k=1}^n \|x_k\|^2 \right)^{1/2}$$

whenever $x_1, \dots, x_n \in X$. More concisely, the preceding inequality reads as follows:

$$\|(Tx_k)|\mathcal{R}_n\| \leq c\, \|(x_k)|l_2^n\|.$$

We refer to

$$\varrho(\mathcal{R}_n, \mathcal{I}_n) : T \longrightarrow \varrho(T|\mathcal{R}_n, \mathcal{I}_n)$$

as a **Rademacher type ideal norm.**

4.2.2 The above definition has the following counterpart:

For $T \in \mathfrak{L}(X,Y)$, we denote by $\varrho(T|\mathcal{I}_n, \mathcal{R}_n)$ the least constant $c \geq 0$ such that

$$\left(\sum_{k=1}^n \|Tx_k\|^2 \right)^{1/2} \leq c \left(\int_0^1 \left\| \sum_{k=1}^n x_k r_k(t) \right\|^2 dt \right)^{1/2}$$

whenever $x_1, \dots, x_n \in X$. More concisely, the preceding inequality reads as follows:

$$\|(Tx_k)|l_2^n\| \leq c\, \|(x_k)|\mathcal{R}_n\|.$$

We refer to

$$\varrho(\mathcal{I}_n, \mathcal{R}_n) : T \longrightarrow \varrho(T|\mathcal{I}_n, \mathcal{R}_n)$$

as a **Rademacher cotype ideal norm.**

4.2.3 The following doubling inequalities are obvious.

PROPOSITION.

$$\varrho(\mathcal{R}_{2n}, \mathcal{I}_{2n}) \leq \sqrt{2}\, \varrho(\mathcal{R}_n, \mathcal{I}_n) \quad \textit{and} \quad \varrho(\mathcal{I}_{2n}, \mathcal{R}_{2n}) \leq \sqrt{2}\, \varrho(\mathcal{I}_n, \mathcal{R}_n).$$

4.2.4 In a sense, type and cotype ideal norms are dual to each other; see 3.7.5.

PROPOSITION.

$$\varrho(\mathcal{I}_n, \mathcal{R}_n) \leq \varrho'(\mathcal{R}_n, \mathcal{I}_n) \leq \varrho(\mathcal{I}_n, \mathcal{R}_n) \circ \boldsymbol{\delta}(\mathcal{R}_n, \mathcal{R}_n).$$

4.2.5 The following observation is a supplement to 3.7.6.

PROPOSITION. *The Rademacher cotype ideal norm $\varrho(\mathfrak{I}_n,\mathfrak{R}_n)$ fails to be surjective.*

PROOF. Since l_q is a quotient of l_1, the assertion follows from

$$\varrho(l_1|\mathfrak{I}_n,\mathfrak{R}_n)\asymp 1 \quad \text{and} \quad \varrho(l_q|\mathfrak{I}_n,\mathfrak{R}_n)=n^{1/2-1/q} \quad \text{if } 2\le q<\infty;$$

see 4.2.7 and 4.2.8.

4.2.6 PROPOSITION. *The sequences $(\varrho(\mathfrak{R}_n,\mathfrak{I}_n))$ and $(\varrho(\mathfrak{I}_n,\mathfrak{R}_n))$ are submultiplicative.*

PROOF. Let $T\in\mathfrak{L}(X,Y)$, $S\in\mathfrak{L}(Y,Z)$ and $x_l\in X$ for $l=1,\dots,mn$. Write $l=(h-1)n+k$ with $h=1,\dots,m$ and $k=1,\dots,n$. The unconditionality of $\mathfrak{R}_{mn}$ yields

$$\|(STx_{(h-1)n+k})|\mathfrak{R}_{mn}\|^2=\|(STx_{(h-1)n+k}r_h(s))|\mathfrak{R}_{mn}\|^2 \tag{1}$$

for $s\in[0,1)$; see 3.5.3. Moreover, since the systems

$$(r_1,\dots,r_n) \quad \text{and} \quad (r_{(h-1)n+1},\dots,r_{(h-1)n+n})$$

have identical distributions for $h=1,\dots,m$, we obtain

$$\int_0^1\Big\|\sum_{k=1}^n Tx_{(h-1)n+k}r_k(t)\Big\|^2 dt=\int_0^1\Big\|\sum_{k=1}^n Tx_{(h-1)n+k}r_{(h-1)n+k}(t)\Big\|^2 dt. \tag{2}$$

Obviously, (1) and (2) imply

$$\begin{aligned}
&\|(STx_{(h-1)n+k})|\mathfrak{R}_{mn}\|^2=\\
&=\int_0^1\int_0^1\Big\|\sum_{h=1}^m\sum_{k=1}^n STx_{(h-1)n+k}r_{(h-1)n+k}(t)r_h(s)\Big\|^2 dt\,ds\\
&\le\varrho(S|\mathfrak{R}_m,\mathfrak{I}_m)^2\int_0^1\sum_{h=1}^m\Big\|\sum_{k=1}^n Tx_{(h-1)n+k}r_{(h-1)n+k}(t)\Big\|^2 dt\\
&=\varrho(S|\mathfrak{R}_m,\mathfrak{I}_m)^2\sum_{h=1}^m\int_0^1\Big\|\sum_{k=1}^n Tx_{(h-1)n+k}r_k(t)\Big\|^2 dt\\
&\le\varrho(S|\mathfrak{R}_m,\mathfrak{I}_m)^2\varrho(T|\mathfrak{R}_n,\mathfrak{I}_n)^2\sum_{h=1}^m\sum_{k=1}^n\|x_{(h-1)n+k}\|^2.
\end{aligned}$$

This proves that

$$\varrho(ST|\mathfrak{R}_{mn},\mathfrak{I}_{mn})\le\varrho(S|\mathfrak{R}_m,\mathfrak{I}_m)\varrho(T|\mathfrak{R}_n,\mathfrak{I}_n).$$

The submultiplicativity of the Rademacher cotype ideal norms can be checked in the same way.

4.2.7 We now state a special case of 3.7.11.

EXAMPLE.

$$\varrho(l_p^n|\mathcal{R}_n,\mathcal{I}_n) = \varrho(L_p|\mathcal{R}_n,\mathcal{I}_n) = n^{1/p-1/2} \quad \text{if } 1 \le p \le 2,$$
$$\varrho(l_q^n|\mathcal{I}_n,\mathcal{R}_n) = \varrho(L_q|\mathcal{I}_n,\mathcal{R}_n) = n^{1/2-1/q} \quad \text{if } 2 \le q \le \infty.$$

4.2.8 The next result follows from 3.8.16.

EXAMPLE.

$$\varrho(L_p|\mathcal{I}_n,\mathcal{R}_n) = K_p'(\mathcal{R}_n) \quad \text{if } 1 \le p \le 2,$$
$$\varrho(L_q|\mathcal{R}_n,\mathcal{I}_n) = K_q(\mathcal{R}_n) \quad \text{if } 2 \le q \le \infty.$$

4.2.9 The preceding examples are summarized in a table:

	$1 \le r \le 2$	$2 \le r < \infty$	$r = \infty$
$\varrho(L_r\|\mathcal{R}_n,\mathcal{I}_n)$	$= n^{1/r-1/2}$ 4.2.7	$= K_r(\mathcal{R}_n)$ 4.2.8	$= \sqrt{n}$ 4.2.8
$\varrho(L_r\|\mathcal{I}_n,\mathcal{R}_n)$	$= K_r'(\mathcal{R}_n)$ 4.2.8	$= n^{1/2-1/r}$ 4.2.7	$= \sqrt{n}$ 4.2.7

Recall that

$1 \le K_r'(\mathcal{R}_n) \le \sqrt{2}$ if $1 \le r \le 2$ and $1 \le K_r(\mathcal{R}_n) \le \sqrt{r-1}$ if $2 \le r < \infty$,

while $K_\infty(\mathcal{R}_n) = \sqrt{n}$.

4.2.10 Let n and N be natural numbers.

EXAMPLE. $\varrho(l_\infty^N|\mathcal{I}_n,\mathcal{R}_n) = \varrho(l_1^N|\mathcal{R}_n,\mathcal{I}_n) = \min\{\sqrt{n},\sqrt{N}\}$.

PROOF. The upper estimate follows from

$$\varrho(l_\infty^N|\mathcal{I}_n,\mathcal{R}_n) \le \varrho(l_1^N|\mathcal{R}_n,\mathcal{I}_n) \le \sqrt{n}$$

and

$$\begin{aligned}\varrho(l_\infty^N|\mathcal{I}_n,\mathcal{R}_n) &\le \varrho(l_1^N|\mathcal{R}_n,\mathcal{I}_n)\\ &\le \|Id: l_1^N \to l_2^N\|\, \varrho(l_2^N|\mathcal{R}_n,\mathcal{I}_n)\, \|Id: l_2^N \to l_1^N\| = \sqrt{N}.\end{aligned}$$

On the other hand, for $m := \min\{n,N\}$, we have

$$\sqrt{m} = \varrho(l_\infty^m|\mathcal{I}_m,\mathcal{R}_m) \le \varrho(l_\infty^N|\mathcal{I}_n,\mathcal{R}_n) \le \varrho(l_1^N|\mathcal{R}_n,\mathcal{I}_n).$$

4.2.11 The next result should be compared with 4.12.13.

EXAMPLE. $\varrho(l_\infty^N|\mathcal{R}_n,\mathcal{I}_n) \asymp \min\{\sqrt{n},\sqrt{1+\log N}\}$.

PROOF. Obviously,

$$\varrho(l_\infty^N|\mathcal{R}_n,\mathcal{I}_n) \le \sqrt{n}.$$

By 3.3.8 and 4.1.2, we have

$$\varrho(l_q^N|\mathcal{R}_n, \mathcal{I}_n) \leq \sqrt{q-1}.$$

Thus the relation

$$\varrho(l_\infty^N|\mathcal{R}_n, \mathcal{I}_n) \prec \sqrt{1+\log N}$$

follows from 1.3.6. This proves the upper estimate.

Let $m := \min\{n, 1 + [\log_2 N]\}$. Then l_1^m can be embedded into l_∞^N uniformly. Hence, by 4.2.7,

$$\sqrt{m} = \varrho(l_1^m|\mathcal{R}_m, \mathcal{I}_m) \prec \varrho(l_\infty^N|\mathcal{R}_n, \mathcal{I}_n).$$

4.2.12 The preceding examples are summarized in the following table:

	l_1^N	l_∞^N
$\varrho(\mathcal{R}_n, \mathcal{I}_n)$	$= \min\{\sqrt{n}, \sqrt{N}\}$ 4.2.10	$\asymp \min\{\sqrt{n}, \sqrt{1+\log N}\}$ 4.2.11
$\varrho(\mathcal{I}_n, \mathcal{R}_n)$	$\asymp 1$ 4.2.8	$= \min\{\sqrt{n}, \sqrt{N}\}$ 4.2.10

4.2.13 Next, assuming that the underlying Banach spaces are complex, we state an easy corollary of 3.6.2.

PROPOSITION.

$$\varrho(\mathcal{P}_n(e^{i\tau}), \mathcal{P}_n(\pm 1)) \leq \tfrac{\pi}{2} \quad \textit{and} \quad \varrho(\mathcal{P}_n(\pm 1), \mathcal{P}_n(e^{i\tau})) \leq \tfrac{\pi}{2}.$$

PROOF. Note that

$$\|(x_k)|\mathcal{P}_n(e^{i\tau})\| = \|(x_k)|\mathcal{P}_n(e^{i\tau}) \otimes \mathcal{P}_n(\pm 1)\| \leq \tfrac{\pi}{2} \, \|(x_k)|\mathcal{P}_n(\pm 1)\|$$

and

$$\|(x_k)|\mathcal{P}_n(\pm 1)\| \leq \tfrac{\pi}{2} \, \|(x_k)|\mathcal{P}_n(e^{i\tau}) \otimes \mathcal{P}_n(\pm 1)\| = \tfrac{\pi}{2} \, \|(x_k)|\mathcal{P}_n(e^{i\tau})\|.$$

4.2.14 Now we are able to show that the sequences of type ideal norms associated with the Bernoulli and the Steinhaus systems, respectively, are uniformly equivalent. The same holds for the cotype ideal norms.

PROPOSITION.

$$\varrho(\mathcal{P}_n(e^{i\tau}), \mathcal{I}_n) \leq \varrho(\mathcal{P}_n(\pm 1), \mathcal{I}_n) \leq \tfrac{\pi}{2} \varrho(\mathcal{P}_n(e^{i\tau}), \mathcal{I}_n),$$

$$\varrho(\mathcal{I}_n, \mathcal{P}_n(e^{i\tau})) \leq \varrho(\mathcal{I}_n, \mathcal{P}_n(\pm 1)) \leq \tfrac{\pi}{2} \varrho(\mathcal{I}_n, \mathcal{P}_n(e^{i\tau})).$$

PROOF. Since $\mathcal{P}_n(e^{i\tau})$ is $\mathbb{R}$-unconditional and $\mathcal{R}_n$ is $\mathbb{R}$-unimodular, the left-hand inequalities follow from 3.7.9. The others are consequences of the preceding proposition.

4.2.15 Recall from 3.1.1 that $\mathcal{A}_n(\mathbb{F})$ is the subsystem of $\mathcal{A}_n=(a_1,\dots,a_n)$ consisting of all functions a_k with $k\in\mathbb{F}$. The following results are due to A. Hinrichs; see [hin 1].

LEMMA. *Let $\{\mathbb{F}_1,\dots,\mathbb{F}_m\}$ be a covering of $\{1,\dots,n\}$ such that every number $k\in\{1,\dots,n\}$ belongs to exactly d of these subsets. Then, for any orthonormal system $\mathcal{A}_n=(a_1,\dots,a_n)$,*

$$\varrho(\mathcal{A}_n,\mathcal{I}_n)\le\sqrt{\tfrac{m}{d}}\max_{1\le\alpha\le m}\varrho(\mathcal{A}_n(\mathbb{F}_\alpha),\mathcal{I}_n(\mathbb{F}_\alpha)).$$

PROOF. We have

$$\|(Tx_k)|\mathcal{A}_n\|=\left(\int\limits_M\Big\|\sum_{k=1}^n Tx_k a_k(s)\Big\|^2 d\mu(s)\right)^{1/2}$$

$$=\left(\int\limits_M\Big\|\tfrac{1}{d}\sum_{\alpha=1}^m\sum_{k\in\mathbb{F}_\alpha} Tx_k a_k(s)\Big\|^2 d\mu(s)\right)^{1/2}$$

$$\le\tfrac{1}{d}\sum_{\alpha=1}^m\|(Tx_k)|\mathcal{A}_n(\mathbb{F}_\alpha)\|\le\tfrac{1}{d}\sum_{\alpha=1}^m\varrho(T|\mathcal{A}_n(\mathbb{F}_\alpha),\mathcal{I}_n(\mathbb{F}_\alpha))\,\|(x_k)|l_2(\mathbb{F}_\alpha)\|$$

$$\le\tfrac{\sqrt{m}}{d}\max_{1\le\alpha\le m}\varrho(\mathcal{A}_n(\mathbb{F}_\alpha),\mathcal{I}_n(\mathbb{F}_\alpha))\left(\sum_{\alpha=1}^m\|(x_k)|l_2(\mathbb{F}_\alpha)\|^2\right)^{1/2}$$

$$=\tfrac{\sqrt{m}}{d}\max_{1\le\alpha\le m}\varrho(\mathcal{A}_n(\mathbb{F}_\alpha),\mathcal{I}_n(\mathbb{F}_\alpha))\left(\sum_{\alpha=1}^m\sum_{k\in\mathbb{F}_\alpha}\|x_k\|^2\right)^{1/2}$$

$$=\sqrt{\tfrac{m}{d}}\max_{1\le\alpha\le m}\varrho(\mathcal{A}_n(\mathbb{F}_\alpha),\mathcal{I}_n(\mathbb{F}_\alpha))\,\|(x_k)|l_2^n\|.$$

4.2.16 We are now prepared to establish **Hinrichs's inequality**.

PROPOSITION. *Let $\mathcal{A}_n$ be the system formed by* **all** *characters of any finite Abelian group $\mathbb{G}_n$. Then, for every subset $\mathbb{F}$ of $\{1,\dots,n\}$,*

$$n^{-1/2}\varrho(\mathcal{A}_n,\mathcal{I}_n)\le|\mathbb{F}|^{-1/2}\varrho(\mathcal{A}_n(\mathbb{F}),\mathcal{I}_n(\mathbb{F})).$$

PROOF. By assumption, the map $\varphi:k\longrightarrow a_k$ defines a bijection from $\{1,\dots,n\}$ onto $\mathbb{G}'_n=\mathcal{A}_n$. Letting $\mathbb{F}_h:=\varphi^{-1}(a_h\varphi(\mathbb{F}))$ yields a covering $\{\mathbb{F}_1,\dots,\mathbb{F}_n\}$ of $\{1,\dots,n\}$ such that every number k belongs to exactly $d:=|\mathbb{F}|$ of these subsets. Moreover, $\mathcal{A}_n(\mathbb{F}_h)=a_h\mathcal{A}_n(\mathbb{F})$ implies that

$$\varrho(\mathcal{A}_n(\mathbb{F}_h),\mathcal{I}_n(\mathbb{F}_h))=\varrho(\mathcal{A}_n(\mathbb{F}),\mathcal{I}_n(\mathbb{F})).$$

Thus the conclusion follows by applying Lemma 4.2.15.

4.2.17 PROPOSITION.

$$2^{-n/2}\,\boldsymbol{\varrho}(\mathcal{I}_{2^n},\mathcal{R}_{2^n}) \le 2^{-n/2}\,\boldsymbol{\varrho}'(\mathcal{R}_{2^n},\mathcal{I}_{2^n}) \le n^{-1/2}\,\boldsymbol{\varrho}(\mathcal{R}_n,\mathcal{I}_n).$$

PROOF. Let $\mathcal{W}_{2^n}$ denote the system of all characters on $\mathbb{E}_2^n$; see 6.1.5. Then, by the preceding proposition,

$$2^{-n/2}\,\boldsymbol{\varrho}(\mathcal{W}_{2^n},\mathcal{I}_{2^n}) \le n^{-1/2}\,\boldsymbol{\varrho}(\mathcal{P}_n(\pm 1),\mathcal{I}_n).$$

Furthermore, we know from 4.2.4, 3.7.9, and 3.8.14 that

$$\boldsymbol{\varrho}(\mathcal{I}_{2^n},\mathcal{P}_{2^n}(\pm 1)) \le \boldsymbol{\varrho}'(\mathcal{P}_{2^n}(\pm 1),\mathcal{I}_{2^n}) \le \boldsymbol{\varrho}'(\mathcal{W}_{2^n},\mathcal{I}_{2^n}) = \boldsymbol{\varrho}(\mathcal{W}_{2^n},\mathcal{I}_{2^n}).$$

REMARK. The above result is optimal in the following sense:
Assume that $\varphi : \mathbb{N} \to \mathbb{N}$ is a function such that

$$\varphi(n)^{-1/2}\,\boldsymbol{\varrho}(\mathcal{I}_{\varphi(n)},\mathcal{R}_{\varphi(n)}) \le c\,n^{-1/2}\,\boldsymbol{\varrho}(\mathcal{R}_n,\mathcal{I}_n).$$

Recall from 4.2.12 that

$$\boldsymbol{\varrho}(l_\infty^N|\mathcal{I}_N,\mathcal{R}_N) = \sqrt{N} \quad\text{and}\quad \boldsymbol{\varrho}(l_\infty^N|\mathcal{R}_n,\mathcal{I}_n) \asymp \min\{\sqrt{n},\sqrt{1+\log N}\}.$$

Hence, letting $N := \varphi(n)$, we obtain

$$1 \le c\,n^{-1/2}\,\boldsymbol{\varrho}(l_\infty^N|\mathcal{R}_n,\mathcal{I}_n) \prec c\,n^{-1/2}\,(1+\log\varphi(n))^{1/2},$$

which means that $a^n \prec \varphi(n)$ for some $a > 1$.

4.3 Operators of Rademacher type

4.3.1 Let $1 < p \le 2$ and $1 \le v < \infty$. For $T \in \mathfrak{L}(X,Y)$, we denote by $\boldsymbol{\varrho}_p^{(v)}(T|\mathcal{R}_n,\mathcal{I}_n)$ the least constant $c \ge 0$ such that

$$\left(\int_0^1 \Big\| \sum_{k=1}^n T x_k r_k(t) \Big\|^v dt\right)^{1/v} \le c \left(\sum_{k=1}^n \|x_k\|^p\right)^{1/p}$$

whenever $x_1,\dots,x_n \in X$. More concisely, the above inequality reads as follows:

$$\|(Tx_k)|\mathcal{R}_n\|_v \le c\,\|(x_k)|l_p^n\|.$$

We refer to

$$\boldsymbol{\varrho}_p^{(v)}(\mathcal{R}_n,\mathcal{I}_n) : T \longrightarrow \boldsymbol{\varrho}_p^{(v)}(T|\mathcal{R}_n,\mathcal{I}_n)$$

as a **Rademacher type ideal norm**. The case $p = v = 2$ has already been treated in the previous section.

Replacing $\|\cdot|l_p^n\|$ by the Lorentz norm $\|\cdot|l_{p,1}^n\|$ yields the **weak Rademacher type ideal norm** $\boldsymbol{\varrho}_{p,1}^{(v)}(\mathcal{R}_n,\mathcal{I}_n)$.

4.3.2 To synchronize the theories of

Rademacher type
(presented in this section)
and
Rademacher cotype
(presented in the next but one section)

we include an

$$\emptyset$$

paragraph.

4.3.3 It follows from the Khintchine–Kahane inequality 4.1.10 that the asymptotic behaviour of $\varrho_p^{(v)}(T|\mathcal{R}_n, \mathcal{I}_n)$ is determined by the exponent p only.

PROPOSITION. *Fix $1 < p \le 2$. Then the sequences of the ideal norms $\varrho_p^{(v)}(\mathcal{R}_n, \mathcal{I}_n)$ are uniformly equivalent whenever $1 \le v < \infty$.*

4.3.4 Let $1 < p \le 2$, $1 \le v < \infty$, and fix any finite subset $\mathbb{F}$ of $\mathbb{N}$. Then, for $T \in \mathfrak{L}(X, Y)$, we denote by $\varrho_p^{(v)}(T|\mathcal{R}(\mathbb{F}), \mathcal{I}(\mathbb{F}))$ the least constant $c \ge 0$ such that

$$\left(\int_0^1 \Big\| \sum_{k \in \mathbb{F}} T x_k r_k(t) \Big\|^v dt \right)^{1/v} \le c \left(\sum_{k \in \mathbb{F}} \|x_k\|^p \right)^{1/p}$$

whenever (x_k) is an $\mathbb{F}$-tuple in X. More concisely, the above inequality reads as follows:

$$\|(Tx_k)|\mathcal{R}(\mathbb{F})\|_v \le c\, \|(x_k)|l_p(\mathbb{F})\|.$$

Obviously,

$$\varrho_p^{(v)}(\mathcal{R}(\mathbb{F}), \mathcal{I}(\mathbb{F})) : T \longrightarrow \varrho_p^{(v)}(T|\mathcal{R}(\mathbb{F}), \mathcal{I}(\mathbb{F}))$$

is an ideal norm. If $|\mathbb{F}| = n$, then $\mathcal{R}(\mathbb{F})$ and $\mathcal{R}_n$ have identical distributions. Hence

$$\varrho_p^{(v)}(\mathcal{R}(\mathbb{F}), \mathcal{I}(\mathbb{F})) = \varrho_p^{(v)}(\mathcal{R}_n, \mathcal{I}_n).$$

REMARK. In view of the last equality, the preceding definition seems to be superfluous. Nevertheless, we have introduced that concept in order to indicate the analogy with the case of trigonometric functions, Walsh functions and Haar functions, where the results are more complicated; see 5.6.15, 6.3.4 and 7.4.20.

4.3.5 We now provide a crucial tool; see also 4.5.5 and 5.6.16.

LEMMA. *Let* $1<p<u\le 2$ *and* $1\le v<\infty$. *Then, for* $T\in\mathfrak{L}(X,Y)$, *the following properties are equivalent:*

(1) *There exists a constant* $c_0\ge 1$ *such that*

$$\varrho_u^{(v)}(T|\mathcal{R}_n,\mathcal{I}_n)\le c_0 n^{1/p-1/u} \quad \textit{for } n=1,2,\ldots.$$

(2) *There exists a constant* $c\ge 1$ *such that*

$$\varrho_{p,1}^{(v)}(T|\mathcal{R}_n,\mathcal{I}_n)\le c \quad \textit{for } n=1,2,\ldots.$$

PROOF. The implication **(2)**$\Rightarrow$**(1)** follows from

$$\|(x_k)|l_{p,1}^n\|\le An^{1/p-1/u}\,\|(x_k)|l_u^n\| \qquad \text{(by 0.5.6)},$$

where $A>0$ is a constant.

Fix $x_1,\ldots,x_n\in X$. By 0.5.8, we can find a partition of $\{1,\ldots,n\}$ into pairwise disjoint subsets $\mathbb{F}_1,\ldots,\mathbb{F}_m$ such that

$$\sum_{h=1}^m|\mathbb{F}_h|^{1/p-1/u}\Big(\sum_{k\in\mathbb{F}_h}\|x_k\|^u\Big)^{1/u}\le 2\,\|(x_k)|l_{p,1}^n\|.$$

Assume that **(1)** holds. It now follows from 4.3.4 that

$$\Big(\int_0^1\Big\|\sum_{k=1}^n Tx_kr_k(t)\Big\|^v dt\Big)^{1/v}\le\sum_{h=1}^m\Big(\int_0^1\Big\|\sum_{k\in\mathbb{F}_h}Tx_kr_k(t)\Big\|^v dt\Big)^{1/v}\le$$

$$\le\sum_{h=1}^m\varrho_u^{(v)}(T|\mathcal{R}(\mathbb{F}_h),\mathcal{I}(\mathbb{F}_h))\Big(\sum_{k\in\mathbb{F}_h}\|x_k\|^u\Big)^{1/u}$$

$$\le c_0\sum_{h=1}^m|\mathbb{F}_h|^{1/p-1/u}\Big(\sum_{k\in\mathbb{F}_h}\|x_k\|^u\Big)^{1/u}\le 2c_0\|(x_k)|l_{p,1}^n\|.$$

Hence $\varrho_{p,1}^{(v)}(T|\mathcal{R}_n,\mathcal{I}_n)\le c:=2c_0$.

4.3.6 Let $1<p\le 2$. An operator T has **Rademacher type** p if

$$\|T|\mathfrak{RT}_p\|:=\sup_n\ \varrho_p^{(2)}(T|\mathcal{R}_n,\mathcal{I}_n)$$

is finite. These operators form the Banach ideal

$$\mathfrak{RT}_p:=\mathfrak{L}[\varrho_p^{(2)}(\mathcal{R}_n,\mathcal{I}_n)].$$

In view of the Khintchine–Kahane inequality 4.1.10, the expressions

$$\|T|\mathfrak{L}[\varrho_p^{(v)}(\mathcal{R}_n,\mathcal{I}_n)]\|:=\sup_n\ \varrho_p^{(v)}(T|\mathcal{R}_n,\mathcal{I}_n)\quad\text{with } 1\le v<\infty$$

yield equivalent norms.

REMARK. Trivially, all operators have Rademacher type 1. Therefore we put $\mathfrak{RT}_1:=\mathfrak{L}$.

4.3.7 Let $1<p<2$. An operator T has **weak Rademacher type** p if

$$\|T|\mathfrak{RT}_p^{weak}\| := \sup_n n^{1/2-1/p}\, \varrho(T|\mathcal{R}_n, \mathcal{I}_n).$$

is finite. These operators form the Banach ideal

$$\mathfrak{RT}_p^{weak} := \mathfrak{L}[n^{1/2-1/p}\, \varrho(\mathcal{R}_n, \mathcal{I}_n)].$$

REMARK. In the case $p=2$, we would get $\mathfrak{RT}_2$.

4.3.8 PROPOSITION. *The operator ideals* $\mathfrak{RT}_p$ *and* $\mathfrak{RT}_p^{weak}$ *are injective, surjective and* l_2*-stable.*

REMARK. In the setting of spaces, the above result implies that RT_p and RT_p^{weak} are stable when passing to subspaces and quotients. We also have stability under the formation of finite direct sums and l_2-multiples.

4.3.9 The following result explains why the attribute *weak* is used in Definition 4.3.7.

PROPOSITION. *If* $1<p<u\le 2$ *and* $1\le v<\infty$, *then*

$$\mathfrak{RT}_p^{weak} = \mathfrak{L}\Big[\varrho_{p,1}^{(v)}(\mathcal{R}_n, \mathcal{I}_n)\Big] = \mathfrak{L}\Big[n^{1/u-1/p}\varrho_u^{(v)}(\mathcal{R}_n, \mathcal{I}_n)\Big].$$

PROOF. The assertion follows at once from 4.3.5 combined with the Khintchine–Kahane inequality 4.1.10.

4.3.10 We are now able to establish the main result of this section.

THEOREM. $\mathfrak{RT}_2 \subset \mathfrak{RT}_{p_0} \subset \mathfrak{RT}_{p_0}^{weak} \subset \mathfrak{RT}_p$ *if* $1<p<p_0<2$.

In particular,

$$\bigcup_{1<p\le 2} \mathfrak{RT}_p = \bigcup_{1<p<2} \mathfrak{RT}_p^{weak}.$$

PROOF. In view of 0.5.4, we have $\|\cdot|l_{p_0}^n\| \le \|\cdot|l_{p_0,1}^n\|$. Hence

$$\varrho_{p_0,1}^{(2)}(\mathcal{R}_n, \mathcal{I}_n) \le \varrho_{p_0}^{(2)}(\mathcal{R}_n, \mathcal{I}_n)$$

which implies the inclusion $\mathfrak{RT}_{p_0} \subseteq \mathfrak{RT}_{p_0}^{weak}$.

On the other hand, $\mathfrak{RT}_{p_0}^{weak} \subseteq \mathfrak{RT}_p$ follows from

$$\|\cdot|l_{p_0,1}^n\| \le A\,\|\cdot|l_p^n\| \qquad \text{(by 0.5.6)},$$

where $A \ge 1$ is a constant.

It will be shown in 4.3.11 that

$$L_p \in \mathsf{RT}_p \setminus \mathsf{RT}_{p_0}^{weak} \quad \text{if } 1<p<p_0<2.$$

Thus the inclusion $\mathfrak{RT}_{p_0}^{weak} \subseteq \mathfrak{RT}_p$ is strict. Finally, Example 4.3.13 tells us that the inclusion $\mathfrak{RT}_{p_0} \subseteq \mathfrak{RT}_{p_0}^{weak}$ is strict as well.

4.3.11 EXAMPLE.

$L_p \in \mathsf{RT}_p \setminus \mathsf{RT}_{p_0}^{weak}$ *if* $1<p<p_0<2$ *and* $L_q \in \mathsf{RT}_2$ *if* $2 \le q < \infty$.

PROOF. In view of 3.7.16,

$$1 \le \varrho_p^{(2)}(L_p|\mathcal{R}_n,\mathcal{I}_n) \le \varrho_p^{(p')}(L_p|\mathcal{R}_n,\mathcal{I}_n) \le 1.$$

Thus $L_p \in \mathsf{RT}_p$. Moreover, we know from 4.2.7 that

$$\varrho(L_p|\mathcal{R}_n,\mathcal{I}_n) = n^{1/p-1/2}.$$

Hence $L_p \notin \mathsf{RT}_{p_0}^{weak}$. Finally, $L_q \in \mathsf{RT}_2$ follows from 4.2.8.

4.3.12 We now evaluate the difference between $\mathfrak{RT}_p$ and $\mathfrak{RT}_p^{weak}$.

PROPOSITION. $\varrho_p^{(v)}(\mathcal{R}_n,\mathcal{I}_n) \le (1+\log n)^{1/p'}\,\varrho_{p,1}^{(v)}(\mathcal{R}_n,\mathcal{I}_n)$.

PROOF. By 0.5.5,

$$\|\cdot|l_{p,1}^n\| \le (1+\log n)^{1/p'}\,\|\cdot|l_p^n\|.$$

4.3.13 For $\lambda \ge 0$, the diagonal operator D_λ is defined by

$$D_\lambda : (\xi_k) \longrightarrow (k^{-\lambda}\xi_k).$$

The following example shows that the preceding estimate is sharp.

EXAMPLE.

$$\varrho_p^{(2)}(D_{1/p'} : l_1 \to l_1|\mathcal{R}_n,\mathcal{I}_n) \asymp (1+\log n)^{1/p'},$$
$$\varrho_{p,1}^{(2)}(D_{1/p'} : l_1 \to l_1|\mathcal{R}_n,\mathcal{I}_n) \asymp 1.$$

PROOF. We know from 3.7.15 that

$$\varrho_p^{(2)}(D_{1/p'} : l_1 \to l_1|\mathcal{R}_n,\mathcal{I}_n) = \left(\sum_{k=1}^n \frac{1}{k}\right)^{1/p'} \asymp (1+\log n)^{1/p'}$$

and

$$\varrho_2^{(2)}(D_{1/p'} : l_1 \to l_1|\mathcal{R}_n,\mathcal{I}_n) = \left(\sum_{k=1}^n k^{-2/p'}\right)^{1/2} \asymp n^{1/p-1/2}.$$

By 4.3.5, the last formula is equivalent to

$$\varrho_{p,1}^{(2)}(D_{1/p'} : l_1 \to l_1|\mathcal{R}_n,\mathcal{I}_n) \asymp 1.$$

4.3.14 The subsequent considerations are due to M. Junge; see [jun].

LEMMA. *Let* $\{\mathbb{F}_1,\dots,\mathbb{F}_n\}$ *be any partition of* $\{1,\dots,N\}$ *into pairwise disjoint subsets. Then*

$$\|(Sy_k)|\mathcal{R}_N\|_p \le \varrho_p^{(p)}(S|\mathcal{R}_n,\mathcal{I}_n)\left(\sum_{h=1}^n \|(y_k)|\mathcal{R}(\mathbb{F}_h)\|_p^p\right)^{1/p}$$

whenever $y_1,\dots,y_N \in Y$ *and* $S \in \mathfrak{L}(Y,Z)$.

PROOF. Since $\mathcal{R}_n$ is $\mathbb{R}$-unconditional,

$$\int_0^1 \Big\| \sum_{h=1}^n \sum_{k\in\mathbb{F}_h} Sy_k r_k(t) \Big\|^p dt = \int_0^1 \Big\| \sum_{h=1}^n \sum_{k\in\mathbb{F}_h} Sy_k r_k(t) r_h(s) \Big\|^p dt.$$

Integrating over $s \in [0,1)$, we obtain

$$\|(Sy_k)|\mathcal{R}_N\|_p^p = \int_0^1\int_0^1 \Big\| \sum_{h=1}^n S\Big(\sum_{k\in\mathbb{F}_h} y_k r_k(t)\Big) r_h(s) \Big\|^p ds\, dt$$

$$\leq \varrho_p^{(p)}(S|\mathcal{R}_n, \mathcal{I}_n)^p \int_0^1 \sum_{h=1}^n \Big\| \sum_{k\in\mathbb{F}_h} y_k r_k(t) \Big\|^p dt.$$

4.3.15 We are now in a position to show that, in the setting of spaces, the estimate established in 4.3.12 can be improved significantly.

PROPOSITION.

$$\varrho_p^{(p)}(\mathcal{R}_{2^n}, \mathcal{I}_{2^n}) \leq 4\, \varrho_p^{(p)}(\mathcal{R}_n, \mathcal{I}_n) \circ \varrho_{p,1}^{(p)}(\mathcal{R}_{2^n}, \mathcal{I}_{2^n}).$$

PROOF. Without loss of generality, we may assume that

$$\|x_1\| \geq \|x_2\| \geq \ldots \geq \|x_{2^n}\|.$$

Let $\mathbb{F}_1 := \{1,2\}$ and $\mathbb{F}_h := \{2^{h-1}+1, \ldots, 2^h\}$ for $h = 2, \ldots, n$. Then

$$\|(x_k)|l_{p,1}(\mathbb{F}_1)\| = \|x_1\| + 2^{1/p-1}\|x_2\| \leq 2\,\|x_1\|$$

and, by 0.10.2,

$$\|(x_k)|l_{p,1}(\mathbb{F}_h)\| \leq \sum_{k=1}^{|\mathbb{F}_h|} k^{1/p-1}\|x_{2^{h-1}}\|$$

$$\leq p|\mathbb{F}_h|^{1/p}\|x_{2^{h-1}}\| \leq 2\cdot 2^{(h-1)/p}\|x_{2^{h-1}}\|$$

for $h = 2, \ldots, n$. Hence 0.10.3 implies that

$$\sum_{h=1}^n \|(x_k)|l_{p,1}(\mathbb{F}_h)\|^p \leq 2^p\|x_1\|^p + 2^p \sum_{h=2}^n 2^{h-1}\|x_{2^{h-1}}\|^p \leq 2\cdot 2^p \sum_{k=1}^{2^n}\|x_k\|^p.$$

In view of the previous lemma, we now obtain

$$\|(STx_k)|\mathcal{R}_{2^n}\|_p^p \leq \varrho_p^{(p)}(S|\mathcal{R}_n, \mathcal{I}_n)^p \sum_{h=1}^n \|(Tx_k)|\mathcal{R}(\mathbb{F}_h)\|_p^p$$

$$\leq \varrho_p^{(p)}(S|\mathcal{R}_n, \mathcal{I}_n)^p \sum_{h=1}^n \varrho_{p,1}^{(p)}(T|\mathcal{R}(\mathbb{F}_h), \mathcal{I}(\mathbb{F}_h))^p \|(x_k)|l_{p,1}(\mathbb{F}_h)\|^p$$

$$\leq 2^{p+1}\varrho_p^{(p)}(S|\mathcal{R}_n, \mathcal{I}_n)^p \varrho_{p,1}^{(p)}(T|\mathcal{R}_{2^n}, \mathcal{I}_{2^n})^p \sum_{k=1}^{2^n}\|x_k\|^p,$$

which completes the proof.

4.3.16 Using the terminology defined in 0.10.6, the previous proposition yields the following result.

PROPOSITION. *If a Banach space X has weak Rademacher type p, then the sequence $(\varrho_p^{(2)}(X|\mathcal{R}_n, \mathcal{I}_n))$ is very slowly growing.*

REMARK. In view of 4.3.13, the gap between $\mathfrak{RT}_p$ and $\mathfrak{RT}_p^{weak}$ is much larger than the gap between RT_p and RT_p^{weak}. In other words, it is hard to find a Banach space $X \in \mathsf{RT}_p^{weak} \setminus \mathsf{RT}_p$. Examples were constructed by L. Tzafriri as a modification of Tsirelson's space; see [tza] and the section on Tirilman's space in [CAS*, pp. 107–116].

4.3.17 Recall that Σ_n is the finite summation operator defined in 0.7.3.

EXAMPLE. $\varrho(\Sigma_n : l_1^n \to l_\infty^n|\mathcal{R}_n, \mathcal{I}_n) \le 2.$

PROOF. By Doob's inequality 7.1.9, we get

$$\Big(\int_0^1 \max_{1\le h\le n}\Big|\sum_{k\le h}\xi_k r_k(t)\Big|^2 dt\Big)^{1/2} \le 2\Big(\int_0^1 \Big|\sum_{k=1}^n \xi_k r_k(t)\Big|^2 dt\Big)^{1/2} = 2\Big(\sum_{k=1}^n |\xi_k|^2\Big)^{1/2}.$$

Since all permutations $(r_{\pi(1)}, \dots, r_{\pi(n)})$ have identical distributions, we conclude from 3.7.14 that $\varrho(\Sigma_n : l_1^n \to l_\infty^n|\mathcal{R}_n, \mathcal{I}_n) \le 2$.

4.3.18 We now pass to the infinite summation operator Σ.

EXAMPLE. $\Sigma \in \mathfrak{RT}_2(l_1, l_\infty) \cap \mathfrak{RC}_2(l_1, l_\infty).$

PROOF. In view of

$$\varrho(\Sigma : l_1 \to l_\infty|\mathcal{R}_n, \mathcal{I}_n) = \lim_{N\to\infty} \varrho(\Sigma_N : l_1^N \to l_\infty^N|\mathcal{R}_n, \mathcal{I}_n)$$

and

$$\varrho(\Sigma_N : l_1^N \to l_\infty^N|\mathcal{R}_n, \mathcal{I}_n) \le 2,$$

we have $\Sigma \in \mathfrak{RT}_2(l_1, l_\infty)$. The rest follows in the same way or by duality.

4.3.19 An operator T has **Rademacher subtype** if

$$\varrho(T|\mathcal{R}_n, \mathcal{I}_n) = o(\sqrt{n}).$$

These operators form the closed ideal

$$\mathfrak{RT} := \mathfrak{L}_0[n^{-1/2}\,\varrho(\mathcal{R}_n, \mathcal{I}_n)].$$

4.3.20 PROPOSITION. *The operator ideal $\mathfrak{RT}$ is injective, surjective, symmetric and l_2-stable.*

PROOF. Injectivity and surjectivity follow from 3.7.6, while symmetry is a consequence of Proposition 4.2.17, and 3.3.4 yields the l_2-stability.

4.4 B-convexity

4.4.1 To begin with, we establish the main result of this section.

THEOREM. $\bigcup_{1<p\le 2} \mathsf{RT}_p = \bigcup_{1<p<2} \mathsf{RT}_p^{weak} = \mathsf{RT}.$

PROOF. If $X \in \mathsf{RT}$, then $\varrho(X|\mathcal{R}_m, \mathcal{I}_m) < \sqrt{m}$ for some $m \ge 2$. Since we know from 3.3.6 and 4.2.6 that the sequence $(\varrho(\mathcal{R}_n, \mathcal{I}_n))$ is non-decreasing and submultiplicative, 1.3.5 provides us with an $\varepsilon > 0$ such that $\varrho(X|\mathcal{R}_n, \mathcal{I}_n) \prec n^{1/2-\varepsilon}$ for $n = 1, 2, \ldots$. Hence X has weak Rademacher type p whenever $1/p' \le \varepsilon$. This proves that

$$\mathsf{RT} \subseteq \bigcup_{1<p<2} \mathsf{RT}_p^{weak}.$$

The reverse inclusion is obvious, and the equality

$$\bigcup_{1<p\le 2} \mathsf{RT}_p = \bigcup_{1<p<2} \mathsf{RT}_p^{weak}$$

was shown in 4.3.10.

REMARK. In view of the above result, Banach spaces in RT are said to have **non-trivial Rademacher type**.

4.4.2 Let $\alpha \ge 0$, and denote by B_α the diagonal operator on $[l_2, l_\infty^{2^k}]$ defined by

$$B_\alpha : (x_k) \longrightarrow ((1 + \log k)^{-\alpha} x_k), \quad \text{where} \quad x_k \in l_\infty^{2^k}.$$

The next example shows that the phenomenon described in the previous paragraph only occurs for spaces, not for operators. That is,

$$\bigcup_{1<p\le 2} \mathfrak{RT}_p = \bigcup_{1<p<2} \mathfrak{RT}_p^{weak} \ne \mathfrak{RT}.$$

EXAMPLE. $\varrho(B_\alpha|\mathcal{R}_n, \mathcal{I}_n) \asymp n^{1/2}(1 + \log n)^{-\alpha}.$

PROOF. Recall that

$$\varrho(l_\infty^N|\mathcal{R}_n, \mathcal{I}_n) \asymp \min\{\sqrt{n}, \sqrt{1+\log N}\} \qquad \text{(by 4.2.11).}$$

The desired relation is now a consequence of $(\triangle)$ in 1.2.12.

REMARK. We stress that $B_\alpha \in \mathfrak{RT}$ does not admit a factorization through a space in RT.

4.4.3 A Banach space X **contains** l_1^n with bound $c \ge 1$ if there exists an operator $A_n \in \mathfrak{L}(l_1^n, X)$ such that

$$\|(\xi_k)|l_1^n\| \le \|A_n(\xi_k)\| \le c\,\|(\xi_k)|l_1^n\|$$

for $\xi_1, \ldots, \xi_n \in \mathbb{K}$. Writing A_n in the form

$$A_n(\xi_k) = \sum_{k=1}^{n} \xi_k x_k$$

yields

$$\|(\xi_k)|l_1^n\| \le \Big\| \sum_{k=1}^{n} \xi_k x_k \Big\| \le c\,\|(\xi_k)|l_1^n\|.$$

In other terms, X has an n-dimensional subspace $E_n := \operatorname{span}\{x_1, \ldots, x_n\}$ with $d(E_n, l_1^n) \le c$.

We say that X **contains the spaces** l_1^n **uniformly** if the above condition is fulfilled for some constant $c \ge 1$ and all $n = 1, 2, \ldots$, simultaneously.

4.4.4 Fix any $n \ge 2$. A Banach space X is said to be $\mathbf{B}_n$**-convex** if there exists $0 < \varepsilon < 1$ such that, whenever $\|x_1\| = \ldots = \|x_n\| = 1$, at least one of the absolutely convex combinations

$$\frac{1}{n} \sum_{k=1}^{n} \zeta_k x_k \quad \text{with} \quad |\zeta_1| = \ldots = |\zeta_n| = 1$$

has norm less than $1 - \varepsilon$.

The class of all B_n-convex Banach spaces is denoted by B_n, and Banach spaces belonging to $\mathsf{B} := \bigcup_{n=2}^{\infty} \mathsf{B}_n$ are called **B-convex**.

REMARKS. R. C. James [jam 1] refers to $X \in \mathsf{B}_n$ as a non-l_1^n space. His terminology is justified by the next criterion.

In the case of complex Banach spaces, we must carefully distinguish between real and complex B_n-convexity. This will be done by using the symbols $\mathsf{B}_n^{\mathbb{R}}$ and $\mathsf{B}_n^{\mathbb{C}}$. However, the concepts of real and complex B-convexity coincide, $\mathsf{B}^{\mathbb{R}} = \mathsf{B}^{\mathbb{C}}$.

4.4.5 PROPOSITION. *For every Banach space X and* **fixed** $n \ge 2$, *the following are equivalent:*

(1) $\varrho(X|\mathcal{P}_n, \mathcal{I}_n) = \sqrt{n}$.

(2) *X contains l_1^n for all bounds $c > 1$.*

(3) *X fails to be B_n-convex.*

PROOF. **(1)⇒(2)**: By 3.7.12, we can find elements $\boldsymbol{x}_1, \ldots, \boldsymbol{x}_n$ in some ultrapower $X^{\mathcal{U}}$ such that $\|\boldsymbol{x}_1\| = \ldots = \|\boldsymbol{x}_n\| = 1$ and

$$\Big\| \sum_{k=1}^{n} \zeta_k \boldsymbol{x}_k \Big\| = n \quad \text{whenever} \quad |\zeta_1| = \ldots = |\zeta_n| = 1.$$

Let $\xi_1, \ldots, \xi_n \in \mathbb{K}$ and $\sum\limits_{k=1}^{n} |\xi_k| = 1$. Then

$$\Big\| \sum_{k=1}^{n} \xi_k \boldsymbol{x}_k \Big\| \leq 1.$$

Choosing $\zeta_1, \ldots, \zeta_n$ such that $\xi_k = \zeta_k |\xi_k|$ and $|\zeta_k| = 1$, we obtain

$$n = \Big\| \sum_{k=1}^{n} \zeta_k \boldsymbol{x}_k \Big\| = \Big\| \sum_{k=1}^{n} \zeta_k (1 - |\xi_k|) \boldsymbol{x}_k + \sum_{k=1}^{n} \xi_k \boldsymbol{x}_k \Big\|$$
$$\leq \sum_{k=1}^{n} (1 - |\xi_k|) \|\boldsymbol{x}_k\| + \Big\| \sum_{k=1}^{n} \xi_k \boldsymbol{x}_k \Big\| = n - 1 + \Big\| \sum_{k=1}^{n} \xi_k \boldsymbol{x}_k \Big\|.$$

Hence

$$1 \leq \Big\| \sum_{k=1}^{n} \xi_k \boldsymbol{x}_k \Big\|.$$

This proves that the linear span of $\boldsymbol{x}_1, \ldots, \boldsymbol{x}_n$ is isometric to l_1^n.

Finally, we use the fact that $X^{\mathcal{U}}$ is finitely representable in X. Thus for every $c > 1$ there exists an n-dimensional subspace E_n of X with $d(E_n, l_1^n) \leq c$.

(2)⇒(3): Taking the unit vectors, we see that l_1^n is not B_n-convex.

(3)⇒(1): Given $0 < \varepsilon < 1$, there exist elements $x_1, \ldots, x_n \in X$ such that $\|x_1\| = \ldots = \|x_n\| = 1$ and

$$\Big\| \frac{1}{n} \sum_{k=1}^{n} \zeta_k x_k \Big\| \geq 1 - \varepsilon \quad \text{whenever} \quad |\zeta_1| = \ldots = |\zeta_n| = 1.$$

Substituting these elements in the defining inequality of $\boldsymbol{\varrho}(\mathcal{P}_n, \mathcal{I}_n)$ yields the lower estimate $(1 - \varepsilon)\sqrt{n} \leq \boldsymbol{\varrho}(X | \mathcal{P}_n, \mathcal{I}_n)$.

REMARK. It follows from $\boldsymbol{\varrho}(l_1^n | \mathcal{P}_n, \mathcal{I}_n) = \boldsymbol{\varrho}(l_1^n | \mathcal{P}_{n+1}, \mathcal{I}_{n+1}) = \sqrt{n}$ that $l_1^n \in \mathsf{B}_{n+1} \setminus \mathsf{B}_n$. Hence being B_n-convex is an isometric property. This fact should be compared with 4.4.8.

4.4.6 In the preceding proposition n was fixed. Now we deal with statements which hold for all n, simultaneously.

PROPOSITION. *For every Banach space X, the following are equivalent:*

(1) $\boldsymbol{\varrho}(X | \mathcal{P}_n, \mathcal{I}_n) = \sqrt{n}$ *for* $n = 1, 2, \ldots$.

(2) *X contains the spaces l_1^n uniformly for* **all** *bounds $c > 1$.*

(3) *X contains the spaces l_1^n uniformly for* **some** *bound $c \geq 1$.*

PROOF. **(1)⇒(2)** follows from 4.4.5, and **(2)⇒(3)** is trivial.

(3)⇒(1): Since $\varrho(l_1^n|\mathcal{P}_n, \mathcal{I}_n) = \sqrt{n}$, we infer from condition **(3)** that $\sqrt{n} \le c\,\varrho(X|\mathcal{P}_n, \mathcal{I}_n)$ for $n = 1, 2, \ldots$. This completes the proof, by 4.4.1.

4.4.7 Combining the above observations, we get a classical result which is due to G. Pisier; see [pis 2] and [pis 3].

THEOREM. *For every Banach space X, the following are equivalent:*

(1) *X has Rademacher subtype.*

(2) *X has non-trivial Rademacher type.*

(3) *X does not contain the spaces l_1^n uniformly.*

(4) *X is* B-*convex.*

4.4.8 PROPOSITION. B-*convexity is an isomorphic property.*

PROOF. Note that conditions **(1)**, **(2)** and **(3)** in the previous theorem are invariant under isomorphisms.

4.4.9 PROPOSITION. *The class of* B-*convex spaces is stable when passing to subspaces, quotients and duals.*

PROOF. Since B = RT, the assertion follows from Proposition 4.3.20. For a direct proof, we refer to [gie].

4.4.10 We mention that both the non-reflexive Banach space l_1 and the reflexive Banach space $[l_2, l_1^n]$ fail to be B-convex.

4.4.11 For $T \in \mathfrak{L}(X,Y)$ and $n = 1, 2, \ldots$, the defining inequality of $\varrho(T|\mathcal{R}_n, \mathcal{I}_n)$ may be considered under the additional condition that the norms of all elements $x_1, \ldots, x_n$ equal 1. This yields

$$\left(\int_0^1 \Big\| \sum_{k=1}^n T x_k r_k(t) \Big\|^2 dt \right)^{1/2} \le c\sqrt{n},$$

and the best possible constant will be denoted by $\varrho^{eq}(T|\mathcal{R}_n, \mathcal{I}_n)$. We refer to

$$\varrho^{eq}(\mathcal{R}_n, \mathcal{I}_n) : T \longrightarrow \varrho^{eq}(T|\mathcal{R}_n, \mathcal{I}_n)$$

as an **equal-norm Rademacher type ideal norm.**

4.4.12 First of all, we state an immediate consequence of the contraction principle 3.5.4.

PROPOSITION.

$$\varrho^{eq}(T|\mathcal{R}_n, \mathcal{I}_n) = \frac{1}{\sqrt{n}} \sup \left\{ \left(\int_0^1 \Big\| \sum_{k=1}^n T x_k r_k(t) \Big\|^2 dt \right)^{1/2} : x_1, \ldots, x_n \in U_X \right\}.$$

4.4.13 The following example shows that the sequence $(\varrho^{eq}(\mathcal{R}_n, \mathcal{I}_n))$ fails to be non-decreasing. We consider the real case.

EXAMPLE.

$$\varrho^{eq}(l_1^n|\mathcal{R}_{n+1}, \mathcal{I}_{n+1}) = \sqrt{\tfrac{n^2+1}{n+1}} < \sqrt{n} = \varrho^{eq}(l_1^n|\mathcal{R}_n, \mathcal{I}_n) \quad \textit{for } n \geq 2.$$

PROOF. Choose extreme points $x_1, \ldots, x_{n+1} \in U_{l_1^n}$ such that

$$\varrho^{eq}(l_1^n|\mathcal{R}_{n+1}, \mathcal{I}_{n+1}) = \frac{1}{\sqrt{n+1}} \left(\int_0^1 \left\| \sum_{k=1}^{n+1} x_k r_k(t) \right\|^2 dt \right)^{1/2}.$$

In the case $\dim[x_1, \ldots, x_{n+1}] = n$, by changes of signs and a permutation, it may be arranged that $x_k = u_k^{(n)}$ for $k = 1, \ldots, n$ and $x_{n+1} = u_n^{(n)}$. Then

$$\left(\int_0^1 \left\| \sum_{k=1}^{n+1} x_k r_k(t) \right\|^2 dt \right)^{1/2} = \left(\int_0^1 \Big(n-1+|r_n(t)+r_{n+1}(t)|\Big)^2 dt \right)^{1/2} = \sqrt{n^2+1}.$$

The case $\dim[x_1, \ldots, x_{n+1}] < n$ cannot occur, since under this assumption

$$\left(\int_0^1 \left\| \sum_{k=1}^{n+1} x_k r_k(t) \right\|^2 dt \right)^{1/2} \leq \sqrt{n-1}\sqrt{n+1} < \sqrt{n^2+1}.$$

4.4.14 The next observation is elementary as well.

PROPOSITION. $\varrho^{eq}(\mathcal{R}_{2n}, \mathcal{I}_{2n}) \leq \sqrt{2}\varrho^{eq}(\mathcal{R}_n, \mathcal{I}_n)$.

4.4.15 We now provide an auxiliary result that shows how Rademacher averages behave under spreading. To this purpose, let

$$\mathbb{L} := \bigcup_{k=1}^{n} \mathbb{L}_k,$$

where $\mathbb{L}_1, \ldots, \mathbb{L}_n$ are pairwise disjoint, finite and non-empty subsets of $\mathbb{N}$. We associate with every n-tuple (x_k) in X the $\mathbb{L}$-tuple that has constant coordinates $|\mathbb{L}_k|^{-1/2} x_k$ on $\mathbb{L}_k$ for $k = 1, \ldots, n$.

LEMMA.

$$\int_0^1 \left\| \sum_{k=1}^{n} x_k r_k(t) \right\|^2 dt \leq 2 \int_0^1 \left\| \sum_{k=1}^{n} \sum_{l \in \mathbb{L}_k} |\mathbb{L}_k|^{-1/2} x_k r_l(u) \right\|^2 du.$$

PROOF. The unconditionality of Rademacher systems implies that

$$\int_0^1\!\!\int_0^1 \Big\| \sum_{k=1}^{n} \Big| \sum_{l\in\mathbb{L}_k} |\mathbb{L}_k|^{-1/2} r_l(u) \Big| x_k r_k(t) \Big\|^2 du\,dt =$$

$$= \int_0^1\!\!\int_0^1 \Big\| \sum_{k=1}^{n} \sum_{l\in\mathbb{L}_k} |\mathbb{L}_k|^{-1/2} x_k r_k(t) r_l(u) \Big\|^2 du\,dt$$

$$= \int_0^1 \Big\| \sum_{k=1}^{n} \sum_{l\in\mathbb{L}_k} |\mathbb{L}_k|^{-1/2} x_k r_l(u) \Big\|^2 du. \tag{1}$$

For $0 \le t < 1$, the triangle inequality for integrals and Hölder's inequality give

$$\Big\| \int_0^1 \sum_{k=1}^{n} \Big| \sum_{l\in\mathbb{L}_k} |\mathbb{L}_k|^{-1/2} r_l(u) \Big| x_k r_k(t)\, du \Big\| \le$$

$$\le \int_0^1 \Big\| \sum_{k=1}^{n} \Big| \sum_{l\in\mathbb{L}_k} |\mathbb{L}_k|^{-1/2} r_l(u) \Big| x_k r_k(t) \Big\| du$$

$$\le \bigg(\int_0^1 \Big\| \sum_{k=1}^{n} \Big| \sum_{l\in\mathbb{L}_k} |\mathbb{L}_k|^{-1/2} r_l(u) \Big| x_k r_k(t) \Big\|^2 du \bigg)^{1/2}. \tag{2}$$

Next, we integrate (2) (squared) over $t \in [0,1)$. The resulting formula combined with (1) yields

$$\int_0^1 \Big\| \sum_{k=1}^{n} \int_0^1 \Big| \sum_{l\in\mathbb{L}_k} |\mathbb{L}_k|^{-1/2} r_l(u) \Big| du\, x_k r_k(t) \Big\|^2 dt \le$$

$$\le \int_0^1\!\!\int_0^1 \Big\| \sum_{k=1}^{n} \Big| \sum_{l\in\mathbb{L}_k} |\mathbb{L}_k|^{-1/2} r_l(u) \Big| x_k r_k(t) \Big\|^2 du\,dt$$

$$= \int_0^1 \Big\| \sum_{k=1}^{n} \sum_{l\in\mathbb{L}_k} |\mathbb{L}_k|^{-1/2} x_k r_l(u) \Big\|^2 du.$$

By Khintchine's inequality,

$$1 \le \sqrt{2} \int_0^1 \Big| \sum_{l\in\mathbb{L}_k} |\mathbb{L}_k|^{-1/2} r_l(u) \Big| du.$$

The conclusion now follows from the principle of contraction 3.5.4.

4.4.16 The following result is taken from [bou*, pp. 160–162].

THEOREM. *The sequences* $(\boldsymbol{\varrho}(\mathcal{R}_n, \mathcal{I}_n))$ *and* $(\boldsymbol{\varrho}^{eq}(\mathcal{R}_n, \mathcal{I}_n))$ *are uniformly equivalent.*

PROOF. Fix n, and let $x_1, \ldots, x_n \in X$. Without loss of generality, we may assume that

$$\sum_{k=1}^{n} \|x_k\|^2 = 1.$$

Define

$$m := \max\{\, h : 3 \cdot 4^h \le 64n \,\}, \quad \mathbb{F}_h := \{\, k : 2^{-h} < \|x_k\| \le 2^{1-h} \,\}$$

and

$$\mathbb{F} := \bigcup_{h=1}^{m} \mathbb{F}_h.$$

Given $k \in \mathbb{F}$, there exists a unique number $h \le m$ such that $k \in \mathbb{F}_h$. Let $\mathbb{L}_1, \ldots, \mathbb{L}_m$ be pairwise disjoint subsets of $\mathbb{N}$ such that $|\mathbb{L}_k| = 4^{m-h}$. Then

$$\sum_{k\in\mathbb{F}} |\mathbb{L}_k| = \sum_{k\in\mathbb{F}} 4^{m-h} < 4^m \sum_{k\in\mathbb{F}} \|x_k\|^2 \le 4^m \le \tfrac{64}{3} n < 32n.$$

Since the norms of all elements $2^{h-1}x_k$ are less than or equal to 1, we obtain

$$\left(\int_0^1 \Big\| \sum_{k\in\mathbb{F}} \sum_{l\in\mathbb{L}_k} 2^{h-1}\, T x_k r_l(u) \Big\|^2 du \right)^{1/2} \le \sqrt{32n}\, \boldsymbol{\varrho}^{eq}(T|\mathcal{R}_{32n}, \mathcal{I}_{32n}). \quad (1)$$

Moreover, by 4.4.15,

$$\int_0^1 \Big\| \sum_{k\in\mathbb{F}} T x_k r_k(t) \Big\|^2 dt \le 2 \int_0^1 \Big\| \sum_{k\in\mathbb{F}} \sum_{l\in\mathbb{L}_k} |\mathbb{L}_k|^{-1/2} T x_k r_l(u) \Big\|^2 du. \quad (2)$$

Because of $2^{h-1} = 2^{m-1}|\mathbb{L}_k|^{-1/2}$, we infer from (1) and (2) that

$$2^{m-1} \left(\int_0^1 \Big\| \sum_{k\in\mathbb{F}} T x_k r_k(t) \Big\|^2 dt \right)^{1/2} \le \sqrt{64n}\, \boldsymbol{\varrho}^{eq}(T|\mathcal{R}_{32n}, \mathcal{I}_{32n}).$$

Since $\boldsymbol{\varrho}^{eq}(T|\mathcal{R}_{32n}, \mathcal{I}_{32n}) \le 4\sqrt{2}\boldsymbol{\varrho}^{eq}(T|\mathcal{R}_n, \mathcal{I}_n)$ and $64n < 3 \cdot 4^{m+1}$, this inequality turns into

$$\begin{aligned} \left(\int_0^1 \Big\| \sum_{k\in\mathbb{F}} T x_k r_k(t) \Big\|^2 dt \right)^{1/2} &\le 8\sqrt{2}\sqrt{\tfrac{64n}{4^m}}\, \boldsymbol{\varrho}^{eq}(T|\mathcal{R}_n, \mathcal{I}_n) \\ &\le 40 \boldsymbol{\varrho}^{eq}(T|\mathcal{R}_n, \mathcal{I}_n). \end{aligned} \quad (3)$$

On the other hand, it follows from

$$\left(\int_0^1 \Big\| \sum_{k\notin\mathbb{F}} Tx_k r_k(t) \Big\|^2 dt\right)^{1/2} \le \boldsymbol{\varrho}(T|\mathcal{R}_n,\mathcal{I}_n)\left(\sum_{k\notin\mathbb{F}} \|x_k\|^2\right)^{1/2}$$

and

$$\sum_{k\notin\mathbb{F}} \|x_k\|^2 = \sum_{h=m+1}^{\infty} \sum_{k\in\mathbb{F}_h} \|x_k\|^2 \le \sum_{h=m+1}^{\infty} \frac{|\mathbb{F}_h|}{4^{h-1}} \le \frac{n}{4^m}\sum_{h=0}^{\infty}\frac{1}{4^h} = \frac{n}{4^m}\frac{4}{3} < \frac{1}{4}$$

that

$$\left(\int_0^1 \Big\| \sum_{k\notin\mathbb{F}} Tx_k r_k(t) \Big\|^2 dt\right)^{1/2} \le \frac{1}{2}\boldsymbol{\varrho}(T|\mathcal{R}_n,\mathcal{I}_n). \tag{4}$$

Next, (3) and (4) yield

$$\left(\int_0^1 \Big\| \sum_{k=1}^{n} Tx_k r_k(t) \Big\|^2 dt\right)^{1/2} \le 40\boldsymbol{\varrho}^{eq}(T|\mathcal{R}_n,\mathcal{I}_n) + \frac{1}{2}\boldsymbol{\varrho}(T|\mathcal{R}_n,\mathcal{I}_n).$$

Hence

$$\boldsymbol{\varrho}(T|\mathcal{R}_n,\mathcal{I}_n) \le 40\boldsymbol{\varrho}^{eq}(T|\mathcal{R}_n,\mathcal{I}_n) + \frac{1}{2}\boldsymbol{\varrho}(T|\mathcal{R}_n,\mathcal{I}_n)$$

which in turn implies that

$$\boldsymbol{\varrho}(T|\mathcal{R}_n,\mathcal{I}_n) \le 80\,\boldsymbol{\varrho}^{eq}(T|\mathcal{R}_n,\mathcal{I}_n).$$

The inequality $\boldsymbol{\varrho}^{eq}(T|\mathcal{R}_n,\mathcal{I}_n) \le \boldsymbol{\varrho}(T|\mathcal{R}_n,\mathcal{I}_n)$ is obvious.

4.4.17 An operator $T \in \mathfrak{L}(X,Y)$ is said to be **uniformly l_1^n-injective** if there exist constants $c > 0$, $a \ge 1$, and a sequence of subspaces E_n such that

$$c\,\|Tx\| \ge \|x\| \quad \text{for all } x \in E_n, \quad \dim(E_n){=}n \quad \text{and} \quad d(E_n, l_1^n) \le a.$$

This is equivalent to the following property:
We can find a sequence of commutative diagrams

$$\begin{array}{ccc} X & \xrightarrow{\quad T \quad} & Y \\ {\scriptstyle A_n}\uparrow & & \uparrow{\scriptstyle B_n} \\ l_1^n & \xrightarrow[\quad I_n \quad]{} & l_1^n \end{array}$$

such that $A_n \in \mathfrak{L}(l_1^n, X)$ and $B_n \in \mathfrak{L}(l_1^n, Y)$ satisfy the conditions

$$\begin{array}{c} \|(\xi_k)|l_1^n\| \le a_0\,\|A_n(\xi_k)\| \le a\,\|(\xi_k)|l_1^n\| \\ \text{and} \qquad\qquad\qquad\qquad\qquad\qquad\qquad\qquad \text{for } \xi_1,\ldots,\xi_n \in \mathbb{K} \\ \|(\xi_k)|l_1^n\| \le b_0\,\|B_n(\xi_k)\| \le b\,\|(\xi_k)|l_1^n\| \end{array}$$

with constants $a, b \ge 1$ and $a_0, b_0 > 0$.

4.4.18 In order to prove the next theorem, we need a deep result due to J. Elton (real case) and A. Pajor (complex case); see [elt] and [PAJ, pp. 60–61 and 77].

THEOREM. *For $0 < \delta \leq 1$, there exist constants $c \geq 1$ and $0 < \theta \leq 1$ such that the following holds:*
If

$$\|(x_k)|\mathcal{R}_N\| \geq \delta N, \quad x_1, \ldots, x_N \in U_X \quad \textit{and} \quad N = 1, 2, \ldots,$$

then we can find a subset $\mathbb{F}$ of $\{1, \ldots, N\}$ such that

$$|\mathbb{F}| \geq \theta N \quad \textit{and} \quad c\Big\|\sum_{k\in\mathbb{F}} \xi_k x_k\Big\| \geq \sum_{k\in\mathbb{F}} |\xi_k| \quad \textit{for } \xi_1, \ldots, \xi_N \in \mathbb{K}.$$

4.4.19 We now extend 4.4.7 to the setting of operators.

THEOREM. *For every operator $T \in \mathfrak{L}(X, Y)$, the following are equivalent:*

(1) *T has Rademacher subtype.*
(2) *T fails to be uniformly l_1^n-injective.*

PROOF. **(1)⇒(2)**: Suppose that $T \in \mathfrak{L}(X, Y)$ is uniformly l_1^n-injective and choose operators $A_n \in \mathfrak{L}(l_1^n, X)$ and $B_n \in \mathfrak{L}(l_1^n, Y)$ as described in 4.4.17. Then the injectivity of $\varrho(\mathcal{R}_n, \mathcal{I}_n)$ and $B_n I_n = T A_n$ implies that

$$\sqrt{n} = \varrho(l_1^n|\mathcal{R}_n, \mathcal{I}_n) \leq b_0 \varrho(TA_n|\mathcal{R}_n, \mathcal{I}_n) \leq a a_0^{-1} b_0 \varrho(T|\mathcal{R}_n, \mathcal{I}_n).$$

(2)⇒(1): Let $\|T\| = 1$ and $\varrho(T|\mathcal{R}_n, \mathcal{I}_n) \neq o(\sqrt{n})$. Then, by 4.4.16, we also have $\varrho^{eq}(T|\mathcal{R}_n, \mathcal{I}_n) \neq o(\sqrt{n})$. Hence there exists δ such that the condition

$$\varrho^{eq}(T|\mathcal{R}_N, \mathcal{I}_N) > \delta\sqrt{N}$$

is satisfied for infinitely many natural numbers N. For every such N, we can find $x_1, \ldots, x_N \in U_X$ with

$$\left(\int_0^1 \Big\|\sum_{k=1}^N T x_k r_k(t)\Big\|^2 dt\right)^{1/2} \geq \delta N.$$

Note that $Tx_1, \ldots, Tx_N \in U_Y$. Consequently, if $c \geq 1$ and $0 < \theta \leq 1$ are chosen according to 4.4.18, then there exists a subset $\mathbb{F}$ of $\{1, \ldots, N\}$ such that

$$|\mathbb{F}| \geq \theta N \quad \text{and} \quad c\Big\|\sum_{k\in\mathbb{F}} \xi_k T x_k\Big\| \geq \sum_{k\in\mathbb{F}} |\xi_k|.$$

Taking N sufficiently large and passing to a subset (if necessary), we may arrange that $|\mathbb{F}| = n$ for any given n. Then

$$A_n(\xi_k) := c \sum_{k \in \mathbb{F}} \xi_k x_k$$

defines an operator $A_n \in \mathfrak{L}(l_1(\mathbb{F}), X)$ such that

$$\sum_{k \in \mathbb{F}} |\xi_k| \leq c \Big\| \sum_{k \in \mathbb{F}} \xi_k T x_k \Big\| = \|TA_n(\xi_k)\| \leq \|A_n(\xi_k)\| \leq c \sum_{k \in \mathbb{F}} |\xi_k|.$$

Thus T is uniformly l_1^n-injective.

4.5 Operators of Rademacher cotype

4.5.1 Let $2 \leq q < \infty$ and $1 \leq u < \infty$. For $T \in \mathfrak{L}(X, Y)$, we denote by $\boldsymbol{\varrho}_u^{(q)}(T|\mathcal{I}_n, \mathcal{R}_n)$ the least constant $c \geq 0$ such that

$$\Big(\sum_{k=1}^{n} \|Tx_k\|^q \Big)^{1/q} \leq c \Big(\int_0^1 \Big\| \sum_{k=1}^{n} x_k r_k(t) \Big\|^u dt \Big)^{1/u}$$

whenever $x_1, \ldots, x_n \in X$. More concisely, the above inequality reads as follows:

$$\|(Tx_k)|l_q^n\| \leq c \, \|(x_k)|\mathcal{R}_n\|_u.$$

We refer to

$$\boldsymbol{\varrho}_u^{(q)}(\mathcal{I}_n, \mathcal{R}_n) : T \longrightarrow \boldsymbol{\varrho}_u^{(q)}(T|\mathcal{I}_n, \mathcal{R}_n)$$

as a **Rademacher cotype ideal norm**. The case $q = u = 2$ has already been treated in Section 4.2.

Replacing $\|\cdot|l_q^n\|$ by the Lorentz quasi-norm $\|\cdot|l_{q,\infty}^n\|$ yields the **weak Rademacher cotype ideal quasi-norm** $\boldsymbol{\varrho}_u^{(q,\infty)}(\mathcal{I}_n, \mathcal{R}_n)$.

REMARK. In order to get an equivalent ideal norm, we can use the Lorentz norms $\|\cdot|l_{q,\infty}^n\|_s$ defined in 0.5.9.

4.5.2 The duality established in 4.2.4 extends as follows.

PROPOSITION.

$$\boldsymbol{\varrho}_{v'}^{(p')}(\mathcal{I}_n, \mathcal{R}_n) \leq \boldsymbol{\varrho}_p^{(v)}{}'(\mathcal{R}_n, \mathcal{I}_n) \leq \boldsymbol{\varrho}_{v'}^{(p')}(\mathcal{I}_n, \mathcal{R}_n) \circ \boldsymbol{\delta}_{v'}^{(v')}(\mathcal{R}_n, \mathcal{R}_n).$$

REMARK. Though the theories of Rademacher type and cotype cannot be derived from each other, there is a strong parallelism which will be emphasized by the correspondence:

$$\textbf{4.3.x} \longleftrightarrow \textbf{4.5.x}.$$

4.5.3 It follows from the Khintchine–Kahane inequality 4.1.10 that the asymptotic behaviour of $\varrho_u^{(q)}(T|\mathcal{I}_n,\mathcal{R}_n)$ is determined by the exponent q only.

PROPOSITION. *Fix $2 \le q < \infty$. Then the sequences of the ideal norms $\varrho_u^{(q)}(\mathcal{I}_n,\mathcal{R}_n)$ are uniformly equivalent whenever $1 \le u < \infty$.*

4.5.4 Let $2 \le q < \infty$, $1 \le u < \infty$, and fix any finite subset $\mathbb{F}$ of $\mathbb{N}$. Then, for $T \in \mathfrak{L}(X,Y)$, we denote by $\varrho_u^{(q)}(T|\mathcal{I}(\mathbb{F}),\mathcal{R}(\mathbb{F}))$ the least constant $c \ge 0$ such that

$$\Big(\sum_{k\in\mathbb{F}} \|Tx_k\|^q\Big)^{1/q} \le c \left(\int_0^1 \Big\|\sum_{k\in\mathbb{F}} x_k r_k(t)\Big\|^u dt\right)^{1/u}$$

whenever (x_k) is an $\mathbb{F}$-tuple in X. More concisely, the preceding inequality reads as follows:

$$\|(Tx_k)|l_q(\mathbb{F})\| \le c\,\|(x_k)|\mathcal{R}(\mathbb{F})\|_u.$$

Obviously,

$$\varrho_u^{(q)}(\mathcal{I}(\mathbb{F}),\mathcal{R}(\mathbb{F})) : T \longrightarrow \varrho_u^{(q)}(T|\mathcal{I}(\mathbb{F}),\mathcal{R}(\mathbb{F}))$$

is an ideal norm. If $|\mathbb{F}| = n$, then $\mathcal{R}(\mathbb{F})$ and $\mathcal{R}_n$ have identical distributions. Hence

$$\varrho_u^{(q)}(\mathcal{I}(\mathbb{F}),\mathcal{R}(\mathbb{F})) = \varrho_u^{(q)}(\mathcal{I}_n,\mathcal{R}_n).$$

4.5.5 We now prove a counterpart of 4.3.5; see also 5.6.16.

LEMMA. *Let $2 \le v < q < \infty$ and $1 \le u < \infty$. Then, for $T \in \mathfrak{L}(X,Y)$, the following properties are equivalent:*

(1) *There exists a constant $c_0 \ge 1$ such that*

$$n^{1/q-1/v}\,\varrho_u^{(v)}(T|\mathcal{I}_n,\mathcal{R}_n) \le c_0 \quad \text{for} \quad n = 1,2,\ldots .$$

(2) *There exists a constant $c \ge 1$ such that*

$$\varrho_u^{(q,\infty)}(T|\mathcal{I}_n,\mathcal{R}_n) \le c \quad \text{for} \quad n = 1,2,\ldots .$$

PROOF. The implication **(2)⇒(1)** follows from

$$\|(x_k)|l_v^n\| \le A\, n^{1/v-1/q}\, \|(x_k)|l_{q,\infty}^n\| \qquad \text{(by 0.5.6)},$$

where $A > 0$ is a constant.

Fix $x_1,\ldots,x_n \in X$. Thanks to 0.5.10, we can find a subset $\mathbb{F}$ of $\{1,\ldots,n\}$ such that

$$\|(Tx_k)|l_{q,\infty}^n\| \le |\mathbb{F}|^{1/q-1/v}\Big(\sum_{k\in\mathbb{F}} \|Tx_k\|^v\Big)^{1/v}.$$

Assume that **(1)** holds. It now follows from 4.5.4 that

$$\|(Tx_k)|l_{q,\infty}^n\| \le |\mathbb{F}|^{1/q-1/v}\Big(\sum_{k\in\mathbb{F}}\|Tx_k\|^v\Big)^{1/v} \le$$
$$\le |\mathbb{F}|^{1/q-1/v}\,\varrho_u^{(v)}(T|\mathfrak{I}(\mathbb{F}),\mathfrak{R}(\mathbb{F}))\bigg(\int_0^1\Big\|\sum_{k\in\mathbb{F}}x_k r_k(t)\Big\|^u dt\bigg)^{1/u}$$
$$\le c_0\bigg(\int_0^1\Big\|\sum_{k=1}^n x_k r_k(t)\Big\|^u dt\bigg)^{1/u}.$$

Hence $\varrho_u^{(q,\infty)}(T|\mathfrak{I}_n,\mathfrak{R}_n) \le c_0$.

4.5.6 Let $2\le q<\infty$. An operator $T\in\mathfrak{L}$ has **Rademacher cotype** q if

$$\|T|\mathfrak{RC}_q\| := \sup_n \varrho_2^{(q)}(T|\mathfrak{I}_n,\mathfrak{R}_n)$$

is finite. These operators form the Banach ideal

$$\mathfrak{RC}_q := \mathfrak{L}[\varrho_2^{(q)}(\mathfrak{I}_n,\mathfrak{R}_n)].$$

In view of the Khintchine–Kahane inequality 4.1.10, the expressions

$$\|T|\mathfrak{L}[\varrho_u^{(q)}(\mathfrak{I}_n,\mathfrak{R}_n)]\| := \sup_n \varrho_u^{(q)}(T|\mathfrak{I}_n,\mathfrak{R}_n) \quad \text{with } 1\le u<\infty$$

yield equivalent norms.

REMARK. Trivially, all operators have Rademacher cotype ∞. Therefore we put $\mathfrak{RC}_\infty := \mathfrak{L}$.

4.5.7 Let $2<q<\infty$. An operator T has **weak Rademacher cotype** q if

$$\|T|\mathfrak{RC}_q^{weak}\| := \sup_n n^{1/q-1/2}\,\varrho(T|\mathfrak{I}_n,\mathfrak{R}_n)$$

is finite. These operators form the Banach ideal

$$\mathfrak{RC}_q^{weak} := \mathfrak{L}[n^{1/q-1/2}\,\varrho(\mathfrak{I}_n,\mathfrak{R}_n)].$$

REMARK. In the case $q=2$, we would get $\mathfrak{RC}_2$.

4.5.8 PROPOSITION. *The operator ideals $\mathfrak{RC}_q$ and $\mathfrak{RC}_q^{weak}$ are injective and l_2-stable, but not surjective.*

PROOF. The non-surjectivity follows from 4.2.5.

REMARK. In the setting of spaces, the preceding result implies that the classes RC_q and RC_q^{weak} are stable when passing to subspaces. We also have stability under the formation of finite direct sums and l_2-multiples.

4.5.9 The following result explains why the attribute *weak* is used in Definition 4.5.7.

PROPOSITION. *If* $2 \le v < q < \infty$ *and* $1 \le u < \infty$, *then*

$$\mathfrak{RC}_q^{weak} = \mathfrak{L}\Big[\varrho_u^{(q,\infty)}(\mathcal{I}_n, \mathcal{R}_n)\Big] = \mathfrak{L}\Big[n^{1/q-1/v}\varrho_u^{(v)}(\mathcal{I}_n, \mathcal{R}_n)\Big].$$

PROOF. The assertion follows at once from 4.5.5 combined with the Khintchine–Kahane inequality 4.1.10.

4.5.10 We are now able to establish the main result of this section.

THEOREM. $\mathfrak{RC}_2 \subset \mathfrak{RC}_{q_0} \subset \mathfrak{RC}_{q_0}^{weak} \subset \mathfrak{RC}_q$ *if* $2 < q_0 < q < \infty$. *In particular,*

$$\bigcup_{2 \le q < \infty} \mathfrak{RC}_q = \bigcup_{2 < q < \infty} \mathfrak{RC}_q^{weak}.$$

PROOF. In view of 0.5.4, we have $\| \cdot | l_{q_0,\infty}^n \| \le \| \cdot | l_{q_0}^n \|$. Hence

$$\varrho_2^{(q_0,\infty)}(\mathcal{I}_n, \mathcal{R}_n) \le \varrho_2^{(q_0)}(\mathcal{I}_n, \mathcal{R}_n)$$

which implies the inclusion $\mathfrak{RC}_{q_0} \subseteq \mathfrak{RC}_{q_0}^{weak}$.

On the other hand, $\mathfrak{RC}_{q_0}^{weak} \subseteq \mathfrak{RC}_{q_0}$ follows from

$$\| \cdot | l_q^n \| \le A \, \| \cdot | l_{q_0,\infty}^n \| \qquad \text{(by 0.5.6)},$$

where $A \ge 1$ is a constant.

It will be shown in 4.5.11 that

$$L_q \in \mathsf{RC}_q \setminus \mathsf{RC}_{q_0}^{weak} \quad \text{if } 2 < q_0 < q < \infty.$$

Thus the inclusion $\mathfrak{RC}_{q_0}^{weak} \subseteq \mathfrak{RC}_q$ is strict. Finally, Example 4.5.13 tells us that the inclusion $\mathfrak{RC}_{q_0} \subseteq \mathfrak{RC}_{q_0}^{weak}$ is strict as well.

4.5.11 EXAMPLE.

$$L_q \in \mathsf{RC}_q \setminus \mathsf{RC}_{q_0}^{weak} \quad \textit{if } 2 < q_0 < q < \infty \quad \textit{and} \quad L_p \in \mathsf{RC}_2 \quad \textit{if } 1 \le p \le 2.$$

PROOF. In view of 3.7.16 and 4.5.2,

$$1 \le \varrho_2^{(q)}(L_q|\mathcal{I}_n, \mathcal{R}_n) \le \varrho_{q'}^{(2)}(L_{q'}|\mathcal{R}_n, \mathcal{I}_n) \le \varrho_{q'}^{(q)}(L_{q'}|\mathcal{R}_n, \mathcal{I}_n) \le 1.$$

Thus $L_q \in \mathsf{RC}_q$. Moreover, we know from 4.2.7 that

$$\varrho(L_q|\mathcal{I}_n, \mathcal{R}_n) = n^{1/2-1/q}.$$

Hence $L_q \notin \mathsf{RC}_{q_0}^{weak}$. Finally, $L_p \in \mathsf{RC}_2$ follows from 4.2.8.

4.5.12 We now evaluate the difference between $\mathfrak{RC}_q$ and $\mathfrak{RC}_q^{weak}$.

PROPOSITION. $\varrho_r^{(q)}(\mathcal{I}_n, \mathcal{R}_n) \le (1 + \log n)^{1/q} \varrho_r^{(q,\infty)}(\mathcal{I}_n, \mathcal{R}_n)$.

PROOF. By 0.5.5,

$$\| \cdot | l_q^n \| \le (1 + \log n)^{1/q} \| \cdot | l_{q,\infty}^n \|.$$

4.5.13 Recall that the diagonal operator D_λ is defined by

$$D_\lambda : (\xi_k) \longrightarrow (k^{-\lambda}\xi_k).$$

The following example shows that the preceding estimate is sharp.

EXAMPLE.

$$\boldsymbol{\varrho}_2^{(q)}(D_{1/q} : l_\infty \to l_\infty | \mathcal{I}_n, \mathcal{R}_n) \asymp (1+\log n)^{1/q}$$
$$\boldsymbol{\varrho}_2^{(q,\infty)}(D_{1/q} : l_\infty \to l_\infty | \mathcal{I}_n, \mathcal{R}_n) \asymp 1.$$

PROOF. We know from 3.7.15 that

$$\boldsymbol{\varrho}_2^{(q)}(D_{1/q} : l_\infty \to l_\infty | \mathcal{I}_n, \mathcal{R}_n) = \Big(\sum_{k=1}^{n} \frac{1}{k}\Big)^{1/q} \asymp (1+\log n)^{1/q}$$

and

$$\boldsymbol{\varrho}_2^{(2)}(D_{1/q} : l_\infty \to l_\infty | \mathcal{I}_n, \mathcal{R}_n) = \Big(\sum_{k=1}^{n} k^{-2/q}\Big)^{1/2} \asymp n^{1/2-1/q}.$$

By 4.5.5, the last formula is equivalent to

$$\boldsymbol{\varrho}_2^{(q,\infty)}(D_{1/q} : l_\infty \to l_\infty | \mathcal{I}_n, \mathcal{R}_n) \asymp 1.$$

4.5.14 In order to prove 4.5.15, an auxiliary result is required.

LEMMA. *Let $\{\mathbb{F}_1, \dots, \mathbb{F}_n\}$ be any partition of $\{1, \dots, N\}$ into pairwise disjoint subsets. Then*

$$\Big(\sum_{h=1}^{n} \|(Tx_k)|\mathcal{R}(\mathbb{F}_h)\|_q^q\Big)^{1/q} \le \boldsymbol{\varrho}_q^{(q)}(T|\mathcal{I}_n, \mathcal{R}_n)\, \|(x_k)|\mathcal{R}_N\|_q$$

whenever $x_1, \dots, x_N \in X$ and $T \in \mathfrak{L}(X,Y)$.

PROOF. Since $\mathcal{R}_n$ is $\mathbb{R}$-unconditional,

$$\int_0^1 \Big\| \sum_{h=1}^{n} \sum_{k\in\mathbb{F}_h} x_k r_k(t) r_h(s) \Big\|^q dt = \int_0^1 \Big\| \sum_{h=1}^{n} \sum_{k\in\mathbb{F}_h} x_k r_k(t) \Big\|^q dt. \qquad (*)$$

Substituting

$$\sum_{k\in\mathbb{F}_h} x_k r_k(t) \quad \text{with} \quad h = 1, \dots, n$$

in the defining inequality of $\boldsymbol{\varrho}_q^{(q)}(T|\mathcal{I}_n, \mathcal{R}_n)$ yields

$$\sum_{h=1}^{n} \Big\| \sum_{k\in\mathbb{F}_h} Tx_k r_k(t) \Big\|^q \le \boldsymbol{\varrho}_q^{(q)}(T|\mathcal{I}_n, \mathcal{R}_n)^q \int_0^1 \Big\| \sum_{h=1}^{n} \Big(\sum_{k\in\mathbb{F}_h} x_k r_k(t) \Big) r_h(s) \Big\|^q ds.$$

Integrating over $t \in [0,1)$ and using $(*)$, we obtain

$$\sum_{h=1}^{n} \|(Tx_k)|\mathcal{R}(\mathbb{F}_h)\|_q^q \le \varrho_q^{(q)}(T|\mathcal{I}_n,\mathcal{R}_n)^q \int_0^1 \Big\|\sum_{h=1}^{n}\sum_{k\in\mathbb{F}_h} x_k r_k(t)\Big\|^q dt$$
$$= \varrho_q^{(q)}(T|\mathcal{I}_n,\mathcal{R}_n)^q \|(x_k)|\mathcal{R}_N\|_q^q.$$

4.5.15 We are now in a position to show that, in the setting of spaces, the estimate established in 4.5.12 can be improved significantly.

PROPOSITION.

$$\varrho_q^{(q)}(\mathcal{I}_{2^n},\mathcal{R}_{2^n}) \le 2\,\varrho_q^{(q,\infty)}(\mathcal{I}_{2^n},\mathcal{R}_{2^n}) \circ \varrho_q^{(q)}(\mathcal{I}_n,\mathcal{R}_n).$$

PROOF. Without loss of generality, we may assume that

$$\|STx_1\| \ge \|STx_2\| \ge \ldots \ge \|STx_{2^n}\|.$$

Let $\mathbb{F}_0 := \{1\}$ and

$$\mathbb{F}_h := \{2^{h-1}+1,\ldots,2^h\} \quad \text{for } h = 1,\ldots,n.$$

Then $\|(STx_k)|l_{q,\infty}(\mathbb{F}_0)\| = \|STx_1\|$ and

$$\|(STx_k)|l_{q,\infty}(\mathbb{F}_h)\| \ge 2^{(h-1)/q}\|STx_{2^h}\| \quad \text{for } h = 1,\ldots,n.$$

In view of the previous lemma, we now obtain

$$\begin{aligned}
\sum_{k=1}^{2^n}\|STx_k\|^q &= \|STx_1\|^q + \sum_{h=1}^{n}\sum_{k\in\mathbb{F}_h}\|STx_k\|^q \\
&\le \|STx_1\|^q + \sum_{h=1}^{n} 2^{h-1}\|STx_{2^{h-1}}\|^q \\
&\le \|(STx_k)|l_{q,\infty}(\mathbb{F}_0)\|^q + 2\sum_{h=0}^{n-1}\|(STx_k)|l_{q,\infty}(\mathbb{F}_h)\|^q \\
&\le 3\sum_{h=0}^{n-1}\|(STx_k)|l_{q,\infty}(\mathbb{F}_h)\|^q \\
&\le 3\sum_{h=0}^{n-1}\varrho_q^{(q,\infty)}(S|\mathcal{I}(\mathbb{F}_h),\mathcal{R}(\mathbb{F}_h))^q\|(Tx_k)|\mathcal{R}(\mathbb{F}_h)\|_q^q \\
&\le 3\,\varrho_q^{(q,\infty)}(S|\mathcal{I}_{2^n},\mathcal{R}_{2^n})^q\sum_{h=0}^{n-1}\|(Tx_k)|\mathcal{R}(\mathbb{F}_h)\|_q^q \\
&\le 3\,\varrho_q^{(q,\infty)}(S|\mathcal{I}_{2^n},\mathcal{R}_{2^n})^q\,\varrho_q^{(q)}(T|\mathcal{I}_n,\mathcal{R}_n)_q^q\,\|(x_k)|\mathcal{R}_{2^{n-1}}\|_q^q.
\end{aligned}$$

This implies the desired inequality, since $\|(x_k)|\mathcal{R}_{2^{n-1}}\|_q \le \|(x_k)|\mathcal{R}_{2^n}\|_q$.

4.5.16 Using the terminology defined in 0.10.6, the previous proposition yields the following result.

PROPOSITION. *If a Banach space X has weak Rademacher cotype q, then the sequence $(\varrho_2^{(q)}(X|\mathfrak{I}_n, \mathfrak{R}_n))$ is very slowly growing.*

REMARK. In view of 4.5.13, the gap between $\mathfrak{RC}_q$ and $\mathfrak{RC}_q^{weak}$ is much larger than the gap between RC_q and RC_q^{weak}. In other words, it is hard to find a Banach space $X \in \mathsf{RC}_q^{weak} \setminus \mathsf{RC}_q$. Examples were constructed by L. Tzafriri as a modification of Tsirelson's space; see [tza] and the section on Tirilman's space in [CAS*, pp. 107–116].

4.5.17 Next, we state a dual version of 4.3.17.

EXAMPLE. $\varrho(\Sigma_n : l_1^n \to l_\infty^n | \mathfrak{I}_n, \mathfrak{R}_n) \leq 2.$

PROOF. Use $\varrho(\mathfrak{I}_n, \mathfrak{R}_n) \leq \varrho'(\mathfrak{R}_n, \mathfrak{I}_n)$ and $\Sigma_n = R_n \Sigma_n' R_n$ from 0.7.4.

4.5.18 We now pass to the infinite summation operator Σ.

EXAMPLE. $\Sigma \in \mathfrak{RC}_2(l_1, l_\infty)$.

4.5.19 An operator T has **Rademacher subcotype** if

$$\varrho(T|\mathfrak{I}_n, \mathfrak{R}_n) = o(\sqrt{n}).$$

These operators form the closed ideal

$$\mathfrak{RC} := \mathfrak{L}_0[n^{-1/2}\, \varrho(\mathfrak{I}_n, \mathfrak{R}_n)].$$

4.5.20 PROPOSITION. *The operator ideal $\mathfrak{RC}$ is injective and l_2-stable, but neither surjective nor symmetric.*

PROOF. Injectivity and l_2-stability are obvious. On the other hand, we have $l_1 \in \mathsf{RC}$ and $c_0, l_\infty \notin \mathsf{RC}$.

4.5.21 We conclude this section with some duality relations which underline once more the parallelism between the theories of Rademacher type and cotype. See also 4.14.11.

THEOREM.

$$(\mathfrak{RT})' \subset \mathfrak{RC}, \quad (\mathfrak{RT}_p)' \subset \mathfrak{RC}_{p'} \quad \textit{and} \quad (\mathfrak{RT}_p^{weak})' \subset \mathfrak{RC}_{p'}^{weak}.$$

PROOF. Use 4.2.4 and 4.5.2.

4.6 MP-convexity

The analogy with Section 4.4 is emphasized by the correspondence

$$\mathbf{4.4.x} \longleftrightarrow \mathbf{4.6.x},$$

which extends that between 4.3 and 4.5.

4.6.1 To begin with, we establish the main result of this section.

THEOREM. $\displaystyle\bigcup_{2\le q<\infty} \mathsf{RC}_q = \bigcup_{2<q<\infty} \mathsf{RC}_q^{weak} = \mathsf{RC}.$

PROOF. If $X \in \mathsf{RC}$, then $\varrho(X|\mathfrak{I}_m, \mathfrak{R}_m) < \sqrt{m}$ for some $m \ge 2$. Since we know from 3.3.6 and 4.2.6 that the sequence $(\varrho(\mathfrak{I}_n, \mathfrak{R}_n))$ is non-decreasing and submultiplicative, 1.3.5 provides us with an $\varepsilon > 0$ such that $\varrho(X|\mathfrak{I}_n, \mathfrak{R}_n) \prec n^{1/2-\varepsilon}$ for $n = 1, 2, \ldots$. Hence X has weak Rademacher cotype q whenever $1/q \le \varepsilon$. This proves that

$$\mathsf{RC} \subseteq \bigcup_{2<q<\infty} \mathsf{RC}_q^{weak}.$$

The reverse inclusion is obvious, and the equality

$$\bigcup_{2\le q<\infty} \mathsf{RC}_q = \bigcup_{2<q<\infty} \mathsf{RC}_q^{weak}$$

was shown in 4.5.10.

REMARK. In view of the above result, Banach spaces in RC are said to have **non-trivial Rademacher cotype.**

4.6.2 Let $\alpha \ge 0$, and denote by C_α the diagonal operator on $[l_2, l_\infty^{2^k}]$ defined by

$$C_\alpha : (x_k) \longrightarrow (k^{-\alpha} x_k), \qquad \text{where } x_k \in l_\infty^{2^k}.$$

The next example shows that the phenomenon described in the previous paragraph only occurs for spaces, not for operators. That is,

$$\bigcup_{2\le q<\infty} \mathfrak{RC}_q = \bigcup_{2<q<\infty} \mathfrak{RC}_q^{weak} \ne \mathfrak{RC}.$$

EXAMPLE. $\varrho(C_\alpha|\mathfrak{I}_n, \mathfrak{R}_n) \asymp n^{1/2}(1 + \log n)^{-\alpha}.$

PROOF. Recall that

$$\varrho(l_\infty^N|\mathfrak{I}_n, \mathfrak{R}_n) \asymp \min\{\sqrt{n}, \sqrt{N}\} \qquad \text{(by 4.2.10)}.$$

The desired relation is now a consequence of ($\triangle$) in 1.2.12.

REMARK. We stress that $C_\alpha \in \mathfrak{RC}$ does not admit a factorization through a space in RC.

4.6.3 A Banach space X **contains** l_∞^n with bound $c \geq 1$ if there exists an operator $A_n \in \mathfrak{L}(l_\infty^n, X)$ such that

$$\|(\xi_k)|l_\infty^n\| \leq \|A_n(\xi_k)\| \leq c\,\|(\xi_k)|l_\infty^n\|$$

for $\xi_1, \ldots, \xi_n \in \mathbb{K}$. Writing A_n in the form

$$A_n(\xi_k) = \sum_{k=1}^{n} \xi_k x_k$$

yields

$$\|(\xi_k)|l_\infty^n\| \leq \Big\| \sum_{k=1}^{n} \xi_k x_k \Big\| \leq c\,\|(\xi_k)|l_\infty^n\|.$$

In other words, X has an n-dimensional subspace $E_n := \text{span}\{x_1, \ldots, x_n\}$ with $d(E_n, l_\infty^n) \leq c$.

We say that X **contains the spaces** l_∞^n **uniformly** if the above condition is fulfilled for some constant $c \geq 1$ and all $n = 1, 2, \ldots$, simultaneously.

4.6.4 Fix any $n \geq 2$. A Banach space X is said to be $\mathbf{MP}_n$**-convex** if there exists $\varepsilon > 0$ such that, whenever $\|x_1\| = \ldots = \|x_n\| = 1$, at least one of the linear combinations

$$\sum_{k=1}^{m} \zeta_k x_k \quad \text{with} \quad |\zeta_1| = \ldots = |\zeta_n| = 1$$

has norm greater than $1+\varepsilon$.

The class of all MP$_n$-convex Banach spaces is denoted by MP_n, and Banach spaces belonging to $\mathsf{MP} := \bigcup_{n=2}^{\infty} \mathsf{MP}_n$ are called **MP-convex.**

REMARKS. The next criterion implies that the closed unit ball of an MP$_n$-convex space does not have any n-dimensional section which is almost a parallelepiped. To some extent, this fact motivates the name *convexity.* The letters MP could be interpreted as a credit to the authors of [mau 1], [pis 1] and [mau*].

In the case of complex Banach spaces, we must carefully distinguish between real and complex MP$_n$-convexity. This will be done by using the symbols $\mathsf{MP}_n^{\mathbb{R}}$ and $\mathsf{MP}_n^{\mathbb{C}}$. However, the concepts of real and complex MP-convexity coincide, $\mathsf{MP}^{\mathbb{R}} = \mathsf{MP}^{\mathbb{C}}$.

4.6.5 PROPOSITION. *For every Banach space X and* **fixed** $n \geq 2$, *the following are equivalent:*

(1) $\varrho(X|\mathcal{I}_n, \mathcal{P}_n) = \sqrt{n}$.

(2) *X contains l_∞^n for all bounds $c > 1$.*

(3) *X fails to be* MP$_n$*-convex.*

PROOF. **(1)⇒(2)**: By 3.7.13, we can find elements $\boldsymbol{x}_1, \dots, \boldsymbol{x}_n$ in some ultrapower $X^{\mathcal{U}}$ such that $\|\boldsymbol{x}_1\| = \dots = \|\boldsymbol{x}_n\| = 1$ and

$$\Big\| \sum_{k=1}^{n} \zeta_k \boldsymbol{x}_k \Big\| = 1 \quad \text{whenever} \quad |\zeta_1| = \dots = |\zeta_n| = 1.$$

Let $\xi_1, \dots, \xi_n \in \mathbb{K}$ and $\max_{1 \le k \le n} |\xi_k| = 1$. An extreme point argument implies that

$$\Big\| \sum_{k=1}^{n} \xi_k \boldsymbol{x}_k \Big\| \le 1.$$

Choosing $h \in \{1, \dots, n\}$ such that $|\xi_h| = 1$, we have

$$2 = 2\|\xi_h \boldsymbol{x}_h\| \le \Big\| \xi_h \boldsymbol{x}_h + \sum_{k \ne h} \xi_k \boldsymbol{x}_k \Big\| + \Big\| \xi_h \boldsymbol{x}_h - \sum_{k \ne h} \xi_k \boldsymbol{x}_k \Big\| \le \Big\| \sum_{k=1}^{n} \xi_k \boldsymbol{x}_k \Big\| + 1.$$

Hence

$$1 \le \Big\| \sum_{k=1}^{n} \xi_k \boldsymbol{x}_k \Big\|.$$

This proves that the linear span of $\boldsymbol{x}_1, \dots, \boldsymbol{x}_n$ is isometric to l_∞^n.

Finally, we use the fact that $X^{\mathcal{U}}$ is finitely representable in X. Thus for every $c > 1$ there exists an n-dimensional subspace E_n of X with $d(E_n, l_\infty^n,) \le c$.

(2)⇒(3): Taking the unit vectors, we see that l_∞^n is not MP$_n$-convex.

(3)⇒(1): Given $\varepsilon > 0$, there exist elements $x_1, \dots, x_n \in X$ such that $\|x_1\| = \dots = \|x_n\| = 1$ and

$$\Big\| \sum_{k=1}^{n} \zeta_k x_k \Big\| \ge 1 + \varepsilon \quad \text{whenever} \quad |\zeta_1| = \dots = |\zeta_n| = 1.$$

Substituting these elements in the defining inequality of $\varrho(\mathfrak{I}_n, \mathfrak{P}_n)$ yields $\sqrt{n} \le (1+\varepsilon)\varrho(X|\mathfrak{I}_n, \mathfrak{P}_n)$.

REMARK. It follows from $\varrho(l_\infty^n|\mathfrak{I}_n, \mathfrak{P}_n) = \varrho(l_\infty^n|\mathfrak{I}_{n+1}, \mathfrak{P}_{n+1}) = \sqrt{n}$ that $l_\infty^n \in \mathsf{MP}_{n+1} \setminus \mathsf{MP}_n$. Hence being MP$_n$-convex is an isometric property.

4.6.6 In the preceding proposition n was fixed. Now we deal with statements which hold for all n, simultaneously.

PROPOSITION. *For every Banach space X, the following are equivalent:*

(1) $\varrho(X|\mathfrak{I}_n, \mathfrak{P}_n) = \sqrt{n}$ *for* $n = 1, 2, \dots$.

(2) *X contains the spaces l_∞^n uniformly for* **all** *bounds $c > 1$.*

(3) *X contains the spaces l_∞^n uniformly for* **some** *bound $c \ge 1$.*

PROOF. **(1)⇒(2)** follows from 4.6.5, and **(2)⇒(3)** is trivial.

(3)⇒(1): Since $\varrho(l_\infty^n|\mathcal{I}_n,\mathcal{P}_n)=\sqrt{n}$, we infer from condition **(3)** that $\sqrt{n}\le c\,\varrho(X|\mathcal{I}_n,\mathcal{P}_n)$ for $n=1,2,\ldots$. This completes the proof, by 4.6.1.

4.6.7 Combining the above observations, we get a counterpart of 4.4.7.

THEOREM. *For every Banach space X, the following are equivalent:*

(1) *X has Rademacher subcotype.*
(2) *X has non-trivial Rademacher cotype.*
(3) *X does not contain the spaces l_∞^n uniformly.*
(4) *X is* MP*-convex.*

4.6.8 PROPOSITION. MP*-convexity is an isomorphic property.*

PROOF. Note that conditions **(1)**, **(2)** and **(3)** in the previous theorem are invariant under isomorphisms.

4.6.9 PROPOSITION. *The class of* MP*-convex spaces is stable when passing to subspaces.*

4.6.10 We mention that both the non-reflexive Banach space c_0 and the reflexive Banach space $[l_2,l_\infty^n]$ fail to be MP-convex. However, $l_1(\mathbb{I})\in$ MP. This implies that MP fails to be stable under the formation of quotients.

4.6.11 At this point the parallelism between Sections 4.4 and 4.6 breaks down. Of course, for $T\in\mathfrak{L}(X,Y)$ and $n=1,2,\ldots$, we may consider the defining inequality of $\varrho(T|\mathcal{I}_n,\mathcal{R}_n)$ under the additional condition that the norms of all elements $Tx_1,\ldots,Tx_n$ equal 1. Unfortunately, it seems to be unknown whether the function $\varrho^{eq}(\mathcal{I}_n,\mathcal{R}_n)$ obtained in this way is an ideal norm. Moreover, A. Hinrichs [hin 6] showed that

$$\limsup_{n\to\infty}\varrho^{eq}(l_\infty^N|\mathcal{I}_n,\mathcal{R}_n)\asymp\sqrt{\frac{N}{1+\log N}}$$

whereas $\varrho(l_\infty^N|\mathcal{I}_n,\mathcal{R}_n)=\sqrt{N}$ for $n\ge N$. Thus $\varrho^{eq}(\mathcal{I}_n,\mathcal{R}_n)$ and $\varrho(\mathcal{I}_n,\mathcal{R}_n)$ cannot be uniformly equivalent.

4.6.12–4.6.16 Now we make a JUMP OF ENUMERATION in order to continue the correspondence with Section 4.4.

4.6.17 An operator $T\in\mathfrak{L}(X,Y)$ is said to be **uniformly l_∞^n-injective** if there exist constants $c>0$, $a\ge 1$, and a sequence of subspaces E_n such that

$$c\,\|Tx\|\ge\|x\|\quad\text{for all } x\in E_n,\quad \dim(E_n)=n\quad\text{and}\quad d(E_n,l_\infty^n)\le a.$$

This is equivalent to the following property:
We can find a sequence of commutative diagrams

$$\begin{array}{ccc} X & \xrightarrow{\quad T \quad} & Y \\ A_n \uparrow & & \uparrow B_n \\ l_\infty^n & \xrightarrow[\quad I_n \quad]{} & l_\infty^n \end{array}$$

such that $A_n \in \mathfrak{L}(l_\infty^n, X)$ and $B_n \in \mathfrak{L}(l_\infty^n, Y)$ satisfy the conditions

$$\|(\xi_k)|l_\infty^n\| \le a_0 \,\|A_n(\xi_k)\| \le a \,\|(\xi_k)|l_\infty^n\|$$

and $\qquad\qquad\qquad\qquad\qquad\qquad\qquad\qquad$ for $\xi_1, \dots, \xi_n \in \mathbb{K}$

$$\|(\xi_k)|l_\infty^n\| \le b_0 \,\|B_n(\xi_k)\| \le b \,\|(\xi_k)|l_\infty^n\|$$

with constants $a, b \ge 1$ and $a_0, b_0 > 0$.

Since l_∞^n has the metric extension property, we can find operators $V_n \in \mathfrak{L}(Y, l_\infty^n)$ such that $V_n B_n = I_n$ and $\|V_n\| \le b_0$. Therefore, an operator $T \in \mathfrak{L}(X, Y)$ is uniformly l_∞^n-injective if and only if it **factors** all spaces l_∞^n with a fixed bound $c > 0$. This means that there are operators $A_n \in \mathfrak{L}(l_\infty^n, X)$ and $V_n \in \mathfrak{L}(X, l_\infty^n)$ such that

$$I_n = V_n T A_n \quad \text{and} \quad \|V_n\| \, \|A_n\| \le c.$$

4.6.18 For a proof of the following theorem, the reader is referred to [bea 2] and a forthcoming paper of A. Hinrichs; see [hin 6].

THEOREM. *For every operator $T \in \mathfrak{L}(X, Y)$, the following are equivalent:*

(1) *T has Rademacher subcotype.*
(2) *T fails to be uniformly l_∞^n-injective.*
(3) *T does not factor the identity map of l_∞^n uniformly.*

REMARK. In the l_∞^n-setting, N. Alon/V. D. Milman proved a certain counterpart of the Elton–Pajor theorem 4.4.18; see [alo*]. Unfortunately, we were not able to find an approach based on this result.

4.6.19 Next, we state a corollary of 4.2.17.

THEOREM. $\mathfrak{RT} \subset \mathfrak{RC}$.

REMARK. Of course, the inclusions $\mathsf{RT} \subset \mathsf{RC}$ and $\mathsf{B} \subset \mathsf{MP}$ are equivalent. We even have $\mathsf{B}_n^{\mathbb{R}} \subset \mathsf{MP}_{2^n}^{\mathbb{R}}$. On the other hand, $l_\infty^N \in \mathsf{B}_2^{\mathbb{C}}$ shows that $\mathsf{B}_n^{\mathbb{C}} \not\subset \mathsf{MP}_N^{\mathbb{C}}$ for any choice of n and N.

4.6.20 THEOREM. *Every Banach space of non-trivial Rademacher type also has non-trivial Rademacher cotype.*

PROOF. The conclusion follows from

$$\bigcup_{1<p\le 2} \mathsf{RT}_p = \mathsf{RT} \quad \text{and} \quad \bigcup_{2\le q<\infty} \mathsf{RC}_q = \mathsf{RC}.$$

REMARK. Since $L_q \in \mathsf{RT}_2 \setminus \mathsf{RC}_{q_0}$ whenever $2 < q_0 < q < \infty$, there cannot exist any function $f : (1,2] \to [2,\infty)$ depending only on p such that $\mathsf{RT}_p \subseteq \mathsf{RC}_{f(p)}$. However, there is a function $\varphi : (1,2] \times [1,\infty) \to [2,\infty)$ such that

$$X \in \mathsf{RT}_p \quad \text{and} \quad q = \varphi(p, \|X|\mathfrak{RT}_p\|) \quad \text{imply} \quad X \in \mathsf{RC}_q.$$

More specifically, we may take $\varphi(p,t) = 2 + (2t)^{p'}$; see [koe*c, p. 92].

4.6.21 We know from 4.6.2 that

$$\varrho(C_\alpha|\mathcal{I}_n, \mathcal{R}_n) \asymp n^{1/2}(1+\log n)^{-\alpha}.$$

Combining this relation with the following example shows that the above theorem fails in the setting of operators. That is,

$$\bigcup_{1<p\le 2} \mathfrak{RT}_p \not\subseteq \bigcup_{2\le q<\infty} \mathfrak{RC}_q.$$

EXAMPLE. *If* $0 \le \alpha \le 1/2$, *then*

$$\varrho(C_\alpha|\mathcal{R}_n, \mathcal{I}_n) \asymp n^{1/2-\alpha}.$$

PROOF. Recall that

$$\varrho(l_\infty^N|\mathcal{R}_n, \mathcal{I}_n) \asymp \min\{\sqrt{n}, \sqrt{1+\log N}\} \qquad \text{(by 4.2.11)}.$$

The desired relation is now a consequence of ($\triangle$) in 1.2.12.

4.7 Gaussian random variables

CONVENTION. In the following, we must carefully distinguish between the real and the complex case. These will be indicated by adding $\mathbb{R}$ and $\mathbb{C}$, respectively, as subscripts or superscripts. But, whenever it is possible to treat both cases simultaneously, we omit this overcharged notation. For example, the real and the complex Gaussian measures are denoted by $\gamma_\mathbb{R}^n$ and $\gamma_\mathbb{C}^n$, respectively, while γ^n stands for both. Recall that $\mathbb{K}^n$ is the synonym of $\mathbb{R}^n$ and $\mathbb{C}^n$.

4.7.1 For every Borel set B of $\mathbb{R}^n$, the **real Gaussian measure** $\gamma_{\mathbb{R}}^n$ is defined by

$$\gamma_{\mathbb{R}}^n(B) := \frac{1}{(2\pi)^{n/2}} \int\limits_B \exp(-\tfrac{1}{2}\sum_{k=1}^n \sigma_k^2)\, d\sigma_1 \ldots d\sigma_n.$$

The **complex Gaussian measure** $\gamma_{\mathbb{C}}^n$ is obtained from $\gamma_{\mathbb{R}}^{2n}$ by the canonical mapping which assigns to every point $(a,b) \in \mathbb{R}^n \times \mathbb{R}^n$ the point $s = a + ib \in \mathbb{C}^n$.

Note that $\gamma_{\mathbb{R}}^n$ and $\gamma_{\mathbb{C}}^n$ are invariant under orthogonal and unitary transformations of $\mathbb{R}^n$ and $\mathbb{C}^n$, respectively.

4.7.2 Let $1 \le r < \infty$. The r-th **absolute Gaussian moments** are defined by

$$M_r^{\mathbb{R}} := \left(\frac{1}{\sqrt{2\pi}} \int\limits_{\mathbb{R}} |\sigma|^r \exp(-\tfrac{1}{2}\sigma^2)\, d\sigma \right)^{1/r}$$

and

$$M_r^{\mathbb{C}} := \left(\frac{1}{2\pi} \int\limits_{\mathbb{R}} \int\limits_{\mathbb{R}} \left|\tfrac{1}{2}(\alpha^2 + \beta^2)\right|^{r/2} \exp(-\tfrac{1}{2}(\alpha^2 + \beta^2))\, d\alpha\, d\beta \right)^{1/r}.$$

It turns out that

$$M_r^{\mathbb{R}} = \sqrt{2} \left[\frac{\Gamma(\frac{r+1}{2})}{\Gamma(\frac{1}{2})} \right]^{1/r} \quad \text{and} \quad M_r^{\mathbb{C}} = \Gamma(\tfrac{r+2}{2})^{1/r}.$$

In particular,

$$M_1^{\mathbb{R}} = \sqrt{\tfrac{2}{\pi}} = 0.7978\ldots \quad \text{and} \quad M_2^{\mathbb{R}} = 1,$$

$$M_1^{\mathbb{C}} = \sqrt{\tfrac{\pi}{4}} = 0.8862\ldots \quad \text{and} \quad M_2^{\mathbb{C}} = 1.$$

By Jessen's inequality 0.4.10, we have

$$M_p^{\mathbb{R}} \le M_p^{\mathbb{C}} \quad \text{if } 1 \le p \le 2 \quad \text{and} \quad M_q^{\mathbb{C}} \le M_q^{\mathbb{R}} \quad \text{if } 2 \le q < \infty.$$

Moreover, it follows from Stirling's formula 0.10.5 that

$$M_q^{\mathbb{R}} \le \sqrt{\tfrac{q+1}{e}} \quad \text{and} \quad \lim_{q\to\infty} \frac{M_q^{\mathbb{R}}}{\sqrt{\frac{q+1}{e}}} = 1,$$

$$M_q^{\mathbb{C}} \ge \sqrt{\tfrac{q}{2e}} \quad \text{and} \quad \lim_{q\to\infty} \frac{M_q^{\mathbb{C}}}{\sqrt{\frac{q}{2e}}} = 1.$$

4.7.3 We define

$$g_{\mathbb{R},k}^{(n)}(s) := \sigma_k \quad \text{for } s = (\sigma_1, \ldots, \sigma_n) \in \mathbb{R}^n$$

and

$$g_{\mathbb{C},k}^{(n)}(s) := \tfrac{1}{\sqrt{2}}\sigma_k \quad \text{for } s = (\sigma_1, \ldots, \sigma_n) \in \mathbb{C}^n.$$

The **Gauss systems**

$$\mathcal{G}_n^{\mathbb{R}} := (g_{\mathbb{R},1}^{(n)}, \dots, g_{\mathbb{R},n}^{(n)}) \quad \text{and} \quad \mathcal{G}_n^{\mathbb{C}} := (g_{\mathbb{C},1}^{(n)}, \dots, g_{\mathbb{C},n}^{(n)})$$

are unconditional and orthonormal in $L_2(\mathbb{R}^n, \gamma_{\mathbb{R}}^n)$ and $L_2(\mathbb{C}^n, \gamma_{\mathbb{C}}^n)$, respectively.

In the complex case, the measure $\gamma_{\mathbb{C}}^n$ is invariant under the transformation $(\sigma_1, \dots, \sigma_n) \to (\overline{\sigma}_1, \dots, \overline{\sigma}_n)$. So we have

$$\|(x_k)|\overline{\mathcal{G}}_n^{\mathbb{C}}\| = \|(x_k)|\mathcal{G}_n^{\mathbb{C}}\|.$$

4.7.4 Note that

$$\|g_k^{(n)}|L_r\| = \Big(\int\limits_{\mathbb{K}^n} |g_k^{(n)}(s)|^r \, d\gamma^n(s) \Big)^{1/r} = M_r.$$

The orthogonal and unitary invariance of $\gamma_{\mathbb{R}}^n$ and $\gamma_{\mathbb{C}}^n$, respectively, imply the **Khintchine equality**

$$\Big(\int\limits_{\mathbb{K}^n} \Big| \sum_{k=1}^n \xi_k g_k^{(n)}(s) \Big|^r d\gamma^n(s) \Big)^{1/r} = M_r \Big(\sum_{k=1}^n |\xi_k|^2 \Big)^{1/2} \quad \text{for } \xi_1, \dots, \xi_n \in \mathbb{K};$$

see Section 3.2. This means that

$$\begin{aligned} K_p'(\mathcal{G}_n) &= M_p(\mathcal{G}_n)^{-1} = M_p^{-1} && \text{if } 1 \le p \le 2, \\ K_q(\mathcal{G}_n) &= M_q(\mathcal{G}_n) \quad\; = M_q && \text{if } 2 \le q < \infty. \end{aligned}$$

Note that these Khintchine constants do not depend on $n = 1, 2, \dots$.

4.7.5 Using the notation from 3.2.3, for $1 \le r < \infty$, we let

$$N_r(\mathcal{G}_n) := \Big(\int\limits_{\mathbb{K}^n} \Big(\frac{1}{n} \sum_{k=1}^n |g_k^{(n)}(s)|^r \Big)^{2/r} d\gamma^n(s) \Big)^{1/2}.$$

In the limiting case $r = \infty$, the usual modification yields

$$N_\infty(\mathcal{G}_n) := \Big(\int\limits_{\mathbb{K}^n} \max_{1 \le k \le n} |g_k^{(n)}(s)|^2 d\gamma^n(s) \Big)^{1/2}.$$

Since $\|\cdot|L_p\| \le \|\cdot|L_2\| \le \|\cdot|L_q\|$, we have

$$N_p(\mathcal{G}_n) \ge M_p \quad \text{if } 1 \le p \le 2, \quad \text{and} \quad N_q(\mathcal{G}_n) \le M_q \quad \text{if } 2 \le q \le \infty.$$

Moreover, it follows from the law of large numbers that

$$\lim_{n \to \infty} N_r(\mathcal{G}_n) = M_r \quad \text{for } 1 \le r < \infty.$$

In the special case $r = 1$, this result is an immediate consequence of

$$N_1(\mathcal{G}_n) = \sqrt{(1 - \tfrac{1}{n}) M_1^2 + \tfrac{1}{n}} \searrow M_1.$$

On the other hand,

$$N_\infty(\mathcal{G}_n) \asymp \sqrt{1+\log n};$$

see [LED*, p. 80].

REMARK. An upper estimate of $N_\infty(\mathcal{G}_n)$ follows from

$$N_\infty(\mathcal{G}_n) \le n^{1/q}\, N_q(\mathcal{G}_n) \le n^{1/q}\, M_q \le n^{1/q} \sqrt{\tfrac{q+1}{e}}$$

by letting $q := \log n > 2$.

4.7.6 For Gaussian random variables we also have a **Khintchine–Kahane inequality** which follows from 4.1.10, by the central limit theorem 4.7.15; see also [LED*, p. 103].

PROPOSITION. *Let $0 < u < v < \infty$. Then there exists a constant $K(u,v) \ge 1$ such that*

$$\left(\int\limits_{\mathbb{K}^n} \Big\| \sum_{k=1}^n x_k g_k^{(n)}(s) \Big\|^v d\gamma^n(s) \right)^{1/v} \le K(u,v) \left(\int\limits_{\mathbb{K}^n} \Big\| \sum_{k=1}^n x_k g_k^{(n)}(s) \Big\|^u d\gamma^n(s) \right)^{1/u}$$

for $x_1, \dots, x_n \in X$ and any Banach space X.

REMARK. The reverse inequality

$$\left(\int\limits_{\mathbb{K}^n} \Big\| \sum_{k=1}^n x_k g_k^{(n)}(s) \Big\|^u d\gamma^n(s) \right)^{1/u} \le \left(\int\limits_{\mathbb{K}^n} \Big\| \sum_{k=1}^n x_k g_k^{(n)}(s) \Big\|^v d\gamma^n(s) \right)^{1/v}$$

is obvious.

4.7.7 We now compare the real and the complex Gauss systems with each other; see also 4.2.13. This only makes sense when the underlying Banach spaces are complex.

PROPOSITION. $\varrho(\mathcal{G}_n^{\mathbb{C}}, \mathcal{G}_n^{\mathbb{R}}) \le \sqrt{2}$ *and* $\varrho(\mathcal{G}_n^{\mathbb{R}}, \mathcal{G}_n^{\mathbb{C}}) \le \sqrt{2}$.

PROOF. Recall that

$$\|(x_k)|\mathcal{G}_n^{\mathbb{C}}\| = \left(\int\limits_{\mathbb{R}^n}\int\limits_{\mathbb{R}^n} \Big\| \sum_{k=1}^n x_k \frac{\alpha_k + i\beta_k}{\sqrt{2}} \Big\|^2 d\gamma^n(a)\, d\gamma^n(b) \right)^{1/2},$$

where $a = (\alpha_k) \in \mathbb{R}^n$ and $b = (\beta_k) \in \mathbb{R}^n$. Associating with every n-tuple $(x_1, \dots, x_n)$ the $2n$-tuple $(x_1, \dots, x_n; ix_1, \dots, ix_n)$ yields

$$\|(x_k)|\mathcal{G}_n^{\mathbb{C}}\| = \tfrac{1}{\sqrt{2}} \|(x_k; ix_k)|\mathcal{G}_{2n}^{\mathbb{R}}\|.$$

The conclusion now follows from

$$\begin{aligned} \|(x_k)|\mathcal{G}_n^{\mathbb{R}}\| &\le \tfrac{1}{2} \left(\|(x_k; +ix_k)|\mathcal{G}_{2n}^{\mathbb{R}}\| + \|(x_k; -ix_k)|\mathcal{G}_{2n}^{\mathbb{R}}\| \right) \\ &= \|(x_k; ix_k)|\mathcal{G}_{2n}^{\mathbb{R}}\| \le 2\|(x_k)|\mathcal{G}_n^{\mathbb{R}}\|. \end{aligned}$$

4.7.8 Next, we prove an auxiliary result about the Gamma function.

LEMMA. *Let*

$$F_a(x) := \frac{\Gamma(x+a)}{x^a\Gamma(x)} \quad \textit{for } 0 < x, a < \infty.$$

Then

$$\lim_{x\to\infty} F_a(x) = 1.$$

Moreover, F_a is increasing if $0 < a \le 1$ and decreasing if $1 \le a < \infty$.

PROOF. The fact that

$$\lim_{x\to\infty} F_a(x) = 1$$

follows at once from Stirling's formula 0.10.5.

The recurrence formula of the Gamma function implies

$$F_a(x) = F_a(x+1)\varphi_a(x) \quad \text{with } \varphi_a(x) := \left(\tfrac{x+1}{x}\right)^a \tfrac{x}{x+a}.$$

By induction, we obtain

$$F_a(x) = F_a(x+n)\prod_{k=0}^{n-1}\varphi_a(x+k).$$

Letting $n \to \infty$ yields

$$F_a(x) = \prod_{k=0}^{\infty}\varphi_a(x+k).$$

The monotonicity of $F_a(x)$ can now be inferred from

$$\varphi_a'(x) = \frac{a(1-a)(x+1)^{a-1}}{x^a(x+a)^2}.$$

4.7.9 Let

$$\mathbb{S}^n_{\mathbb{R}} := \Big\{ s = (\sigma_k) \in \mathbb{R}^n : \sum_{k=1}^{n} |\sigma_k|^2 = 1 \Big\}$$

be the $(n-1)$-dimensional real unit sphere equipped with the orthogonally invariant normalized measure $\omega^n_{\mathbb{R}}$. The complex sphere $\mathbb{S}^n_{\mathbb{C}}$ and the unitarily invariant measure $\omega^n_{\mathbb{C}}$ are defined analogously. Note that $\omega^n_{\mathbb{C}}$ is just the image of $\omega^{2n}_{\mathbb{R}}$ under the canonical mapping between $\mathbb{R}^{2n}$ and $\mathbb{C}^n$.

4.7.10 We conclude from

$$\int_{\mathbb{S}^n} \sum_{h=1}^{n} |\sigma_h|^2 \, d\omega^n(s) = 1$$

that

$$\int_{\mathbb{S}^n} |\sigma_k|^2 \, d\omega^n(s) = \frac{1}{n} \quad \text{for } k = 1, \ldots, n.$$

Hence, letting

$$q_{\mathbb{R},k}^{(n)}(s) := \sqrt{n}\,\sigma_k \quad \text{for } s = (\sigma_1, \ldots, \sigma_n) \in \mathbb{S}_{\mathbb{R}}^n$$

and

$$q_{\mathbb{C},k}^{(n)}(s) := \sqrt{n}\,\sigma_k \quad \text{for } s = (\sigma_1, \ldots, \sigma_n) \in \mathbb{S}_{\mathbb{C}}^n,$$

we obtain the **spherical systems**

$$\mathcal{Q}_n^{\mathbb{R}} := (q_{\mathbb{R},1}^{(n)}, \ldots, q_{\mathbb{R},n}^{(n)}) \quad \text{and} \quad \mathcal{Q}_n^{\mathbb{C}} := (q_{\mathbb{C},1}^{(n)}, \ldots, q_{\mathbb{C},n}^{(n)})$$

which are orthonormal in $L_2(\mathbb{S}_{\mathbb{R}}^n, \omega_{\mathbb{R}}^n)$ and $L_2(\mathbb{S}_{\mathbb{C}}^n, \omega_{\mathbb{C}}^n)$, respectively.

In the complex case, the measure $\omega_{\mathbb{C}}^n$ is invariant under the transformation $(\sigma_1, \ldots, \sigma_n) \to (\overline{\sigma}_1, \ldots, \overline{\sigma}_n)$. So we have

$$\|(x_k)|\overline{\mathcal{Q}_n^{\mathbb{C}}}\| = \|(x_k)|\mathcal{Q}_n^{\mathbb{C}}\|.$$

4.7.11 Using the notation from 3.2.3, for $1 \le r < \infty$, we let

$$M_r(\mathcal{Q}_n) := \left(\int\limits_{\mathbb{S}^n} \left| \frac{1}{\sqrt{n}} \sum_{k=1}^n q_k^{(n)}(s) \right|^r d\omega^n(s) \right)^{1/r} = \left(\int\limits_{\mathbb{S}^n} |q_k^{(n)}(s)|^r d\omega^n(s) \right)^{1/r}.$$

Note that the right-hand expression does not depend on k. In the limiting case $r = \infty$, the usual modification yields

$$M_\infty(\mathcal{Q}_n) := \sqrt{n}.$$

It turns out that

$$M_r(\mathcal{Q}_n^{\mathbb{R}}) = \sqrt{n} \left[\frac{\Gamma(\frac{r+1}{2})}{\Gamma(\frac{1}{2})} \frac{\Gamma(\frac{n}{2})}{\Gamma(\frac{r+n}{2})} \right]^{1/r},$$

$$M_r(\mathcal{Q}_n^{\mathbb{C}}) = \sqrt{n} \left[\Gamma(\tfrac{r+2}{2}) \frac{\Gamma(n)}{\Gamma(\frac{r+2n}{2})} \right]^{1/r}.$$

By 4.7.8, it follows from

$$M_r(\mathcal{Q}_n^{\mathbb{R}}) = \left[\frac{F_{r/2}(\frac{1}{2})}{F_{r/2}(\frac{n}{2})} \right]^{1/r} \quad \text{and} \quad M_r(\mathcal{Q}_n^{\mathbb{C}}) = \left[\frac{F_{r/2}(1)}{F_{r/2}(n)} \right]^{1/r}$$

that

$$\lim_{n\to\infty} M_r(\mathcal{Q}_n) = M_r.$$

The same lemma tells us that the sequence $(M_r(\mathcal{Q}_n))$ is decreasing if $1 \le r \le 2$ and increasing if $2 \le r < \infty$.

4.7.12 Note that

$$\|q_n^{(k)}|L_r\| = \Big(\int_{\mathbb{S}^n} |q_k^{(n)}(s)|^r \, d\omega^n(s) \Big)^{1/r} = M_r(\mathcal{Q}_n).$$

The orthogonal and unitary invariance of $\omega_{\mathbb{R}}^n$ and $\omega_{\mathbb{C}}^n$, respectively, imply the **Khintchine equality**

$$\bigg(\int_{\mathbb{S}^n} \Big| \sum_{k=1}^n \xi_k q_k^{(n)}(s) \Big|^r d\omega^n(s) \bigg)^{1/r} = M_r(\mathcal{Q}_n) \Big(\sum_{k=1}^n |\xi_k|^2 \Big)^{1/2}$$

for $\xi_1, \ldots, \xi_n \in \mathbb{K}$. So, for $n = 1, 2, \ldots$, we have the Khintchine constants

$$\begin{aligned} K_p'(\mathcal{Q}_n) &= M_p(\mathcal{Q}_n)^{-1} && \text{if } 1 \le p \le 2, \\ K_q(\mathcal{Q}_n) &= M_q(\mathcal{Q}_n) && \text{if } 2 \le q < \infty. \end{aligned}$$

4.7.13 Using polar coordinates

$$\sigma := \|s|l_2^n\| \quad \text{and} \quad s_0 := s/\sigma \quad \text{for } s \in \mathbb{K}^n \setminus \{o\},$$

we obtain

$$d\gamma_{\mathbb{R}}^n(s) = \frac{2^{1-n/2}}{\Gamma(\frac{n}{2})} \sigma^{n-1} \exp(-\tfrac{1}{2}\sigma^2) \, d\sigma \, d\omega^n(s_0)$$

and

$$d\gamma_{\mathbb{C}}^n(s) = \frac{2^{1-n}}{\Gamma(n)} \sigma^{2n-1} \exp(-\tfrac{1}{2}\sigma^2) \, d\sigma \, d\omega^{2n}(s_0).$$

By this transformation, it can be shown that

$$\|(x_k)|\mathcal{G}_n\| = \|(x_k)|\mathcal{Q}_n\|$$

for $x_1, \ldots, x_n \in X$ and all Banach spaces X.

4.7.14 For $1 \le r < \infty$, we let

$$N_r(\mathcal{Q}_n) := \bigg(\int_{\mathbb{S}^n} \Big(\frac{1}{n} \sum_{k=1}^n |q_k^{(n)}(s)|^r \Big)^{2/r} d\omega^n(s) \bigg)^{1/2}.$$

In the limiting case $r = \infty$, the usual modification is required. The transformation described in the preceding paragraph yields

$$N_r(\mathcal{G}_n) = N_r(\mathcal{Q}_n).$$

Since $\|\cdot|L_p\| \le \|\cdot|L_2\| \le \|\cdot|L_q\|$, we have

$$N_p(\mathcal{Q}_n) \ge M_p(\mathcal{Q}_n) \quad \text{if } 1 \le p \le 2, \quad \text{and} \quad N_q(\mathcal{Q}_n) \le M_q(\mathcal{Q}_n) \quad \text{if } 2 \le q < \infty.$$

However, we stress that, for fixed n, the gap between $N_q(\mathcal{Q}_n)$ and $M_q(\mathcal{Q}_n)$ increases as $q \to \infty$. More precisely,

$$\lim_{q \to \infty} N_q(\mathcal{Q}_n) \asymp \sqrt{1 + \log n} \quad \text{and} \quad \lim_{q \to \infty} M_q(\mathcal{Q}_n) = \sqrt{n}.$$

4.7.15 Let $C_{exp}(\mathbb{R}^n)$ denote the Banach space of all continuous functions $f : \mathbb{R}^n \to \mathbb{R}$ such that

$$f(s_1, \ldots, s_n) \exp\Big(-\sum_{k=1}^{n} |s_k|\Big) \to 0 \quad \text{as} \sum_{k=1}^{n} |s_k| \to \infty$$

equipped with the norm

$$\|f|C_{exp}(\mathbb{R}^n)\| := \sup\Big\{ |f(s_1, \ldots, s_n)| \exp\Big(-\sum_{k=1}^{n} |s_k|\Big) : s_1, \ldots, s_n \in \mathbb{R} \Big\}.$$

Define the functionals $G^{(n)}, R_1^{(n)}, R_2^{(n)}, \ldots \in C_{exp}(\mathbb{R}^n)'$ by

$$G^{(n)}(f) := \int_{-\infty}^{+\infty} \cdots \int_{-\infty}^{+\infty} f(s_1, \ldots, s_n)\, d\gamma(s_1) \ldots d\gamma(s_n)$$

and

$$R_m^{(n)}(f) := \int_0^1 \cdots \int_0^1 f\Big(\frac{1}{\sqrt{m}} \sum_{h=1}^{m} r_h(t_1), \ldots, \frac{1}{\sqrt{m}} \sum_{h=1}^{m} r_h(t_n)\Big) dt_1 \ldots dt_n.$$

We proceed with a folklore result; the proof is taken from [kwa 2, p. 585].

CENTRAL LIMIT THEOREM (improved version).

$$\lim_{m\to\infty} R_m^{(n)}(f) = G^{(n)}(f) \quad \textit{for} \quad f \in C_{exp}(\mathbb{R}^n).$$

PROOF. Note that

$$1 < (\cosh u)^{1/u^2} < \sqrt{e} \quad \text{whenever } u > 0.$$

Now, for $\|f|C_{exp}(\mathbb{R}^n)\| \leq 1$, we obtain

$$|R_m^{(n)}(f)| = \Big| \int_0^1 \ldots \int_0^1 f\Big(\frac{1}{\sqrt{m}} \sum_{h=1}^{m} r_h(t_1), \ldots, \frac{1}{\sqrt{m}} \sum_{h=1}^{m} r_h(t_n)\Big) dt_1 \ldots dt_n \Big|$$

$$\leq \int_0^1 \ldots \int_0^1 \exp\Big(\sum_{k=1}^{n} \Big| \frac{1}{\sqrt{m}} \sum_{h=1}^{m} r_h(t_k) \Big| \Big) dt_1 \ldots dt_n$$

$$= \Big[\int_0^1 \exp\Big(\Big| \frac{1}{\sqrt{m}} \sum_{h=1}^{m} r_h(t) \Big| \Big) dt \Big]^n$$

$$\leq \Big[\int_0^1 \exp\Big(+\frac{1}{\sqrt{m}} \sum_{h=1}^{m} r_h(t) \Big) dt + \int_0^1 \exp\Big(-\frac{1}{\sqrt{m}} \sum_{h=1}^{m} r_h(t) \Big) dt \Big]^n$$

$$= \left[\left(\int_0^1 \exp\Big(+\frac{1}{\sqrt{m}} r_1(t)\Big) dt \right)^m + \left(\int_0^1 \exp\Big(-\frac{1}{\sqrt{m}} r_1(t)\Big) dt \right)^m \right]^n$$

$$= 2^n \left[\cosh\Big(\frac{1}{\sqrt{m}}\Big) \right]^{mn} \le (2\sqrt{e})^n.$$

Hence

$$\|R_m^{(n)} | C_{exp}(\mathbb{R}^n)'\| \le (2\sqrt{e})^n \quad \text{for} \quad m = 1, 2, \ldots .$$

The common central limit theorem says that

$$\lim_{m\to\infty} R_m^{(n)}(f) = G^{(n)}(f) \tag{$*$}$$

for all bounded continuous functions f. Finally, a density argument shows that $(*)$ even holds for all $f \in C_{exp}(\mathbb{R}^n)$; see [DUN*1, p. 55].

4.8 Gauss versus Rademacher

4.8.1 We now compare the Gauss system with the Rademacher and Steinhaus systems. In view of

$$\varrho(\mathcal{G}_n^{\mathbb{C}}, \mathcal{G}_n^{\mathbb{R}}) \le \sqrt{2} \quad \text{and} \quad \varrho(\mathcal{G}_n^{\mathbb{R}}, \mathcal{G}_n^{\mathbb{C}}) \le \sqrt{2},$$

it would suffice to deal with the real Gauss system only. However, this causes asymmetries which look quite unnatural. Thus we will treat both cases simultaneously. In order to emphasize this parallelism, it is useful to replace the Rademacher system $\mathcal{R}_n$ by the Bernoulli system $\mathcal{P}_n(\pm 1)$ and the Steinhaus system $\mathcal{P}_n(e^{i\tau})$, respectively. Whenever possible, we will use the unified notation $\mathcal{P}_n = (p_1^{(n)}, \ldots, p_n^{(n)})$, where $p_k^{(n)}$ are the coordinate functionals (projections). Moreover, $\mathbb{E}^n$ equipped with the Haar measure λ^n will be a synonym of $\mathbb{E}_2^n$ and $\mathbb{T}^n$.

4.8.2 LEMMA. *Let $\mathcal{A}_n = (a_1, \ldots, a_n)$ and $\mathcal{B}_n = (b_1, \ldots, b_n)$ be orthonormal systems in $L_2(M, \mu)$ and $L_2(N, \nu)$, respectively. If*

$$(a_1, \ldots, a_n) \quad \textit{and} \quad (|a_1| \otimes b_1, \ldots, |a_n| \otimes b_n)$$

have identical distributions and

$$\int_M |a_k(s)|\, d\mu(s) = \alpha \quad \textit{for } k = 1, \ldots, n,$$

then

$$\varrho(\mathcal{B}_n, \mathcal{A}_n) \le \alpha^{-1}.$$

PROOF. For $x_1, \dots, x_n \in X$,

$$\boldsymbol{f}(s) := \sum_{k=1}^{n} |a_k(s)|\, b_k \otimes x_k$$

defines an $[L_2(N,\nu), X]$-valued function on M. By 3.1.7,

$$\Big(\int_M \|\boldsymbol{f}(s)\|^2 d\mu(s)\Big)^{1/2} = \Big(\int_M\int_N \Big\|\sum_{k=1}^{n} x_k\, |a_k(s)|\, b_k(t)\Big\|^2 d\mu(s)\, d\nu(t)\Big)^{1/2}$$

$$= \Big(\int_M \Big\|\sum_{k=1}^{n} x_k a_k(s)\Big\|^2 d\mu(s)\Big)^{1/2} = \|(x_k)|\mathcal{A}_n\|.$$

On the other hand, we have

$$\Big\|\int_M \boldsymbol{f}(s) d\mu(s)\Big| L_2\Big\| = \Big(\int_N \Big\|\int_M \sum_{k=1}^{n} x_k |a_k(s)| b_k(t) d\mu(s)\Big\|^2 d\nu(t)\Big)^{1/2} = \alpha \|(x_k)|\mathcal{B}_n\|.$$

The required inequality is now implied by

$$\Big\|\int_M \boldsymbol{f}(s)\, d\mu(s)\Big| L_2\Big\| \le \int_M \|\boldsymbol{f}(s)|L_2\|\, d\mu(s) \le \Big(\int_M \|\boldsymbol{f}(s)|L_2\|^2 d\mu(s)\Big)^{1/2}.$$

4.8.3 We know from 4.7.2 that $M_1^{\mathbb{R}} = \sqrt{\frac{2}{\pi}}$ and $M_1^{\mathbb{C}} = \sqrt{\frac{\pi}{4}}$.

PROPOSITION. $\varrho(\mathcal{P}_n, \mathcal{G}_n) \le M_1^{-1}$.

PROOF. Note that, in the real and in the complex case,

$$(g_1^{(n)}, \dots, g_n^{(n)}) \quad \text{and} \quad (|g_1^{(n)}| \otimes p_1^{(n)}, \dots, |g_n^{(n)}| \otimes p_n^{(n)})$$

have identical distributions. This can be seen by looking at the Fourier transforms (characteristic functions). The assertion is now a consequence of Lemma 4.8.2 and

$$\int_{\mathbb{K}^n} |g_k^{(n)}(s)|\, d\gamma^n(s) = M_1.$$

REMARK. Applying the central limit theorem 4.7.15, we may conclude from 4.4.15 that $\varrho(\mathcal{R}_n, \mathcal{G}_n^{\mathbb{R}}) \le \sqrt{2}$.

4.8.4 Since

$$\lim_{n\to\infty} N_1(\mathcal{G}_n) = M_1,$$

the following example tells us that the preceding inequality is sharp.

EXAMPLE. $N_1(\mathcal{G}_n)^{-1} \le \varrho(l_1^n|\mathcal{P}_n, \mathcal{G}_n) \le \varrho(L_1|\mathcal{P}_n, \mathcal{G}_n) \le M_1^{-1}$.

PROOF. The lower estimate of $\varrho(l_1^n|\mathcal{P}_n,\mathcal{G}_n)$ follows from

$$\|(u_k^{(n)})|[\mathcal{P}_n,l_1^n]\| \le \varrho(l_1^n|\mathcal{P}_n,\mathcal{G}_n)\,\|(u_k^{(n)})|[\mathcal{G}_n,l_1^n]\|,$$

$$\|(u_k^{(n)})|[\mathcal{P}_n,l_1^n]\| = n \quad\text{and}\quad \|(u_k^{(n)})|[\mathcal{G}_n,l_1^n]\| = N_1(\mathcal{G}_n)\,n.$$

On the other hand, 3.3.7 yields

$$\varrho(l_1^n|\mathcal{P}_n,\mathcal{G}_n) \le \varrho(L_1|\mathcal{P}_n,\mathcal{G}_n) \le K_1'(\mathcal{G}_n) = M_1^{-1}.$$

4.8.5 Because of 4.8.3, the Riemann ideal norms $\varrho(\mathcal{P}_n,\mathcal{G}_n)$ deserve no further attention, since they are uniformly equivalent to the operator norm. Thus, from now on, we concentrate on $\varrho(\mathcal{G}_n,\mathcal{P}_n)$.

PROPOSITION. *The ideal norm $\varrho(\mathcal{G}_n,\mathcal{P}_n)$ is non-surjective and non-symmetric.*

PROOF. Since l_∞ is a quotient of some $l_1(\mathbb{I})$, the non-surjectivity and non-symmetry follow from

$$\varrho(l_1(\mathbb{I})|\mathcal{G}_n,\mathcal{P}_n) \asymp 1 \quad\text{and}\quad \varrho(l_\infty|\mathcal{G}_n,\mathcal{P}_n) \asymp \sqrt{1+\log n};$$

see 4.8.8 and 4.8.10.

4.8.6 The following result can easily be checked.

PROPOSITION. *The sequence $(\varrho(\mathcal{G}_n,\mathcal{P}_n))$ is non-decreasing. Moreover, we have*

$$\varrho(\mathcal{G}_{2n},\mathcal{P}_{2n}) \le \sqrt{2}\,\varrho(\mathcal{G}_n,\mathcal{P}_n).$$

4.8.7 We now state a very important result.

PROPOSITION. $\varrho(\mathcal{G}_n,\mathcal{P}_n) \le N_\infty(\mathcal{G}_n) \asymp \sqrt{1+\log n}.$

PROOF. Since $\mathcal{P}_n$ is unconditional, the contraction principle 3.5.4 implies that

$$\int\limits_{\mathbb{E}^n} \Big\|\sum_{k=1}^n x_k\, g_k^{(n)}(s) p_k^{(n)}(t)\Big\|^2 d\lambda^n(t) \le$$
$$\le \max_{1\le k\le n} |g_k^{(n)}(s)|^2 \int\limits_{\mathbb{E}^n} \Big\|\sum_{k=1}^n x_k\, p_k^{(n)}(t)\Big\|^2 d\lambda^n(t).$$

Integrating over $s\in\mathbb{K}^n$ yields

$$\|(x_k)|\mathcal{G}_n\otimes\mathcal{P}_n\| \le \Big(\int\limits_{\mathbb{K}^n} \max_{1\le k\le n} |g_k^{(n)}(s)|^2 d\gamma^n(s)\Big)^{1/2} \|(x_k)|\mathcal{P}_n\|.$$

Finally, we recall from 4.7.5 that $N_\infty(\mathcal{G}_n) \asymp \sqrt{1+\log n}$.

4.8.8 Next, we show that the preceding result is sharp.

EXAMPLE.

$$\varrho(l_\infty^n|\mathcal{G}_n,\mathcal{P}_n) = \varrho(L_\infty|\mathcal{G}_n,\mathcal{P}_n) = N_\infty(\mathcal{G}_n) \asymp \sqrt{1+\log n}.$$

PROOF. The lower estimate of $\varrho(l_\infty^n|\mathcal{G}_n,\mathcal{P}_n)$ follows from

$$\|(u_k^{(n)})|[\mathcal{G}_n,l_\infty^n]\| \le \varrho(l_\infty^n|\mathcal{G}_n,\mathcal{P}_n)\,\|(u_k^{(n)})|[\mathcal{P}_n,l_\infty^n]\|,$$

$$\|(u_k^{(n)})|[\mathcal{G}_n,l_\infty^n]\| = N_\infty(\mathcal{G}_n) \quad\text{and}\quad \|(u_k^{(n)})|[\mathcal{P}_n,l_\infty^n]\| = 1.$$

Since every infinite dimensional space L_∞ contains a subspace isometric to l_∞^n, we obtain

$$\varrho(l_\infty^n|\mathcal{G}_n,\mathcal{P}_n) \le \varrho(L_\infty|\mathcal{G}_n,\mathcal{P}_n).$$

Finally, in view of the previous proposition,

$$\varrho(L_\infty|\mathcal{G}_n,\mathcal{P}_n) \le N_\infty(\mathcal{G}_n).$$

4.8.9 EXAMPLE. *Let* $2 \le q < \infty$. *Then*

$$N_q(\mathcal{Q}_n) \le \varrho(l_q^n|\mathcal{G}_n,\mathcal{P}_n) \le \varrho(L_q|\mathcal{G}_n,\mathcal{P}_n) \le M_q(\mathcal{Q}_n) \le M_q \le \sqrt{q-1}.$$

PROOF. Since $\mathcal{G}_n$ can be replaced by $\mathcal{Q}_n$, we conclude from 3.3.8 that

$$\varrho(L_q|\mathcal{Q}_n,\mathcal{P}_n) \le M_q(\mathcal{Q}_n).$$

On the other hand, the inequality

$$N_q(\mathcal{Q}_n) \le \varrho(l_q^n|\mathcal{Q}_n,\mathcal{P}_n) \le \varrho(L_q|\mathcal{Q}_n,\mathcal{P}_n)$$

follows as in the above proof.

REMARK. In analogy with 4.8.8, we conjecture that

$$\varrho(l_q^n|\mathcal{G}_n,\mathcal{P}_n) \stackrel{?}{=} \varrho(L_q|\mathcal{G}_n,\mathcal{P}_n) \stackrel{?}{=} N_q(\mathcal{G}_n).$$

In any case,

$$\lim_{n\to\infty} N_q(\mathcal{Q}_n) = \lim_{n\to\infty} M_q(\mathcal{Q}_n) = M_q \qquad \text{(by 4.7.5)}$$

implies that

$$\lim_{n\to\infty} \varrho(L_q|\mathcal{G}_n,\mathcal{P}_n) = M_q.$$

4.8.10 The next fact follows from 3.3.7.

EXAMPLE. *Let* $1 \le p \le 2$. *Then*

$$1 \le \varrho(l_p^n|\mathcal{G}_n,\mathcal{P}_n) \le \varrho(L_p|\mathcal{G}_n,\mathcal{P}_n) \le M_p(\mathcal{Q}_n)^{-1} < 2.$$

4.8.11 We now summarize the results from 4.8.8, 4.8.9 and 4.8.10. Since only asymptotic relations are considered, there is no difference between the real and the complex case. Hence we may return to the Rademacher system.

EXAMPLE.

$$\varrho(L_r|\mathcal{G}_n,\mathcal{R}_n)\asymp 1 \quad \textit{if } 1\le r<\infty \quad \textit{and} \quad \varrho(L_\infty|\mathcal{G}_n,\mathcal{R}_n)\asymp\sqrt{1+\log n}.$$

4.8.12 The above example can be viewed as a special case of a general result, which follows from 4.8.22. A similar situation will be described in 4.12.12.

DICHOTOMY. *For every Banach space X, we have*

$$\textit{either} \quad \varrho(X|\mathcal{G}_n,\mathcal{R}_n)\asymp 1 \quad \textit{or} \quad \varrho(X|\mathcal{G}_n,\mathcal{R}_n)\asymp\sqrt{1+\log n}.$$

4.8.13 In the preceding examples the function spaces are assumed to be infinite dimensional. In order to treat the finite dimensional case, let n and N be natural numbers.

EXAMPLE. $\varrho(l_\infty^N|\mathcal{G}_n,\mathcal{R}_n)\asymp\min\{\sqrt{1+\log n},\sqrt{1+\log N}\}.$

PROOF. By 4.8.7,

$$\varrho(l_\infty^N|\mathcal{G}_n,\mathcal{R}_n)\prec\sqrt{1+\log n}.$$

We know from 4.8.9 that $\varrho(l_q^N|\mathcal{G}_n,\mathcal{R}_n)\le\sqrt{q-1}$. So 1.3.6 gives

$$\varrho(l_\infty^N|\mathcal{G}_n,\mathcal{R}_n)\prec\sqrt{1+\log N}.$$

This proves the upper estimate. Let $m:=\min\{n,N\}$. Then

$$\|(u_k^{(m)})|[\mathcal{G}_m,l_\infty^m]\|\le\varrho(l_\infty^m|\mathcal{G}_m,\mathcal{R}_m)\,\|(u_k^{(m)})|[\mathcal{R}_m,l_\infty^m]\|,$$

$$\|(u_k^{(m)})|[\mathcal{G}_m,l_\infty^m]\|=N_\infty(\mathcal{G}_m) \quad \text{and} \quad \|(u_k^{(m)})|[\mathcal{R}_m,l_\infty^m]\|=1$$

yield

$$\sqrt{1+\log m}\prec\varrho(l_\infty^m|\mathcal{G}_m,\mathcal{R}_m)\le\varrho(l_\infty^N|\mathcal{G}_n,\mathcal{R}_n).$$

Thus the lower estimate holds too.

4.8.14 Let C_α denote the diagonal operator on $[l_2,l_\infty^{2^k}]$ defined by

$$C_\alpha:(x_k)\longrightarrow(k^{-\alpha}x_k), \quad \text{where} \quad x_k\in l_\infty^{2^k}.$$

We are now able to show that the phenomenon described in 4.8.12 only occurs for spaces, not for operators.

EXAMPLE. *If* $0\le\alpha\le 1/2$, *then*

$$\varrho(C_\alpha|\mathcal{G}_n,\mathcal{R}_n)\asymp(1+\log n)^{1/2-\alpha}.$$

PROOF. Recall that

$$\varrho(l_\infty^N|\mathcal{G}_n,\mathcal{R}_n) \asymp \min\{\sqrt{1+\log n}, \sqrt{1+\log N}\} \qquad \text{(by 4.8.13).}$$

The desired relation is now a consequence of $(\triangle)$ in 1.2.12.

4.8.15 Next, we provide some elementary results about the distribution function

$$D(u) := \gamma\{\,\sigma \in \mathbb{R} : |\sigma| \geq u\,\} = \sqrt{\tfrac{2}{\pi}} \int_u^\infty \exp(-\sigma^2/2)d\sigma \quad \text{for} \quad u \geq 0.$$

Obviously, D is decreasing, and $D(0) = 1$. Since

$$D(u) = \sqrt{\tfrac{2}{\pi}} \exp(-u^2/2) \int_0^\infty \exp(-u\tau - \tau^2/2)d\tau,$$

we have

$$D(u) \leq \exp(-u^2/2). \tag{1}$$

Hence

$$\int_0^\infty D(u)^{1/q}\, du \leq \int_0^\infty \exp(-u^2/2q)\, du = \sqrt{\tfrac{\pi q}{2}}\,. \tag{2}$$

Let $v^{-1} := D(u)$. Then it follows from

$$v^{-2}\frac{dv}{du} = \sqrt{\tfrac{2}{\pi}} \exp(-u^2/2) \quad \text{and} \quad \exp(u^2/2) \leq D(u)^{-1} = v$$

that

$$\int_a^\infty \sqrt{D(u)}\, du = \sqrt{\tfrac{\pi}{2}} \int_{D(a)^{-1}}^\infty v^{-5/2} \exp(u^2/2)\, dv$$

$$\leq \sqrt{\tfrac{\pi}{2}} \int_{D(a)^{-1}}^\infty v^{-3/2}\, dv = \sqrt{2\pi D(a)}. \tag{3}$$

Integration by parts yields

$$\int_u^\infty \exp(-\sigma^2/2)d\sigma = \frac{1}{u}\exp(-u^2/2) - \int_u^\infty \sigma^{-2}\exp(-\sigma^2/2)d\sigma$$

and

$$\int_u^\infty \exp(-\sigma^2/2)d\sigma = \left(\frac{1}{u} - \frac{1}{u^3}\right)\exp(-u^2/2) + 3\int_u^\infty \sigma^{-4}\exp(-\sigma^2/2)d\sigma.$$

However, it is simpler to check these formulas by differentiation. Now we get,

$$\Big(\frac{1}{u}-\frac{1}{u^3}\Big)\exp(-u^2/2) < \sqrt{\tfrac{\pi}{2}}\,D(u) < \frac{1}{u}\exp(-u^2/2). \qquad (4)$$

Easy manipulations show that

$$\exp(1-u^2) \le \sqrt{\frac{2}{\pi}}\Big(\frac{1}{u}-\frac{1}{u^3}\Big)\exp(-u^2/2) \quad \text{for } u \ge 3.$$

Hence, in view of (4),

$$\sqrt{\tfrac{\pi}{2}}\,v^{-1}\exp(u^2/2) \le \frac{1}{u} \le (1+\log v)^{-1/2}.$$

Define $0 = u_1 < u_2 < \ldots$ by $D(u_m) = \frac{1}{m}$. Then, for $u_m \ge 3$, we have

$$\int\limits_{u_m}^{u_{m+1}} \sqrt{D(u)}\,du = \sqrt{\tfrac{\pi}{2}}\int\limits_{m}^{m+1} v^{-5/2}\exp(u^2/2)\,dv \le \int\limits_{m}^{m+1} v^{-3/2}(1+\log v)^{-1/2}dv$$

$$\le m^{-3/2}(1+\log m)^{-1/2}.$$

More precisely, $m \ge 29$ suffices. This proves that

$$\int\limits_{u_m}^{u_{m+1}} \sqrt{D(u)}\,du \prec m^{-3/2}(1+\log m)^{-1/2}. \qquad (5)$$

4.8.16 The following inequalities originate from the classical work of B. Maurey and G. Pisier, who used the theory of absolutely summing operators as a main tool; see [mau*], [DIE*a, p. 250] and [TOM, p. 188]. Our proof is taken from [LED*, pp. 251–252] and [pis 18, pp. 187–188]. The basic idea goes back to S. Kwapień, and the final version is due to A. Hinrichs; see [hin 5]. Probabilists may forgive the awkward presentation of the analysts.

THEOREM. *There exists a constant* $c \ge 1$ *such that*

$$\varrho(\mathcal{G}_n, \mathcal{R}_n) \le c\sum_{k=1}^{n} k^{-3/2}(1+\log k)^{-1/2}\varrho(\mathcal{I}_k, \mathcal{R}_k)$$

and

$$\varrho(\mathcal{G}_n, \mathcal{R}_n) \le c\sqrt{q}\,\varrho_2^{(q)}(\mathcal{I}_n, \mathcal{R}_n) \quad \text{if } 2 \le q < \infty.$$

PROOF. By 4.7.7, it is enough to treat the case of the real Gauss system. Let A be a measurable subset of $\mathbb{R}$, and denote its characteristic function by φ. Take n independent copies of φ, viewed as a random variable, by letting

$$\varphi_k(s) := \varphi(\sigma_k) \quad \text{for} \quad s = (\sigma_1, \ldots, \sigma_n) \in \mathbb{R}^n.$$

We are going to show that

$$\left(\int\limits_{\mathbb{R}^n}\int\limits_0^1 \Big\| \sum_{k=1}^n Tx_k\,\varphi_k(s)\,r_k(t)\Big\|^2 d\gamma^n(s)\,dt\right)^{1/2} \le$$

$$\le \sqrt{2\gamma(A)}\,\boldsymbol{\varrho}(T|\mathcal{I}_m,\mathcal{R}_m)\,\|(x_k)|\mathcal{R}_n\| \qquad (1)$$

whenever $\frac{1}{m+1} < \gamma(A) \le \frac{1}{m}$ and $m = 1, 2, \ldots$.

To begin with, let us assume that $\gamma(A) = \frac{1}{m}$. Write $A^{(1)} := A$, and choose measurable subsets $A^{(2)}, \ldots, A^{(m)}$ of $\mathbb{R}$ with $\gamma(A^{(h)}) = \frac{1}{m}$ which are pairwise disjoint. Let $\varphi^{(h)}$ denote the characteristic function of $A^{(h)}$. Take n independent copies of $(\varphi^{(1)}, \ldots, \varphi^{(m)})$, viewed as a vector of random variables, by letting

$$\varphi_k^{(h)}(s) := \varphi^{(h)}(\sigma_k) \quad \text{for} \quad s = (\sigma_1, \ldots, \sigma_n) \in \mathbb{R}^n.$$

We also need an auxiliary Rademacher system $\mathcal{R}_m = (\varrho_1, \ldots, \varrho_m)$ which is independent of the original system $\mathcal{R}_n = (r_1, \ldots, r_n)$. Define

$$x_h(s,t) := \sum_{k=1}^n x_k\,\varphi_k^{(h)}(s)\,r_k(t).$$

Then

$$\sum_{h=1}^m x_h(s,t)\,\varrho_h(\tau) = \sum_{k=1}^n x_k\,r_k(t)\left[\sum_{h=1}^m \varphi_k^{(h)}(s)\,\varrho_h(\tau)\right].$$

Note that $(\varphi_k^{(1)}(s), \ldots, \varphi_k^{(m)}(s))$ is a unit vector, since $(A^{(1)}, \ldots, A^{(m)})$ was chosen to be a partition of $\mathbb{R}$. Hence

$$\left|\sum_{h=1}^m \varphi_k^{(h)}(s)\,\varrho_h(\tau)\right| = 1.$$

The $\mathbb{R}$-unconditionality of $\mathcal{R}_n$ implies that

$$\int\limits_0^1 \Big\|\sum_{h=1}^m x_h(s,t)\,\varrho_h(\tau)\Big\|^2 dt = \int\limits_0^1 \Big\|\sum_{k=1}^n x_k\,r_k(t)\Big\|^2 dt = \|(x_k)|\mathcal{R}_n\|^2.$$

It follows from the previous equality and Fubini's theorem that

$$\left(\int\limits_{\mathbb{R}^n}\int\limits_0^1 \left[\sum_{h=1}^m \|Tx_h(s,t)\|^2\right] d\gamma^n(s)dt\right)^{1/2} \le$$

$$\le \boldsymbol{\varrho}(T|\mathcal{I}_m,\mathcal{R}_m)\left(\int\limits_{\mathbb{R}^n}\int\limits_0^1 \left[\int\limits_0^1 \Big\|\sum_{h=1}^m x_h(s,t)\,\varrho_h(\tau)\Big\|^2 d\tau\right] d\gamma^n(s)\,dt\right)^{1/2}$$

$$= \varrho(T|\mathcal{I}_m, \mathcal{R}_m)\left(\int_{\mathbb{R}^n}\int_0^1\left[\int_0^1\Big\|\sum_{h=1}^m x_h(s,t)\,\varrho_h(\tau)\Big\|^2 dt\right]d\gamma^n(s)\,d\tau\right)^{1/2}$$
$$= \varrho(T|\mathcal{I}_m, \mathcal{R}_m)\,\|(x_k)|\mathcal{R}_n\|.$$

Since $(\varphi_1, \dots, \varphi_n)$ and $(\varphi_1^{(h)}, \dots, \varphi_n^{(h)})$ have identical distributions, we obtain

$$\int_{\mathbb{R}^n}\int_0^1 \|Tx_h(s,t)\|^2\, d\gamma^n(s)\,dt = \int_{\mathbb{R}^n}\int_0^1 \Big\|\sum_{k=1}^n Tx_k\,\varphi_k^{(h)}(s)\,r_k(t)\Big\|^2 d\gamma^n(s)\,dt$$
$$= \int_{\mathbb{R}^n}\int_0^1 \Big\|\sum_{k=1}^n Tx_k\,\varphi_k(s)\,r_k(t)\Big\|^2 d\gamma^n(s)\,dt.$$

Hence the left-hand expression does not depend on $h = 1, \dots, m$. Thus

$$\left(\int_{\mathbb{R}^n}\int_0^1 \Big\|\sum_{k=1}^n Tx_k\,\varphi_k(s)\,r_k(t)\Big\|^2 d\gamma^n(s)\,dt\right)^{1/2} =$$
$$= \left(\frac{1}{m}\sum_{h=1}^m \int_{\mathbb{R}^n}\int_0^1 \|Tx_h(s,t)\|^2\, d\gamma^n(s)\,dt\right)^{1/2}$$
$$\le \sqrt{\tfrac{1}{m}}\,\varrho(T|\mathcal{I}_m, \mathcal{R}_m)\,\|(x_k)|\mathcal{R}_n\|.$$

If $\frac{1}{m+1} < \gamma(A) \le \frac{1}{m}$, then we choose a measurable subset $B \supseteq A$ such that $\gamma(B) = \frac{1}{m}$. Note that

$$\sqrt{\tfrac{1}{m}} \le \sqrt{2\,\gamma(A)}.$$

Let $(\psi_1, \dots, \psi_n)$ be the system of characteristic functions associated with B in the same way as $(\varphi_1, \dots, \varphi_n)$ is associated with A. Then we have $|\varphi_k(s)| \le |\psi_k(s)|$, and the contraction principle 3.5.4 implies that

$$\left(\int_{\mathbb{R}^n}\int_0^1 \Big\|\sum_{k=1}^n Tx_k\,\varphi_k(s)\,r_k(t)\Big\|^2 d\gamma^n(s)\,dt\right)^{1/2} \le$$
$$\le \left(\int_{\mathbb{R}^n}\int_0^1 \Big\|\sum_{k=1}^n Tx_k\,\psi_k(s)\,r_k(t)\Big\|^2 d\gamma^n(s)\,dt\right)^{1/2}$$
$$\le \sqrt{\tfrac{1}{m}}\,\varrho(T|\mathcal{I}_m, \mathcal{R}_m)\,\|(x_k)|\mathcal{R}_n\|$$
$$\le \sqrt{2\,\gamma(A)}\,\varrho(T|\mathcal{I}_m, \mathcal{R}_m)\|(x_k)|\mathcal{R}_n\|,$$

which is the desired inequality (1).

Now we use the notation and results provided in the preceding paragraph. Define $A_u := \{\sigma \in \mathbb{R} : |\sigma| \geq u\}$ for $u \geq 0$, and note that $D(u) = \gamma(A_u)$. Let

$$\varphi_k(s,u) := \varphi(\sigma_k, u) \quad \text{for} \quad s = (\sigma_1, \ldots, \sigma_n) \in \mathbb{R},$$

where $\varphi(\cdot, u)$ is the characteristic function of A_u. Then it follows from (1) and

$$|g_k^{(n)}(s)| = |\sigma_k| = \int_0^\infty \varphi_k(s,u)\, du$$

that

$$\left(\int_{\mathbb{R}^n} \int_0^1 \Big\| \sum_{k=1}^n Tx_k \, |g_k^{(n)}(s)| \, r_k(t) \Big\|^2 d\gamma^n(s)\, dt \right)^{1/2} =$$

$$= \left(\int_{\mathbb{R}^n} \int_0^1 \Big\| \sum_{k=1}^n Tx_k \int_0^\infty \varphi_k(s,u) du \, r_k(t) \Big\|^2 d\gamma^n(s)\, dt \right)^{1/2}$$

$$\leq \int_0^\infty \left(\int_{\mathbb{R}^n} \int_0^1 \Big\| \sum_{k=1}^n Tx_k \, \varphi_k(s,u) \, r_k(t) \Big\|^2 d\gamma^n(s)\, dt \right)^{1/2} du$$

$$= \sum_{m=1}^\infty \int_{u_m}^{u_{m+1}} \left(\int_{\mathbb{R}^n} \int_0^1 \Big\| \sum_{k=1}^n Tx_k \, \varphi_k(s,u) \, r_k(t) \Big\|^2 d\gamma^n(s)\, dt \right)^{1/2} du$$

$$\leq \sqrt{2} \left[\sum_{m=1}^\infty \int_{u_m}^{u_{m+1}} \sqrt{D(u)}\, du \, \varrho(T|\mathfrak{I}_m, \mathfrak{R}_m) \right] \|(x_k)|\mathfrak{R}_n\|.$$

The observation that the systems

$$(g_1^{(n)}, \ldots, g_n^{(n)}) \quad \text{and} \quad (|g_1^{(n)}| \otimes r_1, \ldots, |g_n^{(n)}| \otimes r_n)$$

have identical distributions yields

$$\varrho(T|\mathfrak{G}_n, \mathfrak{R}_n) \leq \sqrt{2} \sum_{m=1}^\infty \int_{u_m}^{u_{m+1}} \sqrt{D(u)}\, du \, \varrho(T|\mathfrak{I}_m, \mathfrak{R}_m). \tag{2}$$

We stress that this inequality holds for $n = 1, 2, \ldots$ uniformly and that the right-hand series may tend to infinity.

If M is any n-dimensional subspace of X, then John's theorem 0.2.4 tells us that $d(M, l_2^n) \leq \sqrt{n}$. Hence

$$\begin{aligned} \varrho(TJ_M^X|\mathfrak{I}_m, \mathfrak{R}_m) &\leq \|T\| \, \varrho(M|\mathfrak{I}_m, \mathfrak{R}_m) \\ &\leq \|T\| \, d(M, l_2^n) \, \varrho(l_2^n|\mathfrak{I}_m, \mathfrak{R}_m) \leq \sqrt{n} \, \|T\| \end{aligned}$$

for $m=1,2,\ldots$. Since $D(u_m)=\frac{1}{m}$, it follows from (2) and formula (3) in the previous paragraph that

$$\begin{aligned}
&\varrho(TJ_M^X|\mathcal{G}_n,\mathcal{R}_n)\leq \\
&\leq \sqrt{2}\Big[\sum_{m=1}^{n}\int_{u_m}^{u_{m+1}}\sqrt{D(u)}\,du+\sum_{m=n+1}^{\infty}\int_{u_m}^{u_{m+1}}\sqrt{D(u)}\,du\Big]\,\varrho(TJ_M^X|\mathcal{I}_m,\mathcal{R}_m)\\
&\leq \sqrt{2}\sum_{m=1}^{n}\int_{u_m}^{u_{m+1}}\sqrt{D(u)}\,du\,\varrho(T|\mathcal{I}_m,\mathcal{R}_m)+\sqrt{2n}\int_{u_{n+1}}^{\infty}\sqrt{D(u)}\,du\,\|T\|\\
&\leq \sqrt{2}\sum_{m=1}^{n}\int_{u_m}^{u_{m+1}}\sqrt{D(u)}\,du\,\varrho(T|\mathcal{I}_m,\mathcal{R}_m)+2\sqrt{\pi}\,\|T\|.
\end{aligned}$$

Taking into account that

$$\varrho(T|\mathcal{G}_n,\mathcal{R}_n)=\sup\{\,\varrho(TJ_M^X|\mathcal{G}_n,\mathcal{R}_n)\,:\,M\in \mathrm{DIM}_{\leq n}(X)\,\},$$

we obtain

$$\varrho(T|\mathcal{G}_n,\mathcal{R}_n)\leq \sqrt{2}\sum_{m=1}^{n}\int_{u_m}^{u_{m+1}}\sqrt{D(u)}\,du\,\varrho(T|\mathcal{I}_m,\mathcal{R}_m)+2\sqrt{\pi}\,\|T\|. \qquad (3)$$

Finally, the first part of the assertion is a consequence of (5) in 4.8.15 and $\|T\|\leq\varrho(T|\mathcal{I}_m,\mathcal{R}_m)$. The second part can be seen as follows: We conclude from

$$\varrho(\mathcal{I}_m,\mathcal{R}_m)\leq m^{1/2-1/q}\varrho_2^{(q)}(\mathcal{I}_m,\mathcal{R}_m)$$

and

$$m^{1/2-1/q}=D(u_m)^{1/q-1/2}\leq D(u)^{1/q-1/2}\quad\text{for}\quad u\geq u_m$$

that

$$D(u)^{1/2}\varrho(\mathcal{I}_m,\mathcal{R}_m)\leq D(u)^{1/q}\varrho_2^{(q)}(\mathcal{I}_m,\mathcal{R}_m)\quad\text{for}\quad u\geq u_m.$$

Therefore

$$\varrho(T|\mathcal{G}_n,\mathcal{R}_n)\leq c_0\int_0^{u_{n+1}}D(u)^{1/q}\,du\,\varrho_2^{(q)}(T|\mathcal{I}_n,\mathcal{R}_n)\leq c\sqrt{q}\,\varrho_2^{(q)}(T|\mathcal{I}_n,\mathcal{R}_n),$$

by (2) in 4.8.15.

REMARK. Substituting $X:=L_q$ in $\varrho(X|\mathcal{G}_n,\mathcal{R}_n)\leq c\sqrt{q}\,\varrho_2^{(q)}(X|\mathcal{I}_n,\mathcal{R}_n)$ shows that this inequality is sharp (up to a constant $c\geq 1$); see 3.7.16 and 4.8.9 (Remark).

4.8.17 We refer to T as a **GR-operator** if

$$\|T|\mathfrak{GR}\| := \sup_n \varrho(T|\mathcal{G}_n, \mathcal{R}_n)$$

is finite. These operators form the Banach ideal

$$\mathfrak{GR} := \mathfrak{L}[\varrho(\mathcal{G}_n, \mathcal{R}_n)].$$

4.8.18 Quasi-GR-operators are defined by the property

$$\varrho(T|\mathcal{G}_n, \mathcal{R}_n) = o(\sqrt{1+\log n}).$$

They form the closed ideal

$$\mathfrak{QGR} := \mathfrak{L}_0[(1+\log n)^{-1/2}\varrho(\mathcal{G}_n, \mathcal{R}_n)].$$

4.8.19 Next, we state an analogue of 4.5.20.

PROPOSITION. *The operator ideals* $\mathfrak{GR}$ *and* $\mathfrak{QGR}$ *are injective and* l_2*-stable, but neither surjective nor symmetric.*

4.8.20 Now we establish the highlight of this section. For a similar result, the reader is referred to 6.4.16.

THEOREM. $$\bigcup_{2\le q<\infty} \mathfrak{RC}_q \subset \mathfrak{GR} \subset \mathfrak{QGR} = \mathfrak{RC}.$$

PROOF. The inclusion

$$\bigcup_{2\le q<\infty} \mathfrak{RC}_q \subseteq \mathfrak{GR} \tag{1}$$

follows from 4.8.16, and

$$\mathfrak{GR} \subseteq \mathfrak{QGR} \tag{2}$$

is trivial. It will be shown in 4.9.6 that

$$\varrho(\mathcal{I}_n, \mathcal{G}_n) \prec \sqrt{\tfrac{n}{1+\log n}}.$$

Hence

$$\varrho(\mathcal{I}_n, \mathcal{R}_n) \le \varrho(\mathcal{I}_n, \mathcal{G}_n) \circ \varrho(\mathcal{G}_n, \mathcal{R}_n) \prec \sqrt{\tfrac{n}{1+\log n}}\, \varrho(\mathcal{G}_n, \mathcal{R}_n),$$

which proves that

$$\mathfrak{QGR} \subseteq \mathfrak{RC}. \tag{3}$$

To check the reverse inclusion, given $T \in \mathfrak{RC}$ and $\varepsilon > 0$, we successively choose k_0 and n_0 such that

$$\varrho(T|\mathcal{I}_k, \mathcal{R}_k) \le \varepsilon\sqrt{k} \quad \text{for } k \ge k_0 \quad \text{and} \quad \frac{1+\log k_0}{1+\log n_0} \le \varepsilon^2.$$

Note that

$$\sum_{k=1}^{N} k^{-1}(1+\log k)^{-1/2} \leq 1 + \int_{1}^{N} t^{-1}(1+\log t)^{-1/2}\, dt < 2\sqrt{1+\log N}.$$

Using the first inequality from 4.8.16 and $\varrho(T|\mathfrak{I}_k, \mathfrak{R}_k) \leq \sqrt{k}\|T\|$, we get

$$\begin{aligned}\varrho(T|\mathfrak{G}_n, \mathfrak{R}_n) &\leq c \sum_{k=1}^{n} k^{-3/2}(1+\log k)^{-1/2} \varrho(T|\mathfrak{I}_k, \mathfrak{R}_k)\\ &\leq c \sum_{k=1}^{k_0} k^{-1}(1+\log k)^{-1/2}\, \|T\| + \varepsilon c \sum_{k=k_0+1}^{n} k^{-1}(1+\log k)^{-1/2}\\ &\leq 2c\sqrt{1+\log k_0}\, \|T\| + 2\varepsilon c \sqrt{1+\log n} \leq 2\varepsilon c(\|T\|+1)\sqrt{1+\log n}\end{aligned}$$

for $n > n_0$. So $T \in \mathfrak{QGR}$.

Recall that

$$\varrho(l_\infty^N|\mathfrak{I}_n, \mathfrak{R}_n) = \min\{\sqrt{n}, \sqrt{N}\} \qquad \text{(by 4.2.10)}$$

and

$$\varrho(l_\infty^N|\mathfrak{G}_n, \mathfrak{R}_n) \asymp \min\{\sqrt{1+\log n}, \sqrt{1+\log N}\} \qquad \text{(by 4.8.13).}$$

Hence ($\triangle$) in 1.2.12 yields

$$\varrho(C_{1/2}|\mathfrak{I}_n, \mathfrak{R}_n) \asymp n^{1/2}(1+\log n)^{-1/2} \quad \text{and} \quad \varrho(C_{1/2}|\mathfrak{G}_n, \mathfrak{R}_n) \asymp 1.$$

Thus inclusion (1) is strict. On the other hand, Example 4.8.14 shows that inclusion (2) is strict as well.

4.8.21 THEOREM. $\displaystyle\bigcup_{2 \leq q < \infty} \mathsf{RC}_q = \mathsf{GR} = \mathsf{QGR} = \mathsf{RC}.$

PROOF. Since

$$\bigcup_{2 \leq q < \infty} \mathsf{RC}_q = \mathsf{RC} \qquad \text{by (4.6.1),}$$

the assertion follows from the above theorem via a sandwich argument.

4.8.22 Finally, we join the previous results with 4.6.7. For an analogous statement, the reader is referred to 4.14.10.

THEOREM. *For every Banach space X, the following are equivalent:*

(1) *X has Rademacher subcotype.*
(2) *X has non-trivial Rademacher cotype.*
(3) *X does not contain the spaces l_∞^n uniformly.*
(4) *X is* MP*-convex.*
(5) $\varrho(X|\mathfrak{G}_n, \mathfrak{R}_n) = O(1)$.
(6) $\varrho(X|\mathfrak{G}_n, \mathfrak{R}_n) = o(\sqrt{1+\log n})$.

4.9 Gauss type and cotype ideal norms

4.9.1 Now the results about Rademacher type and cotype, presented in Section 4.2, will be transferred to the Gauss system. In other words, we study the **Gauss type ideal norm** $\boldsymbol{\varrho}(\mathcal{G}_n, \mathcal{I}_n)$ and the **Gauss cotype ideal norm** $\boldsymbol{\varrho}(\mathcal{I}_n, \mathcal{G}_n)$. Note that, in view of

$$\|(x_k)|\mathcal{G}_n\| = \|(x_k)|\mathcal{Q}_n\|,$$

the orthonormal system $\mathcal{G}_n$ can be replaced by $\mathcal{Q}_n$. In some cases, this trick yields better constants.

4.9.2 First of all, we stress that many results from Section 4.2 hold verbatim for the Gauss system. For instance, the doubling inequalities from 4.2.3 read as follows.

PROPOSITION.

$$\boldsymbol{\varrho}(\mathcal{G}_{2n}, \mathcal{I}_{2n}) \le \sqrt{2}\, \boldsymbol{\varrho}(\mathcal{G}_n, \mathcal{I}_n) \quad \textit{and} \quad \boldsymbol{\varrho}(\mathcal{I}_{2n}, \mathcal{G}_{2n}) \le \sqrt{2}\, \boldsymbol{\varrho}(\mathcal{I}_n, \mathcal{G}_n).$$

4.9.3 We also mention a counterpart of 4.2.4 which can be deduced from 3.7.5 and the fact that $\|(x_k)|\overline{\mathcal{G}}_n\| = \|(x_k)|\mathcal{G}_n\|$.

PROPOSITION.

$$\boldsymbol{\varrho}(\mathcal{I}_n, \mathcal{G}_n) \le \boldsymbol{\varrho}'(\mathcal{G}_n, \mathcal{I}_n) \le \boldsymbol{\varrho}(\mathcal{I}_n, \mathcal{G}_n) \circ \boldsymbol{\delta}(\mathcal{G}_n, \mathcal{G}_n).$$

4.9.4 The next result tells us that there is almost no difference between the type ideal norms $\boldsymbol{\varrho}(\mathcal{R}_n, \mathcal{I}_n)$ and $\boldsymbol{\varrho}(\mathcal{G}_n, \mathcal{I}_n)$. We know from 4.7.2 that $M_1^{\mathbb{R}} = \sqrt{\frac{2}{\pi}}$ and $M_1^{\mathbb{C}} = \sqrt{\frac{\pi}{4}}$.

PROPOSITION. *The sequences of the ideal norms* $\boldsymbol{\varrho}(\mathcal{P}_n, \mathcal{I}_n)$ *and* $\boldsymbol{\varrho}(\mathcal{G}_n, \mathcal{I}_n)$ *are uniformly equivalent. More precisely,*

$$\boldsymbol{\varrho}(\mathcal{G}_n, \mathcal{I}_n) \le \boldsymbol{\varrho}(\mathcal{P}_n, \mathcal{I}_n) \le M_1\, \boldsymbol{\varrho}(\mathcal{G}_n, \mathcal{I}_n).$$

PROOF. The left-hand inequality is a corollary of 3.7.9. Moreover, by 3.3.5 and 4.8.3, we have

$$\boldsymbol{\varrho}(\mathcal{P}_n, \mathcal{I}_n) \le \boldsymbol{\varrho}(\mathcal{P}_n, \mathcal{G}_n) \circ \boldsymbol{\varrho}(\mathcal{G}_n, \mathcal{I}_n) \le M_1\, \boldsymbol{\varrho}(\mathcal{G}_n, \mathcal{I}_n).$$

REMARK. Recall from 4.2.14 that

$$\boldsymbol{\varrho}(\mathcal{P}_n(e^{i\tau}), \mathcal{I}_n) \le \boldsymbol{\varrho}(\mathcal{P}_n(\pm 1), \mathcal{I}_n) \le \tfrac{\pi}{2}\boldsymbol{\varrho}(\mathcal{P}_n(e^{i\tau}), \mathcal{I}_n).$$

4.9.5 For cotype ideal norms only half of the above result remains true, since on the right-hand side there occurs the factor $N_\infty(\mathcal{G}_n) \asymp \sqrt{1+\log n}$.

PROPOSITION. $\boldsymbol{\varrho}(\mathcal{I}_n, \mathcal{G}_n) \le \boldsymbol{\varrho}(\mathcal{I}_n, \mathcal{P}_n) \le N_\infty(\mathcal{G}_n)\, \boldsymbol{\varrho}(\mathcal{I}_n, \mathcal{G}_n)$.

PROOF. The left-hand inequality is a corollary of 3.7.9. Moreover, by 3.3.5 and 4.8.7, we have

$$\varrho(\mathcal{I}_n, \mathcal{P}_n) \le \varrho(\mathcal{I}_n, \mathcal{G}_n) \circ \varrho(\mathcal{G}_n, \mathcal{P}_n) \le N_\infty(\mathcal{G}_n)\, \varrho(\mathcal{I}_n, \mathcal{G}_n).$$

REMARK. Recall from 4.2.14 that

$$\varrho(\mathcal{I}_n, \mathcal{P}_n(e^{i\tau})) \le \varrho(\mathcal{I}_n, \mathcal{P}_n(\pm 1)) \le \tfrac{\pi}{2}\varrho(\mathcal{I}_n, \mathcal{P}_n(e^{i\tau})).$$

4.9.6 The historical roots of the following result go back to a paper of H. König dealing with eigenvalue distributions of γ-summing operators; see [koe 1, pp. 313–315]. The very elegant proof is due to A. Hinrichs. Since $\varrho(L_\infty|\mathcal{I}_n, \mathcal{R}_n) = \sqrt{n}$, it follows that the factor $N_\infty(\mathcal{G}_n) \asymp \sqrt{1+\log n}$ in the previous inequality cannot be avoided.

PROPOSITION. $\varrho(\mathcal{I}_n, \mathcal{G}_n) \le \frac{\sqrt{n}}{N_\infty(\mathcal{G}_n)} \asymp \sqrt{\frac{n}{1+\log n}}.$

PROOF. Let

$$A_h := \Big\{ s = (\sigma_1, \dots, \sigma_n) : |\sigma_h| = \max_{1\le k\le n} |\sigma_k| \Big\} \quad \text{for } h = 1, \dots, n.$$

Then $A_1 \cup \ldots \cup A_n = \mathbb{K}^n$ and $\gamma^n(A_h) = \frac{1}{n}$. By symmetry,

$$n \int_{A_h} |g_h^{(n)}(s)|^2 d\gamma^n(s) = \int_{\mathbb{K}^n} \max_{1\le k\le n} |g_k^{(n)}(s)|^2 d\gamma^n(s) = N_\infty(\mathcal{G}_n)^2.$$

Given $x_1, \dots, x_n \in X$, we have

$$\|x_h\, g_h^{(n)}(s)\| \le \frac{1}{2} \left\{ \begin{array}{c} \Big\| x_h\, g_h^{(n)}(s) + \sum\limits_{k\ne h} x_k\, g_k^{(n)}(s) \Big\| \\ + \\ \Big\| x_h\, g_h^{(n)}(s) - \sum\limits_{k\ne h} x_k\, g_k^{(n)}(s) \Big\| \end{array} \right\}.$$

Hence

$$\begin{aligned} \frac{N_\infty(\mathcal{G}_n)}{\sqrt{n}} \|x_h\| &= \Big(\int_{A_h} \|x_h g_h^{(n)}(s)\|^2 d\gamma^n(s) \Big)^{1/2} \\ &\le \frac{1}{2} \left\{ \begin{array}{c} \Big(\int_{A_h} \Big\| x_h\, g_h^{(n)}(s) + \sum\limits_{k\ne h} x_k\, g_k^{(n)}(s) \Big\|^2 d\gamma^n(s) \Big)^{1/2} \\ + \\ \Big(\int_{A_h} \Big\| x_h\, g_h^{(n)}(s) - \sum\limits_{k\ne h} x_k\, g_k^{(n)}(s) \Big\|^2 d\gamma^n(s) \Big)^{1/2} \end{array} \right\} \\ &= \Big(\int_{A_h} \Big\| \sum_{k=1}^n x_k\, g_k^{(n)}(s) \Big\|^2 d\gamma^n(s) \Big)^{1/2}. \end{aligned}$$

Summing over $h=1,\dots,n$, we finally arrive at

$$\frac{N_\infty(\mathcal{G}_n)^2}{n}\sum_{h=1}^{n}\|x_h\|^2 \le \sum_{h=1}^{n}\int\limits_{A_h}\Big\|\sum_{k=1}^{n} x_k\, g_k^{(n)}(s)\Big\|^2 d\gamma^n(s)$$

$$= \int\limits_{\mathbb{K}^n}\Big\|\sum_{k=1}^{n} x_k\, g_k^{(n)}(s)\Big\|^2 d\gamma^n(s).$$

4.9.7 The next result tells us that the above estimate is sharp.

EXAMPLE.

$$\varrho(l_\infty^n|\mathcal{I}_n,\mathcal{G}_n)=\varrho(L_\infty|\mathcal{I}_n,\mathcal{G}_n)=\frac{\sqrt{n}}{N_\infty(\mathcal{G}_n)}\asymp\sqrt{\frac{n}{1+\log n}}.$$

PROOF. The lower estimate of $\varrho(l_\infty^n|\mathcal{I}_n,\mathcal{G}_n)$ follows from

$$\|(u_k^{(n)})|[\mathcal{I}_n,l_\infty^n]\| \le \varrho(l_\infty^n|\mathcal{I}_n,\mathcal{G}_n)\,\|(u_k^{(n)})|[\mathcal{G}_n,l_\infty^n]\|,$$

$$\|(u_k^{(n)})|[\mathcal{I}_n,l_\infty^n]\| = \sqrt{n} \quad\text{and}\quad \|(u_k^{(n)})|[\mathcal{G}_n,l_\infty^n]\| = N_\infty(\mathcal{G}_n).$$

Since every infinite dimensional space L_∞ contains a subspace isometric to l_∞^n, we obtain

$$\varrho(l_\infty^n|\mathcal{I}_n,\mathcal{G}_n)\le\varrho(L_\infty|\mathcal{I}_n,\mathcal{G}_n).$$

4.9.8 We now establish a counterpart of 4.2.7.

EXAMPLE.

$$N_p(\mathcal{G}_n)\; n^{1/p-1/2}\le\varrho(l_p^n|\mathcal{G}_n,\mathcal{I}_n)\le\varrho(L_p|\mathcal{G}_n,\mathcal{I}_n)\le n^{1/p-1/2} \quad \textit{if } 1\le p\le 2,$$

$$N_q(\mathcal{G}_n)^{-1} n^{1/2-1/q}\le\varrho(l_q^n|\mathcal{I}_n,\mathcal{G}_n)\le\varrho(L_q|\mathcal{I}_n,\mathcal{G}_n)\le n^{1/2-1/q} \quad \textit{if } 2\le q\le\infty.$$

PROOF. In view of 3.7.10, it is enough the check the lower estimates

$$N_p(\mathcal{G}_n)n^{1/p-1/2}\le\varrho(l_p^n|\mathcal{G}_n,\mathcal{I}_n) \quad\text{and}\quad N_q(\mathcal{G}_n)^{-1}n^{1/2-1/q}\le\varrho(l_q^n|\mathcal{I}_n,\mathcal{G}_n),$$

which follow from

$$\|(u_k^{(n)})|[\mathcal{G}_n,l_r^n]\| = N_r(\mathcal{G}_n)\,n^{1/r} \quad\text{and}\quad \|(u_k^{(n)})|[\mathcal{I}_n,l_r^n]\| = \sqrt{n}.$$

4.9.9 In the limiting case $p=1$ the above result can be strengthened.

EXAMPLE. $\varrho(l_1^n|\mathcal{G}_n,\mathcal{I}_n)=\varrho(L_1|\mathcal{G}_n,\mathcal{I}_n)=N_1(\mathcal{G}_n)\,\sqrt{n}.$

PROOF. Fix any natural number N, and choose $x_1,\dots,x_n\in l_1^N$ such that

$$\|(x_k)|[\mathcal{G}_n,l_1^N]\| = \varrho(l_1^N|\mathcal{G}_n,\mathcal{I}_n) \quad\text{and}\quad \|(x_k)|[l_2^n,l_1^N]\| = 1.$$

By the extreme point lemma 0.9.5, we may assume that $x_1,\dots,x_n$ are multiples of unit vectors. Thus there exist a map π from $\{1,\dots,n\}$ into

$\{1,\ldots,N\}$ and coefficients $\xi_1,\ldots,\xi_n \in \mathbb{K}$ such that $x_k = \xi_k u^{(n)}_{\pi(k)}$. Note that $\|(\xi_k)|l_2^n\| = \|(x_k)|[l_2^n, l_1^N]\| = 1$ and

$$\|(x_k)|[\mathcal{G}_n, l_1^N]\|^2 = \int\limits_{\mathbb{K}^n} \Big(\sum_{i=1}^{N} \Big| \sum_{\pi(k)=i} \xi_k g_k^{(n)}(s) \Big| \Big)^2 d\gamma^n(s)$$

$$\leq \int\limits_{\mathbb{K}^n} \Big(\sum_{k=1}^{n} \Big| \xi_k g_k^{(n)}(s) \Big| \Big)^2 d\gamma^n(s).$$

Thus we need to evaluate the maximum of a quadratic form on the positive cone of the unit sphere. The corresponding matrix $A_n = (a_{hk})$ is given by

$$a_{hk} := \begin{cases} 1 & \text{if } h = k, \\ a := M_1^2 & \text{if } h \neq k. \end{cases}$$

Observe that $0 < a < 1$ and $A_n = (1-a)I_n + aU_n$, where

$$U_n = \begin{pmatrix} 1 & \dots & 1 \\ \vdots & 1 & \vdots \\ 1 & \dots & 1 \end{pmatrix}.$$

Since $\operatorname{rank} U_n = 1$, there exist $(n-1)$ eigenvectors to the eigenvalue $1-a$. Hence $1+(n-1)a$ is the largest eigenvalue with $(1,\ldots,1)$ as an eigenvector. This observation shows that

$$\max \Big\{ \sum_{h=1}^{n} \sum_{k=1}^{n} |\xi_h| \, a_{hk} \, |\xi_k| \, : \, |\xi_1|^2 + \ldots + |\xi_n|^2 = 1 \Big\} = 1 + (n-1)a.$$

In view of

$$N_1(\mathcal{G}_n) = \sqrt{(1-\tfrac{1}{n})M_1^2 + \tfrac{1}{n}},$$

we therefore arrive at

$$\varrho(l_1^N|\mathcal{G}_n, \mathcal{I}_n) = \|(x_k)|[\mathcal{G}_n, l_1^N]\| \leq N_1(\mathcal{G}_n)\sqrt{n}.$$

Since L_1 is assumed to be infinite dimensional, a standard approximation argument yields

$$\varrho(L_1|\mathcal{G}_n, \mathcal{I}_n) = \lim_{N\to\infty} \varrho(l_1^N|\mathcal{G}_n, \mathcal{I}_n) \leq N_1(\mathcal{G}_n)\sqrt{n}$$

which is the required upper estimate. The lower estimate was proved in the previous paragraph.

4.9.10 Next, we state a counterpart of 4.2.8.

EXAMPLE.

$$1 \le \varrho(L_p|\mathfrak{I}_n, \mathfrak{G}_n) \le M_1^{-1} \qquad \text{if } 1 \le p \le 2,$$
$$M_1\, M_q(\mathfrak{Q}_n) \le \varrho(L_q|\mathfrak{G}_n, \mathfrak{I}_n) \le M_q(\mathfrak{Q}_n) \quad \text{if } 2 \le q \le \infty.$$

PROOF. In the case $1 \le p \le 2$, the assertion follows from 3.3.7 and $M_1 \le M_p$. In the rest of this proof, we replace $\mathfrak{G}_n$ by $\mathfrak{Q}_n$. The upper estimate $\varrho(L_q|\mathfrak{Q}_n, \mathfrak{I}_n) \le M_q(\mathfrak{Q}_n)$ is a consequence of 3.3.8. Substituting the functions

$$f_k(s) := \begin{cases} |q_k^{(n)}(s)|^{q'-2} q_k^{(n)}(s) & \text{if } \sigma_k \ne 0, \\ 0 & \text{if } \sigma_k = 0, \end{cases} \qquad \text{with } k = 1, \ldots, n$$

in the defining inequality of $\varrho(\mathfrak{Q}_n, \mathfrak{I}_n)$, we obtain the lower estimate

$$M_{q'}(\mathfrak{Q}_n)\, M_q(\mathfrak{Q}_n) \le \varrho(L_q(\mathbb{S}^n)|\mathfrak{Q}_n, \mathfrak{I}_n).$$

This unpleasant computation is left to the willing reader; see also 4.12.10. The general case follows from the fact that $L_q(\mathbb{S}^n)$ can be embedded isometrically into every infinite dimensional space L_q. This completes the proof, in view of $M_1 \le M_{q'} \le M_{q'}(\mathfrak{Q}_n)$.

4.9.11 In the limiting case $q = \infty$ the above result can be strengthened.

EXAMPLE. $\varrho(L_\infty|\mathfrak{G}_n, \mathfrak{I}_n) = N_1(\mathfrak{G}_n)\sqrt{n}.$

PROOF. Using the injectivity and the surjectivity of the type ideal norms, the assertion is an easy consequence of 4.9.9 and the following facts: L_∞ contains some L_1 as a subspace, L_∞ is a quotient space of some L_1.

4.9.12 Throughout, we use the symbol $f(n,r) \overset{n,r}{\asymp} g(n,r)$ to indicate that the estimates $f(n,r) \le c\, g(n,r)$ and $g(n,r) \le c\, f(n,r)$ hold for $n = 1, 2, \ldots$, where $c > 0$ is not only independent of n but also of r. With this notation, the preceding examples are summarized in the following table, which should be compared with that in 4.2.9.

	$1 \le r \le 2$	$2 \le r < \infty$	$r = \infty$
$\varrho(L_r\|\mathfrak{G}_n, \mathfrak{I}_n)$	$\overset{n,r}{\asymp} n^{1/r-1/2}$ 4.9.8	$\overset{n,r}{\asymp} M_r(\mathfrak{Q}_n)$ 4.9.10	$= N_1(\mathfrak{G}_n)\sqrt{n}$ 4.9.11
$\varrho(L_r\|\mathfrak{I}_n, \mathfrak{G}_n)$	$\overset{n,r}{\asymp} 1$ 4.9.10	? ? ?	$\asymp \sqrt{\frac{n}{1+\log n}}$ 4.9.7

REMARK. Certainly, it is only of minor interest to compute the constants $\varrho(L_r|\mathfrak{G}_n, \mathfrak{I}_n)$ and $\varrho(L_r|\mathfrak{I}_n, \mathfrak{G}_n)$ exactly. Nevertheless, it would be desirable to remove the question marks in the above table. Thus we risk some conjectures, which are supported by 4.9.7 and 4.9.9:

$$\varrho(l_p^n|\mathcal{G}_n,\mathcal{I}_n) \overset{?}{=} \varrho(L_p|\mathcal{G}_n,\mathcal{I}_n) \overset{?}{=} N_p(\mathcal{G}_n)\, n^{1/p-1/2} \quad \text{if } 1<p\le 2,$$

$$\varrho(l_q^n|\mathcal{I}_n,\mathcal{G}_n) \overset{?}{=} \varrho(L_q|\mathcal{I}_n,\mathcal{G}_n) \overset{?}{=} N_q(\mathcal{G}_n)^{-1} n^{1/2-1/q} \quad \text{if } 2\le q<\infty.$$

Moreover, in analogy with 3.8.16 and 4.2.8, it seems likely that

$$\varrho(L_p(\mathbb{S}^n)|\mathcal{I}_n,\mathcal{G}_n) \overset{?}{=} \varrho(L_p|\mathcal{I}_n,\mathcal{G}_n) \overset{?}{=} M_p(\mathcal{Q}_n) \quad \text{if } 1\le p\le 2,$$

$$\varrho(L_q(\mathbb{S}^n)|\mathcal{G}_n,\mathcal{I}_n) \overset{?}{=} \varrho(L_q|\mathcal{G}_n,\mathcal{I}_n) \overset{?}{=} M_q(\mathcal{Q}_n) \quad \text{if } 2\le q\le \infty.$$

4.9.13 In the following, let n and N be natural numbers.

EXAMPLE. $\varrho(l_\infty^N|\mathcal{I}_n,\mathcal{G}_n) \asymp \min\Big\{\sqrt{\frac{n}{1+\log n}}, \sqrt{\frac{N}{1+\log N}}\Big\}.$

PROOF. By 4.9.6,

$$\varrho(X|\mathcal{I}_n,\mathcal{G}_n) \prec \sqrt{\frac{n}{1+\log n}}$$

for all Banach spaces X. If $n \ge N$, then we conclude from 9.3.14 that

$$\varrho(l_\infty^N|\mathcal{I}_n,\mathcal{G}_n) \le \sqrt{2}\,\varrho(l_\infty^N|\mathcal{I}_N,\mathcal{G}_N) \prec \sqrt{\frac{N}{1+\log N}}.$$

This yields the upper estimate. On the other hand, for $m := \min\{n, N\}$, we have

$$\sqrt{\frac{m}{1+\log m}} \prec \varrho(l_\infty^m|\mathcal{I}_m,\mathcal{G}_m) \le \varrho(l_\infty^N|\mathcal{I}_n,\mathcal{G}_n) \qquad \text{(by 4.9.7)}.$$

4.9.14 Finally, we present a counterpart of the table given in 4.2.12.

	l_1^N	l_∞^N
$\varrho(\mathcal{G}_n,\mathcal{I}_n)$	$\asymp \min\{\sqrt{n}, \sqrt{N}\}$ 4.9.4	$\asymp \min\{\sqrt{n}, \sqrt{1+\log N}\}$ 4.9.4
$\varrho(\mathcal{I}_n,\mathcal{G}_n)$	$\asymp 1$ 4.9.10	$\asymp \min\{\sqrt{\frac{n}{1+\log n}}, \sqrt{\frac{N}{1+\log N}}\}$ 4.9.13

4.10 Operators of Gauss type and cotype – Kwapień's theorem –

4.10.1 Obviously, the definitions from Section 4.3 can be repeated for the Gauss system. In particular, we get the Banach operator ideals $\mathfrak{GT}_p$, $\mathfrak{GT}_p^{weak}$ and $\mathfrak{GT}$. However, in view of 4.9.4, this is more or less superfluous.

PROPOSITION.

$\mathfrak{GT}_p = \mathfrak{RT}_p$, $\mathfrak{GT}_p^{weak} = \mathfrak{RT}_p^{weak}$ *and* $\mathfrak{GT} = \mathfrak{RT}$.

4.10.2 The same is possible for the definitions from Section 4.5. Then we obtain the Banach operator ideals $\mathfrak{GC}_q$ and $\mathfrak{GC}_q^{weak}$, but now the situation is more complicated.

THEOREM.

$\mathfrak{GC}_{q_0} \subset \mathfrak{RC}_q \subset \mathfrak{GC}_q$ *and* $\mathfrak{GC}_{q_0}^{weak} \subset \mathfrak{RC}_q^{weak} \subset \mathfrak{GC}_q^{weak}$ *if* $2<q_0<q<\infty$.

In particular,

$$\bigcup_{2\leq q<\infty} \mathfrak{RC}_q = \bigcup_{2<q<\infty} \mathfrak{RC}_q^{weak} = \bigcup_{2\leq q<\infty} \mathfrak{GC}_q = \bigcup_{2<q<\infty} \mathfrak{GC}_q^{weak}.$$

PROOF. In the weak case, the inclusions follow from

$$\varrho(\mathcal{I}_n, \mathcal{G}_n) \leq \varrho(\mathcal{I}_n, \mathcal{P}_n) \prec \sqrt{1+\log n}\, \varrho(\mathcal{I}_n, \mathcal{G}_n) \qquad \text{(by 4.9.5)},$$

and the classical case can be treated in the same way. Diagonal operators show that all inclusions are strict; see 4.10.4.

4.10.3 In the setting of spaces, the preceding result can be improved, and we get a full counterpart of 4.10.1. See also 4.10.11.

PROPOSITION. $\mathsf{GC}_q = \mathsf{RC}_q$ *and* $\mathsf{GC}_q^{weak} = \mathsf{RC}_q^{weak}$.

PROOF. Assume that X has weak Gauss cotype q_0. Then it has Rademacher cotype for all $q_1 > q_0$. Thus, by 4.8.16, there exists a constant $c \geq 1$ such that

$$\varrho(X|\mathcal{G}_n, \mathcal{R}_n) \leq c \quad \text{for } n = 1, 2, \ldots .$$

Now it follows from

$$\varrho(X|\mathcal{I}_n, \mathcal{R}_n) \leq \varrho(X|\mathcal{I}_n, \mathcal{G}_n)\, \varrho(X|\mathcal{G}_n, \mathcal{R}_n)$$

that for X the concepts of weak Rademacher cotype and weak Gauss cotype coincide. The classical case can be treated similarly.

4.10.4 Let D_α denote the diagonal operator on $[l_2, l_\infty^{2^k}]$ defined by

$$D_\alpha : (x_k) \longrightarrow (2^{-k\alpha} x_k), \quad \text{where} \quad x_k \in l_\infty^{2^k}.$$

We now show that for operators the Rademacher system and the Gauss system yield different concepts of cotype.

EXAMPLE. *If* $0 \leq \alpha \leq 1/2$, *then*

$$\varrho(D_\alpha|\mathcal{I}_n, \mathcal{R}_n) \asymp n^{1/2-\alpha} \quad \textit{and} \quad \varrho(D_\alpha|\mathcal{I}_n, \mathcal{G}_n) \asymp n^{1/2-\alpha}(1+\log n)^{-1/2}.$$

PROOF. Recall that

$$\varrho(l_\infty^N|\mathcal{I}_n, \mathcal{R}_n) = \min\{\sqrt{n}, \sqrt{N}\} \qquad \text{(by 4.2.10)}$$

and

$$\varrho(l_\infty^N|\mathcal{I}_n,\mathcal{G}_n) \asymp \min\left\{\sqrt{\tfrac{n}{1+\log n}},\sqrt{\tfrac{N}{1+\log N}}\right\} \qquad \text{(by 4.9.13).}$$

The desired relations are now consequences of $(\triangle)$ in 1.2.12.

4.10.5 Next, we establish **Kwapień's inequality**, which is one of the most fundamental results presented in this book; see [kwa 2]. The ideal norm $\boldsymbol{\kappa}_n$ was defined in 2.3.15. The following proof illustrates clearly that, in certain cases, the Gauss system has some advantage over the Rademacher system.

THEOREM. $\boldsymbol{\kappa}_n \le \varrho(\mathcal{I}_n,\mathcal{G}_n)\circ\varrho(\mathcal{G}_n,\mathcal{I}_n) \le \varrho(\mathcal{I}_n,\mathcal{P}_n)\circ\varrho(\mathcal{P}_n,\mathcal{I}_n)$.

PROOF. Since the Gaussian measure γ^n is invariant under transforms associated with matrices $A_n = (\alpha_{hk}) \in \mathbb{U}^n$, we have

$$\left(\sum_{h=1}^n \Big\|\sum_{k=1}^n \alpha_{hk} STx_k\Big\|^2\right)^{1/2} \le$$

$$\le \varrho(S|\mathcal{I}_n,\mathcal{G}_n)\left(\int_{\mathbb{K}^n}\Big\|\sum_{h=1}^n\Big[\sum_{k=1}^n \alpha_{hk}Tx_k\Big]g_h^{(n)}(s)\Big\|^2 d\gamma^n(s)\right)^{1/2}$$

$$= \varrho(S|\mathcal{I}_n,\mathcal{G}_n)\left(\int_{\mathbb{K}^n}\Big\|\sum_{k=1}^n Tx_k g_k^{(n)}(s)\Big\|^2 d\gamma^n(s)\right)^{1/2}$$

$$\le \varrho(S|\mathcal{I}_n,\mathcal{G}_n)\,\varrho(T|\mathcal{G}_n,\mathcal{I}_n)\left(\sum_{k=1}^n \|x_k\|^2\right)^{1/2},$$

which implies that

$$\boldsymbol{\kappa}_n \le \varrho(\mathcal{I}_n,\mathcal{G}_n)\circ\varrho(\mathcal{G}_n,\mathcal{I}_n).$$

Finally, we know from 4.9.4 and 4.9.5 that

$$\varrho(\mathcal{G}_n,\mathcal{I}_n) \le \varrho(\mathcal{P}_n,\mathcal{I}_n) \quad\text{and}\quad \varrho(\mathcal{I}_n,\mathcal{G}_n) \le \varrho(\mathcal{I}_n,\mathcal{P}_n).$$

REMARK. Note that every surjection Q from l_1 onto l_q with $2 < q < \infty$ is the product of a Rademacher cotype 2 operator and a Rademacher type 2 operator, but fails to be Hilbertian. Thus the order of the ideal norms in the inequality

$$\boldsymbol{\kappa}_n \le \varrho(\mathcal{I}_n,\mathcal{G}_n)\circ\varrho(\mathcal{G}_n,\mathcal{I}_n) \le \varrho(\mathcal{I}_n,\mathcal{P}_n)\circ\varrho(\mathcal{P}_n,\mathcal{I}_n)$$

cannot be interchanged.

4.10.6 The following product formula, which goes back to H. König/J. R. Retherford/N. Tomczak-Jaegermann, is an immediate consequence of the preceding inequality; see [koe*b, pp. 114–115].

PROPOSITION.

Let $1 < p < 2 < q < \infty$ *and* $1/r - 1/2 = 1/p - 1/q < 1/2$. *Then*

$$\mathfrak{RC}_q^{weak} \circ \mathfrak{RT}_p^{weak} \subseteq \mathfrak{GC}_q^{weak} \circ \mathfrak{GT}_p^{weak} \subseteq \mathfrak{KT}_r^{weak}.$$

REMARK. At first sight, the assumption $1/p - 1/q < 1/2$ looks quite artificial, but we will see in 5.9.7 (Remark) that, actually, there is no function $f : (1,2] \times [2,\infty) \to (0,1/2)$ with the property that

$$\mathfrak{RC}_q^{weak} \circ \mathfrak{RT}_p^{weak} \subseteq \mathfrak{KR}_{f(p,q)}^{weak}$$

if $1/2 \le 1/p - 1/q$. However, in the setting of Banach lattices, G. Pisier showed that

$$\mathsf{RC}_q^{weak} \cap \mathsf{RT}_p^{weak} \subseteq \mathsf{KT}_r^{weak},$$

where $1/r := \max\{1/p, 1/q'\}$; see [pis 9, p. 14] and [pis 10, p. 275].

4.10.7 In the limiting case $p = q = 2$, we obtain the result that was the historical starting point of the theory presented in this book.

KWAPIEŃ'S THEOREM.

$$\mathfrak{RC}_2 \circ \mathfrak{RT}_2 = \mathfrak{H} \quad \textit{and} \quad \mathsf{RT}_2 \cap \mathsf{RC}_2 = \mathsf{H}.$$

REMARK. It follows from 4.10.5 (Remark) that $\mathfrak{RT}_2 \circ \mathfrak{RC}_2 \not\subseteq \mathfrak{H}$.

4.10.8 In 4.10.2 we introduced the Banach operator ideals $\mathfrak{GC}_q$ and $\mathfrak{GC}_q^{weak}$, but not $\mathfrak{GC}$. The reason for this sin of omission is to be found in Proposition 4.9.6, which says that

$$\varrho(\mathcal{I}_n, \mathcal{G}_n) \prec \sqrt{\tfrac{n}{1+\log n}}.$$

Thus 1.2.8 suggests the following definition.

An operator T has **Gauss subcotype** if

$$\varrho(T|\mathcal{I}_n, \mathcal{G}_n) = o\left(\sqrt{\tfrac{n}{1+\log n}}\right).$$

These operators form the closed ideal

$$\mathfrak{GC} := \mathfrak{L}_0\left[\sqrt{\tfrac{1+\log n}{n}}\, \varrho(\mathcal{I}_n, \mathcal{G}_n)\right].$$

In the rest of this section we show how to overcome the seeming trouble made by the additional term $\sqrt{1+\log n}$. Luckily, it turns out that everything is fine. All results are due to A. Hinrichs.

4.10.9 At first sight, it is surprising that the right-hand expression in the following inequality does not depend on $m = 1, 2, \ldots$.

LEMMA. $\varrho(\mathcal{G}_m \overset{\text{full}}{\otimes} \mathcal{I}_n, \mathcal{G}_{mn}) \le \frac{\sqrt{n}}{N_\infty(\mathcal{G}_n)} \varrho(\mathcal{G}_n, \mathcal{R}_n).$

PROOF. Let

$$A_h := \Big\{ s = (\sigma_1, \ldots, \sigma_n) : |\sigma_h| = \max_{1 \le k \le n} |\sigma_k| \Big\} \quad \text{for } h = 1, \ldots, n.$$

Then $A_1 \cup \ldots \cup A_n = \mathbb{K}^n$ and $\gamma^n(A_h) = \frac{1}{n}$. By symmetry,

$$n \int\limits_{A_h} |g_h^{(n)}(s)|^2 d\gamma^n(s) = \int\limits_{\mathbb{K}^n} \max_{1 \le k \le n} |g_k^{(n)}(s)|^2 d\gamma^n(s) = N_\infty(\mathcal{G}_n)^2.$$

Given $x_1, \ldots, x_{mn} \in X$, we let

$$x_k(w) := \sum_{i=1}^{m} x_{m(k-1)+i} \; g_{m(k-1)+i}^{(mn)}(w) \quad \text{for } k = 1, \ldots, n \text{ and } w \in \mathbb{K}^{mn}.$$

Note that

$$\|Tx_h(w)\, g_h^{(n)}(s)\| \le \frac{1}{2} \left\{ \begin{array}{c} \Big\| Tx_h(w)\, g_h^{(n)}(s) + \sum\limits_{k \ne h} Tx_k(w)\, g_k^{(n)}(s) \Big\| \\ + \\ \Big\| Tx_h(w)\, g_h^{(n)}(s) - \sum\limits_{k \ne h} Tx_k(w)\, g_k^{(n)}(s) \Big\| \end{array} \right\}.$$

Hence

$$\frac{N_\infty(\mathcal{G}_n)}{\sqrt{n}} \Big(\int\limits_{\mathbb{K}^m} \Big\| \sum_{i=1}^{m} Tx_{m(h-1)+i} \; g_i^{(m)}(u) \Big\|^2 d\gamma^m(u) \Big)^{1/2} =$$

$$= \Big(\int\limits_{A_h} \int\limits_{\mathbb{K}^{mn}} \Big\| \sum_{i=1}^{m} Tx_{m(h-1)+i} \; g_{m(h-1)+i}^{(mn)}(w) g_h^{(n)}(s) \Big\|^2 d\gamma^n(s)\, d\gamma^{mn}(w) \Big)^{1/2}$$

$$\le \frac{1}{2} \left\{ \begin{array}{c} \Big(\int\limits_{A_h} \int\limits_{\mathbb{K}^{mn}} \Big\| Tx_h(w)\, g_h^{(n)}(s) + \sum\limits_{k \ne h} Tx_k(w)\, g_k^{(n)}(s) \Big\|^2 d\gamma^n(s)\, d\gamma^{mn}(w) \Big)^{1/2} \\ + \\ \Big(\int\limits_{A_h} \int\limits_{\mathbb{K}^{mn}} \Big\| Tx_h(w)\, g_h^{(n)}(s) - \sum\limits_{k \ne h} Tx_k(w)\, g_k^{(n)}(s) \Big\|^2 d\gamma^n(s)\, d\gamma^{mn}(w) \Big)^{1/2} \end{array} \right\}$$

$$= \Big(\int\limits_{A_h} \int\limits_{\mathbb{K}^{mn}} \Big\| \sum_{k=1}^{n} Tx_k(w)\, g_k^{(n)}(s) \Big\|^2 d\gamma^n(s)\, d\gamma^{mn}(w) \Big)^{1/2}.$$

Summing over $h = 1, \ldots, n$, we finally arrive at

$$\frac{N_\infty(\mathcal{G}_n)^2}{n} \, \|(Tx_{m(h-1)+i}) | \mathcal{G}_m \overset{\text{full}}{\otimes} \mathcal{I}_n\|^2 =$$

$$= \frac{N_\infty(\mathcal{G}_n)^2}{n} \int\limits_{\mathbb{K}^m} \sum_{h=1}^{n} \Big\| \sum_{i=1}^{m} Tx_{m(h-1)+i} \, g_i^{(m)}(u) \Big\|^2 d\gamma^m(u)$$

$$\leq \sum_{h=1}^{n} \int\limits_{A_h} \int\limits_{\mathbb{K}^{mn}} \Big\| \sum_{k=1}^{n} Tx_k(w) \, g_k^{(n)}(s) \Big\|^2 d\gamma^n(s) \, d\gamma^{mn}(w)$$

$$= \int\limits_{\mathbb{K}^n} \int\limits_{\mathbb{K}^{mn}} \Big\| \sum_{k=1}^{n} Tx_k(w) \, g_k^{(n)}(s) \Big\|^2 d\gamma^n(s) \, d\gamma^{mn}(w)$$

$$\leq \varrho(T|\mathcal{G}_n, \mathcal{R}_n)^2 \int\limits_0^1 \int\limits_{\mathbb{K}^{mn}} \Big\| \sum_{k=1}^{n} x_k(w) \, r_k(t) \Big\|^2 dt \, d\gamma^{mn}(w)$$

$$= \varrho(T|\mathcal{G}_n, \mathcal{R}_n)^2 \int\limits_{\mathbb{K}^{mn}} \Big\| \sum_{k=1}^{n} \sum_{i=1}^{m} x_{m(k-1)+i} \, g_{m(k-1)+i}^{(mn)}(w) \Big\|^2 d\gamma^{mn}(w)$$

$$= \varrho(T|\mathcal{G}_n, \mathcal{R}_n)^2 \, \|(x_{m(h-1)+i}) | \mathcal{G}_{mn}\|^2.$$

4.10.10 LEMMA. $\varrho(\mathcal{I}_{mn}, \mathcal{G}_{mn}) \leq \frac{\sqrt{n}}{N_\infty(\mathcal{G}_n)} \, \varrho(\mathcal{I}_m, \mathcal{G}_m) \circ \varrho(\mathcal{G}_n, \mathcal{R}_n).$

PROOF. We conclude from $\mathcal{I}_{mn} = \mathcal{I}_m \overset{\text{full}}{\otimes} \mathcal{I}_n$,

$$\varrho(\mathcal{I}_m \overset{\text{full}}{\otimes} \mathcal{I}_n, \mathcal{G}_m \overset{\text{full}}{\otimes} \mathcal{I}_n) \leq \varrho(\mathcal{I}_m, \mathcal{G}_m)$$

and

$$\varrho(\mathcal{G}_m \overset{\text{full}}{\otimes} \mathcal{I}_n, \mathcal{G}_{mn}) \leq \frac{\sqrt{n}}{N_\infty(\mathcal{G}_n)} \, \varrho(\mathcal{G}_n, \mathcal{R}_n)$$

that

$$\begin{aligned} \varrho(\mathcal{I}_{mn}, \mathcal{G}_{mn}) &= \varrho(\mathcal{I}_m \overset{\text{full}}{\otimes} \mathcal{I}_n, \mathcal{G}_{mn}) \\ &\leq \varrho(\mathcal{I}_m \overset{\text{full}}{\otimes} \mathcal{I}_n, \mathcal{G}_m \overset{\text{full}}{\otimes} \mathcal{I}_n) \circ \varrho(\mathcal{G}_m \overset{\text{full}}{\otimes} \mathcal{I}_n, \mathcal{G}_{mn}) \\ &\leq \frac{\sqrt{n}}{N_\infty(\mathcal{G}_n)} \, \varrho(\mathcal{I}_m, \mathcal{G}_m) \circ \varrho(\mathcal{G}_n, \mathcal{R}_n). \end{aligned}$$

WARNING. Do not mistake $\mathcal{G}_{mn}$ for $\mathcal{G}_m \overset{\text{full}}{\otimes} \mathcal{G}_n$.

4.10.11 Now we are able to show that the concept of Gauss subcotype, introduced in 4.10.8, was indeed the right one.

THEOREM. $\mathfrak{GC} = \mathfrak{RC} = \mathfrak{QGR}$.

PROOF. We know from 4.9.5 that

$$\varrho(\mathcal{I}_n, \mathcal{R}_n) \prec \sqrt{1 + \log n} \, \varrho(\mathcal{I}_n, \mathcal{G}_n).$$

Hence $\mathfrak{GC} \subseteq \mathfrak{RC}$ is trivial. The equality $\mathfrak{RC} = \mathfrak{QGR}$ was proved in 4.8.20. Thus it remains to show that $\mathfrak{QGR} \subseteq \mathfrak{GC}$. Indeed, by 4.9.6 and 4.10.10, we have

$$\varrho(\mathcal{I}_n, \mathcal{G}_n) \prec \sqrt{\tfrac{n}{1+\log n}} \quad \text{and} \quad \varrho(\mathcal{I}_{n^2}, \mathcal{G}_{n^2}) \prec \sqrt{\tfrac{n}{1+\log n}}\, \varrho(\mathcal{I}_n, \mathcal{G}_n) \circ \varrho(\mathcal{G}_n, \mathcal{R}_n).$$

Therefore

$$\varrho(\mathcal{I}_{n^2}, \mathcal{G}_{n^2}) \prec \sqrt{\tfrac{n^2}{1+\log n^2}} \left[\frac{1}{\sqrt{1+\log n}}\, \varrho(\mathcal{G}_n, \mathcal{R}_n) \right].$$

This means that

$$\varrho(T|\mathcal{G}_n, \mathcal{R}_n) = o(\sqrt{1+\log n}) \quad \text{implies} \quad \varrho(\mathcal{I}_{n^2}, \mathcal{G}_{n^2}) = o\left(\sqrt{\tfrac{n^2}{1+\log n^2}} \right).$$

By monotonicity, the last relation does not hold for n^2 only, but for all n.

4.11 Sidon constants

4.11.1 Let $\mathcal{A}_n = (a_1, \dots, a_n)$ be a system of characters defined on a compact Abelian group $\mathbb{G}$. Then there exists a constant $S \geq 1$ such that

$$\sum_{k=1}^{n} |\xi_k| \leq S \sup \left\{ \left| \sum_{k=1}^{n} \xi_k a_k(s) \right| : s \in \mathbb{G} \right\}$$

whenever $\xi_1, \dots, \xi_n \in \mathbb{K}$. We refer to

$$S(\mathcal{A}_n) := \inf S$$

as the **Sidon constant** of $\mathcal{A}_n$.

REMARK. If not otherwise specified, the scalar field is assumed to be complex.

4.11.2 PROPOSITION. $1 \leq S(\mathcal{A}_n) \leq \sqrt{n}$.

PROOF. We have

$$\sum_{k=1}^{n} |\xi_k| \leq \sqrt{n} \Big(\sum_{k=1}^{n} |\xi_k|^2 \Big)^{1/2} = \sqrt{n} \Big(\int_{\mathbb{G}} \Big| \sum_{k=1}^{n} \xi_k a_k(s) \Big|^2 d\mu(s) \Big)^{1/2}$$

$$\leq \sqrt{n} \sup \left\{ \left| \sum_{k=1}^{n} \xi_k a_k(s) \right| : s \in \mathbb{G} \right\}.$$

4.11.3 Obviously, the Steinhaus system has the best possible Sidon constant.

EXAMPLE. $S(\mathcal{P}_n(e^{i\tau})) = 1.$

PROOF. In fact, we even get

$$\sum_{k=1}^{n}|\xi_k| = \sup\left\{\left|\sum_{k=1}^{n}\xi_k e^{i\tau_k}\right| : \tau_1,\dots\tau_n \in \mathbb{R}\right\}.$$

4.11.4 The above observation remains true for the Bernoulli system when we only allow real scalars. The complex case, however, requires some computation.

EXAMPLE. $S(\mathcal{P}_n(\pm 1)) \le \frac{\pi}{2}$.

PROOF. Given $\xi_1,\dots,\xi_n \in \mathbb{C}$, we choose $t_1,\dots,t_n \in (-\pi,+\pi]$ such that $\xi_k = |\xi_k|e^{it_k}$. Next, for every $t \in (-\pi,+\pi]$, the sign $\varepsilon_k(t)$ is defined by $\varepsilon_k(t)\cos(t+t_k) = |\cos(t+t_k)|$. Then

$$\begin{aligned}\sup_{\mathbb{E}_2^n}\left|\sum_{k=1}^{n}\xi_k\varepsilon_k\right| &= \sup_{\mathbb{E}_2^n}\left|\sum_{k=1}^{n}\xi_k e^{it}\varepsilon_k\right| \ge \left|\sum_{k=1}^{n}|\xi_k|e^{i(t+t_k)}\varepsilon_k(t)\right| \\ &= \left|\sum_{k=1}^{n}|\xi_k|\cos(t+t_k)\varepsilon_k(t) + i\sum_{k=1}^{n}|\xi_k|\sin(t+t_k)\varepsilon_k(t)\right| \\ &\ge \sum_{k=1}^{n}|\xi_k|\,|\cos(t+t_k)|.\end{aligned}$$

Integration over t yields

$$\sup_{\mathbb{E}_2^n}\left|\sum_{k=1}^{n}\xi_k\varepsilon_k\right| \ge \sum_{k=1}^{n}|\xi_k|\,\frac{1}{2\pi}\int_{-\pi}^{+\pi}|\cos(t+t_k)|\,dt = \frac{2}{\pi}\sum_{k=1}^{n}|\xi_k|.$$

4.11.5 The following consequence of the previous result was used in the proof of the contraction principle (mixed case); see 3.5.4.

LEMMA. *For $\alpha_1,\dots,\alpha_n \in \mathbb{C}$ with $|\alpha_k| \le 1$ there exists an $\mathbb{E}_2^n$-tuple of complex numbers λ_e such that*

$$\sum_{\mathbb{E}_2^n}|\lambda_e| \le \tfrac{\pi}{2} \quad \textit{and} \quad \alpha_k = \sum_{\mathbb{E}_2^n}\lambda_e\,\varepsilon_k \quad \textit{for } k=1,\dots,n.$$

PROOF. Consider the operator $J : l_1^n \to l_\infty(\mathbb{E}_2^n)$ which transforms $u_k^{(n)}$ into $p_k^{(n)}$. The estimate $S(\mathcal{P}_n(\pm 1)) \le \frac{\pi}{2}$ means that $\|x\| \le \frac{\pi}{2}\|Jx\|$ for all $x \in l_1^n$. Hence, by [PIE 2, p. 27], we have $U_\infty^n \subseteq \frac{\pi}{2}\, J'(U_1(\mathbb{E}_2^n))$, where U_∞^n and $U_1(\mathbb{E}_2^n)$ denote the closed unit balls of l_∞^n and $l_1(\mathbb{E}_2^n)$, respectively. This is the required result, since $J' : l_1(\mathbb{E}_2^n) \to l_\infty^n$ sends the unit vector u_e into e.

4.11.6 The preceding examples and the following result show that, in a sense, the Sidon constant $S(\mathcal{A}_n)$ measures the *lacunarity* of $\mathcal{A}_n$.

PROPOSITION. *Let $\mathcal{A}_n$ be the system constituted by* **all** *characters of a finite Abelian group $\mathbb{G}_n$. Then*

$$\sqrt{\tfrac{n}{2}} \le S(\mathcal{A}_n) \le \sqrt{n}.$$

PROOF. Fix any enumeration $\mathbb{G}_n = \{s_1, \dots, s_n\}$. Substituting

$$\xi_k := \sum_{h=1}^{n} a_k(-s_h)\varepsilon_h \quad \text{with} \quad e = (\varepsilon_1, \dots, \varepsilon_n) \in \mathbb{E}_2^n$$

in

$$\sum_{k=1}^{n} |\xi_k| \le S(\mathcal{A}_n) \sup_{1 \le l \le n} \Big| \sum_{k=1}^{n} \xi_k a_k(s_l) \Big|$$

and using the orthogonality relation

$$\sum_{k=1}^{n} a_k(t) = \begin{cases} n & \text{if } t = 1, \\ 0 & \text{if } t \ne 1, \end{cases}$$

yields

$$\sum_{k=1}^{n} \Big| \sum_{h=1}^{n} a_k(-s_h)\varepsilon_h \Big| \le S(\mathcal{A}_n) \sup_{1 \le l \le n} \Big| \sum_{h=1}^{n} \varepsilon_h \sum_{k=1}^{n} a_k(s_l - s_h) \Big| = n\, S(\mathcal{A}_n).$$

By summation over $e \in \mathbb{E}_2^n$ and Khintchine's inequality, we now obtain

$$\frac{1}{\sqrt{2}} n^{3/2} \le K_1'(\mathcal{P}_n(\pm 1))^{-1} \sum_{k=1}^{n} \Big(\sum_{h=1}^{n} |a_k(-s_h)|^2 \Big)^{1/2} \le n\, S(\mathcal{A}_n).$$

REMARK. The lower estimate can be improved if the Bernoulli system is replaced by the Steinhaus system. Then we get $\sqrt{\frac{\pi n}{4}}$ instead of $\sqrt{\frac{n}{2}}$.

4.11.7 Let $\mathcal{A}_n = (a_1, \dots, a_n)$ and $\mathcal{B}_n = (b_1, \dots, b_n)$ be systems of characters defined on compact Abelian groups $\mathbb{G}$ and $\mathbb{H}$, respectively. Then there exists a constant $S \ge 1$ such that

$$\sup \Big\{ \Big| \sum_{k=1}^{n} \xi_k b_k(t) \Big| : t \in \mathbb{H} \Big\} \le S \sup \Big\{ \Big| \sum_{k=1}^{n} \xi_k a_k(s) \Big| : s \in \mathbb{G} \Big\}$$

whenever $\xi_1, \dots, \xi_n \in \mathbb{K}$. We refer to

$$S(\mathcal{B}_n, \mathcal{A}_n) := \inf S$$

as the **relative Sidon constant** of $\mathcal{A}_n$ and $\mathcal{B}_n$. In this context the original Sidon constant is given by

$$S(\mathcal{A}_n) = S(\mathcal{P}_n(e^{i\tau}), \mathcal{A}_n).$$

On the other hand, we have

$$S(\mathcal{A}_n, \mathcal{P}_n(e^{i\tau})) = 1.$$

4.11.8 The Riesz representation theorem identifies $C(\mathbb{G})'$ with the set of all regular complex-valued Borel measures on $\mathbb{G}$:

$$\langle f, \lambda \rangle = \int\limits_{\mathbb{G}} f(s)\, d\lambda(s) \quad \text{for} \quad f \in C(\mathbb{G}).$$

The total variation $|\lambda|$ of a measure λ is defined by

$$|\lambda|(B) := \sup \sum_{h=1}^{m} |\lambda(B_h)|,$$

the supremum being taken over all finite partitions of the Borel set B into pairwise disjoint Borel subsets $B_1, \ldots, B_m$. Note that $\|\lambda\| = |\lambda|(\mathbb{G})$. We refer to

$$f * \lambda(s) := \int\limits_{\mathbb{G}} f(s-u)\, d\lambda(u)$$

as the convolution of $f \in C(\mathbb{G})$ and $\lambda \in C(\mathbb{G})'$. This concept also makes sense when f is replaced by a continuous X-valued function $\boldsymbol{f}$ on $\mathbb{G}$. In what follows, we only need the elementary case in which

$$\boldsymbol{f} = \sum_{k=1}^{n} a_k \otimes x_k$$

with $a_1, \ldots, a_n \in \mathbb{G}'$ and $x_1, \ldots, x_n \in X$. It follows from

$$\|\boldsymbol{f} * \lambda(s)\| \le \int\limits_{\mathbb{G}} \|\boldsymbol{f}(s-u)\|\, d|\lambda|(u)$$

that

$$\|\boldsymbol{f} * \lambda | L_2\| \le \int\limits_{\mathbb{G}} \Bigg(\int\limits_{\mathbb{G}} \|\boldsymbol{f}(s-u)\|^2\, d\mu(s) \Bigg)^{1/2} d|\lambda|(u).$$

Taking into account the translation invariance of the Haar measure μ, we obtain **Young's inequality**

$$\|\boldsymbol{f} * \lambda | L_2\| \le \|\boldsymbol{f} | L_2\|\, \|\lambda\|.$$

4.11.9 The next result is due to G. Pisier and A. Pełczyński; see [pis 6] and [pel 2].

PROPOSITION.

$$\varrho(\mathcal{A}_n \otimes \mathcal{B}_n, \mathcal{A}_n) \le S(\mathcal{B}_n, \mathcal{A}_n) \quad \textit{and} \quad \varrho(\mathcal{A}_n, \mathcal{A}_n \otimes \mathcal{B}_n) \le S(\mathcal{B}_n, \mathcal{A}_n).$$

PROOF. For $t \in \mathbb{H}$, we define a functional λ_t° on the linear span of $\overline{\mathcal{A}}_n$ by

$$\Big\langle \sum_{k=1}^{n} \xi_k \overline{a_k}, \lambda_t^\circ \Big\rangle := \sum_{k=1}^{n} \xi_k b_k(t).$$

Obviously,

$$\|\lambda_t^\circ\| \le S(\mathcal{B}_n, \overline{\mathcal{A}}_n) = S(\mathcal{B}_n, \mathcal{A}_n).$$

Choose any norm-preserving extension $\lambda_t \in C(\mathbb{G})'$. Given $v \in \mathbb{G}$, we consider the function

$$\boldsymbol{f}_v := \sum_{k=1}^{n} a_k b_k(v) \otimes x_k \in [L_2(\mathbb{G}), X].$$

Then it follows from

$$a_k * \lambda_t(s) = \int_{\mathbb{G}} a_k(s-u)\, d\lambda_t(u) = a_k(s) \int_{\mathbb{G}} \overline{a_k}(u)\, d\lambda_t(u) = a_k(s)\, b_k(t)$$

that $\boldsymbol{f}_v * \lambda_t = \boldsymbol{f}_{v+t}$ for $t, v \in \mathbb{G}$. Thus, in view of Young's inequality and $\|\lambda_t\| \le S(\mathcal{B}_n, \mathcal{A}_n)$,

$$\|\boldsymbol{f}_{v+t}|L_2\| \le S(\mathcal{B}_n, \mathcal{A}_n)\, \|\boldsymbol{f}_v|L_2\|.$$

Letting $v = o$ and $v = -t$, we obtain

$$\int_{\mathbb{G}} \Big\| \sum_{k=1}^{n} x_k a_k(s) b_k(t) \Big\|^2 d\mu(s) \le S(\mathcal{B}_n, \mathcal{A}_n)^2 \int_{\mathbb{G}} \Big\| \sum_{k=1}^{n} x_k a_k(s) \Big\|^2 d\mu(s)$$

and

$$\int_{\mathbb{G}} \Big\| \sum_{k=1}^{n} x_k a_k(s) \Big\|^2 d\mu(s) \le S(\mathcal{B}_n, \mathcal{A}_n)^2 \int_{\mathbb{G}} \Big\| \sum_{k=1}^{n} x_k a_k(s) b_k(-t) \Big\|^2 d\mu(s),$$

respectively. Finally, integration over t yields

$$\|(x_k)|\mathcal{A}_n \otimes \mathcal{B}_n\| \le S(\mathcal{B}_n, \mathcal{A}_n)\, \|(x_k)|\mathcal{A}_n\|$$

and

$$\|(x_k)|\mathcal{A}_n\| \le S(\mathcal{B}_n, \mathcal{A}_n)\, \|(x_k)|\mathcal{A}_n \otimes \mathcal{B}_n\|.$$

4.11.10 We now state some consequences of the previous inequalities.

PROPOSITION.

$$\boldsymbol{\varrho}(\mathcal{B}_n, \mathcal{A}_n) \le S(\mathcal{A}_n, \mathcal{B}_n)\, S(\mathcal{B}_n, \mathcal{A}_n),$$

$$\boldsymbol{\varrho}(\mathcal{A}_n, \mathcal{I}_n) \le S(\mathcal{B}_n, \mathcal{A}_n)\, \boldsymbol{\varrho}(\mathcal{B}_n, \mathcal{I}_n) \text{ and } \boldsymbol{\varrho}(\mathcal{I}_n, \mathcal{A}_n) \le S(\mathcal{B}_n, \mathcal{A}_n)\, \boldsymbol{\varrho}(\mathcal{I}_n, \mathcal{B}_n).$$

In particular,

$$\boldsymbol{\varrho}(\mathcal{P}_n(e^{i\tau}), \mathcal{I}_n) \le \boldsymbol{\varrho}(\mathcal{A}_n, \mathcal{I}_n) \le S(\mathcal{A}_n)\, \boldsymbol{\varrho}(\mathcal{P}_n(e^{i\tau}), \mathcal{I}_n),$$

$$\boldsymbol{\varrho}(\mathcal{I}_n, \mathcal{P}_n(e^{i\tau})) \le \boldsymbol{\varrho}(\mathcal{I}_n, \mathcal{A}_n) \le S(\mathcal{A}_n)\, \boldsymbol{\varrho}(\mathcal{I}_n, \mathcal{P}_n(e^{i\tau})),$$

$$\boldsymbol{\varrho}(\mathcal{A}_n, \mathcal{P}_n(e^{i\tau})) \le S(\mathcal{A}_n) \text{ and } \boldsymbol{\varrho}(\mathcal{P}_n(e^{i\tau}), \mathcal{A}_n) \le S(\mathcal{A}_n).$$

PROOF. Use the triangle inequality 3.3.5. Then, for example, 3.7.8 gives

$$\varrho(\mathcal{A}_n,\mathcal{I}_n) \le \varrho(\mathcal{A}_n,\mathcal{A}_n\otimes\mathcal{B}_n)\circ\varrho(\mathcal{A}_n\otimes\mathcal{B}_n,\mathcal{I}_n) \le S(\mathcal{B}_n,\mathcal{A}_n)\,\varrho(\mathcal{B}_n,\mathcal{I}_n).$$

REMARK. See also 3.7.9 and 4.2.14.

4.11.11 A system of characters $a_1,\dots,a_n$ is said to be **dissociate** if

$$a_1^{\delta_1}\dots a_n^{\delta_n} \ne 1$$

whenever $\delta_1,\dots,\delta_n \in \{+1,0,-1\}$ and $|\delta_1|+\dots+|\delta_n|>0$.

REMARK. This definition, which differs from that in [HEW*2, p. 425], is due to J. Bourgain; see [bou 6].

4.11.12 Obviously, $\mathcal{P}_n(e^{i\tau})$ is dissociate. The most important non-trivial example of a dissociate system is constituted by the characters $e_{2^1},\dots,e_{2^n}$ defined on $\mathbb{T}$; see 3.8.6.

4.11.13 Let $\mathcal{A}_n=(a_1,\dots,a_n)$ be a system of characters on a compact Abelian group $\mathbb{G}$. Fix $\sigma_1,\dots,\sigma_n\in\mathbb{T}$ and $0<\varrho\le 1$. The **Riesz product** $R\in C(\mathbb{G})$ is defined by

$$R := \prod_{k=1}^{n}\big[1+\tfrac{1}{2}(\sigma_k a_k+\sigma_k^{-1}a_k^{-1})\varrho\big].$$

Moreover, the coefficients R_m of the expansion

$$R=\sum_{m=0}^{n} R_m\varrho^m$$

are given by

$$R_m := 2^{-m}\sum(\sigma_1 a_1)^{\delta_1}\dots(\sigma_n a_n)^{\delta_n},$$

where the sum ranges over all exponents $\delta_1,\dots,\delta_n\in\{-1,0,+1\}$ such that $|\delta_1|+\dots+|\delta_n|=m$. In particular,

$$R_0=1 \quad\text{and}\quad R_1=\frac{1}{2}\Big(\sum_{k=1}^{n}\sigma_k a_k+\sum_{k=1}^{n}\sigma_k^{-1}a_k^{-1}\Big).$$

4.11.14 We now state some elementary properties of Riesz products.

LEMMA. *If $\mathcal{A}_n=(a_1,\dots,a_n)$ is dissociate, then*

$$\|R|L_1\|=1 \quad\textit{and}\quad \sum_{m=0}^{n}|(R_m,a)|\le 1 \quad\textit{for } a\in\mathbb{G}'.$$

PROOF. Since

$$1+\tfrac{1}{2}(\sigma_k a_k+\sigma_k^{-1}a_k^{-1})\varrho = 1+\tfrac{1}{2}(\sigma_k a_k+\overline{\sigma_k a_k})\varrho = 1+\mathrm{Re}(\sigma_k a_k)\varrho \ge 0,$$

the function R is non-negative on $\mathbb{G}$, and from $a_1^{\delta_1}\dots a_n^{\delta_n}\neq 1$ we conclude that

$$\int_{\mathbb{G}} a_1(s)^{\delta_1}\dots a_n(s)^{\delta_n}\,d\mu(s)=0$$

whenever $|\delta_1|+\dots+|\delta_n|>0$. Thus

$$\|R|L_1\|=\int_{\mathbb{G}} R(s)\,d\mu(s)=1.$$

In the case when $\sigma_1=\dots=\sigma_n=1$ and $\varrho=1$, we write E and $E_1,\dots,E_n$ instead of R and $R_1,\dots,R_n$, respectively. For $a\in\mathbb{G}'$, we now obtain

$$|(R_m,a)|=\left|\frac{1}{2^m}\sum\sigma_1^{\delta_1}\dots\sigma_n^{\delta_n}\right|\le\frac{1}{2^m}\sum 1^{\delta_1}\dots 1^{\delta_n}=(E_m,a),$$

where both sums range over all $\delta_1,\dots,\delta_n\in\{+1,0,-1\}$ such that

$$|\delta_1|+\dots+|\delta_n|=m \quad\text{and}\quad a_1^{\delta_1}\dots a_n^{\delta_n}=a.$$

Hence

$$\sum_{m=0}^{n}|(R_m,a)|\le\sum_{m=0}^{n}(E_m,a)=(E,a)\le\|E|L_1\|=1.$$

4.11.15 PROPOSITION. *Let $\mathcal{A}_n=(a_1,\dots,a_n)$ be a dissociate system of characters on a compact Abelian group $\mathbb{G}$. Then $S(\mathcal{A}_n)\le 16$.*

PROOF. Write

$$f:=\sum_{k=1}^{n}\xi_k a_k,$$

and define $\sigma_1,\dots,\sigma_n\in\mathbb{T}$ by $\xi_k=\sigma_k|\xi_k|$. Let R and R^* denote the Riesz products associated with $\sigma_1,\dots,\sigma_n$ and $i\sigma_1,\dots,i\sigma_n$, respectively. Then it follows from

$$1\notin\mathcal{A}_n,\quad R_0-iR_0^*=1-i\quad\text{and}\quad R_1-iR_1^*=\sum_{k=1}^{n}\sigma_k a_k$$

that

$$(f,R_0-iR_0^*)=0\quad\text{and}\quad(f,R_1-iR_1^*)=\sum_{k=1}^{n}|\xi_k|.$$

On the other hand, in view of the preceding lemma, we have

$$|(f,R-iR^*)|\le\|f|C\|\,\|R-iR^*|L_1\|\le 2\,\|f|C\|$$

and

$$\begin{aligned}\sum_{m=2}^{n}|(f,R_m-iR_m^*)|\varrho^m&\le\sum_{m=2}^{n}\sum_{k=1}^{n}|(f,a_k)|\big(|(R_m,a_k)|+|(R_m^*,a_k)|\big)\varrho^m\\&=\sum_{k=1}^{n}|(f,a_k)|\sum_{m=2}^{n}\big(|(R_m,a_k)|+|(R_m^*,a_k)|\big)\varrho^m\le 2\varrho^2\sum_{k=1}^{n}|\xi_k|.\end{aligned}$$

Hence

$$2\,\|f|C\| \geq |(f, R - iR^*)| = \Big|\sum_{m=0}^{n} (f, R_m - iR_m^*)\varrho^m\Big|$$

$$\geq |(f, R_1 - iR_1^*)|\varrho - \sum_{m=2}^{n} |(f, R_m - iR_m^*)|\varrho^m \geq \varrho(1 - 2\varrho)\sum_{k=1}^{n} |\xi_k|.$$

Finally, letting $\varrho = \frac{1}{4}$ yields

$$\sum_{k=1}^{n} |\xi_k| \leq 16\,\|f|C\|.$$

4.11.16 PROPOSITION. *Every system of N characters contains a dissociate subsystem of cardinality $n \geq \log_3 N$.*

PROOF (J. Seigner). Let $\mathbb{F}$ be a maximal subset of $\{1, \dots, N\}$ for which $\mathcal{A}_N(\mathbb{F})$ is dissociate. Define

$$\operatorname{hull}(\mathcal{A}_N(\mathbb{F})) := \Big\{ \prod_{k \in \mathbb{F}} a_k^{\delta_k} \;:\; \delta_k = +1, 0, -1 \Big\}.$$

Assume that there is a character $a_{k_0} \in \mathcal{A}_N \setminus \operatorname{hull}(\mathcal{A}_N(\mathbb{F}))$. Then the system $\mathcal{A}_N(\mathbb{F})$ enlarged by a_{k_0} would still be dissociate. This contradiction implies that $\mathcal{A}_N \subseteq \operatorname{hull}(\mathcal{A}_N(\mathbb{F}))$. Consequently,

$$N = |\mathcal{A}_N| \leq |\operatorname{hull}(\mathcal{A}_N(\mathbb{F}))| \leq 3^{|\mathbb{F}|}.$$

REMARK. The above result is best possible, since a dissociate subsystem of $(\mathbb{E}_3^n)'$ contains at most n characters.

4.11.17 Let us fix the following rule:
From every set with n elements, we take away $d(n)$ elements, where $d : \mathbb{N} \to \mathbb{N}$ is a given function with $1 \leq d(n) \leq n$. Repeating this process as long as possible, we denote by $f(n)$ the number of steps needed to exhaust the original set.

LEMMA. *Let $d(n) \geq c \log n$ and $0 < c \leq 1$. Then*

$$f(n) \leq \frac{3}{c}\,\frac{n}{1+\log n}.$$

PROOF. Define

$$F(x) := \frac{x}{1+\log x} \quad \text{for } x \geq 1.$$

Then

$$F'(x) = \frac{\log x}{(1+\log x)^2} \quad \text{and} \quad F''(x) = \frac{1 - \log x}{x(1+\log x)^3}.$$

Hence F is increasing for $x \geq 1$, and F' is decreasing for $x \geq e$. If $n \leq 7 < e^2$, then it follows from $1+\log n < 3$ that

$$f(n) \leq n \leq \frac{n}{c} \leq \frac{3}{c}\,\frac{n}{1+\log n}.$$

In the case $n \geq 8 > e^2$, we have

$$n - c\log n > 8 - \log 8 > e \quad \text{and} \quad \Big(\frac{\log n}{1+\log n}\Big)^2 \geq \Big(\frac{\log 8}{1+\log 8}\Big)^2 > \frac{1}{3}.$$

The mean value theorem implies that

$$F(n) - F(n - c\log n) = c\log n\, F'(n - \theta c\log n) \geq c\log n\, F'(n) > \frac{c}{3},$$

where $0 < \theta < 1$. Now the proof can be completed by induction on n. Indeed, using the monotonicity of F, we have

$$f(n) = f(n-d(n))+1 \leq \tfrac{3}{c}F(n-d(n))+1 \leq \tfrac{3}{c}F(n-c\log n)+1 \leq \tfrac{3}{c}F(n).$$

4.11.18 We now generalize **Hinrichs's inequality** stated in 4.2.16.

PROPOSITION. *Let $\mathcal{A}_{2^n} = (a_1, \ldots, a_{2^n})$ be any system of characters on a compact Abelian group $\mathbb{G}$. Then*

$$2^{-n/2}\,\varrho(\mathcal{A}_{2^n}, \mathcal{I}_{2^n}) \leq 50\, n^{-1/2}\,\varrho(\mathcal{P}_n(e^{i\tau}), \mathcal{I}_n).$$

PROOF. Letting $d(k) := 1 + [\log_3 k]$, we apply the process just described to $\mathcal{A}_{2^n}$ by taking away dissociate subsystems or the singleton $\{1\}$. This yields a partition $\{\mathbb{F}_1, \ldots, \mathbb{F}_m\}$ of $\{1, \ldots, 2^n\}$ into pairwise disjoint subsets. Hence, by 4.2.15,

$$\varrho(\mathcal{A}_{2^n}, \mathcal{I}_{2^n}) \leq \sqrt{m}\, \max_{1\leq\alpha\leq m} \varrho(\mathcal{A}_{2^n}(\mathbb{F}_\alpha), \mathcal{I}_{2^n}(\mathbb{F}_\alpha)).$$

It follows from 4.11.10, 4.11.15 and

$$|\mathbb{F}_\alpha| \leq d(2^n) \leq 1 + \log_3 2^n < 2n$$

that

$$\begin{aligned}\varrho(\mathcal{A}_{2^n}(\mathbb{F}_\alpha), \mathcal{I}_{2^n}(\mathbb{F}_\alpha)) &\leq S(\mathcal{A}_{2^n}(\mathbb{F}_\alpha))\, \varrho(\mathcal{P}_{2n}(e^{i\tau}), \mathcal{I}_{2n})\\ &\leq 16\sqrt{2}\, \varrho(\mathcal{P}_n(e^{i\tau}), \mathcal{I}_n).\end{aligned}$$

On the other hand, in view of $d(k) \geq \log_3 k$, we may apply 4.11.17 with $c := \frac{1}{\log 3}$. Therefore

$$m \leq 3\log 3\, \frac{2^n}{1+\log 2^n} < \frac{\log 27}{\log 2}\,\frac{2^n}{n}.$$

This completes the proof, since $16\sqrt{2\,\frac{\log 27}{\log 2}} < 50$.

4.11.19 Let $\mathcal{A}_n=(a_1,\dots,a_n)$ be a system of characters on a compact Abelian group $\mathbb{G}$. A Banach space X is said to be **$\mathcal{A}_n$-convex** if there exists $0<\varepsilon<1$ such that, whenever $\|x_1\|=\dots=\|x_n\|=1$, at least one of the absolutely convex combinations

$$\frac{1}{n}\sum_{k=1}^{n} x_k a_k(s) \quad \text{with } s\in\mathbb{G}$$

has norm less than $1-\varepsilon$.

4.11.20 We proceed with a geometric corollary of Hinrichs's inequality.

PROPOSITION. *Let $\mathcal{A}_n=(a_1,\dots,a_n)$ be a system of characters on a compact Abelian group $\mathbb{G}$. Then, for every Banach space X, the following are equivalent:*

(1) $\varrho(X|\mathcal{A}_n,\mathfrak{I}_n)=\sqrt{n}$.

(2) *X fails to be $\mathcal{A}_n$-convex.*

PROOF. **(1)⇒(2)**: By 3.7.12, there exist $\boldsymbol{x}_1,\dots,\boldsymbol{x}_n$ in some ultrapower $X^{\mathcal{U}}$ such that

$$\|\boldsymbol{x}_1\|=\dots=\|\boldsymbol{x}_n\|=1 \quad\text{and}\quad \left\|\frac{1}{n}\sum_{k=1}^{n}\boldsymbol{x}_k a_k(s)\right\|=1 \quad\text{whenever } s\in\mathbb{G}.$$

To justify the following notation, we recall from 3.7.12 that $\mathcal{U}$ was chosen to be an ultrafilter on $\mathbb{R}_+$ containing all intervals $(0,\delta)$ with $\delta>0$. As observed in 0.8.2, every equivalence class $\boldsymbol{x}_k$ can be represented in the form $\boldsymbol{x}_k=(x_k^{(\delta)})^{\mathcal{U}}$ with $\|x_k^{(\delta)}\|=1$ for $\delta>0$. Now it follows from

$$\mathcal{U}\text{-}\lim_{\delta}\left\|\frac{1}{n}\sum_{k=1}^{n}x_k^{(\delta)}a_k(s)\right\|=1$$

that X cannot be $\mathcal{A}_n$-convex. The implication **(2)⇒(1)** is trivial.

4.11.21 PROPOSITION. *If $\varrho(X|\mathcal{A}_n,\mathfrak{I}_n)=\sqrt{n}$, then X contains l_1^N with bound 17 whenever $N\le\log_3 n$.*

PROOF. As in the previous proof, we fix elements $\boldsymbol{x}_1,\dots,\boldsymbol{x}_n$ in some ultrapower $X^{\mathcal{U}}$ such that

$$\|\boldsymbol{x}_1\|=\dots=\|\boldsymbol{x}_n\|=1 \quad\text{and}\quad \left\|\sum_{k=1}^{n}\boldsymbol{x}_k a_k(s)\right\|=n \quad\text{whenever } s\in\mathbb{G}.$$

The Hahn–Banach theorem yields $\boldsymbol{x}'_s\in(X^{\mathcal{U}})'$ with

$$\left\langle\sum_{k=1}^{n}\boldsymbol{x}_k a_k(s),\boldsymbol{x}'_s\right\rangle=n \quad\text{and}\quad \|\boldsymbol{x}'_s\|=1.$$

It follows from

$$\sum_{k=1}^{n} \langle \boldsymbol{x}_k a_k(s), \boldsymbol{x}'_s \rangle = n \quad \text{and} \quad |\langle \boldsymbol{x}_k a_k(s), \boldsymbol{x}'_s \rangle| \le \|\boldsymbol{x}_k a_k(s)\| \, \|\boldsymbol{x}'_s\| = 1$$

that $\langle \boldsymbol{x}_k a_k(s), \boldsymbol{x}'_s \rangle = 1$. Hence, for every subset $\mathbb{F} \subseteq \{1, \dots, n\}$, we have

$$\Big| \sum_{k\in\mathbb{F}} \xi_k a_k(-s) \Big| = \Big| \Big\langle \sum_{k\in\mathbb{F}} \xi_k \boldsymbol{x}_k, \boldsymbol{x}'_s \Big\rangle \Big| \le \Big\| \sum_{k\in\mathbb{F}} \xi_k \boldsymbol{x}_k \Big\| \, \|\boldsymbol{x}'_s\|,$$

which in turn implies that

$$\begin{aligned} S(\mathcal{A}_n(\mathbb{F}))^{-1} \sum_{k\in\mathbb{F}} |\xi_k| &\le \sup \Big\{ \Big| \sum_{k\in\mathbb{F}} \xi_k a_k(-s) \Big| : s \in \mathbb{G} \Big\} \\ &\le \Big\| \sum_{k\in\mathbb{F}} \xi_k \boldsymbol{x}_k \Big\| \le \sum_{k\in\mathbb{F}} |\xi_k|. \end{aligned}$$

However, Propositions 4.11.16 and 4.11.15 provide us with some subset $\mathbb{F} \subseteq \{1, \dots, n\}$ such that $|\mathbb{F}| \ge \log_3 n$ and $S(\mathcal{A}_n(\mathbb{F})) \le 16$. Therefore

$$\frac{1}{16} \sum_{k\in\mathbb{F}} |\xi_k| \le \Big\| \sum_{k\in\mathbb{F}} \xi_k \boldsymbol{x}_k \Big\| \le \sum_{k\in\mathbb{F}} |\xi_k|.$$

This completes the proof, since $X^{\mathcal{U}}$ is finitely representable in X.

4.11.22 Finally, we give a very general characterization of B-convexity.

PROPOSITION. *Fix any sequence of systems $\mathcal{A}_1, \mathcal{A}_2, \dots$ consisting of characters on compact Abelian groups. Then, for every Banach space X, the following are equivalent:*

(1) *X is* B*-convex.*

(2) *There exists n_0 such that X is $\mathcal{A}_n$-convex for* **all** $n \ge n_0$.

(3) *X is $\mathcal{A}_n$-convex for* **some** $n \ge 2$.

PROOF. **(1)⇒(2)**: By Hinrichs's inequality and $\varrho(X|\mathcal{R}_n, \mathcal{I}_n) = o(\sqrt{n})$, we can find n_0 such that

$$\varrho(X|\mathcal{A}_n, \mathcal{I}_n) < \sqrt{n} \quad \text{for } n \ge n_0.$$

Condition **(2)** now follows from 4.11.20.

(2)⇒(3) is trivial.

(3)⇒(1): If X is $\mathcal{A}_n$-convex, then 3.7.9 and 4.11.20 imply that

$$\varrho(X|\mathcal{R}_n, \mathcal{I}_n) \le \varrho(X|\mathcal{A}_n, \mathcal{I}_n) < \sqrt{n}.$$

Thus X is B-convex, by 4.4.5.

4.12 The Dirichlet ideal norms $\boldsymbol{\delta}(\mathcal{R}_n, \mathcal{R}_n)$ and $\boldsymbol{\delta}(\mathcal{G}_n, \mathcal{G}_n)$

4.12.1 We now treat the Dirichlet ideal norms associated with the Rademacher and the Gauss systems, $\boldsymbol{\delta}(\mathcal{R}_n, \mathcal{R}_n)$ and $\boldsymbol{\delta}(\mathcal{G}_n, \mathcal{G}_n)$. In order to get parallel results in the real and in the complex case, it is sometimes advisable to replace the Rademacher system by the Bernoulli or the Steinhaus system, respectively.

If the underlying Banach spaces are complex, then we have

$$\boldsymbol{\delta}(\mathcal{P}_n(\pm 1), \mathcal{P}_n(\pm 1)) \asymp \boldsymbol{\delta}(\mathcal{P}_n(e^{i\tau}), \mathcal{P}_n(e^{i\tau})) \quad \text{and} \quad \boldsymbol{\delta}(\mathcal{G}_n^{\mathbb{R}}, \mathcal{G}_n^{\mathbb{R}}) \asymp \boldsymbol{\delta}(\mathcal{G}_n^{\mathbb{C}}, \mathcal{G}_n^{\mathbb{C}});$$

see 3.4.8, 4.2.13 and 4.7.7. Moreover, in view of

$$\|(x_k)|\overline{\mathcal{P}}_n(e^{i\tau})\| = \|(x_k)|\mathcal{P}_n(e^{i\tau})\| \quad \text{and} \quad \|(x_k)|\overline{\mathcal{G}}_n^{\mathbb{C}}\| = \|(x_k)|\mathcal{G}_n^{\mathbb{C}}\|,$$

it makes no difference to use $\overline{\mathcal{P}}_n(e^{i\tau})$ and $\overline{\mathcal{G}}_n^{\mathbb{C}}$ instead of $\mathcal{P}_n(e^{i\tau})$ and $\mathcal{G}_n^{\mathbb{C}}$.

4.12.2 PROPOSITION. *The sequences* $(\boldsymbol{\delta}(\mathcal{R}_n, \mathcal{R}_n))$ *and* $(\boldsymbol{\delta}(\mathcal{G}_n, \mathcal{G}_n))$ *are non-decreasing. Moreover,*

$$\boldsymbol{\delta}(\mathcal{R}_{2n}, \mathcal{R}_{2n}) \le 2\,\boldsymbol{\delta}(\mathcal{R}_n, \mathcal{R}_n) \quad \textit{and} \quad \boldsymbol{\delta}(\mathcal{G}_{2n}, \mathcal{G}_{2n}) \le 2\,\boldsymbol{\delta}(\mathcal{G}_n, \mathcal{G}_n).$$

PROOF. The monotonicity follows from 4.1.4, and the 'moreover' part is a consequence of 3.4.13.

4.12.3 The next result is a corollary of 3.4.6.

PROPOSITION. *The ideal norms* $\boldsymbol{\delta}(\mathcal{R}_n, \mathcal{R}_n)$ *and* $\boldsymbol{\delta}(\mathcal{G}_n, \mathcal{G}_n)$ *are symmetric.*

4.12.4 We now compare the ideal norms $\boldsymbol{\delta}(\mathcal{R}_n, \mathcal{R}_n)$ and $\boldsymbol{\delta}(\mathcal{G}_n, \mathcal{G}_n)$ with each other. In order to get nice constants, the Rademacher system $\mathcal{R}_n$ is replaced by $\mathcal{P}_n$.

PROPOSITION. $\boldsymbol{\delta}(\mathcal{P}_n, \mathcal{P}_n) \le M_1^{-2}\, \boldsymbol{\delta}(\mathcal{G}_n, \mathcal{G}_n)$.

PROOF. The assertion follows from 3.4.8 and 4.8.3.

4.12.5 It is unknown whether a reverse estimate holds.

PROBLEM. *Are the sequences of the Dirichlet ideal norms* $\boldsymbol{\delta}(\mathcal{R}_n, \mathcal{R}_n)$ *and* $\boldsymbol{\delta}(\mathcal{G}_n, \mathcal{G}_n)$ *uniformly equivalent?*

4.12.6 PROPOSITION. $\boldsymbol{\varrho}(\mathcal{G}_n, \mathcal{P}_n) \le M_1^{-1}\, \boldsymbol{\delta}(\mathcal{G}_n, \mathcal{G}_n)$.

PROOF. By 3.4.10, 3.4.8 and 4.8.3, we have

$$\boldsymbol{\varrho}(\mathcal{G}_n, \mathcal{P}_n) \le \boldsymbol{\delta}(\mathcal{G}_n, \mathcal{P}_n) \le \boldsymbol{\delta}(\mathcal{G}_n, \mathcal{G}_n) \circ \boldsymbol{\varrho}'(\mathcal{P}_n, \mathcal{G}_n) \le M_1^{-1}\, \boldsymbol{\delta}(\mathcal{G}_n, \mathcal{G}_n).$$

4.12.7 EXAMPLE. *Let $1 \le p \le 2$. Then*

$$K'_p(\mathcal{P}_n)^{-1} K_{p'}(\mathcal{P}_n) \le \boldsymbol{\delta}(L_p|\mathcal{P}_n, \mathcal{P}_n) = \boldsymbol{\delta}(L_{p'}|\mathcal{P}_n, \mathcal{P}_n) \le K_{p'}(\mathcal{P}_n).$$

PROOF. The upper estimate is a consequence of 3.4.6 and 3.4.15.

Recall from 4.8.1 that $\mathbb{E}^n$ is a synonym of $\mathbb{E}_2^n$ and $\mathbb{T}^n$. The lower estimate for $L_p(\mathbb{E}^n)$ is implied by

$$\boldsymbol{\varrho}(L_{p'}(\mathbb{E}^n)|\mathcal{P}_n, \mathcal{I}_n) \le \boldsymbol{\varrho}(L_p(\mathbb{E}^n)|\mathcal{I}_n, \mathcal{P}_n)\, \boldsymbol{\delta}(L_p(\mathbb{E}^n)|\mathcal{P}_n, \mathcal{P}_n),$$

since

$$\boldsymbol{\varrho}(L_{p'}(\mathbb{E}^n)|\mathcal{P}_n, \mathcal{I}_n) = K_{p'}(\mathcal{P}_n) \quad \text{and} \quad \boldsymbol{\varrho}(L_p(\mathbb{E}^n)|\mathcal{I}_n, \mathcal{P}_n) = K'_p(\mathcal{P}_n);$$

see 3.8.15. The general case follows from the fact that $L_p(\mathbb{E}^n)$ can be embedded isometrically into every infinite dimensional space L_p.

REMARK. Note that

$$1 \le K'_p(\mathcal{P}_n(\pm 1)) \le \sqrt{2} \quad \text{and} \quad 1 \le K'_p(\mathcal{P}_n(e^{i\tau})) \le \sqrt{\tfrac{4}{\pi}}.$$

Hence the dominating quantity in the above inequality is $K_{p'}(\mathcal{P}_n)$. In view of symmetry, it seems plausible that

$$\boldsymbol{\delta}(L_r(\mathbb{E}^n)|\mathcal{P}_n, \mathcal{P}_n) \stackrel{?}{=} M_r(\mathcal{P}_n)\, M_{r'}(\mathcal{P}_n) \quad \text{for } 1 < r < \infty.$$

This conjecture is true in the limiting cases $r = 1$ and $r = \infty$; see 3.8.12.

4.12.8 In order to establish a counterpart of the preceding result for the Gauss system, some preliminaries are required. First of all, we refer to the fact that

$$\boldsymbol{\delta}(\mathcal{G}_n, \mathcal{G}_n) = \boldsymbol{\delta}(\mathcal{Q}_n, \mathcal{Q}_n).$$

Therefore we can replace $\mathcal{G}_n$ by $\mathcal{Q}_n$, which yields better constants. Let

$$Q_n x := \sum_{k=1}^{n} \xi_k q_k^{(n)} \quad \text{for } x = (\xi_k) \in l_2^n,$$

$$Q_n^* f := (\langle f, \overline{q}_k^{(n)} \rangle) \quad \text{for } f \in L_r(\mathbb{S}^n).$$

The next result is related to 3.4.14.

LEMMA. $\|Q_n Q_n^* : L_r(\mathbb{S}^n) \to L_r(\mathbb{S}^n)\| = M_r(\mathcal{Q}_n)\, M_{r'}(\mathcal{Q}_n).$

PROOF. By 4.7.12, the upper estimate follows from

$$\|Q_n Q_n^* : L_r \to L_r\| \le \|Q_n^* : L_r \to l_2^n\|\, \|Q_n : l_2^n \to L_r\| = M_r(\mathcal{Q}_n)\, M_{r'}(\mathcal{Q}_n).$$

To check the lower estimate, we recall that

$$\left(\int_{\mathbb{S}^n} \left| \frac{1}{\sqrt{n}} \sum_{k=1}^{n} q_k^{(n)}(s) \right|^{r'} d\omega^n(s) \right)^{1/r'} = M_{r'}(\mathcal{Q}_n).$$

Hence, given $\varepsilon>0$, there exists $f\in L_r(\mathbb{S}^n)$ such that

$$\int\limits_{\mathbb{S}^n} f(s)\sum_{k=1}^n \overline{q_k^{(n)}(s)}\,d\omega^n(s)=\sqrt{n}\,M_{r'}(\mathcal{Q}_n) \quad\text{and}\quad \|f|L_r\|\le 1+\varepsilon.$$

Then it follows from

$$\sum_{k=1}^n\langle f,\overline{q}_k^{(n)}\rangle\le\sqrt{n}\Big(\sum_{k=1}^n|\langle f,\overline{q}_k^{(n)}\rangle|^2\Big)^{1/2}$$

that

$$M_{r'}(\mathcal{Q}_n)\le\Big(\sum_{k=1}^n|\langle f,\overline{q}_k^{(n)}\rangle|^2\Big)^{1/2}.$$

By Khintchine's equality 4.7.12, we obtain

$$\begin{aligned}M_r(\mathcal{Q}_n)\Big(\sum_{k=1}^n|\langle f,\overline{q}_k^{(n)}\rangle|^2\Big)^{1/2}&=\Big(\int\limits_{\mathbb{S}^n}\Big|\sum_{k=1}^n\langle f,\overline{q}_k^{(n)}\rangle q_k^{(n)}(t)\Big|^r d\omega^n(t)\Big)^{1/r}\\&=\|Q_nQ_n^*f|L_r\|\le(1+\varepsilon)\|Q_nQ_n^*:L_r\to L_r\|.\end{aligned}$$

Combining these facts and letting $\varepsilon\to 0$ yields the lower estimate.

4.12.9 We denote by $\mathbb{U}^n$ the compact (non-Abelian) group of all orthogonal (real case) or unitary (complex case) square matrices of order n. Moreover, χ^n stands for its Haar measure. Note that, for $t\in\mathbb{S}^n$, the measure ω^n is the image of χ^n under the map $A_n\to s=A_nt$. In other terms,

$$\int\limits_{\mathbb{U}^n} f(A_nt)\,d\chi^n(A_n)=\int\limits_{\mathbb{S}^n} f(s)\,d\omega^n(s)\quad\text{for}\quad f\in L_1(\mathbb{S}^n).$$

Next, we stress that $\boldsymbol{f}\in[L_2(\mathbb{S}^n),L_r(\mathbb{U}^n)]$ can be viewed as a measurable function $f:\mathbb{S}^n\times\mathbb{U}^n\to\mathbb{K}$ for which

$$\|\boldsymbol{f}|[L_2,L_r]\|=\Bigg[\int\limits_{\mathbb{S}^n}\Big(\int\limits_{\mathbb{U}^n}|f(t,A_n)|^r d\chi^n(A_n)\Big)^{2/r}d\omega^n(t)\Bigg]^{1/2}$$

is finite. We now define an operator J from $L_r(\mathbb{S}^n)$ into $[L_2(\mathbb{S}^n),L_r(\mathbb{U}^n)]$ by letting

$$Jf(t,A_n):=f(A_nt)\quad\text{for } f\in L_r(\mathbb{S}^n),\ t\in\mathbb{S}^n \text{ and } A_n\in\mathbb{U}^n.$$

LEMMA. *$J\in\mathfrak{L}(L_r(\mathbb{S}^n),[L_2(\mathbb{S}^n),L_r(\mathbb{U}^n)])$ is a metric injection.*

PROOF. For $f\in L_r(\mathbb{S}^n)$ and $t\in\mathbb{S}^n$, we have

$$\int\limits_{\mathbb{U}^n}|f(A_nt)|^r d\chi^n(A_n)=\int\limits_{\mathbb{S}^n}|f(s)|^r d\omega^n(s).$$

Therefore

$$\|Jf|[L_2, L_r]\| = \left[\int\limits_{\mathbb{S}^n}\left(\int\limits_{\mathbb{U}^n} |f(A_n t)|^r d\chi^n(A_n)\right)^{2/r} d\omega^n(t)\right]^{1/2}$$

$$= \left[\int\limits_{\mathbb{S}^n}\left(\int\limits_{\mathbb{S}^n} |f(s)|^r d\omega^n(s)\right)^{2/r} d\omega^n(t)\right]^{1/2} = \|f|L_r\|.$$

4.12.10 We are now in a position to establish a counterpart of 4.12.7.

EXAMPLE. *Let* $1 \le p \le 2$. *Then*

$$M_p(\mathcal{Q}_n)\, M_{p'}(\mathcal{Q}_n) \le \boldsymbol{\delta}(L_p|\mathcal{Q}_n, \mathcal{Q}_n) = \boldsymbol{\delta}(L_{p'}|\mathcal{Q}_n, \mathcal{Q}_n) \le M_{p'}(\mathcal{Q}_n).$$

PROOF. The upper estimate is a consequence of 3.4.6 and 3.4.15.

The following manipulations show that the diagram

$$\begin{array}{ccc} L_p(\mathbb{S}^n) & \xrightarrow{\;Q_n Q_n^*\;} & L_p(\mathbb{S}^n) \\ J\big\downarrow & & \big\downarrow J \\ [L_2(\mathbb{S}^n), L_p(\mathbb{U}^n)] & \xrightarrow[{[Q_n Q_n^*, L_p]}]{} & [L_2(\mathbb{S}^n), L_p(\mathbb{U}^n)] \end{array}$$

is commutative. Indeed, for $f \in L_p(\mathbb{S}^n)$, $t \in \mathbb{S}^n$ and $A_n = (\alpha_{hk}) \in \mathbb{U}^n$, we get

$$JQ_nQ_n^* f(t, A_n) = \sum_{h=1}^{n} \int\limits_{\mathbb{S}^n} f(a)\, \overline{q_h^{(n)}(a)}\, d\omega^n(a)\, q_h^{(n)}(A_n t)$$

$$= \sum_{h=1}^{n}\sum_{k=1}^{n} \int\limits_{\mathbb{S}^n} f(a)\, \overline{q_h^{(n)}(a)}\, d\omega^n(a)\, \alpha_{hk} q_k^{(n)}(t)$$

and, by letting $b = A_n^* a$,

$$[Q_nQ_n^*, L_p] Jf(t, A_n) = \sum_{k=1}^{n} \int\limits_{\mathbb{S}^n} f(A_n b)\, \overline{q_k^{(n)}(b)}\, d\omega^n(b)\, q_k^{(n)}(t)$$

$$= \sum_{k=1}^{n} \int\limits_{\mathbb{S}^n} f(a)\, \overline{q_k^{(n)}(A_n^* a)}\, d\omega^n(a)\, q_k^{(n)}(t)$$

$$= \sum_{h=1}^{n}\sum_{k=1}^{n} \int\limits_{\mathbb{S}^n} f(a)\, \overline{\alpha_{kh}^*\, q_h^{(n)}(a)}\, d\omega^n(a)\, q_k^{(n)}(t).$$

Thus 4.12.8 implies that

$$\begin{aligned} M_p(\mathcal{Q}_n)\, M_{p'}(\mathcal{Q}_n) &= \|Q_n Q_n^* : L_p(\mathbb{S}^n) \to L_p(\mathbb{S}^n)\| \\ &\le \|[Q_n Q_n^*, L_p] : [L_2(\mathbb{S}^n), L_p(\mathbb{U}^n)] \to [L_2(\mathbb{S}^n), L_p(\mathbb{U}^n)]\| \\ &= \boldsymbol{\delta}(L_p(\mathbb{U}^n)|\mathcal{Q}_n, \mathcal{Q}_n). \end{aligned}$$

This proves the lower estimate for $L_p(\mathbb{U}^n)$. The general case follows from the fact that $L_p(\mathbb{U}^n)$ can be embedded isometrically into every infinite dimensional space L_p.

REMARK. In analogy with 4.12.7, we conjecture that

$$\boldsymbol{\delta}(L_r|\mathcal{Q}_n, \mathcal{Q}_n) \stackrel{?}{=} M_r(\mathcal{Q}_n)\, M_{r'}(\mathcal{Q}_n) \quad \text{for } 1 \le r \le \infty.$$

4.12.11 We now summarize the results from 4.12.7 and 4.12.10.

EXAMPLE.

$$\boldsymbol{\delta}(L_r|\mathcal{R}_n, \mathcal{R}_n) \asymp 1 \ \text{ if } 1<r<\infty \text{ and } \ \boldsymbol{\delta}(L_1|\mathcal{R}_n, \mathcal{R}_n) = \boldsymbol{\delta}(L_\infty|\mathcal{R}_n, \mathcal{R}_n) \asymp \sqrt{n},$$
$$\boldsymbol{\delta}(L_r|\mathcal{G}_n, \mathcal{G}_n) \asymp 1 \ \text{ if } 1<r<\infty \text{ and } \ \boldsymbol{\delta}(L_1|\mathcal{G}_n, \mathcal{G}_n) = \boldsymbol{\delta}(L_\infty|\mathcal{G}_n, \mathcal{G}_n) \asymp \sqrt{n}.$$

4.12.12 The preceding example can be viewed as a special case of a general result, which will follow from Pisier's theorem 4.14.10. A similar situation has already been described in 4.8.12.

DICHOTOMY. *For every Banach space X, we have either*

$$\boldsymbol{\delta}(X|\mathcal{R}_n, \mathcal{R}_n) \asymp \boldsymbol{\delta}(X|\mathcal{G}_n, \mathcal{G}_n) \asymp 1 \quad \text{or} \quad \boldsymbol{\delta}(X|\mathcal{R}_n, \mathcal{R}_n) \asymp \boldsymbol{\delta}(X|\mathcal{G}_n, \mathcal{G}_n) \asymp \sqrt{n}.$$

4.12.13 In what follows, let n and N be natural numbers.

EXAMPLE.

$$\begin{aligned} \boldsymbol{\delta}(l_1^N|\mathcal{R}_n, \mathcal{R}_n) &= \boldsymbol{\delta}(l_\infty^N|\mathcal{R}_n, \mathcal{R}_n) \asymp \min\{\sqrt{n}, \sqrt{1+\log N}\}, \\ \boldsymbol{\delta}(l_1^N|\mathcal{G}_n, \mathcal{G}_n) &= \boldsymbol{\delta}(l_\infty^N|\mathcal{G}_n, \mathcal{G}_n) \asymp \min\{\sqrt{n}, \sqrt{1+\log N}\}. \end{aligned}$$

PROOF. By 3.4.6 and 4.12.4, we have

$$\boldsymbol{\delta}(l_1^N|\mathcal{R}_n, \mathcal{R}_n) \prec \boldsymbol{\delta}(l_1^N|\mathcal{G}_n, \mathcal{G}_n) = \boldsymbol{\delta}(l_\infty^N|\mathcal{G}_n, \mathcal{G}_n)$$

and

$$\boldsymbol{\delta}(l_1^N|\mathcal{R}_n, \mathcal{R}_n) = \boldsymbol{\delta}(l_\infty^N|\mathcal{R}_n, \mathcal{R}_n) \prec \boldsymbol{\delta}(l_\infty^N|\mathcal{G}_n, \mathcal{G}_n).$$

Hence it suffices to estimate $\boldsymbol{\delta}(l_1^N|\mathcal{R}_n, \mathcal{R}_n)$ from below and $\boldsymbol{\delta}(l_\infty^N|\mathcal{G}_n, \mathcal{G}_n)$ from above.

Obviously,

$$\boldsymbol{\delta}(l_\infty^N|\mathcal{G}_n, \mathcal{G}_n) \le \sqrt{n}.$$

On the other hand,

$$\boldsymbol{\delta}(l_q^N|\mathcal{G}_n, \mathcal{G}_n) \le \sqrt{q} \qquad \text{(by 4.12.10).}$$

Thus, applying 1.3.6 yields

$$\delta(l_\infty^N|\mathcal{G}_n,\mathcal{G}_n) \prec \sqrt{1+\log N}.$$

This proves the upper estimate.

Letting $m := \min\{n, [\log_2 N]\}$, the required lower estimate can be deduced from 3.8.12, which tells us that

$$\begin{aligned}\sqrt{\tfrac{m}{2}} &\leq M_1(\mathcal{P}_m(\pm1))\sqrt{m} = \delta(L_1(\mathbb{E}_2^m)|\mathcal{P}_m(\pm1),\mathcal{P}_m(\pm1))\\ &\leq \delta(l_1^N|\mathcal{P}_n(\pm1),\mathcal{P}_n(\pm1)).\end{aligned}$$

REMARK. Since $\varrho(l_1^N|\mathcal{I}_n,\mathcal{R}_n)\asymp 1$, it follows from

$$\varrho(l_\infty^N|\mathcal{R}_n,\mathcal{I}_n) \leq \varrho(l_1^N|\mathcal{I}_n,\mathcal{R}_n)\,\delta(l_1^N|\mathcal{R}_n,\mathcal{R}_n)$$

that

$$\varrho(l_\infty^N|\mathcal{R}_n,\mathcal{I}_n) \prec \delta(l_1^N|\mathcal{R}_n,\mathcal{R}_n).$$

Thus half of the proofs in 4.2.11 and in this paragraph could have been omitted.

4.12.14 Let C_α denote the diagonal operator on $[l_2, l_\infty^{2^k}]$ defined by

$$C_\alpha : (x_k) \longrightarrow (k^{-\alpha}x_k), \quad \text{where} \quad x_k \in l_\infty^{2^k}.$$

We are now in a position to show that the dichotomy described in 4.12.12 only occurs for spaces, not for operators.

EXAMPLE. *If* $0 \leq \alpha \leq 1/2$, *then*

$$\delta(C_\alpha|\mathcal{R}_n,\mathcal{R}_n) \asymp \delta(C_\alpha|\mathcal{G}_n,\mathcal{G}_n) \asymp n^{1/2-\alpha}.$$

PROOF. Recall that

$$\delta(l_\infty^N|\mathcal{R}_n,\mathcal{R}_n) \asymp \min\{\sqrt{n}, \sqrt{1+\log N}\} \qquad \text{(by 4.12.13)}.$$

The desired relation is now a consequence of $(\triangle)$ in 1.2.12.

4.13 Inequalities between $\delta(\mathcal{R}_n,\mathcal{R}_n)$ and $\varrho(\mathcal{R}_n,\mathcal{I}_n)$

4.13.1 In this section, we compare the ideal norms $\delta(\mathcal{R}_n,\mathcal{R}_n)$ and $\varrho(\mathcal{R}_n,\mathcal{I}_n)$ with each other.

PROPOSITION. $\delta(\mathcal{R}_n,\mathcal{R}_n) \leq \sqrt{\frac{\pi}{2}}\,\varrho(\mathcal{R}_{2^n},\mathcal{I}_{2^n})$.

PROOF (G. Pisier). We need the following facts:
Recall from 4.8.3 that

$$\varrho(\mathcal{R}_n,\mathcal{G}_n^{\mathbb{R}}) \leq \sqrt{\tfrac{\pi}{2}}. \tag{1}$$

By 4.9.4,

$$\boldsymbol{\varrho}(\mathcal{G}_n^{\mathbb{R}}, \mathcal{I}_n) \leq \boldsymbol{\varrho}(\mathcal{R}_n, \mathcal{I}_n). \tag{2}$$

Let $\mathcal{W}_{2^n}$ denote the system of all characters on $\mathbb{E}_2^n$; see 6.1.5. Since the Gaussian measure $\gamma_{2^n}^{\mathbb{R}}$ is invariant under the associated orthogonal transformation, we obtain

$$\boldsymbol{\varrho}(\mathcal{G}_{2^n}^{\mathbb{R}}, \mathcal{W}_{2^n}) = \boldsymbol{\varrho}(\mathcal{G}_{2^n}^{\mathbb{R}}, \mathcal{I}_{2^n}). \tag{3}$$

Furthermore, in view of 3.4.12

$$\boldsymbol{\delta}(\mathcal{R}_{2^n}, \mathcal{W}_{2^n}) = \boldsymbol{\varrho}(\mathcal{R}_{2^n}, \mathcal{W}_{2^n}). \tag{4}$$

Next, we choose a subset $\mathbb{F}$ of $\{1, \ldots, 2^n\}$ such that $\mathcal{W}_{2^n}(\mathbb{F})$ coincides with $\mathcal{P}_n(\pm 1)$. Thus $\mathcal{W}_{2^n}(\mathbb{F})$, $\mathcal{R}_{2^n}(\mathbb{F})$ and $\mathcal{R}_n$ have identical distributions. Hence

$$\boldsymbol{\delta}(\mathcal{R}_n, \mathcal{R}_n) = \boldsymbol{\delta}(\mathcal{R}_{2^n}(\mathbb{F}), \mathcal{W}_{2^n}(\mathbb{F})). \tag{5}$$

Finally, 4.1.4 implies that

$$\boldsymbol{\delta}(\mathcal{R}_{2^n}(\mathbb{F}), \mathcal{W}_{2^n}(\mathbb{F})) \leq \boldsymbol{\delta}(\mathcal{R}_{2^n}, \mathcal{W}_{2^n}). \tag{6}$$

We are now prepared to conclude as follows:

$$\begin{aligned}
\boldsymbol{\delta}(\mathcal{R}_n, \mathcal{R}_n) &= \boldsymbol{\delta}(\mathcal{R}_{2^n}(\mathbb{F}), \mathcal{W}_{2^n}(\mathbb{F})) \leq \boldsymbol{\delta}(\mathcal{R}_{2^n}, \mathcal{W}_{2^n}) && \text{(by (5) and (6))},\\
&= \boldsymbol{\varrho}(\mathcal{R}_{2^n}, \mathcal{W}_{2^n}) \leq \sqrt{\tfrac{\pi}{2}}\, \boldsymbol{\varrho}(\mathcal{G}_{2^n}^{\mathbb{R}}, \mathcal{W}_{2^n}) && \text{(by (4) and (1))},\\
&= \sqrt{\tfrac{\pi}{2}}\, \boldsymbol{\varrho}(\mathcal{G}_{2^n}^{\mathbb{R}}, \mathcal{I}_{2^n}) \leq \sqrt{\tfrac{\pi}{2}}\, \boldsymbol{\varrho}(\mathcal{R}_{2^n}, \mathcal{I}_{2^n}) && \text{(by (3) and (2))}.
\end{aligned}$$

REMARK. It is unknown whether the above result is optimal in the following sense. Suppose that $\boldsymbol{\delta}(\mathcal{R}_n, \mathcal{R}_n) \leq c\, \boldsymbol{\varrho}(\mathcal{R}_{\varphi(n)}, \mathcal{I}_{\varphi(n)})$, where $\varphi : \mathbb{N} \to \mathbb{N}$. Do we have $a^n \prec \varphi(n)$ for some $a > 1$?

4.13.2 Next, we present a counterpart of the previous inequality.

PROPOSITION. $2^{-n/2}\boldsymbol{\varrho}(\mathcal{R}_{2^n}, \mathcal{I}_{2^n}) \leq \sqrt{2}\, n^{-1/2}\boldsymbol{\delta}(\mathcal{R}_n, \mathcal{R}_n)$.

PROOF. Given $x_1, \ldots, x_{2^n} \in X$, we define $\boldsymbol{f} : [0,1) \times [0,1) \to X$ by

$$\boldsymbol{f}(s,t) := x_k\, r_k(t) \quad \text{for} \quad s \in \Delta_n^{(k)} := \left[\tfrac{k-1}{2^n}, \tfrac{k}{2^n}\right) \quad \text{and} \quad t \in [0,1).$$

Then

$$\int_0^1\!\!\int_0^1 \|\boldsymbol{f}(s,t)\|^2 ds\, dt = \int_0^1 \left[\sum_{k=1}^{2^n} \int_{\Delta_n^{(k)}} \|\boldsymbol{f}(s,t)\|^2 ds\right] dt = \frac{1}{2^n}\sum_{k=1}^{2^n} \|x_k\|^2.$$

By integration over $t \in [0,1)$, it follows from

$$\int_0^1 \left\| \sum_{h=1}^{n} T\langle \boldsymbol{f}(\cdot,t), r_h\rangle\, r_h(u) \right\|^2 du \leq \boldsymbol{\delta}(T|\mathcal{R}_n, \mathcal{R}_n)^2 \int_0^1 \|\boldsymbol{f}(s,t)\|^2 ds$$

that

$$\int_0^1\!\!\int_0^1 \Big\| \sum_{h=1}^{n} T\,\langle \boldsymbol{f}(\cdot,t), r_h\rangle\, r_h(u) \Big\|^2 du\,dt \le \boldsymbol{\delta}(T|\mathcal{R}_n,\mathcal{R}_n)^2 \int_0^1\!\!\int_0^1 \|\boldsymbol{f}(s,t)\|^2 ds\,dt.$$

Hence

$$\int_0^1\!\!\int_0^1 \Big\| \sum_{h=1}^{n} T\,\langle \boldsymbol{f}(\cdot,t), r_h\rangle\; r_h(u) \Big\|^2 du\,dt \le \boldsymbol{\delta}(T|\mathcal{R}_n,\mathcal{R}_n)^2 \frac{1}{2^n}\sum_{k=1}^{2^n} \|x_k\|^2. \quad (1)$$

Moreover, we have

$$\begin{aligned}\sum_{h=1}^{n} \langle \boldsymbol{f}(\cdot,t), r_h\rangle\; r_h &= \sum_{h=1}^{n}\sum_{k=1}^{2^n} x_k r_k(t) \int_{\Delta_n^{(k)}} r_h(s)\,ds\, r_h \\ &= \frac{1}{2^n}\sum_{k=1}^{2^n} x_k r_k(t) \sum_{h=1}^{n} \varepsilon_{hk} r_h, \qquad (2)\end{aligned}$$

where $\varepsilon_{hk} = \pm 1$ is the sign of r_h on $\Delta_n^{(k)}$. Finally, Khintchine's inequality yields

$$\lambda_k := \int_0^1 \Big| \sum_{h=1}^{n} \varepsilon_{hk} r_h(u) \Big|\, du \ge \sqrt{\tfrac{n}{2}} \quad \text{for } k = 1, \ldots, 2^n. \qquad (3)$$

We now conclude as follows:

$$\|(Tx_k)|\mathcal{R}_{2^n}\| = \Big(\int_0^1 \Big\| \sum_{k=1}^{2^n} T x_k r_k(t) \Big\|^2 dt \Big)^{1/2}$$

(by definition)

$$\le \Big(\frac{2}{n} \int_0^1 \Big\| \sum_{k=1}^{2^n} T x_k r_k(t) \lambda_k \Big\|^2 dt \Big)^{1/2}$$

(by (3) and the contraction principle)

$$= \Big(\frac{2}{n} \int_0^1 \Big\| \int_0^1 \sum_{k=1}^{2^n} T x_k r_k(t) \Big| \sum_{h=1}^{n} \varepsilon_{hk} r_h(u) \Big|\, du \Big\|^2 dt \Big)^{1/2}$$

(by definition of λ_k)

$$\le \Big(\frac{2}{n} \int_0^1\!\!\int_0^1 \Big\| \sum_{k=1}^{2^n} T x_k r_k(t) \Big| \sum_{h=1}^{n} \varepsilon_{hk} r_h(u) \Big| \Big\|^2 du\,dt \Big)^{1/2}$$

(by the triangle inequality for integrals)

$$= \Big(\frac{2}{n} \int_0^1\!\!\int_0^1 \Big\| \sum_{k=1}^{2^n} T x_k r_k(t) \sum_{h=1}^{n} \varepsilon_{hk} r_h(u) \Big\|^2 du\,dt \Big)^{1/2}$$

(by the unconditionality of $\mathcal{R}_n$)

$$= \left(\frac{2}{n} \int_0^1 \int_0^1 \left\| 2^n \sum_{h=1}^{n} T \langle f(\cdot, t), r_h \rangle \, r_h(u) \right\|^2 du \, dt \right)^{1/2} \qquad \text{(by (2))}$$

$$\leq \boldsymbol{\delta}(T|\mathcal{R}_n, \mathcal{R}_n) \left(2 \, \tfrac{2^n}{n} \sum_{k=1}^{2^n} \|x_k\|^2 \right)^{1/2} = \sqrt{2 \, \tfrac{2^n}{n}} \, \boldsymbol{\delta}(T|\mathcal{R}_n, \mathcal{R}_n) \, \|(x_k)|\mathcal{I}_{2^n}\|, \qquad \text{(by (1))}$$

which implies the required inequality.

REMARK. Recall from 4.2.10 and 4.12.13 that

$$\boldsymbol{\varrho}(l_1^N|\mathcal{R}_N, \mathcal{I}_N) = \sqrt{N} \quad \text{and} \quad \boldsymbol{\delta}(l_1^N|\mathcal{R}_n, \mathcal{R}_n) \asymp \min\{\sqrt{n}, \sqrt{1+\log N}\}.$$

Therefore, reasoning as in 4.2.17 (Remark), it follows that the above result is optimal.

4.13.3 In this paragraph, all functions are assumed to be real. Recall that a finite measure space (M, μ) is non-atomic if for every measurable set B and $0 < \theta < 1$ we can find a measurable set B_θ in B such that $\mu(B_\theta) = \theta \, \mu(B)$.

SUBLEMMA. *Let (M, μ) be a non-atomic finite measure space. If*

$$g \in L_1(M, \mu) \quad \textit{and} \quad \int_M g(\xi) \, d\mu(\xi) = 0,$$

then there is a measurable subset B for which

$$\int_B g(\xi) \, d\mu(\xi) \geq \tfrac{1}{4} \int_M |g(\xi)| \, d\mu(\xi) \quad \textit{and} \quad \mu(B) = \tfrac{1}{2} \mu(M).$$

PROOF. By homogeneity, it may be assumed that $\mu(M) = 1$. Let

$$M_+ := \Big\{ \xi \in M : g(\xi) \geq 0 \Big\} \quad \text{and} \quad M_- := \Big\{ \xi \in M : g(\xi) \leq 0 \Big\}.$$

Since $M_+ \cup M_- = M$, at least one of the relations $\mu(M_+) \geq \frac{1}{2}$ and $\mu(M_-) \geq \frac{1}{2}$ holds.

To treat the first case, we define

$$M_t := \Big\{ \xi \in M : g(\xi) \geq t \Big\} \quad \text{for } t \geq 0$$

and

$$s := \sup \Big\{ t \geq 0 : \mu(M_t) \geq \tfrac{1}{2} \Big\}.$$

Then it follows from

$$M_s = \bigcap_{t<s} M_t \quad \text{if } s > 0 \quad \text{and} \quad M_s = M_+ \quad \text{if } s = 0$$

that $\mu(M_s) \geq \frac{1}{2}$. On the other hand, letting

$$M_{s+0} := \bigcup_{t>s} M_t,$$

we have $\mu(M_{s+0}) \leq \frac{1}{2}$. Choose a measurable set B such that

$$M_{s+0} \subseteq B \subseteq M_s \subseteq M_+ \quad \text{and} \quad \mu(B) = \tfrac{1}{2}.$$

For $\xi \in B$, we conclude from $g(\xi) \geq s$ that

$$\int_B g(\xi)\, d\mu(\xi) \geq s\,\mu(B) = s/2,$$

while $\xi \notin B$, via $\xi \notin M_{s+0}$, implies that $g(\xi) \leq s$ and, therefore,

$$\int_{M_+ \setminus B} g(\xi)\, d\mu(\xi) \leq s\,\mu(M_+ \setminus B) \leq s/2.$$

Consequently,

$$\int_{M_+ \setminus B} g(\xi)\, d\mu(\xi) \leq \int_B g(\xi)\, d\mu(\xi).$$

By assumption,

$$\int_{M_+} g(\xi)\, d\mu(\xi) + \int_{M_-} g(\xi)\, d\mu(\xi) = 0.$$

Hence

$$\int_M |g(\xi)|\, d\mu(\xi) = \int_{M_+} g(\xi)\, d\mu(\xi) - \int_{M_-} g(\xi)\, d\mu(\xi) = 2 \int_{M_+} g(\xi)\, d\mu(\xi)$$

$$= 2 \int_B g(\xi)\, d\mu(\xi) + 2 \int_{M_+ \setminus B} g(\xi)\, d\mu(\xi) \leq 4 \int_B g(\xi)\, d\mu(\xi).$$

Thus we have indeed

$$\int_B g(\xi)\, d\mu(\xi) \geq \tfrac{1}{4} \int_M |g(\xi)|\, d\mu(\xi).$$

In the second case, it is necessary to pass from g to $-g$.

LEMMA. *Let (M, μ) be a non-atomic probability space. Fix $0 < \delta \leq 1$ and $n = 1, 2, \ldots$. If F is a subset of the closed unit ball in $L_\infty(M, \mu)$ such that*

$$\frac{1}{2^n} \sum_{\mathbb{E}_2^n} \left\| \sum_{k=1}^n f_k\, \varepsilon_k \middle| L_\infty \right\| \leq \tfrac{1}{4} \delta^2 n \quad \text{for } f_1, \ldots, f_n \in F,$$

then there exists an orthogonal projection $P \in \mathfrak{L}(L_2(M,\mu))$ with

$$\operatorname{rank}(P) \le 2^{n-1} \quad \textit{and} \quad \|f - Pf|L_2\| \le \delta \quad \textit{for } f \in F.$$

PROOF. Define the orthogonal projection $P \in \mathfrak{L}(L_2(M,\mu))$ by

$$Pf(\xi) := \int_M f(\eta)\, d\mu(\eta) \quad \text{if } \xi \in M.$$

In the case when

$$\|f - Pf|L_2\| \le \delta \quad \text{for all } f \in F,$$

we are done. Otherwise there is $f \in F$ such that

$$\|f - Pf|L_2\| > \delta.$$

Let $g := f - Pf$. It follows from

$$\|g|L_\infty\| \le \|f|L_\infty\| + \|Pf|L_\infty\| \le 2 \quad \text{and} \quad \delta^2 < \|g|L_2\|^2 \le \|g|L_1\|\, \|g|L_\infty\|$$

that

$$\|g|L_1\| > \tfrac{1}{2}\delta^2.$$

In view of

$$\int_M g(\xi)\, d\mu(\xi) = 0,$$

the preceding sublemma applies to g, and we can find a measurable subset $B_{+1}^{(1)}$ of M such that

$$\int_{B_{+1}^{(1)}} g(\xi)\, d\mu(\xi) \ge \tfrac{1}{4} \int_M |g(\xi)|\, d\mu(\xi) \quad \text{and} \quad \mu(B_{+1}^{(1)}) = \tfrac{1}{2}.$$

Let $B_{-1}^{(1)} := M \setminus B_{+1}^{(1)}$. Then

$$\int_{B_{+1}^{(1)}} f(\xi)\, d\mu(\xi) - \int_{B_{-1}^{(1)}} f(\xi)\, d\mu(\xi) = \int_{B_{+1}^{(1)}} g(\xi)\, d\mu(\xi) - \int_{B_{-1}^{(1)}} g(\xi)\, d\mu(\xi)$$

$$= 2 \int_{B_{+1}^{(1)}} g(\xi)\, d\mu(\xi) > \tfrac{1}{2} \int_M |g(\xi)|\, d\mu(\xi) > \tfrac{1}{4}\delta^2.$$

Since $\mathbb{E}_2^1$ consists of the two elements $e = (\varepsilon_1)$ with $\varepsilon_1 = +1$ and $\varepsilon_1 = -1$, we obtain

$$\sup_{f \in F} \sum_{e \in \mathbb{E}_2^1} \varepsilon_1 \int_{B_e^{(1)}} f(\xi)\, d\mu(\xi) > \tfrac{1}{4}\delta^2. \qquad (I_1)$$

Proceeding by induction, we assume that, for some $m < n$, there exists a partition of M into pairwise disjoint measurable subsets $B_e^{(m)}$ such that $\mu(B_e^{(m)}) = \frac{1}{2^m}$ for $e = (\varepsilon_1, \ldots, \varepsilon_m) \in \mathbb{E}_2^m$ and

$$\sup_{f \in F} \sum_{e \in \mathbb{E}_2^m} \varepsilon_k \int_{B_e^{(m)}} f(\xi)\, d\mu(\xi) > \tfrac{1}{4}\delta^2 \quad \text{for } k = 1, \ldots, m. \qquad (I_m)$$

Define the 2^m-dimensional orthogonal projection $P \in \mathfrak{L}(L_2(M,\mu))$ by

$$Pf(\xi) := \frac{1}{\mu(B_e^{(m)})} \int_{B_e^{(m)}} f(\eta)\, d\mu(\eta) \quad \text{if } \xi \in B_e^{(m)}.$$

In the case when

$$\|f - Pf | L_2\| \le \delta \quad \text{for all } f \in F,$$

we are done. Otherwise there is $f \in F$ such that

$$\|f - Pf | L_2\| > \delta.$$

Let $g := f - Pf$. As above, we get

$$\|g | L_1\| > \tfrac{1}{2}\delta^2.$$

In view of

$$\int_{B_e^{(m)}} g(\xi)\, d\mu(\xi) = 0,$$

the preceding sublemma applies to the restriction of g to $B_e^{(m)}$. So, for every $e \in \mathbb{E}_2^m$, we can find a measurable subset $B_{e,+1}^{(m+1)}$ of $B_e^{(m)}$ such that

$$\int_{B_{e,+1}^{(m+1)}} g(\xi)\, d\mu(\xi) \ge \tfrac{1}{4} \int_{B_e^{(m)}} |g(\xi)|\, d\mu(\xi) \quad \text{and} \quad \mu(B_{e,+1}^{(m+1)}) = \tfrac{1}{2^{m+1}}.$$

Let $B_{e,-1}^{(m+1)} := B_e^{(m)} \setminus B_{e,+1}^{(m+1)}$. Then

$$\int_{B_{e,+1}^{(m+1)}} f(\xi)\, d\mu(\xi) - \int_{B_{e,-1}^{(m+1)}} f(\xi)\, d\mu(\xi) = \int_{B_{e,+1}^{(m+1)}} g(\xi)\, d\mu(\xi) - \int_{B_{e,-1}^{(m+1)}} g(\xi)\, d\mu(\xi)$$

$$= 2 \int_{B_{e,+1}^{(m+1)}} g(\xi)\, d\mu(\xi) > \tfrac{1}{2} \int_{B_e^{(m)}} |g(\xi)|\, d\mu(\xi).$$

Summation over $e \in \mathbb{E}_2^m$ yields

$$\sum_{e \in \mathbb{E}_2^m} \int_{B_{e,+1}^{(m+1)}} f(\xi)\, d\mu(\xi) - \sum_{e \in \mathbb{E}_2^m} \int_{B_{e,-1}^{(m+1)}} f(\xi)\, d\mu(\xi) > \tfrac{1}{2} \int_M |g(\xi)|\, d\mu(\xi) > \tfrac{1}{4}\delta^2.$$

So inequalities $(I_1), \ldots, (I_{m+1})$ hold for the refined partition $\{B^{(m+1)}_{e,\pm 1}\}$, where the index $e, \pm 1$ with $e \in \mathbb{E}_2^m$ ranges over $\mathbb{E}_2^{m+1}$.

Assume that this process can be extended until $m = n$. Then, by (I_n), we can find $f_k \in F$ such that

$$\sum_{e \in \mathbb{E}_2^n} \varepsilon_k \int\limits_{B_e^{(n)}} f_k(\xi)\, d\mu(\xi) > \tfrac{1}{4}\delta^2.$$

Summation over k yields

$$\sum_{e \in \mathbb{E}_2^n} \int\limits_{B_e^{(n)}} \sum_{k=1}^n f_k(\xi)\, \varepsilon_k\, d\mu(\xi) > \tfrac{1}{4}\delta^2\, n.$$

On the other hand, by summation over $e \in \mathbb{E}_2^n$, it follows from

$$\int\limits_{B_e^{(n)}} \sum_{k=1}^n f_k(\xi)\, \varepsilon_k\, d\mu(\xi) \le \mu(B_e^{(n)}) \left\| \sum_{k=1}^n f_k\, \varepsilon_k \middle| L_\infty \right\|$$

that

$$\sum_{e \in \mathbb{E}_2^n} \int\limits_{B_e^{(n)}} \sum_{k=1}^n f_k(\xi)\, \varepsilon_k\, d\mu(\xi) \le \sum_{e \in \mathbb{E}_2^n} \mu(B_e^{(n)}) \left\| \sum_{k=1}^n f_k\, \varepsilon_k \middle| L_\infty \right\| \le \tfrac{1}{4}\delta^2\, n.$$

This contradiction guarantees that the induction terminates before n, which proves the existence of the required projection.

4.13.4 For $T \in \mathfrak{L}(X, Y)$ and $n = 1, 2, \ldots$, the n-th **approximation number** is defined by

$$a_n(T) := \inf\Big\{ \|T - L\| : L \in \mathfrak{L}(X, Y),\ \operatorname{rank}(L) < n \Big\}.$$

More information about these quantities is given in [PIE 2, p. 146] and [PIE 3, pp. 83–90].

4.13.5 The following results and Lemma 4.13.3 are due to J. Bourgain; see [TOM, pp. 214–217 and 232].

PROPOSITION. *Let $T \in \mathfrak{L}(X, Y)$ and $S \in \mathfrak{P}_2(Y, Z)$. Then*

$$a_{2^n}(ST) \le 2\, \|S|\mathfrak{P}_2\| \left[n^{-1/2}\, \boldsymbol{\varrho}(T|\mathfrak{R}_n, \mathfrak{I}_n)\, \|T\| \right]^{1/2}.$$

PROOF. By homogeneity, we may assume that $\|T\| = 1$ and $\|S|\mathfrak{P}_2\| = 1$. Let λ be the Lebesgue measure on $[0, 1]$. Note that, for any probability space (M, μ), the product measure $\lambda \times \mu$ is non-atomic. Indeed, if B is a measurable subset of $[0, 1] \times M$, then $(\lambda \times \mu)(B \cap (0, t) \times M)$ is a continuous function of t on $[0, 1]$ which increases from 0 to $(\lambda \times \mu)(B)$.

Define
$$Uf(t,\xi) := f(\xi) \quad \text{for } f \in L_\infty(M,\mu)$$
and
$$Vg(\xi) := \int_0^1 g(t,\xi)\,dt \quad \text{for } g \in L_2([0,1]\times M, \lambda\times\mu).$$

Then the diagram
$$\begin{array}{ccc} L_\infty(M,\mu) & \xrightarrow{\ Id\ } & L_2(M,\mu) \\ {\scriptstyle U}\big\downarrow & & \big\uparrow{\scriptstyle V} \\ L_\infty([0,1]\times M, \lambda\times\mu) & \xrightarrow[Id_0]{} & L_2([0,1]\times M, \lambda\times\mu), \end{array}$$
in which Id and Id_0 denote identity operators, is commutative. The factorization theorem for 2-summing operators yields a diagram
$$\begin{array}{ccc} Y & \xrightarrow{\ S\ } & Z \\ {\scriptstyle A}\big\downarrow & & \big\uparrow{\scriptstyle B} \\ L_\infty(M,\mu) & \xrightarrow[Id]{} & L_2(M,\mu), \end{array}$$
where $\|A\| = \|B\| = 1$; see [DIE*a, p. 48] or [PIE 3, p. 58]. In view of our preliminary considerations, we may assume that the probability space (M,μ) is non-atomic. Observe that
$$\int_0^1 \Big\| \sum_{k=1}^n ATx_k\, r_k(t) \Big| L_\infty \Big\|\, dt \le \bigg(\int_0^1 \Big\| \sum_{k=1}^n ATx_k\, r_k(t) \Big| L_\infty \Big\|^2 dt \bigg)^{1/2}$$
$$\le \|A\|\, \boldsymbol{\varrho}(T|\mathcal{R}_n, \mathcal{I}_n) \bigg(\sum_{k=1}^n \|x_k\|^2 \bigg)^{1/2} \quad \le \boldsymbol{\varrho}(T|\mathcal{R}_n, \mathcal{I}_n)\, \sqrt{n}$$
for $x_1, \ldots, x_n \in U_X$. Since the Rademacher system can be replaced by the Bernoulli system, Lemma 4.13.3 applies with
$$F := AT(U_X) \quad \text{and} \quad \delta := 2\,n^{-1/4}\, \boldsymbol{\varrho}(T|\mathcal{R}_n, \mathcal{I}_n)^{1/2}$$
and produces an orthogonal projection $P \in \mathfrak{L}(L_2(M,\mu))$ such that
$$\operatorname{rank}(P) \le 2^{n-1} \quad \text{and} \quad \|ATx - PATx|L_2\| \le \delta \quad \text{for } x \in U_X.$$
This means that $\|AT - PAT\| \le \delta$. Hence
$$a_{2^n}(ST) \le \|ST - BPAT\| \le 2\,n^{-1/4}\, \boldsymbol{\varrho}(T|\mathcal{R}_n, \mathcal{I}_n)^{1/2}.$$

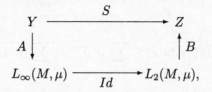

4.13.6 We are now in a position to establish a basic inequality. The universal Dirichlet ideal norm $\boldsymbol{\delta}_n$ was defined in 3.10.1.

PROPOSITION. *There exists a constant $c \geq 1$ such that*

$$2^{-n/2}\boldsymbol{\delta}_{2^n}(T) \leq c\left[n^{-1/2}\,\boldsymbol{\varrho}(T|\mathcal{R}_n,\mathcal{I}_n)\,\|T\|\right]^{1/2}.$$

PROOF. By homogeneity, we may assume that $\|T\| = 1$. In order to estimate $\boldsymbol{\delta}_{2^n}(T)$, we verify property **(4′)** in 3.10.5 and apply 3.10.4. Thus, let $B \in \mathfrak{L}(Y, l_2^{2^n})$ with $\|B|\mathfrak{P}_2\| \leq 1$. In view of 4.13.5 (see the end of the proof), we can find operators $L_1, \ldots, L_n \in \mathfrak{L}(X, l_2^{2^n})$ such that

$$\operatorname{rank}(L_k) \leq 2^{k-1} \quad \text{and} \quad \|BT - L_k\| \leq 2\,k^{-1/4}\,\boldsymbol{\varrho}(T|\mathcal{R}_n,\mathcal{I}_n)^{1/2}.$$

In particular, it is possible to take $L_1 := O$. It follows from

$$BT = (L_2 - L_1) + \ldots + (L_n - L_{n-1}) + (BT - L_n)$$

that

$$BT = S_1 + S_2 + \ldots + S_n,$$

where $\operatorname{rank}(S_k) \leq 2^{k+1}$ for $k = 1, \ldots, n$. Moreover,

$$\|S_k\| \leq \|BT - L_{k+1}\| + \|BT - L_k\| \leq 4\,k^{-1/4}\,\boldsymbol{\varrho}(T|\mathcal{R}_n,\mathcal{I}_n)^{1/2}$$

for $k = 1, \ldots, n-1$. In the case $k = n$, we have

$$\|S_n\| = \|BT - L_n\| \leq 2\,n^{-1/4}\,\boldsymbol{\varrho}(T|\mathcal{R}_n,\mathcal{I}_n)^{1/2}.$$

Hence

$$\sum_{k=1}^{n} 2^{(k+1)/2}\,\|S_k\| \leq \left(4\sum_{k=1}^{n-1}\frac{2^{(k+1)/2}}{k^{1/4}} + 2\,\frac{2^{(n+1)/2}}{n^{1/4}}\right)\boldsymbol{\varrho}(T|\mathcal{R}_n,\mathcal{I}_n)^{1/2}.$$

By induction, it follows that

$$\sum_{k=1}^{n-1}\frac{2^{k/2}}{k^{1/4}} \leq 2.6\cdot\frac{2^{n/2}}{n^{1/4}} \quad \text{for} \quad n \geq 12.$$

The cases when $n \leq 12$ can be checked with the help of a computer. Consequently,

$$4\sum_{k=1}^{n-1}\frac{2^{(k+1)/2}}{k^{1/4}} + 2\,\frac{2^{(n+1)/2}}{n^{1/4}} \leq 18\cdot\frac{2^{n/2}}{n^{1/4}}.$$

Next, we refer to an important inequality about 2-summing operators which says that

$$\|S|\mathfrak{P}_2\| \leq \sqrt{N}\,\|S\| \quad \text{whenever} \quad \operatorname{rank}(S) \leq N;$$

see [PIE 3, p. 45]. Hence

$$\|T'B'|\mathfrak{P}_2\| \leq \sum_{k=1}^{n} \|S'_k|\mathfrak{P}_2\| \leq \sum_{k=1}^{n} 2^{(k+1)/2}\,\|S'_k\| \leq 18 \cdot \frac{2^{n/2}}{n^{1/4}}\,\varrho(T|\mathcal{R}_n,\mathcal{I}_n)^{1/2}.$$

Thus, by 3.10.4 and 3.10.5, we finally arrive at

$$\boldsymbol{\delta}_{2^n}(T) \leq 18 \cdot \frac{2^{n/2}}{n^{1/4}}\,\varrho(T|\mathcal{R}_n,\mathcal{I}_n)^{1/2}$$

which is the desired inequality under the assumption that $\|T\| = 1$.

Until now, we have only dealt with the real case. The complex case, however, follows from 2.3.20 and 3.10.10.

4.13.7 The inequality just proved is sufficient for application in the following section. Nevertheless, compared with 4.13.2 or Hinrichs's inequality 4.11.18, it does not look very aesthetic.

PROBLEM. *Is there a constant* $c \geq 1$ *such that*

$$2^{-n/2}\boldsymbol{\delta}_{2^n} \leq c\,n^{-1/2}\varrho(\mathcal{R}_n,\mathcal{I}_n)\ ?$$

4.14 The vector-valued Rademacher projection – Pisier's theorem –

4.14.1 The first part of this section deals with the smallest non-trivial operator ideals that can be obtained from the Dirichlet ideal norms $\boldsymbol{\delta}(\mathcal{R}_n,\mathcal{R}_n)$ and $\boldsymbol{\delta}(\mathcal{G}_n,\mathcal{G}_n)$.

We refer to T as an **RP-operator** if

$$\|T|\mathfrak{RP}\| := \sup_n \boldsymbol{\delta}(T|\mathcal{R}_n,\mathcal{R}_n)$$

is finite. These operators form the Banach ideal

$$\mathfrak{RP} := \mathfrak{L}[\boldsymbol{\delta}(\mathcal{R}_n,\mathcal{R}_n)].$$

The letters $\mathfrak{RP}$ stand for vector-valued **Rademacher projection**. Replacing the Rademacher system by the Gauss system yields the Banach ideal

$$\mathfrak{GP} := \mathfrak{L}[\boldsymbol{\delta}(\mathcal{G}_n,\mathcal{G}_n)]$$

constituted by the **GP-operators.**

4.14.2 PROPOSITION. *The operator ideal* $\mathfrak{GP}$ *is injective, surjective, symmetric and* l_2*-stable.*

4.14.3 The **Rademacher projection** is defined by

$$P_R : f \longrightarrow \sum_{k=1}^{\infty} \langle f, r_k \rangle \, r_k,$$

and Khintchine's inequality tells us that P_R yields an operator from $L_u[0,1)$ into $L_v[0,1)$ whenever $1 < u, v < \infty$.

4.14.4 An operator $T \in \mathfrak{L}(X,Y)$ is **compatible with the Rademacher projection** P_R from $L_u[0,1)$ into $L_v[0,1)$ if

$$P_R \otimes T : L_u[0,1) \otimes X \longrightarrow L_v[0,1) \otimes Y$$

admits a continuous extension

$$[P_R, T] : [L_u[0,1), X] \longrightarrow [L_v[0,1), Y].$$

In this case, we let

$$\| T | \mathfrak{RP}_u^{(v)} \| := \| [P_R, T] : [L_u[0,1), X] \rightarrow [L_v[0,1), Y] \|.$$

The class of these operators is a Banach ideal, denoted by $\mathfrak{RP}_u^{(v)}$.

4.14.5 THEOREM. *The operator ideal $\mathfrak{RP}_u^{(v)}$ does not depend on the parameters v and u with $1 < u, v < \infty$.*

PROOF. The independence of v is a consequence of the Khintchine–Kahane inequality 4.1.10, and the rest follows from $\big(\mathfrak{RP}_u^{(v)}\big)' = \mathfrak{RP}_{v'}^{(u')}$.

4.14.6 PROPOSITION. $\mathfrak{RP} = \mathfrak{RP}_2^{(2)}$, *and the associated ideal norms coincide.*

PROOF. The unconditionality of the Rademacher system implies that

$$\boldsymbol{\delta}(T | \mathcal{R}_n, \mathcal{R}_n) \leq \| [P_R, T] : [L_2[0,1), X] \rightarrow [L_2[0,1), Y] \|,$$

and, by Fatou's lemma, we have

$$\| [P_R, T] : [L_2[0,1), X] \rightarrow [L_2[0,1), Y] \| \leq \liminf_{n \to \infty} \boldsymbol{\delta}(T | \mathcal{R}_n, \mathcal{R}_n).$$

4.14.7 The remaining part of this section deals with the largest non-trivial operator ideals that can be obtained from the Dirichlet ideal norms $\boldsymbol{\delta}(\mathcal{R}_n, \mathcal{R}_n)$ and $\boldsymbol{\delta}(\mathcal{G}_n, \mathcal{G}_n)$.

Quasi-RP-operators are defined by the property

$$\boldsymbol{\delta}(T | \mathcal{R}_n, \mathcal{R}_n) = o(\sqrt{n}).$$

They form the closed ideal

$$\mathfrak{QRP} := \mathfrak{L}_0[n^{-1/2} \, \boldsymbol{\delta}(\mathcal{R}_n, \mathcal{R}_n)].$$

Replacing the Rademacher system by the Gauss system yields the closed ideal

$$\mathfrak{QGP} := \mathfrak{L}_0[n^{-1/2}\,\delta(\mathcal{G}_n, \mathcal{G}_n)]$$

constituted by the **quasi-GP-operators.**

4.14.8 We now establish the main result of this section.

THEOREM. $\mathfrak{KT} = \mathfrak{GT} = \mathfrak{RT} = \mathfrak{QGP} = \mathfrak{QRP}$.

PROOF. Recall the definitions

$$\mathfrak{KT} := \left\{ T \in \mathfrak{L} : \delta_n(T) = o(\sqrt{n}) \right\} \quad \text{(by 2.3.27, 3.10.10)},$$

$$\mathfrak{RT} := \left\{ T \in \mathfrak{L} : \varrho(T|\mathcal{R}_n, \mathcal{I}_n) = o(\sqrt{n}) \right\} \quad \text{(by 4.3.19)},$$

$$\mathfrak{GT} := \left\{ T \in \mathfrak{L} : \varrho(T|\mathcal{G}_n, \mathcal{I}_n) = o(\sqrt{n}) \right\} \quad \text{(by 4.10.1)},$$

$$\mathfrak{QGP} := \left\{ T \in \mathfrak{L} : \delta(T|\mathcal{G}_n, \mathcal{G}_n) = o(\sqrt{n}) \right\} \quad \text{(by 4.14.7)},$$

$$\mathfrak{QRP} := \left\{ T \in \mathfrak{L} : \delta(T|\mathcal{R}_n, \mathcal{R}_n) = o(\sqrt{n}) \right\} \quad \text{(by 4.14.7)}.$$

Of course, $\mathfrak{GT} = \mathfrak{RT}$. Moreover, $M_1^2\,\delta(\mathcal{R}_n, \mathcal{R}_n) \le \delta(\mathcal{G}_n, \mathcal{G}_n) \le \delta_n$ and 4.13.2 imply that

$$\mathfrak{KT} \subseteq \mathfrak{QGP} \subseteq \mathfrak{QRP} \subseteq \mathfrak{RT}.$$

On the other hand, 4.13.6 yields

$$\mathfrak{RT} \subseteq \mathfrak{KT}.$$

4.14.9 In the setting of spaces, the above chain of equations can be enlarged.

THEOREM. $\mathsf{KT} = \mathsf{GT} = \mathsf{RT} = \mathsf{QGP} = \mathsf{QRP} = \mathsf{GP} = \mathsf{RP}$.

PROOF. In view of $\mathsf{GP} \subseteq \mathsf{RP} \subseteq \mathsf{QRP}$, it is enough to show that $\mathsf{QRP} \subseteq \mathsf{GP}$. Therefore, let $X \in \mathsf{QRP}$. This means that

$$\delta(X|\mathcal{R}_n, \mathcal{R}_n) = o(\sqrt{n}).$$

By duality, we also have

$$\delta(X'|\mathcal{R}_n, \mathcal{R}_n) = o(\sqrt{n}).$$

The injectivity of the Dirichlet ideal norms and

$$\delta(l_1^{2^n}|\mathcal{R}_n, \mathcal{R}_n) = \delta(l_\infty^{2^n}|\mathcal{R}_n, \mathcal{R}_n) \asymp \sqrt{n} \qquad \text{(by 4.12.13)}$$

now imply that neither X nor X' can contain the spaces l_1^n and l_∞^n uniformly. Therefore we may conclude from 4.8.16 that the sequences

$(\varrho(X|\mathcal{G}_n,\mathcal{R}_n))$ and $(\varrho(X'|\mathcal{G}_n,\mathcal{R}_n))$ are bounded. By Pisier's theorem stated in the next paragraph, the sequence $(\delta(X|\mathcal{R}_n,\mathcal{R}_n))$ is bounded as well. Note that

$$\delta(X|\mathcal{G}_n,\mathcal{G}_n) \le \varrho(X|\mathcal{G}_n,\mathcal{R}_n)\,\delta(X|\mathcal{R}_n,\mathcal{R}_n)\,\varrho(X'|\mathcal{G}_n,\mathcal{R}_n),$$

by 3.4.8. Hence $X \in \mathsf{GP}$.

REMARK. For historical and economical reasons, spaces belonging to

$$\mathsf{B} = \mathsf{KT} = \mathsf{GT} = \mathsf{RT} = \mathsf{QGP} = \mathsf{QRP} = \mathsf{GP} = \mathsf{RP}$$

are simply called **B-convex**; see 4.4.4. The attribute **K-convex** is also quite common. G. Pisier says in [pis 16, p. 144] that the reason for the choice of the letter K remains a well kept secret!

4.14.10 We are now able to state one of the most celebrated theorems presented in this book. For an analogous result, the reader is referred to 4.8.22.

PISIER'S THEOREM. *For every Banach space X, the following are equivalent:*

(1) *X has Rademacher subtype.*
(2) *X has non-trivial Rademacher type.*
(3) *X does not contain the spaces l_1^n uniformly.*
(4) *X is* B*-convex.*
(5) $\delta(X|\mathcal{R}_n,\mathcal{R}_n) = O(1)$.
(6) $\delta(X|\mathcal{R}_n,\mathcal{R}_n) = o(\sqrt{n})$.

PROOF. In view of 4.4.7, it remains to show that **(1)⇔(5)⇔(6)**. These equivalences will be proved in 6.4.17 and 6.3.12 with the help of Walsh type ideal norms:

$$\mathsf{RT} = \mathsf{RP} = \mathsf{QRP} = \mathsf{WT}.$$

4.14.11 As a corollary of the preceding theorem we obtain the duality relations between Rademacher type and cotype. See also 4.5.21.

PROPOSITION.

$$(\mathsf{RT}_p)' = \mathsf{RC}_{p'} \cap \mathsf{B} \quad \textit{and} \quad (\mathsf{RC}_q \cap B)' = \mathsf{RT}_{q'}.$$

4.14.12 In view of inequality 4.13.1, we have $\mathfrak{RT}_2 \subset \mathfrak{RP}$. However, the following example shows that

$$\mathfrak{RT}_p \not\subseteq \mathfrak{RP} \quad \text{if } 1 < p < 2.$$

On the other hand, it is obvious that $\mathfrak{RT}_p \subset \mathfrak{RT} = \mathfrak{QRP}$. Thus the difference between the operator ideals $\mathfrak{RP}$ and $\mathfrak{QRP}$ turns out to be considerable.

Let $\alpha \geq 0$, and denote by C_α the diagonal operator on $[l_2, l_\infty^{2^k}]$ defined by

$$C_\alpha : (x_k) \longrightarrow (k^{-\alpha} x_k), \quad \text{where} \quad x_k \in l_\infty^{2^k}.$$

EXAMPLE. *Let $1 < p < 2$. Then the operator $C_{1/p'}$ has Rademacher type p, but*

$$\boldsymbol{\delta}(C_{1/p'} | \mathcal{R}_n, \mathcal{R}_n) \asymp n^{1/p-1/2}.$$

PROOF. We know from 4.2.11 that

$$\boldsymbol{\varrho}_2^{(2)}(l_\infty^N | \mathcal{R}_n, \mathcal{I}_n) \asymp \min\{\sqrt{n}, \sqrt{1+\log N}\}.$$

On the other hand,

$$\boldsymbol{\varrho}_1^{(2)}(l_\infty^N | \mathcal{R}_n, \mathcal{I}_n) = 1.$$

Thus interpolation yields

$$\boldsymbol{\varrho}_p^{(2)}(l_\infty^N | \mathcal{R}_n, \mathcal{I}_n) \prec \min\{n^{1/p'}, (1+\log N)^{1/p'}\}.$$

Now, reasoning as in 1.2.12, we obtain $\boldsymbol{\varrho}_p^{(2)}(C_{1/p'} | \mathcal{R}_n, \mathcal{I}_n) \asymp 1$. The fact that $\boldsymbol{\delta}(C_{1/p'} | \mathcal{R}_n, \mathcal{R}_n) \asymp n^{1/p-1/2}$ was shown in 4.12.14.

4.15 Parseval ideal norms and γ-summing operators – Dvoretzky's theorem –

4.15.1 In what follows, we treat the Parseval ideal norm $\boldsymbol{\pi}(\mathcal{G}_n)$ associated with the Gauss system, which is of special importance. For the general theory, the reader is referred to Sections 2.2 and 3.11.

4.15.2 We denote by $\mathbb{U}^n$ the compact (non-Abelian) group of all orthogonal (real case) or unitary (complex case) square matrices of order n. Moreover, χ^n stands for its Haar measure. Recall from 3.11.7 that $A_n = (\alpha_{hk}) \in \mathbb{U}^n$ transforms every orthonormal system $\mathcal{A}_n = (a_1, \ldots, a_n)$ into the orthonormal system

$$A_n \circ \mathcal{A}_n := \Big(\sum_{k=1}^n \alpha_{1k}\, a_k, \ldots, \sum_{k=1}^n \alpha_{nk}\, a_k \Big).$$

LEMMA. *The averaging formula*

$$\|(x_h) | \mathcal{G}_n\| = \Big(\int_{\mathbb{U}^n} \|(x_h) | A_n \circ \mathcal{A}_n\|^2 \, d\chi^n(A_n) \Big)^{1/2}$$

holds for all orthonormal systems $\mathcal{A}_n = (a_1, \ldots, a_n)$.

PROOF. Fix $t=(\tau_k)\in\mathbb{S}^n$, and note that ω^n, the $\mathbb{U}^n$-invariant measure on $\mathbb{S}^n$, is the image of χ^n under the mapping $\mathcal{A}_n \to s=\mathcal{A}_n t$. Taking into account 4.7.13, we obtain

$$\|(x_h)|\mathcal{G}_n\|^2 = \|(x_h)|\mathcal{Q}_n\|^2 = n\int\limits_{\mathbb{S}^n}\Big\|\sum_{h=1}^n x_h\sigma_h\Big\|^2 d\omega^n(s)$$
$$= n\int\limits_{\mathbb{U}^n}\Big\|\sum_{h=1}^n x_h\sum_{k=1}^n \alpha_{hk}\tau_k\Big\|^2 d\chi^n(\mathcal{A}_n).$$

Hence, by homogeneity,

$$\|(x_h)|\mathcal{G}_n\|^2\sum_{k=1}^n|\tau_k|^2 = n\int\limits_{\mathbb{U}^n}\Big\|\sum_{h=1}^n x_h\sum_{k=1}^n \alpha_{hk}\tau_k\Big\|^2 d\chi^n(\mathcal{A}_n)$$

whenever $\tau_1,\dots,\tau_n\in\mathbb{K}$. Letting $\tau_k := a_k(\xi)$, integration over $\xi\in M$ yields

$$\|(x_h)|\mathcal{G}_n\|^2\int\limits_M\sum_{k=1}^n|a_k(\xi)|^2\,d\mu(\xi) =$$
$$= n\int\limits_{\mathbb{U}^n}\bigg[\int\limits_M\Big\|\sum_{h=1}^n x_h\sum_{k=1}^n \alpha_{hk}a_k(\xi)\Big\|^2 d\mu(\xi)\bigg]\,d\chi^n(\mathcal{A}_n)$$

which implies the required formula, since

$$\int\limits_M\sum_{k=1}^n|a_k(\xi)|^2\,d\mu(\xi)=n.$$

4.15.3 We are now in a position to show the minimality of $\boldsymbol{\pi}(\mathcal{G}_n)$.

THEOREM. *For every orthonormal system $\mathcal{A}_n=(a_1,\dots,a_n)$,*

$$\boldsymbol{\pi}(\mathcal{G}_n)\le\boldsymbol{\pi}(\mathcal{A}_n).$$

PROOF. Recall from 3.11.7 that $\boldsymbol{\pi}(\mathcal{A}_n\circ\mathcal{A}_n)=\boldsymbol{\pi}(\mathcal{A}_n)$. Hence, by the averaging formula,

$$\|(Tx_h)|\mathcal{G}_n\| = \bigg(\int\limits_{\mathbb{U}^n}\|(Tx_h)|\mathcal{A}_n\circ\mathcal{A}_n\|^2\,d\chi^n(\mathcal{A}_n)\bigg)^{1/2}$$
$$\le\bigg(\int\limits_{\mathbb{U}^n}\boldsymbol{\pi}(T|\mathcal{A}_n\circ\mathcal{A}_n)^2\,\|(x_k)|w_2^n\|^2\,d\chi^n(\mathcal{A}_n)\bigg)^{1/2} = \boldsymbol{\pi}(T|\mathcal{A}_n)\,\|(x_k)|w_2^n\|.$$

4.15.4 PROPOSITION. *The sequences $(\pi(\mathcal{G}_n))$ and $(\pi(\mathcal{P}_n))$ are uniformly equivalent. More precisely,*

$$\pi(\mathcal{G}_n) \le \pi(\mathcal{P}_n) \le M_1\, \pi(\mathcal{G}_n).$$

PROOF. By the foregoing observation, 3.11.5 and 4.8.3, we have

$$\pi(\mathcal{G}_n) \le \pi(\mathcal{P}_n) \le \varrho(\mathcal{P}_n, \mathcal{G}_n) \circ \pi(\mathcal{G}_n) \le M_1\, \pi(\mathcal{G}_n).$$

4.15.5 PROPOSITION. *Let $\mathcal{A}_n = (a_1, \dots, a_n)$ be any system of characters on a compact Abelian group $\mathbb{G}$. Then*

$$\pi(\mathcal{A}_n) \le S(\mathcal{A}_n)\, \pi(\mathcal{P}_n(e^{i\tau})).$$

PROOF. Recall from 3.11.5 and 4.11.10 that

$$\pi(\mathcal{A}_n) \le \varrho(\mathcal{A}_n, \mathcal{P}_n(e^{i\tau})) \circ \pi(\mathcal{P}_n(e^{i\tau})) \quad \text{and} \quad \varrho(\mathcal{A}_n, \mathcal{P}_n(e^{i\tau})) \le S(\mathcal{A}_n).$$

4.15.6 For any system of characters on a compact Abelian group, we denote by $S_0(\mathcal{A}_n)$ the least constant $c_0 > 0$ such that

$$\pi(\mathcal{A}_n) \le c_0\, \pi(\mathcal{P}_n(e^{i\tau})).$$

Then the preceding proposition means that $S_0(\mathcal{A}_n) \le S(\mathcal{A}_n)$. Probably, we also have a reverse estimate.

PROBLEM. *Does there exist a constant $c > 0$ such that $S(\mathcal{A}_n) \le c\, S_0(\mathcal{A}_n)$?*

4.15.7 An operator $T \in \mathfrak{L}(X, Y)$ is called **γ-summing** if there exists a constant $c \ge 0$ such that

$$\left(\int\limits_{\mathbb{K}^n} \left\| \sum_{k=1}^{n} T x_k g_k^{(n)}(s) \right\|^2 d\gamma^n(s) \right)^{1/2} \le c \sup \left\{ \left(\sum_{k=1}^{n} |\langle x_k, x' \rangle|^2 \right)^{1/2} : x' \in U_X^\circ \right\}$$

for $x_1, \dots, x_n \in X$ and $n = 1, 2, \dots$. More concisely, the preceding inequality reads as follows:

$$\|(Tx_k)|\mathcal{G}_n\| \le c\, \|(x_k)|w_2^n\|.$$

We let

$$\|T|\mathfrak{P}_\gamma\| := \inf c,$$

where the infimum is taken over all constants $c \ge 0$ with the above property.

The class of γ-summing operators is a Banach operator ideal, denoted by $\mathfrak{P}_\gamma$. Using the notation introduced in 1.2.8 , we have

$$\mathfrak{P}_\gamma = \mathfrak{L}[\pi(\mathcal{G}_n)] \quad \text{and} \quad \|T|\mathfrak{P}_\gamma\| = \sup_n \pi(T|\mathcal{G}_n).$$

4.15.8 LEMMA. $\pi(T|\mathcal{G}_n) = \|(Tu_k^{(n)})|\mathcal{G}_n\|$ *for* $T \in \mathfrak{L}(l_2^n, X)$.

PROOF. We know from 3.11.2 that

$$\pi(T|\mathcal{G}_n) = \sup\left\{ \|(TA_n u_k^{(n)})|\mathcal{G}_n\| \, : \, A_n \in \mathbb{U}^n \right\}.$$

The assertion now follows from the fact that the Gaussian measure is invariant under all transforms $A_n \in \mathbb{U}^n$.

REMARK. In the notation $\ell(T)$, the Parseval ideal norm $\pi(T|\mathcal{G}_n)$ of operators $T \in \mathfrak{L}(l_2^n, X)$ plays a fundamental role in Banach space geometry; see [MIL*, p. 106] and [TOM, pp. 80–85]. The following considerations will show that it is useful to have $\pi(T|\mathcal{G}_n)$ not only in this special case but even for all operators.

4.15.9 LEMMA. $\pi(T|\mathcal{G}_m) = \pi(T|\mathcal{G}_n)$ *for* $T \in \mathfrak{L}(l_2^n, X)$ *and* $m > n$.

PROOF. Define $J \in \mathfrak{L}(l_2^n, l_2^m)$ and $Q \in \mathfrak{L}(l_2^m, l_2^n)$ by

$$J : (\xi_1, \ldots, \xi_n) \longrightarrow (\xi_1, \ldots, \xi_n, 0, \ldots, 0)$$

and

$$Q : (\xi_1, \ldots, \xi_n, \xi_{n+1}, \ldots, \xi_m) \longrightarrow (\xi_1, \ldots, \xi_n).$$

Since $\|J\| = \|Q\| = 1$, it follows from $T = TQJ$ and the preceding lemma that

$$\begin{aligned} \pi(T|\mathcal{G}_m) &= \pi(TQJ|\mathcal{G}_m) \le \pi(TQ|\mathcal{G}_m) = \|(TQu_k^{(m)})|\mathcal{G}_m\| \\ &= \|(Tu_k^{(n)})|\mathcal{G}_n\| = \pi(T|\mathcal{G}_n). \end{aligned}$$

The reverse estimate is obvious.

4.15.10 The above result can be extended to all finite rank operators. In this way, we get an analogue of 2.2.8.

PROPOSITION. *Let* $T \in \mathfrak{F}(X, Y)$. *Then*

$$\|T|\mathfrak{P}_\gamma\| = \pi(T|\mathcal{G}_N) \quad \textit{whenever } \operatorname{rank}(T) = N.$$

PROOF. Fix $\|A_m : l_2^m \to X\| \le 1$, and let $n := \operatorname{rank}(TA_m) \le \operatorname{rank}(T) = N$. Choose any isometry $B_{mn} \in \mathfrak{L}(l_2^n, l_2^m)$ that maps l_2^n onto the orthogonal complement of the null space of TA_m. Then $TA_m = TA_m B_{mn} B_{mn}^*$. Since $TA_m B_{mn} \in \mathfrak{L}(l_2^n, Y)$, we know from 4.15.9 that

$$\pi(TA_m|\mathcal{G}_m) = \pi(TA_m B_{mn}|\mathcal{G}_m) = \pi(TA_m B_{mn}|\mathcal{G}_n) \le \pi(T|\mathcal{G}_n).$$

Thus 3.11.2 implies that $\pi(T|\mathcal{G}_m) \le \pi(T|\mathcal{G}_N)$.

4.15.11 Next, for n-dimensional Banach spaces, we establish two lower estimates of $\|X|\mathfrak{P}_\gamma\| = \pi(X|\mathcal{G}_n)$ in terms of Gauss type and cotype ideal norms; see [fig*a, p. 77] and [MIL*, pp. 54–55] in connection with Theorem 4.15.18.

PROPOSITION. *If* $\dim(X) = n$, *then*

$$\varrho(X|\mathcal{G}_n, \mathcal{I}_n) \le \pi(X|\mathcal{G}_n) \quad \textit{and} \quad \sqrt{\tfrac{\pi}{2}}\, \varrho(X|\mathcal{I}_n, \mathcal{G}_n)^{-1} \le \pi(X|\mathcal{G}_n).$$

PROOF. The left-hand inequality follows from $\|(x_k)|w_2^n\| \le \|(x_k)|l_2^n\|$. On the other hand, for the 2-summing norm we have $\|X|\mathfrak{P}_2\| = \sqrt{n}$; see [DIE*a, p. 87] or [PIE 2, p. 385]. Hence 2.2.8 and 3.11.5 imply that

$$\sqrt{n} = \|X|\mathfrak{P}_2\| \le \sqrt{2}\,\pi(X|\mathcal{I}_n) \le \sqrt{2}\,\varrho(X|\mathcal{I}_n, \mathcal{G}_n)\,\pi(X|\mathcal{G}_n).$$

4.15.12 PROPOSITION. *There exists a constant* $c > 0$ *such that*

$$\pi(X|\mathcal{G}_n) \ge c\sqrt{1+\log n}$$

for all n*-dimensional Banach spaces* X.

PROOF. The assertion can be deduced from 4.9.6 and the previous proposition.

4.15.13 Some γ-summing ideal norms were evaluated by W. Linde/ A. Pietsch; see [lin*a].

EXAMPLE.

$$\begin{aligned} \|l_p^n|\mathfrak{P}_\gamma\| &\asymp \sqrt{n} && \textit{if } 1 \le p \le 2, \\ \|l_q^n|\mathfrak{P}_\gamma\| &\asymp n^{1/q} && \textit{if } 2 \le q < \infty, \\ \|l_\infty^n|\mathfrak{P}_\gamma\| &\asymp \sqrt{1+\log n}. \end{aligned}$$

PROOF. First of all, we recall from 4.15.10 that

$$\|X|\mathfrak{P}_\gamma\| = \pi(X|\mathcal{G}_n) \le \sqrt{n}$$

for all n-dimensional Banach spaces.

On the other hand, 4.9.10 tells us that $\varrho(l_p^n|\mathcal{I}_n, \mathcal{G}_n) \le M_1^{-1}$. Hence, by 4.15.11,

$$M_1\sqrt{n} \le \sqrt{n}\,\varrho(l_p^n|\mathcal{I}_n, \mathcal{G}_n)^{-1} \le \sqrt{2}\,\pi(l_p^n|\mathcal{G}_n).$$

Next, we infer from 3.11.11 that

$$\pi(l_q^n|\mathcal{G}_n) \le M_q\, n^{1/q} \le \sqrt{q}\, n^{1/q}.$$

Since

$$\|(u_k^{(n)})|[\mathcal{G}_n, l_q^n]\| = n^{1/q} N_q(\mathcal{G}_n) \quad \text{and} \quad \|(u_k^{(n)})|[w_2^n, l_q^n]\| = 1,$$

we also have the reverse estimate

$$n^{1/q}\, N_q(\mathcal{G}_n) \le \boldsymbol{\pi}(l_q^n|\mathcal{G}_n).$$

Finally, applying 1.3.6 yields

$$\boldsymbol{\pi}(l_\infty^n|\mathcal{G}_n) \le \sqrt{1+\log n},$$

while the lower estimate was proved in 4.15.12.

REMARK. The following table shows that, for the classical Minkowski spaces l_r^n, the left-hand inequality in 4.15.11 is very coarse, while the right-hand one is sharp.

	$1 \le r \le 2$	$2 \le r < \infty$	$r = \infty$
$\boldsymbol{\varrho}(l_r^n\|\mathcal{G}_n, \mathcal{I}_n)$	$\asymp n^{1/r-1/2}$	$\asymp 1$	$\asymp \sqrt{1+\log n}$
$\sqrt{n}\,\boldsymbol{\varrho}(l_r^n\|\mathcal{I}_n, \mathcal{G}_n)^{-1}$	$\asymp \sqrt{n}$	$\asymp n^{1/r}$	$\asymp \sqrt{1+\log n}$
$\boldsymbol{\pi}(l_r^n\|\mathcal{G}_n)$	$\asymp \sqrt{n}$	$\asymp n^{1/r}$	$\asymp \sqrt{1+\log n}$

4.15.14 LEMMA. $\sqrt{n} \le d(l_2^n, X)\, \boldsymbol{\pi}(X|\mathcal{G}_n)$
for all n-dimensional Banach spaces X.

PROOF. Since $\boldsymbol{\pi}(\mathcal{G}_n)$ is an ideal norm, we have

$$\boldsymbol{\pi}(l_2^n|\mathcal{G}_n) \le d(l_2^n, X)\, \boldsymbol{\pi}(X|\mathcal{G}_n).$$

The assertion now follows from $\boldsymbol{\pi}(l_2^n|\mathcal{G}_n) = \sqrt{n}$.

4.15.15 For every finite dimensional Banach space X and $\varepsilon > 0$, the **Dvoretzky dimension** $D(X, \varepsilon)$ is defined to be the largest n such that there exists an n-dimensional subspace X_n of X with $d(X_n, l_2^n) \le 1 + \varepsilon$.

4.15.16 From 4.15.14, we obtain an elementary, but remarkable fact.

PROPOSITION. $D(X, \varepsilon) \le (1+\varepsilon)^2 \|X|\mathfrak{P}_\gamma\|^2$
for all finite dimensional Banach spaces X.

4.15.17 More important is an estimate in the reverse direction. We omit the proof and refer to [PIS 2, pp. 41–59] and [pis 18, pp. 169–175]. Note that taking the supremum over Pisier's 'dimension' (Varenna version) of all X-valued Gaussian random variables yields $\|X|\mathfrak{P}_\gamma\|^2$.

PROPOSITION. *There exists a constant $c(\varepsilon)>0$, depending only on $\varepsilon > 0$, such that*

$$D(X, \varepsilon) \ge c(\varepsilon)\, \|X|\mathfrak{P}_\gamma\|^2$$

for all finite dimensional Banach spaces X.

REMARKS. One may take $c(\varepsilon) = c\varepsilon^2$, where $c > 0$ is an absolute constant. Because of

$$\boldsymbol{\varrho}(X|\mathfrak{G}_n, \mathfrak{I}_n) \leq \boldsymbol{\pi}(X|\mathfrak{G}_n) \leq \|X|\mathfrak{P}_\gamma\| \qquad \text{(by 4.15.11)},$$

we get Theorem 9.7 in [MIL*, p. 55] for free:

$$D(X, \varepsilon) \geq c(\varepsilon)\, \boldsymbol{\varrho}(X|\mathfrak{G}_n, \mathfrak{I}_n)^2.$$

However, this is an extremely weak estimate, since the above table shows that the difference between $\boldsymbol{\pi}(X|\mathfrak{G}_n)$ and $\boldsymbol{\varrho}(X|\mathfrak{G}_n, \mathfrak{I}_n)$ can be quite large.

4.15.18 We now combine all facts obtained so far.

THEOREM. *There exist constants $a(\varepsilon) > 0$ and $b(\varepsilon) > 0$, depending on $\varepsilon > 0$, such that*

$$a(\varepsilon)\, \|X|\mathfrak{P}_\gamma\|^2 \leq D(X, \varepsilon) \leq b(\varepsilon)\, \|X|\mathfrak{P}_\gamma\|^2$$

for all finite dimensional Banach spaces X.

4.15.19 What follows is the most celebrated result from the local theory of Banach spaces; see [dvo]. The approach via Gaussian methods, presented above, is due to G. Pisier.

DVORETZKY'S THEOREM. *There is a constant $d(\varepsilon)>0$, depending on $\varepsilon > 0$, such that*

$$D(X, \varepsilon) \geq d(\varepsilon)(1 + \log n)$$

for all n-dimensional Banach spaces X.

PROOF. Combine the previous theorem with 4.15.12.

4.15.20 Finally, we restate 4.15.13 in terms of Dvoretzky's dimension. Without using the concept of γ-summing operators, the following result was obtained by T. Figiel/J. Lindenstrauss/V. Milman in their famous paper [fig*a, p. 64]. However, at this time the fascinating interplay between $D(X, \varepsilon)$ and $\|X|\mathfrak{P}_\gamma\|$ was unknown.

EXAMPLE. *For every fixed $\varepsilon > 0$,*

$$\begin{aligned} D(l_p^n, \varepsilon) &\asymp n && \text{if } 1 \leq p \leq 2, \\ D(l_q^n, \varepsilon) &\asymp n^{2/q} && \text{if } 2 \leq q < \infty, \\ D(l_\infty^n, \varepsilon) &\asymp 1 + \log n. \end{aligned}$$

4.16 The Maurey–Pisier theorem

For the sake of completeness, we conclude this chapter by stating the fundamental Maurey–Pisier theorem [mau*, p. 85]. Proofs can be found in [MIL*, pp. 86–97] and, partly, in [DIE*a, pp. 226 and 289–303].

4.16.1 Let $1 \le r \le \infty$. A Banach space X **contains** l_r^n with bound $c \ge 1$ if there exists an operator $A_n \in \mathfrak{L}(l_r^n, X)$ such that

$$\|(\xi_k)|l_r^n\| \le \|A_n(\xi_k)\| \le c\,\|(\xi_k)|l_r^n\|$$

for $\xi_1, \dots, \xi_n \in \mathbb{K}$. Writing A_n in the form

$$A_n(\xi_k) = \sum_{k=1}^{n} \xi_k x_k$$

yields

$$\|(\xi_k)|l_r^n\| \le \left\| \sum_{k=1}^{n} \xi_k x_k \right\| \le c\,\|(\xi_k)|l_r^n\|.$$

In other words, X has an n-dimensional subspace $E_n := \operatorname{span}\{x_1, \dots, x_n\}$ with $d(E_n, l_r^n) \le c$.

We say that X **contains the spaces** l_r^n **uniformly** if the above condition is fulfilled for some constant $c \ge 1$ and all $n = 1, 2, \dots$, simultaneously.

4.16.2 A deep result from [kri] implies that the equivalences **(2)**$\Leftrightarrow$**(3)** established in 4.4.6 and 4.6.6 remain true for $1 < r < \infty$. However, in the general case, the proof is much more complicated; see [GUE, p. 103].

KRIVINE'S THEOREM. *For every Banach space X and $1 \le r \le \infty$, the following are equivalent:*

(1) *X contains the spaces l_r^n uniformly for* **all** *bounds $c > 1$.*

(2) *X contains the spaces l_r^n uniformly for* **some** *bound $c \ge 1$.*

4.16.3 The **Minkowski spectrum** $\operatorname{spec}(X)$ of an infinite dimensional Banach space X consists of all exponents $1 \le r \le \infty$ for which the spaces l_r^n are uniformly contained in X. It easily turns out that $\operatorname{spec}(X)$ is a closed subset of $[1, \infty]$; see [SCHW, p. 57].

REMARK. The Dvoretzky theorem 4.15.19 tells us that the Minkowski spectrum is never empty, since $2 \in \operatorname{spec}(X)$.

4.16.4 For every infinite dimensional Banach space, we define the **Rademacher type index**

$$p(X) := \sup\left\{ p \in [1, 2] \,:\, X \in \mathsf{RT}_p \right\}$$

and the **Rademacher cotype index**

$$q(X) := \inf\left\{ q \in [2, \infty] \,:\, X \in \mathsf{RC}_q \right\}.$$

4.16.5 We are now prepared to state the announced highlight of the local theory of Banach spaces.

MAUREY–PISIER THEOREM. *For every infinite dimensional Banach space X,*

$$p(X) = \min \operatorname{spec}(X) \quad \textit{and} \quad q(X) = \max \operatorname{spec}(X).$$

We even have $\operatorname{spec}(X) \cap [1,2] = [p(X), 2]$.

REMARK. The limiting cases $p(X) = 1$ and $q(X) = \infty$, which are of particular interest, were treated in Sections 4.4 and 4.6, with full proofs.

5

Trigonometric ideal norms

This chapter deals with ideal norms obtained from the classical Fourier systems,

$$\mathcal{E}_n := (e_1, \ldots, e_n), \quad \mathcal{C}_n := (c_1, \ldots, c_n) \quad \text{and} \quad \mathcal{S}_n := (s_1, \ldots, s_n),$$

where $e_k(t) := \exp(ikt)$, $c_k(t) := \sqrt{2}\cos kt$ and $s_k(t) := \sqrt{2}\sin kt$. Quite often, it is advantageous to consider their discrete counterparts, $\mathcal{E}_n^\circ$, $\mathcal{C}_n^\circ$ and $\mathcal{S}_n^\circ$. This is mainly because the elements of $\mathcal{E}_n^\circ$ can be interpreted as characters on the cyclic group $\mathbb{E}_n$.

There are two major blocks of results. One part is devoted to operators compatible with the Hilbert transform. In particular, we characterize those Banach spaces X for which a generalized version of the classical M. Riesz theorem holds. A typical criterion says that this is the case if and only if $\boldsymbol{\delta}(X|\mathcal{E}_n, \mathcal{E}_n) = O(1)$ and/or $\boldsymbol{\varrho}(X|\mathcal{C}_n, \mathcal{S}_n) = O(1)$. We may also use the Kwapień ideals norms $\boldsymbol{\kappa}(H_n)$ associated with the Hilbert matrices H_n.

In another part, the vector-valued Fourier transform is treated. Here the ideal norms $\boldsymbol{\varrho}(\mathcal{E}_n, \mathcal{I}_n)$ and $\boldsymbol{\varrho}(\mathcal{I}_n, \mathcal{E}_n)$ become relevant. Taking the famous Hausdorff–Young inequality as a starting point, we get, among others, the following classes of Banach spaces which are stable when passing to subspaces, quotients and duals:

ET_p	:	Banach spaces of Fourier type p,
ET	:	Banach spaces of Fourier subtype,
EC_q	:	Banach spaces of Fourier cotype q,
EC	:	Banach spaces of Fourier subcotype.

In contrast to the situation for Rademacher functions, there is no difference between type and cotype:

$$\mathsf{ET}_p = \mathsf{EC}_{p'}, \quad \mathsf{EC}_q = \mathsf{ET}_{q'} \quad \text{and} \quad \mathsf{ET} = \mathsf{EC}.$$

We are going to prove that

$$\bigcup_{1<p\leq 2} \mathsf{ET}_p = \mathsf{ET} = \mathsf{B} \quad \text{and} \quad \mathsf{ET}_2 = \mathsf{H}.$$

Unfortunately, no geometric characterization of spaces in ET_p is known.

For more information about trigonometric functions, we refer to the following book:

Zygmund .... *Trigonometric series* [ZYG].

5.1 Trigonometric functions

5.1.1 We write

$$e_k(t) := \exp(ikt) \quad \text{for } t \in \mathbb{R} \text{ and } k \in \mathbb{Z}.$$

The **Fourier system**

$$\mathcal{E}_n := (e_1, \dots, e_n)$$

is orthonormal in $L_2[-\pi, +\pi)$ equipped with the inner product

$$(f, g) = \frac{1}{2\pi} \int_{-\pi}^{+\pi} f(t)\overline{g(t)}\, dt.$$

Note that

$$\|(x_k)|\mathcal{E}_n\|_r = \left(\frac{1}{2\pi} \int_{-\pi}^{+\pi} \Big\| \sum_{k=1}^{n} x_k \exp(ikt) \Big\|^r dt \right)^{1/r}.$$

The substitution $t \to -t$ yields

$$\|(x_k)|\overline{\mathcal{E}}_n\|_r = \|(x_k)|\mathcal{E}_n\|_r.$$

5.1.2 Let

$$e_k^{(n)} := \begin{pmatrix} \varepsilon_{1k}^{(n)} \\ \vdots \\ \varepsilon_{nk}^{(n)} \end{pmatrix} \quad \text{with } \varepsilon_{hk}^{(n)} := n^{-1/2} \exp\left(\tfrac{2\pi i}{n} hk\right)$$

for $h, k = 1, \dots, n$. The **discrete Fourier system**

$$\mathcal{E}_n^\circ := (e_1^{(n)}, \dots, e_n^{(n)})$$

is orthonormal in l_2^n. Normalizing the underlying measure as described in 3.1.8, we get $\mathcal{N}\mathcal{E}_n^\circ$, the system of all characters on the cyclic group $\mathbb{E}_n$. By 3.8.8, the associated **Fourier matrix**

$$E_n^\circ := \left(\varepsilon_{hk}^{(n)}\right)$$

is unitary. Note that

$$\|(x_k)|\mathcal{N}\mathcal{E}_n^\circ\|_r = \left(\frac{1}{n}\sum_{h=1}^{n}\left\|\sum_{k=1}^{n} x_k \exp\left(\tfrac{2\pi i}{n}hk\right)\right\|^r\right)^{1/r}.$$

We obtain the same value if $\sum\limits_{h=1}^{n}$ is replaced by $\sum\limits_{h=0}^{n-1}$. The substitution $h \to n-h$ now yields a special case of 3.8.10,

$$\|(x_k)|\overline{\mathcal{E}}_n^\circ\|_r = \|(x_k)|\mathcal{E}_n^\circ\|_r.$$

5.1.3 We write

$$c_k(t) := \sqrt{2}\cos kt \quad \text{and} \quad s_k(t) := \sqrt{2}\sin kt \quad \text{for } t \in \mathbb{R} \text{ and } k \in \mathbb{N}.$$

The **cosine system** and the **sine system**,

$$\mathcal{C}_n := (c_1, \ldots, c_n) \quad \text{and} \quad \mathcal{S}_n := (s_1, \ldots, s_n),$$

are orthonormal in $L_2[0,\pi)$ equipped with the inner product

$$(f,g) := \frac{1}{\pi}\int_0^\pi f(t)\overline{g(t)}\,dt.$$

Moreover, we have

$$\|(x_k)|\mathcal{C}_n\| = \left(\frac{2}{\pi}\int_0^\pi\left\|\sum_{k=1}^{n} x_k \cos kt\right\|^2 dt\right)^{1/2}$$

and

$$\|(x_k)|\mathcal{S}_n\| = \left(\frac{2}{\pi}\int_0^\pi\left\|\sum_{k=1}^{n} x_k \sin kt\right\|^2 dt\right)^{1/2}.$$

5.1.4 Let $\varphi = \sum\limits_{k\in\mathbb{Z}} \xi_k e_k$ be any trigonometric polynomial. Then

$$\frac{1}{2\pi}\int_{-\pi}^{+\pi} \varphi(t)\,dt = \langle \varphi, \overline{e_0}\rangle = \xi_0.$$

Moreover, whenever $\xi_{\pm n} = \xi_{\pm 2n} = \ldots = 0$, we have

$$\frac{1}{n}\sum_{h=1}^{n}\varphi\left(\tfrac{2\pi h}{n}\right) = \frac{1}{n}\sum_{k\in\mathbb{Z}}\xi_k\sum_{h=1}^{n}\exp\left(\tfrac{2\pi i}{n}hk\right) = \xi_0.$$

This yields the discretization formula

$$\frac{1}{2\pi}\int_{-\pi}^{+\pi}\varphi(t)\,dt = \frac{1}{n}\sum_{h=1}^{n}\varphi\left(\tfrac{2\pi h}{n}\right).$$

5.1.5 We write

$$c_k^{(n)} := \begin{pmatrix} \gamma_{0,k}^{(n)} \\ \vdots \\ \gamma_{n,k}^{(n)} \end{pmatrix} \quad \text{and} \quad s_k^{(n)} := \begin{pmatrix} \sigma_{1,k}^{(n)} \\ \vdots \\ \sigma_{n,k}^{(n)} \end{pmatrix}$$

with

$$\gamma_{0k}^{(n)} := \sqrt{\tfrac{2}{2n+1}},$$

$$\gamma_{hk}^{(n)} := \tfrac{2}{\sqrt{2n+1}} \cos \tfrac{2\pi}{2n+1} hk \quad \text{and} \quad \sigma_{hk}^{(n)} := \tfrac{2}{\sqrt{2n+1}} \sin \tfrac{2\pi}{2n+1} hk$$

for $h, k = 1, \dots, n$. The **discrete cosine system** and the **discrete sine system**,

$$\mathcal{C}_n^\circ := (c_1^{(n)}, \dots, c_n^{(n)}) \quad \text{and} \quad \mathcal{S}_n^\circ := (s_1^{(n)}, \dots, s_n^{(n)}),$$

are orthonormal in the Hilbert spaces l_2^{n+1} and l_2^n, respectively. Indeed, by the discretization formula stated in the preceding paragraph, we have

$$\frac{2}{2n+1} \sum_{|l| \le n} \cos \tfrac{2\pi}{2n+1} hl \cos \tfrac{2\pi}{2n+1} kl = \frac{1}{\pi} \int_{-\pi}^{+\pi} \cos ht \cos kt \, dt$$

and

$$\frac{2}{2n+1} \sum_{|l| \le n} \sin \tfrac{2\pi}{2n+1} hl \sin \tfrac{2\pi}{2n+1} kl = \frac{1}{\pi} \int_{-\pi}^{+\pi} \sin ht \sin kt \, dt.$$

It turns out that

$$\|(x_k)|\mathcal{C}_n^\circ\| = \left(\frac{2}{2n+1} \Big\| \sum_{k=1}^n x_k \Big\|^2 + \frac{4}{2n+1} \sum_{h=1}^n \Big\| \sum_{k=1}^n x_k \cos \tfrac{2\pi}{2n+1} hk \Big\|^2 \right)^{1/2}$$

and

$$\|(x_k)|\mathcal{S}_n^\circ\| = \left(\frac{4}{2n+1} \sum_{h=1}^n \Big\| \sum_{k=1}^n x_k \sin \tfrac{2\pi}{2n+1} hk \Big\|^2 \right)^{1/2},$$

or, in a more symmetric form,

$$\|(x_k)|\mathcal{C}_n^\circ\| = \left(\frac{2}{2n+1} \sum_{|h| \le n} \Big\| \sum_{k=1}^n x_k \cos \tfrac{2\pi}{2n+1} hk \Big\|^2 \right)^{1/2}$$

and

$$\|(x_k)|\mathcal{S}_n^\circ\| = \left(\frac{2}{2n+1} \sum_{|h| \le n} \Big\| \sum_{k=1}^n x_k \sin \tfrac{2\pi}{2n+1} hk \Big\|^2 \right)^{1/2}.$$

The associated **cosine matrices and sine matrices**

$$C_{n+1,n}^\circ := \begin{pmatrix} C_{1,n}^\circ \\ C_n^\circ \end{pmatrix} \quad \text{with} \quad \begin{array}{l} C_{1,n}^\circ := \left(\gamma_{0k}^{(n)}\right), \\ C_n^\circ \;:= \left(\gamma_{hk}^{(n)}\right), \end{array} \quad \text{and} \quad S_n^\circ := \left(\sigma_{hk}^{(n)}\right)$$

are isometric and orthogonal, respectively. If $n \geq 2$, then we can choose $x = (\xi_k) \in l_2^n$ such that $\|x|l_2^n\| = 1$ and $\sum_{k=1}^n \xi_k = 0$. Hence

$$\|C_n^\circ x|l_2^n\| = \|C_{n+1,n}^\circ x|l_2^{n+1}\| = \|x|l_2^n\| = 1.$$

This proves that $\|C_n^\circ\| = 1$. However, $C_1^\circ = \big(-\frac{1}{\sqrt{3}}\big)$.

5.1.6 Next, we recall some classical concepts.

Dirichlet kernel:

$$D_n(t) := 1 + 2\sum_{k=1}^n \cos kt = \sum_{|k|\leq n} \exp(ikt) = \frac{\sin\frac{2n+1}{2}t}{\sin\frac{1}{2}t},$$

conjugate Dirichlet kernel:

$$\widetilde{D}_n(t) := 2\sum_{k=1}^n \sin kt = \frac{\cos\frac{1}{2}t - \cos\frac{2n+1}{2}t}{\sin\frac{1}{2}t},$$

Fejér kernel:

$$F_n(t) := \frac{D_0(t) + \ldots + D_n(t)}{n+1} = \frac{1}{n+1}\left(\frac{\sin\frac{n+1}{2}t}{\sin\frac{1}{2}t}\right)^2 \geq 0,$$

de la Vallée–Poussin kernel:

$$V_n(t) := \frac{D_n(t) + \ldots + D_{2n-1}(t)}{n} = 2F_{2n-1}(t) - F_{n-1}(t).$$

The Fourier coefficients of V_n are given by

$$\langle V_n, \overline{e_k}\rangle = \frac{1}{2\pi}\int_{-\pi}^{+\pi} V_n(t)\overline{e_k(t)}\,dt = \begin{cases} 1 & \text{if } |k| \leq n, \\ \frac{2n-|k|}{n} & \text{if } n < |k| < 2n, \\ 0 & \text{otherwise.} \end{cases}$$

It follows from

$$\frac{1}{2\pi}\int_{-\pi}^{+\pi} |F_n(t)|\,dt = \frac{1}{2\pi}\int_{-\pi}^{+\pi} F_n(t)\,dt = \langle F_n, \overline{e_0}\rangle = 1$$

that

$$\frac{1}{2\pi}\int_{-\pi}^{+\pi} |V_n(t)|\,dt \leq 3.$$

5.1.7 Let

$$d_n(t) := \exp\big(-i\tfrac{n+1}{2}t\big)\sum_{k=1}^n \exp(ikt) = \frac{\sin\frac{n}{2}t}{\sin\frac{1}{2}t}.$$

Note that

$$D_n(t) = d_{2n+1}(t) = 2\cos\tfrac{n+1}{2}t\, d_n(t) + 1 \quad \text{and} \quad \widetilde{D}_n(t) = 2\sin\tfrac{n+1}{2}t\, d_n(t).$$

We have $\|d_n|L_2\| = \sqrt{n}$ and $\|d_n|L_\infty\| = n$. Moreover,

$$|d_n(t)| \le \frac{\pi}{|t|} \quad \text{for } 0 < |t| \le \pi$$

implies that

$$\|d_n|L_{1,\infty}\|^* := \sup_{s>0} s\, d_n^*(s) \le \pi.$$

Thus, by real interpolation 0.6.3,

$$\sqrt{n}\, M_r(\mathcal{E}_n) = \|d_n|L_r\| \asymp n^{1/r'} \quad \text{for } 1 < r \le \infty.$$

In the limiting case $r = 1$ the quantity

$$L(\mathcal{E}_n) := \sqrt{n}\, M_1(\mathcal{E}_n) = \|d_n|L_1\| \asymp 1 + \log n$$

is called the n-th **Lebesgue constant**; see [ZYG, I, p. 67] and 3.8.12.

5.1.8 We write

$$d_h^{(n)}(t) := n^{-1/2} d_n\big(t - \tfrac{2\pi}{n} h\big).$$

The **Dirichlet system**

$$\mathcal{D}_n := (d_1^{(n)}, \ldots, d_n^{(n)})$$

is orthonormal in $L_2[-\pi, +\pi)$, since it is obtained from $\mathcal{E}_n$ by a unitary transformation:

$$d_h^{(n)}(t) = \exp(-i\tfrac{n+1}{2} t) \exp(i\tfrac{\pi}{n} h)(-1)^h \sum_{k=1}^{n} \varepsilon_{hk}^{(n)} e_k(t).$$

REMARKS. Note that $\operatorname{span}\{d_h^{(n)}(t) : h = 1, \ldots, n\}$ coincides with

$$\operatorname{span}\Big\{ \cos \tfrac{2k-1}{2} t,\ \sin \tfrac{2k-1}{2} t : k = 1, \ldots, m \Big\} \text{ when } n = 2m \text{ is even}$$

and with

$$\operatorname{span}\Big\{ 1,\ \cos kt,\ \sin kt : k = 1, \ldots, m \Big\} \text{ when } n = 2m + 1 \text{ is odd.}$$

The latter is the $(2m + 1)$-dimensional subspace of all trigonometric polynomials of degree less than or equal to m.

Moreover, it follows from

$$n^{-1/2} d_h^{(n)}\big(\tfrac{2\pi}{n} k\big) = \begin{cases} 1 & \text{if } h = k, \\ 0 & \text{if } h \ne k, \end{cases}$$

that, for $x_1, \ldots, x_n \in X$,

$$\boldsymbol{f}(t) = n^{-1/2} \sum_{h=1}^{n} x_h d_h^{(n)}(t)$$

yields an interpolating function $\boldsymbol{f}$ with $\boldsymbol{f}\big(\tfrac{2\pi}{n} k\big) = x_k$ for $k = 1, \ldots, n$.

5.2 The Dirichlet ideal norms $\boldsymbol{\delta}(\mathcal{E}_n, \mathcal{E}_n)$

5.2.1 In this section, we treat Dirichlet ideal norms associated with the trigonometric systems and their discrete counterparts. For example,

$$\boldsymbol{\delta}(\mathcal{E}_n, \mathcal{E}_n), \quad \boldsymbol{\delta}(\mathcal{E}_n, \mathcal{E}_n^\circ), \quad \boldsymbol{\delta}(\mathcal{E}_n^\circ, \mathcal{E}_n), \quad \boldsymbol{\delta}(\mathcal{C}_n, \mathcal{C}_n), \quad \text{and} \quad \boldsymbol{\delta}(\mathcal{S}_n, \mathcal{S}_n).$$

Some related Riemann ideal norms will also be considered.

In view of

$$\|(x_k)|\overline{\mathcal{E}}_n\| = \|(x_k)|\mathcal{E}_n\| \quad \text{and} \quad \|(x_k)|\overline{\mathcal{E}}_n^\circ\| = \|(x_k)|\mathcal{E}_n^\circ\|,$$

the orthonormal systems $\mathcal{E}_n$ and $\mathcal{E}_n^\circ$ can be replaced by $\overline{\mathcal{E}}_n$ and $\overline{\mathcal{E}}_n^\circ$, respectively.

5.2.2 It seems to be unknown whether the sequence $(\boldsymbol{\delta}(\mathcal{E}_n, \mathcal{E}_n))$ is non-decreasing. However, we have at least the following information.

PROPOSITION. $\boldsymbol{\delta}(\mathcal{E}_{m+n}, \mathcal{E}_{m+n}) \le \boldsymbol{\delta}(\mathcal{E}_m, \mathcal{E}_m) + \boldsymbol{\delta}(\mathcal{E}_n, \mathcal{E}_n)$,

$$\boldsymbol{\delta}(\mathcal{E}_{n+1}, \mathcal{E}_{n+1}) \le 2\,\boldsymbol{\delta}(\mathcal{E}_n, \mathcal{E}_n) \quad \textit{and} \quad \boldsymbol{\delta}(\mathcal{E}_n, \mathcal{E}_n) \le 2\,\boldsymbol{\delta}(\mathcal{E}_{n+1}, \mathcal{E}_{n+1}),$$

$$\boldsymbol{\delta}(\mathcal{E}_{2n}, \mathcal{E}_{2n}) \le 2\,\boldsymbol{\delta}(\mathcal{E}_n, \mathcal{E}_n) \quad \textit{and} \quad \boldsymbol{\delta}(\mathcal{E}_n, \mathcal{E}_n) \le \boldsymbol{\delta}(\mathcal{E}_{2n}, \mathcal{E}_{2n}).$$

PROOF. We only check the last inequality, since the other ones are special cases of 3.4.13. To this end, extend $\boldsymbol{f} \in [L_2[-\pi, +\pi), X]$ periodically and define $\boldsymbol{g}(t) := \boldsymbol{f}(2t)$. Then

$$\langle \boldsymbol{g}, \overline{e_{2k}} \rangle = \langle \boldsymbol{f}, \overline{e_k} \rangle, \quad \langle \boldsymbol{g}, \overline{e_{2k+1}} \rangle = o \quad \text{and} \quad \|\boldsymbol{g}|L_2\| = \|\boldsymbol{f}|L_2\|.$$

Hence

$$\|(T\,\langle \boldsymbol{g}, \overline{e_k} \rangle)|\mathcal{E}_{2n}\| \le \boldsymbol{\delta}(T|\mathcal{E}_{2n}, \mathcal{E}_{2n})\,\|\boldsymbol{g}|L_2\|$$

passes into

$$\|(T\,\langle \boldsymbol{f}, \overline{e_k} \rangle)|\mathcal{E}_n\| \le \boldsymbol{\delta}(T|\mathcal{E}_{2n}, \mathcal{E}_{2n})\,\|\boldsymbol{f}|L_2\|.$$

This proves that $\boldsymbol{\delta}(\mathcal{E}_n, \mathcal{E}_n) \le \boldsymbol{\delta}(\mathcal{E}_{2n}, \mathcal{E}_{2n})$.

5.2.3 PROPOSITION. *The ideal norm $\boldsymbol{\delta}(\mathcal{E}_n, \mathcal{E}_n)$ is symmetric.*

PROOF. By 3.4.6, we have $\boldsymbol{\delta}'(\mathcal{E}_n, \mathcal{E}_n) = \boldsymbol{\delta}(\overline{\mathcal{E}}_n, \overline{\mathcal{E}}_n)$, which yields the symmetry, since $\overline{\mathcal{E}}_n$ can be replaced by $\mathcal{E}_n$.

5.2.4 We now establish a **Marcinkiewicz inequality**. The following proof is adapted from [ZYG, II, p. 30].

PROPOSITION. *Let $1 \le r < \infty$. Then*

$$\left(\frac{1}{n}\sum_{h=1}^{n}\left\|\sum_{k=1}^{n} x_k \exp(\tfrac{2\pi i}{n}hk)\right\|^r\right)^{1/r} \le 3\left(\frac{1}{2\pi}\int_{-\pi}^{+\pi}\left\|\sum_{k=1}^{n} x_k \exp(ikt)\right\|^r dt\right)^{1/r}$$

for $x_1, \dots, x_n \in X$ and all Banach spaces X.

PROOF. Since the case $n = 1$ is trivial, we may assume that $n \geq 2$. If $m := [\frac{n}{2}]$, then

$$2m \leq n \leq 2m + 1.$$

Letting $V_m^0 := V_m e_{m+1}$, the Fourier coefficients of the m-th de la Vallée–Poussin kernel $V_m = 2F_{2m-1} - F_{m-1}$ are shifted to the right by $m+1$ steps. In particular,

$$\langle V_m^0, \overline{e_k} \rangle = \langle V_m, \overline{e_{k-m-1}} \rangle = 1 \quad \text{for } k = 1, \ldots, 2m+1.$$

Given $x_1, \ldots, x_n \in X$, we write

$$\boldsymbol{f} := \sum_{k=1}^{n} e_k \otimes x_k.$$

It follows from $n \leq 2m + 1$ that

$$\boldsymbol{f}(s) = \frac{1}{2\pi} \int_{-\pi}^{+\pi} V_m^0(s-t) \boldsymbol{f}(t)\, dt.$$

Writing $A_m := 2F_{2m-1} + F_{m-1}$ and applying Hölder's inequality, we obtain

$$\|\boldsymbol{f}(s)\| \leq \frac{1}{2\pi} \int_{-\pi}^{+\pi} |V_m^0(s-t)|\, \|\boldsymbol{f}(t)\|\, dt \leq \frac{1}{2\pi} \int_{-\pi}^{+\pi} A_m(s-t)^{1/r'+1/r} \|\boldsymbol{f}(t)\|\, dt$$

$$\leq \left(\frac{1}{2\pi} \int_{-\pi}^{+\pi} A_m(s-t)\, dt \right)^{1/r'} \left(\frac{1}{2\pi} \int_{-\pi}^{+\pi} A_m(s-t) \|\boldsymbol{f}(t)\|^r\, dt \right)^{1/r}.$$

We see from 5.1.6 that

$$\frac{1}{2\pi} \int_{-\pi}^{+\pi} A_m(s-t)\, dt = 3.$$

Hence

$$\|\boldsymbol{f}(s)\| \leq 3^{1/r'} \left(\frac{1}{2\pi} \int_{-\pi}^{+\pi} A_m(s-t) \|\boldsymbol{f}(t)\|^r\, dt \right)^{1/r}.$$

Summing over $s = \frac{2\pi}{n} h$ with $h = 1, \ldots, n$ yields

$$\sum_{h=1}^{n} \|\boldsymbol{f}(\tfrac{2\pi}{n} h)\|^r \leq 3^{r/r'} \frac{1}{2\pi} \int_{-\pi}^{+\pi} \sum_{h=1}^{n} A_m(\tfrac{2\pi}{n} h - t) \|\boldsymbol{f}(t)\|^r\, dt.$$

Since A_m is a trigonometric polynomial of degree $2m-1<n$, it follows from 5.1.4 that

$$\frac{1}{n}\sum_{h=1}^{n} A_m\left(\tfrac{2\pi}{n}h - t\right) = \frac{1}{2\pi}\int_{-\pi}^{+\pi} A_n(s-t)\,ds = 3.$$

Thus we finally arrive at

$$\frac{1}{n}\sum_{h=1}^{n}\left\|\sum_{k=1}^{n} x_k \exp\left(\tfrac{2\pi i}{n}hk\right)\right\|^r = \frac{1}{n}\sum_{h=1}^{n}\left\|\boldsymbol{f}\left(\tfrac{2\pi}{n}h\right)\right\|^r$$
$$\leq 3^{r/r'+1}\,\frac{1}{2\pi}\int_{-\pi}^{+\pi}\|\boldsymbol{f}(t)\|^r\,dt,$$

which is the desired inequality.

5.2.5 In the language of ideal norms and for $r=2$, the previous inequality reads as follows.

PROPOSITION. $\boldsymbol{\varrho}(\mathcal{E}_n^\circ, \mathcal{E}_n) \leq 3.$

5.2.6 We now establish an easy consequence.

PROPOSITION. $\boldsymbol{\delta}(\mathcal{E}_n^\circ, \mathcal{E}_n) \leq 3\,\boldsymbol{\delta}(\mathcal{E}_n, \mathcal{E}_n).$

PROOF. In view of 3.4.8,

$$\boldsymbol{\delta}(\mathcal{E}_n^\circ, \mathcal{E}_n) \leq \boldsymbol{\varrho}(\mathcal{E}_n^\circ, \mathcal{E}_n)\circ\boldsymbol{\delta}(\mathcal{E}_n, \mathcal{E}_n) \leq 3\,\boldsymbol{\delta}(\mathcal{E}_n, \mathcal{E}_n).$$

5.2.7 PROPOSITION. $\boldsymbol{\delta}(\mathcal{E}_n, \mathcal{E}_n) \leq \boldsymbol{\delta}(\mathcal{E}_n^\circ, \mathcal{E}_n).$

PROOF. By definition,

$$\frac{1}{n}\sum_{h=1}^{n}\left\|\sum_{k=1}^{n} T\,\langle \boldsymbol{f}, \overline{e_k}\rangle \exp\left(\tfrac{2\pi i}{n}hk\right)\right\|^2 \leq \boldsymbol{\delta}(T|\mathcal{E}_n^\circ, \mathcal{E}_n)^2\,\frac{1}{2\pi}\int_{-\pi}^{+\pi}\|\boldsymbol{f}(s)\|^2\,ds$$

for all $\boldsymbol{f} \in [L_2[-\pi, +\pi), X]$. Note that

$$\langle \boldsymbol{f}(\,\cdot + t), \overline{e_k}\rangle = \langle \boldsymbol{f}, \overline{e_k}\rangle\, e_k(t).$$

Therefore, replacing $\boldsymbol{f}(\,\cdot\,)$ by $\boldsymbol{f}(\,\cdot + t)$, we obtain

$$\frac{1}{n}\sum_{h=1}^{n}\left\|\sum_{k=1}^{n} T\,\langle \boldsymbol{f}, \overline{e_k}\rangle \exp\left(ik\left(t+\tfrac{2\pi}{n}h\right)\right)\right\|^2 \leq \boldsymbol{\delta}(T|\mathcal{E}_n^\circ, \mathcal{E}_n)^2\,\frac{1}{2\pi}\int_{-\pi}^{+\pi}\|\boldsymbol{f}(s+t)\|^2 ds.$$

Using the translation invariance of the Lebesgue measure, integration over t yields

$$\frac{1}{2\pi}\int_{-\pi}^{+\pi}\left\|\sum_{k=1}^{n} T\,\langle \boldsymbol{f}, \overline{e_k}\rangle\, e_k(t)\right\|^2 dt \leq \boldsymbol{\delta}(T|\mathcal{E}_n^\circ, \mathcal{E}_n)^2\,\frac{1}{2\pi}\int_{-\pi}^{+\pi}\|\boldsymbol{f}(s)\|^2\,ds.$$

This proves that $\boldsymbol{\delta}(T|\mathcal{E}_n, \mathcal{E}_n) \leq \boldsymbol{\delta}(T|\mathcal{E}_n^\circ, \mathcal{E}_n)$.

5.2.8 PROPOSITION. $\boldsymbol{\delta}(\mathcal{E}_n^\circ, \mathcal{E}_n^\circ) = 1$.

PROOF. The assertion follows from the fact that $\mathcal{E}_n^\circ$ is a basis of l_2^n.

5.2.9 We now combine the preceding results.

THEOREM. *The sequences of the following ideal norms are uniformly equivalent:*

$$\boldsymbol{\delta}(\mathcal{E}_n, \mathcal{E}_n), \quad \boldsymbol{\delta}(\mathcal{E}_n^\circ, \mathcal{E}_n), \quad \boldsymbol{\delta}(\mathcal{E}_n, \mathcal{E}_n^\circ), \quad \boldsymbol{\varrho}(\mathcal{E}_n, \mathcal{E}_n^\circ).$$

PROOF. By 5.2.6 and 5.2.7, we have

$$\boldsymbol{\delta}(\mathcal{E}_n^\circ, \mathcal{E}_n) \asymp \boldsymbol{\delta}(\mathcal{E}_n, \mathcal{E}_n).$$

Moreover, it follows from 3.4.6 and 5.2.1 that

$$\boldsymbol{\delta}(\mathcal{E}_n, \mathcal{E}_n^\circ) = \boldsymbol{\delta}'(\mathcal{E}_n^\circ, \mathcal{E}_n) \asymp \boldsymbol{\delta}'(\mathcal{E}_n, \mathcal{E}_n) = \boldsymbol{\delta}(\mathcal{E}_n, \mathcal{E}_n).$$

Finally, $\boldsymbol{\delta}(\mathcal{E}_n, \mathcal{E}_n^\circ) = \boldsymbol{\varrho}(\mathcal{E}_n, \mathcal{E}_n^\circ)$ is a consequence of 3.4.12 and 5.2.8.

5.2.10 PROPOSITION. $\boldsymbol{\varrho}(\mathcal{C}_n, \mathcal{E}_n) \le \sqrt{2}$ *and* $\boldsymbol{\varrho}(\mathcal{S}_n, \mathcal{E}_n) \le \sqrt{2}$.

PROOF. By Euler's formula, we have

$$c_k = \tfrac{1}{\sqrt{2}}(e_k + \overline{e}_k).$$

Hence

$$\|(x_k)|\mathcal{C}_n\| \le \tfrac{1}{\sqrt{2}}\big(\|(x_k)|\mathcal{E}_n\| + \|(x_k)|\overline{\mathcal{E}}_n\|\big) = \sqrt{2}\,\|(x_k)|\mathcal{E}_n\|.$$

This proves the left-hand inequality. The right-hand inequality can be obtained in the same way.

5.2.11 We now state a corollary of the previous results.

PROPOSITION.

$$\boldsymbol{\delta}(\mathcal{C}_n, \mathcal{C}_n) \le 2\,\boldsymbol{\delta}(\mathcal{E}_n, \mathcal{E}_n), \quad \boldsymbol{\delta}(\mathcal{S}_n, \mathcal{S}_n) \le 2\,\boldsymbol{\delta}(\mathcal{E}_n, \mathcal{E}_n),$$
$$\boldsymbol{\delta}(\mathcal{S}_n, \mathcal{C}_n) \le 2\,\boldsymbol{\delta}(\mathcal{E}_n, \mathcal{E}_n), \quad \boldsymbol{\delta}(\mathcal{C}_n, \mathcal{S}_n) \le 2\,\boldsymbol{\delta}(\mathcal{E}_n, \mathcal{E}_n).$$

PROOF. In view of 3.4.8, we have

$$\boldsymbol{\delta}(\mathcal{C}_n, \mathcal{C}_n) \le \boldsymbol{\varrho}(\mathcal{C}_n, \mathcal{E}_n) \circ \boldsymbol{\delta}(\mathcal{E}_n, \mathcal{E}_n) \circ \boldsymbol{\varrho}'(\mathcal{C}_n, \mathcal{E}_n) \le 2\,\boldsymbol{\delta}(\mathcal{E}_n, \mathcal{E}_n).$$

This proves the first inequality. The other ones can be obtained in the same way.

5.2.12 Next, the estimate from 5.2.5 is extended to the cosine and sine systems by analogous methods.

PROPOSITION. $\boldsymbol{\varrho}(\mathcal{C}_n^\circ, \mathcal{C}_n) \le 3$ *and* $\boldsymbol{\varrho}(\mathcal{S}_n^\circ, \mathcal{S}_n) \le 3$.

PROOF. A simplified version of the proof in 5.2.4 shows that

$$\left(\frac{1}{2n+1}\sum_{|h|\le n}\Big\|\sum_{|k|\le n} x_k \exp(\tfrac{2\pi i}{2n+1}hk)\Big\|^2\right)^{1/2} \le$$
$$\le 3\left(\frac{1}{2\pi}\int_{-\pi}^{+\pi}\Big\|\sum_{|k|\le n} x_k \exp(ikt)\Big\|^2 dt\right)^{1/2}$$

for $x_{-n},\dots,x_0,\dots,x_n \in X$. Given $x_1,\dots,x_n \in X$, we let $x_0 := o$ and $x_{-k} := x_k$ for $k = 1,\dots,n$. Then the above inequality passes into

$$\left(\frac{4}{2n+1}\sum_{|h|\le n}\Big\|\sum_{k=1}^{n} x_k \cos\tfrac{2\pi}{2n+1}hk\Big\|^2\right)^{1/2} \le 3\left(\frac{4}{2\pi}\int_{-\pi}^{+\pi}\Big\|\sum_{k=1}^{n} x_k \cos kt\Big\|^2 dt\right)^{1/2},$$

which proves that $\boldsymbol{\varrho}(X|\mathcal{C}_n^\circ, \mathcal{C}_n) \le 3$. The same result can be obtained by substituting $(x_n,\dots,x_1,o,x_1,\dots,x_n)$ in the defining inequality of $\boldsymbol{\varrho}(X|\mathcal{E}_{2n+1}^\circ, \mathcal{E}_{2n+1})$. Of course, $\boldsymbol{\varrho}(X|\mathcal{S}_n^\circ, \mathcal{S}_n) \le 3$ follows analogously.

5.2.13 We are now in a position to prove a lengthy supplement to 5.2.6.

PROPOSITION.

$$\boldsymbol{\delta}(\mathcal{C}_n^\circ, \mathcal{C}_n) \le 3\,\boldsymbol{\delta}(\mathcal{C}_n, \mathcal{C}_n), \quad \boldsymbol{\delta}(\mathcal{S}_n^\circ, \mathcal{S}_n) \le 3\,\boldsymbol{\delta}(\mathcal{S}_n, \mathcal{S}_n), \tag{1}$$
$$\boldsymbol{\delta}(\mathcal{C}_n, \mathcal{C}_n^\circ) \le 3\,\boldsymbol{\delta}(\mathcal{C}_n, \mathcal{C}_n), \quad \boldsymbol{\delta}(\mathcal{S}_n, \mathcal{S}_n^\circ) \le 3\,\boldsymbol{\delta}(\mathcal{S}_n, \mathcal{S}_n), \tag{2}$$
$$\boldsymbol{\delta}(\mathcal{S}_n^\circ, \mathcal{C}_n) \le 3\,\boldsymbol{\delta}(\mathcal{S}_n, \mathcal{C}_n), \quad \boldsymbol{\delta}(\mathcal{C}_n^\circ, \mathcal{S}_n) \le 3\,\boldsymbol{\delta}(\mathcal{C}_n, \mathcal{S}_n), \tag{3}$$
$$\boldsymbol{\delta}(\mathcal{S}_n, \mathcal{C}_n^\circ) \le 3\,\boldsymbol{\delta}(\mathcal{S}_n, \mathcal{C}_n), \quad \boldsymbol{\delta}(\mathcal{C}_n, \mathcal{S}_n^\circ) \le 3\,\boldsymbol{\delta}(\mathcal{C}_n, \mathcal{S}_n), \tag{4}$$
$$\boldsymbol{\delta}(\mathcal{S}_n^\circ, \mathcal{C}_n^\circ) \le 3\,\boldsymbol{\delta}(\mathcal{S}_n, \mathcal{C}_n^\circ), \quad \boldsymbol{\delta}(\mathcal{C}_n^\circ, \mathcal{S}_n^\circ) \le 3\,\boldsymbol{\delta}(\mathcal{C}_n, \mathcal{S}_n^\circ), \tag{5}$$
$$\boldsymbol{\delta}(\mathcal{S}_n^\circ, \mathcal{C}_n^\circ) \le 3\,\boldsymbol{\delta}(\mathcal{S}_n^\circ, \mathcal{C}_n), \quad \boldsymbol{\delta}(\mathcal{C}_n^\circ, \mathcal{S}_n^\circ) \le 3\,\boldsymbol{\delta}(\mathcal{C}_n^\circ, \mathcal{S}_n), \tag{6}$$
$$\boldsymbol{\varrho}(\mathcal{S}_n^\circ, \mathcal{C}_n) \le 3\,\boldsymbol{\varrho}(\mathcal{S}_n, \mathcal{C}_n), \quad \boldsymbol{\varrho}(\mathcal{C}_n^\circ, \mathcal{S}_n) \le 3\,\boldsymbol{\varrho}(\mathcal{C}_n, \mathcal{S}_n), \tag{7}$$
$$\boldsymbol{\varrho}(\mathcal{S}_n^\circ, \mathcal{C}_n^\circ) \le 3\,\boldsymbol{\varrho}(\mathcal{S}_n, \mathcal{C}_n^\circ), \quad \boldsymbol{\varrho}(\mathcal{C}_n^\circ, \mathcal{S}_n^\circ) \le 3\,\boldsymbol{\varrho}(\mathcal{C}_n, \mathcal{S}_n^\circ), \tag{8}$$
$$\boldsymbol{\varrho}(\mathcal{S}_n^\circ, \mathcal{C}_n) \le 3\,\boldsymbol{\varrho}(\mathcal{S}_n^\circ, \mathcal{C}_n^\circ), \quad \boldsymbol{\varrho}(\mathcal{C}_n^\circ, \mathcal{S}_n) \le 3\,\boldsymbol{\varrho}(\mathcal{C}_n^\circ, \mathcal{S}_n^\circ). \tag{9}$$

PROOF. By 3.4.8, we conclude from $\boldsymbol{\varrho}(\mathcal{C}_n^\circ, \mathcal{C}_n) \le 3$ and $\boldsymbol{\varrho}(\mathcal{S}_n^\circ, \mathcal{S}_n) \le 3$ that

$$\boldsymbol{\delta}(\mathcal{C}_n^\circ, \mathcal{C}_n) \le \boldsymbol{\varrho}(\mathcal{C}_n^\circ, \mathcal{C}_n) \circ \boldsymbol{\delta}(\mathcal{C}_n, \mathcal{C}_n) \le 3\,\boldsymbol{\delta}(\mathcal{C}_n, \mathcal{C}_n)$$

and

$$\boldsymbol{\delta}(\mathcal{S}_n^\circ, \mathcal{S}_n) \le \boldsymbol{\varrho}(\mathcal{S}_n^\circ, \mathcal{S}_n) \circ \boldsymbol{\delta}(\mathcal{S}_n, \mathcal{S}_n) \le 3\,\boldsymbol{\delta}(\mathcal{S}_n, \mathcal{S}_n),$$

which proves (1). Inequalities (3), (5), (7), (8) and (9) can be obtained similarly, while (2), (4) and (6) follow by duality from (1), (3) and (5), respectively.

5.2.14 Let K be the operator on $L_2(\mathbb{R}_+)$ defined by

$$K : f(s) \longrightarrow g(t) := \frac{1}{\pi} \int_0^\infty \frac{f(s)}{s+t} \, ds.$$

Then we have a continuous analogue of 2.4.2; see [HAR*, p. 226].

PROPOSITION. $\|K\| \leq 1$.

5.2.15 The next inequalities are a major step in proving 5.2.16.

PROPOSITION.

$$\boldsymbol{\delta}(\mathcal{C}_n^\circ, \mathcal{C}_n) \leq 12\, \boldsymbol{\delta}(\mathcal{S}_n^\circ, \mathcal{S}_n) \quad \textit{and} \quad \boldsymbol{\delta}(\mathcal{S}_n^\circ, \mathcal{S}_n) \leq 11\, \boldsymbol{\delta}(\mathcal{C}_n^\circ, \mathcal{C}_n).$$

PROOF. Define

$$\boldsymbol{g}(t) := \frac{1}{\pi} \int_0^\pi \boldsymbol{f}(s) D_n(s+t) \, ds \quad \text{for } \boldsymbol{f} \in [L_2[0,\pi), X] \text{ and } 0 \leq t \leq \pi.$$

Write

$$\boldsymbol{g}(t) = \frac{1}{\pi} \int_0^{\pi - t} \boldsymbol{f}(s) D_n(s+t) \, ds + \frac{1}{\pi} \int_{\pi - t}^{\pi} \boldsymbol{f}(s) D_n(s+t) \, ds.$$

Substituting $\pi - s$ for s in the second integral yields

$$\boldsymbol{g}(t) = \frac{1}{\pi} \int_0^{\pi - t} \boldsymbol{f}(s) D_n(s+t) \, ds + \frac{1}{\pi} \int_0^{t} \boldsymbol{f}(\pi - s) D_n(\pi - s + t) \, ds.$$

In view of $D_n(\pi - s + t) = D_n(-\pi + s - t) = D_n(s + \pi - t)$, we have

$$\boldsymbol{g}(t) = \boldsymbol{g}_+(t) + \boldsymbol{g}_-(\pi - t),$$

where

$$\boldsymbol{g}_+(t) := \frac{1}{\pi} \int_0^{\pi - t} \boldsymbol{f}(s) D_n(s+t) \, ds \text{ and } \boldsymbol{g}_-(t) := \frac{1}{\pi} \int_0^{\pi - t} \boldsymbol{f}(\pi - s) D_n(s+t) \, ds.$$

Let

$$\gamma_+(t) := \int_0^\pi \frac{\|\boldsymbol{f}(s)\|}{s+t} \, ds \quad \text{and} \quad \gamma_-(t) := \int_0^\pi \frac{\|\boldsymbol{f}(\pi - s)\|}{s+t} \, ds.$$

It follows from

$$|D_n(s+t)| \leq \frac{\pi}{s+t} \quad \text{for } 0 \leq s+t \leq \pi$$

that

$$\|\boldsymbol{g}_\pm(t)\| \leq \gamma_\pm(t).$$

Since $\gamma_\pm$ is non-increasing, 5.2.14 implies that

$$\frac{1}{2n+1}\sum_{h=1}^{2n+1}\Big\|\boldsymbol{g}_\pm\Big(\tfrac{\pi}{2n+1}h\Big)\Big\|^2 \le \frac{1}{2n+1}\sum_{h=1}^{2n+1}\gamma_\pm\Big(\tfrac{\pi}{2n+1}h\Big)^2$$
$$\le \frac{1}{\pi}\int_0^\pi \gamma_\pm(t)^2\,dt \le \pi\int_0^\pi \|\boldsymbol{f}(s)\|^2\,ds = \pi^2\|\boldsymbol{f}|L_2\|^2.$$

Therefore

$$\left(\frac{1}{2n+1}\sum_{h=1}^{n}\Big\|\boldsymbol{g}(\tfrac{2\pi}{2n+1}h)\Big\|^2\right)^{1/2} \le \left(\frac{1}{2n+1}\sum_{h=1}^{2n+1}\Big\|\boldsymbol{g}(\tfrac{\pi}{2n+1}h)\Big\|^2\right)^{1/2}$$
$$\le \left(\frac{1}{2n+1}\sum_{h=1}^{2n+1}\Big\|\boldsymbol{g}_+(\tfrac{\pi}{2n+1}h)\Big\|^2\right)^{1/2} + \left(\frac{1}{2n+1}\sum_{h=1}^{2n+1}\Big\|\boldsymbol{g}_-(\tfrac{\pi}{2n+1}h)\Big\|^2\right)^{1/2}$$
$$\le 2\pi\,\|\boldsymbol{f}|L_2\|.$$

We conclude from

$$D_n(s+t) = 1 + 2\sum_{k=1}^{n}\cos ks\cos kt - 2\sum_{k=1}^{n}\sin ks\sin kt$$

that

$$\boldsymbol{g}(t) = \frac{1}{\pi}\int_0^\pi \boldsymbol{f}(s)\,ds + \sqrt{2}\sum_{k=1}^{n}\langle\boldsymbol{f}, c_k\rangle\cos kt - \sqrt{2}\sum_{k=1}^{n}\langle\boldsymbol{f}, s_k\rangle\sin kt.$$

The above estimate now yields

$$\left(\frac{4}{2n+1}\sum_{h=1}^{n}\Big\|\sum_{k=1}^{n}T\langle\boldsymbol{f}, c_k\rangle\cos\tfrac{2\pi}{2n+1}hk - \sum_{k=1}^{n}T\langle\boldsymbol{f}, s_k\rangle\sin\tfrac{2\pi}{2n+1}hk\Big\|^2\right)^{1/2} =$$
$$= \sqrt{2}\left(\frac{1}{2n+1}\sum_{h=1}^{n}\Big\|T\boldsymbol{g}\big(\tfrac{2\pi}{2n+1}h\big) - \frac{1}{\pi}\int_0^\pi T\boldsymbol{f}(s)\,ds\Big\|^2\right)^{1/2}$$
$$\le \sqrt{2}\left(2\,\pi + \sqrt{\tfrac{n}{2n+1}}\right)\|T\|\,\|\boldsymbol{f}|L_2\| \le (\pi\sqrt{8}+1)\|T\|\,\|\boldsymbol{f}|L_2\|.$$

Moreover,

$$\sqrt{\tfrac{2}{2n+1}}\Big\|\sum_{k=1}^{n}T\langle\boldsymbol{f}, c_k\rangle\Big\| = \sqrt{\tfrac{2}{2n+1}}\Big\|\Big\langle T\boldsymbol{f}, \sum_{k=1}^{n}c_k\Big\rangle\Big\|$$
$$\le \sqrt{\tfrac{2}{2n+1}}\|T\|\,\|\boldsymbol{f}|L_2\|\,\Big\|\sum_{k=1}^{n}c_k\Big|L_2\Big\| \le \sqrt{\tfrac{2n}{2n+1}}\|T\|\,\|\boldsymbol{f}|L_2\| \le \|T\|\,\|\boldsymbol{f}|L_2\|.$$

Combining the preceding results, we finally arrive at

$$\begin{aligned}
&\|(T\langle \boldsymbol{f}, c_k\rangle)|\mathcal{C}_n^\circ\| = \\
&= \left(\frac{2}{2n+1}\left\|\sum_{k=1}^{n} T\langle \boldsymbol{f}, c_k\rangle\right\|^2 + \frac{4}{2n+1}\sum_{h=1}^{n}\left\|\sum_{k=1}^{n} T\langle \boldsymbol{f}, c_k\rangle \cos\tfrac{2\pi}{2n+1}hk\right\|^2\right)^{1/2} \\
&\le \|T\|\,\|\boldsymbol{f}|L_2\| + \left(\frac{4}{2n+1}\sum_{h=1}^{n}\left\|\sum_{k=1}^{n} T\langle \boldsymbol{f}, s_k\rangle \sin\tfrac{2\pi}{2n+1}hk\right\|^2\right)^{1/2} + \\
&\quad + \left(\frac{4}{2n+1}\sum_{h=1}^{n}\left\|\sum_{k=1}^{n} T\langle \boldsymbol{f}, c_k\rangle \cos\tfrac{2\pi}{2n+1}hk - \sum_{k=1}^{n} T\langle \boldsymbol{f}, s_k\rangle \sin\tfrac{2\pi}{2n+1}hk\right\|^2\right)^{1/2} \\
&\le \left[\boldsymbol{\delta}(T|\mathcal{S}_n^\circ, \mathcal{S}_n) + \left(\pi\sqrt{8}+2\right)\|T\|\right]\|\boldsymbol{f}|L_2\|.
\end{aligned}$$

This proves that

$$\boldsymbol{\delta}(\mathcal{C}_n^\circ, \mathcal{C}_n) \le 12\,\boldsymbol{\delta}(\mathcal{S}_n^\circ, \mathcal{S}_n),$$

since $\pi\sqrt{8}+3 = 11.8857\ldots < 12$.

Analogously, we obtain

$$\begin{aligned}
&\|(T\langle \boldsymbol{f}, s_k\rangle)|\mathcal{S}_n^\circ\| = \left(\frac{4}{2n+1}\sum_{h=1}^{n}\left\|\sum_{k=1}^{n} T\langle \boldsymbol{f}, s_k\rangle \sin\tfrac{2\pi}{2n+1}hk\right\|^2\right)^{1/2} \\
&\le \left(\frac{4}{2n+1}\sum_{h=1}^{n}\left\|\sum_{k=1}^{n} T\langle \boldsymbol{f}, c_k\rangle \cos\tfrac{2\pi}{2n+1}hk\right\|^2\right)^{1/2} + \\
&\quad + \left(\frac{4}{2n+1}\sum_{h=1}^{n}\left\|\sum_{k=1}^{n} T\langle \boldsymbol{f}, s_k\rangle \sin\tfrac{2\pi}{2n+1}hk - \sum_{k=1}^{n} T\langle \boldsymbol{f}, c_k\rangle \cos\tfrac{2\pi}{2n+1}hk\right\|^2\right)^{1/2} \\
&\le \left[\boldsymbol{\delta}(T|\mathcal{C}_n^\circ, \mathcal{C}_n) + \left(\pi\sqrt{8}+1\right)\|T\|\right]\|\boldsymbol{f}|L_2\|,
\end{aligned}$$

which yields

$$\boldsymbol{\delta}(\mathcal{S}_n^\circ, \mathcal{S}_n) \le 11\,\boldsymbol{\delta}(\mathcal{C}_n^\circ, \mathcal{C}_n).$$

5.2.16 PROPOSITION.

$$\boldsymbol{\delta}(\mathcal{E}_n, \mathcal{E}_n) \le 28\,\boldsymbol{\delta}(\mathcal{S}_n^\circ, \mathcal{S}_n) \quad \textit{and} \quad \boldsymbol{\delta}(\mathcal{E}_n, \mathcal{E}_n) \le 26\,\boldsymbol{\delta}(\mathcal{C}_n^\circ, \mathcal{C}_n).$$

PROOF. For every function $\boldsymbol{f} \in [L_2[-\pi, +\pi), X]$, we denote by $\boldsymbol{Cf}$ and $\boldsymbol{Sf}$ the restrictions of the even part $\frac{1}{2}[\boldsymbol{f}(+t) + \boldsymbol{f}(-t)]$ and the odd part $\frac{1}{2}[\boldsymbol{f}(+t) - \boldsymbol{f}(-t)]$ to the interval $[0, \pi)$, respectively. Then

$$\frac{1}{2\pi}\int_{-\pi}^{+\pi} \boldsymbol{f}(t)\cos kt\,dt = \frac{1}{\pi}\int_0^{\pi} \boldsymbol{Cf}(t)\cos kt\,dt$$

and

$$\frac{1}{2\pi}\int_{-\pi}^{+\pi} \boldsymbol{f}(t)\sin kt\,dt = \frac{1}{\pi}\int_0^{\pi} \boldsymbol{Sf}(t)\sin kt\,dt.$$

Note that $\boldsymbol{C}$ and $\boldsymbol{S}$ are metric surjections from $[L_2[-\pi,+\pi),X]$ onto $[L_2[0,\pi),X]$. We conclude from

$$D_n(s-t) = 1 + 2\sum_{k=1}^{n}\cos ks\cos kt + 2\sum_{k=1}^{n}\sin ks\sin kt$$

that

$$\begin{aligned}\boldsymbol{g}(t) &:= \frac{1}{2\pi}\int_{-\pi}^{+\pi} T\boldsymbol{f}(s)D_n(s-t)\,ds \\ &= \frac{1}{2\pi}\int_{-\pi}^{+\pi} T\boldsymbol{f}(s)\,ds + \sqrt{2}\sum_{k=1}^{n}\Big(T\langle \boldsymbol{Cf}, c_k\rangle\cos kt + T\langle \boldsymbol{Sf}, s_k\rangle\sin kt\Big).\end{aligned}$$

Hence

$$\begin{aligned}&\left(\frac{1}{2n+1}\sum_{|h|\le n}\left\|\boldsymbol{g}\Big(\tfrac{2\pi}{2n+1}h\Big)\right\|^2\right)^{1/2} \le \\ &\le \|T\|\,\|\boldsymbol{f}|L_2\| + \left\{\begin{array}{c}\left(\dfrac{2}{2n+1}\displaystyle\sum_{|h|\le n}\left\|\sum_{k=1}^{n}T\langle \boldsymbol{Cf}, c_k\rangle\cos kt\right\|^2\right)^{1/2} \\ + \\ \left(\dfrac{2}{2n+1}\displaystyle\sum_{|h|\le n}\left\|\sum_{k=1}^{n}T\langle \boldsymbol{Sf}, s_k\rangle\sin kt\right\|^2\right)^{1/2}\end{array}\right\} \\ &\le \|T\|\,\|\boldsymbol{f}|L_2\| + \|(T\langle \boldsymbol{Cf}, c_k\rangle)|\mathcal{C}_n^\circ\| + \|(T\langle \boldsymbol{Sf}, s_k\rangle)|\mathcal{S}_n^\circ\| \le c\,\|\boldsymbol{f}|L_2\|,\end{aligned}$$

where $c := \|T\| + \boldsymbol{\delta}(T|\mathcal{C}_n^\circ, \mathcal{C}_n) + \boldsymbol{\delta}(T|\mathcal{S}_n^\circ, \mathcal{S}_n)$.

Given any $\boldsymbol{f}_0 \in [L_2[-\pi,+\pi),X]$, we define $\boldsymbol{f} := \boldsymbol{f}_0\,e_{-n-1}$. Then the substitution $k' := k+n+1$ yields

$$\begin{aligned}\sum_{k'=1}^{2n+1} T\langle \boldsymbol{f}_0, \overline{e_{k'}}\rangle\, e_{k'} &= \sum_{|k|\le n} T\langle \boldsymbol{f}_0, \overline{e_{k+n+1}}\rangle\, e_{k+n+1} \\ &= \sum_{|k|\le n} T\langle \boldsymbol{f}, \overline{e_k}\rangle\, e_k e_{n+1} = \boldsymbol{g}\, e_{n+1}.\end{aligned}$$

Since $g\big(\frac{2\pi}{2n+1}h\big)$ is a $(2n+1)$-periodic function of the index h, we get

$$\|(T\langle f_0, \overline{e_{k'}}\rangle)|\mathcal{E}^\circ_{2n+1}\| = \left(\frac{1}{2n+1}\sum_{h=1}^{2n+1}\left\|g\big(\tfrac{2\pi}{2n+1}h\big)\right\|^2\right)^{1/2}$$
$$\leq c\,\|f|L_2\| = c\,\|f_0|L_2\|.$$

This implies that

$$\delta(T|\mathcal{E}^\circ_{2n+1}, \mathcal{E}_{2n+1}) \leq \|T\| + \delta(T|\mathcal{C}^\circ_n, \mathcal{C}_n) + \delta(T|\mathcal{S}^\circ_n, \mathcal{S}_n).$$

Hence

$$\delta(T|\mathcal{E}^\circ_{2n+1}, \mathcal{E}_{2n+1}) \leq 14\,\delta(T|\mathcal{S}^\circ_n, \mathcal{S}_n)$$

and

$$\delta(T|\mathcal{E}^\circ_{2n+1}, \mathcal{E}_{2n+1}) \leq 13\,\delta(T|\mathcal{C}^\circ_n, \mathcal{C}_n),$$

by 5.2.15. The assertions now follow from

$$\delta(\mathcal{E}_n, \mathcal{E}_n) \leq \delta(\mathcal{E}_{2n}, \mathcal{E}_{2n}) \leq 2\,\delta(\mathcal{E}_{2n+1}, \mathcal{E}_{2n+1}) \leq 2\,\delta(\mathcal{E}^\circ_{2n+1}, \mathcal{E}_{2n+1});$$

see 5.2.2 and 5.2.7.

5.2.17 We now establish a counterpart of 5.2.8.

PROPOSITION. $\delta(\mathcal{S}^\circ_n, \mathcal{S}^\circ_n) = 1$ *and* $1 \leq \delta(\mathcal{C}^\circ_n, \mathcal{C}^\circ_n) \leq 2.$

PROOF. Obviously, $\delta(\mathcal{S}^\circ_n, \mathcal{S}^\circ_n) = 1$ follows from the fact that $\mathcal{S}^\circ_n$ is a basis of l_2^n. Note that $\mathcal{C}^\circ_n = (c_1^{(n)}, \dots, c_n^{(n)})$ together with the vector

$$c_0^{(n)} = \frac{1}{\sqrt{2n+1}}\begin{pmatrix} 1 \\ \sqrt{2} \\ \vdots \\ \sqrt{2} \end{pmatrix}$$

forms a basis of l_2^{n+1}. So the orthogonal projection from l_2^{n+1} onto the 1-dimensional subspace spanned by $c_0^{(n)}$ is given by the positive matrix

$$P_{n+1} := \frac{1}{2n+1}\begin{pmatrix} 1 & \sqrt{2} & \dots & \sqrt{2} \\ \sqrt{2} & 2 & \dots & 2 \\ \vdots & \vdots & \ddots & \vdots \\ \sqrt{2} & 2 & \dots & 2 \end{pmatrix}.$$

Now it follows from 0.4.9 that

$$\delta(X|\mathcal{C}^\circ_n, \mathcal{C}^\circ_n) = \|[I_{n+1} - P_{n+1}, X] : [l_2^{n+1}, X] \to [l_2^{n+1}, X]\|$$
$$\leq \left\{\begin{array}{c} \|[I_{n+1}, X] : [l_2^{n+1}, X] \to [l_2^{n+1}, X]\| \\ + \\ \|[P_{n+1}, X] : [l_2^{n+1}, X] \to [l_2^{n+1}, X]\| \end{array}\right\} = 2.$$

REMARK. Define $\boldsymbol{f} \in [l_2^{n+1}, l_1^n]$ by

$$\boldsymbol{f}(k) := \begin{cases} o & \text{if } k = 0, \\ u_k^{(n)} & \text{if } k = 1, \ldots, n. \end{cases}$$

Then

$$\|\boldsymbol{f} - [P_{n+1}, l_1^n]\boldsymbol{f} \,|[l_2^{n+1}, l_1^n]\| \geq \tfrac{4n-3}{2n+1}\sqrt{n} \quad \text{and} \quad \|\boldsymbol{f}\,|[l_2^{n+1}, l_1^n]\| = \sqrt{n}.$$

Hence, substituting $\boldsymbol{f}$ into the defining inequality of $\boldsymbol{\delta}(l_1^n|\mathcal{C}_n^\circ, \mathcal{C}_n^\circ)$, we obtain

$$\tfrac{4n-3}{2n+1} \leq \boldsymbol{\delta}(l_1^n|\mathcal{C}_n^\circ, \mathcal{C}_n^\circ),$$

which proves that $\lim\limits_{n\to\infty} \boldsymbol{\delta}(l_1^n|\mathcal{C}_n^\circ, \mathcal{C}_n^\circ) = 2$. Thus the upper estimate of $\boldsymbol{\delta}(X|\mathcal{C}_n^\circ, \mathcal{C}_n^\circ)$ by 2 is sharp.

5.2.18 Combining the preceding results, we now establish a supplement to 5.2.9; see [wen 2].

THEOREM. *The sequences of the following ideal norms are uniformly equivalent:*

$$\boldsymbol{\delta}(\mathcal{E}_n, \mathcal{E}_n),$$
$$\boldsymbol{\delta}(\mathcal{C}_n, \mathcal{C}_n),\ \boldsymbol{\delta}(\mathcal{C}_n^\circ, \mathcal{C}_n),\ \boldsymbol{\delta}(\mathcal{C}_n, \mathcal{C}_n^\circ),\ \boldsymbol{\varrho}(\mathcal{C}_n, \mathcal{C}_n^\circ),$$
$$\boldsymbol{\delta}(\mathcal{S}_n, \mathcal{S}_n),\ \boldsymbol{\delta}(\mathcal{S}_n^\circ, \mathcal{S}_n),\ \boldsymbol{\delta}(\mathcal{S}_n, \mathcal{S}_n^\circ),\ \boldsymbol{\varrho}(\mathcal{S}_n, \mathcal{S}_n^\circ).$$

PROOF. Recall that

$$\begin{aligned} \boldsymbol{\delta}(\mathcal{E}_n, \mathcal{E}_n) &\leq 26\,\boldsymbol{\delta}(\mathcal{C}_n^\circ, \mathcal{C}_n) && \text{(by 5.2.16)}, \\ \boldsymbol{\delta}(\mathcal{C}_n^\circ, \mathcal{C}_n) &\leq \ \ 3\,\boldsymbol{\delta}(\mathcal{C}_n, \mathcal{C}_n) && \text{(by 5.2.13)}, \\ \boldsymbol{\delta}(\mathcal{C}_n, \mathcal{C}_n) &\leq \ \ 2\,\boldsymbol{\delta}(\mathcal{E}_n, \mathcal{E}_n) && \text{(by 5.2.11)}. \end{aligned}$$

Hence

$$\boldsymbol{\delta}(\mathcal{E}_n, \mathcal{E}_n) \leq 26\,\boldsymbol{\delta}(\mathcal{C}_n^\circ, \mathcal{C}_n) \leq 78\,\boldsymbol{\delta}(\mathcal{C}_n, \mathcal{C}_n) \leq 156\,\boldsymbol{\delta}(\mathcal{E}_n, \mathcal{E}_n),$$

and, by duality,

$$\boldsymbol{\delta}(\mathcal{E}_n, \mathcal{E}_n) \leq 26\,\boldsymbol{\delta}(\mathcal{C}_n, \mathcal{C}_n^\circ) \leq 78\,\boldsymbol{\delta}(\mathcal{C}_n, \mathcal{C}_n) \leq 156\,\boldsymbol{\delta}(\mathcal{E}_n, \mathcal{E}_n).$$

Finally, we deduce from 3.4.8, 3.4.10, and 5.2.17 that

$$\boldsymbol{\varrho}(\mathcal{C}_n, \mathcal{C}_n^\circ) \leq \boldsymbol{\delta}(\mathcal{C}_n, \mathcal{C}_n^\circ) \leq \boldsymbol{\varrho}(\mathcal{C}_n, \mathcal{C}_n^\circ) \circ \boldsymbol{\delta}(\mathcal{C}_n^\circ, \mathcal{C}_n^\circ) \leq 2\,\boldsymbol{\varrho}(\mathcal{C}_n, \mathcal{C}_n^\circ).$$

This completes the proof for the cosine systems. The sine systems can be treated in the same manner.

5.2.19 We now prepare the proof of Proposition 5.2.22.

LEMMA. *For $s \in \mathbb{R}$ and $x_1, \dots, x_n \in X$,*

$$\|(x_k \cos ks)|\mathfrak{C}_n\| \le \|(x_k)|\mathfrak{C}_n\|, \qquad \|(x_k \sin ks)|\mathfrak{S}_n\| \le \|(x_k)|\mathfrak{C}_n\|,$$
$$\|(x_k \sin ks)|\mathfrak{C}_n\| \le \|(x_k)|\mathfrak{S}_n\|, \qquad \|(x_k \cos ks)|\mathfrak{S}_n\| \le \|(x_k)|\mathfrak{S}_n\|.$$

PROOF. Using the translation invariance of the Lebesgue measure and the addition theorems for trigonometric functions, it follows that

$$\|(x_k \cos ks)|\mathfrak{C}_n\| = \left(\frac{1}{\pi}\int_{-\pi}^{+\pi}\Big\|\sum_{k=1}^{n} x_k \cos ks \cos kt\Big\|^2 dt\right)^{1/2}$$

$$\le \frac{1}{2}\left\{\begin{array}{c}\left(\frac{1}{\pi}\int_{-\pi}^{+\pi}\Big\|\sum_{k=1}^{n} x_k \cos k(t+s)\Big\|^2 dt\right)^{1/2} \\ + \\ \left(\frac{1}{\pi}\int_{-\pi}^{+\pi}\Big\|\sum_{k=1}^{n} x_k \cos k(t-s)\Big\|^2 dt\right)^{1/2}\end{array}\right\} = \|(x_k)|\mathfrak{C}_n\|.$$

This proves the first inequality. The rest can be checked similarly.

5.2.20 LEMMA.

$$\varrho(\mathfrak{C}_n \otimes \mathfrak{C}_n, \mathfrak{C}_n) \le \sqrt{2}, \qquad \varrho(\mathfrak{C}_n^\circ \otimes \mathfrak{C}_n, \mathfrak{C}_n) \le \sqrt{2},$$
$$\varrho(\mathfrak{S}_n \otimes \mathfrak{S}_n, \mathfrak{C}_n) \le \sqrt{2}, \qquad \varrho(\mathfrak{S}_n^\circ \otimes \mathfrak{S}_n, \mathfrak{C}_n) \le \sqrt{2},$$
$$\varrho(\mathfrak{C}_n \otimes \mathfrak{S}_n, \mathfrak{S}_n) \le \sqrt{2}, \qquad \varrho(\mathfrak{C}_n^\circ \otimes \mathfrak{S}_n, \mathfrak{S}_n) \le \sqrt{2},$$
$$\varrho(\mathfrak{S}_n \otimes \mathfrak{C}_n, \mathfrak{S}_n) \le \sqrt{2}, \qquad \varrho(\mathfrak{S}_n^\circ \otimes \mathfrak{C}_n, \mathfrak{S}_n) \le \sqrt{2}.$$

PROOF. The first inequality from 5.2.19 implies that

$$\|(x_k)|\mathfrak{C}_n \otimes \mathfrak{C}_n\|^2 = \frac{1}{2\pi}\int_{-\pi}^{+\pi} \|(x_k\sqrt{2}\cos ks)|\mathfrak{C}_n\|^2 ds \le 2\,\|(x_k)|\mathfrak{C}_n\|^2.$$

Analogously, we get

$$\|(x_k)|\mathfrak{C}_n^\circ \otimes \mathfrak{C}_n\|^2 = \frac{2}{2n+1}\sum_{|h|\le n}\Big\|\Big(x_k \cos \tfrac{2\pi}{2n+1}hk\Big)\Big|\mathfrak{C}_n\Big\|^2 \le 2\,\|(x_k)|\mathfrak{C}_n\|^2.$$

The other inequalities can be obtained in the same way.

5.2.21 LEMMA. *For $x_1, \dots, x_n \in X$,*

$$\|(x_k)|\mathfrak{C}_n\|^2 \le \|(x_k)|\mathfrak{C}_n \otimes \mathfrak{C}_n^\circ\|^2 + \|(x_k)|\mathfrak{S}_n \otimes \mathfrak{S}_n^\circ\|^2,$$
$$\|(x_k)|\mathfrak{S}_n\|^2 \le \|(x_k)|\mathfrak{S}_n \otimes \mathfrak{C}_n^\circ\|^2 + \|(x_k)|\mathfrak{C}_n \otimes \mathfrak{S}_n^\circ\|^2,$$
$$\|(x_k)|\mathfrak{C}_n^\circ\|^2 \le \|(x_k)|\mathfrak{C}_n^\circ \otimes \mathfrak{C}_n^\circ\|^2 + \|(x_k)|\mathfrak{S}_n^\circ \otimes \mathfrak{S}_n^\circ\|^2,$$
$$\|(x_k)|\mathfrak{S}_n^\circ\|^2 \le \|(x_k)|\mathfrak{S}_n^\circ \otimes \mathfrak{C}_n^\circ\|^2 + \|(x_k)|\mathfrak{C}_n^\circ \otimes \mathfrak{S}_n^\circ\|^2.$$

PROOF. For $s \in \mathbb{R}$, the translation invariance of the Lebesgue measure yields

$$\|(x_k)|\mathcal{C}_n\|^2 = \frac{1}{\pi}\int_{-\pi}^{+\pi}\Big\|\sum_{k=1}^{n} x_k \cos k(t-s)\Big\|^2 dt.$$

The addition theorems for trigonometric functions imply that

$$\|(x_k)|\mathcal{C}_n\|^2 \le 2\left\{\begin{array}{c}\displaystyle\frac{1}{\pi}\int_{-\pi}^{+\pi}\Big\|\sum_{k=1}^{n} x_k \cos kt \cos ks\Big\|^2 dt \\ + \\ \displaystyle\frac{1}{\pi}\int_{-\pi}^{+\pi}\Big\|\sum_{k=1}^{n} x_k \sin kt \sin ks\Big\|^2 dt\end{array}\right\}.$$

Averaging this inequality over $s := \frac{2\pi}{2n+1}h$ with $|h| \le n$, we get

$$\|(x_k)|\mathcal{C}_n\|^2 \le \|(x_k)|\mathcal{C}_n \otimes \mathcal{C}_n^\circ\|^2 + \|(x_k)|\mathcal{S}_n \otimes \mathcal{S}_n^\circ\|^2.$$

This proves the first inequality. The second one can be verified similarly.

Note that

$$\|(x_k)|\mathcal{C}_n^\circ\|^2 = \frac{2}{2n+1}\sum_{|h|\le n}\Big\|\sum_{k=1}^{n} x_k \cos \tfrac{2\pi}{2n+1}k(h-l)\Big\|^2$$

for all $l \in \mathbb{Z}$. Using once more the addition theorems for trigonometric functions, we obtain

$$\|(x_k)|\mathcal{C}_n^\circ\|^2 \le 2\left\{\begin{array}{c}\displaystyle\frac{2}{2n+1}\sum_{|h|\le n}\Big\|\sum_{k=1}^{n} x_k \cos \tfrac{2\pi}{2n+1}kh \cos \tfrac{2\pi}{2n+1}kl\Big\|^2 \\ + \\ \displaystyle\frac{2}{2n+1}\sum_{|h|\le n}\Big\|\sum_{k=1}^{n} x_k \sin \tfrac{2\pi}{2n+1}kh \sin \tfrac{2\pi}{2n+1}kl\Big\|^2\end{array}\right\}.$$

Averaging this inequality over l with $|l| \le n$ gives

$$\|(x_k)|\mathcal{C}_n^\circ\|^2 \le \|(x_k)|\mathcal{C}_n^\circ \otimes \mathcal{C}_n^\circ\|^2 + \|(x_k)|\mathcal{S}_n^\circ \otimes \mathcal{S}_n^\circ\|^2.$$

This proves the third inequality. The last one follows in the same way.

REMARK. Since $\|(x_k)|\mathcal{S}_n^\circ \otimes \mathcal{C}_n^\circ\| = \|(x_k)|\mathcal{C}_n^\circ \otimes \mathcal{S}_n^\circ\|$, the fourth inequality means that

$$\varrho(\mathcal{S}_n^\circ, \mathcal{S}_n^\circ \otimes \mathcal{C}_n^\circ) \le \sqrt{2}.$$

5.2.22 We are now in a position to establish one of the most crucial inequalities of this chapter.

PROPOSITION. $\varrho(\mathcal{S}_n^\circ, \mathcal{C}_n) \le 2\,\varrho(\mathcal{C}_n^\circ, \mathcal{S}_n)$.

PROOF. Obviously,

$$\varrho(\mathcal{S}_n^\circ, \mathcal{C}_n) \le \varrho(\mathcal{S}_n^\circ, \mathcal{S}_n^\circ \otimes \mathcal{C}_n^\circ) \circ \varrho(\mathcal{S}_n^\circ \otimes \mathcal{C}_n^\circ, \mathcal{S}_n^\circ \otimes \mathcal{S}_n) \circ \varrho(\mathcal{S}_n^\circ \otimes \mathcal{S}_n, \mathcal{C}_n).$$

From 5.2.21 (Remark) and 5.2.20, we know that

$$\varrho(\mathcal{S}_n^\circ, \mathcal{S}_n^\circ \otimes \mathcal{C}_n^\circ) \le \sqrt{2} \quad \text{and} \quad \varrho(\mathcal{S}_n^\circ \otimes \mathcal{S}_n, \mathcal{C}_n) \le \sqrt{2}.$$

Hence 3.6.3 yields

$$\varrho(\mathcal{S}_n^\circ, \mathcal{C}_n) \le 2\,\varrho(\mathcal{S}_n^\circ \otimes \mathcal{C}_n^\circ, \mathcal{S}_n^\circ \otimes \mathcal{S}_n) \le 2\,\varrho(\mathcal{C}_n^\circ, \mathcal{S}_n).$$

5.2.23 Next, we provide an auxiliary result.

LEMMA. *Let $x_1, \ldots, x_n \in X$ and $x_{-1} = x_0 = x_{n+1} = x_{n+2} = o$. Then, for $t \in \mathbb{R}$,*

$$2 \sin t \sum_{k=1}^{n} x_k \sin kt = \sum_{k=0}^{n+1} (x_{k+1} - x_{k-1}) \cos kt,$$
$$2 \sin t \sum_{k=1}^{n} x_k \cos kt = \sum_{k=1}^{n+1} (x_{k-1} - x_{k+1}) \sin kt.$$

PROOF. The above equations follow from

$$2 \sin t \, \sin kt = \cos(k-1)t - \cos(k+1)t,$$
$$2 \sin t \, \cos kt = \sin(k+1)t - \sin(k-1)t$$

by rearranging the summation.

5.2.24 LEMMA. *Let $\Delta := \{m+1, \ldots, n-m\}$, where $m := \left[\frac{n}{3}\right]$. Then, for $x_1, \ldots, x_n \in X$,*

$$\left(\frac{4}{2n+1} \sum_{h \in \Delta} \Big\| \sum_{k=1}^{n} T x_k \cos \tfrac{2\pi}{2n+1} hk \Big\|^2 \right)^{1/2} \le 5\,\varrho(T|\mathcal{S}_n^\circ, \mathcal{C}_n)\, \|(x_k)|\mathcal{S}_n\|.$$

PROOF. Since $\frac{1}{5}\pi \le \frac{2\pi}{2n+1} h \le \frac{4}{5}\pi$ for $h \in \Delta$, we have $\sin \frac{1}{5}\pi \le \sin \frac{2\pi}{2n+1} h$.

Hence $\|T\| \le \boldsymbol{\varrho}(T|\mathcal{S}_n^\circ, \mathcal{C}_n)$, $\|x_n\| \le \|(x_k)|\mathcal{S}_n\|$, and Lemma 5.2.23 imply

$$\begin{aligned}
2\sin\tfrac{1}{5}\pi\bigg(\frac{4}{2n+1}\sum_{h\in\Delta}\Big\|\sum_{k=1}^{n} Tx_k \cos\tfrac{2\pi}{2n+1}hk\Big\|^2\bigg)^{1/2} &\le \\
\le \bigg(\frac{4}{2n+1}\sum_{h=1}^{n}\Big\|2\sin\tfrac{2\pi}{2n+1}h\sum_{k=1}^{n} Tx_k\cos\tfrac{2\pi}{2n+1}hk\Big\|^2\bigg)^{1/2}& \\
= \bigg(\frac{4}{2n+1}\sum_{h=1}^{n}\Big\|\sum_{k=1}^{n+1}(Tx_{k+1}-Tx_{k-1})\sin\tfrac{2\pi}{2n+1}hk\Big\|^2\bigg)^{1/2}& \\
\le \bigg(\frac{4}{2n+1}\sum_{h=1}^{n}\Big\|\sum_{k=1}^{n}(Tx_{k+1}-Tx_{k-1})\sin\tfrac{2\pi}{2n+1}hk\Big\|^2\bigg)^{1/2} &+ \sqrt{2}\,\|Tx_n\| \\
\le \|(Tx_{k+1}-Tx_{k-1})|\mathcal{S}_n^\circ\| + \sqrt{2}\,\boldsymbol{\varrho}(T|\mathcal{S}_n^\circ,\mathcal{C}_n)\|(x_k)|\mathcal{S}_n\|& \\
\le \boldsymbol{\varrho}(T|\mathcal{S}_n^\circ,\mathcal{C}_n)\big(\|(x_{k+1}-x_{k-1})|\mathcal{C}_n\| + \sqrt{2}\,\|(x_k)|\mathcal{S}_n\|\big).&
\end{aligned}$$

Next, from $\|x_1\| \le \|(x_k)|\mathcal{S}_n\|$, $\|x_n\| \le \|(x_k)|\mathcal{S}_n\|$ and Lemma 5.2.23 we infer that

$$\begin{aligned}
\|(x_{k+1}-x_{k-1})|\mathcal{C}_n\| &\le \bigg(\frac{2}{\pi}\int_0^\pi\Big\|\sum_{k=0}^{n+1}(x_{k+1}-x_{k-1})\cos kt\Big\|^2 dt\bigg)^{1/2} + \sqrt{2}\|x_1\| + \|x_n\| \\
&= \bigg(\frac{2}{\pi}\int_0^\pi\Big\|2\sin t\sum_{k=1}^{n} x_k\sin kt\Big\|^2 dt\bigg)^{1/2} + \sqrt{2}\|x_1\| + \|x_n\| \\
&\le (3+\sqrt{2})\,\|(x_k)|\mathcal{S}_n\|.
\end{aligned}$$

Combining the above inequalities, we obtain

$$\begin{aligned}
2\sin\tfrac{1}{5}\pi\bigg(\frac{4}{2n+1}\sum_{h\in\Delta}\Big\|\sum_{k=1}^{n} Tx_k\cos\tfrac{2\pi}{2n+1}hk\Big\|^2\bigg)^{1/2} &\le \\
\le (3+2\sqrt{2})\,\boldsymbol{\varrho}(T|\mathcal{S}_n^\circ,\mathcal{C}_n)\,\|(x_k)|\mathcal{S}_n\|.&
\end{aligned}$$

The assertion follows, since $(3+2\sqrt{2})/(2\sin\frac{1}{5}\pi) = 4.9580\ldots < 5$.

5.2.25 We now prove a counterpart of 5.2.22.

PROPOSITION. $\boldsymbol{\varrho}(\mathcal{C}_n^\circ, \mathcal{S}_n) \le 10\,\boldsymbol{\varrho}(\mathcal{S}_n^\circ, \mathcal{C}_n)$.

PROOF. Letting $m := \left[\frac{n}{3}\right]$, we define

$$\Delta_- := \{1,\ldots,m\}, \quad \Delta := \{m+1,\ldots,n-m\}, \quad \Delta_+ := \{n-m+1,\ldots,n\}.$$

Then $|\Delta_\pm| = m \le n-2m = |\Delta|$. Hence $\Delta_\pm \subseteq \Delta \pm m$. Since

$$\|(Tx_k)|\mathcal{C}_n^\circ\| = \bigg(\frac{2}{2n+1}\Big\|\sum_{k=1}^{n} Tx_k\Big\|^2 + \frac{4}{2n+1}\sum_{h=1}^{n}\Big\|\sum_{k=1}^{n} Tx_k\cos\tfrac{2\pi}{2n+1}hk\Big\|^2\bigg)^{1/2},$$

we obtain

$$\|(Tx_k)|\mathcal{C}_n^\circ\| \le \sqrt{I_0^2 + I_-^2 + I^2 + I_+^2}, \qquad (I_*)$$

where

$$I_0 := \left(\frac{2}{2n+1}\Big\|\sum_{k=1}^n Tx_k\Big\|^2\right)^{1/2},$$

$$I := \left(\frac{4}{2n+1}\sum_{h\in\Delta}\Big\|\sum_{k=1}^n Tx_k \cos\tfrac{2\pi}{2n+1}hk\Big\|^2\right)^{1/2},$$

$$I_\pm := \left(\frac{4}{2n+1}\sum_{h\in\Delta_\pm}\Big\|\sum_{k=1}^n Tx_k \cos\tfrac{2\pi}{2n+1}hk\Big\|^2\right)^{1/2}.$$

We know from 3.1.5 and 3.4.4 that

$$I_0 \le \|T\|\left(\frac{2}{2n+1}\Big\|\sum_{k=1}^n x_k\Big\|^2\right)^{1/2} \le \left(\frac{2n}{2n+1}\right)^{1/2} \boldsymbol{\delta}(T|\mathcal{S}_n^\circ,\mathcal{C}_n)\,\|(x_k)|\mathcal{S}_n\|.$$

Thus

$$I_0 \le \boldsymbol{\delta}(T|\mathcal{S}_n^\circ,\mathcal{C}_n)\,\|(x_k)|\mathcal{S}_n\|. \qquad (I_0)$$

Moreover, by 5.2.24,

$$I \le 5\,\boldsymbol{\varrho}(T|\mathcal{S}_n^\circ,\mathcal{C}_n)\,\|(x_k)|\mathcal{S}_n\|. \qquad (I)$$

In order to estimate $I_\pm$, we once again exploit the addition theorems for trigonometric functions to get

$$I_\pm \le \left(\frac{4}{2n+1}\sum_{h\in\Delta}\Big\|\sum_{k=1}^n Tx_k \cos\tfrac{2\pi}{2n+1}(h\pm m)k\Big\|^2\right)^{1/2}$$

$$= \left(\frac{4}{2n+1}\sum_{h\in\Delta}\left\|\sum_{k=1}^n Tx_k \begin{pmatrix}\cos\frac{2\pi}{2n+1}hk\cos\frac{2\pi}{2n+1}mk\\ \mp\\ \sin\frac{2\pi}{2n+1}hk\sin\frac{2\pi}{2n+1}mk\end{pmatrix}\right\|^2\right)^{1/2}$$

$$\le \left\{\begin{array}{c}\left(\dfrac{4}{2n+1}\displaystyle\sum_{h\in\Delta}\Big\|\sum_{k=1}^n T\big(x_k\cos\tfrac{2\pi}{2n+1}mk\big)\cos\tfrac{2\pi}{2n+1}hk\Big\|^2\right)^{1/2}\\ +\\ \left(\dfrac{4}{2n+1}\displaystyle\sum_{h\in\Delta}\Big\|\sum_{k=1}^n T\big(x_k\sin\tfrac{2\pi}{2n+1}mk\big)\sin\tfrac{2\pi}{2n+1}hk\Big\|^2\right)^{1/2}\end{array}\right\}.$$

We now estimate the first summand by 5.2.24 and the second summand by applying the defining inequality of $\boldsymbol{\varrho}(T|\mathcal{S}_n^\circ,\mathcal{C}_n)$. This implies that

$$I_\pm \le \boldsymbol{\varrho}(T|\mathcal{S}_n^\circ,\mathcal{C}_n)\Big[5\,\big\|\big(x_k\cos\tfrac{2\pi}{2n+1}mk\big)|\mathcal{S}_n\big\| + \big\|\big(x_k\sin\tfrac{2\pi}{2n+1}mk\big)|\mathcal{C}_n\big\|\Big].$$

Hence, by 5.2.19, we arrive at

$$I_\pm \le 6\, \boldsymbol{\varrho}(T|\mathcal{S}_n^\circ, \mathcal{C}_n)\, \|(x_k)|\mathcal{S}_n\|. \tag{$I_\pm$}$$

Combining estimates (I_*), (I_0), (I) and $(I_\pm)$ yields

$$\|(Tx_k)|\mathcal{C}_n^\circ\| \le \sqrt{1+5^2+6^2+6^2}\, \boldsymbol{\varrho}(T|\mathcal{S}_n^\circ, \mathcal{C}_n)\, \|(x_k)|\mathcal{S}_n\|.$$

5.2.26 In the following, we write

$$\boldsymbol{\mu}_n(T) := \max\Big\{\boldsymbol{\varrho}(T|\mathcal{C}_n, \mathcal{S}_n), \boldsymbol{\varrho}(T|\mathcal{S}_n, \mathcal{C}_n)\Big\}$$

and

$$\boldsymbol{\mu}_n^\circ(T) := \max\Big\{\boldsymbol{\varrho}(T|\mathcal{C}_n^\circ, \mathcal{S}_n), \boldsymbol{\varrho}(T|\mathcal{S}_n^\circ, \mathcal{C}_n)\Big\}.$$

Obviously, the sequences of these ideal norms are non-decreasing.

5.2.27 LEMMA. $\boldsymbol{\mu}_n \le 2\,\boldsymbol{\mu}_n^\circ$.

PROOF. The first inequality in 5.2.21 tells us that

$$\|(Tx_k)|\mathcal{C}_n\|^2 \le \|(Tx_k)|\mathcal{C}_n \otimes \mathcal{C}_n^\circ\|^2 + \|(Tx_k)|\mathcal{S}_n \otimes \mathcal{S}_n^\circ\|^2.$$

Using 3.6.3 and 5.2.20, we get

$$\|(Tx_k)|\mathcal{C}_n\|^2 \le \left\{\begin{array}{c} \boldsymbol{\varrho}(T|\mathcal{C}_n \otimes \mathcal{C}_n^\circ, \mathcal{C}_n \otimes \mathcal{S}_n)^2\, \|(x_k)|\mathcal{C}_n \otimes \mathcal{S}_n\|^2 \\ + \\ \boldsymbol{\varrho}(T|\mathcal{S}_n \otimes \mathcal{S}_n^\circ, \mathcal{S}_n \otimes \mathcal{C}_n)^2 \|(x_k)|\mathcal{S}_n \otimes \mathcal{C}_n\|^2 \end{array}\right\}$$

$$\le 2\, \boldsymbol{\varrho}(T|\mathcal{C}_n^\circ, \mathcal{S}_n)^2\, \|(x_k)|\mathcal{S}_n\|^2 + 2\, \boldsymbol{\varrho}(T|\mathcal{S}_n^\circ, \mathcal{C}_n)^2\, \|(x_k)|\mathcal{S}_n\|^2.$$

This proves that

$$\boldsymbol{\varrho}(\mathcal{C}_n, \mathcal{S}_n) \le 2 \max\{\boldsymbol{\varrho}(\mathcal{C}_n^\circ, \mathcal{S}_n), \boldsymbol{\varrho}(\mathcal{S}_n^\circ, \mathcal{C}_n)\} = 2\, \boldsymbol{\mu}_n^\circ(T).$$

The inequality

$$\boldsymbol{\varrho}(\mathcal{S}_n, \mathcal{C}_n) \le 2 \max\{\boldsymbol{\varrho}(\mathcal{C}_n^\circ, \mathcal{S}_n), \boldsymbol{\varrho}(\mathcal{S}_n^\circ, \mathcal{C}_n)\} = 2\, \boldsymbol{\mu}_n^\circ(T)$$

can be obtained similarly.

5.2.28 In anticipation of 5.4.1, we now deal with finite Riesz projections.

LEMMA. *For $x_{-n}, \ldots, x_0, \ldots, x_{+n} \in X$,*

$$\left(\frac{1}{2\pi}\int_{-\pi}^{+\pi}\Big\|\sum_{k=1}^{n} Tx_k \exp(ikt)\Big\|^2 dt\right)^{1/2} \le 4\, \boldsymbol{\mu}_n(T) \left(\frac{1}{2\pi}\int_{-\pi}^{+\pi}\Big\|\sum_{|k|\le n} x_k \exp(ikt)\Big\|^2 dt\right)^{1/2}.$$

PROOF. Letting $u_k := x_k + x_{-k}$ and $v_k := x_k - x_{-k}$ for $k = 1, \ldots, n$, it follows from $u_k + v_k = 2x_k$ and Euler's formula that

$$x_k \exp(ikt) = \tfrac{1}{2}\big[u_k \cos kt + v_k \cos kt + iu_k \sin kt + iv_k \sin kt\big].$$

Hence

$$\|(Tx_k)|\mathcal{E}_n\| = \left(\frac{1}{2\pi}\int_{-\pi}^{+\pi}\Big\|\sum_{k=1}^{n} Tx_k \exp(ikt)\Big\|^2 dt\right)^{1/2}$$

$$\leq \frac{1}{2}\left\{\begin{array}{l} \left(\frac{1}{2\pi}\int_{-\pi}^{+\pi}\Big\|\sum_{k=1}^{n} Tu_k \cos kt\Big\|^2 dt\right)^{1/2} \\ + \left(\frac{1}{2\pi}\int_{-\pi}^{+\pi}\Big\|\sum_{k=1}^{n} Tv_k \cos kt\Big\|^2 dt\right)^{1/2} \\ + \left(\frac{1}{2\pi}\int_{-\pi}^{+\pi}\Big\|\sum_{k=1}^{n} Tu_k \sin kt\Big\|^2 dt\right)^{1/2} \\ + \left(\frac{1}{2\pi}\int_{-\pi}^{+\pi}\Big\|\sum_{k=1}^{n} Tv_k \sin kt\Big\|^2 dt\right)^{1/2} \end{array}\right\}$$

$$\leq \frac{1}{2}\left\{\begin{array}{ll} \|T\| & \left(\frac{1}{2\pi}\int_{-\pi}^{+\pi}\Big\|\sum_{k=1}^{n} u_k \cos kt\Big\|^2 dt\right)^{1/2} \\ + \boldsymbol{\varrho}(T|\mathcal{C}_n,\mathcal{S}_n) & \left(\frac{1}{2\pi}\int_{-\pi}^{+\pi}\Big\|\sum_{k=1}^{n} v_k \sin kt\Big\|^2 dt\right)^{1/2} \\ + \boldsymbol{\varrho}(T|\mathcal{S}_n,\mathcal{C}_n) & \left(\frac{1}{2\pi}\int_{-\pi}^{+\pi}\Big\|\sum_{k=1}^{n} u_k \cos kt\Big\|^2 dt\right)^{1/2} \\ + \quad \|T\| & \left(\frac{1}{2\pi}\int_{-\pi}^{+\pi}\Big\|\sum_{k=1}^{n} v_k \sin kt\Big\|^2 dt\right)^{1/2} \end{array}\right\}$$

$$\leq \boldsymbol{\mu}_n(T)\left[\left(\frac{1}{2\pi}\int_{-\pi}^{+\pi}\Big\|\sum_{k=1}^{n} u_k \cos kt\Big\|^2 dt\right)^{1/2} + \left(\frac{1}{2\pi}\int_{-\pi}^{+\pi}\Big\|\sum_{k=1}^{n} v_k \sin kt\Big\|^2 dt\right)^{1/2}\right].$$

By the obvious fact that $\|u\| + \|v\| \leq \|u + iv\| + \|u - iv\|$, we obtain

$$\|(Tx_k)|\mathcal{E}_n\| \leq \boldsymbol{\mu}_n(T)\left\{\begin{array}{c} \left(\frac{1}{2\pi}\int_{-\pi}^{+\pi}\Big\|\sum_{k=1}^{n}(u_k \cos kt + iv_k \sin kt)\Big\|^2 dt\right)^{1/2} \\ + \\ \left(\frac{1}{2\pi}\int_{-\pi}^{+\pi}\Big\|\sum_{k=1}^{n}(u_k \cos kt - iv_k \sin kt)\Big\|^2 dt\right)^{1/2} \end{array}\right\}.$$

Substituting $-t$ for t in the lower term on the right-hand side yields

$$\|(Tx_k)|\mathcal{E}_n\| \leq 2\boldsymbol{\mu}_n(T)\Big(\frac{1}{2\pi}\int\limits_{-\pi}^{+\pi}\Big\|\sum_{k=1}^{n}(u_k\cos kt + iv_k\sin kt)\Big\|^2 dt\Big)^{1/2}.$$

Finally, we conclude from

$$\sum_{k=1}^{n}(u_k\cos kt + iv_k\sin kt) = \sum_{|k|\leq n} x_k\exp(ikt) - x_0$$

and 3.1.5 that

$$\begin{aligned}\Big(\frac{1}{2\pi}\int\limits_{-\pi}^{+\pi}\Big\|\sum_{k=1}^{n}(u_k\cos kt + iv_k\sin kt)\Big\|^2 dt\Big)^{1/2} &\leq \\ \leq \Big(\frac{1}{2\pi}\int\limits_{-\pi}^{+\pi}\Big\|\sum_{|k|\leq n} x_k\exp(ikt)\Big\|^2 dt\Big)^{1/2} &+ \|x_0\| \\ \leq 2\Big(\frac{1}{2\pi}\int\limits_{-\pi}^{+\pi}\Big\|\sum_{|k|\leq n} x_k\exp(ikt)\Big\|^2 dt\Big)^{1/2}&.\end{aligned}$$

This proves the desired result.

5.2.29 LEMMA. $\boldsymbol{\delta}(\mathcal{E}_n, \mathcal{E}_n) \leq 96\,\boldsymbol{\mu}_n$.

PROOF. If V_m denotes the m-th de la Vallée–Poussin kernel, then

$$V_m * \boldsymbol{f} = \sum_{k\in\mathbb{Z}} e_k \otimes x_k^{(m)} \quad\text{with}\quad x_k^{(m)} := \langle V_m * \boldsymbol{f}, \overline{e_k}\rangle = \langle V_m, \overline{e_k}\rangle\,\langle \boldsymbol{f}, \overline{e_k}\rangle\,.$$

We know from 5.1.6 that

$$x_k^{(m)} = \begin{cases} \langle \boldsymbol{f}, \overline{e_k}\rangle & \text{if } |k| \leq m, \\ o & \text{if } |k| \geq 2m, \end{cases} \qquad\text{and}\qquad \|V_m * \boldsymbol{f}|L_2\| \leq 3\,\|\boldsymbol{f}|L_2\|.$$

By the triangle inequality,

$$\|(T\langle \boldsymbol{f}, \overline{e_k}\rangle)|\mathcal{E}_m\| = \Big(\frac{1}{2\pi}\int\limits_{-\pi}^{+\pi}\Big\|\sum_{k=1}^{m} Tx_k^{(m)}e_k(t)\Big\|^2 dt\Big)^{1/2} \leq I_1 + I_2,$$

where

$$I_1 := \Big(\frac{1}{2\pi}\int\limits_{-\pi}^{+\pi}\Big\|\sum_{k=1}^{2m-1} Tx_k^{(m)}e_k(t)\Big\|^2 dt\Big)^{1/2}$$

and

$$I_2 := \Big(\frac{1}{2\pi}\int\limits_{-\pi}^{+\pi}\Big\|\sum_{k=m+1}^{2m-1} Tx_k^{(m)}e_k(t)\Big\|^2 dt\Big)^{1/2}.$$

Lemma 5.2.28 implies that

$$I_1 \le 4\boldsymbol{\mu}_{2m-1}(T)\Big(\frac{1}{2\pi}\int_{-\pi}^{+\pi}\Big\|\sum_{|k|\le 2m-1} x_k^{(m)}\exp(ikt)\Big\|^2 dt\Big)^{1/2}$$
$$= 4\,\boldsymbol{\mu}_{2m-1}(T)\,\|V_m * \boldsymbol{f}|L_2\| \le 12\,\boldsymbol{\mu}_{2m-1}(T)\,\|\boldsymbol{f}|L_2\|.$$

To estimate the second term, we recall that $x_k^{(m)} = o$ if $|k| \ge 2m$. Thus

$$I_2 = \Big(\frac{1}{2\pi}\int_{-\pi}^{+\pi}\Big\|\sum_{k=m+1}^{2m-1} Tx_k^{(m)}\exp(ikt)\Big\|^2 dt\Big)^{1/2}$$
$$= \Big(\frac{1}{2\pi}\int_{-\pi}^{+\pi}\Big\|\sum_{k=1}^{3m-1} Tx_{k+m}^{(m)}\exp(ikt)\Big\|^2 dt\Big)^{1/2}$$
$$\le 4\,\boldsymbol{\mu}_{3m-1}(T)\Big(\frac{1}{2\pi}\int_{-\pi}^{+\pi}\Big\|\sum_{|k|\le 3m-1} x_{k+m}^{(m)}\exp(ikt)\Big\|^2 dt\Big)^{1/2}$$
$$= 4\,\boldsymbol{\mu}_{3m-1}(T)\Big(\frac{1}{2\pi}\int_{-\pi}^{+\pi}\Big\|\sum_{|k|\le 2m-1} x_k^{(m)}\exp(ikt)\Big\|^2 dt\Big)^{1/2}$$
$$= 4\,\boldsymbol{\mu}_{3m-1}(T)\,\|V_m * \boldsymbol{f}|L_2\| \le 12\,\boldsymbol{\mu}_{3m-1}(T)\,\|\boldsymbol{f}|L_2\|.$$

Combining the preceding estimates and taking into account the monotonicity of $\boldsymbol{\mu}_n(T)$, we arrive at

$$\boldsymbol{\delta}(T|\mathcal{E}_m, \mathcal{E}_m) \le 24\boldsymbol{\mu}_{3m-1}(T).$$

To complete the proof, choose m such that $3m-1 \le n \le 3m+1$. Then it follows from 5.2.2 that

$$\boldsymbol{\delta}(\mathcal{E}_n, \mathcal{E}_n) = \left\{\begin{array}{ll} \boldsymbol{\delta}(\mathcal{E}_{3m-1}, \mathcal{E}_{3m-1}) & \le 2\,\boldsymbol{\delta}(\mathcal{E}_m, \mathcal{E}_m) + \boldsymbol{\delta}(\mathcal{E}_{m-1}, \mathcal{E}_{m-1}) \\ \boldsymbol{\delta}(\mathcal{E}_{3m}, \mathcal{E}_{3m}) & \le 3\,\boldsymbol{\delta}(\mathcal{E}_m, \mathcal{E}_m) \\ \boldsymbol{\delta}(\mathcal{E}_{3m+1}, \mathcal{E}_{3m+1}) & \le 2\,\boldsymbol{\delta}(\mathcal{E}_m, \mathcal{E}_m) + \boldsymbol{\delta}(\mathcal{E}_{m+1}, \mathcal{E}_{m+1}) \end{array}\right\}$$
$$\le 4\,\boldsymbol{\delta}(\mathcal{E}_m, \mathcal{E}_m).$$

Hence

$$\boldsymbol{\delta}(\mathcal{E}_n, \mathcal{E}_n) \le 4\,\boldsymbol{\delta}(\mathcal{E}_m, \mathcal{E}_m) \le 96\,\boldsymbol{\mu}_{3m-1} \le 96\,\boldsymbol{\mu}_n.$$

5.2.30 PROPOSITION.

$$\boldsymbol{\delta}(\mathcal{E}_n, \mathcal{E}_n) \le 384\,\boldsymbol{\varrho}(\mathcal{C}_n^\circ, \mathcal{S}_n) \quad \textit{and} \quad \boldsymbol{\delta}(\mathcal{E}_n, \mathcal{E}_n) \le 1920\,\boldsymbol{\varrho}(\mathcal{S}_n^\circ, \mathcal{C}_n).$$

PROOF. In view of 5.2.29 and 5.2.27, we have

$$\boldsymbol{\delta}(\mathcal{E}_n, \mathcal{E}_n) \le 96\,\boldsymbol{\mu}_n \le 192\,\boldsymbol{\mu}_n^\circ.$$

On the other hand, 5.2.22 and 5.2.25 imply that

$$\boldsymbol{\mu}_n^\circ \le 2\,\boldsymbol{\varrho}(\mathcal{C}_n^\circ, \mathcal{S}_n) \quad \text{and} \quad \boldsymbol{\mu}_n^\circ \le 10\,\boldsymbol{\varrho}(\mathcal{S}_n^\circ, \mathcal{C}_n),$$

respectively. This yields the assertion.

5.2.31 THEOREM. *The sequences of the following ideal norms are uniformly equivalent:*

$$\begin{array}{cccc} & \boldsymbol{\delta}(\mathcal{E}_n, \mathcal{E}_n), & & \\ \boldsymbol{\delta}(\mathcal{S}_n, \mathcal{C}_n), & \boldsymbol{\varrho}(\mathcal{S}_n, \mathcal{C}_n), & \boldsymbol{\delta}(\mathcal{C}_n, \mathcal{S}_n), & \boldsymbol{\varrho}(\mathcal{C}_n, \mathcal{S}_n), \\ \boldsymbol{\delta}(\mathcal{S}_n^\circ, \mathcal{C}_n), & \boldsymbol{\varrho}(\mathcal{S}_n^\circ, \mathcal{C}_n), & \boldsymbol{\delta}(\mathcal{C}_n^\circ, \mathcal{S}_n), & \boldsymbol{\varrho}(\mathcal{C}_n^\circ, \mathcal{S}_n), \\ \boldsymbol{\delta}(\mathcal{S}_n, \mathcal{C}_n^\circ), & \boldsymbol{\varrho}(\mathcal{S}_n, \mathcal{C}_n^\circ), & \boldsymbol{\delta}(\mathcal{C}_n, \mathcal{S}_n^\circ) = & \boldsymbol{\varrho}(\mathcal{C}_n, \mathcal{S}_n^\circ), \\ \boldsymbol{\delta}(\mathcal{S}_n^\circ, \mathcal{C}_n^\circ), & \boldsymbol{\varrho}(\mathcal{S}_n^\circ, \mathcal{C}_n^\circ), & \boldsymbol{\delta}(\mathcal{C}_n^\circ, \mathcal{S}_n^\circ) = & \boldsymbol{\varrho}(\mathcal{C}_n^\circ, \mathcal{S}_n^\circ). \end{array}$$

PROOF. From 3.4.10, 5.2.11 and 5.2.13 we obtain the following chains of inequalities:

$$\begin{aligned} \boldsymbol{\varrho}(\mathcal{S}_n^\circ, \mathcal{C}_n) &\le 3\,\boldsymbol{\varrho}(\mathcal{S}_n^\circ, \mathcal{C}_n^\circ) \le 3\,\boldsymbol{\delta}(\mathcal{S}_n^\circ, \mathcal{C}_n^\circ) \le 9\,\boldsymbol{\delta}(\mathcal{S}_n^\circ, \mathcal{C}_n) \le 27\,\boldsymbol{\delta}(\mathcal{S}_n, \mathcal{C}_n) \\ &\le 54\,\boldsymbol{\delta}(\mathcal{E}_n, \mathcal{E}_n), \\ \boldsymbol{\varrho}(\mathcal{S}_n^\circ, \mathcal{C}_n) &\le 3\,\boldsymbol{\varrho}(\mathcal{S}_n^\circ, \mathcal{C}_n^\circ) \le 3\,\boldsymbol{\delta}(\mathcal{S}_n^\circ, \mathcal{C}_n^\circ) \le 9\,\boldsymbol{\delta}(\mathcal{S}_n, \mathcal{C}_n^\circ) \le 27\,\boldsymbol{\delta}(\mathcal{S}_n, \mathcal{C}_n) \\ &\le 54\,\boldsymbol{\delta}(\mathcal{E}_n, \mathcal{E}_n), \\ \boldsymbol{\varrho}(\mathcal{S}_n^\circ, \mathcal{C}_n) &\le 3\,\boldsymbol{\varrho}(\mathcal{S}_n^\circ, \mathcal{C}_n^\circ) \le 9\,\boldsymbol{\varrho}(\mathcal{S}_n, \mathcal{C}_n^\circ) \le 27\,\boldsymbol{\delta}(\mathcal{S}_n, \mathcal{C}_n) \le 54\,\boldsymbol{\delta}(\mathcal{E}_n, \mathcal{E}_n), \\ \boldsymbol{\varrho}(\mathcal{S}_n^\circ, \mathcal{C}_n) &\le 3\,\boldsymbol{\varrho}(\mathcal{S}_n, \mathcal{C}_n) \le 3\,\boldsymbol{\delta}(\mathcal{S}_n, \mathcal{C}_n) \le 6\,\boldsymbol{\delta}(\mathcal{E}_n, \mathcal{E}_n). \end{aligned}$$

This proves that, with respect to the order relation $\prec$, the ideal norms

$$\begin{array}{cccc} \boldsymbol{\delta}(\mathcal{S}_n^\circ, \mathcal{C}_n^\circ), & \boldsymbol{\delta}(\mathcal{S}_n^\circ, \mathcal{C}_n), & \boldsymbol{\delta}(\mathcal{S}_n, \mathcal{C}_n^\circ), & \boldsymbol{\delta}(\mathcal{S}_n, \mathcal{C}_n), \\ \boldsymbol{\varrho}(\mathcal{S}_n^\circ, \mathcal{C}_n^\circ), & & \boldsymbol{\varrho}(\mathcal{S}_n, \mathcal{C}_n^\circ), & \boldsymbol{\varrho}(\mathcal{S}_n, \mathcal{C}_n) \end{array}$$

lie between $\boldsymbol{\varrho}(\mathcal{S}_n^\circ, \mathcal{C}_n)$ and $\boldsymbol{\delta}(\mathcal{E}_n, \mathcal{E}_n)$. Hence, by 5.2.30, all of them are uniformly equivalent. The same statement holds when the sine and cosine systems are interchanged.

5.2.32 We now state a *supertheorem* which summarizes all the preceding results. Note that the ideal norms are arranged in the form of two matrices in which the left orthonormal system is used as row index while the right one serves as column index.

THEOREM. *The sequences of the following ideal norms are uniformly equivalent:*

$$\begin{array}{ll} \boldsymbol{\delta}(\mathcal{E}_n, \mathcal{E}_n), & \boldsymbol{\delta}(\mathcal{E}_n, \mathcal{E}_n^\circ), \\ \boldsymbol{\delta}(\mathcal{E}_n^\circ, \mathcal{E}_n), & \text{———}, \end{array}$$

$$\begin{array}{llll} \boldsymbol{\delta}(\mathcal{C}_n, \mathcal{C}_n), & \boldsymbol{\delta}(\mathcal{C}_n, \mathcal{C}_n^\circ), & \boldsymbol{\delta}(\mathcal{C}_n, \mathcal{S}_n), & \boldsymbol{\delta}(\mathcal{C}_n, \mathcal{S}_n^\circ), \\ \boldsymbol{\delta}(\mathcal{C}_n^\circ, \mathcal{C}_n), & \text{----}, & \boldsymbol{\delta}(\mathcal{C}_n^\circ, \mathcal{S}_n), & \boldsymbol{\delta}(\mathcal{C}_n^\circ, \mathcal{S}_n^\circ), \\ \boldsymbol{\delta}(\mathcal{S}_n, \mathcal{C}_n), & \boldsymbol{\delta}(\mathcal{S}_n, \mathcal{C}_n^\circ), & \boldsymbol{\delta}(\mathcal{S}_n, \mathcal{S}_n), & \boldsymbol{\delta}(\mathcal{S}_n, \mathcal{S}_n^\circ), \\ \boldsymbol{\delta}(\mathcal{S}_n^\circ, \mathcal{C}_n), & \boldsymbol{\delta}(\mathcal{S}_n^\circ, \mathcal{C}_n^\circ), & \boldsymbol{\delta}(\mathcal{S}_n^\circ, \mathcal{S}_n), & \text{———}, \end{array}$$

$$\begin{array}{ll} \text{———}, & \boldsymbol{\varrho}(\mathcal{E}_n, \mathcal{E}_n^\circ), \\ \text{----}, & \text{———}, \end{array}$$

$$\begin{array}{llll} \text{———}, & \boldsymbol{\varrho}(\mathcal{C}_n, \mathcal{C}_n^\circ), & \boldsymbol{\varrho}(\mathcal{C}_n, \mathcal{S}_n), & \boldsymbol{\varrho}(\mathcal{C}_n, \mathcal{S}_n^\circ), \\ \text{----}, & \text{———}, & \boldsymbol{\varrho}(\mathcal{C}_n^\circ, \mathcal{S}_n), & \boldsymbol{\varrho}(\mathcal{C}_n^\circ, \mathcal{S}_n^\circ), \\ \boldsymbol{\varrho}(\mathcal{S}_n, \mathcal{C}_n), & \boldsymbol{\varrho}(\mathcal{S}_n, \mathcal{C}_n^\circ), & \text{———}, & \boldsymbol{\varrho}(\mathcal{S}_n, \mathcal{S}_n^\circ), \\ \boldsymbol{\varrho}(\mathcal{S}_n^\circ, \mathcal{C}_n), & \boldsymbol{\varrho}(\mathcal{S}_n^\circ, \mathcal{C}_n^\circ), & \text{----}, & \text{———}. \end{array}$$

PROOF. All collections of ideal norms listed in Theorems 5.2.9, 5.2.18 and 5.2.31 contain $\boldsymbol{\delta}(\mathcal{E}_n, \mathcal{E}_n)$. Hence their union is *connected* with respect to the equivalence relation $\asymp$.

REMARK. The missing ideal norms, indicated by ——— or ----, are equal or uniformly equivalent to the operator norm, respectively.

5.2.33 By means of the Dirichlet system $\mathcal{D}_n := (d_1^{(n)}, \dots, d_n^{(n)})$, it is possible to write $\boldsymbol{\varrho}(\mathcal{E}_n, \mathcal{E}_n^\circ)$ as a type ideal norm.

PROPOSITION. $\boldsymbol{\varrho}(\mathcal{E}_n, \mathcal{E}_n^\circ) = \boldsymbol{\varrho}(\mathcal{D}_n, \mathcal{I}_n)$.

PROOF. Recall from 5.1.8 that

$$d_h^{(n)}(t) = \exp(-i\tfrac{n+1}{2}t)\exp(i\tfrac{\pi}{n}h)(-1)^h \sum_{k=1}^{n} \overline{\varepsilon_{hk}^{(n)}}\, e_k(t).$$

Given $x_1, \dots, x_n \in X$, we define $x_1^\circ \dots, x_n^\circ$ by

$$x_k^\circ := \sum_{h=1}^{n} x_h \exp(-i\tfrac{n+1}{2}t)\exp(i\tfrac{\pi}{n}h)(-1)^h \overline{\varepsilon_{hk}^{(n)}}.$$

Then

$$\sum_{h=1}^{n} x_h\, d_h^{(n)}(t) = \sum_{h=1}^{n} x_h \exp(-i\tfrac{n+1}{2}t)\exp(i\tfrac{\pi}{n}h)(-1)^h \sum_{k=1}^{n} \overline{\varepsilon_{hk}^{(n)}}\, e_k(t)$$

$$= \sum_{k=1}^{n}\left(\sum_{h=1}^{n} x_h \exp(-i\tfrac{n+1}{2}t)\exp(i\tfrac{\pi}{n}h)(-1)^h \overline{\varepsilon_{hk}^{(n)}}\right) e_k(t) = \sum_{k=1}^{n} x_k^\circ e_k(t)$$

and

$$x_h \exp(-i\tfrac{n+1}{2}t)\exp(i\tfrac{\pi}{n}h)(-1)^h = \sum_{k=1}^{n} x_k^\circ \varepsilon_{hk}^{(n)}.$$

Hence

$$\|(Tx_h)|\mathcal{D}_n\| = \|(Tx_k^\circ)|\mathcal{E}_n\| \quad \text{and} \quad \|(x_h)|l_2^n\| = \|(x_k^\circ)|\mathcal{E}_n^\circ\|.$$

Thus the inequalities

$$\|(Tx_h)|\mathcal{D}_n\| \le c\,\|(x_h)|l_2^n\| \quad \text{and} \quad \|(Tx_k^\circ)|\mathcal{E}_n\| \le c\,\|(x_k^\circ)|\mathcal{E}_n^\circ\|$$

are equivalent.

5.2.34 We will show in 5.3.8 that $\boldsymbol{\delta}(\mathcal{E}_n, \mathcal{E}_n) \asymp \boldsymbol{\kappa}(H_n)$. So the following example is equivalent to 2.4.10.

EXAMPLE.

$$\boldsymbol{\delta}(l_1^n|\mathcal{E}_n, \mathcal{E}_n) = \boldsymbol{\delta}(l_\infty^n|\mathcal{E}_n, \mathcal{E}_n) \asymp 1 + \log n,$$
$$\boldsymbol{\delta}(L_1|\mathcal{E}_n, \mathcal{E}_n) = \boldsymbol{\delta}(L_\infty|\mathcal{E}_n, \mathcal{E}_n) \asymp 1 + \log n.$$

REMARK. It follows from 3.8.12 and 5.1.7 that

$$\boldsymbol{\delta}(L_1[-\pi, +\pi)|\mathcal{E}_n, \mathcal{E}_n) = \frac{1}{2\pi} \int_{-\pi}^{+\pi} \left| \sum_{k=1}^{n} e_k(t) \right| dt \asymp 1 + \log n.$$

In the dual case, the lower estimate can be obtained by substituting the L_∞-valued function $\boldsymbol{s}_n(t) := \operatorname{sgn}(d_n(\,\cdot\, - t))$ in the defining inequality of $\boldsymbol{\delta}(L_\infty[-\pi, +\pi)|\mathcal{E}_n, \mathcal{E}_n)$.

5.2.35 The next result can be deduced from 5.4.18 and 5.4.20.

EXAMPLE. $\boldsymbol{\delta}(L_r|\mathcal{E}_n, \mathcal{E}_n) \asymp 1$ *for* $1 < r < \infty$.

REMARK. Using the fact that the Riesz projection 5.4.1 is bounded on $L_r[-\pi, +\pi)$, a direct proof could be obtained from the relation

$$\boldsymbol{\delta}_r^{(r)}(\mathcal{E}_n, \mathcal{E}_n) \asymp \boldsymbol{\varrho}(\mathcal{E}_n, \mathcal{E}_n)$$

which is a corollary of the periodic analogue of 5.4.7.

5.2.36 We refer to T as an **EP-operator** if

$$\|T|\mathfrak{EP}\| := \sup_n \boldsymbol{\delta}(T|\mathcal{E}_n, \mathcal{E}_n)$$

is finite. These operators form the Banach ideal

$$\mathfrak{EP} := \mathfrak{L}[\boldsymbol{\delta}(\mathcal{E}_n, \mathcal{E}_n)].$$

REMARKS. It will be shown in 5.4.18 that $\mathfrak{EP}$ consists of all operators that are compatible with the Riesz projection. Thus the symbol $\mathfrak{RP}$ would be more appropriate. But this combination of letters has already been used in connection with the Rademacher projection; see 4.14.1.

In the definition of $\mathfrak{EP}$, we can replace $\boldsymbol{\delta}(\mathcal{E}_n, \mathcal{E}_n)$ by any ideal norm listed in Theorem 5.2.32.

5.2.37 Quasi-EP-operators are defined by the property

$$\boldsymbol{\delta}(T|\mathcal{E}_n, \mathcal{E}_n) = o(1+\log n).$$

They form the closed ideal

$$\mathfrak{QEP} := \mathfrak{L}_0[(1+\log n)^{-1}\,\boldsymbol{\delta}(\mathcal{E}_n, \mathcal{E}_n)].$$

5.2.38 The final result of this section follows from 3.4.5, 3.4.7, and 5.2.3.

PROPOSITION. *The Banach operator ideals* $\mathfrak{EP}$ *and* $\mathfrak{QEP}$ *are injective, surjective, symmetric and* l_2*-stable.*

5.3 Hilbert matrices and trigonometric systems

5.3.1 Using the cosine and sine matrices C_n°, $C_{1,n}^\circ$, $C_{n+1,n}^\circ$ and S_n°, defined in 5.1.5, we let

$$A_{n+1,n} = \begin{pmatrix} A_{1,n} \\ A_n \end{pmatrix} := C_{n+1,n}^\circ S_n^\circ$$

with

$$\begin{aligned} A_{1,n} &= (\alpha_{0k}^{(n)}) := C_{1,n}^\circ S_n^\circ \quad \text{for } k=1,\dots,n,\\ A_n &= (\alpha_{hk}^{(n)}) := \;C_n^\circ S_n^\circ \quad \text{for } h,k=1,\dots,n. \end{aligned}$$

5.3.2 By means of $A_{n+1,n}$, we write $\boldsymbol{\varrho}(\mathcal{C}_n^\circ, \mathcal{S}_n^\circ)$ as a Kwapień ideal norm.

PROPOSITION. $\boldsymbol{\varrho}(\mathcal{C}_n^\circ, \mathcal{S}_n^\circ) = \boldsymbol{\kappa}(A_{n+1,n})$.

PROOF. The assertion follows from the fact that the inequalities

$$\|(Tx_k)|\mathcal{C}_n^\circ\| \le c\,\|(x_k)|\mathcal{S}_n^\circ\| \quad\text{and}\quad \|(Ty_h)|A_{n+1,n}\| \le c\,\|(y_h)|l_2^n\|$$

pass into each other by the orthogonal transformations

$$y_h = \sum_{k=1}^n \sigma_{hk}^{(n)} x_k \quad\text{and}\quad x_k = \sum_{h=1}^n \sigma_{hk}^{(n)} y_h$$

associated with $S_n^\circ = \big(\sigma_{hk}^{(n)}\big)$.

5.3.3 LEMMA. $\boldsymbol{\kappa}(A_n) \le \boldsymbol{\kappa}(A_{n+1,n}) \le \sqrt{3}\,\boldsymbol{\kappa}(A_n)$.

PROOF. Note that

$$|||A_{1,n}||| = \|(\alpha_{0k}^{(n)})|l_2^n\| \le \|A_{n+1,n}\| \le \|C_{n+1,n}^\circ\|\,\|S_n^\circ\| = 1$$

and $\|A_n\| = 1$ if $n \ge 2$. Hence, by 2.3.2,

$$\boldsymbol{\kappa}(A_{1,n}) \le |||A_{1,n}||| \le \|A_n\| \le \boldsymbol{\kappa}(A_n).$$

We now conclude from 2.3.11 that

$$\boldsymbol{\kappa}(A_n) \le \boldsymbol{\kappa}(A_{n+1,n}) = \boldsymbol{\kappa}\begin{pmatrix} A_{1,n} \\ A_n \end{pmatrix} \le \sqrt{\boldsymbol{\kappa}(A_{1,n})^2 + \boldsymbol{\kappa}(A_n)^2} \le \sqrt{2}\,\boldsymbol{\kappa}(A_n).$$

In the case $n = 1$, we have $\boldsymbol{\kappa}(A_{2,1}) = 1$ and $\boldsymbol{\kappa}(A_1) = \frac{1}{\sqrt{3}}$.

5.3.4 Define

$$\chi_m^{odd} := \begin{cases} \frac{1}{m} & \text{if } m \text{ is odd,} \\ 0 & \text{if } m \text{ is even.} \end{cases}$$

We refer to $H_n^{odd} = \frac{2}{\pi}(\chi_{h-k}^{odd})$ as the n-th **odd Hilbert matrix**. In order to get $H_1^{odd} \ne O$, we let $H_1^{odd} := (\frac{2}{\pi})$.

5.3.5 LEMMA. $\quad |||A_n + H_n^{odd}||| \le 5.$

PROOF. Using a summation formula for the conjugate Dirichlet kernel, stated in 5.1.6, we see that

$$\begin{aligned} \alpha_{hk}^{(n)} &= \frac{2}{2n+1} \sum_{l=1}^{n} \Big(\sin \tfrac{2\pi l}{2n+1}(h+k) - \sin \tfrac{2\pi l}{2n+1}(h-k) \Big) \\ &= \frac{1}{2n+1} \left\{ \begin{array}{c} \dfrac{\cos \frac{\pi}{2n+1}(h+k) - \cos \pi(h+k)}{\sin \frac{\pi}{2n+1}(h+k)} \\ - \\ \dfrac{\cos \frac{\pi}{2n+1}(h-k) - \cos \pi(h-k)}{\sin \frac{\pi}{2n+1}(h-k)} \end{array} \right\}. \end{aligned}$$

In view of

$$1 + \cos 2t = 2\cos^2 t, \quad 1 - \cos 2t = 2\sin^2 t, \quad \sin 2t = 2 \sin t \cos t$$

and

$$\cos \pi(h \pm k) = (-1)^{h \pm k},$$

it follows that

$$\alpha_{hk}^{(n)} = \begin{cases} \dfrac{1}{2n+1}\Big(\cot \frac{\pi}{4n+2}(h+k) - \cot \frac{\pi}{4n+2}(h-k) \Big) & \text{if } h \pm k \text{ is odd,} \\ \dfrac{1}{2n+1}\Big(\tan \frac{\pi}{4n+2}(h-k) - \tan \frac{\pi}{4n+2}(h+k) \Big) & \text{if } h \pm k \text{ is even.} \end{cases}$$

We infer from

$$\Big| \cot t - \tfrac{1}{t} \Big| \le \tfrac{2}{\pi} \quad \text{for} \quad |t| \le \tfrac{\pi}{2}$$

that

$$\Big| \cot \tfrac{\pi}{4n+2}(h \pm k) - \tfrac{4n+2}{\pi(h \pm k)} \Big| \le \tfrac{2}{\pi}.$$

Therefore

$$\left|\alpha_{hk}^{(n)} - \tfrac{2}{\pi}\left(\tfrac{1}{h+k} - \tfrac{1}{h-k}\right)\right| \le \tfrac{4}{\pi(2n+1)} < \tfrac{2}{2n+1} \quad \text{if } h \pm k \text{ is odd.} \qquad (1)$$

Note that

$$\left|\tan \tfrac{\pi}{4n+2}(h-k)\right| < \tan \tfrac{\pi}{4} = 1.$$

Substituting $h' = n - h + 1$ and $k' = n - k + 1$ yields

$$\left|\tan \tfrac{\pi}{4n+2}(h+k) - \tfrac{4n+2}{\pi(h'+k'-1)}\right| = \left|\cot \tfrac{\pi}{4n+2}(h'+k'-1) - \tfrac{4n+2}{\pi(h'+k'-1)}\right| \le \tfrac{2}{\pi}.$$

Hence

$$\left|\alpha_{hk}^{(n)} + \tfrac{2}{\pi}\tfrac{1}{h'+k'-1}\right| \le \tfrac{2}{\pi(2n+1)} + \tfrac{1}{2n+1} < \tfrac{2}{2n+1} \quad \text{if } h \pm k \text{ is even.} \qquad (2)$$

Combining (1) and (2), we arrive at

$$\left|\alpha_{hk}^{(n)} + \tfrac{2}{\pi}\chi_{h-k}^{odd}\right| \le \begin{cases} \tfrac{2}{\pi}\chi_{h+k} + \tfrac{2}{2n+1} & \text{if } h \pm k \text{ is odd,} \\ \tfrac{2}{\pi}\chi_{h'+k'-1} + \tfrac{2}{2n+1} & \text{if } h \pm k \text{ is even.} \end{cases}$$

Thus

$$\left|\alpha_{hk}^{(n)} + \tfrac{2}{\pi}\chi_{h-k}^{odd}\right| \le \tfrac{2}{\pi}\left(\chi_{h+k} + \chi_{h'+k'-1}\right) + \tfrac{2}{2n+1}$$

which implies that

$$|||A_n + H_n^{odd}||| \le 4\,|||K_n||| + \tfrac{2n}{2n+1} < 5,$$

by 2.4.2 and Schur's test 2.1.4.

5.3.6 LEMMA. $\boldsymbol{\kappa}(A_n) \le 9\,\boldsymbol{\kappa}(H_n^{odd}) \le 54\,\boldsymbol{\kappa}(A_n)$.

PROOF. In view of 2.3.10, it follows from

$$\tfrac{2}{\pi} \le \|H_n^{odd}\| \le \boldsymbol{\kappa}(H_n^{odd}), \quad 1 \le \|A_n\| \quad \text{and} \quad |||A_n + H_n^{odd}||| \le 5$$

that

$$\boldsymbol{\kappa}(A_n) \le \left(1 + \frac{|||H_n^{odd} + A_n|||}{\|H_n^{odd}\|}\right)\boldsymbol{\kappa}(H_n^{odd}) \le \left(1 + \tfrac{5\pi}{2}\right)\boldsymbol{\kappa}(H_n^{odd}) \le 9\,\boldsymbol{\kappa}(H_n^{odd})$$

and

$$\boldsymbol{\kappa}(H_n^{odd}) \le \left(1 + \frac{|||A_n + H_n^{odd}|||}{\|A_n\|}\right)\boldsymbol{\kappa}(A_n) \le 6\,\boldsymbol{\kappa}(A_n).$$

5.3.7 LEMMA. $\boldsymbol{\kappa}(H_n^{odd}) \le 2\,\boldsymbol{\kappa}(H_n) \le 4\,\boldsymbol{\kappa}(H_n^{odd})$.

PROOF. Let $H_n^{\bullet} := \tfrac{1}{\pi}\Big((-1)^h \chi_{h-k}(-1)^k\Big)$. Then

$$H_n^{odd} = H_n - H_n^{\bullet} \quad \text{and} \quad \boldsymbol{\kappa}(H_n) = \boldsymbol{\kappa}(H_n^{\bullet}).$$

Hence

$$\boldsymbol{\kappa}(H_n^{odd}) \le \boldsymbol{\kappa}(H_n) + \boldsymbol{\kappa}(H_n^{\bullet}) = 2\,\boldsymbol{\kappa}(H_n).$$

To check the right-hand inequality, we form the following matrices:

H_n^+ : cut the first row and the last column of H_{n+1}^{odd},

H_n^- : cut the last row and the first column of H_{n+1}^{odd}.

Thus, letting

$$\chi_m^+ := \begin{cases} 0 & \text{if } m \text{ is odd,} \\ \frac{1}{m-1} & \text{if } m \text{ is even,} \end{cases} \quad \text{and} \quad \chi_m^- := \begin{cases} 0 & \text{if } m \text{ is odd,} \\ \frac{1}{m+1} & \text{if } m \text{ is even,} \end{cases}$$

we have

$$H_n^+ = \tfrac{2}{\pi}(\chi_{h-k}^+) \quad \text{and} \quad H_n^- = \tfrac{2}{\pi}(\chi_{h-k}^-).$$

Recall that

$$H_n = \tfrac{1}{\pi}(\chi_{h-k}) \quad \text{and} \quad H_n^{odd} = \tfrac{2}{\pi}(\chi_{h-k}^{odd}),$$

where

$$\chi_m := \begin{cases} 0 & \text{if } m = 0, \\ \frac{1}{m} & \text{if } m \neq 0, \end{cases} \quad \text{and} \quad \chi_m^{odd} := \begin{cases} \frac{1}{m} & \text{if } m \text{ is odd,} \\ 0 & \text{if } m \text{ is even.} \end{cases}$$

Consequently, if $L_n := \frac{1}{\pi}(\lambda_{h-k})$ is defined by

$$H_n = \tfrac{1}{2}H_n^{odd} + \tfrac{1}{4}H_n^+ + \tfrac{1}{4}H_n^- + L_n, \tag{1}$$

then

$$\chi_m = \chi_m^{odd} + \tfrac{1}{2}\chi_m^+ + \tfrac{1}{2}\chi_m^- + \lambda_m.$$

Thus

$$\lambda_m = \begin{cases} 0 & \text{if } m = 0, \\ 0 & \text{if } m \text{ is odd,} \\ \frac{1}{m} - \frac{1}{2}\left(\frac{1}{m-1} + \frac{1}{m+1}\right) & \text{if } m \neq 0 \text{ is even.} \end{cases}$$

In the last case,

$$\lambda_m = -\frac{1}{m(m^2-1)}.$$

Hence, for fixed h,

$$\sum_{k=1}^{n} |\lambda_{h-k}| \leq 2 \sum_{l=1}^{\infty} \frac{1}{2l(4l^2-1)} = 2\log 2 - 1;$$

see [BRO]. So, by Schur's test 2.1.4 and

$$\tfrac{2}{\pi} \leq \boldsymbol{\kappa}(H_n^{odd}), \tag{2}$$

we get

$$\boldsymbol{\kappa}(L_n) \leq |||L_n||| \leq \tfrac{1}{\pi}(2\log 2 - 1) \leq (\log 2 - \tfrac{1}{2})\,\boldsymbol{\kappa}(H_n^{odd}). \tag{3}$$

Write H_{n+1}^{odd} in the form

$$H_{n+1}^{odd} = \begin{pmatrix} H_n^{odd} & H_{n,1}^{odd} \\ H_{1,n}^{odd} & O \end{pmatrix}$$

with $H_{n,1}^{odd} := \frac{2}{\pi}(\chi_{h-n-1}^{odd})$ and $H_{1,n}^{odd} := \frac{2}{\pi}(\chi_{n+1-k}^{odd})$. Then

$$\boldsymbol{\kappa}(H_{n,1}^{odd}) = \boldsymbol{\kappa}(H_{1,n}^{odd}) \le \frac{2}{\pi}\bigg(\sum_{l=1}^{\infty} \frac{1}{(2l-1)^2}\bigg)^{1/2} = \frac{1}{\sqrt{2}}, \tag{4}$$

and it follows from 2.3.11 that

$$\boldsymbol{\kappa}(H_n^{\pm}) \le \boldsymbol{\kappa}(H_{n+1}^{odd}) = \boldsymbol{\kappa}\begin{pmatrix} H_n^{odd} & H_{n,1}^{odd} \\ H_{1,n}^{odd} & O \end{pmatrix} \le \sqrt{\boldsymbol{\kappa}(H_n^{odd})^2 + 1}. \tag{5}$$

Combining (1) with inequalities (2) to (5) yields

$$\begin{aligned} \boldsymbol{\kappa}(H_n) &\le \tfrac{1}{2}\boldsymbol{\kappa}(H_n^{odd}) + \tfrac{1}{4}\boldsymbol{\kappa}(H_n^{+}) + \tfrac{1}{4}\boldsymbol{\kappa}(H_n^{-}) + \boldsymbol{\kappa}(L_n) \\ &\le \tfrac{1}{2}\sqrt{\boldsymbol{\kappa}(H_n^{odd})^2 + 1} + \log 2\, \boldsymbol{\kappa}(H_n^{odd}) \\ &\le (\tfrac{1}{2} + \tfrac{\pi}{4} + \log 2)\boldsymbol{\kappa}(H_n^{odd}). \end{aligned}$$

This completes the proof, since $\frac{1}{2} + \frac{\pi}{4} + \log 2 = 1.9785\ldots < 2$.

5.3.8 We are now able to establish the main result of this section, which connects $\boldsymbol{\kappa}(H_n)$ with the ideal norms treated in the previous section.

THEOREM. *The sequences of the ideal norms*

$$\boldsymbol{\delta}(\mathcal{E}_n, \mathcal{E}_n) \quad \textit{and} \quad \boldsymbol{\kappa}(H_n)$$

are uniformly equivalent.

PROOF. We have

$$\begin{aligned} \boldsymbol{\delta}(\mathcal{E}_n, \mathcal{E}_n) &\asymp \boldsymbol{\varrho}(\mathcal{C}_n^\circ, \mathcal{S}_n^\circ) && \text{(by 5.2.32)}, \\ \boldsymbol{\varrho}(\mathcal{C}_n^\circ, \mathcal{S}_n^\circ) &= \boldsymbol{\kappa}(A_{n+1,n}) && \text{(by 5.3.2)}, \\ \boldsymbol{\kappa}(A_{n+1,n}) &\asymp \boldsymbol{\kappa}(A_n) && \text{(by 5.3.3)}, \\ \boldsymbol{\kappa}(A_n) &\asymp \boldsymbol{\kappa}(H_n^{odd}) && \text{(by 5.3.6)}, \\ \boldsymbol{\kappa}(H_n^{odd}) &\asymp \boldsymbol{\kappa}(H_n) && \text{(by 5.3.7)}. \end{aligned}$$

Combining these relations completes the proof.

5.3.9 Recall from 2.4.7, 2.4.9, 5.2.36 and 5.2.37 that

$$\mathfrak{HM} := \mathfrak{L}[\boldsymbol{\kappa}(H_n)] \quad \text{and} \quad \mathfrak{QHM} := \mathfrak{L}_0[(1+\log n)^{-1}\,\boldsymbol{\kappa}(H_n)],$$
$$\mathfrak{EP} := \mathfrak{L}[\boldsymbol{\delta}(\mathcal{E}_n, \mathcal{E}_n)] \quad \text{and} \quad \mathfrak{QEP} := \mathfrak{L}_0[(1+\log n)^{-1}\,\boldsymbol{\delta}(\mathcal{E}_n, \mathcal{E}_n)].$$

The preceding results imply that we actually deal with two ideals only.

THEOREM. $\mathfrak{HM} = \mathfrak{EP}$ *and* $\mathfrak{QHM} = \mathfrak{QEP}$.

5.4 The vector-valued Hilbert transform

The classical theory of the scalar-valued Hilbert transforms is presented, for example, in [BENN*, pp. 126–140, 155, 160], [BUT*b, pp. 305–323, 334–340], [STEI*, pp. 187–188, 217–223] and [TIT, pp. 119–151].

CONVENTION. In this section, we always assume that $1 < r < \infty$.

5.4.1 The **periodic Hilbert transform** is defined by

$$H^{\mathbb{T}} : f(t) \longrightarrow \widetilde{f}^{\mathbb{T}}(s) := \frac{1}{2\pi} \int_{-\pi}^{+\pi} f(t) \cot\left(\frac{s-t}{2}\right) dt.$$

If $f \in L_r[-\pi, +\pi)$, then the right-hand integral exists almost everywhere in the sense of Cauchy's principal value, and the M. Riesz theorem tells us that $H^{\mathbb{T}}$ yields an operator from $L_r[-\pi, +\pi)$ into itself. Note that $H^{\mathbb{T}}$ sends every Fourier series

$$\tfrac{1}{2}\gamma_0 + \sum_{k=1}^{\infty}(\alpha_k \cos kt + \beta_k \sin kt)$$

into its **conjugate Fourier series**

$$\sum_{k=1}^{\infty}(\alpha_k \sin kt - \beta_k \cos kt).$$

Therefore, $H^{\mathbb{T}}$ is often called the **conjugate function operator**.

The (Marcel) **Riesz projection** P_E on $L_r[-\pi, +\pi)$ cuts all Fourier coefficients with negative indices:

$$P_E : \sum_{k=-\infty}^{+\infty} \xi_k \exp(ikt) \longrightarrow \sum_{k=0}^{\infty} \xi_k \exp(ikt).$$

An elementary proof of the fact that P_E is a bounded operator on $L_r[-\pi, +\pi)$ can be found in [LIN*2, pp. 166–167]. We have

$$H^{\mathbb{T}} = i(I - 2P_E + D_0) \quad \text{and} \quad (H^{\mathbb{T}})^2 = -I + D_0,$$

where

$$D_0 : f(t) \longrightarrow \frac{1}{2\pi} \int_{-\pi}^{+\pi} f(t)\, dt$$

is the 1-dimensional projection from $L_r[-\pi, +\pi)$ onto the constants.

5.4.2 The above concept has an analogue on the real line. The **Hilbert transform** is given by

$$H^{\mathbb{R}} : f(t) \longrightarrow \widetilde{f^{\mathbb{R}}}(s) := \frac{1}{\pi} \int_{-\infty}^{+\infty} f(t) \frac{1}{s-t} \, dt.$$

Also in this case, the right-hand integral has to be taken in the principal value sense, it exists almost everywhere for f in $L_r(\mathbb{R})$, and the resulting operator $H^{\mathbb{R}}$ acts continuously on these Banach spaces. Moreover,

$$(H^{\mathbb{R}})^2 = -I.$$

5.4.3 There are different possibilities for defining a **discrete Hilbert transform**. The most standard way is to let

$$H^{\mathbb{Z}} : (\xi_k) \longrightarrow (\widetilde{\xi_h^{\mathbb{Z}}}) := \frac{1}{\pi} \Big(\sum_{k \in \mathbb{Z}}{}' \frac{1}{h-k} \xi_k \Big),$$

where the dash in $\sum{}'$ indicates that the summation extends over all $k \neq h$. We also need the map

$$H_{\Delta}^{\mathbb{Z}} : (\xi_k) \longrightarrow (\widetilde{\xi_h^{\mathbb{Z}}}) := \frac{1}{\pi} \Big(\sum_{k \in \mathbb{Z}} c_{h-k}^{(\Delta)} \, \xi_k \Big)$$

with

$$c_m^{(\Delta)} := \int_{\Delta_m} \int_{\Delta_0} \frac{1}{s-t} \, ds \, dt \quad \text{and} \quad \Delta_k := [\tfrac{2k-1}{2}, \tfrac{2k+1}{2}).$$

Note that $c_0^{(\Delta)} = 0$ and

$$c_{h-k}^{(\Delta)} = \int_{\Delta_h} \int_{\Delta_k} \frac{1}{s-t} \, dsdt.$$

Both of these rules yield operators on $l_r(\mathbb{Z})$, which behave quite similarly.

REMARK. Modifications of the above definition can be found in [kak] and [tit].

5.4.4 The usual operator norms of Hilbert transforms were computed by S. K. Pichorides [pic].

PROPOSITION.

$$\|H^{\mathbb{T}} : L_r[-\pi, +\pi) \to L_r[-\pi, +\pi)\| =$$

$$= \|H^{\mathbb{R}} : L_r(\mathbb{R}) \to L_r(\mathbb{R})\| = \begin{cases} \tan \frac{\pi}{2r} & \text{if } 1 < r \leq 2, \\ \cot \frac{\pi}{2r} & \text{if } 2 \leq r < \infty. \end{cases}$$

REMARK. In the course of this section, we will see that

$$\|H_{\Delta}^{\mathbb{Z}} : l_r(\mathbb{Z}) \to l_r(\mathbb{Z})\| = \|H^{\mathbb{R}} : L_r(\mathbb{R}) \to L_r(\mathbb{R})\|.$$

The exact value of $\|H^{\mathbb{Z}} : l_r(\mathbb{Z}) \to l_r(\mathbb{Z})\|$ seems to be unknown for $r \neq 2$.

5.4.5 An operator $T \in \mathfrak{L}(X, Y)$ is said to be **compatible with the Hilbert transform** $H^{\mathbb{R}}$ on $L_r(\mathbb{R})$ if

$$H^{\mathbb{R}} \otimes T : L_r(\mathbb{R}) \otimes X \longrightarrow L_r(\mathbb{R}) \otimes Y$$

admits a continuous extension

$$[H^{\mathbb{R}}, T] : [L_r(\mathbb{R}), X] \longrightarrow [L_r(\mathbb{R}), Y].$$

In this case, we let

$$\|T|\mathfrak{HT}_r^{\mathbb{R}}\| := \|[H^{\mathbb{R}}, T] : [L_r(\mathbb{R}), X] \to [L_r(\mathbb{R}), Y]\|.$$

The class of these operators is a Banach ideal, denoted by $\mathfrak{HT}_r^{\mathbb{R}}$. The letters $\mathfrak{HT}$ stand for vector-valued **Hilbert transform**.

Operators **compatible with the periodic Hilbert transform** $H^{\mathbb{T}}$ and **compatible with the discrete Hilbert transforms** $H^{\mathbb{Z}}$ and $H_{\Delta}^{\mathbb{Z}}$, respectively, are defined analogously. In this way, we obtain the Banach ideals $\mathfrak{HT}_r^{\mathbb{T}}$, $\mathfrak{HT}_r^{\mathbb{Z}}$ and $\mathfrak{HT}_{\Delta,r}^{\mathbb{Z}}$.

5.4.6 The following facts can be checked by standard techniques.

PROPOSITION.

$$(\mathfrak{HT}_r^{\mathbb{R}})' = \mathfrak{HT}_{r'}^{\mathbb{R}}, \quad (\mathfrak{HT}_r^{\mathbb{T}})' = \mathfrak{HT}_{r'}^{\mathbb{T}}, \quad \textit{and} \quad (\mathfrak{HT}_r^{\mathbb{Z}})' = \mathfrak{HT}_{r'}^{\mathbb{Z}}.$$

5.4.7 Without proof, we state an extrapolation theorem which was independently obtained by J. Schwartz and A. Benedek/A. P. Calderón/R. Panzone; see [ben*], [schw] and [DUN*2, pp. 1165–1171]. Compare with 7.2.10.

THEOREM. *Let $\boldsymbol{K}$ be an $\mathfrak{L}(X, Y)$-valued function on $\mathbb{R}$ which is integrable on all compact sets. Suppose that*

$$\int\limits_{|t| \geq 4|h|} \|\boldsymbol{K}(t+h) - \boldsymbol{K}(t)\| \, dt \leq c \quad \textit{for all } h \in \mathbb{R},$$

where $c > 0$ is a constant. Let

$$\boldsymbol{K} * \boldsymbol{f}(s) := \int\limits_{-\infty}^{+\infty} \boldsymbol{K}(s-t) \boldsymbol{f}(t) \, dt,$$

and consider the property: There exists a constant $c_r > 0$ such that

$$\|\boldsymbol{K} * \boldsymbol{f}|L_r\| \leq c_r \, \|\boldsymbol{f}|L_r\| \tag{C_r}$$

whenever $\boldsymbol{f} : \mathbb{R} \to X$ is a bounded measurable function with compact support.

If (C_r) *holds for* **some** $1 < r < \infty$, *then it holds for* **all** $1 < r < \infty$.

REMARK. Given $c_{r_0} > 0$, we may choose $c_r := C_r c\, c_{r_0}$, where $C_r > 0$ depends only on r.

5.4.8 As shown in [GARC*, p. 492], the preceding theorem remains true if $\boldsymbol{K}$ is only integrable on all compact sets not containing 0. In this form, it applies to the Hilbert kernel $h(t) := \frac{1}{\pi t}$ directly. We prefer, however, an approach via Poisson kernels, which have no singularities at 0.

For $\varepsilon > 0$, the **Poisson transforms** are defined by

$$P_\varepsilon^{\mathbb{R}} : f(t) \longrightarrow f * p_\varepsilon^{\mathbb{R}}(s) := \frac{1}{\pi} \int_{-\infty}^{+\infty} f(t) \frac{\varepsilon}{(s-t)^2 + \varepsilon^2}\, dt$$

and

$$Q_\varepsilon^{\mathbb{R}} : f(t) \longrightarrow f * q_\varepsilon^{\mathbb{R}}(s) := \frac{1}{\pi} \int_{-\infty}^{+\infty} f(t) \frac{s-t}{(s-t)^2 + \varepsilon^2}\, dt.$$

It can be shown that the operators $P_\varepsilon^{\mathbb{R}}$ and $Q_\varepsilon^{\mathbb{R}}$ act from $L_r(\mathbb{R})$ into itself; see [BUT*b, pp. 126 and 314]. The convolution kernels

$$p_\varepsilon^{\mathbb{R}}(t) := \frac{\varepsilon}{\pi(t^2 + \varepsilon^2)} \quad \text{and} \quad q_\varepsilon^{\mathbb{R}}(t) := \frac{t}{\pi(t^2 + \varepsilon^2)}$$

are related by the formula $H^{\mathbb{R}} p_\varepsilon^{\mathbb{R}} = q_\varepsilon^{\mathbb{R}}$. Hence

$$Q_\varepsilon^{\mathbb{R}} f = f * q_\varepsilon^{\mathbb{R}} = f * H^{\mathbb{R}} p_\varepsilon^{\mathbb{R}} = H^{\mathbb{R}} f * p_\varepsilon^{\mathbb{R}} = P_\varepsilon^{\mathbb{R}} H^{\mathbb{R}} f \quad \text{for } f \in L_r(\mathbb{R}). \tag{1}$$

Next, by 0.4.9, we conclude from

$$p_\varepsilon^{\mathbb{R}}(t) \geq 0, \quad \int_{-\infty}^{+\infty} p_\varepsilon^{\mathbb{R}}(t)\, dt = 1, \quad \text{and} \quad \lim_{\varepsilon \to 0} \int_{-a}^{+a} p_\varepsilon^{\mathbb{R}}(t)\, dt = 1 \quad \text{for } a > 0$$

that

$$\big\| [P_\varepsilon^{\mathbb{R}}, X] : [L_r(\mathbb{R}), X] \to [L_r(\mathbb{R}), X] \big\| \leq 1 \tag{2}$$

and

$$\lim_{\varepsilon \to 0} \big\| [P_\varepsilon^{\mathbb{R}}, X]\boldsymbol{f} - \boldsymbol{f} \big| L_r \big\| = 0 \quad \text{for } \boldsymbol{f} \in [L_r(\mathbb{R}), X]. \tag{3}$$

Moreover, (1) and (3) imply

$$\lim_{\varepsilon \to 0} \big\| [Q_\varepsilon^{\mathbb{R}}, T]\boldsymbol{f} - [H^{\mathbb{R}}, T]\boldsymbol{f} \big| L_r \big\| = 0 \tag{4}$$

for $T \in \mathfrak{L}(X, Y)$ and $\boldsymbol{f} \in L_r(\mathbb{R}) \otimes X$.

5.4.9 PROPOSITION. *For every operator $T \in \mathfrak{L}(X, Y)$, the following are equivalent:*

(1) *T is compatible with $H^{\mathbb{R}}$ on $L_r(\mathbb{R})$.*
(2) *T is compatible with $Q_\varepsilon^{\mathbb{R}}$ on $L_r(\mathbb{R})$ for* **all** *$\varepsilon > 0$.*
(3) *T is compatible with $Q_\varepsilon^{\mathbb{R}}$ on $L_r(\mathbb{R})$ for* **some** *$\varepsilon > 0$.*

In this case,

$$\|[H^{\mathbb{R}},T]:[L_r(\mathbb{R}),X]\to[L_r(\mathbb{R}),Y]\| = \|[Q_\varepsilon^{\mathbb{R}},T]:[L_r(\mathbb{R}),X]\to[L_r(\mathbb{R}),Y]\|.$$

PROOF. **(1)⇒(2)**: We know from (1) and (2) in the previous paragraph that

$$\|[P_\varepsilon^{\mathbb{R}},Y]:[L_r(\mathbb{R}),Y]\to[L_r(\mathbb{R}),Y]\|\leq 1 \quad\text{and}\quad [Q_\varepsilon^{\mathbb{R}},T]=[P_\varepsilon^{\mathbb{R}},Y]\,[H^{\mathbb{R}},T].$$

Consequently, if T is compatible with $H^{\mathbb{R}}$, then it is compatible with all $Q_\varepsilon^{\mathbb{R}}$. Moreover,

$$\|[Q_\varepsilon^{\mathbb{R}},T]:[L_r(\mathbb{R}),X]\to[L_r(\mathbb{R}),Y]\| \leq \|[H^{\mathbb{R}},T]:[L_r(\mathbb{R}),X]\to[L_r(\mathbb{R}),Y]\|.$$

(2)⇒(3) is trivial.

(3)⇒(2): Letting $S_a : f(t)\to f(at)$ for $a>0$, we have

$$Q_{a\varepsilon}^{\mathbb{R}} = S_{1/a}Q_\varepsilon^{\mathbb{R}}S_a,$$

and it follows from $\|S_a : L_r(\mathbb{R})\to L_r(\mathbb{R})\| = a^{-1/r}$ that

$$\|[Q_{a\varepsilon}^{\mathbb{R}},T]:[L_r(\mathbb{R}),X]\to[L_r(\mathbb{R}),Y]\| \leq \|[Q_\varepsilon^{\mathbb{R}},T]:[L_r(\mathbb{R}),X]\to[L_r(\mathbb{R}),Y]\|.$$

Thus the compatibility of T with $Q_\varepsilon^{\mathbb{R}}$ does not depend on $\varepsilon>0$, and all operator norms $\|[Q_\varepsilon^{\mathbb{R}},T]:[L_r(\mathbb{R}),X]\to[L_r(\mathbb{R}),Y]\|$ are equal.

(2)⇒(1): As shown in formula (4) of the previous paragraph, we have

$$[H^{\mathbb{R}},T]\boldsymbol{f} = \lim_{\varepsilon\to 0}[Q_\varepsilon^{\mathbb{R}},T]\boldsymbol{f} \quad\text{for } \boldsymbol{f}\in L_r(\mathbb{R})\otimes X.$$

Hence

$$\begin{aligned}\|[H^{\mathbb{R}},T]\boldsymbol{f}|L_r\| &\leq \sup_{\varepsilon>0}\|[Q_\varepsilon^{\mathbb{R}},T]\boldsymbol{f}|L_r\|\\ &\leq \sup_{\varepsilon>0}\|[Q_\varepsilon^{\mathbb{R}},T]:[L_r(\mathbb{R}),X]\to[L_r(\mathbb{R}),Y]\|\,\|\boldsymbol{f}|L_r\|.\end{aligned}$$

However, the right-hand operator norms do not depend on $\varepsilon>0$. Thus

$$\|[H^{\mathbb{R}},T]:[L_r(\mathbb{R}),X]\to[L_r(\mathbb{R}),Y]\| \leq \|[Q_\varepsilon^{\mathbb{R}},T]:[L_r(\mathbb{R}),X]\to[L_r(\mathbb{R}),Y]\|.$$

5.4.10 We now formulate the deepest result of this section, which follows from 5.4.9 and 5.4.7 applied to the $\mathfrak{L}(X,Y)$-valued kernel $q_\varepsilon^{\mathbb{R}}\otimes T$.

THEOREM. *The operator ideal $\mathfrak{HT}_r^{\mathbb{R}}$ is independent of the parameter $1<r<\infty$.*

REMARK. Carefully looking at all estimates, M. Defant [def 2, p. 357] was able to show that

$$\|\cdot|\mathfrak{HT}_2^{\mathbb{R}}\| \leq c_1\,\|\cdot|\mathfrak{HT}_q^{\mathbb{R}}\| \quad\text{and}\quad \|\cdot|\mathfrak{HT}_q^{\mathbb{R}}\| \leq c_2\,q\,\|\cdot|\mathfrak{HT}_2^{\mathbb{R}}\| \quad\text{if } 2<q<\infty,$$

where $c_1>0$ and $c_2>0$ are absolute constants. It seems likely that it is possible to take $c_1=1$.

5.4.11 In this paragraph, we introduce some auxiliary operators.

(1) We let

$$J:(\xi_k)\longrightarrow\sum_{\mathbb{Z}}\xi_k\,\chi_k \quad\text{and}\quad Q:f(t)\longrightarrow\Big(\int_{\Delta_h}f(t)\,dt\Big),$$

where χ_k is the characteristic function of $\Delta_k:=\left[\frac{2k-1}{2},\frac{2k+1}{2}\right)$.

(2) The surjection $R_{2\pi}$ restricts $f:\mathbb{R}\to\mathbb{K}$ to $[-\pi,+\pi)$.

(3) The injection $E_{2\pi}^{(N)}$ multiplies the 2π-periodic extension of any function $f:[-\pi,+\pi)\to\mathbb{K}$ on $\mathbb{R}$ by the characteristic function of $[-N\pi,+N\pi)$.

(4) For $a>0$ and $f:\mathbb{R}\to\mathbb{K}$, the scale transform S_a is defined by

$$S_a : f(t)\longrightarrow f(at).$$

Note that J and Q are dual to each other. Moreover,

$$\|J:l_r(\mathbb{Z})\to L_r(\mathbb{R})\|=1,\quad \|Q:L_r(\mathbb{R})\to l_r(\mathbb{Z})\|=1,$$

$$\|R_{2\pi}:L_r(\mathbb{R})\to L_r[-\pi,+\pi)\|=1,\quad \|E_{2\pi}^{(N)}:L_r[-\pi,+\pi)\to L_r(\mathbb{R})\|=N^{1/r},$$

and

$$\|S_a:L_r(\mathbb{R})\to L_r(\mathbb{R})\|=a^{-1/r}.$$

5.4.12 Next, we provide four lemmas which connect the various Hilbert transforms with each other. In order to avoid cumbersome factors, we temporarily deviate from our custom and equip $[-\pi,\pi)$ with the original Lebesgue measure, rather than with the normalized one.

LEMMA. *Let $f\in L_r[-\pi,+\pi)$ and $g\in L_{r'}[-\pi,+\pi)$. Then*

$$\langle H^{\mathbb{T}}f,g\rangle=\lim_{N\to\infty}\frac{1}{N}\left\langle H^{\mathbb{R}}E_{2\pi}^{(N)}f,E_{2\pi}^{(N)}g\right\rangle.$$

PROOF. Since the trigonometric polynomials are dense and

$$\frac{1}{N}\left|\left\langle H^{\mathbb{R}}E_{2\pi}^{(N)},E_{2\pi}^{(N)}g\right\rangle\right|\le$$
$$\le\|H^{\mathbb{R}}:L_r(\mathbb{R})\to L_r(\mathbb{R})\|\,\|f|L_r[-\pi,+\pi)\|\,\|g|L_{r'}[-\pi,+\pi)\|,$$

it is enough to check the desired formula for $e_k(t)=\exp(ikt)$ and $e_{-h}(s)=\exp(-ihs)$. Applying the coordinate transformations $s=v+u$ and $t=v-u$, we have

$$\frac{1}{N}\left\langle H^{\mathbb{R}}E_{2\pi}^{(N)}e_k,E_{2\pi}^{(N)}e_{-h}\right\rangle=\frac{1}{N\pi}\int_{-N\pi}^{+N\pi}\int_{-N\pi}^{+N\pi}\exp(ikt)\frac{1}{s-t}\exp(-ihs)\,ds\,dt$$
$$=\frac{1}{N\pi}\iint_{|u|+|v|\le N\pi}\exp(-iku-ihu)\frac{1}{u}\exp(ikv-ihv)\,du\,dv.$$

Elementary manipulations now yield

$$\frac{1}{N}\left\langle H^{\mathbb{R}}E_{2\pi}^{(N)}e_k, E_{2\pi}^{(N)}e_{-h}\right\rangle = \begin{cases} -4i\int\limits_0^{N\pi}\frac{\sin 2ku}{u}\,du & \text{if } h=k,\\ \frac{2\exp(iN\pi(k-h))}{iN\pi(k-h)}\int\limits_0^{N\pi}\frac{\cos 2ku-\cos 2hu}{u}\,du & \text{if } h\neq k.\end{cases}$$

On the other hand, it follows from $H^{\mathbb{T}}e_k = -i\,\mathrm{sgn}(k)e_k$ that

$$\langle H^{\mathbb{T}}e_k, e_{-h}\rangle = \begin{cases} -2\pi i\,\mathrm{sgn}(k) & \text{if } h=k,\\ 0 & \text{if } h\neq k.\end{cases}$$

Hence

$$\int\limits_0^{\infty}\frac{\sin ku}{u}\,du = \mathrm{sgn}(k)\frac{\pi}{2}$$

and the convergence of

$$\int\limits_0^{\infty}\frac{\cos ku-\cos hu}{u}\,du$$

imply the assertion.

5.4.13 LEMMA. *Let $f\in L_r(\mathbb{R})$ and $g\in L_{r'}(\mathbb{R})$. Then*

$$\langle H^{\mathbb{R}}f, g\rangle = \lim_{N\to\infty} N\,\langle H^{\mathbb{T}}R_{2\pi}S_Nf, R_{2\pi}S_Ng\rangle.$$

PROOF. Since

$$N\left|\langle H^{\mathbb{T}}R_{2\pi}S_Nf, R_{2\pi}S_Ng\rangle\right| \leq$$
$$\leq \|H^{\mathbb{T}}: L_r[-\pi,+\pi)\to L_r[-\pi,+\pi)\|\,\|f|L_r(\mathbb{R})\|\,\|g|L_{r'}(\mathbb{R})\|,$$

it is enough to check the desired formula on dense subsets. Thus we may assume that the functions $f\in L_r(\mathbb{R})$ and $g\in L_{r'}(\mathbb{R})$ vanish outside an interval $[-N_0\pi,+N_0\pi)$. Let $N\geq 2N_0$. Then

$$\langle H^{\mathbb{R}}f, g\rangle = \frac{1}{2\pi N}\int\limits_{-N_0\pi}^{+N_0\pi}\int\limits_{-N_0\pi}^{+N_0\pi} f(t)\frac{2N}{s-t}g(s)\,ds\,dt$$

and

$$N\,\langle H^{\mathbb{T}}R_{2\pi}S_Nf, R_{2\pi}S_Ng\rangle = \frac{N}{2\pi}\int\limits_{-\pi}^{+\pi}\int\limits_{-\pi}^{+\pi} f(N\tau)\cot\Big(\frac{\sigma-\tau}{2}\Big)g(\sigma)\,d\sigma d\tau$$

$$= \frac{1}{2\pi N}\int\limits_{-N_0\pi}^{+N_0\pi}\int\limits_{-N_0\pi}^{+N_0\pi} f(t)\cot\Big(\frac{s-t}{2N}\Big)g(s)\,ds\,dt.$$

Moreover, since $\left|\frac{s-t}{2N}\right| \le \frac{2\pi N_0}{2N} \le \frac{\pi}{2}$, we know from 0.10.4 that

$$\left|\frac{2N}{s-t} - \cot\left(\frac{s-t}{2N}\right)\right| \le \frac{2}{\pi} < 1.$$

Hence

$$\begin{aligned}
&|\langle H^{\mathbb{R}} f, g\rangle - N\langle H^{\mathbb{T}} R_{2\pi} S_N f, R_{2\pi} S_N g\rangle| \le \\
&\le \frac{1}{2\pi N} \int_{-N_0\pi}^{+N_0\pi} \int_{-N_0\pi}^{+N_0\pi} \left| f(t) \left[\frac{2N}{s-t} - \cot\left(\frac{s-t}{2N}\right)\right] g(s)\right| ds\, dt \\
&\le \frac{1}{2\pi N} \|f|L_1[-N_0\pi, +N_0\pi)\|\, \|g|L_1[-N_0\pi, +N_0\pi)\| \\
&\le \frac{1}{2\pi N} \|f|L_r[-N_0\pi, +N_0\pi)\|\, (2\pi N_0)^{1/r'+1/r}\, \|g|L_{r'}[-N_0\pi, +N_0\pi)\| \\
&\le \frac{N_0}{N} \|f|L_r(\mathbb{R})\|\, \|g|L_{r'}(\mathbb{R})\|.
\end{aligned}$$

Letting $N \to \infty$ completes the proof.

5.4.14 The next formula explains why we have introduced different kinds of discrete Hilbert transforms in 5.4.3.

LEMMA. *Let $x \in l_r(\mathbb{Z})$ and $y \in l_{r'}(\mathbb{Z})$. Then*

$$\langle H_\Delta^{\mathbb{Z}} x, y\rangle = \langle H^{\mathbb{R}} Jx, Jy\rangle.$$

PROOF. The assertion follows from $H_\Delta^{\mathbb{Z}} = QH^{\mathbb{R}}J$ or $c_{h-k}^{(\Delta)} = \langle H^{\mathbb{R}} Ju_k, Ju_h\rangle$.

5.4.15 LEMMA. *Let $f \in L_r(\mathbb{R})$ and $g \in L_{r'}(\mathbb{R})$. Then*

$$\langle H^{\mathbb{R}} f, g\rangle = \lim_{M\to\infty} 2^{-M} \langle H_\Delta^{\mathbb{Z}} QS_{2^{-M}} f, QS_{2^{-M}} g\rangle.$$

PROOF. We conclude from

$$\langle H_\Delta^{\mathbb{Z}} x, y\rangle = \langle H^{\mathbb{R}} Jx, Jy\rangle, \quad \langle S_a^{-1} f, g\rangle = a\,\langle f, S_a g\rangle \quad \text{and} \quad S_a H^{\mathbb{R}} = H^{\mathbb{R}} S_a$$

that

$$\begin{aligned}
\langle H^{\mathbb{R}} S_{2^M} JQS_{2^{-M}} f, S_{2^M} JQS_{2^{-M}} g\rangle &= 2^{-M} \langle H^{\mathbb{R}} JQS_{2^{-M}} f, JQS_{2^{-M}} g\rangle \\
&= 2^{-M} \langle H_\Delta^{\mathbb{Z}} QS_{2^{-M}} f, QS_{2^{-M}} g\rangle.
\end{aligned}$$

Since $S_{2^M} JQS_{2^{-M}}$ is the orthogonal projection from $L_r(\mathbb{R})$ onto the subspace formed by all functions that are constant on the intervals $2^{-M}\Delta_k$, we have

$$\lim_{M\to\infty} \|f - S_{2^M} JQS_{2^{-M}} f|L_r\| = 0 \quad \text{and} \quad \lim_{M\to\infty} \|g - S_{2^M} JQS_{2^{-M}} g|L_{r'}\| = 0.$$

Hence

$$\langle H^{\mathbb{R}} f, g\rangle = \lim_{M\to\infty} \langle H^{\mathbb{R}} S_{2^M} JQS_{2^{-M}} f, S_{2^M} JQS_{2^{-M}} g\rangle.$$

5.4.16 In order to get the next theorem, we need another auxiliary result.

LEMMA. *Let*

$$U \in \mathfrak{L}(L_r(M,\mu)) \quad \textit{and} \quad V \in \mathfrak{L}(L_r(N,\nu)),$$

$$A_N \in \mathfrak{L}(L_r(M,\mu), L_r(N,\nu)) \quad \textit{and} \quad B_N \in \mathfrak{L}(L_{r'}(M,\mu), L_{r'}(N,\nu))$$

be such that $\|[A_N, X]\| \, \|[B_N, Y']\| \le 1$. *Assume that*

$$\langle Uf, g\rangle = \lim_{N\to\infty} \langle VA_N f, B_N g\rangle \quad \textit{for } f \in L_r(M,\mu) \textit{ and } g \in L_{r'}(M,\mu).$$

If $T \in \mathfrak{L}(X,Y)$ *is compatible with* V, *then it is also compatible with* U *and*

$$\|[U,T] : [L_r, X] \to [L_r, Y]\| \le \|[V,T] : [L_r, X] \to [L_r, Y]\|.$$

PROOF. For

$$\sum_{k=1}^{n} f_k \otimes x_k \in L_r(M,\mu) \otimes X \quad \text{and} \quad \sum_{h=1}^{m} g_h \otimes v_h \in L_{r'}(M,\mu) \otimes Y',$$

we have

$$\left|\left\langle \sum_{k=1}^{n} Uf_k \otimes Tx_k, \sum_{h=1}^{m} g_h \otimes v_h \right\rangle\right| =$$

$$= \lim_{N\to\infty} \left|\left\langle \sum_{k=1}^{n} VA_N f_k \otimes Tx_k, \sum_{h=1}^{m} B_N g_h \otimes v_h \right\rangle\right|$$

$$\le \sup_N \left\| \sum_{k=1}^{n} VA_N f_k \otimes Tx_k \middle| L_r \right\| \left\| \sum_{h=1}^{m} B_N g_h \otimes v_h \middle| L_{r'} \right\|$$

$$\le \|[V,T]\| \sup_N \|[A_N, X]\| \|[B_N, Y']\| \left\| \sum_{k=1}^{n} f_k \otimes x_k \middle| L_r \right\| \left\| \sum_{h=1}^{m} g_h \otimes v_h \middle| L_{r'} \right\|$$

$$\le \|[V,T]\| \left\| \sum_{k=1}^{n} f_k \otimes x_k \middle| L_r \right\| \left\| \sum_{h=1}^{m} g_h \otimes v_h \middle| L_{r'} \right\|.$$

5.4.17 Now we show that the various Hilbert transforms lead to one and the same operator ideal which is, moreover, independent of r.

THEOREM. *Let* $T \in \mathfrak{L}(X,Y)$. *Then, for all exponents* r *with* $1 < r < \infty$, *the following are equivalent:*

($\mathbb{R}$) *T is compatible with* $H^{\mathbb{R}}$ *on* $L_r(\mathbb{R})$.

($\mathbb{T}$) *T is compatible with* $H^{\mathbb{T}}$ *on* $L_r[-\pi, +\pi)$.

($\mathbb{P}$) *T is compatible with* P_E *on* $L_r[-\pi, +\pi)$.

($\mathbb{Z}$) *T is compatible with* $H^{\mathbb{Z}}$ *on* $l_r(\mathbb{Z})$.

($\mathbb{Z}$)$_\Delta$ *T is compatible with* $H^{\mathbb{Z}}_\Delta$ *on* $l_r(\mathbb{Z})$.

PROOF. First we assume that the exponent r is fixed.
$(\mathbb{Z}) \Longleftrightarrow (\mathbb{Z})_\Delta$: We have $c_1^{(\Delta)} = 2\log 2$ and

$$c_m^{(\Delta)} = (m+1)\log(m+1) - 2m\log m + (m-1)\log(m-1)$$

for $m = 2, 3, \ldots$. Hence

$$c_m^{(\Delta)} = m\log\Big(1-\frac{1}{m^2}\Big) + \log\frac{1+\frac{1}{m}}{1-\frac{1}{m}} = \frac{1}{m} + \frac{1}{6m^3} + \ldots + \frac{1}{(2k-1)k\,m^{2n-1}} + \ldots,$$

which implies that

$$\Big|c_m^{(\Delta)} - \frac{1}{m}\Big| \le \frac{1}{m^3}.$$

Thus $H_\Delta^{\mathbb{Z}} - H^{\mathbb{Z}}$ is compatible with all operators.

Note that

$$\|[S_a, X] : [L_r(\mathbb{R}), X] \to [L_r(\mathbb{R}), X]\| = a^{-1/r},$$

$$\|[J, X] : [l_r(\mathbb{Z}), X] \to [L_r(\mathbb{R}), X]\| = 1,$$

$$\|[Q, X] : [L_r(\mathbb{R}), X] \to [l_r(\mathbb{Z}), X]\| = 1,$$

$$\|[R_{2\pi}, X] : [L_r(\mathbb{R}), X] \to [L_r[-\pi, +\pi), X]\| = 1,$$

$$\|[E_{2\pi}^{(N)}, X] : [L_r[-\pi, +\pi), X] \to [L_r(\mathbb{R}), X]\| = N^{1/r}.$$

The implications $(\mathbb{R}) \Rightarrow (\mathbb{T})$, $(\mathbb{T}) \Rightarrow (\mathbb{R})$ and $(\mathbb{Z}) \Rightarrow (\mathbb{R})$ now follow by combining 5.4.16 with 5.4.12, 5.4.13 and 5.4.15, respectively.
$(\mathbb{T}) \Longleftrightarrow (\mathbb{P})$: Use $H^{\mathbb{T}} = i(I - 2P_E + D_0)$.
The implication $(\mathbb{R}) \Rightarrow (\mathbb{Z})_\Delta$ is a consequence of 5.4.14.

Finally, we know from 5.4.10 that the compatibility with $H^{\mathbb{R}}$ does not depend on the exponent $1 < r < \infty$.

REMARK. The above theorem tells us that all operator ideals $\mathfrak{H}\mathfrak{T}_r^{\mathbb{R}}$, $\mathfrak{H}\mathfrak{T}_r^{\mathbb{T}}$ $\mathfrak{H}\mathfrak{T}_r^{\mathbb{Z}}$ and $\mathfrak{H}\mathfrak{T}_{\Delta,r}^{\mathbb{Z}}$ with $1 < r < \infty$ coincide, the associated ideal norms being equivalent or even equal. Thus, to simplify matters, we are entitled to use the common symbol $\mathfrak{H}\mathfrak{T}$.

5.4.18 By the preceding theorem, $\mathfrak{H}\mathfrak{T}$ is characterized as the collection of all operators that are **compatible with the Riesz projection P_E** on $L_r[-\pi, +\pi)$. This observation is useful to describe the connection with the Banach ideal $\mathfrak{E}\mathfrak{P} := \mathfrak{L}[\delta(\mathcal{E}_n, \mathcal{E}_n)]$, already defined in 5.2.36.

PROPOSITION. $\mathfrak{E}\mathfrak{P} = \mathfrak{H}\mathfrak{T}$.

PROOF. We introduce the shifted Riesz projection

$$P_E^{(n)} : \sum_{k=-\infty}^{+\infty} \xi_k \exp(ikt) \longrightarrow \sum_{k=n}^{\infty} \xi_k \exp(ikt).$$

Then

$$\begin{aligned}\boldsymbol{\delta}(T|\mathcal{E}_n, \mathcal{E}_n) &= \|[P_E^{(0)} - P_E^{(n+1)}, T] : [L_2[-\pi, +\pi), X] \to [L_2[-\pi, +\pi), Y]\| \\ &\leq 2\,\|[P_E, T] : [L_2[-\pi, +\pi), X] \to [L_2[-\pi, +\pi), Y]\|\end{aligned}$$

for $T \in \mathfrak{HT}$ and $n = 1, 2, \ldots$. Therefore, $\mathfrak{HT} \subseteq \mathfrak{EP}$.

On the other hand, we conclude from Fatou's lemma that

$$\|[P_E, T] : [L_2[-\pi, +\pi), X] \to [L_2[-\pi, +\pi), Y]\| \leq \liminf_{n\to\infty} \boldsymbol{\delta}(T|\mathcal{E}_n, \mathcal{E}_n)$$

for $T \in \mathfrak{EP}$ and $n = 1, 2, \ldots$. Hence $\mathfrak{EP} \subseteq \mathfrak{HT}$.

REMARK. It seems likely that

$$\|[P_E, T] : [L_2[-\pi, +\pi), X] \to [L_2[-\pi, +\pi), Y]\| = \sup_n \boldsymbol{\delta}(T|\mathcal{E}_n, \mathcal{E}_n).$$

However, we do not even know whether the sequence $(\boldsymbol{\delta}(\mathcal{E}_n, \mathcal{E}_n))$ is non-decreasing.

5.4.19 We now establish the connection with Section 2.4 in which the Banach ideal $\mathfrak{HM} := \mathfrak{L}[\boldsymbol{\kappa}(H_n)]$ was introduced.

PROPOSITION. $\mathfrak{HM} = \mathfrak{HT}$.

PROOF. Note that the finite Hilbert matrix H_n is obtained from the infinite Hilbert matrix by cutting all entries with $h > n$ or $k > n$. Therefore

$$\|T|\mathfrak{HM}\| = \lim_{n\to\infty} \boldsymbol{\kappa}(T|H_n) = \|T|\mathfrak{HT}_2^{\mathbb{Z}}\|.$$

REMARK. Recall from 5.3.8 that $\boldsymbol{\delta}(\mathcal{E}_n, \mathcal{E}_n) \asymp \boldsymbol{\kappa}(H_n)$. Hence

$$\mathfrak{L}[\boldsymbol{\delta}(\mathcal{E}_n, \mathcal{E}_n)] = \mathfrak{L}[\boldsymbol{\kappa}_n(H_n)].$$

Thus 5.4.18 and 5.4.19 yield another proof of $\mathfrak{HT}_2^{\mathbb{T}} = \mathfrak{HT}_2^{\mathbb{Z}}$.

5.4.20 We proceed with a corollary of the famous M. Riesz theorem.

EXAMPLE. $\|L_r|\mathfrak{HT}_r^{\mathbb{R}}\| = \|H^{\mathbb{R}} : L_r(\mathbb{R}) \to L_r(\mathbb{R})\|$ *if* $1 < r < \infty$.

PROOF. Note that $\boldsymbol{f} \in [L_r(\mathbb{R}), L_r(M, \mu)]$ can be viewed as an r-integrable function depending on $t \in \mathbb{R}$ and $\xi \in M$. Let

$$\widetilde{f^{\mathbb{R}}}(s, \xi) := \frac{1}{\pi} \int_{-\infty}^{+\infty} f(t, \xi) \frac{1}{s-t}\, dt.$$

Then

$$\int_{-\infty}^{+\infty} |\widetilde{f^{\mathbb{R}}}(s,\xi)|^r \, ds \le \|H^{\mathbb{R}} : L_r(\mathbb{R}) \to L_r(\mathbb{R})\|^r \int_{-\infty}^{+\infty} |f(t,\xi)|^r \, dt.$$

Applying Fubini's theorem, integration over $\xi \in M$ yields

$$\int_{-\infty}^{+\infty}\int_M |\widetilde{f^{\mathbb{R}}}(s,\xi)|^r ds\, d\mu(\xi) \le \|H^{\mathbb{R}} : L_r(\mathbb{R}) \to L_r(\mathbb{R})\|^r \int_{-\infty}^{+\infty}\int_M |f(t,\xi)|^r dt\, d\mu(\xi).$$

This proves that

$$\|L_r|\mathfrak{HT}_r^{\mathbb{R}}\| \le \|H^{\mathbb{R}} : L_r(\mathbb{R}) \to L_r(\mathbb{R})\|.$$

The reverse inequality is obvious.

5.4.21 EXAMPLE. $\|l_1^N|\mathfrak{HT}\| = \|l_\infty^N|\mathfrak{HT}\| \asymp 1 + \log N.$

PROOF. We know from 5.4.4 and 5.4.20 that

$$\|l_q^N|\mathfrak{HT}_q^{\mathbb{R}}\| = \|H^{\mathbb{R}} : L_q(\mathbb{R}) \to L_q(\mathbb{R})\| = \cot \tfrac{\pi}{2q} \le c_0 q,$$

where $c_0 > 0$ is an absolute constant and $2 \le q < \infty$. So it follows from 5.4.10 (Remark) that $\|l_q^N|\mathfrak{HT}_2^{\mathbb{R}}\| \prec q$, and 1.3.6 yields $\|l_\infty^N|\mathfrak{HT}\| \prec 1+\log N$. The lower estimate is implied by 2.4.10 and 5.4.18,

$$1+\log N \asymp \boldsymbol{\kappa}(l_\infty^N|H_N) \le \|l_\infty^N|\mathfrak{HT}_2^{\mathbb{R}}\|.$$

5.4.22 Summarizing the previous facts, we get a new version of 2.4.12.

EXAMPLE. $L_r \in \mathsf{HT}$ *if* $1 < r < \infty$, *and* $L_1, L_\infty \notin \mathsf{HT}$.

5.4.23 Finally, we generalize 2.4.10.

EXAMPLE.

$$\boldsymbol{\kappa}(l_1^N|H_n) = \boldsymbol{\kappa}(l_\infty^N|H_n) \asymp \min\{1+\log n, 1+\log N\}.$$

PROOF. By 2.4.3,

$$\boldsymbol{\kappa}(l_\infty^N|H_n) \le 1+\log n.$$

On the other hand, we conclude from 5.4.21 that

$$\boldsymbol{\kappa}(l_\infty^N|H_n) \le \|l_\infty^N|\mathfrak{HT}_2^{\mathbb{R}}\| \asymp 1 + \log N.$$

This proves the upper estimate. The lower one follows from 2.4.10.

5.5 Fourier type and cotype ideal norms

5.5.1 In this section, we treat the **Fourier type ideal norms** $\varrho(\mathcal{E}_n, \mathcal{I}_n)$ and the **Fourier cotype ideal norms** $\varrho(\mathcal{I}_n, \mathcal{E}_n)$. Recall from 3.7.3 that

$$\delta(\mathcal{E}_n, \mathcal{I}_n) = \varrho(\mathcal{E}_n, \mathcal{I}_n) \quad \text{and} \quad \delta(\mathcal{I}_n, \mathcal{E}_n) = \varrho'(\mathcal{E}_n, \mathcal{I}_n).$$

5.5.2 The following question is related to 3.7.6.

PROBLEM. *Is the Fourier cotype ideal norm $\varrho(\mathcal{I}_n, \mathcal{E}_n)$ non-surjective?*

REMARK. Examples 5.5.20 show that none of the ideal norms $\varrho(\mathcal{E}_n, \mathcal{I}_n)$ and $\varrho(\mathcal{I}_n, \mathcal{E}_n)$ is symmetric.

5.5.3 PROPOSITION.

$$\varrho(\mathcal{E}_{2n}, \mathcal{I}_{2n}) \le \sqrt{2}\, \varrho(\mathcal{E}_n, \mathcal{I}_n) \quad \textit{and} \quad \varrho(\mathcal{I}_{2n}, \mathcal{E}_{2n}) \le \sqrt{2}\, \varrho(\mathcal{I}_n, \mathcal{E}_n).$$

PROOF. We have

$$\|(Tx_k)|\mathcal{E}_{2n}\| = \left(\frac{1}{2\pi}\int_{-\pi}^{+\pi} \Big\| \sum_{k=1}^{2n} Tx_k e_k(t) \Big\|^2 dt\right)^{1/2}$$

$$\le \left(\frac{1}{2\pi}\int_{-\pi}^{+\pi} \Big\| \sum_{k=1}^{n} Tx_k e_k(t) \Big\|^2 dt\right)^{1/2} + \left(\frac{1}{2\pi}\int_{-\pi}^{+\pi} \Big\| \sum_{k=n+1}^{2n} Tx_k e_k(t) \Big\|^2 dt\right)^{1/2}$$

$$= \left(\frac{1}{2\pi}\int_{-\pi}^{+\pi} \Big\| \sum_{k=1}^{n} Tx_k e_k(t) \Big\|^2 dt\right)^{1/2} + \left(\frac{1}{2\pi}\int_{-\pi}^{+\pi} \Big\| \sum_{k=1}^{n} Tx_{k+n} e_k(t) \Big\|^2 dt\right)^{1/2}$$

$$\le \varrho(T|\mathcal{E}_n, \mathcal{I}_n) \left[\Big(\sum_{k=1}^{n} \|x_k\|^2 \Big)^{1/2} + \Big(\sum_{k=1}^{n} \|x_{k+n}\|^2 \Big)^{1/2} \right]$$

$$\le \sqrt{2}\varrho(T|\mathcal{E}_n, \mathcal{I}_n) \|(x_k)|l_2^{2n}\|.$$

This proves the first inequality.

To see the second inequality, note that the substitution $s = t + \pi$ yields

$$\left(\frac{1}{2\pi}\int_{-\pi}^{+\pi} \Big\| \sum_{k=1}^{2n} (-1)^k x_k e_k(t) \Big\|^2 dt\right)^{1/2} = \left(\frac{1}{2\pi}\int_{-\pi}^{+\pi} \Big\| \sum_{k=1}^{2n} x_k e_k(s) \Big\|^2 ds\right)^{1/2}$$
$$= \|(x_k)|\mathcal{E}_{2n}\|.$$

Thus, with $2s = t$, we obtain

$$\|(x_{2k})|\mathcal{E}_n\| = \Big(\frac{1}{2\pi}\int_{-\pi}^{+\pi}\Big\|\sum_{k=1}^{n} x_{2k}e_k(t)\Big\|^2 dt\Big)^{1/2} = \Big(\frac{1}{2\pi}\int_{-\pi}^{+\pi}\Big\|\sum_{k=1}^{n} x_{2k}e_{2k}(s)\Big\|^2 ds\Big)^{1/2}$$

$$\leq \frac{1}{2}\left\{\begin{array}{c}\Big(\frac{1}{2\pi}\int_{-\pi}^{+\pi}\Big\|\sum_{k=1}^{2n}(-1)^k x_k e_k(s)\Big\|^2 ds\Big)^{1/2} \\ + \\ \Big(\frac{1}{2\pi}\int_{-\pi}^{+\pi}\Big\|\sum_{k=1}^{2n} x_k e_k(s)\Big\|^2 ds\Big)^{1/2}\end{array}\right\} = \|(x_k)|\mathcal{E}_{2n}\|.$$

The same estimate can be shown for $\|(x_{2k-1})|\mathcal{E}_n\|$. Hence

$$\begin{aligned}\|(Tx_k)|\mathcal{I}_{2n}\| &= \Big(\sum_{k=1}^{n}\|Tx_{2k}\|^2 + \sum_{k=1}^{n}\|Tx_{2k-1}\|^2\Big)^{1/2} \\ &\leq \varrho(T|\mathcal{I}_n,\mathcal{E}_n)\Big(\|(x_{2k})|\mathcal{E}_n\|^2 + \|(x_{2k-1})|\mathcal{E}_n\|^2\Big)^{1/2} \\ &\leq \sqrt{2}\,\varrho(T|\mathcal{I}_n,\mathcal{E}_n)\,\|(x_k)|\mathcal{E}_{2n}\|.\end{aligned}$$

5.5.4 LEMMA. $\delta(\mathcal{I}_{2m+1},\mathcal{E}_{2m+1}) \leq 3\,\varrho(\mathcal{I}_{4m-1},\mathcal{E}_{4m-1})$.

PROOF. Letting

$$V_m^0 := V_m e_{2m} \quad\text{and}\quad \boldsymbol{f}^0 := \boldsymbol{f} e_{m-1},$$

the Fourier coefficients of the m-th de la Vallée–Poussin kernel V_m and of the function $\boldsymbol{f} \in [L_2[-\pi,+\pi), X]$ are shifted to the right by $2m$ steps and $(m-1)$ steps, respectively. Since the spectrum of $V_m^0 * \boldsymbol{f}^0$ is contained in $\{1,\dots,4m-1\}$, we have

$$V_m^0 * \boldsymbol{f}^0 = \sum_{k=1}^{4m-1}\langle V_m^0 * \boldsymbol{f}^0, \overline{e_k}\rangle\, e_k.$$

If $h = 1,\dots,2m+1$ and $k = m,\dots,3m$ are related by $h = k - m + 1$, then

$$\langle V_m^0 * \boldsymbol{f}^0, \overline{e_k}\rangle = \langle V_m^0, \overline{e_k}\rangle\,\langle \boldsymbol{f}^0, \overline{e_k}\rangle = \langle V_m, \overline{e_{k-2m}}\rangle\,\langle \boldsymbol{f}, \overline{e_{k-m+1}}\rangle = \langle \boldsymbol{f}, \overline{e_h}\rangle.$$

It follows from Young's inequality 4.11.8 and $\|V_m^0|L_1\| \leq 3$ that

$$\begin{aligned}\|\big(\langle V_m^0 * \boldsymbol{f}^0, \overline{e_k}\rangle\big)|\,\mathcal{E}_{4m-1}\| &= \|V_m^0 * \boldsymbol{f}^0|\,L_2\| \leq \|V_m^0|\,L_1\|\,\|\boldsymbol{f}^0|\,L_2\| \\ &\leq 3\,\|\boldsymbol{f}|\,L_2\|.\end{aligned}$$

Hence

$$\|(T\langle \boldsymbol{f}, \overline{e_h}\rangle)|\mathfrak{I}_{2m+1}\| = \Big(\sum_{h=1}^{2m+1} \|T\langle \boldsymbol{f}, \overline{e_h}\rangle\|^2\Big)^{1/2}$$

$$= \Big(\sum_{k=m}^{3m} \big\|T\langle V_m^0 * \boldsymbol{f}^0, \overline{e_k}\rangle\big\|^2\Big)^{1/2} \le \Big(\sum_{k=1}^{4m-1} \big\|T\langle V_m^0 * \boldsymbol{f}^0, \overline{e_k}\rangle\big\|^2\Big)^{1/2}$$

$$= \big\|\big(T\langle V_m^0 * \boldsymbol{f}^0, \overline{e_k}\rangle\big)\big|\mathfrak{I}_{4m-1}\big\|$$

$$\le \boldsymbol{\varrho}(T|\mathfrak{I}_{4m-1}, \mathcal{E}_{4m-1})\,\big\|\big(\langle V_m^0 * \boldsymbol{f}^0, \overline{e_k}\rangle\big)\big|\mathcal{E}_{4m-1}\big\|$$

$$\le 3\,\boldsymbol{\varrho}(T|\mathfrak{I}_{4m-1}, \mathcal{E}_{4m-1})\,\|\boldsymbol{f}|L_2\|.$$

5.5.5 PROPOSITION. $\boldsymbol{\delta}(\mathfrak{I}_n, \mathcal{E}_n) \le 5\,\boldsymbol{\varrho}(\mathfrak{I}_n, \mathcal{E}_n)$.

PROOF. We may assume that $n \ge 2$. Let $m := [\frac{n}{2}]$. Then $n \le 2m+1$ and $4m-1 \le 2n$. Using the monotonicity of $\boldsymbol{\varrho}(\mathfrak{I}_n, \mathcal{E}_n)$ and $\boldsymbol{\delta}(\mathfrak{I}_n, \mathcal{E}_n)$, we conclude from 5.5.3 and 5.5.4 that

$$\boldsymbol{\delta}(\mathfrak{I}_n, \mathcal{E}_n) \le \boldsymbol{\delta}(\mathfrak{I}_{2m+1}, \mathcal{E}_{2m+1}) \le 3\,\boldsymbol{\varrho}(\mathfrak{I}_{4m-1}, \mathcal{E}_{4m-1})$$
$$\le 3\,\boldsymbol{\varrho}(\mathfrak{I}_{2n}, \mathcal{E}_{2n}) \le 3\sqrt{2}\,\boldsymbol{\varrho}(\mathfrak{I}_n, \mathcal{E}_n).$$

5.5.6 Next, we deal with the discrete counterparts of the ideal norms considered above, namely

$$\boldsymbol{\varrho}(\mathcal{E}_n^\circ, \mathfrak{I}_n), \quad \boldsymbol{\delta}(\mathcal{E}_n^\circ, \mathfrak{I}_n), \quad \boldsymbol{\varrho}(\mathfrak{I}_n, \mathcal{E}_n^\circ), \quad \boldsymbol{\delta}(\mathfrak{I}_n, \mathcal{E}_n^\circ).$$

REMARK. Of course, $\boldsymbol{\varrho}(\mathcal{E}_n^\circ, \mathfrak{I}_n)$ is nothing but the **Kwapień ideal norm** $\boldsymbol{\kappa}(E_n^\circ)$ associated with the Fourier matrix

$$E_n^\circ = \big(\varepsilon_{hk}^{(n)}\big) := \Big(n^{-1/2} \exp\big(\tfrac{2\pi i}{n} hk\big)\Big).$$

5.5.7 We know from 3.8.13 that all of these ideal norms coincide.

PROPOSITION.

$$\boldsymbol{\varrho}(\mathcal{E}_n^\circ, \mathfrak{I}_n) = \boldsymbol{\delta}(\mathcal{E}_n^\circ, \mathfrak{I}_n) = \boldsymbol{\varrho}(\mathfrak{I}_n, \mathcal{E}_n^\circ) = \boldsymbol{\delta}(\mathfrak{I}_n, \mathcal{E}_n^\circ).$$

5.5.8 Probably, the answer to the following question is negative.

PROBLEM. *Is the sequence of the ideal norms $\boldsymbol{\varrho}(\mathcal{E}_n^\circ, \mathfrak{I}_n)$ non-decreasing?*

5.5.9 Next, we state a special case of 3.8.14.

PROPOSITION. *The ideal norm $\boldsymbol{\varrho}(\mathcal{E}_n^\circ, \mathfrak{I}_n)$ is symmetric.*

5.5.10 The following result shows another advantage of the discrete Fourier system; see Problem 5.5.2.

PROPOSITION. *The ideal norm $\boldsymbol{\varrho}(\mathcal{E}_n^\circ, \mathfrak{I}_n)$ is injective and surjective.*

5.5.11 The next result is basic for later application in 5.6.27.

PROPOSITION. *The sequence of the ideal norms $\boldsymbol{\varrho}(\mathcal{E}_n^\circ, \mathcal{I}_n)$ is submultiplicative.*

PROOF. In the defining inequality of $\boldsymbol{\varrho}(\mathcal{E}_n^\circ, \mathcal{I}_n)$ we sum over the index set $\{1, \dots, n\}$. Replacing h and k by $n-h$ and $n-k$, respectively, the domain of summation passes into $\{0, \dots, n-1\}$ which is more comfortable for our purpose. Letting $s := pn+h$ and $t := mk+q$ for $p, q = 0, \dots, m-1$ and $h, k = 0, \dots, n-1$, we define bijections between

$$\{0, \dots, m-1\} \times \{0, \dots, n-1\} \quad \text{and} \quad \{0, \dots, mn-1\}.$$

Thus every mn-tuple of elements $x_t \in X$ with $t = 0, \dots, mn-1$ can be indexed by pairs (q, k). It follows from

$$\exp\left(\tfrac{2\pi i}{mn} st\right) = \exp(2\pi i pk) \exp\left(\tfrac{2\pi i}{mn} hq\right) \exp\left(\tfrac{2\pi i}{n} hk\right) \exp\left(\tfrac{2\pi i}{m} pq\right)$$

that

$$\varepsilon_{st}^{(mn)} = \exp\left(\tfrac{2\pi i}{mn} hq\right) \varepsilon_{hk}^{(n)} \varepsilon_{pq}^{(m)}.$$

Hence, for $T \in \mathfrak{L}(X, Y)$ and $S \in \mathfrak{L}(Y, Z)$, we obtain

$$\begin{aligned}
&\sum_{s=0}^{mn-1} \Big\| \sum_{t=0}^{mn-1} S T x_t \varepsilon_{st}^{(mn)} \Big\|^2 = \\
&= \sum_{h=0}^{n-1} \sum_{p=0}^{m-1} \Big\| \sum_{q=0}^{m-1} S \Big[\exp\left(\tfrac{2\pi i}{mn} hq\right) \sum_{k=0}^{n-1} T x_{qk} \varepsilon_{hk}^{(n)} \Big] \varepsilon_{pq}^{(m)} \Big\|^2 \\
&\le \sum_{h=0}^{n-1} \boldsymbol{\varrho}(S|\mathcal{E}_m^\circ, \mathcal{I}_m)^2 \sum_{q=0}^{m-1} \Big\| \exp\left(\tfrac{2\pi i}{mn} hq\right) \sum_{k=0}^{n-1} T x_{qk} \varepsilon_{hk}^{(n)} \Big\|^2 \\
&= \boldsymbol{\varrho}(S|\mathcal{E}_m^\circ, \mathcal{I}_m)^2 \sum_{q=0}^{m-1} \sum_{h=0}^{n-1} \Big\| \sum_{k=0}^{n-1} T x_{qk} \varepsilon_{hk}^{(n)} \Big\|^2 \\
&\le \boldsymbol{\varrho}(S|\mathcal{E}_m^\circ, \mathcal{I}_m)^2 \sum_{q=0}^{m-1} \boldsymbol{\varrho}(T|\mathcal{E}_n^\circ, \mathcal{I}_n)^2 \sum_{k=0}^{n-1} \|x_{qk}\|^2 \\
&= \boldsymbol{\varrho}(S|\mathcal{E}_m^\circ, \mathcal{I}_m)^2 \boldsymbol{\varrho}(T|\mathcal{E}_n^\circ, \mathcal{I}_n)^2 \sum_{t=0}^{mn-1} \|x_t\|^2.
\end{aligned}$$

This completes the proof.

5.5.12 PROPOSITION. $\boldsymbol{\varrho}(\mathcal{E}_n^\circ, \mathcal{I}_n) \le 3\, \boldsymbol{\varrho}(\mathcal{E}_n, \mathcal{I}_n)$.

PROOF. We know from 5.2.5 that $\boldsymbol{\varrho}(\mathcal{E}_n^\circ, \mathcal{E}_n) \le 3$. Thus

$$\boldsymbol{\varrho}(\mathcal{E}_n^\circ, \mathcal{I}_n) \le \boldsymbol{\varrho}(\mathcal{E}_n^\circ, \mathcal{E}_n) \circ \boldsymbol{\varrho}(\mathcal{E}_n, \mathcal{I}_n) \le 3\, \boldsymbol{\varrho}(\mathcal{E}_n, \mathcal{I}_n).$$

5.5.13 LEMMA. $\|(x_k)|\mathcal{E}_n \otimes \mathcal{E}_n^\circ\| = \|(x_k)|\mathcal{E}_n\|$.

PROOF. The translation invariance of the Lebesgue measure yields

$$\begin{aligned}\|(x_k)|\mathcal{E}_n \otimes \mathcal{E}_n^\circ\|^2 &= \frac{1}{n}\sum_{h=1}^{n}\frac{1}{2\pi}\int_{-\pi}^{+\pi}\Big\|\sum_{k=1}^{n} x_k \exp(ikt)\exp(\tfrac{2\pi i}{n}hk)\Big\|^2 dt \\ &= \frac{1}{n}\sum_{h=1}^{n}\frac{1}{2\pi}\int_{-\pi}^{+\pi}\Big\|\sum_{k=1}^{n} x_k \exp\left(ik(t+\tfrac{2\pi}{n}h)\right)\Big\|^2 dt \\ &= \frac{1}{2\pi}\int_{-\pi}^{+\pi}\Big\|\sum_{k=1}^{n} x_k \exp(ikt)\Big\|^2 dt = \|(x_k)|\mathcal{E}_n\|^2.\end{aligned}$$

5.5.14 Next, we state a counterpart of 5.5.12.

PROPOSITION. $\boldsymbol{\varrho}(\mathcal{E}_n, \mathcal{I}_n) \le \boldsymbol{\varrho}(\mathcal{E}_n^\circ, \mathcal{I}_n)$.

PROOF. The preceding lemma implies that $\boldsymbol{\varrho}(\mathcal{E}_n, \mathcal{I}_n) = \boldsymbol{\varrho}(\mathcal{E}_n \otimes \mathcal{E}_n^\circ, \mathcal{I}_n)$. Therefore the assertion follows from 3.7.8.

REMARK. Since $\mathbb{E}_n$ is a subgroup of $\mathbb{T}$, we could also apply 3.8.10.

5.5.15 We are now able to establish the main result of this section.

THEOREM. *The sequences of the following ideal norms are uniformly equivalent:*

$$\boldsymbol{\varrho}(\mathcal{E}_n, \mathcal{I}_n),\ \boldsymbol{\delta}(\mathcal{E}_n, \mathcal{I}_n),\ \boldsymbol{\varrho}(\mathcal{I}_n, \mathcal{E}_n),\ \boldsymbol{\delta}(\mathcal{I}_n, \mathcal{E}_n),$$
$$\boldsymbol{\varrho}(\mathcal{E}_n^\circ, \mathcal{I}_n),\ \boldsymbol{\delta}(\mathcal{E}_n^\circ, \mathcal{I}_n),\ \boldsymbol{\varrho}(\mathcal{I}_n, \mathcal{E}_n^\circ),\ \boldsymbol{\delta}(\mathcal{I}_n, \mathcal{E}_n^\circ).$$

PROOF. We know that

(1) $\boldsymbol{\varrho}(\mathcal{E}_n, \mathcal{I}_n) = \boldsymbol{\delta}(\mathcal{E}_n, \mathcal{I}_n)$ (by 3.7.3),

(2) $\boldsymbol{\varrho}(\mathcal{I}_n, \mathcal{E}_n) \asymp \boldsymbol{\delta}(\mathcal{I}_n, \mathcal{E}_n)$ (by 3.4.10 and 5.5.5),

(3) $\boldsymbol{\varrho}(\mathcal{E}_n, \mathcal{I}_n) \asymp \boldsymbol{\varrho}(\mathcal{E}_n^\circ, \mathcal{I}_n)$ (by 5.5.12 and 5.5.14),

(4) $\boldsymbol{\varrho}(\mathcal{E}_n^\circ, \mathcal{I}_n) = \boldsymbol{\delta}(\mathcal{E}_n^\circ, \mathcal{I}_n) = \boldsymbol{\varrho}(\mathcal{I}_n, \mathcal{E}_n^\circ) = \boldsymbol{\delta}(\mathcal{I}_n, \mathcal{E}_n^\circ)$ (by 5.5.7).

In view of (1) and (4), formula (3) yields

$$\boldsymbol{\delta}(\mathcal{E}_n, \mathcal{I}_n) \asymp \boldsymbol{\delta}(\mathcal{E}_n^\circ, \mathcal{I}_n).$$

By duality, we obtain

$$\boldsymbol{\delta}(\mathcal{I}_n, \mathcal{E}_n) \asymp \boldsymbol{\delta}(\mathcal{I}_n, \mathcal{E}_n^\circ).$$

5.5.16 The following *supertheorem* is an analogue of 5.2.32.

THEOREM. *The sequences of the following ideal norms are uniformly equivalent:*

$$\begin{gathered}\varrho(\mathcal{E}_n, \mathcal{I}_n),\ \delta(\mathcal{E}_n, \mathcal{I}_n),\ \varrho(\mathcal{I}_n, \mathcal{E}_n),\ \delta(\mathcal{I}_n, \mathcal{E}_n),\\ \varrho(\mathcal{E}_n^\circ, \mathcal{I}_n),\ \delta(\mathcal{E}_n^\circ, \mathcal{I}_n),\ \varrho(\mathcal{I}_n, \mathcal{E}_n^\circ),\ \delta(\mathcal{I}_n, \mathcal{E}_n^\circ),\\ \varrho(\mathcal{C}_n, \mathcal{I}_n),\ \delta(\mathcal{C}_n, \mathcal{I}_n),\ \varrho(\mathcal{I}_n, \mathcal{C}_n),\ \delta(\mathcal{I}_n, \mathcal{C}_n),\\ \varrho(\mathcal{C}_n^\circ, \mathcal{I}_n),\ \delta(\mathcal{C}_n^\circ, \mathcal{I}_n),\ \varrho(\mathcal{I}_n, \mathcal{C}_n^\circ),\ \delta(\mathcal{I}_n, \mathcal{C}_n^\circ),\\ \varrho(\mathcal{S}_n, \mathcal{I}_n),\ \delta(\mathcal{S}_n, \mathcal{I}_n),\ \varrho(\mathcal{I}_n, \mathcal{S}_n),\ \delta(\mathcal{I}_n, \mathcal{S}_n),\\ \varrho(\mathcal{S}_n^\circ, \mathcal{I}_n),\ \delta(\mathcal{S}_n^\circ, \mathcal{I}_n),\ \varrho(\mathcal{I}_n, \mathcal{S}_n^\circ),\ \delta(\mathcal{I}_n, \mathcal{S}_n^\circ).\end{gathered}$$

PROOF. We know from 5.2.10 that $\varrho(\mathcal{C}_n, \mathcal{E}_n) \le \sqrt{2}$. Hence

$$\varrho(\mathcal{I}_n, \mathcal{E}_n) \le \varrho(\mathcal{I}_n, \mathcal{C}_n) \circ \varrho(\mathcal{C}_n, \mathcal{E}_n) \le \sqrt{2}\,\varrho(\mathcal{I}_n, \mathcal{C}_n)$$

and

$$\varrho(\mathcal{C}_n, \mathcal{I}_n) \le \varrho(\mathcal{C}_n, \mathcal{E}_n) \circ \varrho(\mathcal{E}_n, \mathcal{I}_n) \le \sqrt{2}\,\varrho(\mathcal{E}_n, \mathcal{I}_n).$$

This implies that

$$\begin{aligned}\varrho(\mathcal{I}_n, \mathcal{E}_n) &\le \sqrt{2}\,\varrho(\mathcal{I}_n, \mathcal{C}_n) \le \sqrt{2}\,\delta(\mathcal{I}_n, \mathcal{C}_n) = \sqrt{2}\,\delta'(\mathcal{C}_n, \mathcal{I}_n)\\ &= \sqrt{2}\,\varrho'(\mathcal{C}_n, \mathcal{I}_n) \le 2\,\varrho'(\mathcal{E}_n, \mathcal{I}_n).\end{aligned}$$

Moreover, by 5.2.12, we have $\varrho(\mathcal{C}_n^\circ, \mathcal{C}_n) \le 3$. Hence

$$\varrho(\mathcal{I}_n, \mathcal{C}_n) \le \varrho(\mathcal{I}_n, \mathcal{C}_n^\circ) \circ \varrho(\mathcal{C}_n^\circ, \mathcal{C}_n) \le 3\,\varrho(\mathcal{I}_n, \mathcal{C}_n^\circ)$$

and

$$\varrho(\mathcal{C}_n^\circ, \mathcal{I}_n) \le \varrho(\mathcal{C}_n^\circ, \mathcal{C}_n) \circ \varrho(\mathcal{C}_n, \mathcal{I}_n) \le 3\,\varrho(\mathcal{C}_n, \mathcal{I}_n).$$

Thus

$$\begin{aligned}\varrho(\mathcal{I}_n, \mathcal{E}_n) \le&\ \sqrt{2}\,\varrho(\mathcal{I}_n, \mathcal{C}_n) \ \le 3\sqrt{2}\,\varrho(\mathcal{I}_n, \mathcal{C}_n^\circ) \ \le 3\sqrt{2}\,\delta(\mathcal{I}_n, \mathcal{C}_n^\circ)\\ &= 3\sqrt{2}\,\delta'(\mathcal{C}_n^\circ, \mathcal{I}_n) = 3\sqrt{2}\,\varrho'(\mathcal{C}_n^\circ, \mathcal{I}_n) \le 9\sqrt{2}\,\varrho'(\mathcal{C}_n, \mathcal{I}_n)\\ &\le 18\,\varrho'(\mathcal{E}_n, \mathcal{I}_n).\end{aligned}$$

Since it follows from 5.5.9 and 5.5.15 that $\varrho(\mathcal{I}_n, \mathcal{E}_n) \asymp \varrho'(\mathcal{E}_n, \mathcal{I}_n)$, all sequences of type and cotype ideal norms associated with $\mathcal{C}_n$ and $\mathcal{C}_n^\circ$ are uniformly equivalent to $(\varrho(\mathcal{I}_n, \mathcal{E}_n))$.

The same holds for the sine systems. In that case, however, we additionally have

$$\varrho(\mathcal{S}_n^\circ, \mathcal{I}_n) = \delta(\mathcal{S}_n^\circ, \mathcal{I}_n) = \varrho(\mathcal{I}_n, \mathcal{S}_n^\circ) = \delta(\mathcal{I}_n, \mathcal{S}_n^\circ).$$

This can be seen as in 3.8.13.

5.5.17 The next result follows immediately from 3.7.11 and 5.5.9.

EXAMPLE.

$$\varrho(l_r^n|\mathcal{E}_n,\mathcal{T}_n) \asymp \varrho(L_r|\mathcal{E}_n,\mathcal{T}_n) \asymp n^{|1/r-1/2|} \quad \textit{if } 1 \le r \le \infty.$$

5.5.18 Without proof, we state a result which extends the previous one.

EXAMPLE.

$$\varrho(l_r^N|\mathcal{E}_n,\mathcal{T}_n) \asymp \min(n^{|1/r-1/2|}, N^{|1/r-1/2|}) \quad \textit{if } 1 \le r \le \infty.$$

5.5.19 In order to illustrate the difference between the ideal norms under consideration, we add some further examples.

EXAMPLE 1. $\quad \varrho(l_\infty^2|\mathcal{E}_2,\mathcal{T}_2) = \left(1+\frac{2}{\pi}\right)^{1/2} = 1.2793\ldots.$

PROOF. We have to find the least constant $c \ge 1$ such that

$$\frac{1}{2\pi}\int_{-\pi}^{+\pi} \|x\exp(it)+y\exp(2it)\|^2\,dt \le c^2\Big[\|x\|^2+\|y\|^2\Big]$$

for all $x, y \in l_\infty^2$. Letting $x=(1,+1)$ and $y=(1,-1)$ yields

$$\frac{1}{2\pi}\int_{-\pi}^{+\pi} \max\Big[\big|\exp(-\tfrac{1}{2}it)+\exp(+\tfrac{1}{2}it)\big|, \big|\exp(-\tfrac{1}{2}it)-\exp(+\tfrac{1}{2}it)\big|\Big]^2 dt \le 2c^2.$$

Hence

$$\frac{1}{2\pi}\int_{-\pi}^{+\pi} \max\Big[\big|\cos\tfrac{1}{2}t\big|, \big|\sin\tfrac{1}{2}t\big|\Big]^2 dt \le \tfrac{1}{2}c^2.$$

Since the left-hand integral has the value $\frac{1}{2}+\frac{1}{\pi}$, we see that

$$c \ge \left(1+\tfrac{2}{\pi}\right)^{1/2}.$$

The upper estimate can be obtained by an extreme point argument and voluminous computations.

EXAMPLE 2. $\quad \varrho(l_1^2|\mathcal{T}_2,\mathcal{E}_2) = \left(\frac{1}{2}+\frac{1}{\pi}\right)^{-1/2} = 1.1054\ldots.$

PROOF. We have to find the least constant $c \ge 1$ such that

$$\|x\|^2+\|y\|^2 \le c^2\frac{1}{2\pi}\int_{-\pi}^{+\pi} \|x\exp(it)+y\exp(2it)\|^2\,dt$$

for all $x, y \in l_1^2$. Letting $x=(1,+1)$ and $y=(1,-1)$ yields

$$8 \le c^2\frac{1}{2\pi}\int_{-\pi}^{+\pi} \Big(\big|\exp(-\tfrac{1}{2}it)+\exp(+\tfrac{1}{2}it)\big| + \big|\exp(-\tfrac{1}{2}it)-\exp(+\tfrac{1}{2}it)\big|\Big)^2\,dt.$$

Hence

$$2c^{-2} \le \frac{1}{2\pi}\int_{-\pi}^{+\pi}\Big(|\cos\tfrac{1}{2}t| + |\sin\tfrac{1}{2}t|\Big)^2\,dt = \frac{1}{2\pi}\int_{-\pi}^{+\pi}\Big(1+|\sin t|\Big)\,dt = 1+\frac{2}{\pi},$$

which implies that

$$c \ge \left(\tfrac{1}{2}+\tfrac{1}{\pi}\right)^{-1/2}.$$

In order to get the upper estimate, one has to solve an optimization problem with 8 real parameters. We omit the horrible details.

EXAMPLE 3. $\varrho(l_\infty^2|\mathcal{I}_2,\mathcal{E}_2)=\sqrt{2}=1.4142\ldots$.

PROOF. This is a special case of 3.7.11.

5.5.20 The above examples and their dual counterparts are summarized in the following table:

	$\varrho(\mathcal{E}_2,\mathcal{I}_2)$	$\varrho(\mathcal{I}_2,\mathcal{E}_2)$	$\delta(\mathcal{I}_2,\mathcal{E}_2)$	discrete case
l_1^2	1.4142	1.1054	1.2793	1.4142
l_∞^2	1.2793	1.4142	1.4142	1.4142

Consequently, among the ideal norms

$$\varrho(\mathcal{E}_n,\mathcal{I}_n)=\delta(\mathcal{E}_n,\mathcal{I}_n),\qquad \varrho(\mathcal{I}_n,\mathcal{E}_n),\qquad \delta(\mathcal{I}_n,\mathcal{E}_n),$$
$$\varrho(\mathcal{E}_n^\circ,\mathcal{I}_n)=\delta(\mathcal{E}_n^\circ,\mathcal{I}_n)=\varrho(\mathcal{I}_n,\mathcal{E}_n^\circ)=\delta(\mathcal{I}_n,\mathcal{E}_n^\circ)$$

there are no equations other than those indicated. This means that we have indeed four different objects.

5.6 Operators of Fourier type – Bourgain's theorem –

5.6.1 Let $1\le u,v<\infty$. For $T\in\mathfrak{L}(X,Y)$, we denote by $\varrho_u^{(v)}(T|\mathcal{E}_n,\mathcal{I}_n)$ the least constant $c\ge 0$ such that

$$\left(\frac{1}{2\pi}\int_{-\pi}^{+\pi}\Big\|\sum_{k=1}^{n}Tx_k\exp(ikt)\Big\|^v dt\right)^{1/v}\le c\left(\sum_{k=1}^{n}\|x_k\|^u\right)^{1/u}$$

whenever $x_1,\ldots,x_n\in X$. In the limiting cases $u=\infty$ and $v=\infty$, the usual modifications are required. More concisely, the above inequality reads as follows:

$$\|(Tx_k)|\mathcal{E}_n\|_v\le c\,\|(x_k)|l_u^n\|.$$

We refer to

$$\varrho_u^{(v)}(\mathcal{E}_n, \mathcal{I}_n) : T \longrightarrow \varrho_u^{(v)}(T|\mathcal{E}_n, \mathcal{I}_n)$$

as a **Fourier type ideal norm**. The case $u = v = 2$ has already been treated in the previous section.

Replacing $\| \cdot |l_u^n\|$ by the Lorentz norm $\| \cdot |l_{u,1}^n\|$, we get the **weak Fourier type ideal norm** $\varrho_{u,1}^{(v)}(\mathcal{E}_n, \mathcal{I}_n)$.

5.6.2 Next, a discrete analogue of the above definition is established. For $T \in \mathfrak{L}(X,Y)$, we denote by $\varrho_u^{(v)}(T|\mathcal{E}_n^\circ, \mathcal{I}_n)$ the least constant $c \geq 0$ such that

$$\left(\sum_{h=1}^{n} \left\| \sum_{k=1}^{n} Tx_k \, n^{-1/2} \exp(\tfrac{2\pi i}{n} hk) \right\|^v \right)^{1/v} \leq c \left(\sum_{k=1}^{n} \|x_k\|^u \right)^{1/u}$$

whenever $x_1, \ldots, x_n \in X$.

Recall that the modified system $\mathcal{N}\mathcal{E}_n^\circ$ consists of all characters on the cyclic group $\mathbb{E}_n$. The associated ideal norm $\varrho_u^{(v)}(T|\mathcal{N}\mathcal{E}_n^\circ, \mathcal{I}_n)$ is defined analogously. In this case, we have the inequality

$$\left(\frac{1}{n} \sum_{h=1}^{n} \left\| \sum_{k=1}^{n} Tx_k \exp(\tfrac{2\pi i}{n} hk) \right\|^v \right)^{1/v} \leq c \left(\sum_{k=1}^{n} \|x_k\|^u \right)^{1/u}.$$

These quantities are related by the formula

$$\varrho_u^{(v)}(T|\mathcal{N}\mathcal{E}_n^\circ, \mathcal{I}_n) = n^{1/2-1/v} \varrho_u^{(v)}(T|\mathcal{E}_n^\circ, \mathcal{I}_n).$$

5.6.3 Each of the ideal norms $\varrho_u^{(v)}(\mathcal{E}_n^\circ, \mathcal{I}_n)$ and $\varrho_u^{(v)}(\mathcal{N}\mathcal{E}_n^\circ, \mathcal{I}_n)$ has certain advantages and disadvantages. In the following, it is preferable to use the normalized version.

PROPOSITION. $\varrho_u^{(v)}(\mathcal{E}_n, \mathcal{I}_n) \leq \varrho_u^{(v)}(\mathcal{N}\mathcal{E}_n^\circ, \mathcal{I}_n) \leq 3\, \varrho_u^{(v)}(\mathcal{E}_n, \mathcal{I}_n)$.

PROOF. By definition,

$$\frac{1}{n} \sum_{h=1}^{n} \left\| \sum_{k=1}^{n} Tx_k \exp(\tfrac{2\pi i}{n} hk) \right\|^v \leq \varrho_u^{(v)}(T|\mathcal{N}\mathcal{E}_n^\circ, \mathcal{I}_n)^v \left(\sum_{k=1}^{n} \|x_k\|^u \right)^{v/u}.$$

Substituting $\exp(ikt)x_k$ for x_k, we obtain

$$\frac{1}{n} \sum_{h=1}^{n} \left\| \sum_{k=1}^{n} Tx_k \exp(ik(t + \tfrac{2\pi}{n} h)) \right\|^v \leq \varrho_u^{(v)}(T|\mathcal{N}\mathcal{E}_n^\circ, \mathcal{I}_n)^v \left(\sum_{k=1}^{n} \|x_k\|^u \right)^{v/u}.$$

Using the translation invariance of the Lebesgue measure, integration over t yields

$$\frac{1}{2\pi} \int_{-\pi}^{+\pi} \left\| \sum_{k=1}^{n} Tx_k \exp(ikt) \right\|^v dt \leq \varrho_u^{(v)}(T|\mathcal{N}\mathcal{E}_n^\circ, \mathcal{I}_n)^v \left(\sum_{k=1}^{n} \|x_k\|^u \right)^{v/u}.$$

This proves the left-hand inequality. The right-hand inequality easily follows from 5.2.4.

5.6.4 As regards duality, the ideal norm $\varrho_u^{(v)}(T|\mathcal{E}_n^\circ, \mathcal{I}_n)$ is more comfortable. The proof can be adapted from 2.3.6.

PROPOSITION. $\varrho_u^{(v)\prime}(\mathcal{E}_n^\circ, \mathcal{I}_n) = \varrho_{v'}^{(u')}(\mathcal{E}_n^\circ, \mathcal{I}_n)$.

5.6.5 The previous result can be transformed as follows.

PROPOSITION. *The sequences of the ideal norms*

$$\varrho_u^{(v)\prime}(\mathcal{E}_n, \mathcal{I}_n) \quad \textit{and} \quad n^{1-1/u-1/v}\, \varrho_{v'}^{(u')}(\mathcal{E}_n, \mathcal{I}_n)$$

are uniformly equivalent. This holds, in particular, for

$$\varrho_p^{(p')\prime}(\mathcal{E}_n, \mathcal{I}_n) \quad \textit{and} \quad \varrho_p^{(p')}(\mathcal{E}_n, \mathcal{I}_n).$$

PROOF. In view of 5.6.2 and 5.6.3, we have

$$\varrho_u^{(v)}(\mathcal{E}_n, \mathcal{I}_n) \asymp \varrho_u^{(v)}(\mathcal{N}\mathcal{E}_n^\circ, \mathcal{I}_n) = n^{1/2-1/v}\, \varrho_u^{(v)}(\mathcal{E}_n^\circ, \mathcal{I}_n)$$

and

$$\varrho_{v'}^{(u')}(\mathcal{E}_n, \mathcal{I}_n) \asymp \varrho_{v'}^{(u')}(\mathcal{N}\mathcal{E}_n^\circ, \mathcal{I}_n) = n^{1/u-1/2}\, \varrho_{v'}^{(u')}(\mathcal{E}_n^\circ, \mathcal{I}_n).$$

Now it follows from 5.6.4 that

$$n^{1/v-1/2}\, \varrho_u^{(v)\prime}(\mathcal{E}_n, \mathcal{I}_n) \asymp n^{1/u'-1/2}\, \varrho_{v'}^{(u')}(\mathcal{E}_n, \mathcal{I}_n).$$

5.6.6 Let $1 < p \le 2$. An operator T has **Fourier type** p if

$$\|T|\mathfrak{F}\mathfrak{T}_p\| := \sup_n \varrho_p^{(p')}(T|\mathcal{E}_n, \mathcal{I}_n)$$

is finite. These operators form the Banach ideal

$$\mathfrak{F}\mathfrak{T}_p := \mathfrak{L}[\varrho_p^{(p')}(\mathcal{E}_n, \mathcal{I}_n)].$$

REMARK. Trivially, all operators have Fourier type 1. Therefore we put $\mathfrak{F}\mathfrak{T}_1 := \mathfrak{L}$.

5.6.7 Let $1 < p < 2$. An operator T has **weak Fourier type** p if

$$\|T|\mathfrak{F}\mathfrak{T}_p^{weak}\| := \sup_n n^{1/2-1/p}\, \varrho(T|\mathcal{E}_n, \mathcal{I}_n)$$

is finite. These operators form the Banach ideal

$$\mathfrak{F}\mathfrak{T}_p^{weak} := \mathfrak{L}[n^{1/2-1/p}\, \varrho(\mathcal{E}_n, \mathcal{I}_n)].$$

REMARK. In the previous definition, $\varrho(\mathcal{E}_n, \mathcal{I}_n)$ can be replaced by any ideal norm occurring in Theorem 5.5.16.

5.6.8 PROPOSITION. *The operator ideals* $\mathfrak{ET}_p$ *and* $\mathfrak{ET}_p^{weak}$ *are injective, surjective, symmetric and* l_2*-stable.*

PROOF. Injectivity and l_2-stability are trivial. The symmetry follows from 5.6.5 and the surjectivity from 1.2.7.

REMARK. In the setting of spaces, the above result implies that the classes ET_p and ET_p^{weak} are stable when passing to subspaces, quotients and duals. We also have stability under the formation of finite direct sums and l_2-multiples.

5.6.9 In order to guarantee that the Banach operator ideal

$$\mathfrak{L}\Big[n^{1/u-1/p}\varrho_u^{(v)}(\mathcal{E}_n,\mathcal{J}_n)\Big]$$

contains non-zero operators, the underlying parameters must satisfy a condition which we now elaborate.

LEMMA. *Let* $1<p<\infty$ *and* $1\le u,v\le\infty$. *Then*

$$\Big\|I_{\mathbb{C}}\Big|\mathfrak{L}\Big[n^{1/u-1/p}\varrho_u^{(v)}(\mathcal{E}_n,\mathcal{J}_n)\Big]\Big\| := \sup_n n^{1/u-1/p}\,\varrho_u^{(v)}(I_{\mathbb{C}}|\mathcal{E}_n,\mathcal{J}_n)$$

is finite if and only if $p\le\min\{2,u,v'\}$.

PROOF. We know from 5.6.3 that

$$\varrho_u^{(v)}(I_{\mathbb{C}}|\mathcal{E}_n,\mathcal{J}_n)\asymp\varrho_u^{(v)}(I_{\mathbb{C}}|\mathcal{N}\mathcal{E}_n^\circ,\mathcal{J}_n)=n^{1/2-1/v}\|E_n^\circ:l_u^n\to l_v^n\|.$$

Using the fact that $E_n^\circ=\big(n^{-1/2}\exp\big(\frac{2\pi i}{n}hk\big)\big)$ is a unitary (n,n)-matrix and the Hausdorff–Young inequality 3.5.2, we obtain

$$\|E_n^\circ:l_u^n\to l_v^n\|\asymp n^\alpha \quad\text{and}\quad n^{1/u-1/p}\,\varrho_u^{(v)}(I_{\mathbb{C}}|\mathcal{E}_n,\mathcal{J}_n)\asymp n^\beta,$$

where the exponents α and $\beta=\alpha+1/2+1/u-1/v-1/p$ take the values indicated in the diagrams:

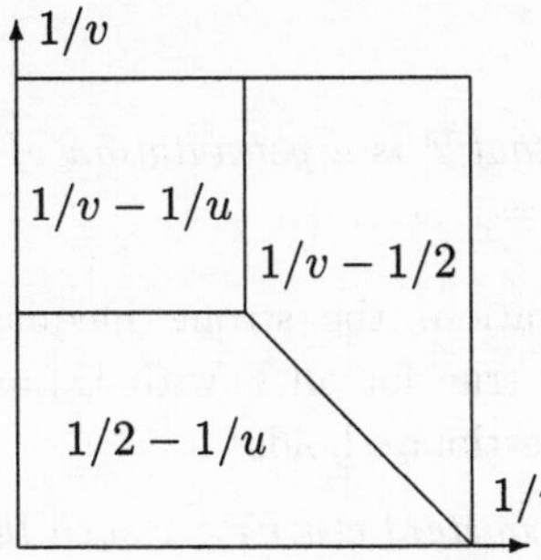

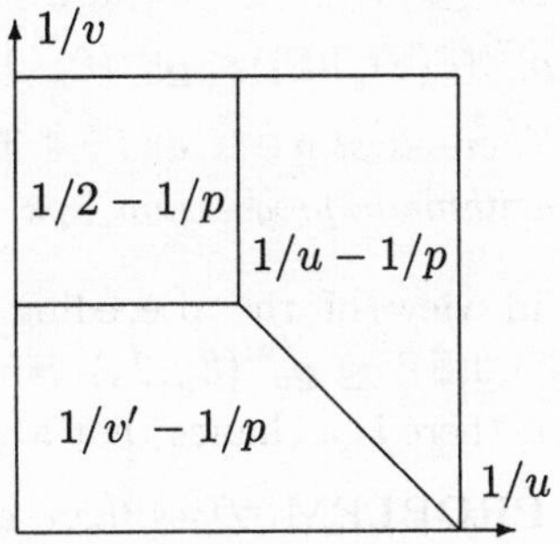

Hence $\beta\le0$ and $p\le\min\{2,u,v'\}$ are equivalent.

5.6.10 Let $1 \le u, v < \infty$, and fix any finite subset $\mathbb{F}$ of $\mathbb{Z}$. Then, for $T \in \mathfrak{L}(X,Y)$, we denote by $\varrho_u^{(v)}(T|\mathcal{E}(\mathbb{F}),\mathcal{I}(\mathbb{F}))$ the least constant $c \ge 0$ such that

$$\left(\frac{1}{2\pi}\int_{-\pi}^{+\pi}\left\|\sum_{k\in\mathbb{F}} Tx_k \exp(ikt)\right\|^v dt\right)^{1/v} \le c\left(\sum_{k\in\mathbb{F}}\|x_k\|^u\right)^{1/u}$$

whenever (x_k) is an $\mathbb{F}$-tuple in X. In the limiting cases $u=\infty$ and $v=\infty$, the usual modifications are required. More concisely, the preceding inequality reads as follows:

$$\|(Tx_k)|\mathcal{E}(\mathbb{F})\|_v \le c\,\|(x_k)|l_u(\mathbb{F})\|.$$

Obviously,

$$\varrho_u^{(v)}(\mathcal{E}(\mathbb{F}),\mathcal{I}(\mathbb{F}))\,:\,T \longrightarrow \varrho_u^{(v)}(T|\mathcal{E}(\mathbb{F}),\mathcal{I}(\mathbb{F}))$$

is an ideal norm.

5.6.11 PROPOSITION. $1 \le \varrho_u^{(v)}(\mathcal{E}(\mathbb{F}),\mathcal{I}(\mathbb{F})) \le |\mathbb{F}|^{1/u'}$.

PROOF. Note that

$$\left(\frac{1}{2\pi}\int_{-\pi}^{+\pi}\left\|\sum_{k\in\mathbb{F}} Tx_k \exp(ikt)\right\|^v dt\right)^{1/v} \le \sum_{k\in\mathbb{F}}\|Tx_k\| \le |\mathbb{F}|^{1/u'}\,\|T\|\left(\sum_{k\in\mathbb{F}}\|x_k\|^u\right)^{1/u}.$$

5.6.12 For $u=v=2$, the next result was obtained in 3.8.26. The general situation can be treated in the same way.

PROPOSITION. *For $1 \le u, v < \infty$ and every finite set $\mathbb{F}$ with $|\mathbb{F}| = n$, the following are equivalent:*

(1) $\varrho_u^{(v)}(\mathcal{E}(\mathbb{F}),\mathcal{I}(\mathbb{F})) = \varrho_u^{(v)}(\mathcal{E}_n,\mathcal{I}_n)$.

(2) $\varrho_u^{(v)}(\mathcal{E}(\mathbb{F}),\mathcal{I}(\mathbb{F})) \le \varrho_u^{(v)}(\mathcal{E}_n,\mathcal{I}_n)$.

(3) *There exist $p \in \mathbb{N}$ and $r \in \mathbb{Z}$ such that $\mathbb{F}$ is a permutation of the arithmetic progression $\{pk + r : k = 1,\dots,n\}$.*

5.6.13 In view of the preceding observation, the strong inequality $\varrho_u^{(v)}(\mathcal{E}(\mathbb{F}),\mathcal{I}(\mathbb{F})) \le \varrho_u^{(v)}(\mathcal{E}_n,\mathcal{I}_n)$ cannot be true for all $\mathbb{F}$ with $|\mathbb{F}| = n$. However, there is a chance that a weaker estimate holds.

PROBLEM. *Does there exist a constant $c(u,v) \ge 1$ such that*

$$\varrho_u^{(v)}(\mathcal{E}(\mathbb{F}),\mathcal{I}(\mathbb{F})) \le c(u,v)\,\varrho_u^{(v)}(\mathcal{E}_n,\mathcal{I}_n) \quad \textit{whenever} \quad |\mathbb{F}| = n\,?$$

5.6.14 Next, we provide an auxiliary result due to J. Bourgain; see [bou 8, p. 244].

LEMMA. *Let $\mathbb{F}$ be any finite subset of $\mathbb{Z}$ and write $m := |\mathbb{F}|$. Then there exists $t_0 \in [-\pi, +\pi)$ such that more than $\frac{1}{30}m$ of the pairwise disjoint arcs*

$$A_h := \Big\{ \exp(it) : \tfrac{\pi}{m}(2h-1) \le t < \tfrac{\pi}{m}(2h+1) \Big\}$$

with $h = 1, \dots, m$ contain at least one of the points $\exp(ikt_0)$ with $k \in \mathbb{F}$.

PROOF. Of course, it is enough to treat the case when $m \ge 30$. Define φ on $[-\pi, +\pi)$ by

$$\varphi(t) := \begin{cases} 1 + \frac{mt}{2\pi} & \text{if } 0 \ge t \ge -\frac{2\pi}{m}, \\ 1 - \frac{mt}{2\pi} & \text{if } 0 \le t \le +\frac{2\pi}{m}, \\ 0 & \text{otherwise,} \end{cases}$$

and extend this peak function 2π-periodically,

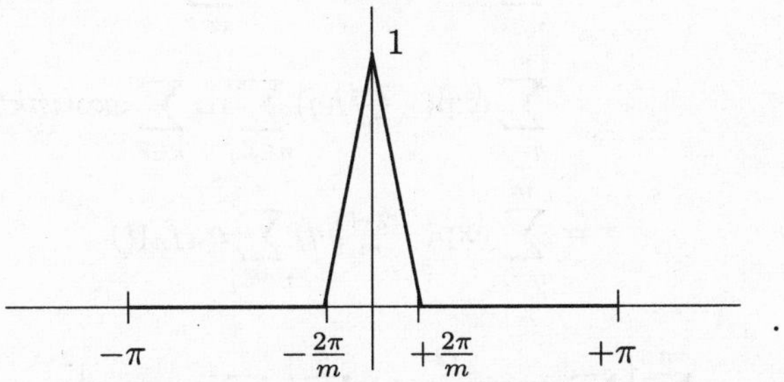

Elementary manipulations show that φ admits the Fourier expansion

$$\varphi(t) = \sum_{n \in \mathbb{Z}} \alpha_n \exp(int),$$

where

$$\alpha_0 = \tfrac{1}{m} \quad \text{and} \quad \alpha_n = \tfrac{1}{m} \left(\frac{\sin \frac{n\pi}{m}}{\frac{n\pi}{m}} \right)^2 \quad \text{for } n \ne 0.$$

Hence

$$0 \le \alpha_n \le \tfrac{1}{m} \quad \text{for } n \in \mathbb{Z} \quad \text{and} \quad 0 \le \alpha_n \le \tfrac{m}{n^2\pi^2} \quad \text{for } n \ne 0.$$

We decompose $\mathbb{Z}$ into the pairwise disjoint subsets

$$\mathbb{Z}_q := \{ pm + q : p \in \mathbb{Z} \} \quad \text{with } q = 1, \dots, m.$$

Then

$$\sum_{n \in \mathbb{Z}_q} \alpha_n = \sum_{p=2}^{\infty} \alpha_{-pm+q} + \alpha_{-m+q} + \alpha_q + \sum_{p=1}^{\infty} \alpha_{pm+q} \le \frac{2}{m}\left(1 + \frac{1}{\pi^2} \sum_{p=1}^{\infty} \frac{1}{p^2}\right).$$

It now follows from

$$\sum_{p=1}^{\infty} \frac{1}{p^2} = \frac{\pi^2}{6}$$

that

$$\sum_{n \in \mathbb{Z}_q} \alpha_n \leq \frac{7}{3m} \quad \text{for } q = 1, \dots, m. \tag{1}$$

Let

$$\varphi_h(t) := \varphi(t - \tfrac{2\pi}{m} h) \quad \text{for } h = 1, \dots, m$$

and

$$f_n(t) := \sum_{k \in \mathbb{F}} \exp(inkt) \quad \text{for } n \in \mathbb{Z}.$$

Since the matrix $m^{-1/2}(\exp(-\frac{2\pi i}{m} hq))$ is unitary, we obtain from

$$\begin{aligned}
\sum_{k \in \mathbb{F}} \varphi_h(kt) &= \sum_{k \in \mathbb{F}} \sum_{n \in \mathbb{Z}} \alpha_n \exp(in(kt - \tfrac{2\pi}{m} h)) \\
&= \sum_{n \in \mathbb{Z}} \exp(-\tfrac{2\pi i}{m} hn)\, \alpha_n \sum_{k \in \mathbb{F}} \exp(inkt) \\
&= \sum_{q=1}^{m} \exp(-\tfrac{2\pi i}{m} hq) \sum_{n \in \mathbb{Z}_q} \alpha_n \sum_{k \in \mathbb{F}} \exp(inkt) \\
&= \sum_{q=1}^{m} \exp(-\tfrac{2\pi i}{m} hq) \sum_{n \in \mathbb{Z}_q} \alpha_n f_n(t)
\end{aligned}$$

that

$$\sum_{h=1}^{m} \Big| \sum_{k \in \mathbb{F}} \varphi_h(kt) \Big|^2 = m \sum_{q=1}^{m} \Big| \sum_{n \in \mathbb{Z}_q} \alpha_n f_n(t) \Big|^2.$$

Integrating over $t \in [-\pi, +\pi)$ and applying the triangle inequality yields

$$\begin{aligned}
\frac{1}{2\pi} \int_{-\pi}^{+\pi} \sum_{h=1}^{m} \Big| \sum_{k \in \mathbb{F}} \varphi_h(kt) \Big|^2 dt &= m \sum_{q=1}^{m} \frac{1}{2\pi} \int_{-\pi}^{+\pi} \Big| \sum_{n \in \mathbb{Z}_q} \alpha_n f_n(t) \Big|^2 dt \\
&\leq m \sum_{q=1}^{m} \Big(\sum_{n \in \mathbb{Z}_q} \alpha_n \|f_n | L_2\| \Big)^2.
\end{aligned}$$

Obviously,

$$\|f_0 | L_2\| = m \quad \text{and} \quad \|f_n | L_2\| = \sqrt{m} \quad \text{if } n \neq 0.$$

We now conclude from (1) that

$$\sum_{q=1}^{m} \Big(\sum_{n \in \mathbb{Z}_q} \alpha_n \|f_n | L_2\| \Big)^2 =$$

$$= \sum_{q=1}^{m-1} \Big(\sum_{n \in \mathbb{Z}_q} \alpha_n \|f_n|L_2\| \Big)^2 + \Big(\alpha_0 \|f_0|L_2\| + \sum_{n \in \mathbb{Z}_m \setminus \{0\}} \alpha_n \|f_n|L_2\| \Big)^2$$

$$\leq (m-1)\left(\tfrac{7}{3}m^{-1/2}\right)^2 + \left(1+\tfrac{7}{3}m^{-1/2}\right)^2 = 1 + (\tfrac{7}{3})^2 + \tfrac{14}{3}m^{-1/2}.$$

Since $m \geq 30$, we have

$$1 + (\tfrac{7}{3})^2 + \tfrac{14}{3}m^{-1/2} \leq 1 + \tfrac{49}{9} + \frac{14}{3\sqrt{30}} < 7.5.$$

Therefore

$$\frac{1}{2\pi}\int_{-\pi}^{+\pi} \sum_{h=1}^{m} \Big| \sum_{k \in \mathbb{F}} \varphi_h(kt) \Big|^2 dt < 7.5\, m. \tag{2}$$

For $t \in [-\pi, +\pi)$, we denote by $\lambda_h(t)$ the number of indices $k \in \mathbb{F}$ such that $\exp(ikt) \in A_h$. Since

$$\varphi_h(kt) \geq \tfrac{1}{2} \quad \text{whenever} \quad \exp(ikt) \in A_h,$$

we have

$$\lambda_h(t) \leq 2 \sum_{k \in \mathbb{F}} \varphi_h(kt).$$

In view of (2), this implies that

$$\frac{1}{2\pi}\int_{-\pi}^{+\pi} \sum_{h=1}^{m} \lambda_h(t)^2\, dt \leq \frac{4}{2\pi}\int_{-\pi}^{+\pi} \sum_{h=1}^{m} \Big| \sum_{k \in \mathbb{F}} \varphi_h(kt) \Big|^2 dt < 30\, m.$$

Consequently, there exists $t_0 \in [-\pi, +\pi)$ such that

$$\sum_{h=1}^{m} \lambda_h(t_0)^2 < 30\, m.$$

Letting $\mathbb{F}_0 := \{h : \lambda_h(t_0) > 0\}$, we have

$$m = \sum_{h=1}^{m} \lambda_h(t) = \sum_{h \in \mathbb{F}_0} \lambda_h(t_0) \leq |\mathbb{F}_0|^{1/2} \Big(\sum_{h \in \mathbb{F}_0} \lambda_h(t_0)^2 \Big)^{1/2} < \sqrt{30m\, |\mathbb{F}_0|}.$$

Hence $|\mathbb{F}_0| > \frac{1}{30}m$, which completes the proof.

5.6.15 We are now in a position to establish an inequality, also due to J. Bourgain [bou 8, pp. 243–244], which is a weak version of that conjectured in 5.6.13.

PROPOSITION. *Let $1 \leq u < \infty$ and $1 \leq v < \infty$. Then there exists a constant $c(u,v) \geq 1$ such that*

$$|\mathbb{F}|^{-1/u} \varrho_\infty^{(v)}(\mathcal{E}(\mathbb{F}), \mathcal{I}(\mathbb{F})) \leq c(u,v)\, \varrho_u^{(v)}(\mathcal{E}_n, \mathcal{I}_n) \quad \textit{whenever} \quad |\mathbb{F}| \leq n.$$

PROOF. For brevity, we write

$$\boldsymbol{\alpha}_n(T) := \sup_{|\mathbb{F}|\le n} |\mathbb{F}|^{-1/u} \boldsymbol{\varrho}_\infty^{(v)}(T|\mathcal{E}(\mathbb{F}), \mathcal{I}(\mathbb{F})).$$

Fix any subset $\mathbb{F}$ of $\mathbb{Z}$ with $m := |\mathbb{F}| \le n$. By 5.6.14, we can find sets

$$\mathbb{F}_0 := \{k_1', \dots, k_{m_0}'\} \subseteq \mathbb{F} \quad \text{and} \quad \{k_1'', \dots, k_{m_0}''\} \subseteq \{1, \dots, m\}$$

such that $m_0 := |\mathbb{F}_0| \ge \frac{1}{30} m$ and

$$\exp(ik_\alpha' t_0) \in A_{k_\alpha''} \quad \text{for } \alpha = 1, \dots, m_0 \text{ and some } t_0 \in [-\pi, +\pi).$$

Since any chord of the unit circle is shorter than the corresponding arc, we have

$$|\exp(ik_\alpha' t_0) - \exp(\tfrac{2\pi i}{m} k_\alpha'')| < |k_\alpha' t_0 - \tfrac{2\pi}{m} k_\alpha''| \le \tfrac{\pi}{m}.$$

Note that

$$|a^h - b^h| \le |a-b|\,|a^{h-1} + a^{h-2}b + \dots + ab^{h-2} + b^{h-1}| \le |a-b|h$$

whenever $|a| = |b| = 1$ and $h = 1, 2, \dots$. Hence

$$|\exp(ihk_\alpha' t_0) - \exp(\tfrac{2\pi i}{m} hk_\alpha'')| \le \tfrac{\pi}{m} h. \tag{1}$$

Let

$$I(\mathbb{A}) := \left(\frac{1}{2\pi} \int_{-\pi}^{+\pi} \Big\| \sum_{k\in\mathbb{A}} Tx_k \exp(ikt) \Big\|^v dt \right)^{1/v}.$$

Obviously,

$$I(\mathbb{F}) \le I(\mathbb{F}_0) + I(\mathbb{F} \setminus \mathbb{F}_0).$$

Now we fix a natural number h_0 which will be specified later. Then

$$\begin{aligned} I(\mathbb{F}_0) &= \left(\frac{1}{2\pi} \int_{-\pi}^{+\pi} \Big\| \sum_{\alpha=1}^{m_0} Tx_{k_\alpha'} \exp(ik_\alpha' t) \Big\|^v dt \right)^{1/v} \\ &= \left(\frac{1}{h_0} \sum_{h=1}^{h_0} \frac{1}{2\pi} \int_{-\pi}^{+\pi} \Big\| \sum_{\alpha=1}^{m_0} Tx_{k_\alpha'} \exp(ik_\alpha'(t + ht_0)) \Big\|^v dt \right)^{1/v} \le I_1 + I_2, \end{aligned}$$

where

$$I_1 := \left(\frac{1}{h_0} \sum_{h=1}^{h_0} \frac{1}{2\pi} \int_{-\pi}^{+\pi} \left\| \sum_{\alpha=1}^{m_0} Tx_{k_\alpha'} \begin{bmatrix} \exp(ihk_\alpha' t_0) \\ - \\ \exp(\frac{2\pi i}{m} hk_\alpha'') \end{bmatrix} \exp(ik_\alpha' t) \right\|^v dt \right)^{1/v}$$

and

$$I_2 := \left(\frac{1}{h_0} \sum_{h=1}^{h_0} \frac{1}{2\pi} \int_{-\pi}^{+\pi} \Big\| \sum_{\alpha=1}^{m_0} Tx_{k_\alpha'} \exp(ik_\alpha' t) \exp(\tfrac{2\pi i}{m} hk_\alpha'') \Big\|^v dt \right)^{1/v}.$$

In view of (1), it follows that

$$I_1 \le \varrho_\infty^{(v)}(T|\mathcal{E}(\mathbb{F}_0),\mathcal{J}(\mathbb{F}_0)) \max_{1\le\alpha\le m_0} |\exp(ihk'_\alpha t_0) - \exp(\tfrac{2\pi i}{m} hk''_\alpha)| \, \|x_{k'_\alpha}\|$$
$$\le \pi \tfrac{h_0}{m} m_0^{1/u} \boldsymbol{\alpha}_n(T) \max_{\mathbb{F}_0} \|x_k\|. \quad (2)$$

Next, by integration over $t \in [-\pi, +\pi)$, we obtain from

$$\left(\frac{1}{m}\sum_{h=1}^{h_0}\left\|\sum_{\alpha=1}^{m_0} Tx_{k'_\alpha}\exp(ik'_\alpha t)\exp(\tfrac{2\pi i}{m}hk''_\alpha)\right\|^v\right)^{1/v} \le$$
$$\le \varrho_u^{(v)}(T|\mathcal{N}\mathcal{E}^\circ_m,\mathcal{J}_m)\left(\sum_{\alpha=1}^{m_0}\|\exp(ik'_\alpha t)x_{k'_\alpha}\|^u\right)^{1/u}$$
$$\le m_0^{1/u}\varrho_u^{(v)}(T|\mathcal{N}\mathcal{E}^\circ_m,\mathcal{J}_m)\max_{k\in\mathbb{F}_0}\|x_k\|$$

and 5.6.3 that

$$I_2 \le 3\left(\tfrac{m}{h_0}\right)^{1/v} m_0^{1/u} \varrho_u^{(v)}(T|\mathcal{E}_m,\mathcal{J}_m) \max_{k\in\mathbb{F}_0}\|x_k\|. \quad (3)$$

Since $\varrho_\infty^{(v)}(T|\mathcal{E}(\mathbb{F}\setminus\mathbb{F}_0),\mathcal{J}(\mathbb{F}\setminus\mathbb{F}_0)) \le |\mathbb{F}\setminus\mathbb{F}_0|^{1/u}\,\boldsymbol{\alpha}_n(T)$, we have

$$I(\mathbb{F}\setminus\mathbb{F}_0) \le (m-m_0)^{1/u}\boldsymbol{\alpha}_n(T) \max_{k\in\mathbb{F}\setminus\mathbb{F}_0}\|x_k\|. \quad (4)$$

Combining (2), (3) and (4) now yields

$$I(\mathbb{F}) \le I(\mathbb{F}_0) + I(\mathbb{F}\setminus\mathbb{F}_0) \le I_1 + I_2 + I(\mathbb{F}\setminus\mathbb{F}_0)$$
$$\le \begin{bmatrix} \pi\frac{h_0}{m} m_0^{1/u}\boldsymbol{\alpha}_n(T) \\ + \\ 3\left(\frac{m}{h_0}\right)^{1/v} m_0^{1/u}\varrho_u^{(v)}(T|\mathcal{E}_m,\mathcal{J}_m) \\ + \\ (m-m_0)^{1/u}\boldsymbol{\alpha}_n(T) \end{bmatrix} \max_{k\in\mathbb{F}}\|x_k\|.$$

Hence, by $m_0 \le m \le n$,

$$|\mathbb{F}|^{-1/u}\varrho_\infty^{(v)}(T|\mathcal{E}(\mathbb{F}),\mathcal{J}(\mathbb{F})) \le$$
$$\le 3\left(\tfrac{m}{h_0}\right)^{1/v}\varrho_u^{(v)}(T|\mathcal{E}_n,\mathcal{J}_n) + \left[\pi\tfrac{h_0}{m} + \left(1-\tfrac{m_0}{m}\right)^{1/u}\right]\boldsymbol{\alpha}_n(T). \quad (5)$$

Define

$$\delta := \tfrac{1}{2\pi}\left(1 - \left(\tfrac{29}{30}\right)^{1/u}\right) < \tfrac{1}{2\pi}.$$

In the case when $|\mathbb{F}| = m \ge \frac{1}{\delta}$, we let $h_0 := [\delta m] \ge 1$. Then

$$\tfrac{1}{2}\delta \le \tfrac{h_0}{m} \le \delta.$$

Moreover, it follows from $m_0 \ge \frac{1}{30}m$ that

$$\pi\tfrac{h_0}{m} + \left(1-\tfrac{m_0}{m}\right)^{1/u} \le \pi\delta + \left(\tfrac{29}{30}\right)^{1/u} = 1 - \pi\delta.$$

Thus, by (5),

$$|\mathbb{F}|^{-1/u}\boldsymbol{\varrho}_\infty^{(v)}(T|\mathcal{E}(\mathbb{F}),\mathcal{I}(\mathbb{F}))\le 3\,(2/\delta)^{1/v}\boldsymbol{\varrho}_u^{(v)}(T|\mathcal{E}_n,\mathcal{I}_n)+(1-\pi\delta)\boldsymbol{\alpha}_n(T). \quad (6)$$

In order to treat the case when $\mathbb{F}$ is small, we conclude from

$$\begin{aligned}\left(\frac{1}{2\pi}\int_{-\pi}^{+\pi}\Big\|\sum_{k\in\mathbb{F}}Tx_k\exp(ikt)\Big\|^v dt\right)^{1/v} &\le |\mathbb{F}|\,\|T\|\max_{k\in\mathbb{F}}\|x_k\| \\ &\le |\mathbb{F}|\,\boldsymbol{\varrho}_u^{(v)}(T|\mathcal{E}_n,\mathcal{I}_n)\max_{k\in\mathbb{F}}\|x_k\|\end{aligned}$$

that

$$\boldsymbol{\varrho}_\infty^{(v)}(T|\mathcal{E}(\mathbb{F}),\mathcal{I}(\mathbb{F}))\le|\mathbb{F}|\,\boldsymbol{\varrho}_u^{(v)}(T|\mathcal{E}_n,\mathcal{I}_n).$$

Consequently, if $|\mathbb{F}|=m\le\frac{1}{\delta}$, then

$$|\mathbb{F}|^{-1/u}\,\boldsymbol{\varrho}_\infty^{(v)}(T|\mathcal{E}(\mathbb{F}),\mathcal{I}(\mathbb{F}))\le(1/\delta)^{1/u'}\,\boldsymbol{\varrho}_u^{(v)}(T|\mathcal{E}_n,\mathcal{I}_n). \quad (7)$$

Letting

$$c_0(u,v):=\max\left\{(1/\delta)^{1/u'},3\cdot(2/\delta)^{1/v}\right\},$$

we obtain from (6) and (7) that

$$|\mathbb{F}|^{-1/u}\boldsymbol{\varrho}_\infty^{(v)}(T|\mathcal{E}(\mathbb{F}),\mathcal{I}(\mathbb{F}))\le c_0(u,v)\,\boldsymbol{\varrho}_u^{(v)}(T|\mathcal{E}_n,\mathcal{I}_n)+(1-\pi\delta)\,\boldsymbol{\alpha}_n(T)$$

for all subsets $\mathbb{F}$ of $\mathbb{Z}$ with $|\mathbb{F}|\le n$. Therefore, by definition of $\boldsymbol{\alpha}_n$,

$$\boldsymbol{\alpha}_n(T)\le c_0(u,v)\,\boldsymbol{\varrho}_u^{(v)}(T|\mathcal{E}_n,\mathcal{I}_n)+(1-\pi\delta)\,\boldsymbol{\alpha}_n(T)$$

which implies that

$$\pi\delta\,\boldsymbol{\alpha}_n(T)\le c_0(u,v)\,\boldsymbol{\varrho}_u^{(v)}(T|\mathcal{E}_n,\mathcal{I}_n),$$

and we are done.

5.6.16 We now establish an analogue of 4.3.5 and 4.5.5.

LEMMA. *Let $1<p<u\le 2$ and $1\le v<\infty$. Then, for $T\in\mathfrak{L}(X,Y)$, the following properties are equivalent:*

(1) *There exists a constant $c_0\ge 1$ such that*

$$\boldsymbol{\varrho}_u^{(v)}(T|\mathcal{E}_n,\mathcal{I}_n)\le c_0 n^{1/p-1/u} \quad \text{for} \quad n=1,2,\dots.$$

(2) *There exists a constant $c\ge 1$ such that*

$$\boldsymbol{\varrho}_{p,1}^{(v)}(T|\mathcal{E}_n,\mathcal{I}_n)\le c \quad \text{for} \quad n=1,2,\dots.$$

PROOF. The implication **(2)⇒(1)** follows from

$$\|(x_k)|l_{p,1}^n\|\le A\,n^{1/p-1/u}\,\|(x_k)|l_u^n\| \qquad \text{(by 0.5.6)},$$

where $A>0$ is a constant.

Fix $x_1, \dots, x_n \in X$. By 0.5.8, we can find a partition of $\{1, \dots, n\}$ into pairwise disjoint subsets $\mathbb{F}_h$ such that

$$\sum_{h=1}^{m} |\mathbb{F}_h|^{1/p} \max_{k \in \mathbb{F}_h} \|x_k\| \le 2\, \|(x_k)|l_{p,1}^n\|.$$

Assume that **(1)** holds. Then it follows from 5.6.15 that

$$\varrho_\infty^{(v)}(T|\mathcal{E}(\mathbb{F}_h), \mathcal{I}(\mathbb{F}_h)) \le c(u,v)\, |\mathbb{F}_h|^{1/u} \varrho_u^{(v)}(T|\mathcal{E}_{|\mathbb{F}_h|}, \mathcal{I}_{|\mathbb{F}_h|}) \le c(u,v) c_0\, |\mathbb{F}_h|^{1/p}.$$

Hence

$$\left(\frac{1}{2\pi}\int_{-\pi}^{+\pi} \Big\| \sum_{k=1}^{n} T x_k e_k(t) \Big\|^v dt\right)^{1/v} \le \sum_{h=1}^{m} \left(\frac{1}{2\pi}\int_{-\pi}^{+\pi} \Big\| \sum_{k \in \mathbb{F}_h} T x_k e_k(t) \Big\|^v dt\right)^{1/v}$$

$$\le \sum_{h=1}^{m} \varrho_\infty^{(v)}(T|\mathcal{E}(\mathbb{F}_h), \mathcal{I}(\mathbb{F}_h)) \max_{k \in \mathbb{F}_h} \|x_k\|$$

$$\le c(u,v)\, c_0 \sum_{h=1}^{m} |\mathbb{F}_h|^{1/p} \max_{k \in \mathbb{F}_h} \|x_k\| \le 2\, c(u,v)\, c_0\, \|(x_k)|l_{p,1}^n\|.$$

This proves that $\varrho_{p,1}^{(v)}(T|\mathcal{E}_n, \mathcal{I}_n) \le c := 2\, c(u,v)\, c_0$.

5.6.17 Next, we restate 5.6.5 and 5.6.16 in terms of operator ideals.

PROPOSITION. *If* $1 < p \le \min\{2, u, v'\}$, *then*

$$\mathfrak{L}'\big[n^{1/u-1/p} \varrho_u^{(v)}(\mathcal{E}_n, \mathcal{I}_n)\big] = \mathfrak{L}\big[n^{1/v'-1/p} \varrho_{v'}^{(u')}(\mathcal{E}_n, \mathcal{I}_n)\big].$$

Moreover, assuming in addition that $p < u$, *we have*

$$\mathfrak{L}\big[n^{1/u-1/p} \varrho_u^{(v)}(\mathcal{E}_n, \mathcal{I}_n)\big] = \mathfrak{L}\big[\varrho_{p,1}^{(v)}(\mathcal{E}_n, \mathcal{I}_n)\big].$$

5.6.18 Apart from some limiting cases, $\mathfrak{L}\big[n^{1/u-1/p} \varrho_u^{(v)}(\mathcal{E}_n, \mathcal{I}_n)\big]$ depends only on the parameter p, not on u and v.

PROPOSITION. *If* $1 < p < \min\{2, u, v'\}$, *then*

$$\mathfrak{E}\mathfrak{T}_p^{weak} = \mathfrak{L}\big[\varrho_{p,1}^{(v)}(\mathcal{E}_n, \mathcal{I}_n)\big] = \mathfrak{L}\big[n^{1/u-1/p} \varrho_u^{(v)}(\mathcal{E}_n, \mathcal{I}_n)\big].$$

PROOF. We conclude from 5.6.17 that

$$\mathfrak{L}'\big[\varrho_{p,1}^{(v)}(\mathcal{E}_n, \mathcal{I}_n)\big] = \mathfrak{L}'\big[n^{1/u-1/p} \varrho_u^{(v)}(\mathcal{E}_n, \mathcal{I}_n)\big]$$

$$= \mathfrak{L}\big[n^{1/v'-1/p} \varrho_{v'}^{(u')}(\mathcal{E}_n, \mathcal{I}_n)\big] = \mathfrak{L}\big[\varrho_{p,1}^{(u')}(\mathcal{E}_n, \mathcal{I}_n)\big].$$

In this way, parameters have been separated: the left-hand ideal does not depend on u, and the right-hand ideal does not depend on v. So all ideals are independent of u and v. In particular, taking $u = v = 2$ proves that the ideals in question coincide with $\mathfrak{E}\mathfrak{T}_p^{weak}$.

REMARK. If $p = u$ and $p < v'$, then

$$\mathfrak{CT}_p \subseteq \mathfrak{L}[\varrho_p^{(v)}(\mathcal{E}_n, \mathcal{I}_n)] \subseteq \mathfrak{CT}_p^{weak}.$$

This is all that we know about the middle ideal, which may or may not depend on v.

5.6.19 It is an open problem whether the second equality in the above proposition also holds in the limiting case $p = v'$. We only know the following inclusion.

PROPOSITION. $\mathfrak{L}[\varrho_{p,1}^{(p')}(\mathcal{E}_n, \mathcal{I}_n)] \subseteq \mathfrak{CT}_p^{weak}$.

PROOF. Indeed, $p' > 2$ and 5.6.18 imply that

$$\mathfrak{L}[\varrho_{p,1}^{(p')}(\mathcal{E}_n, \mathcal{I}_n)] \subseteq \mathfrak{L}[\varrho_{p,1}^{(2)}(\mathcal{E}_n, \mathcal{I}_n)] = \mathfrak{CT}_p^{weak}.$$

5.6.20 However, allowing a small change of parameters, we get a reverse inclusion.

PROPOSITION. $\mathfrak{CT}_{p_0}^{weak} \subseteq \mathfrak{L}[\varrho_{p,1}^{(p')}(\mathcal{E}_n, \mathcal{I}_n)]$ *if* $1 < p < p_0 < 2$.

PROOF. In view of 0.5.6, 5.6.17 and 5.6.18,

$$\begin{aligned}(\mathfrak{CT}_{p_0}^{weak})' &= \mathfrak{L}'[\varrho_{p_0,1}^{(2)}(\mathcal{E}_n, \mathcal{I}_n)] \subseteq \mathfrak{L}'[\varrho_p^{(2)}(\mathcal{E}_n, \mathcal{I}_n)] \\ &= \mathfrak{L}[n^{1/2-1/p}\varrho_2^{(p')}(\mathcal{E}_n, \mathcal{I}_n)] = \mathfrak{L}[\varrho_{p,1}^{(p')}(\mathcal{E}_n, \mathcal{I}_n)],\end{aligned}$$

which implies the desired inclusion, since $\mathfrak{CT}_{p_0}^{weak}$ is symmetric.

5.6.21 PROPOSITION. *Let* $1 < p < p_0 \le 2$. *Then there exists a constant* $c(p, p_0) \ge 1$ *such that*

$$\varrho_p^{(p')}(\mathcal{E}_n, \mathcal{I}_n) \le c(c, p_0)\, \varrho_{p_0,1}^{(p_0')}(\mathcal{E}_n, \mathcal{I}_n).$$

PROOF. Consider the operator

$$[E_n, T] : (x_k) \longrightarrow \sum_{k=1}^{n} T x_k e_k.$$

Define θ by $1/p = (1-\theta)/p_0 + \theta/1$. Then, using the real interpolation method, we have the formulas

$$([l_{p_0,1}^n, X], [l_1^n, X])_{\theta,p} = [l_p^n, X]$$

and

$$([L_{p_0'}[-\pi, +\pi), Y], [C[-\pi, +\pi), Y])_{\theta,p} = [L_{p',p}[-\pi, +\pi), Y]$$

which indicate isomorphisms (but not isometries). Moreover,

$$[L_{p',p}[-\pi, +\pi), Y] \subseteq [L_{p'}[-\pi, +\pi), Y],$$

since $p \le p'$. Hence it follows from

$$\begin{aligned}\|[E_n, T] : [l_{p_0,1}^n, X] \to [L_{p_0'}[-\pi, +\pi), Y]\| &= \varrho_{p_0,1}^{(p_0')}(T|\mathcal{E}_n, \mathcal{I}_n), \\ \|[E_n, T] : [l_1^n, X] \to [C[-\pi, +\pi), Y]\| &= \|T\|\end{aligned}$$

and

$$\|[E_n, T] : [l_p^n, X] \to [L_{p'}[-\pi, +\pi), Y]\| = \varrho_p^{(p')}(T|\mathcal{E}_n, \mathcal{I}_n)$$

that

$$\varrho_p^{(p')}(T|\mathcal{E}_n, \mathcal{I}_n) \le c(p, p_0)\, \varrho_{p_0,1}^{(p'_0)}(T|\mathcal{E}_n, \mathcal{I}_n)^{1-\theta} \|T\|^\theta.$$

This completes the proof, because of $\|T\| \le \varrho_{p_0,1}^{(p'_0)}(T|\mathcal{E}_n, \mathcal{I}_n)$.

5.6.22 In terms of operator ideals, the preceding result reads as follows.

PROPOSITION. $\mathfrak{L}\big[\varrho_{p_0,1}^{(p'_0)}(\mathcal{E}_n, \mathcal{I}_n)\big] \subseteq \mathfrak{ET}_p$ *if* $1 < p < p_0 \le 2$.

5.6.23 We are now in a position to establish an analogue of 4.3.10.

THEOREM. $\mathfrak{ET}_2 \subset \mathfrak{ET}_{p_0} \subset \mathfrak{ET}_{p_0}^{weak} \subset \mathfrak{ET}_p$ *if* $1 < p < p_0 < 2$.
In particular,

$$\bigcup_{1<p\le 2} \mathfrak{ET}_p = \bigcup_{1<p<2} \mathfrak{ET}_p^{weak}.$$

PROOF. The inclusion $\mathfrak{ET}_{p_0} \subseteq \mathfrak{ET}_{p_0}^{weak}$ follows from the trivial estimate

$$\varrho(\mathcal{E}_n, \mathcal{I}_n) \le n^{1/p_0 - 1/2} \varrho_{p_0}^{(p'_0)}(\mathcal{E}_n, \mathcal{I}_n).$$

Choosing p_1 such that $p < p_1 < p_0$, we conclude from 5.6.20 and 5.6.22 that

$$\mathfrak{ET}_{p_0}^{weak} \subseteq \mathfrak{L}\big[\varrho_{p_1,1}^{(p'_1)}(\mathcal{E}_n, \mathcal{I}_n)\big] \subseteq \mathfrak{ET}_p.$$

It will be shown in 5.6.24 that

$$L_p \in \mathsf{ET}_p \setminus \mathsf{ET}_{p_0}^{weak} \quad \text{if } 1 < p < p_0 < 2.$$

Thus the inclusion $\mathfrak{ET}_{p_0}^{weak} \subseteq \mathfrak{ET}_p$ is strict. Finally, Example 5.6.25 tells us that the inclusion $\mathfrak{ET}_{p_0} \subseteq \mathfrak{ET}_{p_0}^{weak}$ is strict as well.

5.6.24 EXAMPLE.

$$L_p,\ L_{p'} \in \mathsf{ET}_p \setminus \mathsf{ET}_{p_0}^{weak} \quad \textit{if } 1 < p < p_0 < 2.$$

PROOF. In view of 3.7.16,

$$\varrho_p^{(p')}(L_p|\mathcal{E}_n, \mathcal{I}_n) = 1.$$

Hence $L_p \in \mathsf{ET}_p$. By symmetry, we also have $L_{p'} \in \mathsf{ET}_p$. Moreover, we know from 5.5.17 that

$$\varrho(L_p|\mathcal{E}_n, \mathcal{I}_n) \asymp \varrho(L_{p'}|\mathcal{E}_n, \mathcal{I}_n) \asymp n^{1/p-1/2},$$

which yields $L_p,\ L_{p'} \notin \mathsf{ET}_{p_0}^{weak}$.

5.6.25 Recall that the diagonal operator D_λ is defined by

$$D_\lambda : (\xi_k) \longrightarrow (k^{-\lambda}\xi_k).$$

EXAMPLE. $D_{1/p'} \in \mathfrak{ET}_p^{weak}(l_1) \setminus \mathfrak{ET}_p(l_1)$ *if* $1<p<2$.

PROOF. We know from 3.7.15 that

$$\varrho_p^{(p')}(D_{1/p'} : l_1 \to l_1 | \mathcal{E}_n, \mathcal{I}_n) = \Big(\sum_{k=1}^{n} \frac{1}{k} \Big)^{1/p'} \asymp (1+\log n)^{1/p'}$$

and

$$\varrho_2^{(2)}(D_{1/p'} : l_1 \to l_1 | \mathcal{E}_n, \mathcal{I}_n) = \Big(\sum_{k=1}^{n} k^{-2/p'} \Big)^{1/2} \asymp n^{1/p-1/2}.$$

Hence $D_{1/p'}$ has the required properties.

5.6.26 An operator T has **Fourier subtype** if

$$\varrho(T|\mathcal{E}_n, \mathcal{I}_n) = o(\sqrt{n}).$$

These operators form the closed ideal

$$\mathfrak{ET} := \mathfrak{L}_0[n^{-1/2}\, \varrho(\mathcal{E}_n, \mathcal{I}_n)].$$

REMARK. ET consists just of all B-convex Banach spaces; see 5.6.30.

5.6.27 The following result is an analogue of 4.4.1.

THEOREM. $\bigcup_{1<p\le 2} \mathsf{ET}_p = \bigcup_{1<p<2} \mathsf{ET}_p^{weak} = \mathsf{ET}$.

PROOF. If $X \in \mathsf{ET}$, then $\varrho(X|\mathcal{E}_m, \mathcal{I}_m) < \sqrt{m}$ for some $m \ge 2$. We know from 3.3.6 that the sequence $(\varrho(\mathcal{E}_n, \mathcal{I}_n))$ is non-decreasing, while $(\varrho(\mathcal{E}_n^\circ, \mathcal{I}_n))$ is submultiplicative, by 5.5.11. Thus 1.3.5 provides us with an $\varepsilon > 0$ such that $\varrho(X|\mathcal{E}_n, \mathcal{I}_n) \prec n^{1/2-\varepsilon}$ for $n = 1, 2, \ldots$. This means that X has weak Fourier type p whenever $1/p' \le \varepsilon$. Hence

$$\mathsf{ET} \subseteq \bigcup_{1<p<2} \mathsf{ET}_p^{weak}.$$

The reverse inclusion is obvious, and the equality

$$\bigcup_{1<p\le 2} \mathsf{ET}_p = \bigcup_{1<p<2} \mathsf{ET}_p^{weak}$$

was shown in 5.6.23.

REMARK. In view of the above result, Banach spaces in ET are said to have **non-trivial Fourier type**.

5.6.28 Let $\alpha \geq 0$, and denote by C_α the diagonal operator on $[l_2, l_\infty^{2^k}]$ defined by

$$C_\alpha : (x_k) \longrightarrow (k^{-\alpha} x_k), \quad \text{where} \quad x_k \in l_\infty^{2^k}.$$

The next example shows that the phenomenon described in the previous paragraph only occurs for spaces, not for operators. That is,

$$\bigcup_{1<p\leq 2} \mathfrak{ET}_p = \bigcup_{1<p<2} \mathfrak{ET}_p^{weak} \neq \mathfrak{ET}.$$

EXAMPLE. $\varrho(C_\alpha|\mathcal{E}_n, \mathcal{I}_n) \asymp n^{1/2}(1+\log n)^{-\alpha}$.

PROOF. Recall that

$$\varrho(l_\infty^N|\mathcal{E}_n, \mathcal{I}_n) \asymp \min\{\sqrt{n}, \sqrt{N}\} \qquad \text{(by 5.5.18).}$$

The desired relation is now a consequence of ($\triangle$) in 1.2.12.

5.6.29 Next, we establish one of the deepest results presented in this book; see [bou 1], [bou 2] and [bou 8]. We hope that the machinery of ideal norms has made the proof more transparent.

BOURGAIN'S THEOREM. $\mathfrak{ET} = \mathfrak{RT}$.

PROOF. In view of 3.7.9 and 4.2.14, we have $\varrho(\mathcal{R}_n, \mathcal{I}_n) \leq \frac{\pi}{2}\, \varrho(\mathcal{E}_n, \mathcal{I}_n)$. Hence $\mathfrak{ET} \subseteq \mathfrak{RT}$.

We now give two proofs of the reverse inclusion.

1st VERSION: Use Hinrichs's inequality 4.11.18

$$2^{-n/2}\, \varrho(\mathcal{E}_{2^n}, \mathcal{I}_{2^n}) \leq 50\, n^{-1/2}\, \varrho(\mathcal{P}_n(e^{i\tau}), \mathcal{I}_n)$$

and $\varrho(\mathcal{P}_n(e^{i\tau}), \mathcal{I}_n) \asymp \varrho(\mathcal{R}_n, \mathcal{I}_n)$.

2nd VERSION: It follows from 3.10.10 that

$$\mathfrak{KT} = \left\{ T \in \mathfrak{L} : \boldsymbol{\tau}_n(T) = o(\sqrt{n}) \right\},$$

where $\boldsymbol{\tau}_n$ denotes the n-th universal type ideal norm. So $\varrho(\mathcal{E}_n, \mathcal{I}_n) \leq \boldsymbol{\tau}_n$ implies that $\mathfrak{KT} \subseteq \mathfrak{ET}$. In other terms, $\mathfrak{ET}$ belongs between $\mathfrak{KT}$ and $\mathfrak{RT}$, which coincide by 4.14.8.

5.6.30 In literary style and in the setting of spaces, the preceding result reads as follows.

THEOREM. *A Banach space has non-trivial Fourier type if and only if it has non-trivial Rademacher type.*

PROOF. Combining the previous result with 4.4.1 and 5.6.27 yields

$$\bigcup_{1<p\leq 2} \mathsf{RT}_p = \mathsf{RT} = \mathsf{ET} = \bigcup_{1<p\leq 2} \mathsf{ET}_p.$$

REMARK. We have a similar situation to that described in 4.6.20 (Remark). This means that there exists a function $\varphi\colon (1,2]\times[1,\infty)\to(1,2]$ such that

$$X\in \mathsf{RT}_p \quad\text{and}\quad p_0=\varphi(p,\|X|\mathfrak{RT}_p\|) \quad\text{imply}\quad X\in \mathsf{ET}_{p_0}.$$

More specifically, we may take $\varphi(p,t)=1+(ct)^{-p'}$, where $c>1$ is a constant; see [bou 2].

5.7 Operators of Fourier cotype

The main purpose of this section is to show that the concept of Fourier cotype is more or less superfluous; see 5.5.15.

5.7.1 To begin with, we transfer Definition 5.6.1 to the case of cotype.

Let $1\le u,v<\infty$. For $T\in\mathfrak{L}(X,Y)$, we denote by $\boldsymbol{\varrho}_u^{(v)}(T|\mathcal{I}_n,\mathcal{E}_n)$ the least constant $c\ge 0$ such that

$$\left(\sum_{k=1}^n\|Tx_k\|^v\right)^{1/v}\le c\left(\frac{1}{2\pi}\int_{-\pi}^{+\pi}\left\|\sum_{k=1}^n x_k\exp(ikt)\right\|^u dt\right)^{1/u}$$

whenever $x_1,\dots,x_n\in X$. In the limiting cases $u=\infty$ and $v=\infty$, the usual modifications are required. More concisely, the above inequality reads as follows:

$$\|(Tx_k)|l_v^n\|\le c\,\|(x_k)|\mathcal{E}_n\|_u.$$

We refer to

$$\boldsymbol{\varrho}_u^{(v)}(\mathcal{I}_n,\mathcal{E}_n) : T\longrightarrow \boldsymbol{\varrho}_u^{(v)}(T|\mathcal{I}_n,\mathcal{E}_n)$$

as a **Fourier cotype ideal norm**. The case $u=v=2$ has already been treated in Section 5.5.

5.7.2 Next, we generalize 5.5.5.

PROPOSITION. $\boldsymbol{\varrho}_u^{(v)}(\mathcal{I}_n,\mathcal{E}_n)\le\boldsymbol{\delta}_u^{(v)}(\mathcal{I}_n,\mathcal{E}_n)\le 6\,\boldsymbol{\varrho}_u^{(v)}(\mathcal{I}_n,\mathcal{E}_n)$.

PROOF. The same reasoning as in 5.5.3 and 5.5.4 yields

$$\boldsymbol{\delta}_u^{(v)}(\mathcal{I}_{2n},\mathcal{E}_{2n})\le 2^{1/v}\,\boldsymbol{\delta}_u^{(v)}(\mathcal{I}_n,\mathcal{E}_n)$$

and

$$\boldsymbol{\delta}_u^{(v)}(\mathcal{I}_{2m+1},\mathcal{E}_{2m+1})\le 3\,\boldsymbol{\varrho}_u^{(v)}(\mathcal{I}_{4m-1},\mathcal{E}_{4m-1}).$$

Note that, in the first inequality, $\sqrt{2}$ is replaced by $2^{1/v}\le 2$. Finally, letting $m:=[\frac{n}{2}]$, we may adapt the proof of 5.5.5 to get

$$\begin{aligned}\boldsymbol{\delta}_u^{(v)}(\mathcal{I}_n,\mathcal{E}_n) &\le \boldsymbol{\delta}_u^{(v)}(\mathcal{I}_{2m+1},\mathcal{E}_{2m+1})\le 3\,\boldsymbol{\varrho}_u^{(v)}(\mathcal{I}_{4m-1},\mathcal{E}_{4m-1})\\ &\le 3\,\boldsymbol{\varrho}_u^{(v)}(\mathcal{I}_{2n},\mathcal{E}_{2n})\le 3\cdot 2^{1/v}\,\boldsymbol{\varrho}_u^{(v)}(\mathcal{I}_n,\mathcal{E}_n).\end{aligned}$$

5.7.3 PROPOSITION. *The sequences of the ideal norms*

$$\varrho_u^{(v)}(\mathfrak{I}_n, \mathfrak{E}_n) \quad and \quad n^{1/u+1/v-1}\, \varrho_u^{(v)}(\mathfrak{E}_n, \mathfrak{I}_n)$$

are uniformly equivalent. This holds, in particular, for

$$\varrho_{q'}^{(q)}(\mathfrak{I}_n, \mathfrak{E}_n) \quad and \quad \varrho_{q'}^{(q)}(\mathfrak{E}_n, \mathfrak{I}_n).$$

PROOF. We have

$$\varrho_u^{(v)}(\mathfrak{I}_n, \mathfrak{E}_n) \asymp \boldsymbol{\delta}_u^{(v)}(\mathfrak{I}_n, \mathfrak{E}_n) \qquad \text{(by 5.7.2)},$$

$$\boldsymbol{\delta}_u^{(v)}(\mathfrak{I}_n, \mathfrak{E}_n) = \varrho_{v'}^{(u')}{}'(\mathfrak{E}_n, \mathfrak{I}_n) \qquad \text{(generalize 3.4.6)},$$

$$\varrho_{v'}^{(u')}{}'(\mathfrak{E}_n, \mathfrak{I}_n) \asymp n^{1/u+1/v-1}\, \varrho_u^{(v)}(\mathfrak{E}_n, \mathfrak{I}_n) \qquad \text{(by 5.6.5)}.$$

5.7.4 Obviously, the definitions from Section 5.6 can be repeated in the case of Fourier cotype. In particular, we get the Banach operator ideals $\mathfrak{FC}_q$, $\mathfrak{FC}_q^{weak}$ and $\mathfrak{FC}$. However, in view of the previous results, this is a matter of pedantry.

PROPOSITION.

$$\mathfrak{FC}_q = \mathfrak{FT}_{q'}, \quad \mathfrak{FC}_q^{weak} = \mathfrak{FT}_{q'}^{weak} \quad and \quad \mathfrak{FC} = \mathfrak{FT}.$$

5.8 The vector-valued Fourier transform

5.8.1 In this section and only in this section, $\mathbb{G}$ is assumed to be a **locally compact Abelian group**, and $\mathbb{G}'$ denotes its character group, which is locally compact and Abelian as well. Moreover, μ and μ' stand for the associated Haar measures, respectively,

5.8.2 Let $[C_0(\mathbb{G}'), X]$ denote the Banach space of all continuous functions $\boldsymbol{f} : \mathbb{G}' \to X$ vanishing at infinity. This means that, outside a suitable compact subset, $\|\boldsymbol{f}(a)\|$ is as small as we please. The norm is given by

$$\|\boldsymbol{f}|C_0(\mathbb{G}')\| := \sup_{\mathbb{G}'} \|\boldsymbol{f}(a)\|.$$

If X is the scalar field $\mathbb{K}$, we write $C_0(\mathbb{G}')$ instead of $[C_0(\mathbb{G}'), X]$.

5.8.3 The **Fourier transform** $F^{\mathbb{G}}$ from $L_1(\mathbb{G})$ into $C_0(\mathbb{G}')$ is defined by

$$F^{\mathbb{G}} : f(t) \longrightarrow \widehat{f^{\mathbb{G}}}(a) := \int_{\mathbb{G}} f(t)\, a(t)\, d\mu(t).$$

We will be interested in the following cases:

$$F^{\mathbb{R}} : f(t) \longrightarrow \widehat{f}(s) \quad := \int_{-\infty}^{+\infty} f(t)\exp(ist)\,dt \qquad \text{(real line } \mathbb{R}\text{)},$$

$$F^{\mathbb{T}} : f(t) \longrightarrow \big(\langle f, e_k\rangle\big) := \Big(\frac{1}{2\pi}\int_{-\pi}^{+\pi} f(t)\exp(ikt)\,dt\Big) \qquad \text{(circle group } \mathbb{T}\text{)},$$

$$F^{\mathbb{Z}} : (\xi_k) \longrightarrow f(s) \quad := \sum_{\mathbb{Z}} \xi_k \exp(iks) \qquad \text{(discrete group } \mathbb{Z}\text{)}.$$

Note that $F^{\mathbb{T}}$ and $F^{\mathbb{Z}}$ are dual to each other, while $F^{\mathbb{R}}$ is self-dual.

5.8.4 By Plancherel's theorem, the Haar measures μ and μ' can be adjusted such that

$$\|\widehat{f^{\mathbb{G}}}|L_2(\mathbb{G}')\| = \|f|L_2(\mathbb{G})\| \quad \text{for } f \in L_1(\mathbb{G}) \cap L_2(\mathbb{G});$$

see [pla] and [RUD, p. 26]. Consequently, there exists an isometric extension of the Fourier transform which maps $L_2(\mathbb{G})$ onto $L_2(\mathbb{G}')$.

Next, we conclude by complex interpolation that the Fourier transform yields an operator $F^{\mathbb{G}}$ acting from $L_p(\mathbb{G})$ into $L_{p'}(\mathbb{G}')$ for $1<p<2$. This is the famous Hausdorff–Young theorem; see [hau] and [you].

5.8.5 For $T \in \mathfrak{L}(X,Y)$ and $\boldsymbol{f} \in [L_1(\mathbb{G}), X]$, we let

$$[F^{\mathbb{G}}, T] : \boldsymbol{f}(t) \longrightarrow T\widehat{\boldsymbol{f}^{\mathbb{G}}}(a) := \int_{\mathbb{G}} T\boldsymbol{f}(t)\,a(t)\,d\mu(t).$$

The Riemann–Lebesgue theorem says that $[F^{\mathbb{G}}, T]\boldsymbol{f} \in [C_0(\mathbb{G}'), Y]$. Moreover,

$$\|[F^{\mathbb{G}}, T] : [L_1(\mathbb{G}), X] \to [C_0(\mathbb{G}'), Y]\| = \|T\|.$$

5.8.6 Let us fix any infinite locally compact Abelian group $\mathbb{G}$, and let $1<p\leq 2$.

An operator $T \in \mathfrak{L}(X,Y)$ is said to be **compatible with the Fourier transform** $F^{\mathbb{G}}$ from $L_p(\mathbb{G})$ into $L_{p'}(\mathbb{G}')$ if

$$F^{\mathbb{G}} \otimes T : L_p(\mathbb{G}) \otimes X \longrightarrow L_{p'}(\mathbb{G}') \otimes Y$$

admits a continuous extension

$$[F^{\mathbb{G}}, T] : [L_p(\mathbb{G}), X] \longrightarrow [L_{p'}(\mathbb{G}'), Y].$$

In this case, we let

$$\|T|\mathfrak{FT}_p^{\mathbb{G}}\| := \|[F^{\mathbb{G}}, T] : [L_p(\mathbb{G}), X] \to [L_{p'}(\mathbb{G}'), Y]\|.$$

The class of these operators is a Banach ideal, denoted by $\mathfrak{FT}_p^{\mathbb{G}}$. The letters $\mathfrak{FT}$ stand for vector-valued **Fourier transform**.

Replacing the space $[L_p(\mathbb{G}), X]$ by $[L_{p,1}(\mathbb{G}), X]$ and $[L_{p'}(\mathbb{G}'), Y]$ by $[L_{p',\infty}(\mathbb{G}'), Y]$ yields the corresponding weak concept. The Banach ideal obtained in this way will be denoted by $\mathfrak{F}\mathfrak{T}_p^{\mathbb{G},weak}$.

REMARK. By the Riemann–Lebesgue theorem mentioned in the previous paragraph, all operators are compatible with the Fourier transform $F^{\mathbb{G}}$ from $L_1(\mathbb{G})$ into $C_0(\mathbb{G}')$.

5.8.7 Now we ask a fundamental question, which should be compared with 6.5.2.

PROBLEM. *Do the operator ideals* $\mathfrak{F}\mathfrak{T}_p^{\mathbb{G}}$ *and* $\mathfrak{F}\mathfrak{T}_p^{\mathbb{G},weak}$ *depend on the underlying locally compact Abelian group* $\mathbb{G}$ *?*

5.8.8 The following elementary properties are stated without proof.

PROPOSITION. *The operator ideals* $\mathfrak{F}\mathfrak{T}_p^{\mathbb{G}}$ *and* $\mathfrak{F}\mathfrak{T}_p^{\mathbb{G},weak}$ *are injective, surjective and* l_2*-stable. Moreover,*

$$(\mathfrak{F}\mathfrak{T}_p^{\mathbb{G}})' = \mathfrak{F}\mathfrak{T}_p^{\mathbb{G}'} \quad \textit{and} \quad (\mathfrak{F}\mathfrak{T}_p^{\mathbb{G},weak})' = \mathfrak{F}\mathfrak{T}_p^{\mathbb{G}',weak}.$$

5.8.9 Next, we introduce some auxiliary operators; see 5.4.11.

(1) For $a > 0$ and $f : \mathbb{R} \to \mathbb{C}$, the scale transform S_a is defined by

$$S_a : f(t) \longrightarrow f(at).$$

(2) We let

$$J : (\xi_k) \longrightarrow \sum_{\mathbb{Z}} \xi_k \, \chi_k,$$

where χ_k is the characteristic function of $\Delta_k := [\frac{2k-1}{2}, \frac{2k+1}{2})$.

(3) The surjection $R_{2\pi}$ restricts $f : \mathbb{R} \to \mathbb{C}$ to $[-\pi, +\pi)$.

(4) The injection $E_{2\pi}$ extends $f : [-\pi, +\pi) \to \mathbb{C}$ to a 2π-periodic function on $\mathbb{R}$.

(5) The map M multiplies $f : \mathbb{R} \to \mathbb{C}$ by $\frac{\sin s/2}{s/2}$.

(6) The map $M_{2\pi}^{-1}$ multiplies $f : [-\pi, +\pi) \to \mathbb{C}$ by $\frac{s/2}{\sin s/2}$.

5.8.10 We now list two formulas which connect the operators just defined with each other.

LEMMA. $\quad ME_{2\pi}F^{\mathbb{Z}} = F^{\mathbb{R}}J \quad$ *and* $\quad F^{\mathbb{Z}} = M_{2\pi}^{-1}R_{2\pi}F^{\mathbb{R}}J.$

PROOF. Use the elementary fact that

$$\widehat{\chi}_k(s) = \exp(iks)\,\frac{\sin \frac{1}{2}s}{\frac{1}{2}s}.$$

5.8.11 LEMMA. *If $2 \le q < \infty$, then*

$$\sum_{\mathbb{Z}} \left| \frac{\sin s}{s + h\pi} \right|^q \le 1 \quad \textit{whenever } s \notin \pi\mathbb{Z}.$$

PROOF. Apply the partial fraction decomposition

$$\frac{1}{\sin^2 s} = \sum_{\mathbb{Z}} \frac{1}{(s + h\pi)^2};$$

see [BENK*, p. 232]. Hence

$$\left(\sum_{\mathbb{Z}} \left| \frac{\sin s}{s + h\pi} \right|^q \right)^{1/q} \le \left(\sum_{\mathbb{Z}} \left| \frac{\sin s}{s + h\pi} \right|^2 \right)^{1/2} = 1.$$

5.8.12 LEMMA. *Let $2 \le q < \infty$. Then*

$$\| [ME_{2\pi}, X] : [L_q[-\pi, +\pi), X] \to [L_q(\mathbb{R}), X] \| \le (2\pi)^{1/q},$$
$$\| [M_{2\pi}^{-1} R_{2\pi}, X] : [L_q(\mathbb{R}), X] \to [L_q[-\pi, +\pi), X] \| \le \tfrac{\pi}{2} \tfrac{1}{(2\pi)^{1/q}}.$$

PROOF. By 5.8.11, we have

$$\| [ME_{2\pi}, X] \boldsymbol{f} | L_q(\mathbb{R}) \|^q = \int\limits_{-\infty}^{+\infty} \| E_{2\pi} \boldsymbol{f}(s) \|^q \left| \frac{\sin \frac{1}{2} s}{\frac{1}{2} s} \right|^q ds$$

$$= \sum_{\mathbb{Z}} \int\limits_{(2h-1)\pi}^{(2h+1)\pi} \| \boldsymbol{f}(s - 2h\pi) \|^q \left| \frac{\sin \frac{1}{2} s}{\frac{1}{2} s} \right|^q ds$$

$$= \int\limits_{-\pi}^{+\pi} \| \boldsymbol{f}(s) \|^q \sum_{\mathbb{Z}} \left| \frac{\sin \frac{s + 2h\pi}{2}}{\frac{s + 2h\pi}{2}} \right|^q ds$$

$$= \int\limits_{-\pi}^{+\pi} \| \boldsymbol{f}(s) \|^q \sum_{\mathbb{Z}} \left| \frac{\sin \frac{s}{2}}{\frac{s + 2h\pi}{2}} \right|^q ds \le \int\limits_{-\pi}^{+\pi} \| \boldsymbol{f}(s) \|^q ds = 2\pi \, \| \boldsymbol{f} | L_q[-\pi, +\pi) \|^q.$$

This proves the first inequality. The second inequality follows from $|\sin s| \ge \frac{2}{\pi} |s|$ for $|s| \le \frac{\pi}{2}$.

5.8.13 We are now able to give a very partial answer to Problem 5.8.7. The following proof goes back to S. Kwapień [kwa 2]; see also [koe 3].

THEOREM. $\mathfrak{FT}_p^{\mathbb{R}} = \mathfrak{FT}_p^{\mathbb{T}} = \mathfrak{FT}_p^{\mathbb{Z}}$ *for* $1 < p \le 2$.

PROOF. In view of 5.8.10, we have the commutative diagram

$$\begin{array}{ccc}
[L_p(\mathbb{R}), X] & \xrightarrow{\ [F^{\mathbb{R}}, T]\ } & [L_{p'}(\mathbb{R}), Y] \\
{\scriptstyle [J, X]} \uparrow & & \downarrow {\scriptstyle [M_{2\pi}^{-1} R_{2\pi}, Y]} \\
[l_p(\mathbb{Z}), X] & \xrightarrow[\ [F^{\mathbb{Z}}, T]\ ]{} & [L_{p'}[-\pi, +\pi), Y].
\end{array}$$

Hence

$$\|T|\mathfrak{F}\mathfrak{T}_p^{\mathbb{Z}}\| = \|[F^{\mathbb{Z}},T] : [l_p(\mathbb{Z}),X] \to [L_{p'}[-\pi,+\pi),Y]\|$$
$$\le \tfrac{\pi}{2}\tfrac{1}{(2\pi)^{1/p'}}\,\|[F^{\mathbb{R}},T] : [L_p(\mathbb{R}),X] \to [L_{p'}(\mathbb{R}),Y]\| = \tfrac{\pi}{2}\tfrac{1}{(2\pi)^{1/p'}}\,\|T|\mathfrak{F}\mathfrak{T}_p^{\mathbb{R}}\|.$$

This proves that $\mathfrak{F}\mathfrak{T}_p^{\mathbb{R}} \subseteq \mathfrak{F}\mathfrak{T}_p^{\mathbb{Z}}$.

In order to check the reverse inclusion, we let

$$\boldsymbol{f} := [J,X](x_k) = \sum_{\mathbb{Z}} x_k\chi_k, \tag{$*$}$$

where $(x_k) \in [l_p(\mathbb{Z}),X]$. Then it follows from 5.8.10 that

$$[F^{\mathbb{R}},T]\boldsymbol{f} = [ME_{2\pi}F^{\mathbb{Z}},T](x_k).$$

Thus, by 5.8.12,

$$\big\|[F^{\mathbb{R}},T]\boldsymbol{f}|L_{p'}(\mathbb{R})\big\| \le$$
$$\le \big\|[ME_{2\pi},Y]:[L_{p'}[-\pi,+\pi),Y]\to[L_{p'}(\mathbb{R}),Y]\big\|\,\big\|[F^{\mathbb{Z}},T](x_k)|L_{p'}[-\pi,+\pi)\big\|$$
$$\le (2\pi)^{1/p'}\big\|[F^{\mathbb{Z}},T] : [l_p(\mathbb{Z}),X] \to [L_{p'}[-\pi,+\pi),Y]\big\|\,\|(x_k)|l_p(\mathbb{Z})\|$$
$$= (2\pi)^{1/p'}\,\|T|\mathfrak{F}\mathfrak{T}_p^{\mathbb{Z}}\|\|\boldsymbol{f}|L_p(\mathbb{R})\|$$

whenever $\boldsymbol{f}$ can be written in the form $(*)$. Note that

$$\|[S_a,X]\boldsymbol{f}|L_p(\mathbb{R})\| = a^{-1/p}\,\|\boldsymbol{f}|L_p(\mathbb{R})\|$$

and

$$\|[F^{\mathbb{R}},T][S_a,X]\boldsymbol{f}|L_{p'}(\mathbb{R})\| = a^{-1/p}\,\|[F^{\mathbb{R}},T]\boldsymbol{f}|L_{p'}(\mathbb{R})\|.$$

The first formula is elementary, and the second formula follows from $F^{\mathbb{R}}S_a = a^{-1}S_{a^{-1}}F^{\mathbb{R}}$. Consequently, the above estimate remains true for all functions $[S_a,X]\boldsymbol{f}$ produced by scaling. Since these functions are dense in $[L_p(\mathbb{R}),X]$, we have

$$\|[F^{\mathbb{R}},T]\boldsymbol{f}|L_{p'}(\mathbb{R})\| \le (2\pi)^{1/p'}\,\|T|\mathfrak{F}\mathfrak{T}_p^{\mathbb{Z}}\|\,\|\boldsymbol{f}|L_p(\mathbb{R})\| \quad \text{for } \boldsymbol{f} \in [L_p(\mathbb{R}),X].$$

Hence

$$\|T|\mathfrak{F}\mathfrak{T}_p^{\mathbb{R}}\| \le (2\pi)^{1/p'}\,\|T|\mathfrak{F}\mathfrak{T}_p^{\mathbb{Z}}\|.$$

This proves that $\mathfrak{F}\mathfrak{T}_p^{\mathbb{Z}} \subseteq \mathfrak{F}\mathfrak{T}_p^{\mathbb{R}}$. The results obtained so far imply that

$$\mathfrak{F}\mathfrak{T}_p^{\mathbb{R}} = \mathfrak{F}\mathfrak{T}_p^{\mathbb{Z}}.$$

Finally, 5.8.8 tells us that

$$\mathfrak{F}\mathfrak{T}_p^{\mathbb{T}} = (\mathfrak{F}\mathfrak{T}_p^{\mathbb{Z}})' = (\mathfrak{F}\mathfrak{T}_p^{\mathbb{R}})' = \mathfrak{F}\mathfrak{T}_p^{\mathbb{R}}.$$

REMARK. From now on, we denote the ideals $\mathfrak{F}\mathfrak{T}_p^{\mathbb{R}} = \mathfrak{F}\mathfrak{T}_p^{\mathbb{T}} = \mathfrak{F}\mathfrak{T}_p^{\mathbb{Z}}$ simply by $\mathfrak{F}\mathfrak{T}_p$.

5.8.14 PROPOSITION. $\mathfrak{FT}_p = \mathfrak{ET}_p$ *if* $1 < p \leq 2$.

PROOF. Obviously, the Banach operator ideals $\mathfrak{FT}_p^{\mathbb{Z}}$ and $\mathfrak{ET}_p$ coincide isometrically.

REMARK. In view of $\mathfrak{FT}_p^{\mathbb{Z}} = \mathfrak{ET}_p$ and $\mathfrak{FT}_p^{\mathbb{T}} = \mathfrak{EC}_{p'}$, the equality $\mathfrak{FT}_p^{\mathbb{T}} = \mathfrak{FT}_p^{\mathbb{Z}}$ also follows from 5.7.4.

5.8.15 We now transfer 5.8.13 to the *weak* setting.

THEOREM. $\mathfrak{FT}_p^{\mathbb{R},weak} = \mathfrak{FT}_p^{\mathbb{T},weak} = \mathfrak{FT}_p^{\mathbb{Z},weak}$ *if* $1 < p \leq 2$.

PROOF. By real interpolation, Lemma 5.8.12 can be extended to Lorentz spaces. Hence the proof of 5.8.13 also works in that case. For example, we have the diagram

$$\begin{array}{ccc}
[L_{p,1}(\mathbb{R}), X] & \xrightarrow{[F^{\mathbb{R}}, T]} & [L_{p',\infty}(\mathbb{R}), Y] \\
{\scriptstyle [J,X]} \uparrow & & \downarrow {\scriptstyle [M_{2\pi}^{-1} R_{2\pi}, Y]} \\
[l_{p,1}(\mathbb{Z}), X] & \xrightarrow[{[F^{\mathbb{Z}}, T]}]{} & [L_{p',\infty}[-\pi,+\pi), Y].
\end{array}$$

REMARK. As in the case of $\mathfrak{FT}_p$, from now on, we denote the ideals $\mathfrak{FT}_p^{\mathbb{R},weak} = \mathfrak{FT}_p^{\mathbb{T},weak} = \mathfrak{FT}_p^{\mathbb{Z},weak}$ by $\mathfrak{FT}_p^{weak}$.

5.8.16 The following result is analogous to half of 5.8.14.

PROPOSITION. $\mathfrak{FT}_p^{weak} \subseteq \mathfrak{ET}_p^{weak}$ *if* $1 < p < 2$.

PROOF. By 0.5.6,

$$\|(x_k)|l_{p,1}^n\| \leq A\, n^{1/p-1/2} \|(x_k)|l_2^n\|,$$

where $A > 0$ is a constant. Similarly, the continuity of the embedding $[L_{p',\infty}[-\pi,+\pi), Y] \subset [L_2[-\pi,+\pi), Y]$ yields another constant $B > 0$ such that

$$\|(Tx_k)|\mathcal{E}_n\| = \left\| \sum_{k=1}^{n} Tx_k e_k \middle| L_2 \right\| \leq B \left\| \sum_{k=1}^{n} Tx_k e_k \middle| L_{p',\infty} \right\|.$$

Hence

$$\varrho(T|\mathcal{E}_n, \mathcal{I}_n) \leq AB\, n^{1/p-1/2} \|T|\mathfrak{FT}_p^{\mathbb{Z},weak}\|.$$

5.8.17 It is unknown whether 5.8.14 extends to the *weak* setting.

PROBLEM. *Do the ideals* $\mathfrak{FT}_p^{weak}$ *and* $\mathfrak{ET}_p^{weak}$ *coincide?*

5.8.18 We are now in a position to establish an analogue of 5.6.23.

THEOREM. $\mathfrak{FT}_2 \subset \mathfrak{FT}_{p_0} \subset \mathfrak{FT}_{p_0}^{weak} \subset \mathfrak{FT}_p$ *if* $1<p<p_0<2$.

In particular,

$$\bigcup_{1<p\leq 2} \mathfrak{FT}_p = \bigcup_{1<p<2} \mathfrak{FT}_p^{weak}.$$

PROOF. The inclusion $\mathfrak{FT}_{p_0} \subseteq \mathfrak{FT}_{p_0}^{weak}$ is obvious, since

$$[L_{p_0,1}(\mathbb{R}), X] \subset [L_{p_0}(\mathbb{R}), X] \quad \text{and} \quad [L_{p_0'}(\mathbb{R}), Y] \subset [L_{p_0',\infty}(\mathbb{R}), Y].$$

On the other hand, we know from 5.8.16, 5.6.23 and 5.8.14 that

$$\mathfrak{FT}_{p_0}^{weak} \subseteq \mathfrak{ET}_{p_0}^{weak} \subseteq \mathfrak{ET}_p = \mathfrak{FT}_p.$$

In order to give a direct proof of the inclusion $\mathfrak{FT}_{p_0}^{weak} \subseteq \mathfrak{FT}_p$, define θ by $1/p = (1-\theta)/p_0 + \theta/1$. Then, using the real interpolation method, we obtain the formulas

$$([L_{p_0,1}(\mathbb{R}), X], [L_1(\mathbb{R}), X])_{\theta,p} = [L_p(\mathbb{R}), X]$$

and

$$([L_{p_0',\infty}(\mathbb{R}), Y], [C_0(\mathbb{R}), Y])_{\theta,p} = [L_{p',p_0}(\mathbb{R}), Y].$$

Moreover,

$$[L_{p',p}(\mathbb{R}), Y] \subseteq [L_{p',p'}(\mathbb{R}), Y],$$

since $p \leq p'$. By definition, $T \in \mathfrak{FT}_{p_0}^{weak}(X, Y)$ means that the operator

$$[F^{\mathbb{R}}, T] : [L_{p_0,1}(\mathbb{R}), X] \longrightarrow [L_{p_0',\infty}(\mathbb{R}), Y]$$

is continuous. As observed in 5.8.5, we always have

$$[F^{\mathbb{R}}, T] : [L_1(\mathbb{R}), X] \longrightarrow [C_0(\mathbb{R}), Y].$$

Thus it follows by interpolation that

$$[F^{\mathbb{R}}, T] : [L_p(\mathbb{R}), X] \longrightarrow [L_{p'}(\mathbb{R}), Y].$$

Hence $T \in \mathfrak{FT}_p(X, Y)$.

It will be shown in 5.8.19 that

$$L_p \in \mathsf{FT}_p \setminus \mathsf{FT}_{p_0}^{weak}.$$

Thus the inclusion $\mathfrak{FT}_{p_0}^{weak} \subseteq \mathfrak{FT}_p$ is strict. Finally, 5.8.21 shows that the inclusion $\mathfrak{FT}_{p_0} \subseteq \mathfrak{FT}_{p_0}^{weak}$ is strict as well.

5.8.19 EXAMPLE. L_p, $L_{p'} \in \mathsf{FT}_p \setminus \mathsf{FT}_{p_0}^{weak}$ *if* $1 < p < p_0 < 2$.

PROOF. Recall that

$$L_p,\, L_{p'} \in \mathsf{ET}_p \setminus \mathsf{ET}_{p_0}^{weak} \qquad \text{(by 5.6.24)},$$
$$\mathsf{FT}_p = \mathsf{ET}_p \qquad \text{(by 5.8.14)},$$
$$\mathsf{FT}_{p_0}^{weak} \subseteq \mathsf{ET}_{p_0}^{weak} \qquad \text{(by 5.8.16)}.$$

5.8.20 To establish the next example, we need a result from interpolation theory which goes back to J. Lions and J. Peetre [lio*, p. 48]. For a proof, the reader is referred to [koe 2].

PROPOSITION. *Let $0<\theta<1$ and $1 \le u,v,w \le \infty$ such that $1/w = 1/u + 1/v - 1$. Assume that B is a bilinear map which transforms $(X_0+X_1)\times(Y_0+Y_1)$ into Z_0+Z_1, where (X_0,X_1), (Y_0,Y_1) and (Z_0,Z_1) denote interpolation couples. If the induced maps*

$$B_0 : X_0 \times Y_0 \xrightarrow{B} Z_0 \quad \textit{and} \quad B_1 : X_1 \times Y_1 \xrightarrow{B} Z_1$$

are bounded, then so is

$$B_{\theta,u,v} : X_{\theta,u} \times Y_{\theta,v} \xrightarrow{B} Z_{\theta,w}.$$

5.8.21 For $t = (\tau_k)$, we define

$$D_t : (\xi_k) \longrightarrow (\tau_k \xi_k).$$

In particular, the diagonal operator associated with $(k^{-1/p'})$ will be denoted by $D_{1/p'}$.

EXAMPLE. $D_{1/p'} \in \mathfrak{FT}_p^{weak}(l_1) \setminus \mathfrak{FT}_p(l_1)$ *if* $1<p<2$.

PROOF. Consider the bilinear operator $B(t,\boldsymbol{f}) := [F^{\mathbb{R}}, D_t]\boldsymbol{f}$, and note that

$$B_0 : l_\infty \times [L_1(\mathbb{R}), l_1] \xrightarrow{B} [C_0(\mathbb{R}), l_1]$$

and

$$B_1 : l_2 \times [L_2(\mathbb{R}), l_1] \xrightarrow{B} [L_2(\mathbb{R}), l_1]$$

are bounded. If $u := \infty$, $v := 1$ and $w := \infty$, then it follows from the preceding proposition that

$$B_{\theta,\infty,1} : l_{p',\infty} \times [L_{p,1}(\mathbb{R}), l_1] \xrightarrow{B} [L_{p',\infty}(\mathbb{R}), l_1]$$

with $1/p' = (1-\theta)/\infty + \theta/2$ is bounded as well. So $D_{1/p'} \in \mathfrak{FT}_p^{weak}(l_1)$, since $(k^{-1/p'}) \in l_{p',\infty}$. On the other hand, $D_{1/p'} \notin \mathfrak{FT}_p(l_1)$ follows from 5.6.25 and 5.8.14.

5.8.22 We do not know whether Tzafriri's example, quoted in 4.3.16 and 4.5.16, yields a Banach space in $\mathsf{FT}_p^{weak} \setminus \mathsf{FT}_p$.

PROBLEM. *Is* FT_p *strictly smaller than* FT_p^{weak} *whenever* $1<p<2$ *?*

5.8.23 The following question is raised because of the analogy with the situation for Rademacher type and cotype; see [jam 7, p. 2].

PROBLEM. *Is it true that* $\mathfrak{FT}_2 = \mathfrak{FT}_2^{weak}$ *?*

5.9 Fourier versus Gauss and Rademacher

5.9.1 We now compare ideal norms associated with the Fourier system, on the one hand, and the Gauss, Rademacher or Steinhaus systems, on the other hand. Since $\mathcal{E}_n$ is complex, there are some advantages in working with the systems $\mathcal{G}_n^{\mathbb{C}}$ and $\mathcal{P}_n(e^{i\tau})$. However, all asymptotic relations also hold for $\mathcal{G}_n^{\mathbb{R}}$ and $\mathcal{R}_n$. To simplify matters, we will formulate most of the statements in terms of $\mathcal{G}_n$ and $\mathcal{R}_n$.

5.9.2 To begin with, we treat type and cotype ideal norms.

PROPOSITION.

$$\varrho_u^{(v)}(\mathcal{P}_n(e^{i\tau}),\mathcal{I}_n) \le \varrho_u^{(v)}(\mathcal{E}_n,\mathcal{I}_n) \quad \textit{and} \quad \varrho_u^{(v)}(\mathcal{I}_n,\mathcal{P}_n(e^{i\tau})) \le \varrho_u^{(v)}(\mathcal{I}_n,\mathcal{E}_n).$$

PROOF. Since $\mathcal{E}_n$ is $\mathbb{C}$-unimodular and $\mathcal{P}_n(e^{i\tau})$ is $\mathbb{C}$-unconditional, the desired inequalities are obtained by using the tensor product technique from 3.7.9.

5.9.3 In the language of operator ideals the previous result, combined with 5.7.3, reads as follows.

THEOREM.

$$\mathfrak{E}\mathfrak{T}_p \subseteq \mathfrak{R}\mathfrak{T}_p \cap \mathfrak{R}\mathfrak{C}_{p'} \quad \textit{and} \quad \mathfrak{E}\mathfrak{T}_p^{weak} \subseteq \mathfrak{R}\mathfrak{T}_p^{weak} \cap \mathfrak{R}\mathfrak{C}_{p'}^{weak}.$$

5.9.4 For p=2, we get a famous theorem due to S. Kwapień [kwa 2].

THEOREM. $\mathsf{ET}_2 = \mathsf{EC}_2 = \mathsf{H}$.

PROOF. Obviously, 4.10.7 and 5.7.4 imply that

$$\mathsf{ET}_2 = \mathsf{EC}_2 \subseteq \mathsf{RT}_2 \cap \mathsf{RC}_2 = \mathsf{H}.$$

The reverse inclusion is trivial.

5.9.5 It seems to be unknown whether the preceding result still holds in the setting of operators.

PROBLEM. *Is it true that* $\mathfrak{E}\mathfrak{T}_2 = \mathfrak{H}$ *?*

REMARK. We suspect that the answer might be negative.

5.9.6 In consequence of 4.10.6 we obtain a counterpart of 5.9.3.

PROPOSITION.
Let $1 < p < 2 < q < \infty$ *and* $1/r - 1/2 = 1/p - 1/q < 1/2$. *Then*

$$\mathfrak{R}\mathfrak{C}_q^{weak} \circ \mathfrak{R}\mathfrak{T}_p^{weak} \subseteq \mathfrak{G}\mathfrak{C}_q^{weak} \circ \mathfrak{G}\mathfrak{T}_p^{weak} \subseteq \mathfrak{E}\mathfrak{T}_r^{weak}.$$

5.9.7 It turns out that the condition $1/r-1/2=1/p-1/q<1/2$ posed in the above proposition cannot be dropped. This follows from an example which goes back to J. Bourgain [bou 6, p. 117]; see also [TOM, p. 212].

EXAMPLE. *For* $1<r<2$ *and* $0<\theta<1$, *there exists a Banach space* $X_{r,\theta}$ *with the following properties:*

$$\varrho(X_{r,\theta}|\mathcal{E}_n,\mathcal{I}_n)\asymp n^{1/r-1/2},$$
$$\varrho(X_{r,\theta}|\mathcal{R}_n,\mathcal{I}_n)\asymp n^{(1-\theta)(1/r-1/2)},$$
$$\varrho(X_{r,\theta}|\mathcal{I}_n,\mathcal{R}_n)\asymp n^{\theta(1/r-1/2)}.$$

PROOF. As observed in 5.8.4, the discrete Fourier transform

$$F^{\mathbb{Z}}:(\xi_k)\longrightarrow\sum_{\mathbb{Z}}\xi_k\exp(ikt)$$

defines an operator from $l_r(\mathbb{Z})$ into $L_{r'}[-\pi,+\pi)$. In what follows, to simplify notation, we write F instead of $F^{\mathbb{Z}}$. In view of 5.1.7,

$$\|(Fu_k)|[\mathcal{E}_n,L_{r'}]\|=\|(e_k)|[\mathcal{E}_n,L_{r'}]\|=\|d_n|L_{r'}\|\asymp n^{1/r}.$$

Moreover, $\|(u_k)|[\mathcal{I}_n,l_r]\|=\sqrt{n}$. Hence it follows from

$$\|(Fu_k)|[\mathcal{E}_n,L_{r'}]\|\leq\varrho(F:l_r\to L_{r'}|\mathcal{E}_n,\mathcal{I}_n)\,\|(u_k)|[\mathcal{I}_n,l_r]\|$$

that

$$n^{1/r-1/2}\prec\varrho(F:l_r\to L_{r'}|\mathcal{E}_n,\mathcal{I}_n). \tag{1}$$

We now build a new Banach space, which is continuously embedded in $L_{r'}[-\pi,+\pi)$, by transferring the norm from $l_r(\mathbb{Z})$ to $F(l_r(\mathbb{Z}))$:

$$\|f|F(l_r(\mathbb{Z}))\|:=\|F^{-1}f|l_r(\mathbb{Z})\|\quad\text{for } f\in F(l_r(\mathbb{Z})).$$

Letting $X_{r,\theta}:=[F(l_r(\mathbb{Z})),L_{r'}[-\pi,+\pi)]_\theta$ yields the factorization

$$F:l_r(\mathbb{Z})\xrightarrow{F}F(l_r(\mathbb{Z}))\xrightarrow{A}X_{r,\theta}\xrightarrow{B}L_{r'}[-\pi,+\pi).$$

By definition of $F(l_r(\mathbb{Z}))$, the first operator is an isometry, while A and B are embedding maps with $\|A\|=\|B\|=1$. Thus (1) and

$$n^{1/r-1/2}\prec\varrho(F:l_r\to L_{r'}|\mathcal{E}_n,\mathcal{I}_n)\leq\|B\|\,\varrho(X_{r,\theta}|\mathcal{E}_n,\mathcal{I}_n)\,\|AF\|$$

imply

$$n^{1/r-1/2}\prec\varrho(X_{r,\theta}|\mathcal{E}_n,\mathcal{I}_n). \tag{2}$$

We know that

$$\varrho(F(l_r(\mathbb{Z}))|\mathcal{R}_n,\mathcal{I}_n)=\varrho(l_r(\mathbb{Z})|\mathcal{R}_n,\mathcal{I}_n)=n^{1/r-1/2}\quad\text{(by 4.2.7)}$$

and

$$\varrho(L_{r'}[-\pi,+\pi)|\mathcal{R}_n,\mathcal{I}_n)\asymp 1\quad\text{(by 4.2.8).}$$

Now interpolation gives

$$\varrho(X_{r,\theta}|\mathcal{R}_n, \mathcal{I}_n) \prec n^{(1-\theta)(1/r-1/2)}. \qquad (3)$$

The duality theorem of complex interpolation says that

$$\Big[F(l_r(\mathbb{Z})), L_{r'}[-\pi,+\pi)\Big]'_\theta \quad \text{and} \quad \Big[F(l_r(\mathbb{Z}))', L_{r'}[-\pi,+\pi)'\Big]_\theta$$

can be identified isometrically; see [BER*, p. 98] or [KRE*, p. 231]. Hence, again by interpolation,

$$\varrho(X'_{r,\theta}|\mathcal{R}_n, \mathcal{I}_n) \prec n^{\theta(1/r-1/2)}.$$

In view of

$$\varrho(X_{r,\theta}|\mathcal{I}_n, \mathcal{R}_n) \le \varrho(X'_{r,\theta}|\mathcal{R}_n, \mathcal{I}_n),$$

we obtain

$$\varrho(X_{r,\theta}|\mathcal{I}_n, \mathcal{R}_n) \prec n^{\theta(1/r-1/2)}. \qquad (4)$$

Finally, (2), (3) and (4) combined with 4.10.5 yield

$$\begin{aligned} n^{1/r-1/2} &\prec \varrho(X_{r,\theta}|\mathcal{E}_n, \mathcal{I}_n) \le \boldsymbol{\kappa}_n(X_{r,\theta}) \\ &\le \varrho(X_{r,\theta}|\mathcal{I}_n, \mathcal{R}_n)\, \varrho(X_{r,\theta}|\mathcal{R}_n, \mathcal{I}_n) \prec n^{1/r-1/2}. \end{aligned}$$

Thus we even have $\asymp$ everywhere. This fact is, in particular, true for (2), (3) and (4).

REMARKS. Let $1 < p \le 2 \le q < \infty$ and $1/r - 1/2 = 1/p - 1/q$.

(a) Assume that $1/p - 1/q < 1/2$. Define $\theta \in (0,1)$ by

$$1/p - 1/2 = (1-\theta)(1/r - 1/2) \quad \text{and} \quad 1/2 - 1/q = \theta(1/r - 1/2).$$

Then

$$X_{r,\theta} \in \mathsf{RT}_p^{weak} \cap \mathsf{RC}_q^{weak} \quad \text{and} \quad X_{r,\theta} \notin \mathsf{ET}_{r_0}^{weak} \quad \text{whenever } r < r_0 \le 2.$$

This shows that the inclusion stated in 5.9.6 is best possible.

(b) Assume that $1/p - 1/q \ge 1/2$ and $1 < r_0 < r \le 2$. Define $\lambda \in (0,1)$ by

$$1/r_0 - 1/2 = \lambda(1/p - 1/q).$$

Let

$$1/p_0 := \lambda/p + (1-\lambda)/2 \quad \text{and} \quad 1/q_0 := \lambda/q + (1-\lambda)/2.$$

Then

$$1 < p < p_0 \le 2, \quad 2 \le q_0 < q < \infty \quad \text{and} \quad 1/r_0 - 1/2 = 1/p_0 - 1/q_0.$$

Thus we have found a Banach space

$$X_0 \in \mathsf{RT}_{p_0}^{weak} \cap \mathsf{RC}_{q_0}^{weak} \subset \mathsf{RT}_p \cap \mathsf{RC}_q$$

that does not belong to ET_r^{weak}. This proves that

$$\mathsf{RT}_p \cap \mathsf{RC}_q \not\subseteq \mathsf{ET}_r \quad \text{whenever } 1/p - 1/q \ge 1/2 \text{ and } 1 < r \le 2.$$

5.9.8 We now begin to prepare the proof of Theorem 5.9.14.

PROPOSITION. *Assume that the orthonormal system $\mathcal{A}_n$ is $\mathbb{C}$-unconditional. Then*

$$\varrho(\mathcal{A}_n,\mathcal{E}_n) \asymp \varrho(\mathcal{A}_n,\mathcal{E}_n^\circ) \quad \textit{and} \quad \varrho(\mathcal{E}_n,\mathcal{A}_n) \asymp \varrho(\mathcal{E}_n^\circ,\mathcal{A}_n).$$

PROOF. Recall that

$$\begin{aligned} \|(x_k)|\mathcal{E}_n\| &= \|(x_k)|\mathcal{A}_n \otimes \mathcal{E}_n\| && \text{(by 3.6.2)},\\ \|(x_k)|\mathcal{E}_n\| &= \|(x_k)|\mathcal{E}_n \otimes \mathcal{E}_n^\circ\| && \text{(by 5.5.13)},\\ \varrho(\mathcal{E}_n^\circ,\mathcal{E}_n) &\le 3 && \text{(by 5.2.5)}. \end{aligned}$$

Hence, applying 3.6.3 and the triangle inequality for Riemann ideal norms, we get

$$\varrho(\mathcal{A}_n,\mathcal{E}_n) = \varrho(\mathcal{A}_n\otimes\mathcal{E}_n,\mathcal{E}_n^\circ\otimes\mathcal{E}_n) \le \varrho(\mathcal{A}_n,\mathcal{E}_n^\circ) \le 3\,\varrho(\mathcal{A}_n,\mathcal{E}_n).$$

This proves the left-hand relation. The right-hand one can be checked similarly.

5.9.9 We proceed with another consequence of $\varrho(\mathcal{E}_n^\circ,\mathcal{E}_n) \le 3$.

PROPOSITION. *Let $\mathcal{A}_n$ be any orthonormal system. Then*

$$\delta(\mathcal{E}_n^\circ,\mathcal{A}_n) \le 3\,\delta(\mathcal{E}_n,\mathcal{A}_n) \quad \textit{and} \quad \delta(\mathcal{A}_n,\mathcal{E}_n^\circ) \le 3\,\delta(\mathcal{A}_n,\mathcal{E}_n).$$

5.9.10 Next, we state a corollary of 3.5.5.

PROPOSITION. *The sequences $(\delta(\mathcal{G}_n,\mathcal{E}_n))$ and $(\delta(\mathcal{R}_n,\mathcal{E}_n))$ are non-decreasing.*

5.9.11 The following inequalities are analogues of 5.5.3.

LEMMA.

$$\delta(\mathcal{G}_{2n},\mathcal{E}_{2n}) \le 2\,\delta(\mathcal{G}_n,\mathcal{E}_n) \quad \textit{and} \quad \delta(\mathcal{R}_{2n},\mathcal{E}_{2n}) \le 2\,\delta(\mathcal{R}_n,\mathcal{E}_n).$$

PROOF. Let

$$\mathcal{G}_{2n} = (\overbrace{g_1^{(2n)},\dots,g_n^{(2n)}}^{\mathcal{G}_n},\overbrace{g_{n+1}^{(2n)},\dots,g_{2n}^{(2n)}}^{\mathcal{G}_n^*})$$

and

$$\mathcal{E}_{2n} = (\overbrace{e_1,\dots,e_n}^{\mathcal{E}_n},\overbrace{e_{n+1},\dots,e_{2n}}^{\mathcal{E}_n^*}).$$

Since $\delta(\mathcal{G}_n^*,\mathcal{E}_n^*) = \delta(\mathcal{G}_n,\mathcal{E}_n)$, the left-hand inequality follows from 3.4.13. The right-hand one can be proved in the same way.

5.9.12 The next inequality looks a little artificial. But it is the crucial tool in the proof of Theorem 5.9.14.

LEMMA. *Let*

$$\mathcal{A}_{4m-1} = (a_1, \dots, a_{m-1}, \overbrace{a_m, \; \dots, a_{3m}}^{\mathcal{A}_{2m+1}}, a_{3m+1}, \dots, a_{4m-1})$$

be $\mathbb{R}$*-unconditional. Then*

$$\boldsymbol{\delta}(\mathcal{A}_{2m+1}, \mathcal{E}_{2m+1}) \le 3\, \boldsymbol{\varrho}(\mathcal{A}_{4m-1}, \mathcal{E}_{4m-1}).$$

PROOF. The first part of the proof in 5.5.4 can be adopted word for word, and the rest has to be modified as follows:
Hence

$$\begin{aligned}
\|(T\langle \boldsymbol{f}, \overline{e_h}\rangle)|\mathcal{A}_{2m+1}\| &= \left(\int\limits_M \Big\| \sum_{h=1}^{2m+1} T\langle \boldsymbol{f}, \overline{e_h}\rangle\, a_{h+m-1}(s) \Big\|^2 d\mu(s) \right)^{1/2} \\
&= \left(\int\limits_M \Big\| \sum_{k=m}^{3m} T\langle V_m^0 * \boldsymbol{f}^0, \overline{e_k}\rangle\, a_k(s) \Big\|^2 d\mu(s) \right)^{1/2} \\
&\le \left(\int\limits_M \Big\| \sum_{k=1}^{4m-1} T\langle V_m^0 * \boldsymbol{f}^0, \overline{e_k}\rangle\, a_k(s) \Big\|^2 d\mu(s) \right)^{1/2} \\
&\qquad\qquad\qquad\qquad\qquad\qquad\text{(by } \mathbb{R}\text{-unconditionality)} \\
&= \big\| \big(T\langle V_m^0 * \boldsymbol{f}^0, \overline{e_k}\rangle\big) \big| \mathcal{A}_{4m-1} \big\| \\
&\le \boldsymbol{\varrho}(T|\mathcal{A}_{4m-1}, \mathcal{E}_{4m-1})\, \big\| \langle V_m^0 * \boldsymbol{f}^0, \overline{e_k}\rangle \big| \mathcal{E}_{4m-1} \big\| \\
&\le 3\, \boldsymbol{\varrho}(T|\mathcal{A}_{4m-1}, \mathcal{E}_{4m-1})\, \|\boldsymbol{f}|L_2\|,
\end{aligned}$$

which yields the desired result.

5.9.13 PROPOSITION.

$$\boldsymbol{\delta}(\mathcal{G}_n, \mathcal{E}_n) \le 6\, \boldsymbol{\varrho}(\mathcal{G}_n, \mathcal{E}_n) \quad \textit{and} \quad \boldsymbol{\delta}(\mathcal{R}_n, \mathcal{E}_n) \le 6\, \boldsymbol{\varrho}(\mathcal{R}_n, \mathcal{E}_n).$$

PROOF. We may assume that $n \ge 2$. Let $m := [\frac{n}{2}]$. Then $n \le 2m+1$ and $4m-1 \le 2n$. Using the monotonicity of $\boldsymbol{\varrho}(\mathcal{G}_n, \mathcal{E}_n)$ and $\boldsymbol{\delta}(\mathcal{G}_n, \mathcal{E}_n)$, we conclude from Lemmas 5.9.11 and 5.9.12 that

$$\begin{aligned}
\boldsymbol{\delta}(\mathcal{G}_n, \mathcal{E}_n) &\le \boldsymbol{\delta}(\mathcal{G}_{2m+1}, \mathcal{E}_{2m+1}) \le 3\, \boldsymbol{\varrho}(\mathcal{G}_{4m-1}, \mathcal{E}_{4m-1}) \\
&\le 3\, \boldsymbol{\varrho}(\mathcal{G}_{2n}, \mathcal{E}_{2n}) \le 3 \cdot 2\, \boldsymbol{\varrho}(\mathcal{G}_n, \mathcal{E}_n).
\end{aligned}$$

This proves the left-hand inequality. The right-hand one can be obtained in the same manner.

5.9.14 We are now in a position to establish a basic result of this section.

THEOREM. *In each of the rows below, the sequences of ideal norms are uniformly equivalent.*

$$
\begin{array}{lllll}
(\mathbf{1})_G & \varrho(\mathcal{G}_n,\mathcal{E}_n), & \varrho(\mathcal{G}_n,\mathcal{E}_n^\circ), & \delta(\mathcal{G}_n,\mathcal{E}_n), & \delta(\mathcal{G}_n,\mathcal{E}_n^\circ),\\
(\mathbf{2})_G & \varrho(\mathcal{E}_n,\mathcal{G}_n), & \varrho(\mathcal{E}_n^\circ,\mathcal{G}_n), & & \\
(\mathbf{3})_G & & & \delta(\mathcal{E}_n,\mathcal{G}_n), & \delta(\mathcal{E}_n^\circ,\mathcal{G}_n),
\end{array}
$$

$$
\begin{array}{lllll}
(\mathbf{1})_R & \varrho(\mathcal{R}_n,\mathcal{E}_n), & \varrho(\mathcal{R}_n,\mathcal{E}_n^\circ), & \delta(\mathcal{R}_n,\mathcal{E}_n), & \delta(\mathcal{R}_n,\mathcal{E}_n^\circ),\\
(\mathbf{2})_R & \varrho(\mathcal{E}_n,\mathcal{R}_n), & \varrho(\mathcal{E}_n^\circ,\mathcal{R}_n), & & \\
(\mathbf{3})_R & & & \delta(\mathcal{E}_n,\mathcal{R}_n), & \delta(\mathcal{E}_n^\circ,\mathcal{R}_n).
\end{array}
$$

PROOF.

$(\mathbf{1})_G$: It follows from 5.9.8 (proof), 3.4.12, 5.9.9 and 5.9.13 that

$$\varrho(\mathcal{G}_n^{\mathbb{C}},\mathcal{E}_n) \le \varrho(\mathcal{G}_n^{\mathbb{C}},\mathcal{E}_n^\circ) = \delta(\mathcal{G}_n^{\mathbb{C}},\mathcal{E}_n^\circ) \le 3\,\delta(\mathcal{G}_n^{\mathbb{C}},\mathcal{E}_n) \le 18\,\varrho(\mathcal{G}_n^{\mathbb{C}},\mathcal{E}_n).$$

$(\mathbf{2})_G$: Use 5.9.8.

$(\mathbf{3})_G$: Dualize $\delta(\mathcal{G}_n^{\mathbb{C}},\mathcal{E}_n) \asymp \delta(\mathcal{G}_n^{\mathbb{C}},\mathcal{E}_n^\circ)$ from $(\mathbf{1})_G$.

The case of the Rademacher system can be treated analogously.

REMARK. Since

$$\varrho(L_1|\mathcal{E}_n,\mathcal{R}_n) \asymp \varrho(L_1|\mathcal{E}_n,\mathcal{G}_n) \asymp 1 \quad\text{and}\quad \delta(L_1|\mathcal{E}_n,\mathcal{R}_n) \asymp \delta(L_1|\mathcal{E}_n,\mathcal{G}_n) \asymp \sqrt{n},$$

we have $\varrho(\mathcal{E}_n,\mathcal{G}_n) \not\asymp \delta(\mathcal{E}_n,\mathcal{G}_n)$ and $\varrho(\mathcal{E}_n,\mathcal{R}_n) \not\asymp \delta(\mathcal{E}_n,\mathcal{R}_n)$.

5.9.15 The following fact tells us that the new ideal norms derived from $\mathcal{G}_n$ and $\mathcal{E}_n$ are uniformly equivalent to the Gauss type or Gauss cotype ideal norms; see [gei*].

PROPOSITION.

$$\varrho(\mathcal{G}_n,\mathcal{E}_n) \asymp \varrho(\mathcal{G}_n,\mathcal{I}_n) \quad \textit{and} \quad \varrho(\mathcal{E}_n,\mathcal{G}_n) \asymp \varrho(\mathcal{I}_n,\mathcal{G}_n).$$

PROOF. Using the invariance of the complex Gaussian measure $\gamma_{\mathbb{C}}^n$ under unitary transformations, we obtain

$$\varrho(\mathcal{G}_n^{\mathbb{C}},\mathcal{E}_n^\circ) = \varrho(\mathcal{G}_n^{\mathbb{C}},\mathcal{I}_n) \quad\text{and}\quad \varrho(\mathcal{E}_n^\circ,\mathcal{G}_n^{\mathbb{C}}) = \varrho(\mathcal{I}_n,\mathcal{G}_n^{\mathbb{C}}).$$

5.9.16 The situation is not clear for the Rademacher system.

PROBLEM. *Is it true that*

$$\varrho(\mathcal{R}_n,\mathcal{E}_n) \asymp \varrho(\mathcal{R}_n,\mathcal{I}_n) \quad \textit{and} \quad \varrho(\mathcal{E}_n,\mathcal{R}_n) \asymp \varrho(\mathcal{I}_n,\mathcal{R}_n)\ ?$$

REMARK. By 4.8.22, the answer is certainly affirmative in the setting of MP-convex Banach spaces.

5.9.17 In the following, we exclude the limiting case $q = \infty$. Thus, in view of the preceding observations, the next example is an immediate consequence of 4.2.7. We prefer, however, to give a direct proof.

EXAMPLE.

$$\varrho(l_p^n|\mathcal{R}_n, \mathcal{E}_n) \asymp \varrho(L_p|\mathcal{R}_n, \mathcal{E}_n) \asymp n^{1/p-1/2} \quad \textit{if } 1 \le p \le 2,$$
$$\varrho(l_q^n|\mathcal{E}_n, \mathcal{R}_n) \asymp \varrho(L_q|\mathcal{E}_n, \mathcal{R}_n) \asymp n^{1/2-1/q} \quad \textit{if } 2 \le q < \infty.$$

The same relations hold if $\mathcal{R}_n$ *is replaced by* $\mathcal{G}_n$.

PROOF. By 5.5.17 and 4.2.8, we have

$$\varrho(l_p^n|\mathcal{I}_n, \mathcal{E}_n) \asymp n^{1/p-1/2} \quad \text{and} \quad \varrho(l_p^n|\mathcal{I}_n, \mathcal{R}_n) \asymp 1.$$

Hence

$$\varrho(l_p^n|\mathcal{I}_n, \mathcal{E}_n) \le \varrho(l_p^n|\mathcal{I}_n, \mathcal{R}_n)\, \varrho(l_p^n|\mathcal{R}_n, \mathcal{E}_n)$$

implies that $n^{1/p-1/2} \prec \varrho(l_p^n|\mathcal{R}_n, \mathcal{E}_n)$. The upper estimate follows from 3.4.17. This proves the relations in the first line. The rest can be checked analogously.

REMARK. Denoting by C the Banach space of all 2π-periodic continuous complex-valued functions on $\mathbb{R}$, we get $\|(e_k)|[\mathcal{E}_n, C]\| = n$ and

$$\|(e_k)|[\mathcal{R}_n, C]\|^2 = \int_0^1 \sup_{|t|\le\pi} \left| \sum_{k=1}^n r_k(s) \exp(ikt) \right|^2 ds \asymp n(1 + \log n).$$

Hence

$$\sqrt{\tfrac{n}{1+\log n}} \prec \varrho(C|\mathcal{E}_n, \mathcal{R}_n);$$

see [hal]. Consequently, we may conjecture that

$$\varrho(l_\infty^n|\mathcal{E}_n, \mathcal{R}_n) \overset{?}{\asymp} \varrho(C|\mathcal{E}_n, \mathcal{R}_n) \overset{?}{\asymp} \sqrt{\tfrac{n}{1+\log n}}.$$

5.9.18 We now state a counterpart of the above example which follows at once from 3.3.7 and 3.3.8.

EXAMPLE.

$$\varrho(l_p^n|\mathcal{E}_n, \mathcal{R}_n) \asymp \varrho(L_p|\mathcal{E}_n, \mathcal{R}_n) \asymp 1 \quad \textit{if } 1 \le p \le 2,$$
$$\varrho(l_q^n|\mathcal{R}_n, \mathcal{E}_n) \asymp \varrho(L_q|\mathcal{R}_n, \mathcal{E}_n) \asymp 1 \quad \textit{if } 2 \le q < \infty.$$

The same relations hold if $\mathcal{R}_n$ *is replaced by* $\mathcal{G}_n$.

REMARK. In the limiting case $q = \infty$, we have

$$\varrho(l_\infty^N|\mathcal{R}_n, \mathcal{E}_n) \asymp \min\{\sqrt{n}, \sqrt{1+\log N}\}.$$

This can be seen as in 4.2.11.

5.9.19 The relationship between the Dirichlet ideal norms $\boldsymbol{\delta}(\mathcal{E}_n, \mathcal{E}_n)$ and $\boldsymbol{\delta}(\mathcal{R}_n, \mathcal{R}_n)$ is quite unclear. At least, for the closed operator ideals

$$\mathfrak{Q}\mathfrak{E}\mathfrak{P} := \Big\{ T \in \mathfrak{L} : \boldsymbol{\delta}(T|\mathcal{E}_n, \mathcal{E}_n) = o(1+\log n) \Big\}$$

and

$$\mathfrak{Q}\mathfrak{R}\mathfrak{P} := \Big\{ T \in \mathfrak{L} : \boldsymbol{\delta}(T|\mathcal{R}_n, \mathcal{R}_n) = o(\sqrt{n}) \Big\}$$

we have the following inclusion.

PROPOSITION. $\mathfrak{Q}\mathfrak{E}\mathfrak{P} \subseteq \mathfrak{Q}\mathfrak{R}\mathfrak{P}$.

PROOF. Recall from 4.14.8 that $\mathfrak{Q}\mathfrak{R}\mathfrak{P} = \mathfrak{R}\mathfrak{T}$. So, if $T \in \mathfrak{L}(X,Y)$ does not belong to $\mathfrak{Q}\mathfrak{R}\mathfrak{P}$, it follows from 4.4.19 that there exists a diagram as described in 4.4.17. Hence, taking into account the injectivity of $\mathfrak{Q}\mathfrak{E}\mathfrak{P}$ and

$$\boldsymbol{\delta}(l_1^n|\mathcal{E}_n, \mathcal{E}_n) \asymp 1 + \log n \qquad \text{(by 5.2.34)},$$

we obtain $T \notin \mathfrak{Q}\mathfrak{E}\mathfrak{P}(X,Y)$.

REMARK. It will be shown in 7.6.13 that operators in $\mathfrak{Q}\mathfrak{E}\mathfrak{P} = \mathfrak{Q}\mathfrak{H}\mathfrak{M}$ are super weakly compact. Thus the above inclusion is strict.

5.9.20 Finally, we observe that the Rademacher system $\mathcal{R}_n$ is closely related to the **lacunary Fourier system** $\mathcal{E}(\mathbb{L}_n)$ with $\mathbb{L}_n := \{2^1, \ldots, 2^n\}$.

PROPOSITION. $\boldsymbol{\varrho}(\mathcal{R}_n, \mathcal{E}(\mathbb{L}_n)) \le 8\pi$ *and* $\boldsymbol{\varrho}(\mathcal{E}(\mathbb{L}_n), \mathcal{R}_n) \le 8\pi$.

PROOF. Since $\mathcal{E}(\mathbb{L}_n)$, viewed as a system of characters on $\mathbb{T}$, is dissociate, 4.11.15 tells us that $S(\mathcal{E}(\mathbb{L}_n)) \le 16$. Hence 4.11.10 yields

$$\boldsymbol{\varrho}(\mathcal{E}(\mathbb{L}_n), \mathcal{P}_n(e^{i\tau})) \le 16 \quad \text{and} \quad \boldsymbol{\varrho}(\mathcal{P}_n(e^{i\tau}), \mathcal{E}(\mathbb{L}_n)) \le 16.$$

Moreover, we know from 4.2.13 that

$$\boldsymbol{\varrho}(\mathcal{P}_n(e^{i\tau}), \mathcal{P}_n(\pm 1)) \le \tfrac{\pi}{2} \quad \text{and} \quad \boldsymbol{\varrho}(\mathcal{P}_n(\pm 1), \mathcal{P}_n(e^{i\tau})) \le \tfrac{\pi}{2}.$$

REMARK. Inequalities of the same kind hold for the lacunary sine and cosine systems; see [pis 6].

6

Walsh ideal norms

In this chapter, we treat ideal norms associated with the Walsh system. Among others, the following classes of Banach spaces will be considered:

WT_p	:	Banach spaces of Walsh type p,
WT	:	Banach spaces of Walsh subtype,
WC_q	:	Banach spaces of Walsh cotype q,
WC	:	Banach spaces of Walsh subcotype.

There is a strong analogy with the situation for trigonometric functions. We do not even know whether WT_p and ET_p are different.

It turns out that

$$\mathsf{WT}_p = \mathsf{WC}_{p'}, \quad \mathsf{WC}_q = \mathsf{WT}_{q'} \text{ and } \mathsf{WT} = \mathsf{WC},$$

$$\bigcup_{1<p\leq 2} \mathsf{WT}_p = \mathsf{WT} = \mathsf{B} \text{ and } \mathsf{WT}_2 = \mathsf{H}.$$

A further highlight of this chapter is the relation

$$\bigcup_{1<p\leq 2} \mathfrak{WT}_p \subset \mathfrak{RP} \subset \mathfrak{QRP} = \mathfrak{WT},$$

proved in 6.4.16, which gives new insight into Pisier's theory about the equivalence of B-convexity and K-convexity.

The investigation of the Dirichlet ideal norms $\boldsymbol{\delta}(\mathcal{W}_n, \mathcal{W}_n)$ will be postponed until Section 8.7, where we describe the relationship with the UMD-property.

For more information concerning Walsh functions, we refer to the following books:

Golubov/Efimov/Skvortsov... *Walsh series and transforms* [GOL*],
Schipp/Wade/Simon .... *Walsh series* [SCHI*].

6.1 Walsh functions

6.1.1 In what follows, $\mathcal{F}(\mathbb{N})$ denotes the collection of all finite subsets of $\mathbb{N} := \{1, 2, \dots\}$. Note that $\mathcal{F}(\mathbb{N})$ is an Abelian group under the operation $(\mathbb{A}, \mathbb{B}) \to \mathbb{A} \triangle \mathbb{B}$ (symmetric difference). The symbol $\mathbb{A} \leq n$ indicates that $\mathbb{A} \subseteq \{1, \dots, n\}$.

6.1.2 For every $\mathbb{A} \in \mathcal{F}(\mathbb{N})$, the **Walsh function** $w_{\mathbb{A}} : [0,1) \to \{-1, +1\}$ is defined by

$$w_{\mathbb{A}} := \prod_{h \in \mathbb{A}} r_h,$$

where $r_1, r_2, \dots$ are the Rademacher functions. In particular, $w_\emptyset \equiv 1$. The functions $w_{\mathbb{A}}$ constitute the orthonormal **Walsh system**, denoted by $\mathcal{W}$. Obviously, we have $w_{\mathbb{A}}\, w_{\mathbb{B}} = w_{\mathbb{A} \triangle \mathbb{B}}$.

6.1.3 There are several ways to enumerate the Walsh system. It is most common to give $w_{\mathbb{A}}$ the index

$$\varphi(\mathbb{A}) := \sum_{h \in \mathbb{A}} 2^{h-1} \in \mathbb{N}_0 := \{0, 1, \dots\}$$

with the understanding that $\varphi(\emptyset) := 0$. Then $w_0 \equiv 1$ and $w_{2^{k-1}} = r_k$.

6.1.4 Define

$$\frac{f \oplus g}{2}(t) := \begin{cases} \;\; f(2t) & \text{if } 0 \leq t < 1/2, \\ +g(2t-1) & \text{if } 1/2 \leq t < 1 \end{cases}$$

and

$$\frac{f \ominus g}{2}(t) := \begin{cases} \;\; f(2t) & \text{if } 0 \leq t < 1/2, \\ -g(2t-1) & \text{if } 1/2 \leq t < 1 \end{cases}$$

whenever f and g are functions on $[0,1)$. The Walsh functions can now be obtained by induction. Starting with $w_0^\circ \equiv 1$, we let

$$w_{2k}^\circ := \begin{cases} \frac{w_k^\circ \oplus w_k^\circ}{2} & \text{if } k \text{ is even}, \\ \frac{w_k^\circ \ominus w_k^\circ}{2} & \text{if } k \text{ is odd}, \end{cases} \quad \text{and} \quad w_{2k+1}^\circ(t) := \begin{cases} \frac{w_k^\circ \ominus w_k^\circ}{2} & \text{if } k \text{ is even}, \\ \frac{w_k^\circ \oplus w_k^\circ}{2} & \text{if } k \text{ is odd}. \end{cases}$$

This construction yields the original enumeration $\{w_0^\circ, w_1^\circ, \dots\}$ with the property that the index n just indicates the number of jumps (the so-called frequency) of w_n°.

REMARK. The enumerations described above pass into each other by permutations that leave the subsystems $\{ w_k : 2^{n-1} \leq n < 2^n \}$ invariant.

6.1.5 The Walsh system $\mathcal{W}_{2^n} := \{ w_0, \ldots, w_{2^n-1} \}$ can be identified with the character group of $\mathbb{E}_2^n$. Indeed, we may define

$$w_{\mathbb{A}}(e) := \prod_{h \in \mathbb{A}} \varepsilon_h \quad \text{for } e = (\varepsilon_1, \ldots, \varepsilon_n) \in \mathbb{E}_2^n.$$

Then $w_{\mathbb{A}}(e^{-1}) = w_{\mathbb{A}}(e)$ and $w_{\mathbb{A}}(de) = w_{\mathbb{A}}(d)\, w_{\mathbb{A}}(e)$ for $d, e \in \mathbb{E}_2^n$.

Starting with $W_1 := (1)$, the orthogonal **Walsh matrices** are obtained by induction:

$$W_{2^{n+1}} := \frac{1}{\sqrt{2}} \begin{pmatrix} +W_{2^n} & +W_{2^n} \\ +W_{2^n} & -W_{2^n} \end{pmatrix}.$$

There exists a map π_n from $\{1, \ldots, 2^n\}$ onto $\{0, \ldots, 2^n - 1\}$ such that the entries of the symmetric matrix W_{2^n} are given by

$$w_{hk}^{(2^n)} = 2^{-n/2}\, w_{\pi_n(h)}(\Delta_n^{(k)}),$$

where $w_h(\Delta_n^{(k)})$ stands for the constant value of w_h on the dyadic interval $\Delta_n^{(k)} := \left[\frac{k-1}{2^n}, \frac{k}{2^n}\right)$; see [SCHI*, p. 23].

6.2 Walsh type and cotype ideal norms

6.2.1 In this section, we treat the **Walsh type ideal norms** $\varrho(\mathcal{W}_n, \mathcal{I}_n)$ and the **Walsh cotype ideal norms** $\varrho(\mathcal{I}_n, \mathcal{W}_n)$.

Remark. Since there is a strong analogy with the theory of Fourier type and cotype developed in Section 5.5, several proofs will be omitted.

6.2.2 We first show that the asymptotic behaviour of the sequence $(\varrho(\mathcal{W}_m, \mathcal{I}_m))$ is fully specified by that of the subsequence $(\varrho(\mathcal{W}_{2^n}, \mathcal{I}_{2^n}))$ which can be treated much more easily.

PROPOSITION. *Let $2^n \le m \le 2^{n+1}$. Then*

$$\varrho(\mathcal{W}_{2^n}, \mathcal{I}_{2^n}) \le \varrho(\mathcal{W}_m, \mathcal{I}_m) \le \varrho(\mathcal{W}_{2^{n+1}}, \mathcal{I}_{2^{n+1}}) \le \sqrt{2}\, \varrho(\mathcal{W}_{2^n}, \mathcal{I}_{2^n}).$$

Proof. We have

$$\|(Tx_{\mathbb{A}})|\mathcal{W}_{2^{n+1}}\| = \left(\int_0^1 \Big\| \sum_{\mathbb{A} \le n+1} Tx_{\mathbb{A}} w_{\mathbb{A}}(t) \Big\|^2 dt \right)^{1/2}$$

$$\le \left(\int_0^1 \Big\| \sum_{\mathbb{A} \le n} Tx_{\mathbb{A}} w_{\mathbb{A}}(t) \Big\|^2 dt \right)^{1/2} + \left(\int_0^1 \Big\| \sum_{\mathbb{A} \le n} Tx_{\mathbb{A} \cup \{n+1\}} w_{\mathbb{A}}(t) r_{n+1}(t) \Big\|^2 dt \right)^{1/2}$$

$$\le \varrho(T|\mathcal{W}_{2^n}, \mathcal{I}_{2^n}) \left[\left(\sum_{\mathbb{A} \le n} \|x_{\mathbb{A}}\|^2 \right)^{1/2} + \left(\sum_{\mathbb{A} \le n} \|x_{\mathbb{A} \cup \{n+1\}}\|^2 \right)^{1/2} \right]$$

$$\leq \sqrt{2}\,\varrho(T|\mathcal{W}_{2^n},\mathcal{I}_{2^n})\Big(\sum_{\mathbb{A}\leq n+1}\|x_{\mathbb{A}}\|^2\Big)^{1/2} = \sqrt{2}\,\varrho(T|\mathcal{W}_{2^n},\mathcal{I}_{2^n})\,\|(x_{\mathbb{A}})|\mathcal{I}_{2^{n+1}}\|.$$

This proves the right-hand inequality. The rest is trivial.

6.2.3 The following relation says that $\varrho(\mathcal{W}_{2^n},\mathcal{I}_{2^n})$ is the Kwapień ideal norm associated with the character matrix W_{2^n}; see 6.1.5.

PROPOSITION. $\varrho(\mathcal{W}_{2^n},\mathcal{I}_{2^n}) = \boldsymbol{\kappa}(W_{2^n})$.

6.2.4 Next, a special case of 3.8.13 is stated.

PROPOSITION.

$$\varrho(\mathcal{W}_{2^n},\mathcal{I}_{2^n}) = \boldsymbol{\delta}(\mathcal{W}_{2^n},\mathcal{I}_{2^n}) = \varrho(\mathcal{I}_{2^n},\mathcal{W}_{2^n}) = \boldsymbol{\delta}(\mathcal{I}_{2^n},\mathcal{W}_{2^n}).$$

6.2.5 We proceed with a corollary of the previous equality and 3.4.6.

PROPOSITION. *The ideal norm $\varrho(\mathcal{W}_{2^n},\mathcal{I}_{2^n})$ is symmetric.*

6.2.6 The main result of this section is an analogue of 5.5.15.

THEOREM. *The sequences of the following ideal norms are uniformly equivalent:*

$$\varrho(\mathcal{W}_n,\mathcal{I}_n),\ \boldsymbol{\delta}(\mathcal{W}_n,\mathcal{I}_n),\ \varrho(\mathcal{I}_n,\mathcal{W}_n),\ \boldsymbol{\delta}(\mathcal{I}_n,\mathcal{W}_n).$$

6.2.7 The next property is basic for later application in 6.3.10.

PROPOSITION. *The sequence $(\varrho(\mathcal{W}_{2^n},\mathcal{I}_{2^n}))$ is submultiplicative,*

$$\varrho(\mathcal{W}_{2^{m+n}},\mathcal{I}_{2^{m+n}}) \leq \varrho(\mathcal{W}_{2^m},\mathcal{I}_{2^m}) \circ \varrho(\mathcal{W}_{2^n},\mathcal{I}_{2^n}).$$

PROOF. Interpreting Walsh functions as characters on Cantor groups, we may identify $\mathcal{W}_{2^{m+n}}$ with the full tensor product of $\mathcal{W}_{2^m}$ and $\mathcal{W}_{2^n}$. Thus the assertion can be deduced from 3.6.5.

6.2.8 We conclude this section with some immediate consequences of 3.7.11 and 6.2.6.

EXAMPLE.

$$\varrho(l_r^n|\mathcal{W}_n,\mathcal{I}_n) \asymp \varrho(L_r|\mathcal{W}_n,\mathcal{I}_n) \asymp n^{|1/r-1/2|} \quad \textit{if } 1 \leq r \leq \infty.$$

6.2.9 The final result is an analogue of 5.5.18.

EXAMPLE.

$$\varrho(l_r^N|\mathcal{W}_n,\mathcal{I}_n) \asymp \min(n^{|1/r-1/2|}, N^{|1/r-1/2|}) \quad \textit{if } 1 \leq r \leq \infty.$$

6.3 Operators of Walsh type

6.3.1 We now deal with the **Walsh type ideal norms** $\varrho_u^{(v)}(\mathcal{W}_n, \mathcal{I}_n)$. The quadratic case $u = v = 2$ has already been treated in the preceding section, and, as in this special setting, the asymptotic behaviour of the sequence $(\varrho_u^{(v)}(\mathcal{W}_n, \mathcal{I}_n))$ is fully specified by that of the subsequence $(\varrho_u^{(v)}(\mathcal{W}_{2^n}, \mathcal{I}_{2^n}))$. The **weak Walsh type ideal norm** $\varrho_{u,1}^{(v)}(\mathcal{W}_n, \mathcal{I}_n)$ will also be considered.

REMARK. Since there is a strong analogy with the theory of Fourier type developed in Section 5.6, we omit several proofs.

6.3.2 First, we observe that the **Walsh cotype ideal norms** are even more superfluous than the Fourier cotype ideal norms; see 5.7.3.

PROPOSITION. $\varrho_u^{(v)}(\mathcal{I}_{2^n}, \mathcal{W}_{2^n}) = 2^{n(1/u+1/v-1)} \varrho_u^{(v)}(\mathcal{W}_{2^n}, \mathcal{I}_{2^n})$.

PROOF. By definition,

$$\|(x_k)|\mathcal{W}_{2^n}\|_r := \left(\int_0^1 \Big\| \sum_{k=0}^{2^n-1} x_k w_k(t) \Big\|^r dt \right)^{1/r}$$

and

$$\|(x_k)|W_{2^n}\|_r := \left(\sum_{h=1}^{2^n} \Big\| \sum_{k=1}^{2^n} x_k w_{hk}^{(2^n)} \Big\|^r \right)^{1/r},$$

where $W_{2^n} = (w_{hk}^{(2^n)})$ is the orthogonal and symmetric Walsh matrix.

Since the Walsh functions w_k with $k < 2^n$ are constant on the dyadic intervals $\Delta_n^{(h)}$, we can form the matrix $2^{-n/2}(w_k(\Delta_n^{(h)}))$, which passes into W_{2^n} by a rearrangement of rows; see 6.1.5. Hence

$$\int_0^1 \Big\| \sum_{k=0}^{2^n-1} x_k w_k(t) \Big\|^r dt = \frac{1}{2^n} \sum_{h=1}^{2^n} \Big\| \sum_{k=1}^{2^n} x_k 2^{n/2} w_{hk}^{(2^n)} \Big\|^r.$$

This means that

$$\|(x_k)|\mathcal{W}_{2^n}\|_r = 2^{n(1/2-1/r)} \|(x_k)|W_{2^n}\|_r. \qquad (*)$$

If we relate (x_k) and (x_h°) by

$$x_h^\circ = \sum_{k=1}^{2^n} w_{hk}^{(2^n)} x_k \quad \text{and} \quad x_k = \sum_{h=1}^{2^n} w_{kh}^{(2^n)} x_h^\circ,$$

then the inequalities

$$\|(Tx_k)|l_v^{2^n}\| \le c\,\|(x_k)|W_{2^n}\|_u \quad \text{and} \quad \|(Tx_h^\circ)|W_{2^n}\|_v \le c\,\|(x_h^\circ)|l_u^{2^n}\|$$

are equivalent. It follows from $(*)$ that the same holds for

$$\|(Tx_k)|l_v^{2^n}\| \le 2^{n(1/u-1/2)} c \, \|(x_k)|\mathcal{W}_{2^n}\|_u$$

and

$$\|(Tx_h^\circ)|\mathcal{W}_{2^n}\|_v \le 2^{n(1/2-1/v)} c \, \|(x_h^\circ)|l_u^{2^n}\|.$$

Thus we have

$$2^{n(1/2-1/u)} \varrho_u^{(v)}(\mathcal{J}_{2^n}, \mathcal{W}_{2^n}) = \inf c = 2^{n(1/v-1/2)} \varrho_u^{(v)}(\mathcal{W}_{2^n}, \mathcal{J}_{2^n}).$$

6.3.3 As a corollary of the preceding result, we get a useful duality formula; see 5.6.5.

PROPOSITION. $\varrho_u^{(v)}{}'(\mathcal{W}_{2^n},\mathcal{J}_{2^n}) = 2^{n(1-1/u-1/v)} \varrho_{v'}^{(u')}(\mathcal{W}_{2^n},\mathcal{J}_{2^n})$.

PROOF. The assertion is implied by the following chain of equalities:

$$\begin{aligned}\varrho_u^{(v)}{}'(\mathcal{W}_{2^n}, \mathcal{J}_{2^n}) &= \boldsymbol{\delta}_u^{(v)}{}'(\mathcal{W}_{2^n}, \mathcal{J}_{2^n}) = \boldsymbol{\delta}_{v'}^{(u')}(\mathcal{J}_{2^n}, \mathcal{W}_{2^n}) \\ &= \varrho_{v'}^{(u')}(\mathcal{J}_{2^n}, \mathcal{W}_{2^n}) = 2^{n(1-1/u-1/v)} \varrho_{v'}^{(u')}(\mathcal{W}_{2^n}, \mathcal{J}_{2^n}).\end{aligned}$$

6.3.4 We now provide an auxiliary result which should be compared with Problem 5.6.13. Its proof is due to J. Bourgain [bou 8, pp. 242–243] and has become a polished diamond.

LEMMA.

$$\varrho_2^{(q)}(\mathcal{W}(\mathbb{F}), \mathcal{J}(\mathbb{F})) \le 2 \, \varrho_2^{(q)}(\mathcal{W}_{2^n}, \mathcal{J}_{2^n}) \quad \textit{whenever} \quad |\mathbb{F}| = 2^n.$$

PROOF. For the present, let $A_N = (a_1, \dots, a_N)$ be any system of characters on a compact Abelian group $\mathbb{G}$. Define

$$f_k(e|t_1, \dots, t_n) := \prod_{\varepsilon_h=-1} a_k(t_h) \tag{1}$$

for $t_1, \dots, t_n \in \mathbb{G}$ and $e = (\varepsilon_1, \dots, \varepsilon_n) \in \mathbb{E}_2^n$. Then the coefficients of the Walsh series

$$f_k(e|t_1, \dots, t_n) = \frac{1}{2^n} \sum_{\mathbb{A} \le n} c_{k,\mathbb{A}}(t_1, \dots, t_n) w_{\mathbb{A}}(e) \tag{2}$$

are given by

$$c_{k,\mathbb{A}}(t_1, \dots, t_n) := \langle f_k(\cdot|t_1, \dots, t_n), w_{\mathbb{A}} \rangle = \frac{1}{2^n} \sum_{e \in \mathbb{E}_2^n} \prod_{\varepsilon_h=-1} a_k(t_h) \prod_{h \in \mathbb{A}} \varepsilon_h. \tag{3}$$

Note that

$$c_{k,\mathbb{A}}(t_1, \dots, t_n) = \frac{1}{2^n} \prod_{h \notin \mathbb{A}} \Big(1 + a_k(t_h)\Big) \prod_{h \in \mathbb{A}} \Big(1 - a_k(t_h)\Big). \tag{4}$$

Indeed, the n-fold product on the right-hand side can be expanded in a sum of 2^n terms indexed by $e = (\varepsilon_1, \dots, \varepsilon_n) \in \mathbb{E}_2^n$. Taking from the factor $1 \pm a_k(t_h)$ the first summand 1 if $\varepsilon_h = +1$ and the second summand $\pm a_k(t_h)$ if $\varepsilon_h = -1$, we get $\prod\limits_{\varepsilon_h=-1} a_k(t_h)$ multiplied by $\prod\limits_{h \in \mathbb{A}} \varepsilon_h$. Let

$$x_{\mathbb{A}}(s|t_1, \dots, t_n) := \sum_{k=1}^{N} x_k c_{k,\mathbb{A}}(t_1, \dots, t_n) a_k(s). \tag{5}$$

Fix $e = (\varepsilon_1, \ldots, \varepsilon_n) \in \mathbb{E}_2^n$. Then the invariance of the Haar measure implies that

$$\|(Tx_k)|\mathcal{A}_N\|_q = \left(\int\limits_{\mathbb{G}} \left\| \sum_{k=1}^{N} Tx_k a_k(s) \right\|^q d\mu(s) \right)^{1/q}$$
$$= \left(\int\limits_{\mathbb{G}} \left\| \sum_{k=1}^{N} Tx_k \Big[\prod_{\varepsilon_h = -1} a_k(t_h) \Big] a_k(s) \right\|^q d\mu(s) \right)^{1/q}.$$

Using (1), (2) and (5), we see that

$$\sum_{k=1}^{N} Tx_k \Big[\prod_{\varepsilon_h=-1} a_k(t_h) \Big] a_k(s) = \frac{1}{2^n} \sum_{k=1}^{N} Tx_k \sum_{\mathbb{A} \le n} c_{k,\mathbb{A}}(t_1, \ldots, t_n) w_{\mathbb{A}}(e) a_k(s)$$
$$= \frac{1}{2^n} \sum_{\mathbb{A} \le n} Tx_{\mathbb{A}}(s|t_1, \ldots, t_n) w_{\mathbb{A}}(e).$$

Hence

$$\|(Tx_k)|\mathcal{A}_N\|_q = \left(\int\limits_{\mathbb{G}} \left\| \frac{1}{2^n} \sum_{\mathbb{A} \le n} Tx_{\mathbb{A}}(s|t_1, \ldots, t_n) w_{\mathbb{A}}(e) \right\|^q d\mu(s) \right)^{1/q}.$$

Averaging over $e = (\varepsilon_1, \ldots, \varepsilon_n) \in \mathbb{E}_2^n$ gives

$$\|(Tx_k)|\mathcal{A}_N\|_q = \left(\frac{1}{2^n} \sum_{e \in \mathbb{E}_2^n} \int\limits_{\mathbb{G}} \left\| \frac{1}{2^n} \sum_{\mathbb{A} \le n} Tx_{\mathbb{A}}(s|t_1, \ldots, t_n) w_{\mathbb{A}}(e) \right\|^q d\mu(s) \right)^{1/q}. \quad (6)$$

From now on, we assume that the characters $a_1, \ldots, a_N$ only take the values ± 1. Then it follows from (4) that $c_{k,\mathbb{A}}(t_1, \ldots, t_n) \ge 0$. Thus, for the auxiliary elements $x_{\mathbb{A}}(s|t_1, \ldots, t_n)$ defined in (5), we have

$$\|x_{\mathbb{A}}(s|t_1, \ldots, t_n)\| \le \sum_{k=1}^{N} \|x_k\| \, c_{k,\mathbb{A}}(t_1, \ldots, t_n).$$

Hence

$$\left(\frac{1}{2^n} \sum_{e \in \mathbb{E}_2^n} \left\| \frac{1}{2^n} \sum_{\mathbb{A} \le n} Tx_{\mathbb{A}}(s|t_1, \ldots, t_n) w_{\mathbb{A}}(e) \right\|^q \right)^{1/q} \le$$
$$\le \varrho_2^{(q)}(T|\mathcal{W}_{2^n}, \mathcal{I}_{2^n}) \left(\sum_{\mathbb{A} \le n} \|x_{\mathbb{A}}(s|t_1, \ldots, t_n)\|^2 \right)^{1/2}$$
$$\le \varrho_2^{(q)}(T|\mathcal{W}_{2^n}, \mathcal{I}_{2^n}) \left(\sum_{\mathbb{A} \le n} \left[\sum_{k=1}^{N} \|x_k\| \, c_{k,\mathbb{A}}(t_1, \ldots, t_n) \right]^2 \right)^{1/2}.$$

Since the last line does not depend on $s \in \mathbb{G}$, integration yields

$$\left(\frac{1}{2^n}\sum_{e\in\mathbb{E}_2^n}\int_{\mathbb{G}}\left\|\frac{1}{2^n}\sum_{\mathbb{A}\leq n} Tx_{\mathbb{A}}(s|t_1,\ldots,t_n)w_{\mathbb{A}}(e)\right\|^q d\mu(s)\right)^{1/q} \leq$$

$$\leq \boldsymbol{\varrho}_2^{(q)}(T|\mathcal{W}_{2^n},\mathcal{I}_{2^n})\left(\sum_{\mathbb{A}\leq n}\left[\sum_{k=1}^N \|x_k\|\, c_{k,\mathbb{A}}(t_1,\ldots,t_n)\right]^2\right)^{1/2}$$

which means, by (6), that

$$\|(Tx_k)|\mathcal{A}_N\|_q \leq \boldsymbol{\varrho}_2^{(q)}(T|\mathcal{W}_{2^n},\mathcal{I}_{2^n})\left(\sum_{\mathbb{A}\leq n}\left[\sum_{k=1}^N \|x_k\|\, c_{k,\mathbb{A}}(t_1,\ldots,t_n)\right]^2\right)^{1/2}.$$

Taking into account that the left-hand side is a constant, we obtain by integration over $t_1,\ldots,t_n \in \mathbb{G}$ that

$$\|(Tx_k)|\mathcal{A}_N\|_q \leq \tag{7}$$

$$\leq \boldsymbol{\varrho}_2^{(q)}(T|\mathcal{W}_{2^n},\mathcal{I}_{2^n})\left(\sum_{\mathbb{A}\leq n}\int_{\mathbb{G}}\ldots\int_{\mathbb{G}}\left[\sum_{k=1}^N \|x_k\|\, c_{k,\mathbb{A}}(t_1,\ldots,t_n)\right]^2 d\mu(t_1)\ldots d\mu(t_n)\right)^{1/2}.$$

Let us further assume that $\mathcal{A}_N = (a_1,\ldots,a_N)$ does not contain the trivial character. Then

$$\Big\{\prod_{\varepsilon_h=-1} a_k(t_h) \,:\, k=1,\ldots,N,\ e\in\mathbb{E}_2^n,\ e\neq(1,\ldots,1)\Big\}$$

is an orthonormal system in $L_2(\mathbb{G}^n)$. Hence

$$\sum_{k=1}^N \|x_k\|\, c_{k,\mathbb{A}}(t_1,\ldots,t_n) = \frac{1}{2^n}\sum_{k=1}^N\sum_{e\in\mathbb{E}_2^n}\|x_k\|\prod_{\varepsilon_h=-1} a_k(t_h)\prod_{h\in\mathbb{A}}\varepsilon_h$$

$$= \frac{1}{2^n}\left[\sum_{k=1}^N \|x_k\| + \sum_{k=1}^N\sum_{\substack{e\in\mathbb{E}_2^n\\ e\neq(1,\ldots,1)}}\|x_k\|\prod_{h\in\mathbb{A}}\varepsilon_h\prod_{\varepsilon_h=-1} a_k(t_h)\right]$$

implies that

$$\left(\int_{\mathbb{G}}\cdots\int_{\mathbb{G}}\left[\sum_{k=1}^N \|x_k\|\, c_{k,\mathbb{A}}(t_1,\ldots,t_n)\right]^2 d\mu(t_1)\ldots d\mu(t_n)\right)^{1/2} \leq$$

$$\leq \frac{1}{2^n}\left[\sum_{k=1}^N \|x_k\| + \left((2^n-1)\sum_{k=1}^N \|x_k\|^2\right)^{1/2}\right]$$

$$\leq \frac{1}{2^n}(\sqrt{N}+\sqrt{2^n-1})\left(\sum_{k=1}^N \|x_k\|^2\right)^{1/2}. \tag{8}$$

Therefore it follows from (7) and (8) that

$$\varrho_2^{(q)}(T|\mathcal{A}_N,\mathfrak{I}_N) \leq 2\,\varrho_2^{(q)}(T|\mathcal{W}_{2^n},\mathfrak{I}_{2^n}) \quad \text{whenever} \quad N \leq 2^n.$$

Finally, we may take for $\mathcal{A}_N$ any Walsh system $\mathcal{W}(\mathbb{F})$ with $|\mathbb{F}| = 2^n$ not containing the function $w_0 \equiv 1$. In fact, should w_0 occur, we simply multiply $\mathcal{W}(\mathbb{F})$ by some Walsh function w_m, where $m \notin \mathbb{F}$.

6.3.5 Let $1 < p \leq 2$. An operator T has **Walsh type** p if

$$\|T|\mathfrak{WT}_p\| := \sup_n \varrho_p^{(p')}(T|\mathcal{W}_n,\mathfrak{I}_n)$$

is finite. These operators form the Banach ideal

$$\mathfrak{WT}_p := \mathfrak{L}[\varrho_p^{(p')}(\mathcal{W}_n,\mathfrak{I}_n)].$$

Moreover, $\mathfrak{WC}_q$ is the ideal of operators of **Walsh cotype** q. However, in view of 6.3.2, we get $\mathfrak{WC}_q = \mathfrak{WT}_{q'}$.

REMARK. Trivially, all operators have Walsh type 1. Therefore we put $\mathfrak{WT}_1 := \mathfrak{L}$.

6.3.6 Let $1 < p < 2$. An operator T has **weak Walsh type** p if

$$\|T|\mathfrak{WT}_p^{weak}\| := \sup_n n^{1/2-1/p}\,\varrho(T|\mathcal{W}_n,\mathfrak{I}_n)$$

is finite. These operators form the Banach ideal

$$\mathfrak{WT}_p^{weak} := \mathfrak{L}[n^{1/2-1/p}\,(\mathcal{W}_n,\mathfrak{I}_n)].$$

Moreover, $\mathfrak{WC}_q^{weak}$ is the ideal of operators of **weak Walsh cotype** q. However, in view of 6.3.2, we get $\mathfrak{WC}_q^{weak} = \mathfrak{WT}_{q'}^{weak}$.

6.3.7 We now establish the main result of this section, which is an analogue of 4.3.10 and 5.6.23. Using 6.3.4 as a substitute for Lemmas 4.3.5 and 5.6.16, the previous proofs can easily be adapted.

THEOREM. $\mathfrak{WT}_2 \subset \mathfrak{WT}_{p_0} \subset \mathfrak{WT}_{p_0}^{weak} \subset \mathfrak{WT}_p$ *if* $1<p<p_0<2$.

In particular,

$$\bigcup_{1<p\leq 2} \mathfrak{WT}_p = \bigcup_{1<p<2} \mathfrak{WT}_p^{weak}.$$

6.3.8 Next, we state an analogue of 5.6.24.

EXAMPLE.

$$L_p,\ L_{p'} \in \mathsf{WT}_p \setminus \mathsf{WT}_{p_0}^{weak} \quad \textit{if } 1 < p < p_0 < 2.$$

6.3.9 The end of this section is parallel to the end of Section 5.6.

An operator T has **Walsh subtype** if

$$\varrho(T|\mathcal{W}_n,\mathcal{I}_n)=o(\sqrt{n}).$$

These operators form the closed ideal

$$\mathfrak{WT}:=\mathfrak{L}_0[n^{-1/2}\,\varrho(\mathcal{W}_n,\mathcal{I}_n)].$$

Moreover, $\mathfrak{WC}$ is the ideal of operators of **Walsh subcotype**. However, in view of 6.3.2, we get $\mathfrak{WC}=\mathfrak{WT}$.

REMARK. WT consists just of all B-convex Banach spaces; see 6.3.13.

6.3.10 The following result is an analogue of 4.4.1 and 5.6.27.

THEOREM. $$\bigcup_{1<p\le 2}\mathsf{WT}_p=\bigcup_{1<p<2}\mathsf{WT}_p^{weak}=\mathsf{WT}.$$

PROOF. We know from 3.3.6 and 6.2.7 that the sequence $(\varrho(\mathcal{W}_{2^n},\mathcal{I}_{2^n}))$ is non-decreasing and submultiplicative. Hence the assertion follows from 1.3.5 and 6.3.7.

REMARK. In view of the above result, Banach spaces in WT are said to have **non-trivial Walsh type**.

6.3.11 Let $\alpha\ge 0$, and denote by C_α the diagonal operator on $[l_2,l_\infty^{2^k}]$ defined by

$$C_\alpha:(x_k)\longrightarrow(k^{-\alpha}x_k),\quad\text{where}\quad x_k\in l_\infty^{2^k}.$$

The next example shows that the phenomenon described in the previous paragraph only occurs for spaces, not for operators. That is,

$$\bigcup_{1<p\le 2}\mathfrak{WT}_p=\bigcup_{1<p<2}\mathfrak{WT}_p^{weak}\neq\mathfrak{WT}.$$

EXAMPLE. $$\varrho(C_\alpha|\mathcal{W}_n,\mathcal{I}_n)\asymp n^{1/2}(1+\log n)^{-\alpha}.$$

PROOF. Recall from 6.2.9 that

$$\varrho(l_\infty^N|\mathcal{W}_n,\mathcal{I}_n)\asymp\min\{\sqrt{n},\sqrt{N}\}.$$

The desired relation is now a consequence of $(\triangle)$ in 1.2.12.

6.3.12 We proceed with a supplement to 4.14.8 and 5.6.29 which goes back to J. Bourgain; see [bou 8].

THEOREM. $\mathfrak{WT}=\mathfrak{RT}$.

PROOF. By 3.7.9, we have $\varrho(\mathcal{R}_n,\mathcal{I}_n)\le\varrho(\mathcal{W}_n,\mathcal{I}_n)$. Hence $\mathfrak{WT}\subseteq\mathfrak{RT}$. The reverse inclusion follows as in 5.6.29.

6.3.13 In literary style and in the setting of spaces, the preceding result reads as follows.

THEOREM. *A Banach space has non-trivial Walsh type if and only if it has non-trivial Rademacher type.*

6.4 Walsh versus Rademacher

6.4.1 For completeness, we begin with a counterpart of 5.9.2.

PROPOSITION.

$$\varrho_u^{(v)}(\mathcal{R}_n, \mathcal{I}_n) \le \varrho_u^{(v)}(\mathcal{W}_n, \mathcal{I}_n) \quad and \quad \varrho_u^{(v)}(\mathcal{I}_n, \mathcal{R}_n) \le \varrho_u^{(v)}(\mathcal{I}_n, \mathcal{W}_n).$$

6.4.2 All results from Section 5.9 can be carried over to Walsh systems. For example, in the language of operator ideals, the previous result combined with 6.3.2 reads as follows.

THEOREM.

$$\mathfrak{WT}_p \subseteq \mathfrak{RT}_p \cap \mathfrak{RC}_{p'} \quad and \quad \mathfrak{WT}_p^{weak} \subseteq \mathfrak{RT}_p^{weak} \cap \mathfrak{RC}_{p'}^{weak}.$$

6.4.3 Next, we state an analogue of 5.9.4.

THEOREM. $\mathsf{WT}_2 = \mathsf{WC}_2 = \mathsf{H}$.

6.4.4 In the remaining part of this section, we prepare and complete the proof of Theorem 6.4.15, which yields an important extension of Pisier's theory of K-convexity to the setting of operators. The following approach was obtained by A. Hinrichs, who successfully combined the original ideas of G. Pisier and the main steps in the proof of the Beurling–Kato theorem with the technique of ideal norms; see [hin 4]. Concerning the classical literature, the reader is referred to [beu], [kat], [pis 12], [pis 15] and [pis 16].

6.4.5 In this paragraph, all operators under consideration act on the 2^n-dimensional Hilbert space $L_2(\mathbb{E}_2^n)$ with $n = 1, 2 \dots$. In particular, I_n stands for the identity map.

For $h = 0, \dots, n$, we denote by $W_n^{(h)}$ the orthogonal projection from $L_2(\mathbb{E}_2^n)$ onto the span of all Walsh functions $w_{\mathbb{A}}$ with $\mathbb{A} \le n$ and $|\mathbb{A}| = h$. In other terms, $W_n^{(h)}$ is the Dirichlet projection associated with the orthonormal system $\mathcal{W}(\mathbb{F}_n^{(h)})$, where $\mathbb{F}_n^{(h)} := \{ \mathbb{A} \le n : |\mathbb{A}| = h \}$.

Let

$$S_n(z) := \sum_{h=0}^{n} z^h W_n^{(h)} \quad \text{for } z \in \mathbb{C}. \tag{1}$$

Since $W_n^{(0)}, \dots, W_n^{(n)}$ are mutually orthogonal, the right-hand side of (1) is a spectral decomposition. This observation is quite useful for computing various functions of $S_n(z)$. In particular, it follows that

$$\|S_n(t)\| = 1 \quad \text{for } 0 \le t \le 1. \tag{2}$$

Obviously, the map $z \to S_n(z)$ defines a holomorphic semigroup with respect to multiplication. Its infinitesimal generator is given by

$$A_n := S_n'(1) = \sum_{h=0}^{n} h W_n^{(h)}, \tag{3}$$

and we have

$$(wI_n - A_n)^{-1} = \sum_{h=0}^{n} \frac{1}{w-h} W_n^{(h)} \quad \text{for } w \in \mathbb{C} \setminus \{0, 1, \dots, n\}. \tag{4}$$

Hence it follows from

$$\frac{1}{w-h} = \frac{t^w}{t^w - t^h} \int_1^t \tau^{-w-1} \tau^h \, d\tau \quad \text{and} \quad S_n(t) = t^{A_n}$$

that

$$(wI_n - A_n)^{-1} = (t^w I_n - S_n(t))^{-1} t^w \int_1^t \tau^{-w-1} S_n(\tau) d\tau \tag{5}$$

whenever $0 < t < 1$ and $w \ne 0, 1, \dots, n$.

Next, let

$$R_n(z) := \frac{1}{2}\Big(I_n - 3S_n(z)\Big). \tag{6}$$

Although this definition looks quite artificial, it is crucial for the following considerations. By the operational calculus, we have

$$R_n(t)^m = \sum_{h=0}^{n} \Big(\frac{1-3t^h}{2}\Big)^m W_n^{(h)} \tag{7}$$

which implies that

$$\|R_n(t)^m\| = 1 \quad \text{for } 0 \le t \le 1 \text{ and } m = 1, 2, \dots. \tag{8}$$

For $k = 1, \dots, n$, we denote by $P_n^{(k)}$ the orthogonal projection from $L_2(\mathbb{E}_2^n)$ onto the span of all Walsh functions $w_{\mathbb{A}}$ with $\mathbb{A} \le n$ and $k \notin \mathbb{A}$. By evaluating the action on all Walsh functions $w_{\mathbb{A}}$, it follows that $S_n(z)$ can be written in the form

$$S_n(z) = \prod_{k=1}^{n} \Big[P_n^{(k)} + z\big(I_n - P_n^{(k)}\big)\Big]. \tag{9}$$

Indeed, we have

$$\Big[P_n^{(k)} + z\big(I_n - P_n^{(k)}\big)\Big] w_{\mathbb{A}} = \begin{cases} z w_{\mathbb{A}} & \text{if } k \in \mathbb{A}, \\ w_{\mathbb{A}} & \text{if } k \notin \mathbb{A}. \end{cases}$$

Finally, let

$$P_n^{(\mathbb{B})} := \prod_{k\in\mathbb{B}} P_n^{(k)}. \tag{10}$$

This is the orthogonal projection from $L_2(\mathbb{E}_2^n)$ onto the span of all Walsh functions $w_{\mathbb{A}}$ with $\mathbb{A} \leq n$ and $\mathbb{A} \cap \mathbb{B} = \emptyset$.

6.4.6 In what follows, we successively estimate the quantities

$$\|[R_n(t)^m, T]\|, \quad \|[(\varrho I_n - R_n(t))^{-1}, T]\|, \quad \|[(\sigma I_n + S_n(t))^{-1}, T]\|,$$

$$\|[(wI_n - A_n)^{-1}, T]\|, \quad \|[S_n(z), T]\|, \quad \text{and} \quad \|[W_n^{(1)}, T]\| = \boldsymbol{\delta}(T|\mathcal{R}_n, \mathcal{R}_n).$$

The constants involved depend neither on the operator $T \in \mathfrak{L}(X, Y)$ nor on $m, n = 1, 2, \ldots$ and the parameters t, w, z, ϱ, σ, but may depend on the exponent p with $1 < p \leq 2$. The final inequality, to be established in Theorem 6.4.15, reads as follows:

$$\boldsymbol{\delta}(\mathcal{R}_n, \mathcal{R}_n) \leq c_p\, \boldsymbol{\varrho}_p^{(p')}(\mathcal{W}_{2^n}, \mathcal{I}_{2^n}).$$

6.4.7 Given $e = (\varepsilon_1, \ldots, \varepsilon_n) \in \mathbb{E}_2^n$, we denote by $u_e^{(n)}$ the $\mathbb{E}_2^n$-tuple with coordinate 1 at index e and 0 otherwise.

LEMMA. *The operators $P_n^{(k)}$ and $S_n(t)$ with $0 \leq t \leq 1$ are positive.*

PROOF. For $d = (\delta_1, \ldots, \delta_n)$ and $e = (\varepsilon_1, \ldots, \varepsilon_n)$ in $\mathbb{E}_2^n$, we have

$$\begin{aligned}\langle P_n^{(k)} u_d^{(n)}, u_e^{(n)} \rangle &= \sum_{k\notin\mathbb{A}} \langle u_d^{(n)}, w_{\mathbb{A}} \rangle \langle w_{\mathbb{A}}, u_e^{(n)} \rangle \\ &= \sum_{k\notin\mathbb{A}} \Big[\frac{1}{2^n} \prod_{h\in\mathbb{A}} \delta_h\Big]\Big[\frac{1}{2^n} \prod_{h\in\mathbb{A}} \varepsilon_h\Big] = \frac{1}{4^n} \prod_{h\neq k} (1 + \delta_h \varepsilon_h) \geq 0\end{aligned}$$

and

$$\begin{aligned}\langle S_n(t) u_d^{(n)}, u_e^{(n)} \rangle &= \sum_{h=0}^{n} t^h \langle W_n^{(h)} u_d^{(n)}, u_e^{(n)} \rangle \\ &= \sum_{h=0}^{n} t^h \sum_{|\mathbb{A}|=h} \langle u_d^{(n)}, w_{\mathbb{A}} \rangle \langle w_{\mathbb{A}}, u_e^{(n)} \rangle \\ &= \sum_{h=0}^{n} t^h \sum_{|\mathbb{A}|=h} \Big[\frac{1}{2^n} \prod_{k\in\mathbb{A}} \delta_k\Big]\Big[\frac{1}{2^n} \prod_{k\in\mathbb{A}} \varepsilon_k\Big] = \frac{1}{4^n} \prod_{k=1}^{n} (1 + t\delta_k \varepsilon_k) \geq 0.\end{aligned}$$

6.4.8 Next we show how Walsh type ideal norms come into play.

LEMMA. *Let $Q_1, \dots, Q_m \in \mathfrak{L}(L_2(\mathbb{E}_2^n))$ be positive contractions. Then, for $\boldsymbol{f} \in [L_2, X]$,*

$$\frac{1}{2^m} \sum_{e \in \mathbb{E}_2^m} \left\| \left[\prod_{j=1}^{m} (I_n + \varepsilon_j Q_j), T \right] \boldsymbol{f} \right\| \leq 2^{m/2} \varrho(T|\mathcal{W}_{2^m}, \mathcal{I}_{2^m}) \, \|\boldsymbol{f}\|.$$

PROOF. In view of 0.4.9, the assumption implies that $\|[Q_j, X]\| \leq 1$. Moreover, the l_2-stability of type ideal norms means that we have $\varrho([L_2, T]|\mathcal{W}_{2^m}, \mathcal{I}_{2^m}) = \varrho(T|\mathcal{W}_{2^m}, \mathcal{I}_{2^m})$. Since

$$\left[\prod_{j=1}^{m} (I_n + \varepsilon_j Q_j), T \right] = \sum_{\mathbb{J} \leq m} [L_2, T] \left[\prod_{j \in \mathbb{J}} Q_j, X \right] w_{\mathbb{J}}(e)$$

for $e = (\varepsilon_1, \dots, \varepsilon_m) \in \mathbb{E}_2^m$, it follows that

$$\begin{aligned}
&\frac{1}{2^m} \sum_{e \in \mathbb{E}_2^m} \left\| \left[\prod_{j=1}^{m} (I_n + \varepsilon_j Q_j), T \right] \boldsymbol{f} \right\| \leq \\
&\quad \leq \left(\frac{1}{2^m} \sum_{e \in \mathbb{E}_2^m} \left\| \left[\prod_{j=1}^{m} (I_n + \varepsilon_j Q_j), T \right] \boldsymbol{f} \right\|^2 \right)^{1/2} \\
&\quad = \left(\frac{1}{2^m} \sum_{e \in \mathbb{E}_2^m} \left\| \sum_{\mathbb{J} \leq m} [L_2, T] \left[\prod_{j \in \mathbb{J}} Q_j, X \right] \boldsymbol{f} w_{\mathbb{J}}(e) \right\|^2 \right)^{1/2} \\
&\quad \leq \varrho([L_2, T]|\mathcal{W}_{2^m}, \mathcal{I}_{2^m}) \left(\sum_{\mathbb{J} \leq m} \left\| \left[\prod_{j \in \mathbb{J}} Q_j, X \right] \boldsymbol{f} \right\|^2 \right)^{1/2} \\
&\quad \leq 2^{m/2} \varrho(T|\mathcal{W}_{2^m}, \mathcal{I}_{2^m}) \, \|\boldsymbol{f}\|.
\end{aligned}$$

6.4.9 The following estimate will be used in the proof of 6.4.11.

LEMMA.

$\|[R_n(t)^m, T]\| \leq 2^{m/2} \varrho(T|\mathcal{W}_{2^m}, \mathcal{I}_{2^m})$ *for $0 \leq t \leq 1$ and $m = 1, 2, \dots$.*

PROOF. By (9) in 6.4.5,

$$S_n(t) = \prod_{k=1}^{n} \left[t I_n + (1-t) P_n^{(k)} \right].$$

Hence

$$S_n(t) = \sum_{\mathbb{B} \leq n} t^{n-|\mathbb{B}|} (1-t)^{|\mathbb{B}|} P_n^{(\mathbb{B})}. \tag{1}$$

In view of

$$\sum_{\mathbb{B} \leq n} t^{n-|\mathbb{B}|} (1-t)^{|\mathbb{B}|} = 1,$$

we also have

$$I_n - S_n(t) = \sum_{\mathbb{B}\leq n} t^{n-|\mathbb{B}|}(1-t)^{|\mathbb{B}|}(I_n - P_n^{(\mathbb{B})}). \tag{2}$$

Define

$$\mathbb{J}_e^+ := \{j : \varepsilon_j = +1\} \quad \text{and} \quad \mathbb{J}_e^- := \{j : \varepsilon_j = -1\} \quad \text{for } e=(\varepsilon_1,\ldots,\varepsilon_m)\in\mathbb{E}_2^m.$$

Obviously, $P_n(I_n + P_n) = 2P_n$ for every projection P_n. We now conclude from (1) and (2) that

$$R_n(t)^m = \frac{1}{2^m}\Big(I_n - S_n(t) - 2S_n(t)\Big)^m$$

$$= \frac{1}{2^m}\sum_{e\in\mathbb{E}_2^m}\Big(-2S_n(t)\Big)^{|\mathbb{J}_e^+|}\Big(I_n - S_n(t)\Big)^{|\mathbb{J}_e^-|}$$

$$= \frac{1}{2^m}\sum_{e\in\mathbb{E}_2^m}\sum_{\mathbb{B}_1,\ldots,\mathbb{B}_m\leq n}\Big(\prod_{j=1}^m t^{n-|\mathbb{B}_j|}(1-t)^{|\mathbb{B}_j|}\Big)\prod_{j\in\mathbb{J}_e^+}(-2P_n^{(\mathbb{B}_j)})\prod_{j\in\mathbb{J}_e^-}(I_n - P_n^{(\mathbb{B}_j)})$$

$$= \sum_{\mathbb{B}_1,\ldots,\mathbb{B}_m\leq n}\Big(\prod_{j=1}^m t^{n-|\mathbb{B}_j|}(1-t)^{|\mathbb{B}_j|}\Big)\frac{1}{2^m}\sum_{e\in\mathbb{E}_2^m}\prod_{j\in\mathbb{J}_e^+}(-P_n^{(\mathbb{B}_j)})\prod_{j=1}^m(I_n + \varepsilon_j P_n^{(\mathbb{B}_j)}).$$

Since $0 \leq t \leq 1$, we have

$$\sum_{\mathbb{B}_1,\ldots,\mathbb{B}_m\leq n}\Big|\prod_{j=1}^m t^{n-|\mathbb{B}_j|}(1-t)^{|\mathbb{B}_j|}\Big| = \prod_{j=1}^m\Big(\sum_{\mathbb{B}_j\leq n} t^{n-|\mathbb{B}_j|}(1-t)^{|\mathbb{B}_j|}\Big) = 1. \tag{3}$$

Taking into account the positivity of $P_n^{(\mathbb{B})}$, which is a consequence of 6.4.7, yields

$$\|[P_n^{(\mathbb{B})}, Y]\| = 1 \quad \text{whenever } \mathbb{B} \leq n, \tag{4}$$

and 6.4.8 implies that, for $\boldsymbol{f} \in [L_2, X]$,

$$\frac{1}{2^m}\sum_{e\in\mathbb{E}_2^m}\Big\|\Big[\prod_{j=1}^m(I_n + \varepsilon_j P_n^{(\mathbb{B}_j)}), T\Big]\boldsymbol{f}\Big\| \leq 2^{m/2}\varrho(T|\mathcal{W}_{2^m}, \mathcal{I}_{2^m})\,\|\boldsymbol{f}\|. \tag{5}$$

The required inequality now follows from

$$\|[R_n(t)^m, T]\boldsymbol{f}\| \leq$$

$$\leq \sum_{\mathbb{B}_1,\ldots,\mathbb{B}_m\leq n}\left[\begin{array}{c}\Big|\prod_{j=1}^m t^{n-|\mathbb{B}_j|}(1-t)^{|\mathbb{B}_j|}\Big| \\ \times \\ \frac{1}{2^m}\sum_{e\in\mathbb{E}_2^m}\Big\|\Big[\prod_{j\in\mathbb{J}_e^+}(-P_n^{(\mathbb{B}_j)}), Y\Big]\Big\|\Big\|\Big[\prod_{j=1}^m(I_n + \varepsilon_j P_n^{(\mathbb{B}_j)}), T\Big]\boldsymbol{f}\Big\|\end{array}\right].$$

6.4.10 Next, a supplement to the previous lemma will be established.

LEMMA.

$$\|[R_n(t)^m, T]\| \le \sqrt{(n+1)2^n}\, \|T\| \quad \textit{for } 0 \le t \le 1 \textit{ and } m = 1, 2, \ldots.$$

PROOF. Recall from (7) in 6.4.5 that

$$R_n(t)^m = \sum_{h=0}^{n} \Big(\frac{1-3t^h}{2}\Big)^m W_n^{(h)}.$$

By 3.4.2 and 3.4.4,

$$\|[W_n^{(h)}, T]\| = \boldsymbol{\delta}(T|\mathcal{W}(\mathbb{F}_n^{(h)}), \mathcal{W}(\mathbb{F}_n^{(h)})) \le |\mathbb{F}_n^{(h)}|^{1/2}\|T\| = \tbinom{n}{h}^{1/2}\|T\|.$$

Moreover, we infer from Hölder's inequality that

$$\Big(\sum_{h=0}^{n} \binom{n}{h}^{1/2}\Big)^2 \le (n+1)\sum_{h=0}^{n}\binom{n}{h} = (n+1)2^n.$$

Combining these facts yields

$$\begin{aligned}\|[R_n(t)^m, T]\| &\le \Big\|\sum_{h=0}^{n} \Big(\frac{1-3t^h}{2}\Big)^m [W_n^{(h)}, T]\Big\| \\ &\le \sum_{h=0}^{n}\binom{n}{h}^{1/2}\|T\| \le \sqrt{(n+1)2^n}\|T\|.\end{aligned}$$

6.4.11 We now begin to prove a series of four lemmas which step by step lead us to the desired result.

LEMMA. *There exist $c_p^{(1)} > 0$ and $d_p \in (2^{1/p}, 2)$ such that*

$$\|[(\varrho I_n - R_n(t))^{-1}, T]\| \le \frac{c_p^{(1)}}{\varrho}\, \boldsymbol{\varrho}_p^{(p')}(T|\mathcal{W}_{2^n}, \mathcal{I}_{2^n}) \quad \textit{for } 0 \le t \le 1 \textit{ and } \varrho \ge d_p.$$

PROOF. By Hölder's inequality,

$$\boldsymbol{\varrho}(\mathcal{W}_{2^m}, \mathcal{I}_{2^m}) \le 2^{m(1/p-1/2)}\, \boldsymbol{\varrho}_p^{(p')}(\mathcal{W}_{2^m}, \mathcal{I}_{2^m}) \quad \text{if } 1 < p \le 2.$$

We know from (8) in 6.4.5 that

$$\|R_n(t)^m\| = 1 \quad \text{for } 0 \le t \le 1 \text{ and } m = 1, 2, \ldots.$$

Hence

$$(\varrho I_n - R_n(t))^{-1} = \frac{1}{\varrho}\sum_{m=0}^{\infty} \frac{R_n(t)^m}{\varrho^m} \quad \text{for } 0 \le t \le 1 \text{ and } \varrho > 1.$$

Consequently, in view of 6.4.9 and 6.4.10,

$$\begin{aligned}\|[(\varrho I_n - R_n(t))^{-1}, T]\| &\le \frac{1}{\varrho}\sum_{m=0}^{\infty}\frac{\|[R_n(t)^m, T]\|}{\varrho^m}\\ &\le \frac{1}{\varrho}\Bigg(\sum_{m=0}^{n} 2^{m/2}\frac{\varrho(T|\mathcal{W}_{2^m},\mathcal{I}_{2^m})}{\varrho^m} + \sum_{m=n+1}^{\infty}\frac{\sqrt{(n+1)2^n}\,\|T\|}{\varrho^m}\Bigg)\\ &\le \frac{1}{\varrho}\,\varrho_p^{(p')}(\mathcal{W}_{2^n},\mathcal{I}_{2^n})\Bigg(\sum_{m=0}^{n}\Big(\frac{2^{1/p}}{\varrho}\Big)^m + \sum_{m=n+1}^{\infty}\sqrt{m}\Big(\frac{2^{1/2}}{\varrho}\Big)^m\Bigg).\end{aligned}$$

In order to guarantee that

$$c_p^{(1)} := \sum_{m=0}^{\infty}\Big(\frac{2^{1/p}}{d_p}\Big)^m + \sum_{m=1}^{\infty}\sqrt{m}\Big(\frac{2^{1/2}}{d_p}\Big)^m$$

is finite, we choose $d_p > 2^{1/p} \ge 2^{1/2}$. In addition, and this will turn out to be crucial, we can arrange that $d_p < 2$.

6.4.12 The next step is trivial.

LEMMA. *There exist $c_p^{(2)} > 0$ and $d_p \in (2^{1/p}, 2)$ such that*

$$\|[(\sigma I_n - S_n(t))^{-1}, T]\| \le \frac{c_p^{(2)}}{1-3\sigma}\,\varrho_p^{(p')}(T|\mathcal{W}_{2^n},\mathcal{I}_{2^n})$$

for $0 \le t \le 1$ and $\frac{1-3\sigma}{2} \ge d_p$.

PROOF. In view of $R_n(t) = \frac{1}{2}\big(I_n - 3S_n(t)\big)$, we have

$$\sigma I_n - S_n(t) = -\tfrac{2}{3}\big(\tfrac{1-3\sigma}{2}I_n - R_n(t)\big).$$

Hence

$$\Big\|\Big[\big(\sigma I_n - S_n(t)\big)^{-1}, T\Big]\Big\| = \tfrac{3}{2}\Big\|\Big[\big(\tfrac{1-3\sigma}{2}I_n - R_n(t)\big)^{-1}, T\Big]\Big\|.$$

The assertion now follows from the previous lemma.

6.4.13 For $a > 0$, we consider the sector

$$\Omega_a := \{\, w = u + iv : u, v \in \mathbb{R},\ |v| \le a\,u \,\}.$$

LEMMA. *There exist $c_p^{(3)} > 0$ and $a_p > 0$ such that*

$$\|[(wI_n - A_n)^{-1}, T]\| \le \frac{c_p^{(3)}}{|w|}\,\varrho_p^{(p')}(T|\mathcal{W}_{2^n},\mathcal{I}_{2^n}) \quad \textit{for } w \in \mathbb{C}\setminus\Omega_{a_p}.$$

PROOF. Let $w = u + iv \in \mathbb{C}\setminus\Omega_a$ and assume that $u \ne 0$ and $v \ne 0$. Putting $t := e^{-\pi/|v|}$, we have $0 < t < 1$ and

$$t^w = e^{-\pi u/|v| - i\pi v/|v|} = -e^{-\pi u/|v|}. \tag{1}$$

Formula (5) in 6.4.5 implies that

$$[(wI_n - A_n)^{-1}, T] = (-t^w)[(t^w I_n - S_n(t))^{-1}, T] \int_t^1 \tau^{-w-1}[S_n(\tau), X]\, d\tau. \quad (2)$$

We conclude from (2) in 6.4.5, 6.4.7 and 0.4.9 that $\|[S_n(\tau), X]\| \leq 1$ for $0 \leq \tau \leq 1$. Hence

$$\Big\| \int_t^1 \tau^{-w-1}[S_n(\tau), X] d\tau \Big\| \leq \int_t^1 \tau^{-u-1} d\tau = \frac{t^{-u} - 1}{u} = \frac{e^{\pi u/|v|} - 1}{u}. \quad (3)$$

Since $d_p < 2$ and

$$\frac{1 - 3t^w}{2} \geq \begin{cases} \frac{1+3e^{-\pi/a}}{2} & \text{if } u > 0, \\ 2 & \text{if } u < 0, \end{cases}$$

choosing the opening $a > 0$ large enough, it can be arranged that $\frac{1-3t^w}{2} \geq d_p$. Therefore the preceding lemma applies, and we get

$$\|[(t^w I_n - S_n(t))^{-1}, T]\| \leq \frac{c_p^{(2)}}{1 + 3e^{-\pi u/|v|}}\, \varrho_p^{(p')}(T|\mathcal{W}_{2^n}, \mathcal{I}_{2^n}). \quad (4)$$

Combing (1), (2), (3) and (4) yields

$$\begin{aligned} &\|[(wI_n - A_n)^{-1}, T]\| \leq \\ &\quad \leq |t^w| \|[(t^w I_n - S_n(t))^{-1}, T]\| \Big\| \int_1^t \tau^{-w-1}[S_n(\tau), X]\, d\tau \Big\| \\ &\quad \leq c_p^{(2)} \frac{1}{u} \frac{1 - e^{-\pi u/|v|}}{1 + 3e^{-\pi u/|v|}}\, \varrho_p^{(p')}(T|\mathcal{W}_{2^n}, \mathcal{I}_{2^n}). \end{aligned}$$

In view of 0.10.4,

$$\frac{1}{u} \frac{1 - e^{-\pi u/|v|}}{1 + 3e^{-\pi u/|v|}} \leq \min\left\{ \frac{1}{|u|}, \frac{\pi}{2|v|} \right\} = \frac{1}{\max\{|u|, \frac{2}{\pi}|v|\}} \leq \frac{\sqrt{1 + (\frac{\pi}{2})^2}}{|w|}.$$

Hence

$$\|[(wI_n - A_n)^{-1}, T]\| \leq 2\, \frac{c_p^{(2)}}{|w|}\, \varrho_p^{(p')}(T|\mathcal{W}_{2^n}, \mathcal{I}_{2^n}) \quad \text{for} \quad w \in \mathbb{C} \setminus \Omega_a$$

provided that $u \neq 0$ and $v \neq 0$. This restriction, however, can be removed, since both sides are continuous functions of w.

6.4.14 LEMMA. *There exist $c_p^{(4)} > 0$ and $r_p > 0$ such that*

$$\|[S_n(z), T]\| \leq c_p^{(4)}\, \varrho_p^{(p')}(T|\mathcal{W}_{2^n}, \mathcal{I}_{2^n}) \quad \textit{for } |z| \leq r_p.$$

PROOF. Given $z = re^{i\varphi}$ with $r > 0$ and $-\pi \leq \varphi < +\pi$, we define the holomorphic function

$$w \longrightarrow z^w := e^{w(\log r + i\varphi)}.$$

If $w = u + iv$, then
$$|z^w| = e^{u\log r - v\varphi}.$$

Cauchy's formula says that
$$z^h = \frac{1}{2\pi i}\int_\Gamma \frac{z^w}{w-h}dw \quad \text{for } h = 0, \ldots, n,$$

where Γ denotes any closed rectifiable path oriented counterclockwise and enclosing $\{0, 1, \ldots, n\}$, the spectrum of A_n. In particular, given $a > 0$ and $N > n$, we may take the triangle Γ_a formed by the lines
$$\Gamma^{\pm}_{a,N} := \Big\{(u, v) : v = \pm a(u+1),\ -1 \le u \le N\Big\}$$
and
$$\Gamma_{a,N} := \Big\{(N, v) : |v| \le a(N+1)\Big\}.$$

The situation is illustrated in the following figure:

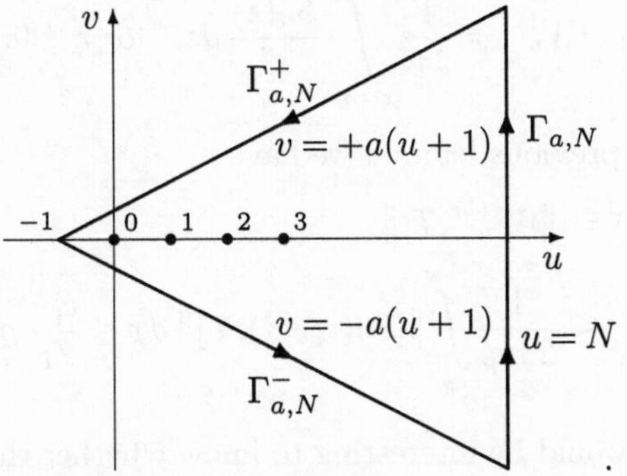

As $N \to \infty$, we get the path $<_a$ consisting of the rays
$$\Gamma^{\pm}_a := \Big\{ (u, v) : v = \pm a(u+1),\ -1 \le u < \infty \Big\}.$$

Since
$$\int_{-a(N+1)}^{+a(N+1)} \left|\frac{z^w}{w-h}\right| dv = \int_{-a(N+1)}^{+a(N+1)} \frac{e^{N\log r - v\varphi}}{\sqrt{(N-h)^2 + v^2}}\, dv = O(e^{N(\log r + \pi a)}),$$
we have
$$z^h = \frac{1}{2\pi i}\int_{<_a} \frac{z^w}{w-h}dw$$

whenever r is so small that $\log r + \pi a < 0$. Hence there exists $r_a > 0$ such that
$$S_n(z) = z^{A_n} = \frac{1}{2\pi i}\int_{<_a} z^w (wI_n - A_n)^{-1} dw \quad \text{for} \quad |z| \le r_a.$$

Moreover,

$$d_a := \sup\left\{ \frac{1}{2\pi} \int\limits_{<_a} \left|\frac{z^w}{w}\right| d|w| \, : \, |z| \le r_a \right\}$$

is finite. Taking the $a_p > 0$ obtained in the previous lemma, we finally arrive at

$$\big\|[S_n(z), T]\big\| \le c_p^{(3)} d_{a_p} \varrho_p^{(p')}(T|\mathcal{W}_{2^n}, \mathcal{I}_{2^n}) \quad \text{for } |z| \le r_{a_p}.$$

6.4.15 We are now able to establish the main result of this section.

THEOREM. *Let $1 < p \le 2$. Then there exists a constant $c_p \ge 1$ such that*

$$\boldsymbol{\delta}(\mathcal{R}_n, \mathcal{R}_n) \le c_p \, \varrho_p^{(p')}(\mathcal{W}_{2^n}, \mathcal{I}_{2^n}).$$

PROOF. Cauchy's formula and (1) in 6.4.5 imply that

$$W_n^{(1)} = \frac{1}{2\pi i} \int\limits_{|z|=r} \frac{S_n(z)}{z^2}\, dz \quad \text{for } r > 0.$$

Thus, by the previous lemma, we have

$$\begin{aligned} \boldsymbol{\delta}(T|\mathcal{R}_n, \mathcal{R}_n) &= \big\|[W_n^{(1)}, T]\big\| \\ &\le \frac{1}{2\pi r_p} \int\limits_{-\pi}^{+\pi} \big\|[S_n(r_p e^{i\varphi}), T]\big\|\, d\varphi \le \frac{c_p^{(4)}}{r_p} \varrho_p^{(p')}(T|\mathcal{W}_{2^n}, \mathcal{I}_{2^n}). \end{aligned}$$

REMARK. It would be interesting to know whether the above inequality remains true if the Walsh system is replaced by the Fourier system.

6.4.16 THEOREM. $\displaystyle\bigcup_{1<p\le 2} \mathfrak{WT}_p \subset \mathfrak{RP} \subset \mathfrak{QRP} = \mathfrak{WT}.$

PROOF. The left-hand inclusion follows from the preceding theorem, and $\mathfrak{RP} \subseteq \mathfrak{QRP}$ is trivial. Moreover, by 4.14.8 and 6.3.12, we have

$$\mathfrak{QRP} = \mathfrak{RT} = \mathfrak{WT}.$$

Next, 4.12.14 and 6.3.11 tell us that

$$\boldsymbol{\delta}(C_{1/2}|\mathcal{R}_n, \mathcal{R}_n) \asymp 1 \quad \text{and} \quad \varrho(C_{1/2}|\mathcal{W}_n, \mathcal{I}_n) \asymp \sqrt{\tfrac{n}{1+\log n}}.$$

Hence

$$C_{1/2} \in \mathfrak{RP} \setminus \bigcup_{1<p<2} \mathfrak{WT}_p^{weak} = \bigcup_{1<p\le 2} \mathfrak{WT}_p.$$

On the other hand, it was shown in 4.14.12 that $\mathfrak{RP}$ is strictly smaller than $\mathfrak{QRP}$.

6.4.17 THEOREM. $\bigcup\limits_{1<p\leq 2} \mathsf{WT}_p = \mathsf{RP} = \mathsf{QRP} = \mathsf{WT}.$

PROOF. We know from 6.3.10 that

$$\bigcup_{1<p\leq 2} \mathsf{WT}_p = \mathsf{WT}.$$

The assertion now follows from the preceding theorem by a sandwich argument.

REMARK. In particular, we have shown that $\mathsf{RP} = \mathsf{RT}$. This is Pisier's theorem:

K-*convexity* = B-*convexity.*

6.5 Walsh versus Fourier

6.5.1 The MAIN PROBLEM arising from this treatise is to illuminate

the similarities and the dissimilarities

of

the Fourier system and the Walsh system.

(the circle group) (the Cantor group)

These orthonormal systems possess many properties in common. However, in some respects they behave quite differently.

6.5.2 It follows from 5.5.17 and 6.2.8 that

$$\boldsymbol{\varrho}(L_r|\mathcal{E}_n,\mathfrak{I}_n) \asymp \boldsymbol{\varrho}(L_r|\mathcal{W}_n,\mathfrak{I}_n).$$

Moreover, combining 5.9.4 and 6.4.3 yields that, for every Banach space X,

$$\boldsymbol{\varrho}(X|\mathcal{E}_n,\mathfrak{I}_n) = O(1) \iff \boldsymbol{\varrho}(X|\mathcal{W}_n,\mathfrak{I}_n) = O(1).$$

Similarly, by 5.6.29 and 6.3.12,

$$\boldsymbol{\varrho}(X|\mathcal{E}_n,\mathfrak{I}_n) = o(\sqrt{n}) \iff \boldsymbol{\varrho}(X|\mathcal{W}_n,\mathfrak{I}_n) = o(\sqrt{n}).$$

Nevertheless, it seems unlikely that both sequences of type ideal norms always show the same asymptotic behaviour.

PROBLEM. *Is it true that* $\boldsymbol{\varrho}(\mathcal{E}_n,\mathfrak{I}_n) \asymp \boldsymbol{\varrho}(\mathcal{W}_n,\mathfrak{I}_n)$ *?*

REMARK. A negative answer would suggest that there exists a huge variety of different types, and that is what we expect. On the other hand, an affirmative answer could even lead to the result that

$$\boldsymbol{\varrho}(\mathcal{E}_n,\mathfrak{I}_n) \asymp \boldsymbol{\varrho}(\mathcal{W}_n,\mathfrak{I}_n) \asymp \boldsymbol{\tau}_n,$$

where $\boldsymbol{\tau}_n$ is the universal type ideal norm defined in 3.10.1.

The most hopeful candidate for a counterexample is the canonical operator

$$E_n x := \sum_{k=1}^{n} \xi_k e_k \quad \text{for } x = (\xi_k) \in l_1^n;$$

see 3.8.17. Then $\varrho(E_n : l_1^n \to C|\mathcal{E}_n, \mathcal{I}_n) = \sqrt{n}$, where C denotes the Banach space of all 2π-periodic continuous complex-valued functions on $\mathbb{R}$. Unfortunately, we were not able to determine the asymptotic behaviour of $\varrho(E_n : l_1^n \to C|\mathcal{W}_n, \mathcal{I}_n)$ as $n \to \infty$.

6.5.3 The situation for the associated Dirichlet ideal norms is also unclear. In this case, we know that

$$\boldsymbol{\delta}(L_1|\mathcal{E}_{2^n}, \mathcal{E}_{2^n}) \asymp n \quad \text{and} \quad \boldsymbol{\delta}(L_1|\mathcal{W}_{2^n}, \mathcal{W}_{2^n}) \asymp 1;$$

see 5.2.34 and 8.7.1. However, it follows that

$$\max_{1 \le m \le n} \boldsymbol{\delta}(L_r|\mathcal{E}_m, \mathcal{E}_m) \asymp \max_{1 \le m \le n} \boldsymbol{\delta}(L_r|\mathcal{W}_m, \mathcal{W}_m).$$

As in the setting of type ideal norms, we do not conjecture that uniform equivalence holds in general.

PROBLEM. *Is it true that*

$$\max_{1 \le m \le n} \boldsymbol{\delta}(\mathcal{E}_m, \mathcal{E}_m) \asymp \max_{1 \le m \le n} \boldsymbol{\delta}(\mathcal{W}_m, \mathcal{W}_m)\ ?$$

In the limiting cases, the following questions arise:

$$\max_{1 \le m \le n} \boldsymbol{\delta}(X|\mathcal{E}_m, \mathcal{E}_m) = O(1) \overset{Burkholder-Bourgain}{\Longleftrightarrow} \max_{1 \le m \le n} \boldsymbol{\delta}(X|\mathcal{W}_m, \mathcal{W}_m) = O(1),$$

$$\max_{1 \le m \le n} \boldsymbol{\delta}(X|\mathcal{E}_m, \mathcal{E}_m) = o(1+\log n) \overset{?}{\underset{\Longleftarrow}{\Longrightarrow}} \max_{1 \le m \le n} \boldsymbol{\delta}(X|\mathcal{W}_m, \mathcal{W}_m) = o(1+\log n)$$

for all Banach spaces X, and

$$\max_{1 \le m \le n} \boldsymbol{\delta}(T|\mathcal{E}_m, \mathcal{E}_m) = O(1) \overset{?}{\Longleftrightarrow} \max_{1 \le m \le n} \boldsymbol{\delta}(T|\mathcal{W}_m, \mathcal{W}_m) = O(1),$$

$$\max_{1 \le m \le n} \boldsymbol{\delta}(T|\mathcal{E}_m, \mathcal{E}_m) = o(1+\log n) \overset{?}{\underset{\Longleftarrow}{\Longrightarrow}} \max_{1 \le m \le n} \boldsymbol{\delta}(T|\mathcal{W}_m, \mathcal{W}_m) = o(1+\log n)$$

for all operators T.

REMARKS. The relation $\Longleftrightarrow$ without question mark is equivalent to the celebrated Burkholder–Bourgain theorem 8.8.1. Concerning the lower direction in $\overset{?}{\underset{\Longleftarrow}{\Longrightarrow}}$, we refer to 8.8.2, where the same problems are stated in terms of other ideal norms.

Since the sequence $(\boldsymbol{\kappa}(H_n))$ is non-decreasing and $\boldsymbol{\delta}(\mathcal{E}_n,\mathcal{E}_n) \asymp \boldsymbol{\kappa}(H_n)$, it follows that

$$\max_{1\le m\le n} \boldsymbol{\delta}(\mathcal{E}_m,\mathcal{E}_m) \asymp \boldsymbol{\delta}(\mathcal{E}_n,\mathcal{E}_n).$$

6.5.4 Despite our lack of knowledge described so far, we have at least the following result which says that the Fourier system and the Walsh system can be distinguished with the help of Banach spaces.

EXAMPLE. $\sqrt{1+\log n} \prec \boldsymbol{\varrho}(l_\infty^n|\mathcal{E}_n,\mathcal{W}_n)$.

PROOF. Let Σ_n be the finite summation operator defined in 0.7.3. Then

$$\boldsymbol{\delta}(\Sigma_n : l_1^n \to l_\infty^n|\mathcal{E}_n,\mathcal{E}_n) \le \boldsymbol{\varrho}(l_\infty^n|\mathcal{E}_n,\mathcal{W}_n)\,\boldsymbol{\delta}(\Sigma_n : l_1^n \to l_\infty^n|\mathcal{W}_n,\mathcal{E}_n)$$

and

$$\boldsymbol{\delta}(\Sigma_n : l_1^n \to l_\infty^n|\mathcal{E}_n,\mathcal{W}_n) \le \boldsymbol{\varrho}(l_\infty^n|\mathcal{E}_n,\mathcal{W}_n)\,\boldsymbol{\delta}(\Sigma_n : l_1^n \to l_\infty^n|\mathcal{W}_n,\mathcal{W}_n).$$

Note that $\Sigma_n' = R_n\Sigma_nR_n$ with $R_n(\xi_k) := (\xi_{n-k+1})$. Therefore it follows from $\|R_n : l_1^n \to l_1^n\| = 1$ that

$$\boldsymbol{\delta}(\Sigma_n : l_1^n \to l_\infty^n|\mathcal{W}_n,\mathcal{E}_n) = \boldsymbol{\delta}(\Sigma_n' : l_1^n \to l_\infty^n|\mathcal{W}_n,\mathcal{E}_n) = \boldsymbol{\delta}(\Sigma_n : l_1^n \to l_\infty^n|\mathcal{E}_n,\mathcal{W}_n).$$

Combining the above facts yields

$$\boldsymbol{\delta}(\Sigma_n : l_1^n \to l_\infty^n|\mathcal{E}_n,\mathcal{E}_n) \le \boldsymbol{\varrho}(l_\infty^n|\mathcal{E}_n,\mathcal{W}_n)^2\,\boldsymbol{\delta}(\Sigma_n : l_1^n \to l_\infty^n|\mathcal{W}_n,\mathcal{W}_n).$$

The assertion is now an immediate consequence of

$$\boldsymbol{\delta}(\Sigma_n : l_1^n \to l_\infty^n|\mathcal{E}_n,\mathcal{E}_n) \asymp 1+\log n \quad\text{and}\quad \boldsymbol{\delta}(\Sigma_{2^n} : l_1^{2^n} \to l_\infty^{2^n}|\mathcal{W}_{2^n},\mathcal{W}_{2^n}) = 1;$$

see 2.4.4 and 5.3.8.

REMARK. On the basis of computer experiments, A. Hinrichs arrived at the conjecture that

$$\boldsymbol{\varrho}(l_\infty^n|\mathcal{E}_n,\mathcal{W}_n) \asymp n^\lambda,$$

where $\lambda = \log_2 \frac{1+\sqrt{3}}{2} = 0.44998\ldots$. So the difference between $\mathcal{E}_n$ and $\mathcal{W}_n$ should be quite large.

7

Haar ideal norms

This chapter is devoted to type and cotype ideal norms associated with the Haar system. The main tools are Burkholder–Gundy–Davis inequalities for vector-valued martingales. Our final goal is James's theory of superreflexivity. We present a new approach which is based on the submultiplicativity of martingale type ideal norms. In this way, B-convexity, MP-convexity and J-convexity (=superreflexivity) can be treated more or less in parallel.

Among others, the following classes of Banach spaces will be considered:

AT_p	:	Banach spaces of Haar type p,
AT	:	Banach spaces of Haar subtype,
SR	:	superreflexive Banach spaces,
AC_q	:	Banach spaces of Haar cotype q,
AC	:	Banach spaces of Haar subcotype.

We are going to prove that

$$\bigcup_{1<p\le 2} \mathsf{AT}_p = \mathsf{AT} = \mathsf{SR} = \mathsf{AC} = \bigcup_{2\le q<\infty} \mathsf{AC}_q$$

and

$$\mathsf{AT}_2 \cap \mathsf{AC}_2 = \mathsf{H}.$$

The duality between Haar type and cotype is described by

$$(\mathsf{AT}_p)' = \mathsf{AC}_{p'} \quad \text{and} \quad (\mathsf{AC}_q)' = \mathsf{AT}_{q'}.$$

Another highlight is Enflo's theorem about uniformly convex and uniformly smooth renormings of superreflexive spaces.

We refer to the following books concerned with Haar functions, martingales, and superreflexivity:

Beauzamy .... *Introduction to Banach spaces and their geometry* [BEA 2],
Deville/Godefroy/Zizler... *Smoothness and renorming in Banach spaces* [DEV*],
van Dulst................. *Reflexive and superreflexive Banach spaces* [DUL],
Garsia..................... *Martingale inequalities* [GARS],
Neveu..................... *Martingales à temps discret* [NEV],
Novikov/Semenov .... *Haar series and linear operators* [NOV*],
Williams .... *Probability with martingales* [WIL].

7.1 Martingales

7.1.1 Let μ be a probability measure on a set M. As usual, $\{f \le g\}$ denotes the collection of all $t \in M$ with $f(t) \le g(t)$; etc.

7.1.2 We now consider any algebra $\mathcal{F}$ formed by a finite number of measurable sets.

An **atom** $A \in \mathcal{F}$ is a set not containing any $A_0 \in \mathcal{F}$ such that $0 < \mu(A_0) < \mu(A)$. Clearly, $\mathcal{F}$-measurable X-valued functions can be characterized by the property that they are constant on the atoms of $\mathcal{F}$. For $1 \le r \le \infty$, these functions form a finite dimensional subspace of $[L_r(M,\mu), X]$, denoted by $[L_r(M,\mathcal{F},\mu), X]$.

7.1.3 The **conditional expectation** of $\boldsymbol{f} \in [L_1(M,\mu), X]$ is the $\mathcal{F}$-measurable function $[E(\mathcal{F}), X]\boldsymbol{f}$ that takes the constant value

$$\frac{1}{\mu(A)} \int_A \boldsymbol{f}(t)\, d\mu(t)$$

on every atom $A \in \mathcal{F}$. For the trivial subalgebra $\{\emptyset, M\}$, we get the **expectation** (mean)

$$[E, X]\boldsymbol{f} := \int_M \boldsymbol{f}(t)\, d\mu(t).$$

Note that, in the scalar-valued case, the mapping $E(\mathcal{F})$ restricted to $L_2(M,\mu)$ is nothing but the orthogonal projection onto the finite dimensional subspace $L_2(M,\mathcal{F},\mu)$.

7.1.4 Fix algebras $\mathcal{F}_0, \dots, \mathcal{F}_n$ formed by a finite number of measurable subsets of M. In particular, let $\mathcal{F}_0 := \{\emptyset, M\}$. We say that functions $\boldsymbol{f}_0, \dots, \boldsymbol{f}_n : M \to X$ are **adapted** to the **filtration** $\mathcal{F}_0 \subseteq \dots \subseteq \mathcal{F}_n$ if $\boldsymbol{f}_k$ is $\mathcal{F}_k$-measurable for $k = 0, \dots, n$.

7.1.5 An X-valued **martingale** of length n is an adapted $(n+1)$-tuple $(\boldsymbol{f}_k)$ such that

$$\boldsymbol{f}_k = [E(\mathcal{F}_k), X]\boldsymbol{f}_h \quad \text{for } 0 \le k \le h \le n.$$

If $\boldsymbol{f}_0 = \boldsymbol{o}$, then $(\boldsymbol{f}_k)$ is said to have mean zero.

The **martingale differences** $\boldsymbol{d}_k$ are defined by

$$\boldsymbol{d}_0 := \boldsymbol{f}_0 \quad \text{and} \quad \boldsymbol{d}_k := \boldsymbol{f}_k - \boldsymbol{f}_{k-1} \quad \text{for } k = 1, \dots, n.$$

Hence

$$\boldsymbol{f}_k = \sum_{h=0}^{k} \boldsymbol{d}_h \quad \text{for } k = 0, \dots, n.$$

REMARK. In view of $\boldsymbol{f}_k = [E(\mathcal{F}_k), X]\boldsymbol{f}_n$, every X-valued martingale of length n is completely specified by $\boldsymbol{f}_n$. So we are justified in writing $\boldsymbol{f}_n$ instead of $(\boldsymbol{f}_k)$. It depends on the underlying situation which symbol is more appropriate.

7.1.6 By a **stopping time** we mean a map $\sigma : M \to \{0, \dots, n\}$ with the property that $\{\sigma = k\} \in \mathcal{F}_k$ for $k = 0, \dots, n$.

For every X-valued martingale $(\boldsymbol{f}_k)$ and every stopping time σ, we define the **stopped martingale**

$$\boldsymbol{f}_k^\sigma(t) := \boldsymbol{f}_{k\wedge\sigma(t)}(t) = \sum_{h=0}^{k\wedge\sigma(t)} \boldsymbol{d}_h(t) \quad \text{for } t \in M \text{ and } k = 0, \dots, n.$$

That $(\boldsymbol{f}_k^\sigma)$ is indeed a martingale follows from the formula

$$\boldsymbol{f}_k^\sigma = \boldsymbol{d}_0 + \sum_{h=1}^{k} \chi_h \boldsymbol{d}_h,$$

where χ_h is the characteristic function of $\{\sigma \ge h\} = M \setminus \bigcup_{k=0}^{h-1} \{\sigma = k\} \in \mathcal{F}_{h-1}$.

7.1.7 Given $T_1, \dots, T_n \in \mathfrak{L}(X, Y)$, we define the **martingale transform**

$$\boldsymbol{T} = (T_k) : \boldsymbol{f}_n(t) = \sum_{k=1}^{n} \boldsymbol{d}_k(t) \longrightarrow \boldsymbol{g}_n(t) = \sum_{k=1}^{n} T_k \boldsymbol{d}_k(t)$$

for every X-valued martingale $\boldsymbol{f}_n$ of length n and of mean zero.

7.1.8 The **maximal function** is defined by

$$\boldsymbol{f}_n^\star(t) := \max_{0 \le k \le n} \|\boldsymbol{f}_k(t)\| \quad \text{for } t \in M.$$

REMARK. Omitting parentheses, we use the symbol $\boldsymbol{T}\boldsymbol{f}_n^\star$ to denote the maximal function of the transformed martingale $\boldsymbol{T}\boldsymbol{f}_n$.

7.1.9 Doob's inequality says that

$$\|\boldsymbol{f}_n^\star|L_r\| \le \tfrac{r}{r-1}\|\boldsymbol{f}_n|L_r\|$$

for all X-valued martingales $\boldsymbol{f}_n$ and $1<r<\infty$. Replacing $|\cdot|$ by $\|\cdot\|$, the proof of the scalar-valued case can be carried over verbatim to vector-valued martingales; see [DOO, pp. 311–318], [GARS, pp. 15 and 83], [LIN*2, p. 153], [NEV, p. 68] and [WIL, p. 143].

7.2 Dyadic martingales

7.2.1 In this section, the underlying probability space is the unit interval $[0,1)$ equipped with the Lebesgue measure λ.

7.2.2 For $k=0,1,2,\dots$ and $j=1,\dots,2^n$, put

$$\Delta_k^{(j)} := \big[\tfrac{j-1}{2^k}, \tfrac{j}{2^k}\big).$$

We refer to these sets as **dyadic intervals**. Let $\mathcal{D}_n$ denote the algebra generated by $\Delta_n^{(j)}$ with $j=1,\dots,2^n$. In particular, $\mathcal{D}_0 := \{\emptyset, [0,1)\}$.

7.2.3 An X-valued martingale of length n adapted to the **dyadic filtration** $\mathcal{D}_0 \subseteq \dots \subseteq \mathcal{D}_n$ is called **dyadic**. The name **Walsh–Paley martingale** is also common. Quite often, we will assume that $\boldsymbol{f}_0 = \boldsymbol{o}$, and $[WP_n^0, X]$ denotes the collection of such martingales.

7.2.4 For $\boldsymbol{f}_n \in [WP_n^0, X]$, the BMO-norm is defined by

$$\|\boldsymbol{f}_n|BMO\| := \sup\left\{\frac{1}{\lambda(A)}\int\limits_A \|\boldsymbol{f}_n(t)-\boldsymbol{f}_{h-1}(t)\|dt \,:\, A\in\mathcal{D}_h,\ 1\le h\le n\right\}.$$

Obviously, it suffices to take the supremum over all $\Delta_h^{(j)}$ with $1\le h\le n$ and $j=1,\dots,2^h$.

7.2.5 In what follows, U and V denote two auxiliary Banach spaces.

LEMMA. *Let $S_1,\dots,S_n \in \mathfrak{L}(X,U)$ and $T_1,\dots,T_n\in\mathfrak{L}(X,V)$. If*

$$\|\boldsymbol{T}\boldsymbol{f}_n|L_1\| \le \|\boldsymbol{S}\boldsymbol{f}_n^\star|L_\infty\| \quad \text{for } \boldsymbol{f}_n\in[WP_n^0,X],$$

then

$$\|\boldsymbol{T}\boldsymbol{f}_n|BMO\| \le 4\,\|\boldsymbol{S}\boldsymbol{f}_n^\star|L_\infty\| \quad \text{for } \boldsymbol{f}_n\in[WP_n^0,X].$$

PROOF. Fix any dyadic interval $\Delta_h^{(j)}$ with $h=1,\dots,n$, and let $\Delta_{h-1}^{(i)}$ be its predecessor, $\Delta_h^{(j)}\subset\Delta_{h-1}^{(i)}$. Note that

$$\gamma(t) := \frac{i-1+t}{2^{h-1}}$$

maps $[0,1)$ onto $\Delta_{h-1}^{(i)}$. Taking into account that $\boldsymbol{g}_n := (\boldsymbol{f}_n - \boldsymbol{f}_{h-1}) \circ \gamma$ is even $\mathcal{D}_{n-h+1}$-measurable, $\boldsymbol{g}_n$ can be viewed as a martingale in $[WP_n^0, X]$. By

$$\lambda(\Delta_{h-1}^{(i)}) = 2\lambda(\Delta_h^{(j)}) \quad \text{and} \quad \|\boldsymbol{Sg}_n^\star|L_\infty\| \le 2\,\|\boldsymbol{Sf}_n^\star|L_\infty\|,$$

we obtain

$$\begin{aligned}\frac{1}{\lambda(\Delta_h^{(j)})} \int\limits_{\Delta_h^{(j)}} \|\boldsymbol{Tf}_n(t) - \boldsymbol{Tf}_{h-1}(t)\|\,dt &\le \\ \le \frac{2}{\lambda(\Delta_{h-1}^{(i)})} &\int\limits_{\Delta_{h-1}^{(i)}} \|\boldsymbol{Tf}_n(t) - \boldsymbol{Tf}_{h-1}(t)\|\,dt \\ = 2\,\|\boldsymbol{Tg}_n|L_1\| &\le 2\,\|\boldsymbol{Sg}_n^\star|L_\infty\| \le 4\,\|\boldsymbol{Sf}_n^\star|L_\infty\|.\end{aligned}$$

7.2.6 As a counterpart of the maximal function defined in 7.1.8, given any dyadic X-valued martingale of length n and of mean zero, we let

$$\boldsymbol{f}_n^\times(t) := \max_{1\le k\le n} \{\|\boldsymbol{d}_k(t)\| + \|\boldsymbol{f}_{k-1}(t)\|\} \quad \text{for } t \in [0,1).$$

It follows from

$$\|\boldsymbol{f}_k(t)\| \le \|\boldsymbol{d}_k(t)\| + \|\boldsymbol{f}_{k-1}(t)\| \le \|\boldsymbol{f}_k(t)\| + 2\,\|\boldsymbol{f}_{k-1}(t)\|$$

that

$$\boldsymbol{f}_n^\star \le \boldsymbol{f}_n^\times \le 3\,\boldsymbol{f}_n^\star.$$

So we can easily pass from $\boldsymbol{f}_n^\star$ to $\boldsymbol{f}_n^\times$, and back.

REMARK. Omitting parentheses, we use the symbol $\boldsymbol{Sf}_n^\times$ to denote the modified maximal function of the transformed martingale $\boldsymbol{Sf}_n$.

7.2.7 LEMMA. *Let $S_1,\dots,S_n \in \mathfrak{L}(X,U)$ and $T_1,\dots,T_n \in \mathfrak{L}(X,V)$. If*

$$\|\boldsymbol{Tf}_n|BMO\| \le \|\boldsymbol{Sf}_n^\times|L_\infty\| \qquad \textit{for } \boldsymbol{f}_n \in [WP_n^0, X],$$

then

$$\lambda\{\,\boldsymbol{Tf}_n^\star > \beta u,\ \boldsymbol{Sf}_n^\times \le \alpha u\} \le \tfrac{\alpha}{\beta-1}\lambda\{\,\boldsymbol{Tf}_n^\star > u\} \quad \textit{for } \boldsymbol{f}_n \in [WP_n^0, X],$$

whenever $u>0$, $\alpha>0$ and $\beta>1$.

PROOF. Thanks to the $\mathcal{D}_{h-1}$-measurability of $\boldsymbol{Sf}_h^\times$,

$$\sigma(t) := \begin{cases} \min\{h-1 : \boldsymbol{Sf}_h^\times(t) > \alpha u\} & \text{if } \boldsymbol{Sf}_n^\times(t) > \alpha u, \\ n & \text{if } \boldsymbol{Sf}_n^\times(t) \le \alpha u, \end{cases}$$

defines a stopping time. Since

$$\boldsymbol{f}_k^\sigma(t) = \boldsymbol{f}_k(t) \quad \text{if } \boldsymbol{Sf}_n^\times(t) \le \alpha u,$$

we have

$$\lambda\{\,\boldsymbol{Tf}_n^\star > \beta u,\ \boldsymbol{Sf}_n^\times \le \alpha u\} \le \lambda\{(\boldsymbol{Tf}_n^\sigma)^\star > \beta u\,\}. \tag{1}$$

Let
$$B := \{(\boldsymbol{Tf}_n^\sigma)^\star > \beta u\}, \quad C := \{(\boldsymbol{Tf}_n^\sigma)^\star > u\},$$
and
$$C_h := \{\|\boldsymbol{Tf}_h^\sigma\| > u,\ \|\boldsymbol{Tf}_k^\sigma\| \le u \text{ if } k < h\} \quad \text{for } h = 1, \ldots, n.$$
Obviously,
$$B \subseteq C = \bigcup_{h=1}^{n} C_h \quad \text{and} \quad C_h \in \mathcal{D}_h.$$

We need another stopping time given by
$$\tau(t) := \begin{cases} \min\{h : \|\boldsymbol{Tf}_h^\sigma(t)\| > \beta u\} & \text{if } (\boldsymbol{Tf}_n^\sigma)^\star(t) > \beta u \\ n & \text{if } (\boldsymbol{Tf}_n^\sigma)^\star(t) \le \beta u. \end{cases}$$
Note that $\tau(t) \ge h$ for $t \in C_h$. In view of
$$\|\boldsymbol{Tf}_n^{\sigma\wedge\tau}(t)\| = \|\boldsymbol{Tf}_{\tau(t)}^\sigma(t)\| > \beta u \quad \text{for } t \in B,$$
$$\|\boldsymbol{Tf}_{h-1}^{\sigma\wedge\tau}(t)\| = \|\boldsymbol{Tf}_{h-1}^\sigma(t)\| \le u \quad \text{for } t \in C_h,$$
and
$$\|\boldsymbol{Tf}_n^{\sigma\wedge\tau}(t)\| \le \|\boldsymbol{Tf}_n^{\sigma\wedge\tau}(t) - \boldsymbol{Tf}_{h-1}^{\sigma\wedge\tau}(t)\| + \|\boldsymbol{Tf}_{h-1}^{\sigma\wedge\tau}(t)\|,$$
we obtain
$$(\beta - 1)u \le \|\boldsymbol{Tf}_n^{\sigma\wedge\tau}(t) - \boldsymbol{Tf}_{h-1}^{\sigma\wedge\tau}(t)\| \quad \text{for } t \in B \cap C_h,$$
which in turn implies that
$$(\beta - 1)u\,\lambda(B \cap C_h) \le \int\limits_{B\cap C_h} \|\boldsymbol{Tf}_n^{\sigma\wedge\tau}(t) - \boldsymbol{Tf}_{h-1}^{\sigma\wedge\tau}(t)\|\,dt$$
$$\le \int\limits_{C_h} \|\boldsymbol{Tf}_n^{\sigma\wedge\tau}(t) - \boldsymbol{Tf}_{h-1}^{\sigma\wedge\tau}(t)\|\,dt \le \lambda(C_h)\,\|\boldsymbol{Tf}_n^{\sigma\wedge\tau}|BMO\|. \quad (2)$$

Moreover, applying the assumption to the stopped martingale $\boldsymbol{f}_n^{\sigma\wedge\tau}$ and using the fact that $(\boldsymbol{Sf}_n^{\sigma\wedge\tau})^\times \le (\boldsymbol{Sf}_n^\sigma)^\times$, we get
$$\|\boldsymbol{Tf}_n^{\sigma\wedge\tau}|BMO\| \le \|(\boldsymbol{Sf}_n^{\sigma\wedge\tau})^\times|L_\infty\| \le \|(\boldsymbol{Sf}_n^\sigma)^\times|L_\infty\| \le \alpha u. \quad (3)$$
Inequalities (2) and (3) yield
$$(\beta - 1)u\,\lambda(B \cap C_h) \le \alpha u\,\lambda(C_h).$$
Thus
$$\lambda(B) = \sum_{h=1}^{n} \lambda(B \cap C_h) \le \frac{\alpha}{\beta - 1} \sum_{h=1}^{n} \lambda(C_h) = \frac{\alpha}{\beta - 1}\lambda(C).$$
This proves that
$$\lambda\{(\boldsymbol{Tf}_n^\sigma)^\star > \beta u\} \le \frac{\alpha}{\beta - 1}\lambda\{\boldsymbol{Tf}_n^\star > u\}. \quad (4)$$
Finally, we combine (1) and (4).

7.2.8 The following lemma is due to D. L. Burkholder [bur 2].

LEMMA. *Let f and g be non-negative Lebesgue measurable functions on $[0,1)$. Assume that there exist constants $\alpha > 0$, $\beta > 1$ and $c > 0$ such that*

$$\lambda\{\, g > \beta u,\ f \le \alpha u\,\} \le c\,\lambda\{\, g > u\,\} \quad \textit{whenever} \quad u > 0.$$

Then

$$\|g|L_r\| \le \alpha^{-1}(\beta^{-r} - c)^{-1/r}\,\|f|L_r\| \quad \textit{for } 1 \le r < \infty$$

provided that $\beta^r c < 1$.

PROOF. Integration by parts yields

$$\int_0^1 g(t)^r\,dt = r\int_0^\infty \lambda\{g>u\}\,u^{r-1}du = r\beta^r\int_0^\infty \lambda\{g>\beta u\}u^{r-1}du$$

$$= r\beta^r\left\{\begin{matrix}\int_0^\infty \lambda\{g>\beta u,\ f\le\alpha u\}u^{r-1}du\\ +\\ \int_0^\infty \lambda\{g>\beta u,\ f>\alpha u\}u^{r-1}du\end{matrix}\right\} \le r\beta^r\left\{\begin{matrix}c\int_0^\infty \lambda\{g>u\}u^{r-1}du\\ +\\ \int_0^\infty \lambda\{f>\alpha u\}u^{r-1}du\end{matrix}\right\}$$

$$= \beta^r\left\{c\int_0^1 g(t)^r dt + \alpha^{-r}\int_0^1 f(t)^r dt\right\}.$$

Hence

$$(1-\beta^r c)\int_0^1 g(t)^r\,dt \le \alpha^{-r}\beta^r\int_0^1 f(t)^r\,dt.$$

7.2.9 We are now prepared to establish the main result of this section, which concerns extrapolation of martingale inequalities.

THEOREM. *Let $S_1,\dots,S_n \in \mathfrak{L}(X,U)$ and $T_1,\dots,T_n \in \mathfrak{L}(X,V)$. If*

$$\|\boldsymbol{T}\boldsymbol{f}_n|L_1\| \le \|\boldsymbol{S}\boldsymbol{f}_n^\star|L_\infty\| \qquad \textit{for } \boldsymbol{f}_n \in [WP_n^0, X],$$

then

$$\|\boldsymbol{T}\boldsymbol{f}_n^\star|L_r\| \le C_r\,\|\boldsymbol{S}\boldsymbol{f}_n^\star|L_r\| \quad \textit{for } \boldsymbol{f}_n \in [WP_n^0, X],$$

whenever $1 \le r < \infty$, where $C_r > 0$ is a constant depending only on r.

PROOF. Since

$$\|\boldsymbol{T}\boldsymbol{f}_n|L_1\| \le \|\boldsymbol{S}\boldsymbol{f}_n^\star|L_\infty\|$$

for all $\boldsymbol{f}_n \in [WP_n^0, X]$, we successively obtain

$$\|\boldsymbol{Tf}_n|BMO\| \le \|4\boldsymbol{Sf}_n^\star|L_\infty\| \qquad \text{(by 7.2.5)},$$

$$\lambda\{\boldsymbol{Tf}_n^\star > \beta u,\ 4\boldsymbol{Sf}_n^\times \le \alpha u\} \le \tfrac{\alpha}{\beta-1}\lambda\{\boldsymbol{Tf}_n^\star > u\} \qquad \text{(by 7.2.7)},$$

$$\|\boldsymbol{Tf}_n^\star|L_r\| \le \alpha^{-1}(\beta^{-r} - \tfrac{\alpha}{\beta-1})^{-1/r}\,\|4\boldsymbol{Sf}_n^\times|L_r\| \qquad \text{(by 7.2.8)},$$

$$\|\boldsymbol{Tf}_n^\star|L_r\| \le 12\,\alpha^{-1}(\beta^{-r} - \tfrac{\alpha}{\beta-1})^{-1/r}\,\|\boldsymbol{Sf}_n^\star|L_r\| \qquad \text{(by 7.2.6)}.$$

Finally, let $\alpha = 3^{-r}$ and $\beta = 2$. Since $(2^{-r} - 3^{-r})^{-1/r} \le 6$ it follows that

$$C_r := 12\,\alpha^{-1}(\beta^{-r} - \frac{\alpha}{\beta-1})^{-1/r} \le 72 \cdot 3^r.$$

REMARK. The constant C_r can be improved significantly; see [gei 1]. However, this is irrelevant for the applications we have in mind.

7.2.10 We proceed with a corollary of the previous theorem. Compare with 5.4.7.

THEOREM. *Let $S_1, \ldots, S_n \in \mathfrak{L}(X,U)$ and $T_1, \ldots, T_n \in \mathfrak{L}(X,V)$. Consider the property: There exists a constant $c_r > 0$ such that*

$$\|\boldsymbol{Tf}_n|L_r\| \le c_r\,\|\boldsymbol{Sf}_n|L_r\| \quad \textit{whenever } \boldsymbol{f}_n \in [WP_n^0, X]. \qquad (\mathrm{M}_r)$$

If (M_r) *holds for* **some** $1 < r < \infty$, *then it holds for* **all** $1 < r < \infty$.

PROOF. Assuming (M_{r_0}), we get

$$\|\boldsymbol{Tf}_n|L_1\| \le \|\boldsymbol{Tf}_n|L_{r_0}\| \le c_{r_0}\,\|\boldsymbol{Sf}_n|L_{r_0}\| \le \|c_{r_0}\boldsymbol{Sf}_n|L_\infty\| \le \|c_{r_0}\boldsymbol{Sf}_n^\star|L_\infty\|$$

for $\boldsymbol{f}_n \in [WP_n^0, X]$. Hence Theorem 7.2.9 and Doob's inequality 7.1.9 imply that

$$\|\boldsymbol{Tf}_n|L_r\| \le \|\boldsymbol{Tf}_n^\star|L_r\| \le C_r\,\|c_{r_0}\boldsymbol{Sf}_n^\star|L_r\| \le \tfrac{r}{r-1}C_r c_{r_0}\,\|\boldsymbol{Sf}_n|L_r\|.$$

REMARK. Given $c_{r_0} > 0$, we may choose $c_r := \frac{r}{r-1}C_r c_{r_0}$, where $C_r > 0$ is taken from 7.2.9.

7.2.11 The above theorems will be applied to the following martingale transforms:

(1) The canonical transform

$$\boldsymbol{I}_T : \sum_{k=1}^n \boldsymbol{d}_k(t) \longrightarrow \sum_{k=1}^n T\boldsymbol{d}_k(t),$$

where $T \in \mathfrak{L}(X, Y)$.

(2) The change of signs transform

$$\boldsymbol{I}_e : \sum_{k=1}^{n} \boldsymbol{d}_k(t) \longrightarrow \sum_{k=1}^{n} \varepsilon_k \boldsymbol{d}_k(t)$$

induced by $e = (\varepsilon_1, \dots, \varepsilon_n) \in \mathbb{E}_2^n$.

(3) The random transform

$$\boldsymbol{R} : \sum_{k=1}^{n} \boldsymbol{d}_k(t) \longrightarrow \sum_{k=1}^{n} r_k(s)\, \boldsymbol{d}_k(t)$$

which maps X-valued into $[L_r[0,1), X]$-valued martingales.

(4) The spreading transform

$$\boldsymbol{S} : \sum_{k=1}^{n} \boldsymbol{d}_k(t) \longrightarrow (\boldsymbol{d}_1(t), \dots, \boldsymbol{d}_n(t))$$

which maps X-valued into $[l_r^n, X]$-valued martingales. Note that

$$\| \boldsymbol{S}\boldsymbol{f}_n(t) | l_r^n \| = \left(\sum_{k=1}^{n} \| \boldsymbol{d}_k(t) \|^r \right)^{1/r}.$$

For $r = 2$, we obtain the well-known *square function.*

7.2.12 To get the classical **Burkholder–Gundy–Davis inequalities**, we finally consider a scalar-valued dyadic martingale (f_k). Then, by Parseval's equality, the orthogonality of $d_1, \dots, d_n$ implies that

$$\| f_n | L_2 \| = \left(\sum_{k=1}^{n} \| d_k | L_2 \|^2 \right)^{1/2} = \left\| \left(\sum_{k=1}^{n} |d_k|^2 \right)^{1/2} \middle| L_2 \right\|.$$

We further note that $\left(\sum_{k=1}^{n} |d_k|^2 \right)^{1/2}$ is the n-th maximal function of the l_2^n-valued martingale $((d_1, \dots, d_k, 0, \dots, 0))$. Hence 7.2.9 yields

$$\| f_n | L_r \| \le a_r \left\| \left(\sum_{k=1}^{n} |d_k|^2 \right)^{1/2} \middle| L_r \right\|$$

for $1 \le r < \infty$. Except for the limiting case $r = 1$, the converse inequality

$$\left\| \left(\sum_{k=1}^{n} |d_k|^2 \right)^{1/2} \middle| L_r \right\| \le b_r \| f_n | L_r \|$$

follows from 7.2.10.

REMARK. Rather precise estimates of the constants $a_r > 0$ and $b_r > 0$ are to be found in [GARS, p. 168]. For example,

$$\| f_n | L_1 \| \le \sqrt{10} \left\| \left(\sum_{k=1}^{n} |d_k|^2 \right)^{1/2} \middle| L_1 \right\|.$$

7.3 Haar functions

7.3.1 The infinite **dyadic tree** $\mathbb{D}_1^\infty$ consists of all pairs (k, j) with $k=1,2,\ldots$ and $j=1,\ldots,2^{k-1}$. Adding on top the pair $(0,1)$, we get $\mathbb{D}_0^\infty$. The k-th **level** is defined by

$$\mathbb{D}_0 := \{(0,1)\} \quad \text{and} \quad \mathbb{D}_k := \{(k,j) : j=1,\ldots,2^{k-1}\} \quad \text{if } k=1,2,\ldots.$$

Let

$$\mathbb{D}_m^n := \bigcup_{k=m}^{n} \mathbb{D}_k \quad \text{for } 0 \le m \le n.$$

WARNING. When dealing with dyadic intervals $\Delta_k^{(j)}$, the index j runs from 1 to 2^k. However, for trees and Haar functions (see below) we have $j=1,\ldots,2^{k-1}$, and the case $k=0$ is exceptional.

7.3.2 A subset $\mathbb{B}$ of $\mathbb{D}_1^\infty$ is called a **branch** if it contains precisely one element of every level $\mathbb{D}_k$ and $(k,j)\in\mathbb{B}$ implies either $(k+1,2j-1)\in\mathbb{B}$ or $(k+1,2j)\in\mathbb{B}$ for $k=1,2,\ldots$. We refer to $\mathbb{B}\cap\mathbb{D}_m^n$ as a branch in $\mathbb{D}_m^n$. Every such branch can be written in the form

$$\mathbb{D}_m^n(t) := \{(k,j)\in\mathbb{D}_m^n : t\in\Delta_{k-1}^{(j)}\} \quad \text{with some } t\in[0,1).$$

In order to get all different branches in $\mathbb{D}_m^n$, it is enough to take the midpoints $t=\frac{2h-1}{2^n}$ of the intervals $\Delta_{n-1}^{(h)}$ with $h=1,\ldots,2^{n-1}$.

In the following example, the bold points form a branch in $\mathbb{D}_1^4$:

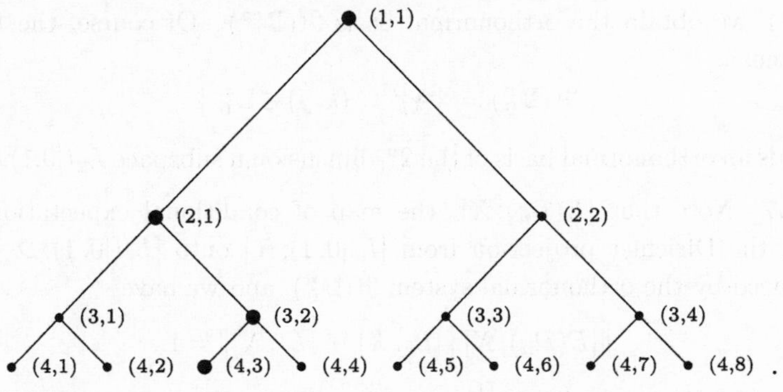

7.3.3 For $k=1,2,\ldots$ and $j=0,\pm1,\pm2,\ldots$, the **Haar functions** $\chi_k^{(j)}$ are defined by

$$\chi_k^{(j)}(t) := \begin{cases} +2^{(k-1)/2} & \text{if } t\in\Delta_k^{(2j-1)}, \\ -2^{(k-1)/2} & \text{if } t\in\Delta_k^{(2j)}, \\ 0 & \text{otherwise.} \end{cases}$$

Note that $\chi_k^{(j)}$ has support equal to $\Delta_{k-1}^{(j)}$. Moreover,

$$\chi_k^{(j)}(t-\tfrac{1}{2^{k-1}})=\chi_k^{(j+1)}(t) \quad\text{and}\quad \chi_{k+1}^{(j)}(t)=\sqrt{2}\,\chi_k^{(j)}(2t).$$

In what follows, we restrict the Haar functions with $k=1,2,\ldots$ and $j=1,2,\ldots,2^{k-1}$ to the interval $[0,1)$.

7.3.4 A **Haar polynomial**

$$f_n=\sum_{\mathbb{D}_0^n} x_k^{(j)}\chi_k^{(j)}$$

with coefficients $x_k^{(j)}\in X$ is an X-valued function taking constant values on all dyadic intervals $\Delta_n^{(1)},\ldots,\Delta_n^{(2^n)}$, the atoms of $\mathcal{D}_n$. Often, we write

$$f_n=\sum_{k=0}^{n} d_k \quad\text{with } d_0:=x_0^{(1)} \quad\text{and}\quad d_k:=\sum_{j=1}^{2^{k-1}} x_k^{(j)}\chi_k^{(j)} \quad\text{for } k=1,\ldots,n.$$

7.3.5 The **Rademacher functions** can be represented in the form

$$r_k=2^{-(k-1)/2}\sum_{j=1}^{2^{k-1}}\chi_k^{(j)}.$$

7.3.6 The **Haar system**

$$\mathcal{H}(\mathbb{D}_1^\infty):=\{\,\chi_k^{(j)} : (k,j)\in\mathbb{D}_1^\infty\,\}$$

is orthonormal in $L_2[0,1)$. Adding $\chi_0^{(1)}$, the characteristic function of $[0,1)$, we obtain the orthonormal basis $\mathcal{H}(\mathbb{D}_0^\infty)$. Of course, the Haar system

$$\mathcal{H}(\mathbb{D}_0^n):=\{\,\chi_k^{(j)} : (k,j)\in\mathbb{D}_0^n\,\}$$

yields an orthonormal basis of the 2^n-dimensional subspace $L_2([0,1),\mathcal{D}_n)$.

7.3.7 Note that $[E(\mathcal{D}_n),X]$, the map of conditional expectation, is just the Dirichlet projection from $[L_r[0,1),X]$ onto $[L_r([0,1),\mathcal{D}_n),X]$ induced by the orthonormal system $\mathcal{H}(\mathbb{D}_0^n)$, and we have

$$\|[E(\mathcal{D}_n),X]:[L_r,X]\to[L_r,X]\|=1.$$

Since the Haar functions $\chi_n^{(1)},\ldots,\chi_n^{(2^{n-1})}$ have disjoint supports,

$$([E(\mathcal{D}_n),X]-[E(\mathcal{D}_{n-1}),X])f=\sum_{j=1}^{2^{n-1}}\left\langle f,\chi_n^{(j)}\right\rangle\chi_n^{(j)}$$

implies that

$$\big\|[E(\mathcal{D}_n),X]-[E(\mathcal{D}_{n-1}),X] : [L_r,X]\to[L_r,X]\big\|=1.$$

7.3.8 Letting

$$f_k := [E(\mathcal{D}_k), X]f_n \quad \text{for } k = 0, \dots, n,$$

yields a one-to-one correspondence between X-valued Haar polynomials and X-valued dyadic martingales. Thus, at first glance, it seems that martingale theory replaces ordinary functions by a more complicated concept. However, we now have at our disposal the powerful inequalities proved in the previous section. The main trick was the use of stopping times which allow us to select the important terms of a Haar polynomial.

7.4 Haar type and cotype ideal norms

7.4.1 Together with the Haar system

$$\mathcal{H}(\mathbb{D}_0^\infty) = \{\, \chi_k^{(j)} : (k,j) \in \mathbb{D}_0^\infty \,\}$$

in $L_2[0,1)$, we consider the canonical orthonormal system

$$\mathcal{I}(\mathbb{D}_0^\infty) = \{\, u_k^{(j)} : (k,j) \in \mathbb{D}_0^\infty \,\}$$

consisting of the unit $\mathbb{D}_0^\infty$-tuples $u_k^{(j)}$ in $l_2(\mathbb{D}_0^\infty)$ that have the coefficient 1 for the index (k,j) and the coefficients 0 otherwise. As usual, given any finite subset $\mathbb{F} \subset \mathbb{D}_0^\infty$, we let

$$\mathcal{H}(\mathbb{F}) = \{\, \chi_k^{(j)} : (k,j) \in \mathbb{F} \,\} \quad \text{and} \quad \mathcal{I}(\mathbb{F}) = \{\, u_k^{(j)} : (k,j) \in \mathbb{F} \,\}.$$

7.4.2 We now treat the **Haar type ideal norm** $\varrho(\mathcal{H}(\mathbb{F}), \mathcal{I}(\mathbb{F}))$ and the **Haar cotype ideal norm** $\varrho(\mathcal{I}(\mathbb{F}), \mathcal{H}(\mathbb{F}))$. The most interesting cases occur when $\mathbb{F}$ is either $\mathbb{D}_1^n$ or $\mathbb{D}_0^n$.

Recall from 3.4.12 that

$$\delta(\mathcal{H}(\mathbb{F}), \mathcal{I}(\mathbb{F})) = \varrho(\mathcal{H}(\mathbb{F}), \mathcal{I}(\mathbb{F}))$$

for any finite subset $\mathbb{F} \subset \mathbb{D}_0^\infty$. Moreover, we have observed in 7.3.7 that

$$\delta(X|\mathcal{H}(\mathbb{D}_0^n), \mathcal{H}(\mathbb{D}_0^n)) = \|[E(\mathcal{D}_n), X] : [L_2, X] \to [L_2, X]\| = 1.$$

Hence

$$\delta(\mathcal{I}(\mathbb{D}_0^n), \mathcal{H}(\mathbb{D}_0^n)) = \varrho(\mathcal{I}(\mathbb{D}_0^n), \mathcal{H}(\mathbb{D}_0^n)).$$

In view of these facts, it is superfluous to deal with Dirichlet ideal norms.

7.4.3 It depends on the underlying situation which of the index sets $\mathbb{D}_0^n$ and $\mathbb{D}_1^n$ is more convenient. Nevertheless, in view of the following elementary observation, which is a special case of 3.3.6, we may easily pass from one to the other.

PROPOSITION.

$$\varrho(\mathcal{H}(\mathbb{D}_1^n),\mathcal{I}(\mathbb{D}_1^n)) \le \varrho(\mathcal{H}(\mathbb{D}_0^n),\mathcal{I}(\mathbb{D}_0^n)) \le 3\,\varrho(\mathcal{H}(\mathbb{D}_1^n),\mathcal{I}(\mathbb{D}_1^n)),$$
$$\varrho(\mathcal{I}(\mathbb{D}_1^n),\mathcal{H}(\mathbb{D}_1^n)) \le \varrho(\mathcal{I}(\mathbb{D}_0^n),\mathcal{H}(\mathbb{D}_0^n)) \le 3\,\varrho(\mathcal{I}(\mathbb{D}_1^n),\mathcal{H}(\mathbb{D}_1^n)).$$

REMARK. At least in the first inequality, the factor 3 can be improved.

7.4.4 For the index set $\mathbb{D}_0^n$, the concepts of Haar type and Haar cotype ideal norms are dual to each other.

PROPOSITION.

$$\varrho'(\mathcal{H}(\mathbb{D}_0^n),\mathcal{I}(\mathbb{D}_0^n)) = \varrho(\mathcal{I}(\mathbb{D}_0^n),\mathcal{H}(\mathbb{D}_0^n)),$$
$$\varrho'(\mathcal{I}(\mathbb{D}_0^n),\mathcal{H}(\mathbb{D}_0^n)) = \varrho(\mathcal{H}(\mathbb{D}_0^n),\mathcal{I}(\mathbb{D}_0^n)).$$

PROOF. By 3.4.6, we have

$$\begin{aligned}\varrho'(\mathcal{H}(\mathbb{D}_0^n),\mathcal{I}(\mathbb{D}_0^n)) &= \delta'(\mathcal{H}(\mathbb{D}_0^n),\mathcal{I}(\mathbb{D}_0^n))\\ &= \delta(\mathcal{I}(\mathbb{D}_0^n),\mathcal{H}(\mathbb{D}_0^n)) = \varrho(\mathcal{I}(\mathbb{D}_0^n),\mathcal{H}(\mathbb{D}_0^n)).\end{aligned}$$

The second equation can be obtained in the same way.

7.4.5 PROPOSITION.

$$\varrho(\mathcal{R}_n,\mathcal{I}_n) \le \varrho(\mathcal{H}(\mathbb{D}_1^n),\mathcal{I}(\mathbb{D}_1^n)) \quad \textit{and} \quad \varrho(\mathcal{I}_n,\mathcal{R}_n) \le \varrho(\mathcal{I}(\mathbb{D}_1^n),\mathcal{H}(\mathbb{D}_1^n)).$$

PROOF. Given $x_1,\ldots,x_n \in X$, we let

$$x_k^{(j)} := 2^{-(k-1)/2} x_k \quad \text{for } k = 1,\ldots,n.$$

Then it follows from

$$r_k = 2^{-(k-1)/2} \sum_{j=1}^{2^{k-1}} \chi_k^{(j)}$$

that

$$\begin{aligned}\|(x_k^{(j)})|\mathcal{H}(\mathbb{D}_1^n)\| &= \left(\int_0^1 \Big\| \sum_{k=1}^{n}\sum_{j=1}^{2^{k-1}} x_k^{(j)}\chi_k^{(j)}(t)\Big\|^2 dt\right)^{1/2}\\ &= \left(\int_0^1 \Big\| \sum_{k=1}^{n}\sum_{j=1}^{2^{k-1}} 2^{-(k-1)/2} x_k\chi_k^{(j)}(t)\Big\|^2 dt\right)^{1/2}\\ &= \left(\int_0^1 \Big\| \sum_{k=1}^{n} x_k r_k(t)\Big\|^2 dt\right)^{1/2} = \|(x_k)|\mathcal{R}_n\|.\end{aligned}$$

Moreover, we have

$$\|(x_k^{(j)})|\mathcal{I}(\mathbb{D}_1^n)\| = \Big(\sum_{k=1}^{n}\sum_{j=1}^{2^{k-1}}\|x_k^{(j)}\|^2\Big)^{1/2} = \Big(\sum_{k=1}^{n}\sum_{j=1}^{2^{k-1}}\|2^{-(k-1)/2}x_k\|^2\Big)^{1/2}$$

$$= \Big(\sum_{k=1}^{n}\|x_k\|^2\Big)^{1/2} = \|(x_k)|\mathcal{I}_n\|.$$

Now the desired inequalities are easily obtained.

7.4.6 PROPOSITION.

$$1 \le \varrho(\mathcal{H}(\mathbb{D}_1^n), \mathcal{I}(\mathbb{D}_1^n)) \le \sqrt{n} \quad \textit{and} \quad 1 \le \varrho(\mathcal{I}(\mathbb{D}_1^n), \mathcal{H}(\mathbb{D}_1^n)) \le \sqrt{n}.$$

PROOF. Obviously,

$$\|(x_k^{(j)})|\mathcal{H}(\mathbb{D}_1^n)\| = \Big(\int_0^1\Big\|\sum_{k=1}^{n}\sum_{j=1}^{2^{k-1}}x_k^{(j)}\chi_k^{(j)}(t)\Big\|^2 dt\Big)^{1/2}$$

$$\le \sum_{k=1}^{n}\Big(\int_0^1\Big\|\sum_{j=1}^{2^{k-1}}x_k^{(j)}\chi_k^{(j)}(t)\Big\|^2 dt\Big)^{1/2} = \sum_{k=1}^{n}\Big(\sum_{j=1}^{2^{k-1}}\|x_k^{(j)}\|^2\Big)^{1/2}$$

$$\le \sqrt{n}\Big(\sum_{k=1}^{n}\sum_{j=1}^{2^{k-1}}\|x_k^{(j)}\|^2\Big)^{1/2} = \sqrt{n}\,\|(x_k^{(j)})|\mathcal{I}(\mathbb{D}_1^n)\|.$$

Hence

$$\varrho(X|\mathcal{H}(\mathbb{D}_1^n), \mathcal{I}(\mathbb{D}_1^n)) \le \sqrt{n}.$$

As observed in 7.3.7, we have

$$\|[E(\mathcal{D}_h), X] - [E(\mathcal{D}_{h-1}), X] : [L_2, X] \to [L_2, X]\| = 1.$$

This means that

$$\Big(\int_0^1\Big\|\sum_{j=1}^{2^{h-1}}x_h^{(j)}\chi_h^{(j)}(t)\Big\|^2 dt\Big)^{1/2} \le \Big(\int_0^1\Big\|\sum_{k=1}^{n}\sum_{j=1}^{2^{k-1}}x_k^{(j)}\chi_k^{(j)}(t)\Big\|^2 dt\Big)^{1/2}$$

for $h = 1, \ldots, n$. Consequently,

$$\|(x_h^{(j)})|\mathcal{I}(\mathbb{D}_1^n)\| = \Big(\sum_{h=1}^{n}\sum_{j=1}^{2^{h-1}}\|x_h^{(j)}\|^2\Big)^{1/2} = \Big(\sum_{h=1}^{n}\int_0^1\Big\|\sum_{j=1}^{2^{h-1}}x_h^{(j)}\chi_h^{(j)}(t)\Big\|^2 dt\Big)^{1/2}$$

$$\le \Big(n\int_0^1\Big\|\sum_{k=1}^{n}\sum_{j=1}^{2^{k-1}}x_k^{(j)}\chi_k^{(j)}(t)\Big\|^2 dt\Big)^{1/2} = \sqrt{n}\,\|(x_k^{(j)})|\mathcal{H}(\mathbb{D}_1^n)\|.$$

Therefore

$$\varrho(X|\mathcal{I}(\mathbb{D}_1^n), \mathcal{H}(\mathbb{D}_1^n)) \le \sqrt{n}.$$

7.4.7 Next, we show that the preceding estimates cannot be improved.

EXAMPLE.

$$\varrho(l_1^n|\mathcal{H}(\mathbb{D}_1^n),\mathcal{I}(\mathbb{D}_1^n)) = \varrho(L_1|\mathcal{H}(\mathbb{D}_1^n),\mathcal{I}(\mathbb{D}_1^n)) = \sqrt{n}.$$

PROOF. It follows from 4.2.7, 7.4.5 and 7.4.6 that

$$\sqrt{n} = \varrho(l_1^n|\mathcal{R}_n,\mathcal{I}_n) \le \varrho(l_1^n|\mathcal{H}(\mathbb{D}_1^n),\mathcal{I}(\mathbb{D}_1^n)) \le \varrho(L_1|\mathcal{H}(\mathbb{D}_1^n),\mathcal{I}(\mathbb{D}_1^n)) \le \sqrt{n}.$$

7.4.8 The above result admits the following generalization.

EXAMPLE.

$$\varrho(l_p^n|\mathcal{H}(\mathbb{D}_1^n),\mathcal{I}(\mathbb{D}_1^n)) = \varrho(L_p|\mathcal{H}(\mathbb{D}_1^n),\mathcal{I}(\mathbb{D}_1^n)) = n^{1/p-1/2} \quad \textit{if } 1 \le p \le 2.$$

PROOF. Since

$$\varrho(\mathcal{H}(\mathbb{D}_1^n),\mathcal{I}(\mathbb{D}_1^n)) = \delta(\mathcal{H}(\mathbb{D}_1^n),\mathcal{I}(\mathbb{D}_1^n)),$$

we can interpolate; see 3.4.9. Hence

$$\begin{aligned}\varrho(L_p|\mathcal{H}(\mathbb{D}_1^n),\mathcal{I}(\mathbb{D}_1^n)) &\le \varrho(L_1|\mathcal{H}(\mathbb{D}_1^n),\mathcal{I}(\mathbb{D}_1^n))^{1-\theta}\varrho(L_2|\mathcal{H}(\mathbb{D}_1^n),\mathcal{I}(\mathbb{D}_1^n))^{\theta} \\ &\le n^{(1-\theta)/2} = n^{1/p-1/2},\end{aligned}$$

where $1/p = (1-\theta)/1 + \theta/2$. This yields the upper estimate. The lower estimate follows as in the previous proof.

7.4.9 In order to get the next example, we use a deep result to be proved later.

EXAMPLE. *For* $2 \le q < \infty$

$$1 \le \varrho(l_q^n|\mathcal{H}(\mathbb{D}_1^n),\mathcal{I}(\mathbb{D}_1^n)) \le \varrho(L_q|\mathcal{H}(\mathbb{D}_1^n),\mathcal{I}(\mathbb{D}_1^n)) \le \sqrt{q-1}.$$

PROOF. By 7.4.3, 7.9.21 and 7.9.17, we have

$$\begin{aligned}\varrho(L_q|\mathcal{H}(\mathbb{D}_1^n),\mathcal{I}(\mathbb{D}_1^n)) &\le \varrho(L_q|\mathcal{H}(\mathbb{D}_0^n),\mathcal{I}(\mathbb{D}_0^n)) \le \|L_q|\mathfrak{L}\circ\mathfrak{UT}_2\| \\ &\le \|L_q|\mathfrak{UT}_2\| = \sqrt{q-1}.\end{aligned}$$

7.4.10 EXAMPLE. $\varrho(L_\infty|\mathcal{H}(\mathbb{D}_1^n),\mathcal{I}(\mathbb{D}_1^n)) = \sqrt{n}$.

PROOF. Note that L_∞ contains isometric copies of l_1^n, and use 7.4.7.

7.4.11 If n and N are natural numbers, then we have the following results, which are analogous to 4.2.10 and 4.2.11.

EXAMPLE.

$$\begin{aligned}\varrho(l_1^N|\mathcal{H}(\mathbb{D}_1^n),\mathcal{I}(\mathbb{D}_1^n)) &= \min\{\sqrt{n},\sqrt{N}\}, \\ \varrho(l_\infty^N|\mathcal{H}(\mathbb{D}_1^n),\mathcal{I}(\mathbb{D}_1^n)) &\asymp \min\{\sqrt{n},\sqrt{1+\log N}\}.\end{aligned}$$

7.4.12 The remaining part of this section is devoted to the proof of Proposition 7.4.20, which requires extensive preparation.

LEMMA.

$$\varrho(\mathcal{H}(\mathbb{D}_{m+1}^{n+1}), \mathcal{I}(\mathbb{D}_{m+1}^{n+1})) \le \varrho(\mathcal{H}(\mathbb{D}_m^n), \mathcal{I}(\mathbb{D}_m^n)) \quad \textit{for } n \ge m \ge 1.$$

PROOF. Let

$$\boldsymbol{f} = \sum_{\mathbb{D}_{m+1}^{n+1}} x_k^{(j)} \chi_k^{(j)}$$

and write

$$\|[L_2, T]\boldsymbol{f}|L_2\|^2 = I_0 + I_1,$$

where

$$I_0 := \int_0^{1/2} \|T\boldsymbol{f}(t)\|^2 \, dt \quad \text{and} \quad I_1 := \int_{1/2}^1 \|T\boldsymbol{f}(t)\|^2 \, dt.$$

Note that $\chi_k^{(j)}$ vanishes on $[0, 1/2)$ for $j = 2^{k-2}+1, \dots, 2^{k-1}$. Moreover,

$$\chi_k^{(j)}(s/2) = \sqrt{2}\, \chi_{k-1}^{(j)}(s).$$

Substituting $s := 2t$ and $h := k-1$, the integral I_0 can be estimated as follows:

$$I_0 = \int_0^{1/2} \Big\| \sum_{k=m+1}^{n+1} \sum_{j=1}^{2^{k-2}} T x_k^{(j)} \chi_k^{(j)}(t) \Big\|^2 dt = \frac{1}{2} \int_0^1 \Big\| \sum_{k=m+1}^{n+1} \sum_{j=1}^{2^{k-2}} T x_k^{(j)} \chi_k^{(j)}(s/2) \Big\|^2 ds$$

$$= \int_0^1 \Big\| \sum_{h=m}^{n} \sum_{j=1}^{2^{h-1}} T x_{h+1}^{(j)} \chi_h^{(j)}(s) \Big\|^2 ds$$

$$\le \varrho(T|\mathcal{H}(\mathbb{D}_m^n), \mathcal{I}(\mathbb{D}_m^n))^2 \sum_{h=m}^{n} \sum_{j=1}^{2^{h-1}} \|x_{h+1}^{(j)}\|^2$$

$$= \varrho(T|\mathcal{H}(\mathbb{D}_m^n), \mathcal{I}(\mathbb{D}_m^n))^2 \sum_{k=m+1}^{n+1} \sum_{j=1}^{2^{k-2}} \|x_k^{(j)}\|^2.$$

Hence

$$I_0 \le \varrho(T|\mathcal{H}(\mathbb{D}_m^n), \mathcal{I}(\mathbb{D}_m^n))^2 \sum_{k=m+1}^{n+1} \sum_{j=1}^{2^{k-2}} \|x_k^{(j)}\|^2.$$

Replacing t by $1-t$ and $x_k^{(j)}$ by $x_k^{(2^{k-1}-j+1)}$, the estimate just obtained passes into

$$I_1 \le \varrho(T|\mathcal{H}(\mathbb{D}_m^n), \mathcal{I}(\mathbb{D}_m^n))^2 \sum_{k=m+1}^{n+1} \sum_{j=2^{k-2}+1}^{2^{k-1}} \|x_k^{(j)}\|^2.$$

Finally, we arrive at

$$\|Tf|L_2\|^2 = I_0 + I_1 \le \varrho(T|\mathcal{H}(\mathbb{D}_m^n), \mathcal{I}(\mathbb{D}_m^n))^2 \sum_{k=m+1}^{n+1} \sum_{j=1}^{2^{k-1}} \|x_k^{(j)}\|^2,$$

which implies the desired inequality.

7.4.13 In what follows, $\mathbb{F}$ denotes a finite subset of $\mathbb{D}_1^n$. The **height** of $\mathbb{F}$ is defined as the maximal cardinality of those subsets that are contained in some finite branch $\mathbb{B}$. That is, we let

$$\text{height}(\mathbb{F}) := \max_{\mathbb{B}} |\mathbb{F} \cap \mathbb{B}|.$$

7.4.14 For $(h,i) \in \mathbb{D}_1^\infty$, we denote by $\varphi_h^{(i)}$ the **fork transform** of $[0,1)$ that interchanges the intervals

$$\Delta_{h+1}^{(4i-2)} \quad \text{and} \quad \Delta_{h+1}^{(4i-1)}.$$

More formally,

$$\varphi_h^{(i)}(t) := \begin{cases} t + \frac{1}{2^{h+1}} & \text{for } t \in \Delta_{h+1}^{(4i-2)}, \\ t - \frac{1}{2^{h+1}} & \text{for } t \in \Delta_{h+1}^{(4i-1)}, \\ t & \text{otherwise.} \end{cases}$$

We now determine $\chi_k^{(j)} \circ \varphi_h^{(i)}$ for $(k,j) \in \mathbb{D}_1^n$. The most interesting behaviour occurs on the fork $\mathbb{F}_h^{(i)}$:

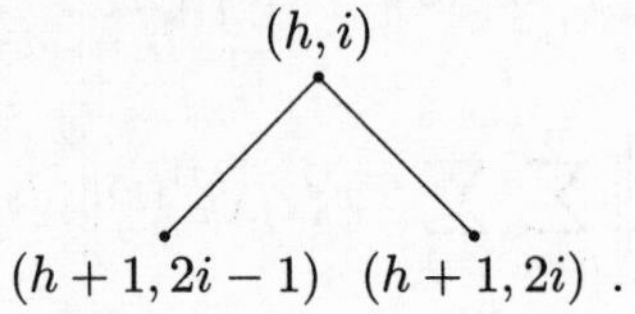

Looking at

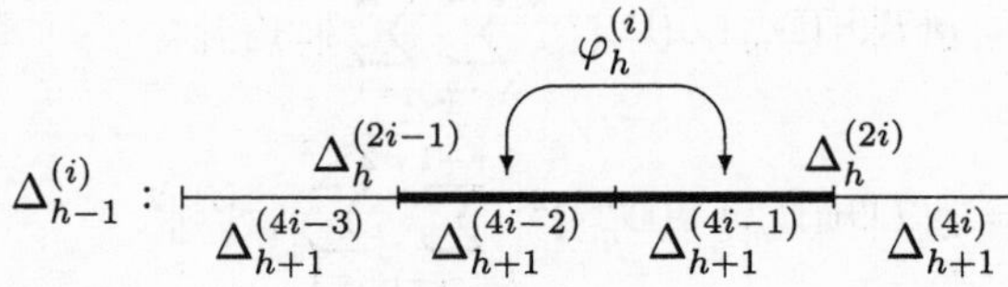

we easily see that

$$\chi_h^{(i)} \circ \varphi_h^{(i)} = \frac{1}{\sqrt{2}} \left(\chi_{h+1}^{(2i-1)} + \chi_{h+1}^{(2i)} \right). \tag{1}$$

Moreover, it follows from

$$\left(\chi_{h+1}^{(2i-1)} + \chi_{h+1}^{(2i)} \right) \circ \varphi_h^{(i)} = \sqrt{2}\, \chi_h^{(i)}$$

and

$$\left(\chi_{h+1}^{(2i-1)} - \chi_{h+1}^{(2i)} \right) \circ \varphi_h^{(i)} = \chi_{h+1}^{(2i-1)} - \chi_{h+1}^{(2i)}$$

that

$$\chi_{h+1}^{(2i-1)} \circ \varphi_h^{(i)} = \frac{1}{\sqrt{2}}\chi_h^{(i)} + \frac{1}{2}\left(\chi_{h+1}^{(2i-1)} - \chi_{h+1}^{(2i)}\right) \tag{2}$$

and

$$\chi_{h+1}^{(2i)} \circ \varphi_h^{(i)} = \frac{1}{\sqrt{2}}\chi_h^{(i)} - \frac{1}{2}\left(\chi_{h+1}^{(2i-1)} - \chi_{h+1}^{(2i)}\right). \tag{3}$$

Next, we consider those Haar functions $\chi_k^{(j)}$ that have support equal to $\Delta_{h+1}^{(4i-2)}$ or $\Delta_{h+1}^{(4i-1)}$. This means that their indices belong to the subtrees

$$\mathbb{S}_{h+2}^{(4i-2)} := \left\{ (k,j) \in \mathbb{D}_1^\infty : \Delta_{k-1}^{(j)} \subseteq \Delta_{h+1}^{(4i-2)} \right\}$$

or

$$\mathbb{S}_{h+2}^{(4i-1)} := \left\{ (k,j) \in \mathbb{D}_1^\infty : \Delta_{k-1}^{(j)} \subseteq \Delta_{h+1}^{(4i-1)} \right\}.$$

In this case, we have

$$\chi_k^{(j)} \circ \varphi_h^{(i)} = \begin{cases} \chi_k^{(j+2^{k-h-2})} & \text{if } (k,j) \in \mathbb{S}_{h+2}^{(4i-2)}, \\ \chi_k^{(j-2^{k-h-2})} & \text{if } (k,j) \in \mathbb{S}_{h+2}^{(4i-1)}. \end{cases}$$

The other Haar functions remain invariant.

7.4.15 Let

$$\Phi_h^{(i)} : (k,j) \to (k,j^*)$$

denote the permutation of $\mathbb{D}_1^\infty \setminus \mathbb{F}_h^{(i)}$ determined by the property that

$$\chi_k^{(j)} \circ \varphi_h^{(i)} = \chi_k^{(j^*)}.$$

Obviously, $\Phi_h^{(i)}$ interchanges the subtrees $\mathbb{S}_{h+2}^{(4i-2)}$ and $\mathbb{S}_{h+2}^{(4i-1)}$, is undefined on $\mathbb{F}_h^{(i)}$, and leaves the rest invariant:

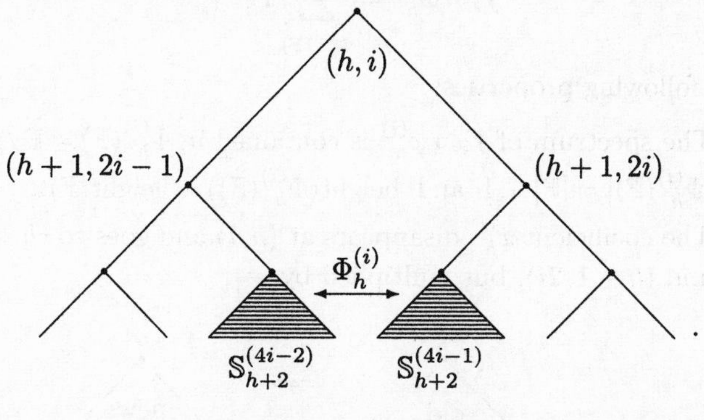

7.4.16 An index $(h,i) \in \mathbb{F}$ is called $\mathbb{F}$-admissible if both its successors $(h+1, 2i-1)$ and $(h+1, 2i)$ are outside $\mathbb{F}$. In this case, $\Phi_h^{(i)}(\mathbb{F})$ is defined as follows:

We cancel (h,i), but add $(h+1, 2i-1)$ and $(h+1, 2i)$, while the remaining part $\mathbb{F}_0 := \mathbb{F} \setminus \{(h,i)\}$ is transformed by $\Phi_h^{(i)}$.

7.4.17 LEMMA. *Let $(h,i) \in \mathbb{F}$ be $\mathbb{F}$-admissible. If*

$$\boldsymbol{f}_{\mathbb{F}} = \sum_{\mathbb{F}} x_k^{(j)} \chi_k^{(j)} \quad \textit{and} \quad \boldsymbol{f}_{\mathbb{F}} \circ \varphi_h^{(i)} = \sum_{\Phi_h^{(i)}(\mathbb{F})} \mathring{x}_k^{(j)} \chi_k^{(j)},$$

then

$$\sum_{\mathbb{F}} \|x_k^{(j)}\|^2 = \sum_{\Phi_h^{(i)}(\mathbb{F})} \|\mathring{x}_k^{(j)}\|^2 \quad \textit{and} \quad \|\boldsymbol{f}_{\mathbb{F}} \circ \varphi_h^{(i)} | L_2\| = \|\boldsymbol{f}_{\mathbb{F}} | L_2\|.$$

PROOF. Apart from $x_h^{(i)}$, $\mathring{x}_{h+1}^{(2i-1)}$ and $\mathring{x}_{h+1}^{(2i)}$, the $\mathbb{F}$-tuple $(x_k^{(j)})$ and the $\Phi_h^{(i)}(\mathbb{F})$-tuple $(\mathring{x}_k^{(j)})$ are rearrangements of each other. Instead of $x_h^{(i)}$ there occurs $\mathring{x}_{h+1}^{(2i-1)} = \frac{1}{\sqrt{2}} x_h^{(i)}$ and $\mathring{x}_{h+1}^{(2i)} = \frac{1}{\sqrt{2}} x_h^{(i)}$. This proves the first equation. The second one follows from the fact that $\varphi_h^{(i)}$ is measure preserving.

7.4.18 In this paragraph, a summary of facts concerning the **fork transform** $\varphi_h^{(i)}$ will be given.

Consider a Haar polynomial

$$\boldsymbol{f}_{\mathbb{F}} = \sum_{\mathbb{F}} x_k^{(j)} \chi_k^{(j)}$$

with spectrum contained in $\mathbb{F}$. Let $(h,i) \in \mathbb{F}$ be $\mathbb{F}$-admissible, which means that its successors $(h+1, 2i-1)$ and $(h+1, 2i)$ are outside $\mathbb{F}$. Then the transformed Haar polynomial

$$\boldsymbol{f}_{\mathbb{F}} \circ \varphi_h^{(i)} = \sum_{\Phi_k^{(i)}(\mathbb{F})} \mathring{x}_k^{(j)} \chi_k^{(j)}$$

has the following properties:

(1) The spectrum of $\boldsymbol{f}_{\mathbb{F}} \circ \varphi_h^{(i)}$ is contained in $\Phi_h^{(i)}(\mathbb{F}) = \mathbb{F} \,\triangle\, \mathbb{F}_h^{(i)}$.

(2) $|\Phi_h^{(i)}(\mathbb{F})| = |\mathbb{F}| + 1$ and $\text{height}(\Phi_h^{(i)}(\mathbb{F})) = \text{height}(\mathbb{F})$.

(3) The coefficient $x_h^{(i)}$ disappears at (h,i) and goes to $(h+1, 2i-1)$ and $(h+1, 2i)$, but multiplied by $\frac{1}{\sqrt{2}}$,

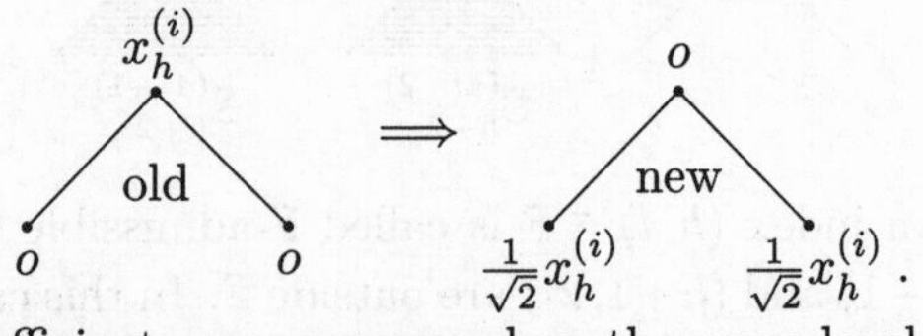

(4) The other coefficients are rearranged on the same level.

(5) $\|(\mathring{x}_k^{(j)}) | l_2(\Phi_h^{(i)}(\mathbb{F}))\| = \|(x_k^{(j)}) | l_2(\mathbb{F})\|$.

(6) $\|\boldsymbol{f}_{\mathbb{F}} \circ \varphi_h^{(i)} | L_2\| = \|\boldsymbol{f}_{\mathbb{F}} | L_2\|$.

7.4.19 Next, we show how to squeeze an index set $\mathbb{F}$ with the help of fork transforms. The idea is to replace an index $(h,i) \in \mathbb{F}$ by its successors $(h+1, 2i-1)$ and $(h+1, 2i)$ provided that the latter do not already belong to $\mathbb{F}$.

LEMMA. *Let* $n = \text{height}(\mathbb{F})$. *Then there exists* $m \in \mathbb{N}$ *and a finite sequence of admissible fork transforms*

$$\mathbb{F} =: \mathbb{F}_0 \longrightarrow \mathbb{F}_1 \longrightarrow \mathbb{F}_2 \longrightarrow \ldots \longrightarrow \mathbb{F}_N$$

such that

$$\mathbb{F}_N \subseteq \mathbb{D}_{m+1}^{m+n}.$$

PROOF. Choose any $m \in \mathbb{N}$ with $\mathbb{F} \subseteq \mathbb{D}_1^{m+n}$. As often as possible, we successively apply admissible fork transforms

$$\mathbb{F}_{l-1} \xrightarrow{\Phi_{h_l}^{(i_l)}} \mathbb{F}_l.$$

In order to guarantee that $\mathbb{F}_l$ is still contained in $\mathbb{D}_1^{m+n}$, we always assume that $h_l < m+n$. Since

$$|\mathbb{F}_l| = |\mathbb{F}_{l-1}| + 1 \quad \text{and} \quad |\mathbb{F}_l| \le |\mathbb{D}_1^{m+n}|,$$

this process terminates after a finite number of steps. Hence, for the last index set $\mathbb{F}_N$, there does not exist any admissible fork $\mathbb{F}_h^{(i)}$ with $h < m+n$.

Choose k such that

$$\mathbb{F}_N \subseteq \mathbb{D}_k^{m+n} \quad \text{and} \quad \mathbb{F}_N \not\subseteq \mathbb{D}_{k+1}^{m+n}.$$

If $k = m+n$, then $n = 1$ and $\mathbb{F}_N \subseteq \mathbb{D}_{m+1}^{m+n}$. Thus we are done.

Otherwise, take any index $(k, j_0) \in \mathbb{F}_N$. Since $k < m+n$, at least one of its successors $(k+1, j_1)$ must belong to $\mathbb{F}_N$. In this way, we find

$$(k, j_0), (k+1, j_1), \ldots, (m+n, j_{m+n-k}) \in \mathbb{F}_N$$

which form a branch in $\mathbb{D}_{k+1}^{m+n}$. Hence

$$m+n-k+1 \le n = \text{height}(\mathbb{F}).$$

Finally, we conclude from $m+1 \le k$ that

$$\mathbb{F}_N \subseteq \mathbb{D}_k^{m+n} \subseteq \mathbb{D}_{m+1}^{m+n}.$$

7.4.20 We are now in a position to establish a very important result which will be crucial for the proof of 7.5.12.

PROPOSITION.

$$\varrho(\mathcal{H}(\mathbb{F}), \mathcal{I}(\mathbb{F})) \le \varrho(\mathcal{H}(\mathbb{D}_1^n), \mathcal{I}(\mathbb{D}_1^n)) \quad \textit{whenever } \text{height}(\mathbb{F}) = n.$$

PROOF. Pick admissible fork transforms $\Phi_{h_1}^{(i_1)}, \ldots, \Phi_{h_N}^{(i_N)}$ as constructed in the above lemma. Let $\varphi_{h_1}^{(i_1)}, \ldots, \varphi_{h_N}^{(i_N)}$ denote the associated bijections defined in 7.4.14. Given

$$\boldsymbol{f}_{\mathbb{F}} = \sum_{\mathbb{F}} x_k^{(j)} \chi_k^{(j)},$$

we let

$$\boldsymbol{f}_{\mathbb{F}_l} := \boldsymbol{f}_{\mathbb{F}_{l-1}} \circ \varphi_{h_l}^{(i_l)} \quad \text{and} \quad \boldsymbol{f}_{\mathbb{F}_0} := \boldsymbol{f}_{\mathbb{F}}.$$

Write

$$\boldsymbol{f}_{\mathbb{F}_l} = \sum_{\mathbb{F}_l} x_k^{(j,l)} \chi_k^{(j)}.$$

By Lemma 7.4.17,

$$\|(x_k^{(j)})|\mathfrak{I}(\mathbb{F})\| = \|(x_k^{(j,0)})|\mathfrak{I}(\mathbb{F}_0)\| = \ldots = \|(x_k^{(j,N)})|\mathfrak{I}(\mathbb{F}_N)\|$$

and

$$\|(Tx_k^{(j)})|\mathcal{H}(\mathbb{F})\| = \|(Tx_k^{(j,0)})|\mathcal{H}(\mathbb{F}_0)\| = \ldots = \|(Tx_k^{(j,N)})|\mathcal{H}(\mathbb{F}_N)\|.$$

Since $\mathbb{F}_N \subseteq \mathbb{D}_{m+1}^{m+n}$, we have

$$\|(Tx_k^{(j,N)})|\mathcal{H}(\mathbb{F}_N)\| \le \boldsymbol{\varrho}(T|\mathcal{H}(\mathbb{D}_{m+1}^{m+n}), \mathfrak{I}(\mathbb{D}_{m+1}^{m+n}))\, \|(x_k^{(j,N)})|\mathfrak{I}(\mathbb{F}_N)\|.$$

Hence

$$\|(Tx_k^{(j)})|\mathcal{H}(\mathbb{F})\| \le \boldsymbol{\varrho}(T|\mathcal{H}(\mathbb{D}_{m+1}^{m+n}), \mathfrak{I}(\mathbb{D}_{m+1}^{m+n}))\, \|(x_k^{(j)})|\mathfrak{I}(\mathbb{F})\|.$$

This proves that

$$\boldsymbol{\varrho}(T|\mathcal{H}(\mathbb{F}), \mathfrak{I}(\mathbb{F})) \le \boldsymbol{\varrho}(T|\mathcal{H}(\mathbb{D}_{m+1}^{m+n}), \mathfrak{I}(\mathbb{D}_{m+1}^{m+n})).$$

The final conclusion follows from

$$\boldsymbol{\varrho}(T|\mathcal{H}(\mathbb{D}_{m+1}^{m+n}), \mathfrak{I}(\mathbb{D}_{m+1}^{m+n})) \le \boldsymbol{\varrho}(T|\mathcal{H}(\mathbb{D}_1^n), \mathfrak{I}(\mathbb{D}_1^n)) \quad \text{(by 7.4.12)}.$$

7.5 Operators of Haar type

7.5.1 Assume that $1 < p \le 2$ and $1 < v < \infty$. For $T \in \mathfrak{L}(X, Y)$, we denote by $\boldsymbol{\varrho}_p^{(v)}(T|\mathcal{H}(\mathbb{D}_1^n), \mathfrak{I}(\mathbb{D}_1^n))$ the least constant $c \ge 0$ such that

$$\left(\int_0^1 \Big\| \sum_{\mathbb{D}_1^n} T x_k^{(j)} \chi_k^{(j)}(t) \Big\|^v dt \right)^{1/v} \le c \left(\int_0^1 \left[\sum_{k=1}^n \Big\| \sum_{j=1}^{2^{k-1}} x_k^{(j)} \chi_k^{(j)}(t) \Big\|^p \right]^{v/p} dt \right)^{1/v}$$

for all $\mathbb{D}_1^n$-tuples $(x_k^{(j)})$ in X. In other words, it is required that

$$\|[L_v, T]\boldsymbol{f}_n|L_v\| \le c\, \|(\boldsymbol{d}_k)|[L_v, l_p^n]\|$$

for all X-valued Haar polynomials $\boldsymbol{f}_n = \sum_{k=1}^{n} \boldsymbol{d}_k$. We refer to

$$\varrho_p^{(v)}(\mathcal{H}(\mathbb{D}_1^n), \mathcal{I}(\mathbb{D}_1^n)) : T \longrightarrow \varrho_p^{(v)}(T|\mathcal{H}(\mathbb{D}_1^n), \mathcal{I}(\mathbb{D}_1^n))$$

as a **Haar type ideal norm**. The case $p = v = 2$ has already been treated in 7.4.2.

The quantity $\varrho_p^{(v)}(T|\mathcal{H}(\mathbb{D}_0^n), \mathcal{I}(\mathbb{D}_0^n))$ can be defined analogously by dropping the condition that $\boldsymbol{f}_n$ has mean zero.

REMARK. If $v=p$, then the defining inequality of $\varrho_p^{(p)}(T|\mathcal{H}(\mathbb{D}_1^n), \mathcal{I}(\mathbb{D}_1^n))$ can be written in the form

$$\left(\int_0^1 \left\| \sum_{\mathbb{D}_1^n} T x_k^{(j)} \chi_k^{(j)}(t) \right\|^p dt \right)^{1/p} \le c \left(\sum_{\mathbb{D}_1^n} \left\| 2^{(k-1)(1/p-1/2)} x_k^{(j)} \right\|^p \right)^{1/p}.$$

So $\varrho_p^{(p)}(\mathcal{H}(\mathbb{D}_1^n), \mathcal{I}(\mathbb{D}_1^n))$ is a *weighted* type ideal norm. This fact justifies the above notation and terminology. However, there is no consistency with that given in 3.7.1.

7.5.2 We now repeat the preceding definitions in the case of cotype.

Assume that $2 \le q < \infty$ and $1 < u < \infty$. For $T \in \mathfrak{L}(X, Y)$, we denote by $\varrho_u^{(q)}(T|\mathcal{I}(\mathbb{D}_1^n), \mathcal{H}(\mathbb{D}_1^n))$ the least constant $c \ge 0$ such that

$$\left(\int_0^1 \left[\sum_{k=1}^{n} \left\| \sum_{j=1}^{2^{k-1}} T x_k^{(j)} \chi_k^{(j)}(t) \right\|^q \right]^{u/q} dt \right)^{1/u} \le c \left(\int_0^1 \left\| \sum_{\mathbb{D}_1^n} x_k^{(j)} \chi_k^{(j)}(t) \right\|^u dt \right)^{1/u}$$

for all $\mathbb{D}_1^n$-tuples $(x_k^{(j)})$ in X. In other words, it is required that

$$\|([L_u, T]\boldsymbol{d}_k)|[L_u, l_q^n]\| \le c\, \|\boldsymbol{f}_n|L_u\|$$

for all X-valued Haar polynomials $\boldsymbol{f}_n = \sum_{k=1}^{n} \boldsymbol{d}_k$. We refer to

$$\varrho_u^{(q)}(\mathcal{I}(\mathbb{D}_1^n), \mathcal{H}(\mathbb{D}_1^n)) : T \longrightarrow \varrho_u^{(q)}(T|\mathcal{I}(\mathbb{D}_1^n), \mathcal{H}(\mathbb{D}_1^n))$$

as a **Haar cotype ideal norm**. The case $q = u = 2$ has already been treated in 7.4.2.

The quantity $\varrho_u^{(q)}(T|\mathcal{I}(\mathbb{D}_0^n), \mathcal{H}(\mathbb{D}_0^n))$ can be defined analogously by dropping the condition that $\boldsymbol{f}_n$ has mean zero.

7.5.3 The following observation is trivial; see 7.4.3.

PROPOSITION.

$$\varrho_p^{(v)}(T|\mathcal{H}(\mathbb{D}_1^n), \mathcal{I}(\mathbb{D}_1^n)) \le \varrho_p^{(v)}(T|\mathcal{H}(\mathbb{D}_0^n), \mathcal{I}(\mathbb{D}_0^n)) \le 3\varrho_p^{(v)}(T|\mathcal{H}(\mathbb{D}_1^n), \mathcal{I}(\mathbb{D}_1^n)),$$
$$\varrho_u^{(q)}(T|\mathcal{I}(\mathbb{D}_1^n), \mathcal{H}(\mathbb{D}_1^n)) \le \varrho_u^{(q)}(T|\mathcal{I}(\mathbb{D}_0^n), \mathcal{H}(\mathbb{D}_0^n)) \le 3\varrho_u^{(q)}(T|\mathcal{I}(\mathbb{D}_1^n), \mathcal{H}(\mathbb{D}_1^n)).$$

7.5.4 Haar type and Haar cotype ideal norms are dual to each other; see 3.4.6 and 7.4.2.

PROPOSITION.

$$\varrho_p^{(v)'}(\mathcal{H}(\mathbb{D}_0^n), \mathcal{I}(\mathbb{D}_0^n)) = \varrho_{v'}^{(p')}(\mathcal{I}(\mathbb{D}_0^n), \mathcal{H}(\mathbb{D}_0^n)),$$

$$\varrho_u^{(q)'}(\mathcal{I}(\mathbb{D}_0^n), \mathcal{H}(\mathbb{D}_0^n)) = \varrho_{q'}^{(u')}(\mathcal{H}(\mathbb{D}_0^n), \mathcal{I}(\mathbb{D}_0^n)).$$

7.5.5 We now establish a very important result which states that the asymptotic behaviour of $\varrho_p^{(v)}(\mathcal{H}(\mathbb{D}_1^n), \mathcal{I}(\mathbb{D}_1^n))$ is determined by the exponent p only.

PROPOSITION. *Fix* $1 < p \leq 2$. *The sequences of the ideal norms* $\varrho_p^{(v)}(\mathcal{H}(\mathbb{D}_1^n), \mathcal{I}(\mathbb{D}_1^n))$ *are uniformly equivalent whenever* $1<v<\infty$.

PROOF. Let $T \in \mathfrak{L}(X, Y)$, and consider the martingale transforms

$$S : \sum_{h=1}^{k} \boldsymbol{d}_h(t) \longrightarrow (\boldsymbol{d}_1(t), \ldots, \boldsymbol{d}_k(t), o, \ldots, o)$$

from $[WP_n^0, X]$ into $[WP_n^0, U]$ with $U := [l_p^n, X]$ and

$$\boldsymbol{I}_T : \sum_{h=1}^{k} \boldsymbol{d}_h(t) \longrightarrow \sum_{h=1}^{k} T\boldsymbol{d}_h(t)$$

from $[WP_n^0, X]$ into $[WP_n^0, Y]$. Then the assertion follows from 7.2.10.

7.5.6 The preceding result can be dualized as follows.

PROPOSITION. *Fix* $2 \leq q < \infty$. *The sequences of the ideal norms* $\varrho_u^{(q)}(\mathcal{I}(\mathbb{D}_1^n), \mathcal{H}(\mathbb{D}_1^n))$ *are uniformly equivalent whenever* $1<u<\infty$.

7.5.7 Let $1 < p \leq 2$. An operator T has **Haar type** p if

$$\|T|\mathfrak{AT}_p\| := \sup_n \varrho_p^{(p)}(T|\mathcal{H}(\mathbb{D}_1^n), \mathcal{I}(\mathbb{D}_1^n))$$

is finite. These operators form the Banach ideal $\mathfrak{AT}_p$.

REMARK. Since $\mathfrak{HT}$ has already been used to denote the ideal related to the vector-valued Hilbert transform, we have decided to use the symbol $\mathfrak{AT}$, where $\mathfrak{A}$ refers to $\mathfrak{A}$lfred H$\mathfrak{aa}$r.

7.5.8 Let $2 \leq q < \infty$. An operator T has **Haar cotype** q if

$$\|T|\mathfrak{AC}_q\| := \sup_n \varrho_q^{(q)}(T|\mathcal{I}(\mathbb{D}_1^n), \mathcal{H}(\mathbb{D}_1^n))$$

is finite. These operators form the Banach ideal $\mathfrak{AC}_q$.

7.5.9 Let $1 < p < 2$. An operator T has **weak Haar type** p if

$$\|T|\mathfrak{AT}_p^{weak}\| := \sup_n n^{1/p-1/2}\, \varrho(T|\mathcal{H}(\mathbb{D}_1^n), \mathcal{I}(\mathbb{D}_1^n))$$

is finite. These operators from the Banach ideal $\mathfrak{AT}_p^{weak}$.

REMARK. In the case $p = 2$, we would get $\mathfrak{AT}_2$.

7.5.10 Let $2 < q < \infty$. An operator T has **weak Haar cotype** q if

$$\|T|\mathfrak{AC}_q^{weak}\| := \sup_n n^{1/2-1/q}\, \varrho(T|\mathcal{I}(\mathbb{D}_1^n), \mathcal{H}(\mathbb{D}_1^n))$$

is finite. These operators form the Banach ideal $\mathfrak{AC}_q^{weak}$.

REMARK. In the case $q = 2$, we would get $\mathfrak{AC}_2$.

7.5.11 The next formulas are corollaries of 7.5.4 and 7.4.4.

PROPOSITION.

$$(\mathfrak{AT}_p)' = \mathfrak{AC}_{p'} \quad \text{and} \quad (\mathfrak{AT}_p^{weak})' = \mathfrak{AC}_{p'}^{weak},$$
$$(\mathfrak{AC}_q)' = \mathfrak{AT}_{q'} \quad \text{and} \quad (\mathfrak{AC}_q^{weak})' = \mathfrak{AT}_{q'}^{weak}.$$

REMARK. The corresponding ideal norms coincide when they are generated with respect to the index sets $\mathbb{D}_0^n$.

7.5.12 THEOREM. $\mathfrak{AT}_2 \subset \mathfrak{AT}_{p_0} \subset \mathfrak{AT}_{p_0}^{weak} \subset \mathfrak{AT}_p$ *if* $1 < p < p_0 < 2$. *In particular,*

$$\bigcup_{1<p\le 2} \mathfrak{AT}_p = \bigcup_{1<p<2} \mathfrak{AT}_p^{weak}.$$

PROOF. Applying Hölder's inequality yields

$$\Big(\int_0^1 \Big\|\sum_{\mathbb{D}_1^n} T x_k^{(j)} \chi_k^{(j)}(t)\Big\|^2 dt\Big)^{1/2} \le$$

$$\le \varrho_{p_0}^{(2)}(T|\mathcal{H}(\mathbb{D}_1^n), \mathcal{I}(\mathbb{D}_1^n)) \Bigg(\int_0^1 \Bigg[\sum_{k=1}^n \Big\|\sum_{j=1}^{2^{k-1}} x_k^{(j)} \chi_k^{(j)}(t)\Big\|^{p_0}\Bigg]^{2/p_0} dt\Bigg)^{1/2}$$

$$\le n^{1/p_0-1/2} \varrho_{p_0}^{(2)}(T|\mathcal{H}(\mathbb{D}_1^n), \mathcal{I}(\mathbb{D}_1^n)) \Bigg(\int_0^1 \sum_{k=1}^n \Big\|\sum_{j=1}^{2^{k-1}} x_k^{(j)} \chi_k^{(j)}(t)\Big\|^2 dt\Bigg)^{1/2}.$$

Hence

$$\varrho(T|\mathcal{H}(\mathbb{D}_1^n), \mathcal{I}(\mathbb{D}_1^n)) \le n^{1/p_0-1/2} \varrho_{p_0}^{(2)}(T|\mathcal{H}(\mathbb{D}_1^n), \mathcal{I}(\mathbb{D}_1^n))$$

which, by 7.5.5, implies that $\mathfrak{AT}_{p_0} \subseteq \mathfrak{AT}_{p_0}^{weak}$.

We now come to the hard part of the proof. Fix any X-valued Haar polynomial

$$f_n = \sum_{\mathbb{D}_1^n} x_k^{(j)} \chi_k^{(j)},$$

write

$$d_k := \sum_{j=1}^{2^{k-1}} x_k^{(j)} \chi_k^{(j)} \quad \text{for } k = 1, \ldots, n,$$

and let

$$\sigma := \sup_{0 \le t < 1} \left(\sum_{k=1}^{n} \|d_k(t)\|^p \right)^{1/p}.$$

Recall from 7.3.2 that

$$\mathbb{D}_1^n(t) = \{(k,j) \in \mathbb{D}_1^n : t \in \Delta_{k-1}^{(j)}\},$$

and define

$$\mathbb{F}_h := \left\{(k,j) \in \mathbb{D}_1^n : 2^{-h/p}\sigma < \|2^{(k-1)/2} x_k^{(j)}\| \le 2^{-(h-1)/p}\sigma\right\}$$

for $h = 1, 2, \ldots$. Since

$$\sigma^p \ge \sum_{k=1}^{n} \|d_k(t)\|^p = \sum_{\mathbb{D}_1^n} \left\| x_k^{(j)} \chi_k^{(j)}(t) \right\|^p$$
$$\ge \sum_{\mathbb{F}_h \cap \mathbb{D}_1^n(t)} \left\| 2^{(k-1)/2} x_k^{(j)} \right\|^p \ge 2^{-h} \sigma^p \, |\mathbb{F}_h \cap \mathbb{D}_1^n(t)|,$$

we have

$$|\mathbb{F}_h \cap \mathbb{D}_1^n(t)| \le 2^h \quad \text{for } 0 \le t < 1. \tag{1}$$

Hence

$$\text{height}(\mathbb{F}_h) \le 2^h \quad \text{for } h = 1, 2, \ldots. \tag{2}$$

Next, we conclude from

$$\sum_{\mathbb{F}_h} \|x_k^{(j)}\|^2 = \int_0^1 \sum_{\mathbb{F}_h} \|x_k^{(j)} \chi_k^{(j)}(t)\|^2 \, dt \le \left[2^{-(h-1)/p}\sigma\right]^2 \int_0^1 |\mathbb{F}_h \cap \mathbb{D}_1^n(t)| \, dt$$

and (1) that

$$\sum_{\mathbb{F}_h} \|x_k^{(j)}\|^2 \le 2^h \left[2^{-(h-1)/p}\sigma\right]^2 \quad \text{for } h = 1, 2, \ldots. \tag{3}$$

If $T \in \mathfrak{AT}_{p_0}^{weak}(X, Y)$, then it follows from 7.4.20 and (2) that

$$\varrho(T|\mathcal{H}(\mathbb{F}_h), \mathcal{I}(\mathbb{F}_h)) \le 2^{(1/p_0 - 1/2)h} \|T|\mathfrak{AT}_p^{weak}\| \quad \text{for } h = 1, 2, \ldots. \tag{4}$$

By (3), (4) and

$$f_n = \sum_{h=1}^{\infty} \sum_{\mathbb{F}_h} x_k^{(j)} \chi_k^{(j)},$$

we now obtain

$$\begin{aligned} \|[L_2,T]f_n|L_2\| &\le \sum_{h=1}^{\infty} \Big\| \sum_{\mathbb{F}_h} Tx_k^{(j)} \chi_k^{(j)} \Big| L_2 \Big\| \\ &\le \sum_{h=1}^{\infty} \varrho(T|\mathcal{H}(\mathbb{F}_h), \mathcal{I}(\mathbb{F}_h)) \Big(\sum_{\mathbb{F}_h} \|x_k^{(j)}\|^2 \Big)^{1/2} \\ &\le 2^{1/p}\sigma \Big(\sum_{h=1}^{\infty} 2^{(1/p_0-1/p)h} \Big) \|T|\mathfrak{AT}_{p_0}^{weak}\|. \end{aligned}$$

Since $\|[L_1,T]f_n|L_1\| \le \|[L_2,T]f_n|L_2\|$, this proves that

$$\|[L_1,T]f_n|L_1\| \le c(p,p_0)\, \|T|\mathfrak{AT}_{p_0}^{weak}\| \, \Big\| \Big(\sum_{k=1}^{n} \|d_k\|^p \Big)^{1/p} \Big| L_\infty \Big\|$$

for all Haar polynomials f_n with mean zero. Note that $c(p,p_0) > 0$ depends only on p and p_0. Applying 7.2.12 yields

$$\|[L_p,T]f_n|L_p\| \le c_0(p,p_0)\, \|T|\mathfrak{AT}_{p_0}^{weak}\| \, \Big\| \Big(\sum_{k=1}^{n} \|d_k\|^p \Big)^{1/p} \Big| L_p \Big\|,$$

where $c_0(p,p_0) > 0$ is another constant. So $\mathfrak{AT}_{p_0}^{weak} \subseteq \mathfrak{AT}_p$.

In the next paragraph we will show that

$$L_p \in \mathsf{AT}_p \setminus \mathsf{AT}_{p_0}^{weak} \quad \text{if } 1 < p < p_0 < 2.$$

Thus the inclusion $\mathfrak{AT}_{p_0}^{weak} \subseteq \mathfrak{AT}_p$ is strict. Finally, the inclusion $\mathfrak{AT}_{p_0} \subseteq \mathfrak{AT}_{p_0}^{weak}$ is strict as well. Indeed, using 7.4.11, it can be shown that the diagonal operator $D_{1/p_0'}$ satisfies

$$\varrho(D_{1/p_0'} : l_1 \to l_1 | \mathcal{H}(\mathbb{D}_1^n), \mathcal{I}(\mathbb{D}_1^n)) \asymp n^{1/p_0-1/2},$$

while 4.3.13 tells us that it even fails to have Rademacher type p_0.

7.5.13 We now present a counterpart of 4.3.11.

EXAMPLE.

$$L_p \in \mathsf{AT}_p \setminus \mathsf{AT}_{p_0}^{weak} \quad \textit{if } 1 < p < p_0 < 2 \quad \textit{and} \quad L_q \in \mathsf{AT}_2 \quad \textit{if } 2 \le q < \infty.$$

PROOF. It will be shown in 7.9.21 that $\mathsf{UT}_p \subseteq \mathsf{AT}_p$. Therefore the relations $L_p \in \mathsf{AT}_p$ and $L_q \in \mathsf{AT}_2$ are consequences of 7.9.19. Moreover, we know from 7.4.8 that

$$\varrho(L_p|\mathcal{H}(\mathbb{D}_1^n), \mathcal{I}(\mathbb{D}_1^n)) = n^{1/p-1/2}.$$

Hence $L_p \notin \mathsf{AT}_{p_0}^{weak}$.

7.5.14 PROBLEM. *Is it true that* $\mathsf{AT}_p \neq \mathsf{AT}_p^{weak}$?

7.5.15 We now generalize the inequalities stated in 7.4.5. The proofs can be adapted.

PROPOSITION.

$$\varrho_p^{(v)}(\mathcal{R}_n, \mathcal{I}_n) \leq \varrho_p^{(v)}(\mathcal{H}(\mathbb{D}_1^n), \mathcal{I}(\mathbb{D}_1^n)),$$
$$\varrho_u^{(q)}(\mathcal{I}_n, \mathcal{R}_n) \leq \varrho_u^{(q)}(\mathcal{I}(\mathbb{D}_1^n), \mathcal{H}(\mathbb{D}_1^n)).$$

7.5.16 The above inequalities imply the following relations.

PROPOSITION.

$$\mathfrak{AT}_p \subset \mathfrak{RT}_p \quad \textit{and} \quad \mathfrak{AT}_p^{weak} \subset \mathfrak{RT}_p^{weak},$$
$$\mathfrak{AC}_q \subset \mathfrak{RC}_q \quad \textit{and} \quad \mathfrak{AC}_q^{weak} \subset \mathfrak{RC}_q^{weak}.$$

REMARK. All inclusions are strict, since there exist non-reflexive spaces with Rademacher type 2 or cotype 2; see [jam 7], [pis*1] and 7.6.15. Another example is given by the summation operator; see 4.3.18 and 7.6.14 (Remark).

7.5.17 From 7.4.5 we obtain a counterpart of Kwapień's theorem 4.10.7 which is due to T. Figiel and G. Pisier; see [fig*b].

THEOREM. $\mathfrak{AC}_2 \circ \mathfrak{AT}_2 = \mathfrak{H}$ *and* $\mathsf{AT}_2 \cap \mathsf{AC}_2 = \mathsf{H}$.

7.5.18 For the proofs of 7.5.19 and 7.7.5, we provide an auxiliary result.

LEMMA. *Assume that* $V_1, \dots, V_{2^n} \in \mathfrak{L}(L_2(M,\mu))$ *are totally compatible, and put*

$$c := \sup\Big\{ \, \|[V_m, X]\| \, : \, 1 \leq m \leq 2^n, \; X \in \mathsf{L} \Big\}.$$

Let $\boldsymbol{a}_1, \dots, \boldsymbol{a}_{2^n} \in [L_2(M,\mu), X]$ *such that*

$$[V_m, X]\boldsymbol{a}_i = \begin{cases} \boldsymbol{a}_i & \text{if } i \leq m, \\ \boldsymbol{o} & \text{if } i > m. \end{cases}$$

Then

$$2^{-n/2}\Big\| \sum_{i=1}^{2^n} [L_2, T]\boldsymbol{a}_i \Big| L_2 \Big\| \leq$$
$$\leq (2c+1)n^{-1/2}\varrho(T|\mathcal{I}(\mathbb{D}_0^n), \mathcal{H}(\mathbb{D}_0^n)) \Big(\sum_{i=1}^{2^n} \|\boldsymbol{a}_i | L_2\|^2 \Big)^{1/2}.$$

PROOF. Define $\boldsymbol{f}_n : [0,1) \times M \to X$ by

$$\boldsymbol{f}_n(s,\xi) := \boldsymbol{a}_i(\xi) \quad \text{for } s \in \Delta_n^{(i)} \text{ and } \xi \in M.$$

Obviously,

$$\int_0^1 \int_M \|\boldsymbol{f}_n(s,\xi)\|^2 ds\, d\mu(\xi) = \sum_{i=1}^{2^n} \int_{\Delta_n^{(i)}} \int_M \|\boldsymbol{f}_n(s,\xi)\|^2 ds\, d\mu(\xi)$$

$$= \frac{1}{2^n} \sum_{i=1}^{2^n} \|\boldsymbol{a}_i|L_2\|^2. \tag{1}$$

Viewing $\boldsymbol{f}_n(\cdot,\xi)$ as a Haar polynomial, we have

$$\boldsymbol{f}_n(s,\xi) = \sum_{\mathbb{D}_0^n} \boldsymbol{x}_k^{(j)}(\xi)\chi_k^{(j)}(s) \quad \text{with} \quad \boldsymbol{x}_k^{(j)}(\xi) := \int_0^1 \boldsymbol{f}_n(t,\xi)\chi_k^{(j)}(t)\, dt. \tag{2}$$

By definition,

$$\sum_{\mathbb{D}_0^n} \|T\boldsymbol{x}_k^{(j)}(\xi)\|^2 \le \varrho(T|\mathcal{I}(\mathbb{D}_0^n), \mathcal{H}(\mathbb{D}_0^n))^2 \int_0^1 \Big\| \sum_{\mathbb{D}_0^n} \boldsymbol{x}_k^{(j)}(\xi)\chi_k^{(j)}(s) \Big\|^2 ds.$$

Integration over $\xi \in M$ yields

$$\int_M \sum_{\mathbb{D}_0^n} \|T\boldsymbol{x}_k^{(j)}(\xi)\|^2 d\mu(\xi) \le$$

$$\le \varrho(T|\mathcal{I}(\mathbb{D}_0^n), \mathcal{H}(\mathbb{D}_0^n))^2 \int_0^1 \int_M \Big\| \sum_{\mathbb{D}_0^n} \boldsymbol{x}_k^{(j)}(\xi)\chi_k^{(j)}(s) \Big\|^2 ds\, d\mu(\xi).$$

Thus, by (1) and (2),

$$\sum_{\mathbb{D}_0^n} \big\|[L_2, T]\boldsymbol{x}_k^{(j)}|L_2\big\|^2 \le \varrho(T|\mathcal{I}(\mathbb{D}_0^n), \mathcal{H}(\mathbb{D}_0^n))^2 \, \frac{1}{2^n} \sum_{i=1}^{2^n} \|\boldsymbol{a}_i|L_2\|^2. \tag{3}$$

Let

$$\mathbb{N}_k^{(j)} := \{\, i : \Delta_n^{(i)} \subseteq \Delta_k^{(j)} \,\},$$

and note that

$$\mathbb{N}_{k-1}^{(j)} = \mathbb{N}_k^{(2j-1)} \cup \mathbb{N}_k^{(2j)}.$$

We now get

$$\boldsymbol{x}_k^{(j)}(\xi) = \int_0^1 \boldsymbol{f}_n(t,\xi)\chi_k^{(j)}(t) dt$$

$$= 2^{(k-1)/2} \Bigg(\int_{\Delta_k^{(2j-1)}} \boldsymbol{f}_n(t,\xi) dt - \int_{\Delta_k^{(2j)}} \boldsymbol{f}_n(t,\xi) dt \Bigg)$$

$$= 2^{(k-1)/2-n} \Bigg(\sum_{i \in \mathbb{N}_k^{(2j-1)}} \boldsymbol{a}_i(\xi) - \sum_{i \in \mathbb{N}_k^{(2j)}} \boldsymbol{a}_i(\xi) \Bigg),$$

which implies that

$$\big\|[L_2,T]\boldsymbol{x}_k^{(j)}\big|L_2\big\| = 2^{(k-1)/2-n}\Big\|\sum_{i\in\mathbb{N}_k^{(2j-1)}}[L_2,T]\boldsymbol{a}_i - \sum_{i\in\mathbb{N}_k^{(2j)}}[L_2,T]\boldsymbol{a}_i\Big|L_2\Big\|. \quad (4)$$

Let $m{:=}2^{n-k}(2j-1)$ denote the largest index in $\mathbb{N}_k^{(2j-1)}$, and note that

$$[V_m,Y][L_2,T] = [V_m,T] = [L_2,T][V_m,X].$$

Then

$$\sum_{i\in\mathbb{N}_k^{(2j-1)}}[L_2,T]\boldsymbol{a}_i = [V_m,Y]\Big(\sum_{i\in\mathbb{N}_k^{(2j-1)}}[L_2,T]\boldsymbol{a}_i - \sum_{i\in\mathbb{N}_k^{(2j)}}[L_2,T]\boldsymbol{a}_i\Big).$$

Hence

$$\begin{aligned}\Big\|\sum_{i\in\mathbb{N}_{k-1}^{(j)}}[L_2,T]\boldsymbol{a}_i\Big|L_2\Big\| &\le \Big\|\sum_{i\in\mathbb{N}_k^{(2j-1)}}[L_2,T]\boldsymbol{a}_i\Big|L_2\Big\| + \Big\|\sum_{i\in\mathbb{N}_k^{(2j)}}[L_2,T]\boldsymbol{a}_i\Big|L_2\Big\| \\ &\le 2\Big\|\sum_{i\in\mathbb{N}_k^{(2j-1)}}[L_2,T]\boldsymbol{a}_i\Big|L_2\Big\| + \Big\|\sum_{i\in\mathbb{N}_k^{(2j-1)}}[L_2,T]\boldsymbol{a}_i - \sum_{i\in\mathbb{N}_k^{(2j)}}[L_2,T]\boldsymbol{a}_i\Big|L_2\Big\| \\ &\le (2c+1)\Big\|\sum_{i\in\mathbb{N}_k^{(2j-1)}}[L_2,T]\boldsymbol{a}_i - \sum_{i\in\mathbb{N}_k^{(2j)}}[L_2,T]\boldsymbol{a}_i\Big|L_2\Big\|.\end{aligned}$$

In view of (4), it follows that

$$\Big\|\sum_{i\in\mathbb{N}_{k-1}^{(j)}}[L_2,T]\boldsymbol{a}_i\Big|L_2\Big\| \le (2c+1)\,2^{n-(k-1)/2}\big\|[L_2,T]\boldsymbol{x}_k^{(j)}\big|L_2\big\|. \quad (5)$$

Averaging over $k=1,\ldots,2^n$ and using the Cauchy–Schwarz inequality, we finally obtain from (5) and (3)

$$\begin{aligned}\Big\|\sum_{i=1}^{2^n}[L_2,T]\boldsymbol{a}_i\Big|L_2\Big\| &= \Big(\frac{1}{n}\sum_{k=1}^{n}\Big\|\sum_{j=1}^{2^{k-1}}\sum_{i\in\mathbb{N}_{k-1}^{(j)}}[L_2,T]\boldsymbol{a}_i\Big|L_2\Big\|^2\Big)^{1/2} \\ &\le \Big(\frac{1}{n}\sum_{k=1}^{n}\Big[\sum_{j=1}^{2^{k-1}}\Big\|\sum_{i\in\mathbb{N}_{k-1}^{(j)}}[L_2,T]\boldsymbol{a}_i\Big|L_2\Big\|\Big]^2\Big)^{1/2} \\ &\le \Big(\frac{1}{n}\sum_{k=1}^{n}2^{k-1}\sum_{j=1}^{2^{k-1}}\Big\|\sum_{i\in\mathbb{N}_{k-1}^{(j)}}[L_2,T]\boldsymbol{a}_i\Big|L_2\Big\|^2\Big)^{1/2} \\ &\le (2c+1)\Big(\frac{2^{2n}}{n}\sum_{k=1}^{n}\sum_{j=1}^{2^{k-1}}\Big\|[L_2,T]\boldsymbol{x}_k^{(j)}\Big|L_2\Big\|^2\Big)^{1/2} \\ &\le (2c+1)\,\boldsymbol{\varrho}(T|\mathcal{I}(\mathbb{D}_0^n),\mathcal{H}(\mathbb{D}_0^n))\Big(\frac{2^n}{n}\sum_{i=1}^{2^n}\|\boldsymbol{a}_i|L_2\|^2\Big)^{1/2}.\end{aligned}$$

7.5.19 Applying the above lemma to dyadic martingales, $V_m := E(\mathcal{D}_m)$, we get an analogue of 4.2.17.

PROPOSITION.

$$2^{-n/2}\varrho(\mathcal{H}(\mathbb{D}_1^{2^n}),\mathcal{I}(\mathbb{D}_1^{2^n})) \le 3\,n^{-1/2}\varrho(\mathcal{I}(\mathbb{D}_0^n),\mathcal{H}(\mathbb{D}_0^n)).$$

7.5.20 An operator T has **Haar subtype** if

$$\varrho(T|\mathcal{H}(\mathbb{D}_1^n),\mathcal{I}(\mathbb{D}_1^n)) = o(\sqrt{n}).$$

These operators form the closed ideal

$$\mathfrak{AT} := \mathfrak{L}_0[n^{-1/2}\,\varrho(\mathcal{H}(\mathbb{D}_1^n),\mathcal{I}(\mathbb{D}_1^n))].$$

The ideal $\mathfrak{AC}$ constituted by operators of **Haar subcotype** is defined analogously.

7.5.21 THEOREM. $\mathfrak{AT} = (\mathfrak{AT})' = (\mathfrak{AC})' = \mathfrak{AC}.$

PROOF. We conclude from 7.4.4 that $(\mathfrak{AT})' = \mathfrak{AC}$ and $(\mathfrak{AC})' = \mathfrak{AT}$. Moreover, 7.5.19 yields $\mathfrak{AC} \subseteq \mathfrak{AT}$. Hence

$$\mathfrak{AC} \subseteq \mathfrak{AT} = (\mathfrak{AC})' \subseteq (\mathfrak{AT})' = \mathfrak{AC}.$$

7.6 Super weakly compact operators

7.6.1 An operator $T \in \mathfrak{L}(X,Y)$ is called **weakly compact** if the image of the closed unit ball U_X is relatively compact with respect to the weak topology of Y induced by Y'.

These operators constitute a closed ideal, usually denoted by $\mathfrak{W}$. However, we will use the symbol $\mathfrak{R}$, since $\mathsf{R} = \{\, X : I_X \in \mathfrak{R}\,\}$ is just the class of **reflexive spaces**.

7.6.2 Recall that K_X denotes the natural embedding from a Banach space X into its bidual X'',

$$\langle K_X x, x'\rangle = \langle x, x'\rangle \quad \text{for } x \in X \text{ and } x' \in X'.$$

Consider the quotient space $X^\pi := X''/K_X(X)$. Reflexivity means that $X^\pi = \{o\}$. In this case, X and X'' can be identified.

With $T \in \mathfrak{L}(X,Y)$, we associate the operator $T^\pi \in \mathfrak{L}(X^\pi, Y^\pi)$ defined by

$$T^\pi : x'' + K_X(X) \longrightarrow T''x'' + K_Y(Y).$$

A criterion proved in [DUN*1, p. 482] says that $T \in \mathfrak{L}(X,Y)$ is weakly compact if and only if T'' maps X'' into $K_Y(Y)$. That is, $T^\pi = O^\pi$. Thus the size of T^π yields a measure of weak non-compactness.

7.6.3 Recapitulate 0.7.1 and 0.7.2.

EXAMPLE. *The infinite summation operators* $\Sigma \in \mathfrak{L}(l_1, l_\infty)$ *and* $\Sigma^t \in \mathfrak{L}(l_1, l_\infty)$ *fail to be weakly compact.*

PROOF. Assume that $\Sigma^t \in \mathfrak{R}(l_1, l_\infty)$. Then Σ^t is also weakly compact as an operator from l_1 into c_0 and, with respect to the weak topology induced by l_1, the sequence formed by

$$\Sigma^t u_k = (\underbrace{1, \ldots, 1}_{k}, 0, \ldots)$$

must have an accumulation point $x = (\xi_h) \in c_0$. Thus, for some non-trivial ultrafilter $\mathcal{U}$ on $\mathbb{N}$, we get

$$\xi_h = \langle x, u_h \rangle = \mathcal{U}\text{-}\lim_k \langle \Sigma^t u_k, u_h \rangle = 1.$$

Hence $x = (1, 1, 1, \ldots) \notin c_0$. This contradiction shows that $\Sigma^t \notin \mathfrak{R}(l_1, l_\infty)$. Finally, since $\mathfrak{R}$ is an ideal, $\Sigma^t = \Sigma U$ implies that $\Sigma \notin \mathfrak{R}(l_1, l_\infty)$.

7.6.4 We say that $T \in \mathfrak{L}(X, Y)$ **factors the infinite summation operator** if there are operators $A \in \mathfrak{L}(l_1, X)$ and $B \in \mathfrak{L}(Y, l_\infty)$ such that

$$\begin{array}{ccc} X & \xrightarrow{\quad T \quad} & Y \\ {\scriptstyle A}\uparrow & & \downarrow{\scriptstyle B} \\ l_1 & \xrightarrow[\quad \Sigma \quad]{} & l_\infty \, . \end{array}$$

Since $\Sigma^t = \Sigma U$ and $\Sigma = U^t \Sigma^t$, factorization of Σ^t is an equivalent property.

Writing

$$A(\xi_k) = \sum_{k=1}^{\infty} \xi_k x_k \quad \text{and} \quad By = (\langle y, y_h' \rangle)$$

shows that the above condition is also equivalent to the existence of bounded sequences (x_k) in X and (y_h') in Y' such that

$$\langle T x_k, y_h' \rangle = \begin{cases} 1 & \text{if } k \le h, \\ 0 & \text{if } k > h. \end{cases} \tag{Σ}$$

In the transposed case, we get the condition

$$\langle T x_k, y_h' \rangle = \begin{cases} 1 & \text{if } k \ge h, \\ 0 & \text{if } k < h. \end{cases} \tag{Σ^t}$$

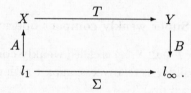

REMARK. Note that $\|U : l_1 \to l_1\| = 2$ and $\|U^t : l_\infty \to l_\infty\| = 2$. In other words, if (x_k) and (y'_h) satisfy (Σ^t), then (x_k) and $(y'_1 - y'_h)$ satisfy (Σ). Hence, when changing between Σ and Σ^t, the norms of the underlying operators, elements and functionals may increase by the factor 2.

7.6.5 The following construction goes back to R. C. James; see [jam 2].

PROPOSITION. *Operators that fail to be weakly compact factor the infinite summation operator. More precisely, if $T \notin \mathfrak{R}(X,Y)$ and $\varepsilon > 0$, then there exist $A \in \mathfrak{L}(l_1, X)$ and $B \in \mathfrak{L}(Y, l_\infty)$ such that $\Sigma^t = BTA$ and $\|B\|\,\|T^\pi\|\,\|A\| < 1+\varepsilon$.*

PROOF. Let $c := \|T^\pi\|^{-1}$, and choose $x'' \in X''$ and $y''' \in Y'''$ such that

$$\|x''\| < c\sqrt{1+\varepsilon}, \quad \|y'''\| = 1,$$
$$\langle T''x'', y'''\rangle = 1 \quad \text{and} \quad \langle K_Y y, y'''\rangle = 0 \quad \text{for } y \in Y.$$

We now proceed by induction. Assume that $x_1, \dots, x_{n-1} \in X$ and $y'_1, \dots, y'_{n-1} \in Y'$ with

$$\|x_k\| < c\sqrt{1+\varepsilon}, \quad \|y'_h\| < \sqrt{1+\varepsilon},$$

$$\langle Tx_k, y'_h\rangle = \begin{cases} 1 & \text{if } k \geq h, \\ 0 & \text{if } k < h, \end{cases} \quad \text{and} \quad \langle T''x'', y'_h\rangle = 1$$

are found. Applying Helly's lemma 0.1.8 to the elements $y''' \in Y'''$ and $T''x'', K_Y T x_1, \dots, K_Y T x_{n-1} \in Y''$ yields a functional $y'_n \in Y'$ such that

$$\|y'_n\| < \sqrt{1+\varepsilon},$$
$$\langle T''x'', y'_n\rangle = \langle T''x'', y'''\rangle \ \text{ and } \ \langle K_Y T x_k, y'_n\rangle = \langle K_Y T x_k, y'''\rangle \ \text{ if } k < n.$$

Again by Helly's lemma, but now applied to the elements $x'' \in X''$ and $T'y'_1, \dots, T'y'_n \in X'$, we find $x_n \in X$ such that

$$\|x_n\| < c\sqrt{1+\varepsilon} \quad \text{and} \quad \langle x_n, T'y'_h\rangle = \langle x'', T'y'_h\rangle \quad \text{if } h \leq n.$$

Since this process does not terminate, we finally obtain sequences (x_k) and (y'_h) such that

$$\|x_k\| < c\sqrt{1+\varepsilon}, \quad \|y'_h\| < \sqrt{1+\varepsilon},$$

and

$$\langle Tx_k, y'_h\rangle = \begin{cases} \langle K_Y T x_k, y'''\rangle = 0 & \text{if } k < h, \\ \langle T''x'', y'''\rangle \ = 1 & \text{if } k \geq h. \end{cases}$$

7.6.6 Combing the preceding results yields a fundamental criterion which is due to J. Lindenstrauss/A. Pełczyński; see [lin*b, p. 322].

THEOREM. *An operator is weakly compact if and only if it does not factor the infinite summation operator.*

PROOF. Since $\mathfrak{R}$ is an operator ideal, it follows from $\Sigma = BTA$ and $\Sigma \notin \mathfrak{R}(l_1, l_\infty)$ that $T \notin \mathfrak{R}(X, Y)$. The reverse implication was shown in the previous paragraph.

7.6.7 An operator $T \in \mathfrak{L}(X, Y)$ is called **super weakly compact** if all of its ultrapowers are weakly compact. These operators constitute a closed ideal, denoted by $\mathfrak{SR}$. Banach spaces contained in SR are named **superreflexive**.

7.6.8 We say that $T \in \mathfrak{L}(X, Y)$ **factors the summation operator** Σ_n with bound $c \geq 1$ if there are operators $A_n \in \mathfrak{L}(l_1^n, X)$ and $B_n \in \mathfrak{L}(X, l_\infty^n)$ such that

$$\Sigma_n = B_n T A_n \quad \text{and} \quad \|B_n\| \, \|A_n\| \leq c.$$

Recall from 0.7.4 that $\Sigma_n^t = R_n \Sigma_n R_n$ and $\Sigma_n = R_n \Sigma_n^t R_n$. Thus factorization of Σ_n^t is an equivalent property. Furthermore, in view of $\|R_n : l_1 \to l_1\| = 1$ and $\|R_n : l_\infty \to l_\infty\| = 1$, the bound remains unchanged.

Writing

$$A_n(\xi_k) = \sum_{k=1}^{n} \xi_k x_k \quad \text{and} \quad B_n y = (\langle y, y_h' \rangle),$$

it follows from $\langle T x_k, y_h' \rangle = \langle \Sigma_n u_k^{(n)}, u_h^{(n)} \rangle$ and

$$\Sigma_n : (\underbrace{0, \ldots, 0}_{k-1}, 1, 0, \ldots, 0) \to (\underbrace{0, \ldots, 0}_{k-1}, 1, \ldots, 1),$$

that the above condition is also equivalent to the existence of elements $x_1, \ldots, x_n$ in X and functionals $y_1', \ldots, y_n'$ in Y' such that

$$\|x_k\| \leq c, \quad \|y_h'\| \leq 1 \quad \text{and} \quad \langle T x_k, y_h' \rangle = \begin{cases} 1 & \text{if } k \leq h, \\ 0 & \text{if } k > h. \end{cases} \qquad (\Sigma_n)$$

In the transposed case, we get the condition

$$\|x_k\| \leq c, \quad \|y_h'\| \leq 1 \quad \text{and} \quad \langle T x_k, y_h' \rangle = \begin{cases} 1 & \text{if } k \geq h, \\ 0 & \text{if } k < h. \end{cases} \qquad (\Sigma_n^t)$$

Observe that, if $x_1, \ldots, x_n$ and $y_1', \ldots, y_n'$ satisfy (Σ_n^t), then $x_n, \ldots, x_1$ and $y_n', \ldots, y_1'$ satisfy (Σ_n).

An operator **factors the finite summation operators** Σ_n **uniformly** if the above (equivalent) conditions are fulfilled for some constant $c \geq 1$ and all $n = 1, 2, \ldots$, simultaneously.

This terminology will also be used in the setting of spaces.

7.6.9 We now establish a counterpart of 7.6.5.

PROPOSITION. *Operators that fail to be super weakly compact factor the finite summation operators uniformly.*

PROOF. We illustrate the following proof by a diagram:

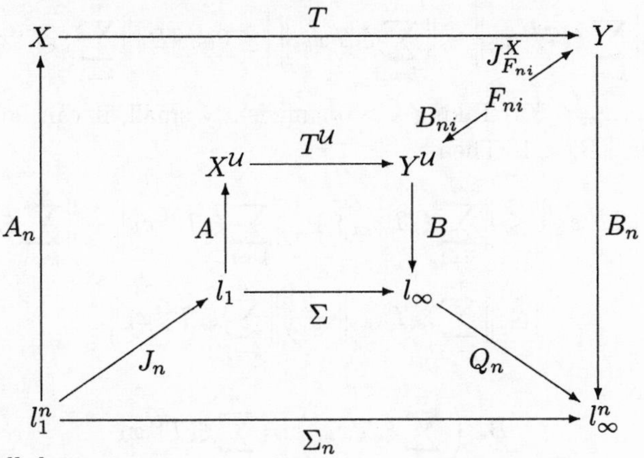

First of all, let

$$J_n : (\xi_1, \dots, \xi_n) \to (\xi_1, \dots, \xi_n, 0, \dots),$$
$$Q_n : (\xi_1, \dots, \xi_n, \xi_{n+1}, \dots) \to (\xi_1, \dots, \xi_n).$$

Without loss of generality, it can be assumed that $T \notin \mathfrak{S}\mathfrak{R}(X, Y)$ has norm 1. Taking any ultrapower $T^{\mathcal{U}}$ which fails to be weakly compact, we find a factorization $\Sigma = BT^{\mathcal{U}}A$ with $\|A\| = 1$. Let $(x_{ki})^{\mathcal{U}} = \boldsymbol{x}_k := Au_k$ such that $\|x_{ki}\| \leq 1$. Then

$$A_n(\xi_k) := \sum_{k=1}^{n} \xi_k x_{ki}$$

defines an operator with $\|A_n\| \leq 1$.

Next, for fixed n, we choose a finite ε-net N in the unit sphere

$$S_1^n := \Big\{ (\xi_k) \in \mathbb{K}^n : \sum_{k=1}^{n} |\xi_k| = 1 \Big\}.$$

If $\mathbb{I}$ is the index set on which the ultrafilter $\mathcal{U}$ lives, then there exists $i \in \mathbb{I}$ such that

$$\left| \Big\| \sum_{k=1}^{n} \xi_k T^{\mathcal{U}} \boldsymbol{x}_k \Big\| - \Big\| \sum_{k=1}^{n} \xi_k T x_{ki} \Big\| \right| \leq \varepsilon \quad \text{for } (\xi_k) \in N.$$

Hence, by $\|T^{\mathcal{U}}\boldsymbol{x}_k\| \leq 1$ and $\|Tx_{ki}\| \leq 1$, we get

$$\left| \Big\| \sum_{k=1}^{n} \xi_k T^{\mathcal{U}} \boldsymbol{x}_k \Big\| - \Big\| \sum_{k=1}^{n} \xi_k T x_{ki} \Big\| \right| \leq 3\varepsilon \quad \text{for } (\xi_k) \in S_1^n.$$

Now it follows from

$$\frac{1}{2n}\sum_{k=1}^{n}|\xi_k| \leq \Big\|\Sigma\Big(\sum_{k=1}^{n}\xi_k u_k\Big)\Big\| \leq \|B\|\Big\|\sum_{k=1}^{n}\xi_k T^{\mathcal{U}}\boldsymbol{x}_k\Big\|$$

that

$$\Big|\Big\|\sum_{k=1}^{n}\xi_k T^{\mathcal{U}}\boldsymbol{x}_k\Big\| - \Big\|\sum_{k=1}^{n}\xi_k T x_{ki}\Big\|\Big| \leq 6n\varepsilon\,\|B\|\Big\|\sum_{k=1}^{n}\xi_k T^{\mathcal{U}}\boldsymbol{x}_k\Big\|$$

for $\xi_1,\dots,\xi_n \in \mathbb{K}$. Taking $\varepsilon > 0$ sufficiently small, it can be arranged that $12n\varepsilon\,\|B\| < 1$. Then

$$\begin{aligned}\Big\|\sum_{k=1}^{n}\xi_k T^{\mathcal{U}}\boldsymbol{x}_k\Big\| &\leq \Big\|\sum_{k=1}^{n}\xi_k T x_{ki}\Big\| + \Big|\Big\|\sum_{k=1}^{n}\xi_k T^{\mathcal{U}}\boldsymbol{x}_k\Big\| - \Big\|\sum_{k=1}^{n}\xi_k T x_{ki}\Big\|\Big| \\ &\leq \Big\|\sum_{k=1}^{n}\xi_k T x_{ki}\Big\| + \frac{1}{2}\Big\|\sum_{k=1}^{n}\xi_k T^{\mathcal{U}}\boldsymbol{x}_k\Big\|.\end{aligned}$$

Therefore

$$B_{ni}\Big(\sum_{k=1}^{n}\xi_k T x_{ki}\Big) := \sum_{k=1}^{n}\xi_k T^{\mathcal{U}}\boldsymbol{x}_k$$

defines an operator from $F_{ni} := \operatorname{span}\{Tx_{1i},\dots,Tx_{ni}\}$ into $Y^{\mathcal{U}}$ with $\|B_{ni}\| \leq 2$. Since l_∞ has the metric extension property, there exists B_n such that $B_n J_{F_{ni}}^{Y} = Q_n B B_{ni}$ and $\|B_n\| \leq 2\|B\|$. A final look at the above diagram completes the proof.

7.6.10 With the help of ultraproduct techniques we can build infinite dimensional objects from finite dimensional ones.

PROPOSITION. *If an operator factors the finite summation operators uniformly, then it has an ultrapower which factors the infinite summation operator.*

PROOF. Fix any non-trivial ultrafilter $\mathcal{U}$ on $\mathbb{N}$. Let J denote the canonical map from l_∞ into $l_\infty^{\mathcal{U}}$ given by $Jx := (x,x,\dots)^{\mathcal{U}}$. We know from the Alaoglu–Bourbaki theorem that the closed unit ball of l_∞ is compact in the weak topology induced by l_1. Taking the limit in this topology, $Q(x_n)^{\mathcal{U}} := \mathcal{U}\text{-}\lim_n x_n$ yields a well-defined operator from $l_\infty^{\mathcal{U}}$ onto l_∞. Obviously, $\|J\| = \|Q\| = 1$.

Choose operators $A_n \in \mathfrak{L}(l_1^n, X)$ and $B_n \in \mathfrak{L}(Y, l_\infty^n)$ such that

$$\|A_n\| \leq 1, \quad \|B_n\| \leq c \quad \text{and} \quad \Sigma_n = B_n T A_n \quad \text{for } n = 1,2,\dots.$$

Then the required factorization of Σ is given by

$$\Sigma = Q(J_n \Sigma_n Q_n)^{\mathcal{U}} J = Q(J_n B_n T A_n Q_n)^{\mathcal{U}} J = Q(J_n B_n)^{\mathcal{U}} T^{\mathcal{U}} (A_n Q_n)^{\mathcal{U}} J.$$

7.6.11 Combing the preceding results, we obtain an analogue of the Lindenstrauss–Pełczyński criterion 7.6.6.

THEOREM. *An operator is super weakly compact if and only if it does not factor the finite summation operators uniformly.*

7.6.12 We now make a simple but useful observation.

LEMMA. *If an operator $T \in \mathfrak{L}(X,Y)$ factors the finite summation operators uniformly, then there exists a constant $c > 0$ such that*

$$\boldsymbol{\alpha}(\Sigma_n : l_1^n \to l_\infty^n) \le c\,\boldsymbol{\alpha}(T)$$

for all ideal norms $\boldsymbol{\alpha}$ and $n = 1, 2, \ldots$.

PROOF. By assumption, we find $A_n \in \mathfrak{L}(l_1^n, X)$ and $B_n \in \mathfrak{L}(X, l_\infty^n)$ such that

$$\Sigma_n = B_n T A_n \quad \text{and} \quad \|B_n\|\,\|A_n\| \le c,$$

where the constant $c > 0$ depends only on T. Hence

$$\boldsymbol{\alpha}(\Sigma_n : l_1^n \to l_\infty^n) \le \|B_n\|\boldsymbol{\alpha}(T)\|A_n\| \le c\,\boldsymbol{\alpha}(X).$$

7.6.13 We proceed with a first consequence of the previous lemma, which is concerned with the QHM-operators defined in 2.4.9.

PROPOSITION. $\mathfrak{QHM} \subseteq \mathfrak{SR}$.

PROOF. Recall from 2.4.4 that

$$\boldsymbol{\kappa}(\Sigma_n : l_1^n \to l_\infty^n | H_n) \asymp 1 + \log n.$$

Hence operators belonging to

$$\mathfrak{QHM} := \Big\{ T \in \mathfrak{L} : \boldsymbol{\kappa}(T|H_n) = o(1 + \log n) \Big\}$$

cannot factor the finite summation operators uniformly.

7.6.14 EXAMPLE. $\boldsymbol{\varrho}(\Sigma_{2^n} : l_1^{2^n} \to l_\infty^{2^n} | \mathcal{I}(\mathbb{D}_0^n), \mathcal{H}(\mathbb{D}_0^n)) \asymp \sqrt{n}$.

PROOF (S. Geiss). The upper estimate follows from 7.4.6. In order to check the lower estimate, we define the $l_1^{2^n}$-valued Haar polynomial

$$\boldsymbol{f}_n = \sum_{\mathbb{D}_0^n} x_k^{(j)} \chi_k^{(j)} \quad \text{by} \quad \boldsymbol{f}_n(t) := u_i^{(2^n)} \quad \text{if } t \in \Delta_n^{(i)}.$$

Write

$$\mathbb{N}_k^{(j)} := \{ i : \Delta_n^{(i)} \subseteq \Delta_k^{(j)} \}.$$

Then we conclude from

$$x_k^{(j)} = 2^{(k-1)/2-n}(0, \ldots, 0, \underbrace{+1, \ldots, +1}_{\mathbb{N}_k^{(2j-1)}}, \underbrace{-1, \ldots, -1}_{\mathbb{N}_k^{(2j)}}, 0, \ldots, 0)$$

that

$$\Sigma_{2^n} x_k^{(j)} = 2^{(k-1)/2-n}(0,\ldots,0,\underbrace{1,\ldots,2^{n-k}}_{\mathbb{N}_k^{(2j-1)}},\underbrace{2^{n-k}-1,\ldots,1}_{\mathbb{N}_k^{(2j)}},0,0,\ldots,0),$$

which in turn yields $\|\Sigma_{2^n} x_k^{(j)} | l_\infty^{2^n}\| = 2^{-(k+1)/2}$. Hence

$$\|(\Sigma_{2^n} x_k^{(j)}) | \mathfrak{I}(\mathbb{D}_0^n)\| \asymp \sqrt{n}.$$

This completes the proof, in view of $\|(x_k^{(j)}) | \mathcal{H}(\mathbb{D}_0^n)\| = \|f_n | L_2\| = 1$.

Remark. As an immediate consequence, we get a supplement to 4.3.18:

$$\Sigma \notin \mathfrak{AT}(l_1, l_\infty) = \mathfrak{AC}(l_1, l_\infty).$$

Hence $\mathfrak{RT}_2 \not\subseteq \mathfrak{AT}_p$ for $1 < p \le 2$ and $\mathfrak{RC}_2 \not\subseteq \mathfrak{AC}_q$ for $2 \le q < \infty$.

7.6.15 The previous example implies another corollary of 7.6.12. Later, we will show that the following inclusion is even an equality; see 7.10.25.

PROPOSITION. $\mathfrak{AT} \subseteq \mathfrak{SR}$.

Remark. Recall from 7.5.21 that $\mathfrak{AT} = \mathfrak{AC}$.

7.7 Martingale type ideal norms

7.7.1 It seems to be unknown whether the sequence of Haar type ideal norms is submultiplicative. The following concept enables us to overcome this lack of knowledge.

For $T \in \mathfrak{L}(X, Y)$ and $n = 1, 2, \ldots$, we denote by $\varrho(T|\mathcal{M}_n, \mathcal{I}_n)$ the least constant $c \ge 0$ such that

$$\Big\| \sum_{k=1}^{n} [L_2, T] \boldsymbol{d}_k \Big| L_2 \Big\| \le c \Big(\sum_{k=1}^{n} \|\boldsymbol{d}_k | L_2\|^2 \Big)^{1/2}$$

for all X-valued martingale differences $\boldsymbol{d}_1, \ldots, \boldsymbol{d}_n$ adapted to any filtration $\mathcal{F}_1 \subseteq \ldots \subseteq \mathcal{F}_n$ on the unit interval $[0,1)$. The algebras $\mathcal{F}_k$ are supposed to be finite. We refer to

$$\varrho(\mathcal{M}_n, \mathcal{I}_n) : T \longrightarrow \varrho(T|\mathcal{M}_n, \mathcal{I}_n)$$

as a **martingale type ideal norm**.

Remark. Note that $\varrho(\mathcal{M}_n, \mathcal{I}_n)$ is not a type ideal norm in the sense of 3.7.1. However, there is a strong analogy and, so far as we can see, this notation will not cause any confusion.

7.7.2 PROPOSITION. $\varrho(\mathcal{M}_n, \mathcal{I}_n) \le \sqrt{n}$.

PROOF. For $\boldsymbol{d}_1, \ldots, \boldsymbol{d}_n \in [L_2[0,1), X]$, we have

$$\Big\| \sum_{k=1}^{n} \boldsymbol{d}_k \Big| L_2 \Big\| \le \sum_{k=1}^{n} \|\boldsymbol{d}_k | L_2\| \le \sqrt{n} \Big(\sum_{k=1}^{n} \|\boldsymbol{d}_k | L_2\|^2 \Big)^{1/2}.$$

7.7.3 PROPOSITION. *The sequence $(\varrho(\mathcal{M}_n, \mathcal{I}_n))$ is non-decreasing and submultiplicative.*

7.7.4 PROPOSITION. $\varrho(\mathcal{H}(\mathbb{D}_1^n), \mathcal{I}(\mathbb{D}_1^n)) \le \varrho(\mathcal{M}_n, \mathcal{I}_n)$.

REMARK. We conjecture that the sequences $(\varrho(\mathcal{H}(\mathbb{D}_1^n), \mathcal{I}(\mathbb{D}_1^n)))$ and $(\varrho(\mathcal{M}_n, \mathcal{I}_n))$ fail to be uniformly equivalent.

7.7.5 Applying 7.5.18 to arbitrary martingales, $V_m := E(\mathcal{F}_m)$, we get a generalization of 7.5.19.

PROPOSITION. $2^{-n/2}\varrho(\mathcal{M}_{2^n}, \mathcal{I}_{2^n}) \le 3n^{-1/2}\varrho(\mathcal{I}(\mathbb{D}_0^n), \mathcal{H}(\mathbb{D}_0^n))$.

7.7.6 An operator T has **martingale subtype** if

$$\varrho(T|\mathcal{M}_n, \mathcal{I}_n) = o(\sqrt{n}).$$

These operators form the closed ideal

$$\mathfrak{MT} := \mathfrak{L}_0[n^{-1/2}\, \varrho(\mathcal{M}_n, \mathcal{I}_n)].$$

7.7.7 In view of the following observation, the preceding definition seems to be superfluous. However, we now have submultiplicativity at our disposal.

THEOREM. $\mathfrak{MT} = \mathfrak{AT}$.

PROOF. By 7.7.4, we get $\mathfrak{MT} \subseteq \mathfrak{AT}$. On the other hand, 7.7.5 implies that $\mathfrak{AC} \subseteq \mathfrak{MT}$. The conclusion now follows from 7.5.21.

7.7.8 Next, we establish an analogue of 4.4.1, 5.6.27 and 6.3.10.

THEOREM. $\displaystyle\bigcup_{1<p\le 2} \mathsf{AT}_p = \bigcup_{1<p<2} \mathsf{AT}_p^{weak} = \mathsf{AT}$.

PROOF. If $X \in \mathsf{AT} = \mathsf{MT}$, then $\varrho(X|\mathcal{M}_m, \mathcal{I}_m) < \sqrt{m}$ for some $m \ge 2$. Since the sequence $(\varrho(\mathcal{M}_n, \mathcal{I}_n))$ is non-decreasing and submultiplicative, 1.3.5 provides us with an $\varepsilon > 0$ such that

$$\varrho(X|\mathcal{H}(\mathbb{D}_1^n), \mathcal{I}(\mathbb{D}_1^n)) \le \varrho(X|\mathcal{M}_n, \mathcal{I}_n) \prec n^{1/2-\varepsilon}$$

for $n = 1, 2, \ldots$. Hence X has weak Haar type p whenever $1/p' \le \varepsilon$. This proves that

$$\mathsf{AT} \subseteq \bigcup_{1<p<2} \mathsf{AT}_p^{weak}.$$

The reverse inclusion is obvious, and the equality

$$\bigcup_{1<p\leq 2} \mathsf{AT}_p = \bigcup_{1<p<2} \mathsf{AT}_p^{weak}$$

was shown in 7.5.12.

REMARKS. In view of the above result, Banach spaces in AT are said to have **non-trivial Haar type.**

By 7.5.11, we also get

$$\bigcup_{2\leq q<\infty} \mathsf{AC}_q = \bigcup_{2<q<\infty} \mathsf{AC}_q^{weak} = \mathsf{AC}.$$

7.7.9 The following counterpart of 4.4.2 and 4.6.2 shows that the phenomenon described in the previous paragraph only occurs for spaces, not for operators. That is,

$$\bigcup_{1<p\leq 2} \mathfrak{AT}_p \neq \mathfrak{AT} \quad \text{and} \quad \bigcup_{2\leq q<\infty} \mathfrak{AC}_q \neq \mathfrak{AC}.$$

EXAMPLE.

$$\varrho(B_\alpha|\mathcal{H}(\mathbb{D}_1^n), \mathcal{I}(\mathbb{D}_1^n)) \asymp n^{1/2}(1+\log n)^{-\alpha},$$
$$\varrho(C_\alpha|\mathcal{I}(\mathbb{D}_1^n), \mathcal{H}(\mathbb{D}_1^n)) \asymp n^{1/2}(1+\log n)^{-\alpha}.$$

PROOF. We know from 7.4.11 and 7.4.4 that

$$\varrho(l_1^N|\mathcal{H}(\mathbb{D}_1^n), \mathcal{I}(\mathbb{D}_1^n)) = \min\{\sqrt{n}, \sqrt{N}\}$$

and

$$\varrho(l_1^N|\mathcal{I}(\mathbb{D}_1^n), \mathcal{H}(\mathbb{D}_1^n)) \asymp \min\{\sqrt{n}, \sqrt{1+\log N}\}.$$

The desired relations are now consequences of ($\triangle$) in 1.2.12.

REMARK. We stress that $B_\alpha, C_\alpha \in \mathfrak{AT} = \mathfrak{AC}$ do not admit factorizations through spaces in $\mathsf{AT} = \mathsf{AC}$. In view of 7.10.25, this observation carries over to $\mathfrak{SR}$. At first sight, this is quite surprising, since a famous theorem of W. J. Davis/T. Figiel/W. B. Johnson/A. Pełczyński asserts that every weakly compact operator can be factored through a reflexive space; see [dav*] and [PIE 2, p. 55]. Thus passing from $\mathfrak{R}$ to $\mathfrak{SR}$ changes the situation significantly.

7.7.10 LEMMA. *Assume that Y is finitely representable in X. If Y factors the finite summation operators uniformly with bound $c \geq 1$, then so does X with bound $(1+\varepsilon)c$ for all $\varepsilon > 0$.*

PROOF. We choose $y_1, \ldots, y_n \in Y$ and $y'_1, \ldots, y'_n \in Y'$ such that

$$\|y_k\| \leq c, \quad \|y'_h\| \leq 1 \quad \text{and} \quad \langle y_k, y'_h\rangle = \begin{cases} 1 \text{ if } k \leq h, \\ 0 \text{ if } k > h. \end{cases} \qquad (\Sigma_n)$$

Let $F := \text{span}\{y_1, \ldots, y_n\}$. Then there exist a subspace E of X with the same dimension and a bijection $T \in \mathfrak{L}(E, F)$ such that $\|T\| = 1$ and $\|T^{-1}\| \le 1 + \varepsilon$. Defining $x_k := T^{-1} y_k$ and taking any norm-preserving extension x'_h of $T' y'_h$, we are done.

7.7.11 PROPOSITION. *Assume that $\varrho(X|\mathcal{M}_n, \mathcal{I}_n) = \sqrt{n}$ for some fixed n. Then $[l_2, X]$ factors the summation operator Σ_n with every bound $c > 1$.*

PROOF. By assumption, for $\delta > 0$, there are X-valued martingale differences $\boldsymbol{d}_1^{(\delta)}, \ldots, \boldsymbol{d}_n^{(\delta)}$ adapted to a filtration $\mathcal{F}_1^{(\delta)} \subseteq \ldots \subseteq \mathcal{F}_n^{(\delta)}$ on $[0, 1)$ such that

$$\left\| \sum_{k=1}^{n} \boldsymbol{d}_k^{(\delta)} \middle| L_2 \right\| \ge \frac{\sqrt{n}}{1+\delta} \left(\sum_{k=1}^{n} \|\boldsymbol{d}_k^{(\delta)} | L_2\|^2 \right)^{1/2}.$$

By scaling, we can arrange that

$$\left\| \sum_{k=1}^{n} \boldsymbol{d}_k^{(\delta)} \middle| L_2 \right\| \ge \frac{n}{1+\delta} \quad \text{and} \quad \sum_{k=1}^{n} \|\boldsymbol{d}_k^{(\delta)} | L_2\|^2 = n. \tag{1}$$

Let $A_i^{(\delta)}$ with $i = 1, \ldots, N^{(\delta)}$ be the atoms of $\mathcal{F}_n^{(\delta)}$, and assign to the normalized characteristic function $\lambda(A_i^{(\delta)})^{-1/2} \chi_i^{(\delta)}$ the i-th unit sequence u_i in l_2. Via this correspondence, $[L_2([0, 1), \mathcal{F}_n^{(\delta)}), X]$ can be identified with the $N^{(\delta)}$-th section of $Y := [l_2, X]$. In particular, we obtain elements $y_1^{(\delta)}, \ldots, y_n^{(\delta)} \in Y$ such that

$$\left\| \sum_{k=1}^{n} y_k^{(\delta)} \right\| \ge \frac{n}{1+\delta} \quad \text{and} \quad \sum_{k=1}^{n} \|y_k^{(\delta)}\|^2 = n. \tag{2}$$

Moreover, the operators of conditional expectation are mapped into $V_1^{(\delta)}, \ldots, V_n^{(\delta)}$. The fact that $\boldsymbol{d}_1, \ldots, \boldsymbol{d}_n$ are martingale differences yields

$$\|V_1^{(\delta)}\| = \ldots = \|V_n^{(\delta)}\| = 1 \quad \text{and} \quad V_h^{(\delta)} y_k^{(\delta)} = \begin{cases} y_k^{(\delta)} & \text{if } k \le h, \\ o & \text{if } k > h. \end{cases} \tag{3}$$

Fix an ultrafilter $\mathcal{U}$ on $\mathbb{R}_+$ containing all intervals $(0, \delta)$ with $\delta > 0$. Now we pass to the equivalence classes $\boldsymbol{y}_k := (y_k^{(\delta)})^{\mathcal{U}}$ in the ultrapower $Y^{\mathcal{U}}$ and conclude from (2) that

$$\left\| \sum_{k=1}^{n} \boldsymbol{y}_k \right\| \ge n \quad \text{and} \quad \sum_{k=1}^{n} \|\boldsymbol{y}_k\|^2 = n. \tag{4}$$

For the ultraproducts $\boldsymbol{V}_1 = (V_1^{(\delta)})^{\mathcal{U}}, \ldots, \boldsymbol{V}_n = (V_n^{(\delta)})^{\mathcal{U}}$ acting on $Y^{\mathcal{U}}$, property (3) turns into

$$\| V_1 \| = \ldots = \| V_n \| = 1 \quad \text{and} \quad V_h y_k = \begin{cases} y_k & \text{if } k \le h, \\ o & \text{if } k > h. \end{cases} \tag{5}$$

Since the inequalities

$$n \le \Big\| \sum_{k=1}^{n} y_k \Big\| \le \sum_{k=1}^{n} \| y_k \| \le \sqrt{n} \Big(\sum_{k=1}^{n} \| y_k \|^2 \Big)^{1/2} = n$$

are even equalities, we get

$$\Big\| \sum_{k=1}^{n} y_k \Big\| = n \quad \text{and} \quad \| y_1 \| = \ldots = \| y_n \| = 1. \tag{6}$$

By the Hahn–Banach theorem, there exist $y' \in (Y^{\mathcal{U}})'$ with

$$\Big\langle \sum_{k=1}^{n} y_k, y' \Big\rangle = n \quad \text{and} \quad \| y' \| = 1.$$

It follows from

$$\sum_{k=1}^{n} \langle y_k, y' \rangle = n \quad \text{and} \quad |\langle y_k, y' \rangle| \le 1$$

that

$$\langle y_k, y' \rangle = 1. \tag{7}$$

Letting $y'_h := V'_h y'$, we obtain $\| y'_h \| \le 1$ and

$$\langle y_k, y'_h \rangle = \langle V_h y_k, y' \rangle = \begin{cases} \langle y_k, y' \rangle = 1 & \text{if } k \le h, \\ \langle o, y' \rangle \ = 0 & \text{if } k > h. \end{cases} \tag{Σ_n}$$

As shown in 7.7.10, at the cost of a factor $1 + \varepsilon$, the finite systems $\{y_1, \ldots, y_n\}$ and $\{y'_1, \ldots, y'_n\}$ can be transferred to $Y = [l_2, X]$ and $Y' = [l_2, X]'$, respectively. This completes the proof.

7.7.12 We know from 7.6.15 that $\mathsf{AT} \subseteq \mathsf{SR}$. The next result is a first step in proving the reverse inclusion. The missing implication, namely $X \in \mathsf{SR} \Rightarrow [l_2, X] \in \mathsf{SR}$, will be established in 7.8.15.

PROPOSITION. *If* $[l_2, X] \in \mathsf{SR}$, *then* $X \in \mathsf{AT}$.

PROOF. Assume that $\varrho(X|\mathcal{M}_n, \mathcal{I}_n) = \sqrt{n}$ for all $n = 1, 2, \ldots$. Then, by the preceding proposition, $[l_2, X]$ factors the summation operators Σ_n uniformly with every bound $c > 1$. Hence $[l_2, X] \notin \mathsf{SR}$. This contradiction shows that $\varrho(X|\mathcal{M}_m, \mathcal{I}_m) < \sqrt{m}$ for some $m \ge 2$. As in the proof of 7.7.8, we now obtain that X has weak Haar type p whenever $p > 1$ is sufficiently small.

7.7.13 For $T \in \mathfrak{L}(X,Y)$ and $n = 1, 2, \ldots$, the defining inequality of $\varrho(T|\mathcal{M}_n, \mathcal{I}_n)$ may be considered under the additional condition that the L_2-norms of all X-valued martingale differences $\boldsymbol{d}_1, \ldots, \boldsymbol{d}_n$ equal 1. This yields

$$\Big\| \sum_{k=1}^{n} [L_2, T]\boldsymbol{d}_k \Big| L_2 \Big\| \le c\sqrt{n},$$

and the best possible constant will be denoted by $\varrho^{eq}(T|\mathcal{M}_n, \mathcal{I}_n)$. We refer to

$$\varrho^{eq}(\mathcal{M}_n, \mathcal{I}_n) : T \longrightarrow \varrho^{eq}(T|\mathcal{M}_n, \mathcal{I}_n)$$

as an **equal-norm martingale type ideal norm.**

7.7.14 There is a strong analogy with some results in Section 4.4. For example, the following fact corresponds to 4.4.12.

PROPOSITION.

$$\varrho^{eq}(T|\mathcal{M}_n, \mathcal{I}_n) = \frac{1}{\sqrt{n}} \sup\left\{ \Big\| \sum_{k=1}^{n} [L_2, T]\boldsymbol{d}_k \Big| L_2 \Big\| : \|\boldsymbol{d}_k|L_2\| \le 1 \right\}.$$

PROOF. If $\|\boldsymbol{d}_k|L_2\| = 1$, then we have

$$\Big\| \sum_{k=1}^{n} \zeta_k [L_2, T]\boldsymbol{d}_k \Big| L_2 \Big\| \le \varrho^{eq}(T|\mathcal{M}_n, \mathcal{I}_n)\sqrt{n} \quad \text{whenever } |\zeta_k| = 1.$$

Thus an extreme point argument yields

$$\Big\| \sum_{k=1}^{n} c_k [L_2, T]\boldsymbol{d}_k \Big| L_2 \Big\| \le \varrho^{eq}(T|\mathcal{M}_n, \mathcal{I}_n)\sqrt{n} \quad \text{whenever } 0 \le c_k \le 1.$$

7.7.15 We now describe a process of spreading martingale differences. Fix $m = 1, 2, \ldots$, and consider the partition

$$[0,1) = [0, \tfrac{1}{m}) \cup \ldots \cup [\tfrac{l-1}{m}, \tfrac{l}{m}) \cup \ldots \cup [\tfrac{m-1}{m}, 1).$$

Note that the function

$$\varphi_l^{[m]}(t) := mt - l + 1 \quad \text{with } l = 1, \ldots, m$$

maps $[\frac{l-1}{m}, \frac{l}{m})$ onto $[0,1)$. For every filtration $\mathcal{F}_1 \subseteq \ldots \subseteq \mathcal{F}_n$, we let $\mathcal{F}_{kl}^{[m]}$ denote the algebra on $[0,1)$ generated by all subsets A of $[\frac{l-1}{m}, \frac{l}{m})$ with $\varphi_l^{[m]}(A) \in \mathcal{F}_k$. Next, $\widetilde{\mathcal{F}}_{kl}^{[m]}$ is defined to be the smallest algebra that contains $\mathcal{F}_{kl}^{[m]}$ as well as all predecessors with respect to lexicographical ordering.

Given X-valued martingale differences $\boldsymbol{d}_1, \ldots, \boldsymbol{d}_n$ adapted to $\mathcal{F}_1 \subseteq \ldots \subseteq \mathcal{F}_n$, the m-fold duplicated differences are obtained by

$$\boldsymbol{d}_{kl}^{[m]}(t) := \begin{cases} \boldsymbol{d}_k(\varphi_l^{[m]}(t)) & \text{if } \frac{l-1}{m} \le t < \frac{l}{m}, \\ o & \text{otherwise.} \end{cases}$$

This definition transfers $\boldsymbol{d}_1, \ldots, \boldsymbol{d}_n$ to each of the intervals $[\frac{l-1}{m}, \frac{l}{m})$, and we get an mn-tuple of X-valued martingale differences

$$\underbrace{\boldsymbol{d}_{11}^{[m]}, \ldots, \boldsymbol{d}_{1m}^{[m]}}_{\boldsymbol{d}_1}, \ldots, \underbrace{\boldsymbol{d}_{k1}^{[m]}, \ldots, \boldsymbol{d}_{km}^{[m]}}_{\boldsymbol{d}_k}, \ldots, \underbrace{\boldsymbol{d}_{n1}^{[m]}, \ldots, \boldsymbol{d}_{nm}^{[m]}}_{\boldsymbol{d}_n}$$

adapted to the filtration

$$\widetilde{\mathcal{F}}_{11}^{[m]} \subseteq \ldots \subseteq \widetilde{\mathcal{F}}_{1m}^{[m]} \subseteq \ldots \subseteq \widetilde{\mathcal{F}}_{k1}^{[m]} \subseteq \ldots \subseteq \widetilde{\mathcal{F}}_{km}^{[m]} \subseteq \ldots \subseteq \widetilde{\mathcal{F}}_{n1}^{[m]} \subseteq \ldots \subseteq \widetilde{\mathcal{F}}_{nm}^{[m]}.$$

Note that

$$\Big\| \sum_{k=1}^{n} \sum_{l=1}^{m} \boldsymbol{d}_{kl}^{[m]} \Big| L_2 \Big\| = \Big\| \sum_{k=1}^{n} \boldsymbol{d}_k \Big| L_2 \Big\| \quad \text{and} \quad \|\boldsymbol{d}_{kl}^{[m]} | L_2\| = \tfrac{1}{\sqrt{m}} \|\boldsymbol{d}_k | L_2\|.$$

Moreover, all differences $\boldsymbol{d}_{kl}^{[m]}$ in any block of length at most m have disjoint supports.

7.7.16 PROPOSITION. *The sequence* $(\boldsymbol{\varrho}^{eq}(\mathcal{M}_n, \mathcal{I}_n))$ *is non-decreasing.*

PROOF. Let $\boldsymbol{d}_1, \ldots, \boldsymbol{d}_n$ be X-valued martingale differences such that $\|\boldsymbol{d}_k | L_2\| = 1$. For $m := n+1$, the construction described in the previous paragraph yields $(n+1)n$ differences $\boldsymbol{d}_{kl}^{[n+1]}$. Define $\boldsymbol{b}_1, \ldots, \boldsymbol{b}_{n+1}$ by adding up n consecutive terms,

$$\boldsymbol{b}_h := \underbrace{\boldsymbol{d}_{h-1,n-h+3}^{[n+1]} + \ldots + \boldsymbol{d}_{h-1,n+1}^{[n+1]}}_{h-1} + \underbrace{\boldsymbol{d}_{h,1}^{[n+1]} + \ldots + \boldsymbol{d}_{h,n-h+1}^{[n+1]}}_{n-h+1}.$$

Then it follows from

$$\|\boldsymbol{b}_h | L_2\| = \sqrt{\tfrac{n}{n+1}}$$

that

$$\Big\| \sum_{k=1}^{n} [L_2, T] \boldsymbol{d}_k \Big| L_2 \Big\| = \Big\| \sum_{h=1}^{n+1} [L_2, T] \boldsymbol{b}_h \Big| L_2 \Big\| \le \boldsymbol{\varrho}^{eq}(T | \mathcal{M}_{n+1}, \mathcal{I}_{n+1}) \sqrt{n+1} \sqrt{\tfrac{n}{n+1}}.$$

7.7.17 The following doubling inequality is obvious.

PROPOSITION. $\quad \boldsymbol{\varrho}^{eq}(\mathcal{M}_{2n}, \mathcal{I}_{2n}) \le \sqrt{2}\, \boldsymbol{\varrho}^{eq}(\mathcal{M}_n, \mathcal{I}_n).$

7.7.18 We now establish a counterpart of 4.4.16.

THEOREM. *The sequences of the ideal norms $\varrho(\mathcal{M}_n, \mathcal{I}_n)$ and $\varrho^{eq}(\mathcal{M}_n, \mathcal{I}_n)$ are uniformly equivalent.*

PROOF. Let $\boldsymbol{d}_1, \ldots, \boldsymbol{d}_n$ be X-valued martingale differences. Without loss of generality, we may assume that

$$\sum_{k=1}^{n} \|\boldsymbol{d}_k|L_2\|^2 = 1.$$

Define

$$m := \max\{\, h : 4^h \le 16n \,\}, \quad \mathbb{F}_h := \{\, k : 2^{-h} < \|\boldsymbol{d}_k|L_2\| \le 2^{1-h} \,\}$$

and

$$\mathbb{F} := \bigcup_{h=1}^{m} \mathbb{F}_h.$$

Given $k \in \mathbb{F}$, there exists a unique number $h \le m$ such that $k \in \mathbb{F}_h$. Let

$$\mathbb{L}_{hi}^{[4^m]} := \{\, 4^h(i-1)+1, \ldots, 4^h i \,\} \quad \text{and} \quad N_k := 4^{m-h}.$$

Then

$$\bigcup_{i=1}^{N_k} \mathbb{L}_{hi}^{[4^m]} = \{1, \ldots, 4^m\}.$$

We now employ the construction described in 7.7.15 to create an X-valued martingale of length

$$N := \sum_{k \in \mathbb{F}} N_k < 4^m \sum_{k \in \mathbb{F}} \|\boldsymbol{d}_k|L_2\|^2 \le 4^m \le 16n < 4^{m+1}$$

by defining the differences

$$\boldsymbol{b}_{ki}^{[4^m]} := \sum_{l \in \mathbb{L}_{hi}^{[4^m]}} \boldsymbol{d}_{kl}^{[4^m]}$$

for $k \in \mathbb{F}$ and $i = 1, \ldots, N_k$. Since $|\mathbb{L}_{hi}^{[4^m]}| = 4^h$, we get

$$\|\boldsymbol{b}_{ki}^{[4^m]}|L_2\| = \Big\| \sum_{l \in \mathbb{L}_{hi}^{[4^m]}} \boldsymbol{d}_{kl}^{[4^m]} \Big| L_2 \Big\| = 2^{h-m} \|\boldsymbol{d}_k|L_2\| \le 2^{1-m}.$$

Hence, by 7.7.14 and $N \le 4^m$,

$$\begin{aligned} \Big\| \sum_{k \in \mathbb{F}} [L_2, T]\boldsymbol{d}_k \Big| L_2 \Big\| &= \Big\| \sum_{k \in \mathbb{F}} \sum_{i=1}^{N_k} [L_2, T]\boldsymbol{b}_{ki}^{[4^m]} \Big| L_2 \Big\| \\ &\le \varrho^{eq}(T|\mathcal{M}_N, \mathcal{I}_N)\sqrt{N}\, 2^{1-m} \\ &\le 2\, \varrho^{eq}(T|\mathcal{M}_N, \mathcal{I}_N). \end{aligned} \tag{1}$$

On the other hand, for the X-valued martingale constituted by the differences $\boldsymbol{d}_k$ with $k \notin \mathbb{F}$, it follows from

$$\sum_{k\notin\mathbb{F}} \|\boldsymbol{d}_k|L_2\|^2 \le 4^{-m} n < \tfrac{1}{4}$$

and

$$\Big\|\sum_{k\notin\mathbb{F}} [L_2,T]\boldsymbol{d}_k \Big| L_2\Big\| \le \varrho(T|\mathcal{M}_n,\mathcal{I}_n)\Big(\sum_{k\notin\mathbb{F}} \|\boldsymbol{d}_k|L_2\|^2\Big)^{1/2}$$

that

$$\Big\|\sum_{k\notin\mathbb{F}} [L_2,T]\boldsymbol{d}_k \Big| L_2\Big\| \le \tfrac{1}{2}\,\varrho(T|\mathcal{M}_n,\mathcal{I}_n). \tag{2}$$

Combining (1) and (2) yields

$$\begin{aligned}\Big\|\sum_{k=1}^{n} [L_2,T]\boldsymbol{d}_k \Big| L_2\Big\| &\le \\ &\le \Big\|\sum_{k\in\mathbb{F}} [L_2,T]\boldsymbol{d}_k \Big| L_2\Big\| + \Big\|\sum_{k\notin\mathbb{F}} [L_2,T]\boldsymbol{d}_k \Big| L_2\Big\| \\ &\le 2\,\varrho^{eq}(T|\mathcal{M}_N,\mathcal{I}_N) + \tfrac{1}{2}\,\varrho(T|\mathcal{M}_n,\mathcal{I}_n).\end{aligned}$$

Hence

$$\varrho(T|\mathcal{M}_n,\mathcal{I}_n) \le 2\,\varrho^{eq}(T|\mathcal{M}_N,\mathcal{I}_N) + \tfrac{1}{2}\,\varrho(T|\mathcal{M}_n,\mathcal{I}_n).$$

Finally, taking into account the estimate $N \le 16n$, the monotonicity of $\varrho^{eq}(\mathcal{M}_n,\mathcal{I}_n)$ and the doubling inequality 7.7.17, we arrive at

$$\varrho(T|\mathcal{M}_n,\mathcal{I}_n) \le 4\varrho^{eq}(T|\mathcal{M}_N,\mathcal{I}_N) \le 4\varrho^{eq}(T|\mathcal{M}_{16n},\mathcal{I}_{16n}) \le 16\varrho^{eq}(T|\mathcal{M}_n,\mathcal{I}_n).$$

The inequality

$$\varrho^{eq}(T|\mathcal{M}_n,\mathcal{I}_n) \le \varrho(T|\mathcal{M}_n,\mathcal{I}_n)$$

is obvious.

7.7.19 PROPOSITION. *$[l_2,T]$ factors the summation operator Σ_n with bound*

$$\frac{6\sqrt{n}}{\varrho^{eq}(T|\mathcal{M}_{2n},\mathcal{I}_{2n})}.$$

PROOF. Given $0 < \delta < 1$, we choose martingale differences $\boldsymbol{d}_1, \ldots, \boldsymbol{d}_{2n}$ adapted to a filtration $\mathcal{F}_1 \subseteq \ldots \subseteq \mathcal{F}_{2n}$ such that

$$\Big\|\sum_{i=1}^{2n} [L_2,T]\boldsymbol{d}_i \Big| L_2\Big\| \ge \delta\,\varrho^{eq}(T|\mathcal{M}_{2n},\mathcal{I}_{2n})\sqrt{2n} \quad\text{and}\quad \|\boldsymbol{d}_i|L_2\| \le 1.$$

In what follows, L_2 denotes the finite dimensional Hilbert space formed by all $\mathcal{F}_{2n}$-measurable functions on $[0,1)$. There exists $\boldsymbol{v}' \in [L_2, Y']$ with

$$\Big\| \sum_{i=1}^{2n} [L_2, T]\boldsymbol{d}_i \Big| L_2 \Big\| = \Big| \Big\langle \sum_{i=1}^{2n} [L_2, T]\boldsymbol{d}_i, \boldsymbol{v}' \Big\rangle \Big| \quad \text{and} \quad \|\boldsymbol{v}'|L_2\| = 1.$$

Let

$$\mathbb{F} := \Big\{ i = 1, \ldots, 2n : 4\sqrt{2n}\, |\langle [L_2, T]\boldsymbol{d}_i, \boldsymbol{v}'\rangle| \ge \delta\, \boldsymbol{\varrho}^{eq}(T|\mathfrak{M}_{2n}, \mathfrak{I}_{2n}) \Big\}$$

and $m := |\mathbb{F}| \le 2n$. Since $\boldsymbol{\varrho}^{eq}(T|\mathfrak{M}_m, \mathfrak{I}_m) \le \boldsymbol{\varrho}^{eq}(T|\mathfrak{M}_{2n}, \mathfrak{I}_{2n})$, we obtain from

$$\begin{aligned}
\delta\, \boldsymbol{\varrho}^{eq}(T|\mathfrak{M}_{2n}, \mathfrak{I}_{2n})\sqrt{2n} &\le \Big| \Big\langle \sum_{i=1}^{2n} [L_2, T]\boldsymbol{d}_i, \boldsymbol{v}' \Big\rangle \Big| \\
&\le \Big| \Big\langle \sum_{i\in\mathbb{F}} [L_2, T]\boldsymbol{d}_i, \boldsymbol{v}' \Big\rangle \Big| + \Big| \Big\langle \sum_{i\notin\mathbb{F}} [L_2, T]\boldsymbol{d}_i, \boldsymbol{v}' \Big\rangle \Big| \\
&\le \boldsymbol{\varrho}^{eq}(T|\mathfrak{M}_m, \mathfrak{I}_m)\sqrt{m} + \tfrac{1}{4}\delta\, \boldsymbol{\varrho}^{eq}(T|\mathfrak{M}_{2n}, \mathfrak{I}_{2n})\sqrt{2n}
\end{aligned}$$

that $\frac{3}{4}\delta\sqrt{2n} \le \sqrt{m}$. Setting $\delta := \frac{2}{3}\sqrt{2}$ yields $n \le m$. So there are $i_1 < \ldots < i_n$ in $\mathbb{F}$. Define $\boldsymbol{u}_k \in [L_2, X]$ and $\boldsymbol{v}_h \in [L_2, Y']$ by

$$\boldsymbol{u}_k := \frac{\boldsymbol{d}_{i_k}}{\langle [L_2, T]\boldsymbol{d}_{i_k}, \boldsymbol{v}'\rangle} \quad \text{and} \quad \boldsymbol{v}'_h := [E(\mathcal{F}_{i_h}), Y']\boldsymbol{v}'.$$

Then

$$\|\boldsymbol{u}_k|L_2\| \le c_n := \frac{6\sqrt{n}}{\boldsymbol{\varrho}^{eq}(T|\mathfrak{M}_{2n}, \mathfrak{I}_{2n})} \quad \text{and} \quad \|\boldsymbol{v}'_h|L_2\| \le 1.$$

Moreover,

$$\begin{aligned}
\langle [L_2, T]\boldsymbol{u}_k, \boldsymbol{v}'_h\rangle &= \frac{\langle [L_2, T]\boldsymbol{d}_{i_k}, [E(\mathcal{F}_{i_h}), Y']\boldsymbol{v}'\rangle}{\langle [L_2, T]\boldsymbol{d}_{i_k}, \boldsymbol{v}'\rangle} \\
&= \frac{\langle [L_2, T][E(\mathcal{F}_{i_h}), X]\boldsymbol{d}_{i_k}, \boldsymbol{v}'\rangle}{\langle [L_2, T]\boldsymbol{d}_{i_k}, \boldsymbol{v}'\rangle} = \begin{cases} 1 & \text{if } k \le h, \\ 0 & \text{if } k > h. \end{cases}
\end{aligned}$$

Thus $[L_2, T]$ factors Σ_n with bound c_n. This completes the proof, since $L_2 = L_2([0,1), \mathcal{F}_{2n})$ can be identified with a finite section of l_2, in a canonical way.

7.7.20 Finally, we combine the preceding results with 7.6.11.

PROPOSITION. *If $[l_2, T]$ is super weakly compact, then T has martingale subtype.*

REMARK. In order to get the inclusion $\mathfrak{SR} \subseteq \mathfrak{MT}$, we need to show that $T \in \mathfrak{SR}$ implies $[l_2, T] \in \mathfrak{SR}$. This will be done in 7.10.24.

7.8 J-convexity – James's theorem

7.8.1 The finite **James operator** J_n is defined by

$$J_n : (\xi_k) \longrightarrow \Big(\sum_{k \le h} \xi_k - \sum_{k > h} \xi_k \Big).$$

The associated (n, n)-matrix has the form

$$J_n = (\iota_{hk}), \quad \text{where} \quad \iota_{hk} := \begin{cases} +1 & \text{if } k \le h, \\ -1 & \text{if } k > h, \end{cases} \quad \text{and} \quad h, k = 1, \ldots, n.$$

Mostly, J_n will be viewed as an operator from l_1^n into l_∞^n. In this case,

$$\| J_n : l_1^n \to l_\infty^n \| = 1.$$

7.8.2 We say that $T \in \mathfrak{L}(X, Y)$ **factors the James operator** J_n with bound $c \ge 1$ if there exist operators $A_n \in \mathfrak{L}(l_1^n, X)$ and $B_n \in \mathfrak{L}(Y, l_\infty^n)$ such that

$$J_n = B_n T A_n \quad \text{and} \quad \|B_n\| \, \|A_n\| \le c.$$

Writing

$$A_n(\xi_k) = \sum_{k=1}^{n} \xi_k x_k \quad \text{and} \quad B_n y = (\langle y, y_h' \rangle),$$

it follows from $\langle T x_k, y_h' \rangle = \langle J_n u_k^{(n)}, u_h^{(n)} \rangle$ and

$$J_n : (\underbrace{0, \ldots, 0}_{k-1}, 1, 0, \ldots, 0) \longrightarrow (\underbrace{-1, \ldots, -1}_{k-1}, +1, \ldots, +1),$$

that the above condition is equivalent to the existence of elements $x_1, \ldots, x_n$ in X and functionals $y_1', \ldots, y_n'$ in Y' such that

$$\|x_k\| \le c, \quad \|y_h'\| \le 1 \quad \text{and} \quad \langle T x_k, x_h' \rangle = \begin{cases} +1 \text{ if } k \le h, \\ -1 \text{ if } k > h. \end{cases} \qquad (J_n)$$

An operator **factors the James operators** J_n **uniformly** if the above (equivalent) conditions are fulfilled for some constant $c \ge 1$ and all $n = 1, 2, \ldots$, simultaneously.

This terminology will also be used in the setting of spaces.

7.8.3 LEMMA. *There exists* $P_n \in \mathfrak{L}(l_1)$ *such that*

$$2\Sigma_n = J_n(I_n + P_n), \quad J_n = \Sigma_n(2I_n - P_n) \quad \textit{and} \quad \|P_n : l_1 \to l_1\| = 1.$$

PROOF. Use the projection P_n induced by the (n, n)-matrix with entries equal to 1 in the first row and 0 otherwise.

REMARK. If an operator factors J_n, then it factors Σ_n with the same bound $c \geq 1$. We also have an implication in the reverse direction. But the bound may get worse. An example is given by the 2-dimensional real Banach space with the regular hexagon as its unit ball; see 2.2.15.

7.8.4 It follows from

$$\|J_n\| := \|(\iota_{hk}) : l_2^n \to l_2^n\| \asymp n \quad \text{and} \quad |||J_n||| := \|(|\iota_{hk}|) : l_2^n \to l_2^n\| = n$$

that $\boldsymbol{\kappa}(T|J_n) \asymp n$ whenever $T \neq O$; see 2.3.2. Thus the Kwapień ideal norms $\boldsymbol{\kappa}(J_n)$ do not give rise to any non-trivial operator ideal. Nevertheless, we are going to show that the limiting case $\boldsymbol{\kappa}(X|J_n) = n$ is of considerable interest.

PROPOSITION. *For every Banach space X and fixed $n \geq 2$, the following are equivalent:*

(1) $\boldsymbol{\kappa}(X|J_n) = n$.

(2) *X factors the James operator J_n with all bounds $c > 1$.*

PROOF. **(1)⇒(2)**: For $\delta > 0$, there exist $x_1^{(\delta)}, \ldots, x_n^{(\delta)} \in X$ such that

$$\Big(\sum_{h=1}^{n} \Big\| \sum_{k \leq h} x_k^{(\delta)} - \sum_{k > h} x_k^{(\delta)} \Big\|^2 \Big)^{1/2} \geq \frac{n}{1+\delta} \Big(\sum_{k=1}^{n} \|x_k^{(\delta)}\|^2 \Big)^{1/2}.$$

By scaling, it can be arranged that

$$\sum_{h=1}^{n} \Big\| \sum_{k \leq h} x_k^{(\delta)} - \sum_{k > h} x_k^{(\delta)} \Big\|^2 \geq \frac{n^3}{(1+\delta)^2} \quad \text{and} \quad \sum_{k=1}^{n} \|x_k^{(\delta)}\|^2 = n.$$

We consider an ultrafilter $\mathcal{U}$ on $\mathbb{R}_+$ containing all intervals $(0, \delta)$ with $\delta > 0$. Passing to the equivalence classes $\boldsymbol{x}_k := (x_k^{(\delta)})^{\mathcal{U}}$ in the ultrapower $X^{\mathcal{U}}$ yields

$$\sum_{h=1}^{n} \Big\| \sum_{k \leq h} \boldsymbol{x}_k - \sum_{k > h} \boldsymbol{x}_k \Big\|^2 \geq n^3 \quad \text{and} \quad \sum_{k=1}^{n} \|\boldsymbol{x}_k\|^2 = n.$$

From

$$\Big\| \sum_{k \leq h} \boldsymbol{x}_k - \sum_{k > h} \boldsymbol{x}_k \Big\| \leq \sum_{k=1}^{n} \|\boldsymbol{x}_k\| \leq \sqrt{n} \Big(\sum_{k=1}^{n} \|\boldsymbol{x}_k\|^2 \Big)^{1/2} \qquad (*)$$

we obtain

$$n^3 \leq \sum_{h=1}^{n} \Big\| \sum_{k \leq h} \boldsymbol{x}_k - \sum_{k > h} \boldsymbol{x}_k \Big\|^2 \leq n \Big(\sum_{k=1}^{n} \|\boldsymbol{x}_k\| \Big)^2 \leq n^2 \sum_{k=1}^{n} \|\boldsymbol{x}_k\|^2 = n^3.$$

Therefore equality holds in $(*)$. In view of 0.4.6, this implies that

$$\|\boldsymbol{x}_1\| = \ldots = \|\boldsymbol{x}_n\| = 1 \quad \text{and} \quad \Big\| \sum_{k \leq h} \boldsymbol{x}_k - \sum_{k > h} \boldsymbol{x}_k \Big\| = n \quad \text{for } h = 1, \ldots, n.$$

The Hahn–Banach theorem yields $\boldsymbol{x}'_1, \ldots, \boldsymbol{x}'_n \in (X^{\mathcal{U}})'$ such that

$$\Big\langle \sum_{k\le h} \boldsymbol{x}_k - \sum_{k>h} \boldsymbol{x}_k, \boldsymbol{x}'_h \Big\rangle = n \quad \text{and} \quad \|\boldsymbol{x}'_h\| = 1.$$

We now conclude from

$$\sum_{k\le h} \langle \boldsymbol{x}_k, \boldsymbol{x}'_h \rangle - \sum_{k>h} \langle \boldsymbol{x}_k, \boldsymbol{x}'_h \rangle = n \quad \text{and} \quad |\langle \boldsymbol{x}_k, \boldsymbol{x}'_h \rangle| \le 1$$

that

$$\langle \boldsymbol{x}_k, \boldsymbol{x}'_h \rangle = \begin{cases} +1 & \text{if } k \le h, \\ -1 & \text{if } k > h. \end{cases}$$

So $X^{\mathcal{U}}$ factors the James operator J_n with all bounds $c > 1$. By an analogue of Lemma 7.7.10, this property carries over to X.

(2)⇒(1): For $c > 1$, there exist $x_1, \ldots, x_n \in X$ and $x'_1, \ldots, x'_n \in X'$ such that

$$\|x_k\| \le c, \quad \|x'_h\| \le 1 \quad \text{and} \quad \langle x_k, x'_h \rangle = \begin{cases} +1 \text{ if } k \le h, \\ -1 \text{ if } k > h. \end{cases} \tag{J_n}$$

Now it follows from

$$n = \Big\langle \sum_{k\le h} x_k - \sum_{k>h} x_k, x'_h \Big\rangle \le \Big\| \sum_{k\le h} x_k - \sum_{k>h} x_k \Big\|$$

that

$$n^3 \le \sum_{h=1}^{n} \Big\| \sum_{k\le h} x_k - \sum_{k>h} x_k \Big\|^2 \le \boldsymbol{\kappa}(X|J_n)^2 \sum_{k=1}^{n} \|x_k\|^2 \le \boldsymbol{\kappa}(X|J_n)^2 n c^2.$$

Letting $c \to 1$ yields $n \le \boldsymbol{\kappa}(X|J_n)$.

7.8.5 For every finite set $\mathbb{H}$ and $m \le |\mathbb{H}|$, let $\mathbb{H}^{[m]}$ denote the collection of all subsets in $\mathbb{H}$ with m elements.

In what follows, we need a celebrated result from combinatorics; see [GRA*, pp. 7–9].

RAMSEY'S THEOREM. *For $n \ge m$ and every finite set $\mathbb{I}$, there exists a number $R(n, m, |\mathbb{I}|)$ such that the following holds whenever $|\mathbb{H}| \ge R(n, m, |\mathbb{I}|)$:*
Given any function $f : \mathbb{H}^{[m]} \to \mathbb{I}$, we can find a subset $\mathbb{H}_0$ in $\mathbb{H}$ with $|\mathbb{H}_0| = n$ and an element $i_0 \in \mathbb{I}$ such that

$$f(\mathbb{F}) = i_0 \quad \textit{for all} \quad \mathbb{F} \in \mathbb{H}_0^{[m]}.$$

7.8.6 The symbol $\{H_1,\dots,H_{2m}\}_<$ means that $H_1 < \dots < H_{2m}$. In what follows, subsets $\mathbb{F}$ of $\mathbb{N}$ with $|\mathbb{F}| = 2m$ will be written in this form.

Fix $x_1,\dots,x_N \in X$ and $x'_1,\dots,x'_N \in X'$ such that

$$\|x_K\| \le c, \quad \|x'_H\| \le 1 \quad \text{and} \quad \langle x_K, x'_H\rangle = \begin{cases} 1 \text{ if } K \le H, \\ 0 \text{ if } K > H, \end{cases} \qquad (\Sigma_N)$$

where $c > 1$ is a constant. For $p = 1,\dots,m$, we now define

$$S_p(\mathbb{F}) := \Big\{ x \in X : \langle x, x'_H\rangle = (-1)^i \text{ if } H_{2i-1} \le H < H_{2i} \text{ and } i = 1,\dots,p \Big\}$$

and

$$\sigma_p(\mathbb{F}) := \inf\{\, \|x\| : x \in S_p(\mathbb{F}) \,\}.$$

LEMMA. $1 \le \sigma_1(\mathbb{F}) \le \dots \le \sigma_m(\mathbb{F}) \le 2mc.$

PROOF. Note that

$$x^\circ := \sum_{j=1}^{m} (-1)^j (x_{H_{2j-1}} - x_{H_{2j}}) \in S_p(\mathbb{F}).$$

Indeed, if $H_{2i-1} \le H < H_{2i}$, then

$$\langle x^\circ, x'_H\rangle = \sum_{j<i} (-1)^j (1-1) + (-1)^i (1-0) + \sum_{j>i} (-1)^j (0-0) = (-1)^i.$$

Since $\|x^\circ\| \le 2mc$, we obtain $\sigma_m(\mathbb{F}) \le 2mc$. On the other hand, it follows from

$$1 = \big|\langle x, x'_{H_1}\rangle\big| \le \|x\| \quad \text{for } x \in S_p(\mathbb{F})$$

that $1 \le \sigma_p(\mathbb{F})$.

7.8.7 First of all, we produce an $(m+1, 2n)$-matrix by splitting the set $\{1,\dots,2(m+1)n\}$ into successive parts of length $2n$. Letting

$$c_{2i}^{(h)} := 2in + 2h - 1 \quad \text{and} \quad c_{2i+1}^{(h)} := 2in + 2h$$

yields

$c_0^{(1)}$	$c_1^{(1)}$	$c_0^{(2)}$	$c_1^{(2)}$	$\cdots$	$\cdots$	$c_0^{(n)}$	$c_1^{(n)}$
$c_2^{(1)}$	$c_3^{(1)}$	$c_2^{(2)}$	$c_3^{(2)}$	$\cdots$	$\cdots$	$c_2^{(n)}$	$c_3^{(n)}$
$\vdots$	$\vdots$	$\vdots$	$\vdots$			$\vdots$	$\vdots$
$c_{2i-2}^{(1)}$	$c_{2i-1}^{(1)}$	$c_{2i-2}^{(2)}$	$c_{2i-1}^{(2)}$	$\cdots$	$\cdots$	$c_{2i-2}^{(n)}$	$c_{2i-1}^{(n)}$
$c_{2i}^{(1)}$	$c_{2i+1}^{(1)}$	$c_{2i}^{(2)}$	$c_{2i+1}^{(2)}$	$\cdots$	$\cdots$	$c_{2i}^{(n)}$	$c_{2i+1}^{(n)}$
$\vdots$	$\vdots$	$\vdots$	$\vdots$			$\vdots$	$\vdots$
$c_{2m-2}^{(1)}$	$c_{2m-1}^{(1)}$	$c_{2m-2}^{(2)}$	$c_{2m-1}^{(2)}$	$\cdots$	$\cdots$	$c_{2m-2}^{(n)}$	$c_{2m-1}^{(n)}$
$c_{2m}^{(1)}$	$c_{2m+1}^{(1)}$	$c_{2m}^{(2)}$	$c_{2m+1}^{(2)}$	$\cdots$	$\cdots$	$c_{2m}^{(n)}$	$c_{2m+1}^{(n)}$

Finally, we form the sets

$$\mathbb{A}^{(k)} := \{c_1^{(k)}, c_2^{(k)}, \dots, c_{2m-1}^{(k)}, c_{2m}^{(k)}\}_<$$

and

$$\mathbb{B}^{(h)} := \{c_3^{(h)}, c_2^{(h+1)}, c_5^{(h)}, c_4^{(h+1)}, \dots, c_{2m-1}^{(h)}, c_{2m-2}^{(h+1)}\}_<,$$

which contain $2m$ and $2m-2$ elements, respectively. Obviously, we have $\mathbb{A}^{(k+1)} = \mathbb{A}^{(k)} + 2$ and $\mathbb{B}^{(h+1)} = \mathbb{B}^{(h)} + 2$. The reader is encouraged to locate $\mathbb{A}^{(1)}$ and $\mathbb{B}^{(1)}$ in the above scheme. Note that

$$c_{2i}^{(h+1)} = c_{2i+1}^{(h)} + 1 \quad \text{and} \quad c_j^{(h+n)} = c_{j+2}^{(h)}.$$

The following inequalities can easily be checked:

$$\begin{array}{ll} c_{2i+1}^{(k)} \le c_{2i+1}^{(h)} < c_{2i}^{(h+1)} \le c_{2i+2}^{(k)} & \text{if } 1 \le k \le h \le n, \\ c_{2i-1}^{(k)} \le c_{2i+1}^{(h)} < c_{2i}^{(h+1)} \le c_{2i}^{(k)} & \text{if } 1 \le h < k \le n. \end{array} \qquad (*)$$

7.8.8 We are now prepared to prove a vital result.

PROPOSITION. *If a Banach space X factors the summation operators Σ_n uniformly for* **some** *bound $c_0 \ge 1$, then it factors the James operators J_n uniformly for* **all** *bounds $c > 1$.*

PROOF. Let N be a huge natural number which will be specified later, with the help of Ramsey's theorem 7.8.5. Choose $x_1, , \dots, x_N \in X$ and $x'_1, \dots, x'_N \in X'$ such that

$$\|x_K\| \le c_0, \quad \|x'_H\| \le 1 \quad \text{and} \quad \langle x_K, x'_H \rangle = \begin{cases} 1 \text{ if } K \le H, \\ 0 \text{ if } K < H. \end{cases} \qquad (\Sigma_N)$$

Given $\varepsilon > 0$, we fix m so large that

$$2mc_0 < (1+\varepsilon)^{m-1}. \qquad (1)$$

Of course, the intervals

$$I_q := \big[(1+\varepsilon)^{q-1}, (1+\varepsilon)^q\big) \quad \text{with } q = 1, \dots, m-1$$

cover $[1, 2mc_0]$.

Let $\mathbb{F}$ be any subset of $\{1, \dots, N\}$ with $|\mathbb{F}| = 2m$. Then we know from 7.8.6 that $1 \le \sigma_1(\mathbb{F}) \le \dots \le \sigma_m(\mathbb{F}) \le 2mc_0$. Thus, by the pigeon-hole principle, there exists $p \in \{2, \dots, m\}$ such that $\sigma_{p-1}(\mathbb{F})$ and $\sigma_p(\mathbb{F})$ belong to one and the same interval I_q. Define $f(\mathbb{F})$ as the smallest p with this property. Applying Ramsey's theorem to the function

$$f : \{1, \dots, N\}^{[2m]} \longrightarrow \{2, \dots, m\}$$

gives a subset $\mathbb{P}$ of $\{1,\ldots,N\}$ and an index $p_0 \in \{2,\ldots,m\}$ such that

$$f(\mathbb{F}) = p_0 \quad \text{for all } \mathbb{F} \in \mathbb{P}^{[2m]}.$$

Next, for $\mathbb{F} \in \mathbb{P}^{[2m]}$, we denote by $g(\mathbb{F})$ the unique index $q \in \{1,\ldots,m-1\}$ with $\sigma_{p_0}(\mathbb{F}) \in I_q$. Now Ramsey's theorem applied to

$$g : \mathbb{P}^{[2m]} \longrightarrow \{1,\ldots,m-1\}$$

yields a subset $\mathbb{Q}$ of $\mathbb{P}$ and an index $q_0 \in \{1,\ldots,m-1\}$ such that

$$\sigma_{p_0-1}(\mathbb{F}) \in I_{q_0} \quad \text{and} \quad \sigma_{p_0}(\mathbb{F}) \in I_{q_0} \quad \text{for all } \mathbb{F} \in \mathbb{Q}^{[2m]}. \tag{2}$$

Moreover, given m and n, the number N (which has not been specified yet) can be chosen so large that we achieve $|\mathbb{Q}| = 2(m+1)n$.

Let φ denote the monotone mapping from $\{1,\ldots,2(m+1)n\}$ onto $\mathbb{Q}$. With the help of this correspondence, the construction performed in 7.8.7 carries over to $\mathbb{Q}$.

Pick $x_k^\circ \in S_{p_0}(\varphi(\mathbb{A}^{(k)}))$ such that

$$\|x_k^\circ\| \le (1+\varepsilon)\sigma_{p_0}(\varphi(\mathbb{A}^{(k)})).$$

Since $\varphi(\mathbb{A}^{(k)}) \in \mathbb{Q}^{[2m]}$, we conclude from (2) that

$$\|x_k^\circ\| \le (1+\varepsilon)^{q_0+1}. \tag{3}$$

Fix $h = 1,\ldots,n$, and let

$$\varphi(c_{2i+1}^{(h)}) \le H < \varphi(c_{2i}^{(h+1)}).$$

Then $(*)$ in 7.8.7 tells us that

$$\varphi(c_{2i+1}^{(k)}) \le H < \varphi(c_{2i+2}^{(k)}) \text{ if } k \le h \quad \text{and} \quad \varphi(c_{2i-1}^{(k)}) \le H < \varphi(c_{2i}^{(k)}) \text{ if } k > h.$$

Hence

$$\langle x_k^\circ, x_H' \rangle = (-1)^{i+1} \quad \text{if } k \le h \quad \text{and} \quad \langle x_k^\circ, x_H' \rangle = (-1)^i \quad \text{if } k > h.$$

Therefore we have

$$\frac{1}{n}\Big\langle \sum_{k\le h} x_k^\circ - \sum_{k>h} x_k^\circ, x_H' \Big\rangle = (-1)^{i+1},$$

which implies that

$$-\frac{1}{n}\Big(\sum_{k\le h} x_k^\circ - \sum_{k>h} x_k^\circ \Big) \in S_m(\varphi(\mathbb{B}^{(h)})) \subseteq S_{p_0-1}(\varphi(\mathbb{B}^{(h)})).$$

Note that $\varphi(\mathbb{B}^{(h)}) \in \mathbb{Q}^{[2m-2]}$. But the use of (2) requires a set in $\mathbb{Q}^{[2m]}$. To satisfy this formal condition, assuming $n \ge 2$, we add to $\mathbb{B}^{(h)}$ two more elements, namely $2nm+2n-1$ and $2nm+2n$. If $\widetilde{\mathbb{B}}^{(h)}$ denotes this new set, then $S_{p_0-1}(\varphi(\mathbb{B}^{(h)})) = S_{p_0-1}(\varphi(\widetilde{\mathbb{B}}^{(h)}))$. Thus

$$\frac{1}{n}\Big\|\sum_{k\leq h} x_k^\circ - \sum_{k>h} x_k^\circ\Big\| \geq (1+\varepsilon)^{q_0-1}. \tag{4}$$

Finally, in view of (3) and (4), it follows from

$$\bigg(\sum_{h=1}^{n}\Big\|\sum_{k\leq h} x_k^\circ - \sum_{k>h} x_k^\circ\Big\|^2\bigg)^{1/2} \leq \boldsymbol{\kappa}(X|J_n)\bigg(\sum_{k=1}^{n}\|x_k^\circ\|^2\bigg)^{1/2}$$

that

$$(1+\varepsilon)^{q_0-1} n \leq (1+\varepsilon)^{q_0+1}\boldsymbol{\kappa}(X|J_n).$$

Letting $\varepsilon \to 0$ yields $\boldsymbol{\kappa}(X|J_n) = n$, which completes the proof, by 7.8.4.

REMARK. The above considerations show the existence of a function $N = f(n, c, c_0)$ such that the following holds:
If X factors Σ_N with bound $c_0 > 1$, then it factors J_n with bound $c > 1$.
No reasonable explicit expression for f seems to be known.

7.8.9 We now summarize the preceding results.

PROPOSITION. *For every Banach space X, the following are equivalent:*

(1) *X factors the summation operators Σ_n uniformly for* **some** *bound $c \geq 1$.*
(2) *X factors the summation operators Σ_n uniformly for* **all** *bounds $c > 1$.*
(3) *X factors the James operators J_n uniformly for* **some** *bound $c \geq 1$.*
(4) *X factors the James operators J_n uniformly for* **all** *bounds $c > 1$.*

PROOF. **(2)⇒(1)** and **(4)⇒(3)** are trivial, **(3)⇒(1)** and **(4)⇒(2)** follow from 7.8.3, and **(1)⇒(4)** was proved in the previous paragraph.

REMARK. The implication **(1)⇒(2)** can be deduced more directly. Indeed, by 7.6.10, it follows from **(1)** that some ultrapower $X^{\mathcal{U}}$ factors the infinite summation operator. Thus $X^{\mathcal{U}}$ is non-reflexive. Taking into account that $\|I_X^\pi\| = 1$, we may apply 7.6.5. This yields uniform factorizations of Σ_n with bound $1+\varepsilon$, in view of 0.8.3 and 7.7.10.

7.8.10 Fix any $n \geq 2$. A Banach space X is said to be **$\mathbf{J}_n$-convex** if there exists $0 < \varepsilon < 1$ such that, whenever $\|x_1\| = \ldots = \|x_n\| = 1$, at least one of the absolutely convex combinations

$$\frac{1}{n}(x_1 + \ldots + x_h - x_{h+1} - \ldots - x_n) \quad \text{with } h = 1, \ldots, n$$

has norm less than $1-\varepsilon$.

The class of all J_n-convex Banach spaces is denoted by J_n, and Banach spaces belonging to $\mathsf{J} := \bigcup_{n=2}^{\infty} \mathsf{J}_n$ are called **J-convex**.

7.8.11 PROPOSITION. *For every Banach space X, the following are equivalent:*

(1) $\kappa(X|J_n) = n$ *for* $n = 1, 2, \dots$.

(2) *X fails to be* J*-convex.*

PROOF. **(1)⇒(2)**: As in the proof of 7.8.4, we find elements $\boldsymbol{x}_1, \dots, \boldsymbol{x}_n$ in some ultrapower $X^{\mathcal{U}}$ such that

$$\|\boldsymbol{x}_1\| = \dots = \|\boldsymbol{x}_n\| = 1 \quad \text{and} \quad \Big\| \sum_{k \le h} \boldsymbol{x}_k - \sum_{k > h} \boldsymbol{x}_k \Big\| = n \quad \text{for } h = 1, \dots, n.$$

To justify the following notation, we recall that $\mathcal{U}$ was chosen to be an ultrafilter on $\mathbb{R}_+$ containing all intervals $(0, \delta)$ with $\delta > 0$. By 0.8.2, every equivalence class $\boldsymbol{x}_k$ can be represented in the form $\boldsymbol{x}_k = (x_k^{(\delta)})^{\mathcal{U}}$ with $\|x_k^{(\delta)}\| = 1$ for $\delta > 0$. Now it follows from

$$\mathcal{U}\text{-}\lim_{\delta} \Big\| \sum_{k \le h} x_k^{(\delta)} - \sum_{k > h} x_k^{(\delta)} \Big\| = n$$

that X cannot be J-convex.

The reverse implication **(2)⇒(1)** is trivial.

7.8.12 Summarizing the preceding results, we now state a fundamental criterion of superreflexivity; see [jam 3], [jam 4] and [jam 5].

JAMES'S THEOREM. *For every Banach space X, the following are equivalent:*

(1) *X is superreflexive.*

(2) *X does not factor the summation operators Σ_n uniformly.*

(3) *X does not factor the James operators J_n uniformly.*

(4) *X is* J*-convex.*

(5) *All Banach spaces finitely representable in X are reflexive.*

PROOF. The implications **(1)⇔(2)⇔(3)** follow from 7.6.11 and 7.8.9.

(3)⇔(4) can be deduced from 7.8.4 and 7.8.11.

(1)⇔(5) is a consequence of 0.8.3 and the fact that subspaces of reflexive spaces are reflexive.

7.8.13 PROPOSITION. *Superreflexivity is an isomorphic invariant.*

PROOF. Note that conditions **(2)** and **(3)** in the previous theorem are invariant under isomorphisms.

7.8.14 PROPOSITION. *The class of superreflexive spaces is stable when passing to subspaces, quotients and duals.*

PROOF. The invariance under formation of subspaces follows directly from the fact that the relation of being finitely representable is transitive.

We now assume that X' fails to be superreflexive. Then there are $x'_1, \dots, x'_n \in X'$ and $x''_1, \dots, x''_n \in X''$ such that

$$\|x'_k\| \leq c, \quad \|x''_h\| \leq 1 \quad \text{and} \quad \langle x''_h, x'_k \rangle = \begin{cases} 1 & \text{if } k \leq h, \\ 0 & \text{if } k > h. \end{cases} \qquad (\Sigma_n)$$

Applying Helly's lemma 0.1.8 to $x''_h \in X''$ and $x'_1, \dots, x'_n \in X'$ yields elements $x_h \in X$ with $\langle x_h, x'_k \rangle = \langle x''_h, x'_k \rangle$ for $h, k = 1, \dots, n$. Hence X also fails to be superreflexive. Thus $X \in \mathsf{SR}$ implies $X' \in \mathsf{SR}$. The reverse is now obvious by viewing X as a subspace of X''.

Quotient spaces can be treated by duality.

7.8.15 PROPOSITION. *If a Banach space X is superreflexive, then so is $[l_2, X]$.*

PROOF. Without the Kwapień ideal norm $\boldsymbol{\kappa}(J_n)$, which was introduced just for the purpose of this proof, our task would be very hard. But now the implication is a trivial consequence of l_2-stability; see 2.3.8.

REMARK. We stress that superreflexivity is inherited from X by all Banach spaces $[L_r(M, \mu), X]$ with $1 < r < \infty$. Indeed, it follows from 7.10.26 that X admits a uniformly convex renorming. Next, by a classical and simple result, $[L_r(M, \mu), X]$ is uniformly convex with respect to this new norm; see 7.10.23. Thus the fact that $X \in \mathsf{SR}$ implies $[l_2, X] \in \mathsf{SR}$ becomes elementary in view of Enflo's theorem 7.10.26. However, we may also go the converse way. That is, Enflo's theorem follows from 7.8.16, 7.7.8, 7.9.22 and 7.10.7.

7.8.16 We are now prepared to establish a fundamental result.

THEOREM. $\mathsf{SR} = \mathsf{AT} = \mathsf{AC}$.

PROOF. We know from 7.5.21 and 7.6.15 that $\mathsf{AT} = \mathsf{AC} \subseteq \mathsf{SR}$. On the other hand, it follows from 7.8.15 that $X \in \mathsf{SR}$ implies $[l_2, X] \in \mathsf{SR}$. Hence $X \in \mathsf{AT}$, by 7.7.12.

7.9 Uniform q-convexity and uniform p-smoothness

7.9.1 To begin with, we prove **Clarkson's inequalities.**

LEMMA. *Let* $\alpha, \beta \in \mathbb{K}$. *Then*

$$\frac{|\alpha+\beta|^p + |\alpha-\beta|^p}{2} \le |\alpha|^p + |\beta|^p \quad \textit{if } 1 \le p \le 2$$

and

$$|\alpha|^q + |\beta|^q \le \frac{|\alpha+\beta|^q + |\alpha-\beta|^q}{2} \quad \textit{if } 2 \le q < \infty.$$

PROOF. Obviously,

$$\left(\frac{|\alpha+\beta|^p + |\alpha-\beta|^p}{2}\right)^{1/p} \le \left(\frac{|\alpha+\beta|^2 + |\alpha-\beta|^2}{2}\right)^{1/2}$$
$$= \left(|\alpha|^2 + |\beta|^2\right)^{1/2} \le \left(|\alpha|^p + |\beta|^p\right)^{1/p},$$

and

$$\left(|\alpha|^q + |\beta|^q\right)^{1/q} \le \left(|\alpha|^2 + |\beta|^2\right)^{1/2}$$
$$= \left(\frac{|\alpha+\beta|^2 + |\alpha-\beta|^2}{2}\right)^{1/2} \le \left(\frac{|\alpha+\beta|^q + |\alpha-\beta|^q}{2}\right)^{1/q}.$$

7.9.2 Let $2 \le q < \infty$. An operator $T \in \mathfrak{L}(X, Y)$ is called **uniformly q-convex** if there exists a constant $c \ge 0$ such that

$$\left\|\frac{Tx_+ - Tx_-}{2}\right\| \le c\left(\frac{\|x_+\|^q + \|x_-\|^q}{2} - \left\|\frac{x_+ + x_-}{2}\right\|^q\right)^{1/q} \qquad (\mathrm{UC}_q)$$

for all $x_\pm \in X$. Substituting $x_\pm := x_0 \pm x$ in (UC_q) yields the condition

$$\|Tx\| \le c\left(\frac{\|x_0 + x\|^q + \|x_0 - x\|^q}{2} - \|x_0\|^q\right)^{1/q}$$

for all $x, x_0 \in X$, which is sometimes more comfortable to handle. We let

$$\|T|\mathfrak{UC}_q\| := \inf c,$$

where the infimum is taken over all $c \ge 0$ with the above property. The class of uniformly q-convex operators is denoted by $\mathfrak{UC}_q$.

7.9.3 The following terminology should be self-explanatory, and the proof is straightforward.

THEOREM. *$\mathfrak{UC}_q$ is a left-sided Banach operator ideal.*

REMARK. Letting $x_\pm = \pm x$ in (UC_q) yields that $\|T\| \le \|T|\mathfrak{UC}_q\|$. Moreover, we obtain from 7.9.1 that $\|I_{\mathbb{K}}|\mathfrak{UC}_q\| = 1$. Note that $\mathfrak{UC}_q$ cannot be a two-sided ideal, since the identity map of l_∞^2 fails to be uniformly q-convex. This follows by substituting $x_\pm = (1, \pm 1)$ into (UC_q).

7.9.4 Let $1 < p \le 2$. An operator $T \in \mathfrak{L}(X,Y)$ is called **uniformly p-smooth** if there exists a constant $c \ge 0$ such that

$$\left(\frac{\|y+Tx\|^p + \|y-Tx\|^p}{2} - \|y\|^p \right)^{1/p} \le c\|x\| \qquad (\mathrm{UT}_p)$$

for all $x \in X$ and $y \in Y$. We let

$$\|T|\mathfrak{UT}_p\| := \inf c,$$

where the infimum is taken over all $c \ge 0$ with the above property. The class of uniformly p-smooth operators is denoted by $\mathfrak{UT}_p$.

REMARKS. The concept of uniform p-smoothness makes also sense for p=1. In this case, however, we get $\|T|\mathfrak{UT}_1\| = \|T\|$ for all operators T. The reason for the choice of the letter $\mathfrak{T}$ in the symbol $\mathfrak{UT}_p$ will become clear in 7.9.21.

7.9.5 THEOREM. *$\mathfrak{UT}_p$ is a right-sided Banach operator ideal.*

PROOF. Although the assertion follows from 7.9.3 and 7.9.6, we give a direct proof of the triangle inequality.

Let $S, T \in \mathfrak{UT}_p(X,Y)$, $x \in X$ and $y \in Y$. Write

$$\alpha := \|S|\mathfrak{UT}_p\| \quad \text{and} \quad \beta := \|T|\mathfrak{UT}_p\|.$$

Then

$$\left(\frac{\|y+(S+T)x\|^p + \|y-(S+T)x\|^p}{2} \right)^{1/p} \le$$

$$\le \left(\frac{\big[\|\frac{\alpha}{\alpha+\beta}y+Sx\| + \|\frac{\beta}{\alpha+\beta}y+Tx\|\big]^p + \big[\|\frac{\alpha}{\alpha+\beta}y-Sx\| + \|\frac{\beta}{\alpha+\beta}y-Tx\|\big]^p}{2} \right)^{1/p}$$

$$\le \left(\frac{\|\frac{\alpha}{\alpha+\beta}y+Sx\|^p + \|\frac{\alpha}{\alpha+\beta}y-Sx\|^p}{2} \right)^{1/p} + \left(\frac{\|\frac{\alpha}{\alpha+\beta}y+Tx\|^p + \|\frac{\alpha}{\alpha+\beta}y-Tx\|^p}{2} \right)^{1/p}$$

$$\le \left(\|\alpha x\|^p + \|\tfrac{\alpha}{\alpha+\beta}y\|^p \right)^{1/p} + \left(\|\beta x\|^p + \|\tfrac{\beta}{\alpha+\beta}y\|^p \right)^{1/p}$$

$$= (\alpha+\beta) \left(\|x\|^p + \|\tfrac{1}{\alpha+\beta}y\|^p \right)^{1/p},$$

which implies that

$$\left(\frac{\|y+(S+T)x\|^p + \|y-(S+T)x\|^p}{2} - \|y\|^p \right)^{1/p} \le$$

$$\le \left(\|S|\mathfrak{UT}_p\| + \|T|\mathfrak{UT}_p\| \right) \|x\|.$$

REMARK. Letting $y = o$ in (UT_p) yields that $\|T\| \le \|T|\mathfrak{UT}_p\|$. Moreover, we obtain from 7.9.1 that $\|I_{\mathbb{K}}|\mathfrak{UT}_p\| = 1$. Note that the identity map of l_∞^2 fails to be uniformly p-smooth. This follows by substituting $x = (\varrho, \varrho)$ and $y = (1, -1)$ into (UT_p) as $\varrho \to 0$.

7.9.6 We now show that the concepts of uniform p-smoothness and uniform q-convexity are dual to each other; see [BEA 2, pp. 311–312].

PROPOSITION. $(\mathfrak{UT}_p)' = \mathfrak{UC}_{p'}$ *and* $(\mathfrak{UC}_q)' = \mathfrak{UT}_{q'}$. *The corresponding ideal norms coincide.*

PROOF. To get $\mathfrak{UC}_{p'} \subseteq (\mathfrak{UT}_p)'$, we let $T \in \mathfrak{UC}_{p'}(X,Y)$ and $c := \|T|\mathfrak{UC}_{p'}\|$. Given $y' \in Y'$ and $x' \in X'$, for every $\varepsilon > 0$ there exist $x_\pm \in X$ such that

$$\|x' \pm T'y'\|^p = \langle x_\pm, x' \pm T'y'\rangle \quad \text{and} \quad \|x_\pm\| \le (1+\varepsilon)\|x' \pm T'y'\|^{p-1}.$$

Since

$$\Big\|\frac{Tx_+ - Tx_-}{2}\Big\|^{p'} \le c^{p'}\Big(\frac{\|x_+\|^{p'} + \|x_-\|^{p'}}{2} - \Big\|\frac{x_+ + x_-}{2}\Big\|^{p'}\Big),$$

we obtain

$$\begin{aligned}
&\|x' + T'y'\|^p + \|x' - T'y'\|^p = \langle x_+, x' + T'y'\rangle + \langle x_-, x' - T'y'\rangle \\
&= \langle Tx_+ - Tx_-, y'\rangle + \langle x_+ + x_-, x'\rangle \\
&\le \|Tx_+ - Tx_-\|\,\|y'\| + \|x_+ + x_-\|\,\|x'\| \\
&\le \Big(c^{-p'}\|Tx_+ - Tx_-\|^{p'} + \|x_+ + x_-\|^{p'}\Big)^{1/p'}\Big(c^p\|y'\|^p + \|x'\|^p\Big)^{1/p} \\
&\le 2^{(p'-1)/p'}\Big(\|x_+\|^{p'} + \|x_-\|^{p'}\Big)^{1/p'}\Big(c^p\|y'\|^p + \|x'\|^p\Big)^{1/p} \\
&\le 2^{1/p}(1+\varepsilon)\Big(\|x' + T'y'\|^p + \|x' - T'y'\|^p\Big)^{1/p'}\Big(c^p\|y'\|^p + \|x'\|^p\Big)^{1/p}.
\end{aligned}$$

Letting $\varepsilon \to 0$ yields

$$\Big(\frac{\|x' + T'y\|^p + \|x' - T'y'\|^p}{2}\Big)^{1/p} \le \Big(c^p\|y'\|^p + \|x'\|^p\Big)^{1/p}.$$

So

$$\Big(\frac{\|x' + T'y'\|^p + \|x' - T'y'\|^p}{2} - \|x'\|^p\Big)^{1/p} \le c\|y'\|.$$

This proves that

$$T' \in \mathfrak{UT}_p(Y', X') \quad \text{and} \quad \|T'|\mathfrak{UT}_p\| \le c = \|T|\mathfrak{UC}_{p'}\|.$$

To get $\mathfrak{UT}_{q'} \subseteq (\mathfrak{UC}_q)'$, we let $T \in \mathfrak{UT}_{q'}(X,Y)$ and $c := \|T|\mathfrak{UT}_{q'}\|$. Given $y', y_0' \in Y'$, for every $\varepsilon > 0$ there exist $x \in X$ and $y \in Y$ such that

$$\begin{aligned}
\|T'y'\|^q &= \langle x, T'y'\rangle \quad &&\text{and} \quad \|x\| \le (1+\varepsilon)\|T'y'\|^{q-1}, \\
\|y_0'\|^q &= \langle y, y_0'\rangle \quad &&\text{and} \quad \|y\| \le (1+\varepsilon)\|y_0'\|^{q-1}.
\end{aligned}$$

By assumption, we have

$$\left(\frac{\|c^q y+Tx\|^{q'}+\|c^q y-Tx\|^{q'}}{2}-\|c^q y\|^{q'}\right)^{1/q'}\leq c\,\|x\|,$$

which implies that

$$\begin{aligned}\left(\frac{\|c^q y+Tx\|^{q'}+\|c^q y-Tx\|^{q'}}{2}\right)^{1/q'} &\leq c\left(\|x\|^{q'}+c^q\|y\|^{q'}\right)^{1/q'}\\ &\leq (1+\varepsilon)c\left(\|T'y'\|^q+c^q\|y_0'\|^q\right)^{1/q'}.\end{aligned}$$

Therefore

$$\begin{aligned}&\|T'y'\|^q+c^q\|y_0'\|^q=\langle x,T'y'\rangle+c^q\,\langle y,y_0'\rangle\\ &=\tfrac{1}{2}\Big[\,\langle c^q y+Tx,y_0'+y'\rangle+\langle c^q y-Tx,y_0'-y'\rangle\Big]\\ &\leq\tfrac{1}{2}\Big[\|c^q y+Tx\|\,\|y_0'+y'\|+\|c^q y-Tx\|\,\|y_0'-y'\|\Big]\\ &\leq\left(\frac{\|c^q y+Tx\|^{q'}+\|c^q y-Tx\|^{q'}}{2}\right)^{1/q'}\left(\frac{\|y_0'+y'\|^q+\|y_0'-y'\|^q}{2}\right)^{1/q}\\ &\leq(1+\varepsilon)c\left(\|T'y'\|^q+c^q\|y_0'\|^q\right)^{1/q'}\left(\frac{\|y_0'+y'\|^q+\|y_0'-y'\|^q}{2}\right)^{1/q}.\end{aligned}$$

Letting $\varepsilon\to 0$ yields

$$\left(\|T'y'\|^q+c^q\|y_0'\|^q\right)^{1/q}\leq c\left(\frac{\|y_0'+y'\|^q+\|y_0'-y'\|^q}{2}\right)^{1/q}.$$

So

$$\|T'y'\|\leq c\left(\frac{\|y_0'+y'\|^q+\|y_0'-y'\|^q}{2}-\|y_0'\|^q\right)^{1/q}.$$

This proves that

$$T'\in\mathfrak{UC}_q(Y',X')\quad\text{and}\quad\|T'|\mathfrak{UC}_q\|\leq c=\|T|\mathfrak{UT}_{q'}\|.$$

So far, we have shown that

$$\mathfrak{UC}_{p'}\subseteq(\mathfrak{UT}_p)'\quad\text{and}\quad\mathfrak{UT}_{q'}\subseteq(\mathfrak{UC}_q)'.$$

If $q=p'$ and $p=q'$, then passing to the dual classes yields

$$(\mathfrak{UC}_q)'\subseteq(\mathfrak{UT}_{q'})''\quad\text{and}\quad(\mathfrak{UT}_p)'\subseteq(\mathfrak{UC}_{p'})''.$$

Hence

$$\mathfrak{UC}_{p'}\subseteq(\mathfrak{UT}_p)'\subseteq(\mathfrak{UC}_{p'})''\subseteq\mathfrak{UC}_{p'}$$

and

$$\mathfrak{UT}_{q'}\subseteq(\mathfrak{UC}_q)'\subseteq(\mathfrak{UT}_{q'})''\subseteq\mathfrak{UT}_{q'}.$$

This completes the proof.

7.9.7 We say that $T \in \mathfrak{L}(X,Y)$ is **uniformly q-convexifiable** if it can be represented in the form $T = BA$ such that $A \in \mathfrak{L}(X, X_0)$ and $B \in \mathfrak{UC}_q(X_0, Y)$. Let

$$\|T|\mathfrak{UC}_q \circ \mathfrak{L}\| := \inf \|B|\mathfrak{UC}_q\| \, \|A\|,$$

where the infimum is taken over all possible factorizations; see 1.2.4. The class of uniformly q-convexifiable operators is denoted by $\mathfrak{UC}_q \circ \mathfrak{L}$.

7.9.8 The term *convexifiable* is justified by the following result.

PROPOSITION. *If $T \in \mathfrak{L}(X,Y)$ is uniformly q-convexifiable, then X admits a renorming such that T becomes uniformly q-convex.*

PROOF. Choose any factorization $T = BA$ with $A \in \mathfrak{L}(X, X_0)$ and $B \in \mathfrak{UC}_q(X_0, Y)$. Let $\|A\| = 1$ and $c := \|B|\mathfrak{UC}_q\|$. Then

$$|||x||| := \big(\|x\|^q + \|Ax\|^q\big)^{1/q} \quad \text{for } x \in X$$

yields an equivalent norm such that $\|x\| \le |||x||| \le 2^{1/q}\|x\|$. Now, for $x_\pm \in X$, it follows from

$$\left\|\frac{BAx_+ - BAx_-}{2}\right\|^q \le c^q \left(\frac{\|Ax_+\|^q + \|Ax_-\|^q}{2} - \left\|\frac{Ax_+ + Ax_-}{2}\right\|^q\right)$$

and

$$0 \le \frac{\|x_+\|^q + \|x_-\|^q}{2} - \left\|\frac{x_+ + x_-}{2}\right\|^q$$

that

$$\left\|\frac{Tx_+ - Tx_-}{2}\right\|^q \le c^q \left(\frac{|||x_+|||^q + |||x_-|||^q}{2} - \left|\left|\left|\frac{x_+ + x_-}{2}\right|\right|\right|^q\right).$$

7.9.9 We say that $T \in \mathfrak{L}(X,Y)$ is **uniformly p-smoothable** if it can be represented in the form $T = BA$ such that $A \in \mathfrak{UT}_p(X, Y_0)$ and $B \in \mathfrak{L}(Y_0, Y)$. Let

$$\|T|\mathfrak{L} \circ \mathfrak{UT}_p\| := \inf \|B\| \, \|A|\mathfrak{UT}_p\|,$$

where the infimum is taken over all possible factorizations; see 1.2.4. The class of uniformly p-smoothable operators is denoted by $\mathfrak{L} \circ \mathfrak{UT}_p$.

7.9.10 We now present a dual counterpart of 7.9.8.

PROPOSITION. *If $T \in \mathfrak{L}(X,Y)$ is uniformly p-smoothable, then Y admits a renorming such that T becomes uniformly p-smooth.*

PROOF. Choose any factorization $T = BA$ with $A \in \mathfrak{UT}_p(X, Y_0)$ and $B \in \mathfrak{L}(Y_0, Y)$. Let $c := \|A|\mathfrak{UT}_p\|$ and $\|B\| = 1$. Then

$$|||y||| := \inf\Big\{ \big(\|y - By_0\|^p + \|y_0\|^p\big)^{1/p} : y_0 \in Y_0 \Big\} \quad \text{for } y \in Y$$

yields an equivalent norm such that $2^{-1/p'}\|y\| \le |||y||| \le \|y\|$. Now, step by step, we conclude from

$$\frac{\|y_0 + Ax\|^p + \|y_0 - Ax\|^p}{2} \le \|y_0\|^p + c^p\|x\|^p \quad \text{for } x \in X \text{ and } y_0 \in Y_0$$

that

$$\left.\begin{cases} \dfrac{\|y+Tx-B(y_0+Ax)\|^p+\|y_0+Ax\|^p}{2} \\ + \\ \dfrac{\|y-Tx-B(y_0-Ax)\|^p+\|y_0-Ax\|^p}{2} \end{cases}\right\} \le \|y-By_0\|^p+\|y_0\|^p+c^p\|x\|^p,$$

next

$$\frac{|||y+Tx|||^p + |||y-Tx|||^p}{2} \le \|y - By_0\|^p + \|y_0\|^p + c^p\|x\|^p,$$

and finally

$$\frac{|||y+Tx|||^p + |||y-Tx|||^p}{2} \le |||y|||^p + c^p\|x\|^p.$$

7.9.11 THEOREM. *$\mathfrak{UC}_q \circ \mathfrak{L}$ and $\mathfrak{L} \circ \mathfrak{UT}_p$ are Banach operator ideals.*

PROOF. For $\mathfrak{UC}_q \circ \mathfrak{L}$, the assertion is a straightforward consequence of the characterization established in 7.9.20, and the class $\mathfrak{L} \circ \mathfrak{UT}_p$ can be treated by duality; see 7.9.12.

7.9.12 We now state a corollary of 7.9.6.

PROPOSITION.

$$(\mathfrak{L} \circ \mathfrak{UT}_p)' = \mathfrak{UC}_{p'} \circ \mathfrak{L} \quad \text{and} \quad (\mathfrak{UC}_q \circ \mathfrak{L})' = \mathfrak{L} \circ \mathfrak{UT}_{q'}.$$

The corresponding ideal norms coincide.

7.9.13 Next, an elementary exercise is performed.

LEMMA. *Let $\alpha, \beta, \gamma \ge 0$ such that $\frac{\alpha+\beta}{2} \ge \gamma$. Then*

$$\left(\frac{\alpha^r + \beta^r}{2} - \gamma^r\right)^{1/r}$$

is a non-decreasing function of r for $1 \le r < \infty$.

PROOF. It follows from

$$\left[\left(\frac{\alpha+\beta}{2} - \gamma\right)^r + \gamma^r\right]^{1/r} \le \left(\frac{\alpha+\beta}{2} - \gamma\right) + \gamma = \frac{\alpha+\beta}{2}$$

that

$$\left(\frac{\alpha+\beta}{2} - \gamma\right)^r + \gamma^r \le \left(\frac{\alpha+\beta}{2}\right)^r \le \frac{\alpha^r + \beta^r}{2}.$$

Hence

$$\frac{\alpha+\beta}{2} - \gamma \le \left(\frac{\alpha^r + \beta^r}{2} - \gamma^r\right)^{1/r}. \qquad (*)$$

If $1 \le r_1 < r_2 < \infty$, then we let $r := r_2/r_1$. Since

$$\frac{\alpha^{r_1}+\beta^{r_1}}{2} \ge \Big(\frac{\alpha+\beta}{2}\Big)^{r_1} \ge \gamma^{r_1},$$

the inequality $(*)$ applies to $\alpha^{r_1}, \beta^{r_1}, \gamma^{r_1} \ge 0$. Thus

$$\frac{\alpha^{r_1}+\beta^{r_1}}{2} - \gamma^{r_1} \le \Big(\frac{\alpha^{r_2}+\beta^{r_2}}{2} - \gamma^{r_2}\Big)^{r_1/r_2}.$$

7.9.14 The preceding lemma implies monotonicity.

PROPOSITION.

$$\mathfrak{UC}_{q_1} \subseteq \mathfrak{UC}_{q_2} \;\; \textit{and} \;\; \mathfrak{UC}_{q_1} \circ \mathfrak{L} \subseteq \mathfrak{UC}_{q_2} \circ \mathfrak{L} \quad \textit{if } 2 \le q_1 \le q_2 < \infty,$$
$$\mathfrak{UT}_{p_1} \subseteq \mathfrak{UT}_{p_2} \;\; \textit{and} \;\; \mathfrak{L} \circ \mathfrak{UT}_{p_1} \subseteq \mathfrak{L} \circ \mathfrak{UT}_{p_2} \quad \textit{if } 1 < p_2 \le p_1 \le 2.$$

7.9.15 The following observation is an immediate consequence of 7.9.1.

EXAMPLE.

$$\|L_p|\mathfrak{UT}_p\| = 1 \;\; \textit{if } 1 < p \le 2 \;\; \textit{and} \;\; \|L_q|\mathfrak{UC}_q\| = 1 \;\; \textit{if } 2 \le q < \infty.$$

7.9.16 We now prove an inequality which goes back to W. Beckner; see [LIN*2, pp. 78–79]. We also refer to [mei].

LEMMA. *If* $1 < p \le 2$, *then*

$$\Big(\frac{|1+z|^p + |1-z|^p}{2}\Big)^{1/p} \ge \sqrt{|z|^2 + p - 1} \quad \textit{for } z \in \mathbb{C}.$$

PROOF. Write $z \in \mathbb{C}$ in the form $z = re^{it}$ with $r \ge 0$ and $-\pi \le t < +\pi$. Then

$$|1+z|^p + |1-z|^p = (1 + r^2 + 2r\cos t)^{p/2} + (1 + r^2 - 2r\cos t)^{p/2}.$$

Regarding the right-hand expression as a function of t, we easily see that the minimum is attained at 0. Hence

$$|1+z|^p + |1-z|^p \ge |1+r|^p + |1-r|^p.$$

Thus it is enough to check the desired inequality for $0 < r < \infty$.

In the case $0 < r < 1$, writing $(1+r)^p$ and $(1-r)^p$ as binomial series, we obtain

$$\frac{(1+r)^p + (1-r)^p}{2} = \sum_{k=0}^{\infty} \binom{p}{2k} r^2 > 1 + \frac{p(p-1)}{2} r^2.$$

By Bernoulli's inequality,

$$\Big(1 + \frac{p(p-1)}{2} r^2\Big)^{2/p} \ge 1 + (p-1)r^2.$$

Hence

$$\left(\frac{(1+r)^p+(1-r)^p}{2}\right)^{2/p} \geq 1+(p-1)r^2 \qquad (*)$$

which implies that

$$\left(\frac{(1+r)^p+(1-r)^p}{2}\right)^{2/p} \geq r^2+1+(p-2)r^2 \geq r^2+p-1.$$

In the case $1<r<\infty$, we conclude from $(*)$ that

$$\left(\frac{(1+\frac{1}{r})^p+(1-\frac{1}{r})^p}{2}\right)^{2/p} \geq 1+(p-1)\frac{1}{r^2}.$$

Multiplying by r^2 yields

$$\left(\frac{(r+1)^p+(r-1)^p}{2}\right)^{2/p} \geq r^2+p-1.$$

This completes the proof.

7.9.17 We stress that the following result holds in the real and in the complex case, simultaneously.

EXAMPLE.

$$\begin{aligned} \|L_p|\mathfrak{UC}_2\| &= \sqrt{p'-1} \quad \text{if } 1<p\leq 2, \\ \|L_q|\mathfrak{UT}_2\| &= \sqrt{q-1} \quad \text{if } 2\leq q<\infty. \end{aligned}$$

PROOF. We know from the preceding lemma that

$$\sqrt{|z|^2+c^2} \leq \left(\frac{|1+z|^p+|1-z|^p}{2}\right)^{1/p},$$

where $c:=\sqrt{p-1}$. Replacing z by z_0/z yields

$$\left(|cz|^2+|z_0|^2\right)^{p/2} \leq \frac{|z_0+z|^p+|z_0-z|^p}{2}.$$

For $f, f_0 \in L_p(\Omega,\omega)$, we obtain

$$\left(\int_\Omega \left[|cf(\xi)|^2+|f_0(\xi)|^2\right]^{p/2} d\omega(\xi)\right)^{1/p} \leq$$

$$\leq \left(\int_\Omega \frac{|f_0(\xi)+f(\xi)|^p+|f_0(\xi)-f(\xi)|^p}{2} d\omega(\xi)\right)^{1/p}.$$

By Jessen's inequality,

$$\left(\left[\int_\Omega |cf(\xi)|^p \, d\omega(\xi) \right]^{2/p} + \left[\int_\Omega |f_0(\xi)|^p \, d\omega(\xi) \right]^{2/p} \right)^{1/2} \leq$$
$$\leq \left(\int_\Omega \left[|cf(\xi)|^2 + |f_0(\xi)|^2 \right]^{p/2} d\omega(\xi) \right)^{1/p}.$$

Therefore

$$\left(\|cf|L_p\|^2 + \|f_0|L_p\|^2 \right)^{1/2} \leq \left(\frac{\|f_0 + f|L_p\|^p + \|f_0 - f|L_p\|^p}{2} \right)^{1/p}$$
$$\leq \left(\frac{\|f_0 + f|L_p\|^2 + \|f_0 - f|L_p\|^2}{2} \right)^{1/2}.$$

Thus we have

$$\|f|L_p\| \leq c^{-1} \left(\frac{\|f_0 + f|L_p\|^2 + \|f_0 - f|L_p\|^2}{2} - \|f_0|L_p\|^2 \right)^{1/2}$$

which proves that

$$\|L_p|\mathfrak{UC}_2\| \leq c^{-1} = (p-1)^{-1/2} = \sqrt{p'-1}.$$

In order to show the reverse inequality, we consider the two elements $x = (\varrho, \varrho)$ and $x_0 = (1, -1)$ in l_p^2. Since

$$\|x|l_p^2\| = 2^{1/p}|\varrho|, \quad \|x_0|l_p^2\| = 2^{1/p} \quad \text{and} \quad \|x \pm x_0|l_p^2\| = (|1+\varrho|^p + |1-\varrho|^p)^{1/p},$$

it follows from the definition of the $\mathfrak{UC}_2$-norm that

$$|\varrho| \leq \left(\left[\frac{|1+\varrho|^p + |1-\varrho|^p}{2} \right]^{2/p} - 1 \right)^{1/2} \|l_p^2|\mathfrak{UC}_2\|.$$

Letting $\varrho \to 0$ yields

$$1 \leq \sqrt{p-1}\, \|l_p^2|\mathfrak{UC}_2\|.$$

This proves that $\|L_p|\mathfrak{UC}_2\| = \sqrt{p'-1}$, because every infinite dimensional Banach space L_p contains isometric copies of l_p^2. The other case follows from $\|L_q|\mathfrak{UT}_2\| = \|L_{q'}|\mathfrak{UC}_2\|$.

7.9.18 The classes of Banach spaces associated with the one-sided operator ideals $\mathfrak{UC}_q$ and $\mathfrak{UT}_p$ are denoted as follows:

UC_q : uniformly q-convex Banach spaces,

UT_p : uniformly p-smooth Banach spaces.

The same terminology will be used when we refer to the underlying norms.

7.9.19 The next observation is closely related to 7.5.13.

EXAMPLE.

$$L_p \in \mathsf{UT}_p \setminus \mathsf{UT}_{p_0} \textit{ if } 1<p<p_0<2 \textit{ and } L_q \in \mathsf{UT}_2 \textit{ if } 2\le q<\infty,$$
$$L_q \in \mathsf{UC}_q \setminus \mathsf{UC}_{q_0} \textit{ if } 2<q_0<q<\infty \textit{ and } L_p \in \mathsf{UC}_2 \textit{ if } 1<p\le 2.$$

PROOF. See 7.9.15 and 7.9.17. Substituting $x_\pm := (\varrho, \pm 1)$ in (UC_{q_0}) and letting $\varrho \to 0$ yields $l_q^2 \notin \mathsf{UC}_{q_0}$.

7.9.20 We now establish the main result of this section, which is due to G. Pisier; see [pis 5]. In order to get isometry, $\mathfrak{AC}_q$ must be equipped with the ideal norm $\boldsymbol{\varrho}_q^{(q)}(T|\mathcal{I}(\mathbb{D}_0^n), \mathcal{H}(\mathbb{D}_0^n))$.

THEOREM. *The Banach operator ideals* $\mathfrak{AC}_q$ *and* $\mathfrak{UC}_q \circ \mathfrak{L}$ *coincide isometrically.*

PROOF. Assume that $T \in \mathfrak{AC}_q(X,Y)$ and write $c := \|T|\mathfrak{AC}_q\|$. Then, for $x \in X$, a semi-norm s is defined by

$$s(x) := \inf \left(c^q \|\boldsymbol{f}_n|L_q\|^q - \sum_{k=1}^n \|[L_q, T]\boldsymbol{d}_k|L_q\|^q \right)^{1/q},$$

where the infimum ranges over all X-valued Haar polynomials

$$\boldsymbol{f}_n = x + \sum_{k=1}^n \boldsymbol{d}_k \quad \text{with } n = 0, 1, \dots .$$

It follows from

$$\|Tx\|^q + \sum_{k=1}^n \|[L_q, T]\boldsymbol{d}_k|L_q\|^q \le c^q \|\boldsymbol{f}_n|L_q\|^q$$

that

$$\|Tx\| \le s(x) \quad \text{for } x \in X. \tag{1}$$

Letting $\boldsymbol{f}_0(t) \equiv x$ on $[0,1)$ yields

$$s(x) \le c\,\|x\| \quad \text{for } x \in X. \tag{2}$$

Fix $x_\pm \in X$ such that $x = \frac{x_+ + x_-}{2}$. Given $\varepsilon > 0$, we choose X-valued Haar polynomials

$$\boldsymbol{f}_n^\pm = x_\pm + \sum_{k=1}^n \boldsymbol{d}_k^\pm$$

such that

$$\|\boldsymbol{f}_n^\pm|L_q\|^q - c^{-q} \sum_{k=1}^n \|[L_q, T]\boldsymbol{d}_k^\pm|L_q\|^q \le s(x_\pm)^q + \varepsilon.$$

Define

$$\boldsymbol{f}_{n+1} = x + \sum_{k=1}^{n+1} \boldsymbol{d}_k$$

by

$$\boldsymbol{f}_{n+1}(t) = \frac{\boldsymbol{f}_n^+ \oplus \boldsymbol{f}_n^-}{2}(t) := \begin{cases} \boldsymbol{f}_n^+(2t) & \text{if } 0 \le t < 1/2, \\ \boldsymbol{f}_n^-(2t-1) & \text{if } 1/2 \le t < 1. \end{cases} \tag{3}$$

Then

$$\boldsymbol{f}_0(t) = x \quad \text{if } 0 \le t < 1 \quad \text{and} \quad \boldsymbol{f}_1(t) = \begin{cases} x_+ & \text{if } 0 \le t < 1/2, \\ x_- & \text{if } 1/2 \le t < 1. \end{cases}$$

Hence

$$\boldsymbol{d}_1(t) = \begin{cases} \frac{x_+ - x_-}{2} & \text{if } 0 \le t < 1/2, \\ \frac{x_- - x_+}{2} & \text{if } 1/2 \le t < 1. \end{cases}$$

Easy manipulations show that

$$\boldsymbol{d}_{k+1} = \frac{\boldsymbol{d}_k^+ \oplus \boldsymbol{d}_k^-}{2} \quad \text{for } k = 1, \ldots, n,$$

where the right-hand side is defined as in (3). Thus

$$s\Big(\frac{x_+ + x_-}{2}\Big)^q \le c^q \|\boldsymbol{f}_{n+1}|L_q\|^q - \sum_{k=1}^{n+1} \|[L_q, T]\boldsymbol{d}_k|L_q\|^q$$

$$\le \Big\| c^q \frac{\boldsymbol{f}_n^+ \oplus \boldsymbol{f}_n^-}{2} \Big| L_q \Big\|^q - \left\{ \begin{array}{c} \|[L_q, T]\boldsymbol{d}_1|L_q\|^q \\ + \\ \displaystyle\sum_{k=1}^{n} \Big\| \frac{[L_q, T]\boldsymbol{d}_k^+ \oplus [L_q, T]\boldsymbol{d}_k^-}{2} \Big| L_q \Big\|^q \end{array} \right\}.$$

Note that

$$\Big\| \frac{\boldsymbol{f}_n^+ \oplus \boldsymbol{f}_n^-}{2} \Big| L_q \Big\|^q = \frac{\|\boldsymbol{f}_n^+|L_q\|^q + \|\boldsymbol{f}_n^-|L_q\|^q}{2},$$

$$\|[L_q, T]\boldsymbol{d}_1|L_q\|^q = \Big\| \frac{Tx_+ - Tx_-}{2} \Big\|^q,$$

and

$$\Big\| \frac{[L_q, T]\boldsymbol{d}_k^+ \oplus [L_q, T]\boldsymbol{d}_k^-}{2} \Big| L_q \Big\|^q = \frac{\|[L_q, T]\boldsymbol{d}_k^+|L_q\|^q + \|[L_q, T]\boldsymbol{d}_k^-|L_q\|^q}{2}$$

for $k = 1, \ldots, n$. We now obtain

$$s\Big(\frac{x_+ + x_-}{2}\Big)^q \le \frac{1}{2} \left\{ \begin{array}{c} c^q \|\boldsymbol{f}_n^+|L_q\|^q - \displaystyle\sum_{k=1}^{n} \|[L_q, T]\boldsymbol{d}_k^+|L_q\|^q \\ + \\ c^q \|\boldsymbol{f}_n^-|L_q\|^q - \displaystyle\sum_{k=1}^{n} \|[L_q, T]\boldsymbol{d}_k^-|L_q\|^q \end{array} \right\} - \Big\| \frac{Tx_+ - Tx_-}{2} \Big\|^q$$

$$\le \frac{s(x_+)^q + s(x_-)^q + 2\varepsilon}{2} - \Big\| \frac{Tx_+ - Tx_-}{2} \Big\|^q.$$

Letting $\varepsilon \to 0$ yields

$$s\Big(\frac{x_+ + x_-}{2}\Big)^q \le \frac{s(x_+)^q + s(x_-)^q}{2} - \Big\|\frac{Tx_+ - Tx_-}{2}\Big\|^q \quad \text{for } x_\pm \in X. \tag{4}$$

Obviously, we have

$$s(\lambda x) = |\lambda| \|x\| \quad \text{for } \lambda \in \mathbb{K} \text{ and } x \in X.$$

Thus s is the Minkowski functional of an open set

$$B := \{\, x \in X : s(x) < 1 \,\}.$$

It follows from (4) that B is midpoint convex. Hence, for $x_\pm \in B$ and every dyadic fraction $\lambda \in (0,1)$, we get

$$x := \lambda x_+ + (1-\lambda)x_- \in B.$$

To prove that B is even convex, write x in the form

$$x = \mu\alpha_+ x_+ + (1-\mu)\alpha_- x_-$$

such that $\alpha_\pm x_\pm \in B$ and $\mu \in (0,1)$ becomes dyadic. The convexity of B means that s is a semi-norm.

Passing to equivalence classes $\widetilde{x} = x + N$ modulo the null space N of s, we get a linear space X_0 equipped with the norm $\|\widetilde{x}\|_0 := s(x)$. Let A denote the canonical map from X into $\widetilde{X}_0$, the completion of X_0. In view of (2), we have $\|A\| \le c$. It follows from (1) that T admits a unique extension $\widetilde{T}_0 \in \mathfrak{L}(\widetilde{X}_0, Y)$ such that $T = \widetilde{T}_0 A$. Moreover, the estimate (4) yields

$$\Big\|\frac{\widetilde{x}_+ + \widetilde{x}_-}{2}\Big\|_0^q \le \frac{\|\widetilde{x}_+\|_0^q + \|\widetilde{x}_-\|_0^q}{2} - \Big\|\frac{\widetilde{T}_0\,\widetilde{x}_+ - \widetilde{T}_0\,\widetilde{x}_-}{2}\Big\|^q \quad \text{for } \widetilde{x}_\pm \in \widetilde{X}_0.$$

Thus $\widetilde{T}_0$ is uniformly q-convex and $\|\widetilde{T}_0|\mathfrak{UC}_q\| \le 1$, which implies that

$$\|T|\mathfrak{UC}_q \circ \mathfrak{L}\| \le \varrho_q^{(q)}(T|\mathfrak{I}(\mathbb{D}_0^n), \mathcal{H}(\mathbb{D}_0^n))$$

for $T \in \mathfrak{AC}_q(X,Y)$ and $n = 1, 2, \ldots$.

To verify the reverse implication, we let $T \in \mathfrak{UC}_q(X,Y)$ and write $c := \|T|\mathfrak{UC}_q\|$. If

$$\boldsymbol{f}_n = \sum_{k=0}^{n} \boldsymbol{d}_k$$

is any X-valued Haar polynomial, then

$$\boldsymbol{d}_k(t) = \begin{cases} +2^{(k-1)/2}\left\langle \boldsymbol{f}_n, \chi_k^{(j)} \right\rangle & \text{if } t \in \Delta_k^{(2j-1)}, \\ -2^{(k-1)/2}\left\langle \boldsymbol{f}_n, \chi_k^{(j)} \right\rangle & \text{if } t \in \Delta_k^{(2j)}, \end{cases}$$

for $k=1,\dots,n$. Since $\boldsymbol{f}_{k-1}$ is constant on $\Delta_{k-1}^{(j)}=\Delta_k^{(2j-1)}\cup\Delta_k^{(2j)}$, we have

$$\int\limits_{\Delta_{k-1}^{(j)}} \|\boldsymbol{f}_{k-1}(t)+\boldsymbol{d}_k(t)\|^q\,dt = \int\limits_{\Delta_{k-1}^{(j)}} \|\boldsymbol{f}_{k-1}(t)-\boldsymbol{d}_k(t)\|^q\,dt.$$

Hence

$$\|\boldsymbol{f}_k|L_q\| = \|\boldsymbol{f}_{k-1}+\boldsymbol{d}_k|L_q\| = \|\boldsymbol{f}_{k-1}-\boldsymbol{d}_k|L_q\|.$$

Integrating

$$\|T\boldsymbol{d}_k(t)\|^q \le c^q\Big(\frac{\|\boldsymbol{f}_{k-1}(t)+\boldsymbol{d}_k(t)\|^q+\|\boldsymbol{f}_{k-1}(t)-\boldsymbol{d}_k(t)\|^q}{2}-\|\boldsymbol{f}_{k-1}(t)\|^q\Big)$$

over t yields

$$\begin{aligned}\|[L_q,T]\boldsymbol{d}_k|L_q\|^q &\le c^q\Big(\frac{\|\boldsymbol{f}_{k-1}+\boldsymbol{d}_k|L_q\|^q+\|\boldsymbol{f}_{k-1}-\boldsymbol{d}_k|L_q\|^q}{2}-\|\boldsymbol{f}_{k-1}|L_q\|^q\Big)\\ &= c^q\Big(\|\boldsymbol{f}_k|L_q\|^q-\|\boldsymbol{f}_{k-1}|L_q\|^q\Big).\end{aligned}$$

Consequently,

$$\sum_{k=1}^n \|[L_q,T]\boldsymbol{d}_k|L_q\|^q \le c^q\Big(\|\boldsymbol{f}_n|L_q\|^q-\|\boldsymbol{f}_0|L_q\|^q\Big).$$

Thus, in view of

$$\|[L_q,T]\boldsymbol{d}_0|L_q\| \le \|T\|\,\|\boldsymbol{d}_0|L_q\| \le c\,\|\boldsymbol{f}_0|L_q\|,$$

we finally arrive at

$$\sum_{k=0}^n \|[L_q,T]\boldsymbol{d}_k|L_q\|^q \le c^q\|\boldsymbol{f}_n|L_q\|^q.$$

This proves that

$$\boldsymbol{\varrho}_q^{(q)}(T|\mathcal{I}(\mathbb{D}_0^n),\mathcal{H}(\mathbb{D}_0^n)) \le \|T|\mathfrak{UC}_q\|$$

for $T\in\mathfrak{UC}_q(X,Y)$ and $n=1,2,\dots$. Obviously, this inequality extends to uniformly q-convexifiable operators.

7.9.21 The preceding result can be dualized, by 7.5.11 and 7.9.6. In this case, we must use the ideal norm $\boldsymbol{\varrho}_p^{(p)}(T|\mathcal{H}(\mathbb{D}_0^n),\mathcal{I}(\mathbb{D}_0^n))$ on $\mathfrak{AT}_p$.

THEOREM. *The Banach operator ideals $\mathfrak{AT}_p$ and $\mathfrak{L}\circ\mathfrak{UT}_p$ coincide isometrically.*

7.9.22 In the setting of spaces, the above theorems read as follows.

THEOREM.

A Banach space belongs to AC_q if and only if it admits an equivalent uniformly q-convex norm.

A Banach space belongs to AT_p if and only if it admits an equivalent uniformly p-smooth norm.

7.10 Uniform convexity and uniform smoothness

7.10.1 An operator $T \in \mathfrak{L}(X,Y)$ is called **uniformly convex** if for every $\varepsilon > 0$ there exists $\delta > 0$ such that

$$\left\| \frac{Tx_+ - Tx_-}{2} \right\| \geq \varepsilon \quad \text{and} \quad \|x_\pm\| = 1 \quad \text{imply} \quad \left\| \frac{x_+ + x_-}{2} \right\| \leq 1 - \delta.$$

The class of these operators is denoted by $\mathfrak{UC}$.

7.10.2 For completeness, we state a trivial result.

THEOREM. *$\mathfrak{UC}$ is a left-sided closed operator ideal.*

7.10.3 In the definition of uniform convexity the elements x_+ and x_- are supposed to have norm 1. Without this restriction, we get a condition which is more flexible.

LEMMA. *For $T \in \mathfrak{L}(X,Y)$, $\varepsilon > 0$ and $1 < r < \infty$, we consider the following properties:*

(UC_ε) *There exists $\delta > 0$ such that*

$$\left\| \frac{Tx_+ - Tx_-}{2} \right\| \geq \varepsilon \quad \textit{and} \quad \|x_\pm\| = 1$$

imply

$$\left\| \frac{x_+ + x_-}{2} \right\| \leq 1 - \delta.$$

($\mathrm{UC}_\varepsilon^r$) *There exists $\delta > 0$ such that*

$$\left\| \frac{Tx_+ - Tx_-}{2} \right\| \geq \varepsilon \left(\frac{\|x_+\|^r + \|x_-\|^r}{2} \right)^{1/r}$$

implies

$$\left\| \frac{x_+ + x_-}{2} \right\| \leq (1-\delta) \left(\frac{\|x_+\|^r + \|x_-\|^r}{2} \right)^{1/r}.$$

Then ($\mathrm{UC}_\varepsilon^r$)$\Rightarrow$($\mathrm{UC}_\varepsilon$)$\Rightarrow$($\mathrm{UC}_{\varepsilon_0}^r$) *whenever $\varepsilon < \varepsilon_0$.*

PROOF. The left-hand implication is trivial. Assume that ($\mathrm{UC}_{\varepsilon_0}^r$) fails. Then, by homogeneity and symmetry, we can find sequences (x_n^+) and (x_n^-) such that

$$\left\| \frac{Tx_n^+ - Tx_n^-}{2} \right\| \geq \varepsilon_0 \left(\frac{\|x_n^+\|^r + \|x_n^-\|^r}{2} \right)^{1/r}, \quad \|x_n^+\| = 1, \ \|x_n^-\| \leq 1, \tag{1}$$

and

$$\lim_{n\to\infty} \frac{\left\| \dfrac{x_n^+ + x_n^-}{2} \right\|^r}{\dfrac{\|x_n^+\|^r + \|x_n^-\|^r}{2}} = 1. \tag{2}$$

Passing to a subsequence, we may assume that $(\|x_n^-\|)$ is convergent. Now, by Hölder's inequality,

$$1 \geq \lim_{n\to\infty} \frac{\left(\dfrac{1+\|x_n^-\|}{2}\right)^r}{\dfrac{1+\|x_n^-\|^r}{2}} \geq \lim_{n\to\infty} \frac{\left\|\dfrac{x_n^+ + x_n^-}{2}\right\|^r}{\dfrac{\|x_n^+\|^r + \|x_n^-\|^r}{2}} = 1.$$

This implies that $\lim\limits_{n\to\infty} \|x_n^-\| = 1$. Let

$$u_n^+ := x_n^+ \quad \text{and} \quad u_n^- := \frac{x_n^-}{\|x_n^-\|}.$$

Clearly, $\lim\limits_{n\to\infty} \|u_n^- - x_n^-\| = 0$. Next, we conclude from (1) that

$$\liminf_{n\to\infty} \left\|\frac{Tu_n^+ - Tu_n^-}{2}\right\| = \liminf_{n\to\infty} \left\|\frac{Tx_n^+ - Tx_n^-}{2}\right\| \geq \varepsilon_0 > \varepsilon,$$

and (2) yields

$$\lim_{n\to\infty} \left\|\frac{u_n^+ + u_n^-}{2}\right\|^r = \lim_{n\to\infty} \frac{\left\|\dfrac{x_n^+ + x_n^-}{2}\right\|^r}{\dfrac{\|x_n^+\|^r + \|x_n^-\|^r}{2}} = 1.$$

Thus $(\mathrm{UC}_\varepsilon)$ fails as well.

7.10.4 An operator $T \in \mathfrak{L}(X,Y)$ is called **uniformly smooth** if for every $\varepsilon > 0$ there exists $\delta > 0$ such that

$$\frac{\|y+Tx\| + \|y-Tx\|}{2} - 1 \leq \varepsilon\|x\| \quad \text{whenever} \quad \|y\| = 1 \quad \text{and} \quad \|x\| \leq \delta.$$

The class of these operators is denoted by $\mathfrak{UT}$.

7.10.5 THEOREM. *$\mathfrak{UT}$ is a right-sided closed operator ideal.*

7.10.6 The duality between $\mathfrak{UC}$ and $\mathfrak{UT}$ can be deduced as in 7.9.6.

PROPOSITION. $(\mathfrak{UC})' = \mathfrak{UT}$ *and* $(\mathfrak{UT})' = \mathfrak{UC}$.

7.10.7 In a sense, $\mathfrak{UC}$ and $\mathfrak{UT}$ are limiting cases of $\mathfrak{UC}_q$ and $\mathfrak{UT}_p$, respectively.

PROPOSITION.

$$\mathfrak{UC}_q \subset \mathfrak{UC} \quad \text{if } 2 \leq q < \infty \quad \text{and} \quad \mathfrak{UT}_p \subset \mathfrak{UT} \quad \text{if } 1 < p \leq 2.$$

PROOF. The first inclusion is trivial, and the second one follows by duality. We give, however, a direct argument. Indeed, note that $(1+s)^{1/p} \leq 1+s$ for $s \geq 0$. Then, whenever $\|y\| = 1$, it follows that

$$\frac{\|y+Tx\| + \|y-Tx\|}{2} - 1 \leq \left(\frac{\|y+Tx\|^p + \|y-Tx\|^p}{2}\right)^{1/p} - 1$$

$$\leq \Big(1 + \|T|\mathfrak{UT}_p\|^p \, \|x\|^p\Big)^{1/p} - 1 \leq \|T|\mathfrak{UT}_p\|^p \, \|x\|^p.$$

7.10.8 An operator $T \in \mathfrak{L}(X,Y)$ is called **uniformly convexifiable** if it can be written in the form $T = BA$ such that $A \in \mathfrak{L}(X, X_0)$ and $B \in \mathfrak{UC}(X_0, Y)$. The class of these operators is denoted by $\mathfrak{UC} \circ \mathfrak{L}$.

7.10.9 An operator $T \in \mathfrak{L}(X,Y)$ is called **uniformly smoothable** if it can be written in the form $T = BA$ such that $A \in \mathfrak{UT}(X, Y_0)$ and $B \in \mathfrak{L}(Y_0, Y)$. The class of these operators is denoted by $\mathfrak{L} \circ \mathfrak{UT}$.

7.10.10 THEOREM. *$\mathfrak{UC} \circ \mathfrak{L}$ and $\mathfrak{L} \circ \mathfrak{UT}$ are closed operator ideals.*

7.10.11 We now establish a counterpart of 7.9.8.

PROPOSITION. *If $T \in \mathfrak{L}(X,Y)$ is uniformly convexifiable, then X admits a renorming such that T becomes uniformly convex. It can even be arranged that the new norm is arbitrarily close to the old norm.*

PROOF. Choose any factorization $T = BA$ with $A \in \mathfrak{L}(X, X_0)$ and $B \in \mathfrak{UC}(X_0, Y)$. Given $\varrho > 0$,

$$|||x||| := \Big(\|x\|^2 + \varrho^2 \|Ax\|^2\Big)^{1/2} \qquad \text{for } x \in X$$

defines an equivalent norm such that $\|x\| \le |||x||| \le \sqrt{1 + \varrho^2 \|A\|^2}\, \|x\|$. For $x_\pm \in X$, it follows from

$$\left\| \frac{Tx_+ - Tx_-}{2} \right\|^2 \ge \varepsilon^2 \frac{|||x_+|||^2 + |||x_-|||^2}{2} \tag{1}$$

that

$$\left\| \frac{BAx_+ - BAx_-}{2} \right\|^2 \ge \varepsilon^2 \varrho^2 \frac{\|Ax_+\|^2 + \|Ax_-\|^2}{2}.$$

Hence, by 7.10.3,

$$\left\| \frac{Ax_+ + Ax_-}{2} \right\|^2 \le (1-\delta)^2 \frac{\|Ax_+\|^2 + \|Ax_-\|^2}{2}. \tag{2}$$

We also have

$$\left\| \frac{x_+ + x_-}{2} \right\|^2 \le \frac{\|x_+\|^2 + \|x_-\|^2}{2} \tag{3}$$

and

$$\left\| \frac{BAx_+ - BAx_-}{2} \right\|^2 \le \|B\|^2 \frac{\|Ax_+\|^2 + \|Ax_-\|^2}{2}. \tag{4}$$

Now (1) and (4) imply

$$\varepsilon^2 \frac{|||x_+|||^2 + |||x_-|||^2}{2} \le \|B\|^2 \frac{\|Ax_+\|^2 + \|Ax_-\|^2}{2}. \tag{5}$$

Finally, combining (2), (3) and (5) yields

$$\begin{aligned}\left|\!\left|\!\left|\frac{x_+ + x_-}{2}\right|\!\right|\!\right|^2 &= \left\|\frac{x_+ + x_-}{2}\right\|^2 + \varrho^2\left\|\frac{Ax_+ + Ax_-}{2}\right\|^2 \\ &\le \frac{\|x_+\|^2 + \|x_-\|^2}{2} + (1-\delta)^2\frac{\varrho^2\|Ax_+\|^2 + \varrho^2\|Ax_-\|^2}{2} \\ &= \frac{|\!|\!|x_+|\!|\!|^2 + |\!|\!|x_-|\!|\!|^2}{2} - \left[1-(1-\delta)^2\right]\frac{\varrho^2\|Ax_+\|^2 + \varrho^2\|Ax_-\|^2}{2} \\ &\le \left(1 - \frac{\varepsilon^2\varrho^2}{\|B\|^2}\left[1-(1-\delta)^2\right]\right)\frac{|\!|\!|x_+|\!|\!|^2 + |\!|\!|x_-|\!|\!|^2}{2}.\end{aligned}$$

REMARK. A dual result holds for uniformly smoothable operators.

7.10.12 The next proof is classic; see [LIN*2, p. 61].

PROPOSITION. *Uniformly convex operators are weakly compact.*

PROOF. Let $x'' \in X''$ and $\|x''\| = 1$. By the Goldstine theorem [DUN*1, p. 424], there exists a directed family (x_α) in the unit sphere of X such that

$$\lim_\alpha \langle x_\alpha, x'\rangle = \langle x'', x'\rangle \quad \text{for } x' \in X'.$$

Hence $\lim\limits_{\alpha,\beta} \|x_\alpha + x_\beta\| = 2$. It now follows from $T \in \mathfrak{UC}(X,Y)$ that $\lim\limits_{\alpha,\beta} \|Tx_\alpha - Tx_\beta\| = 0$. Thus (Tx_α) is a directed Cauchy family in Y which has a limit $y \in Y$. Consequently,

$$\langle T''x'', y'\rangle = \langle x'', T'y'\rangle = \lim_\alpha \langle x_\alpha, T'y'\rangle = \lim_\alpha \langle Tx_\alpha, y'\rangle = \langle y, y'\rangle = \langle K_Y y, y'\rangle$$

for $y' \in Y'$. This proves that T is weakly compact, since T'' maps X'' into $K_Y(Y)$; see 7.6.2.

7.10.13 PROPOSITION. *Uniformly convexifiable operators are super weakly compact.*

PROOF. Obviously, uniform convexity is stable under the formation of ultraproducts. Thus $\mathfrak{UC} \subseteq \mathfrak{R}$ yields $\mathfrak{UC} \subseteq \mathfrak{SR}$ and, since $\mathfrak{SR}$ is an ideal, we get $\mathfrak{UC} \circ \mathfrak{L} \subseteq \mathfrak{SR}$.

7.10.14 Elements $x_k^{(j)} \in X$ form an ***n*-tree** if

$$x_k^{(j)} = \frac{x_{k+1}^{(2j-1)} + x_{k+1}^{(2j)}}{2} \quad \text{for} \quad j = 1, \ldots, 2^k \quad \text{and} \quad k = 0, \ldots, n-1.$$

Clearly, n-trees are uniquely determined by their n-th level,

$$x_n^{(1)}, \ldots, x_n^{(2^n)}.$$

The following diagram shows a 3-tree:

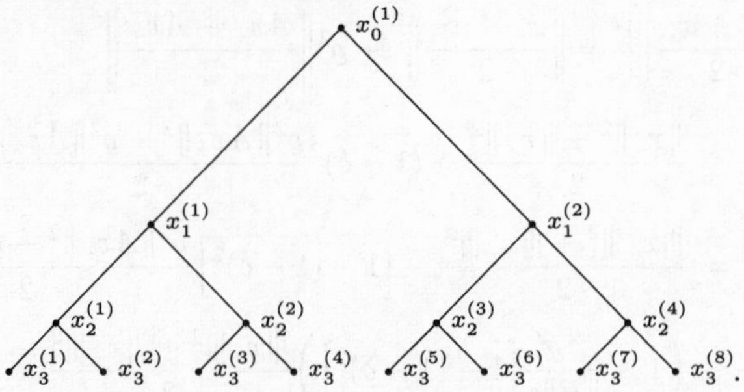

REMARK. Putting $\boldsymbol{f}_k(t) := x_k^{(j)}$ for $t \in \Delta_k^{(j)}$, every n-tree can be viewed as an X-valued dyadic martingale. Conversely, given $(\boldsymbol{f}_k)$ such that $\boldsymbol{f}_k = [E(\mathcal{D}_k), X]\,\boldsymbol{f}_n$, the constant values $x_k^{(j)}$ of $\boldsymbol{f}_k$ on the dyadic intervals $\Delta_k^{(j)}$ form an n-tree.

7.10.15 Let $T \in \mathfrak{L}(X,Y)$ and $\varepsilon > 0$. We call $(x_k^{(j)})$ a **(T, n, ε)-tree** if

$$\|x_{k+1}^{(2j-1)}\| \geq \|x_k^{(j)}\|, \ \|x_{k+1}^{(2j)}\| \geq \|x_k^{(j)}\| \ \text{and} \ \left\| \frac{Tx_{k+1}^{(2j-1)} - Tx_{k+1}^{(2j)}}{2} \right\| \geq \varepsilon \|x_k^{(j)}\|$$

for $j = 1, \ldots, 2^k$ and $k = 0, \ldots, n-1$.

7.10.16 An operator $T \in \mathfrak{L}(X,Y)$ is said to have the **growing tree property** if, given $\varepsilon > 0$, there exists $n_\varepsilon \in \mathbb{N}_0$ such that

$$\max_{1 \leq j \leq 2^{n_\varepsilon}} \|x_{n_\varepsilon}^{(j)}\| \geq 2\|x_0^{(1)}\| \quad \text{for all } (T, n, \varepsilon)\text{-trees } (x_k^{(j)}) \text{ with } n \geq n_\varepsilon.$$

7.10.17 We now prepare the proof of the next proposition. First of all, let us point out a simple fact.

SUBLEMMA. *Define* $f : [0,1] \times X \times X$ *by*

$$f(\lambda, x_0, x_1) := (1-\lambda)x_0 + \lambda x_1.$$

If W_0 and W_1 are weakly compact and convex subsets of X, then so is

$$W_{[\alpha,\beta]} := f([\alpha,\beta], W_0, W_1) \quad \textit{with } 0 \leq \alpha \leq \beta \leq 1.$$

PROOF. Since f is continuous with respect to the weak topology on X, weak compactness carries over from $[\alpha,\beta] \times W_0 \times W_1$ to the image $W_{[\alpha,\beta]}$.

Given $x := (1-\lambda)x_0 + \lambda x_1$ and $y := (1-\mu)y_0 + \mu y_1$, we have

$$(1-\varrho)x+\varrho y = \left\{ \begin{array}{c} \Big((1-\varrho)(1-\lambda) + \varrho(1-\mu)\Big) \dfrac{(1-\varrho)(1-\lambda)x_0 + \varrho(1-\mu)y_0}{(1-\varrho)(1-\lambda) + \varrho(1-\mu)} \\ + \\ \Big((1-\varrho)\lambda + \varrho\mu\Big) \dfrac{(1-\varrho)\lambda x_1 + \varrho\mu y_1}{(1-\varrho)\lambda + \varrho\mu} \end{array} \right\}.$$

This implies the convexity of $W_{[\alpha,\beta]}$.

It is well-known that weakly compact subsets in Banach spaces are dentable; see [DIE, p. 204] and [DIE*b, p. 138]. In what follows, we only need this deep result in a particular and much simpler case which was treated by J. Lindenstrauss, [lin 2]. The proof of the following lemma is due to I. Namioka/E. Asplund, [nam*]. The diameter of a bounded set W is defined by $\operatorname{diam}(W) := \sup \{ \|x_1 - x_2\| : x_1, x_2 \in W \}$.

LEMMA. *Let W be a separable and weakly compact convex set. Then for every $\varepsilon > 0$ there exists a closed convex subset C such that $C \neq W$ and* $\operatorname{diam}(W \setminus C) \leq \varepsilon$.

PROOF. We denote by E the collection of all extreme points of W, and F stands for its weak closure. By assumption, for $\delta > 0$ there exist $x_1, x_2, \ldots \in X$ such that

$$F \subseteq W \subseteq \bigcup_{n=1}^{\infty} \{x_n + \delta U_X\}.$$

Because of separability and weak compactness, we know that W is a complete metrizable space in the weak topology; see [DUN*1, p. 434]. Thus, by Baire's category theorem [DUN*1, p. 20], at least one of the weakly closed sets $F \cap \{x_n + \delta U_X\}$, say $F \cap \{x_{n_0} + \delta U_X\}$, has an interior point relative to F. So we can find in X a weakly open set G such that

$$\emptyset \neq F \cap G \subseteq F \cap \{x_{n_0} + \delta U_X\}.$$

Let

$$W_0 := \overline{\operatorname{conv}}(F \cap G) \quad \text{and} \quad W_1 := \overline{\operatorname{conv}}(F \setminus G).$$

By the preceding sublemma

$$W_{[0,1]} := f([0,1], W_0, W_1) = \operatorname{conv}(W_0 \cup W_1)$$

is weakly compact. Hence $\operatorname{conv}(W_0 \cup W_1) = \overline{\operatorname{conv}}(W_0 \cup W_1)$. Thus it follows from $E \subseteq F \subseteq W_0 \cup W_1 \subseteq \overline{\operatorname{conv}}(W_0 \cup W_1)$ and the Kreĭn–Milman theorem [DUN*1, p. 440] that

$$W = \overline{\operatorname{conv}}(E) = \operatorname{conv}(W_0 \cup W_1).$$

By the preceding sublemma, $W_{[\varrho,1]} := f([\varrho,1], W_0, W_1)$ with $0 < \varrho \leq 1$

is weakly compact and convex. In order to show that $W_{[\varrho,1]} \neq W$, we assume the contrary. Then every $x \in E$ admits a representation

$$x = (1-\lambda)x_0 + \lambda x_1 \quad \text{with } \varrho \leq \lambda \leq 1,\ x_0 \in W_0 \text{ and } x_1 \in W_1.$$

Hence $x = x_1 \in W_1 = \overline{\text{conv}}(F \setminus G)$. Consequently, $F \setminus G$ must contain all extreme points of W; see [DUN*1, p. 440]. However, this is impossible, since there are extreme points in $F \cap G \neq \emptyset$. Thus we have indeed $W_{[\varrho,1]} \neq W$.

By definition, every $x \in W_{[0,\varrho]}$ is of the form

$$x = (1-\lambda)x_0 + \lambda x_1 \quad \text{with } x_0 \in W_0,\ x_1 \in W_1 \text{ and } 0 \leq \lambda \leq \varrho.$$

So

$$\|x - x_0\| \leq \lambda \|x_1 - x_0\| \leq \varrho \operatorname{diam}(W).$$

This implies that

$$\operatorname{diam}(W_{[0,\varrho]}) \leq 2\varrho \operatorname{diam}(W) + \operatorname{diam}(W_0).$$

Since $W_0 \subseteq x_{n_0} + \delta U_X$, we have $\operatorname{diam}(W_0) \leq 2\delta$. Therefore

$$\operatorname{diam}(W \setminus W_{[\varrho,1]}) \leq \operatorname{diam}(W_{[0,\varrho]}) \leq 2\varrho \operatorname{diam}(W) + 2\delta.$$

Choosing ϱ and δ sufficiently small, we are done.

7.10.18 PROPOSITION. *Super weakly compact operators have the growing tree property.*

PROOF. Assume that $T \in \mathfrak{SR}(X,Y)$ does not possess the desired property. Then there exists $\varepsilon > 0$ such that for all $n \in \mathbb{N}$ we can find a (T,n,ε)-tree $(x_{nk}^{(j)})$ with $\|x_{nk}^{(j)}\| < 2\|x_{n0}^{(1)}\|$. By scaling, we may arrange that $\|x_{n0}^{(1)}\| = 1$.

Fix any non-trivial ultrafilter $\mathcal{U}$ on $\mathbb{N}$, and let $x_{nk}^{(j)} := o$ if $k > n$. It follows from $\|x_{nk}^{(j)}\| < 2$ and $T^{\mathcal{U}} \in \mathfrak{R}(X^{\mathcal{U}}, Y^{\mathcal{U}})$ that $\boldsymbol{W}$, the closed convex hull of all $\boldsymbol{y}_k^{(j)} := (Tx_{nk}^{(j)})^{\mathcal{U}}$, is weakly compact. Applying the previous lemma yields a proper closed convex subset $\boldsymbol{C} \subset \boldsymbol{W}$ such that $\operatorname{diam}(\boldsymbol{W} \setminus \boldsymbol{C}) \leq \varepsilon/2$. We conclude from

$$\frac{\boldsymbol{y}_{k+1}^{(2j-1)} + \boldsymbol{y}_{k+1}^{(2j)}}{2} = \boldsymbol{y}_k^{(j)} \quad \text{and} \quad \left\| \frac{\boldsymbol{y}_{k+1}^{(2j)} - \boldsymbol{y}_{k+1}^{(2j-1)}}{2} \right\| \geq \varepsilon \tag{1}$$

that

$$\|\boldsymbol{y}_k^{(j)} - \boldsymbol{y}_{k+1}^{(2j-1)}\| \geq \varepsilon \quad \text{and} \quad \|\boldsymbol{y}_k^{(j)} - \boldsymbol{y}_{k+1}^{(2j)}\| \geq \varepsilon. \tag{2}$$

Of course, there exists $\boldsymbol{y}_k^{(j)} \in \boldsymbol{W} \setminus \boldsymbol{C}$. Then (2) implies that $\boldsymbol{y}_{k+1}^{(2j-1)} \in \boldsymbol{C}$ and $\boldsymbol{y}_{k+1}^{(2j)} \in \boldsymbol{C}$. So it follows from (1) that $\boldsymbol{y}_k^{(j)} \in \boldsymbol{C}$. This contradiction proves that T must have the growing tree property.

7.10.19 We associate with every $\varepsilon > 0$ a constant $\frac{3}{4} \le w_\varepsilon < 1$, which will be specified later; see 7.10.21 (Remark). Furthermore, let

$$w_{\varepsilon,n} := w_\varepsilon + \tfrac{1}{3^n}(1 - w_\varepsilon).$$

Then

$$1 = w_{\varepsilon,0} > w_{\varepsilon,1} > \ldots > w_{\varepsilon,n} > \ldots > w_\varepsilon$$

and

$$w_{\varepsilon,n+1} - w_{\varepsilon,m} < \tfrac{1}{3^{n+1}}(1 - w_\varepsilon) = \tfrac{1}{2}(w_{\varepsilon,n} - w_{\varepsilon,n+1}) \quad \text{if } m > n.$$

Next, we fix an operator $T \in \mathfrak{L}(X, Y)$ and define for $x \in X$ the following quantities:

$$\|x\|_\varepsilon := \inf \frac{w_{\varepsilon,n}}{2^n} \sum_{j=1}^{2^n} \|x_n^{(j)}\|, \tag{1_ε}$$

where the infimum ranges over all (T, n, ε)-trees $(x_k^{(j)})$ with $x_0^{(1)} = x$ and of arbitrary length $n = 0, 1, \ldots$,

$$|||x|||_\varepsilon := \inf \sum_{i=1}^{N} \|x_i\|_\varepsilon, \tag{2_ε}$$

the infimum being taken over all finite decompositions $x = \sum_{i=1}^{N} x_i$ with $N = 1, 2, \ldots$,

$$|||x||| := \left(\sum_{m=1}^{\infty} \frac{1}{2^m} |||x|||_{1/2^m}^2 \right)^{1/2}. \tag{3}$$

7.10.20 LEMMA.

$$w_\varepsilon \|x\| \le |||x|||_\varepsilon \le \|x\|_\varepsilon \le \|x\| \quad \textit{and} \quad \tfrac{3}{4}\|x\| \le |||x||| \le \|x\|.$$

Moreover, $\|\cdot\|_\varepsilon$ is a quasi-norm, whereas $|||\cdot|||_\varepsilon$ and $|||\cdot|||$ are norms.

PROOF. Substituting the trivial tree $x_0^{(1)} := x$ in the defining expression of $\|x\|_\varepsilon$ yields $\|x\|_\varepsilon \le \|x\|$. Moreover, the trivial decomposition $x = x$ implies that $|||x|||_\varepsilon \le \|x\|_\varepsilon$. Next, we conclude from $w_\varepsilon < w_{\varepsilon,n}$ and

$$\|x\| \le \frac{1}{2^n} \sum_{j=1}^{2^n} \|x_n^{(j)}\| \quad \text{for} \quad x = \frac{1}{2^n} \sum_{j=1}^{2^n} x_n^{(j)}$$

that $w_\varepsilon \|x\| \le \|x\|_\varepsilon$. Thus, given any decomposition $x = \sum_{i=1}^{N} x_i$, we have

$$w_\varepsilon \|x\| \le w_\varepsilon \sum_{i=1}^{N} \|x_i\| \le \sum_{i=1}^{N} \|x_i\|_\varepsilon.$$

Hence $\frac{3}{4}\|x\| \le w_\varepsilon \|x\| \le |||x|||_\varepsilon$. Of course,

$$\tfrac{3}{4}\|x\| \le \left(\sum_{m=1}^{\infty} \frac{1}{2^m} |||x|||_{1/2^m}^2 \right)^{1/2} \le \|x\|,$$

$\|\lambda x\|_\varepsilon = |\lambda|\|x\|_\varepsilon$, $|||\lambda x|||_\varepsilon = |\lambda| |||x|||_\varepsilon$ and $|||\lambda x||| = |\lambda| |||x|||$. The quasi-triangle inequality for $\|\cdot\|_\varepsilon$ follows from $\frac{3}{4}\|x\| \le \|x\|_\varepsilon \le \|x\|$, and the definition of $|||\cdot|||_\varepsilon$ is made in order to get the triangle inequality, which carries over to $|||\cdot|||$.

7.10.21 The following proof, which goes back to P. Enflo, is quite lengthy. Using the quantities $\|\cdot\|_\varepsilon$, $|||\cdot|||_\varepsilon$ and $|||\cdot|||$, the required renorming will be obtained step by step.

PROPOSITION. *Operators with the growing tree property are uniformly convexifiable.*

PROOF. Assume that $T \in \mathfrak{L}(X, Y)$ has the growing tree property. Since the case $T = O$ is trivial, we may arrange by scaling that $\|T\| = 1$.

Preliminary step: *For $\varepsilon > 0$ there exist $\frac{3}{4} \le w_\varepsilon < 1$ and $n_\varepsilon \in \mathbb{N}_0$ such that*

$$\|x\|_\varepsilon = \inf \frac{w_{\varepsilon,n}}{2^n} \sum_{j=1}^{2^n} \|x_n^{(j)}\| \quad \text{for } x \in X,$$

where the infimum only ranges over all (T, n, ε)-trees $(x_j^{(k)})$ with $x_0^{(1)} = x$ that have length less than n_ε.

We find a level $n_\varepsilon \in \mathbb{N}_0$ such that all (T, n, ε)-trees $(x_k^{(j)})$ with $n \ge n_\varepsilon$ contain at least one element $x_{n_\varepsilon}^{(j_0)}$ with $\|x_{n_\varepsilon}^{(j_0)}\| \ge 2\|x_0^{(1)}\|$. Moreover, $\|x_{n_\varepsilon}^{(j)}\| \ge \|x_0^{(1)}\|$ for $j \ne j_0$. If $x_0^{(1)} = x \ne o$, then

$$\frac{1}{2^n} \sum_{j=1}^{2^n} \|x_n^{(j)}\| \ge \frac{1}{2^{n_\varepsilon}} \sum_{j=1}^{2^{n_\varepsilon}} \|x_{n_\varepsilon}^{(j)}\| \ge (1 + \tfrac{1}{2^{n_\varepsilon}})\|x\|.$$

Choosing $\frac{3}{4} \le w_\varepsilon < 1$ such that $w_\varepsilon(1 + \frac{1}{2^{n_\varepsilon}}) \ge 1 + \frac{1}{2^{n_\varepsilon+1}}$, we get

$$\frac{w_{\varepsilon,n}}{2^n} \sum_{j=1}^{2^n} \|x_n^{(j)}\| \ge (1 + \tfrac{1}{2^{n_\varepsilon+1}})\|x\|_\varepsilon.$$

So (T, n, ε)-trees with $n \ge n_\varepsilon$ are superfluous for approximating $\|x\|_\varepsilon$.

First step: *For $\varepsilon > 0$ there exists $\delta_1 > 0$ such that*

$$\left\| \frac{Tx_+ - Tx_-}{2} \right\| \ge \varepsilon \quad \text{and} \quad \|x_\pm\| = 1$$

imply

$$\left\| \frac{x_+ + x_-}{2} \right\|_\varepsilon \le \frac{\|x_+\|_\varepsilon + \|x_-\|_\varepsilon}{2} - \delta_1.$$

To avoid inconvenient indices, we write $u := x_+$, $v := x_-$ and $x := \frac{x_+ + x_-}{2}$. Fix w_ε and n_ε as described above. For the present, we let

$$\delta_1 := \delta_0 \min_{0 \le n < n_\varepsilon} (w_{\varepsilon,n} - w_{\varepsilon,n+1}) \le \delta_0 \le \tfrac{1}{2}. \tag{1}$$

Choose a (T, m, ε)-tree $(u_h^{(i)})$ with $u_0^{(1)} = u$ and a (T, n, ε)-tree $(v_k^{(j)})$ with $v_0^{(1)} = v$ such that

$$\frac{w_{\varepsilon,m}}{2^m} \sum_{i=1}^{2^m} \|u_m^{(i)}\| \le \|u\|_\varepsilon + \delta_1 \quad \text{and} \quad \frac{w_{\varepsilon,n}}{2^n} \sum_{j=1}^{2^n} \|v_n^{(j)}\| \le \|v\|_\varepsilon + \delta_1. \tag{2}$$

In addition, we may assume that $m \ge n$ and $m, n < n_\varepsilon$. Putting

$$x_{n+1}^{(j)} := u_n^{(j)} \quad \text{and} \quad x_{n+1}^{(j+2^n)} := v_n^{(j)} \quad \text{for } j = 1, \ldots, 2^n$$

yields a $(T, n+1, \varepsilon)$-tree $(x_k^{(j)})$ with $x_0^{(1)} = x$,

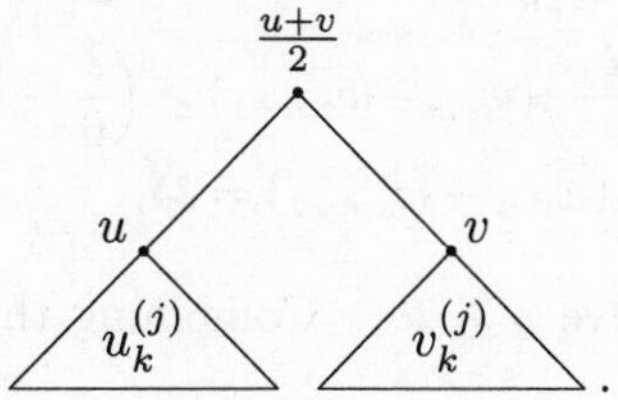

Indeed, we have

$$\|u\| = 1 \ge \|x\|, \quad \|v\| = 1 \ge \|x\| \quad \text{and} \quad \left\| \frac{Tu - Tv}{2} \right\| \ge \varepsilon \ge \varepsilon \|x\|.$$

By definition,

$$\begin{aligned} \|x\|_\varepsilon &\le \frac{w_{\varepsilon,n+1}}{2^{n+1}} \left(\sum_{j=1}^{2^n} \|u_n^{(j)}\| + \sum_{j=1}^{2^n} \|v_n^{(j)}\| \right) \\ &\le \frac{w_{\varepsilon,n+1}}{2^{m+1}} \sum_{i=1}^{2^m} \|u_m^{(i)}\| + \frac{w_{\varepsilon,n+1}}{2^{n+1}} \sum_{j=1}^{2^n} \|v_n^{(j)}\| \\ &= \frac{w_{\varepsilon,m}}{2^{m+1}} \sum_{i=1}^{2^m} \|u_m^{(i)}\| + \frac{w_{\varepsilon,n}}{2^{n+1}} \sum_{j=1}^{2^n} \|v_n^{(j)}\| - \varrho, \end{aligned} \tag{3}$$

where

$$\varrho := \frac{w_{\varepsilon,m} - w_{\varepsilon,n+1}}{2^{m+1}} \sum_{i=1}^{2^m} \|u_m^{(i)}\| + \frac{w_{\varepsilon,n} - w_{\varepsilon,n+1}}{2^{n+1}} \sum_{j=1}^{2^n} \|v_n^{(j)}\|.$$

Note that

$$1 = \|u\| \le \frac{1}{2^m} \sum_{i=1}^{2^m} \|u_m^{(i)}\|, \quad 1 = \|v\| \le \frac{1}{2^n} \sum_{j=1}^{2^n} \|v_n^{(j)}\|, \tag{4}$$

and

$$\frac{w_{\varepsilon,m}}{2^m}\sum_{i=1}^{2^m}\|u_m^{(i)}\| \le \|u\|_\varepsilon + \delta_1 \le \|u\| + \delta_1 = 1 + \delta_1. \tag{5}$$

If $m = n$, then (1) and (4) imply

$$\varrho \ge w_{\varepsilon,n} - w_{\varepsilon,n+1} \ge 2\delta_1.$$

In the case $m > n$, we know from 7.10.19 that

$$0 \le w_{\varepsilon,n+1} - w_{\varepsilon,m} < \tfrac{1}{2}(w_{\varepsilon,n} - w_{\varepsilon,n+1}).$$

Moreover, $w_{\varepsilon,m} > \frac{3}{4}$. Thus, choosing δ_0 sufficiently small, (4) (right part) and (5) yield

$$\begin{aligned}\varrho &\ge \frac{(1+\delta_1)(w_{\varepsilon,m} - w_{\varepsilon,n+1})}{2w_{\varepsilon,m}} + \frac{w_{\varepsilon,n} - w_{\varepsilon,n+1}}{2}\\ &\ge \Big(\frac{1}{2} - \frac{1+\delta_1}{3}\Big)(w_{\varepsilon,n} - w_{\varepsilon,n+1}) \ge \Big(\frac{1}{6} - \frac{\delta_0}{3}\Big)(w_{\varepsilon,n} - w_{\varepsilon,n+1})\\ &\ge 2\delta_0 \min_{0\le n<n_\varepsilon}(w_{\varepsilon,n} - w_{\varepsilon,n+1}) = 2\delta_1.\end{aligned}$$

Hence we always have $\varrho \ge 2\delta_1$. Combining this fact with (2) and (3) gives

$$\|x\|_\varepsilon \le \frac{\|u\|_\varepsilon + \|v\|_\varepsilon}{2} + \delta_1 - \varrho \le \frac{\|x_+\|_\varepsilon + \|x_-\|_\varepsilon}{2} - \delta_1.$$

Second step: *For $\varepsilon > 0$ there exists $\delta_2 > 0$ such that*

$$\left\|\frac{Tx_+ - Tx_-}{2}\right\| \ge 4\varepsilon \quad \textit{and} \quad \|x_\pm\| = 1$$

imply

$$\left|\!\left|\!\left|\frac{x_+ + x_-}{2}\right|\!\right|\!\right|_\varepsilon \le \frac{|\!|\!|x_+|\!|\!|_\varepsilon + |\!|\!|x_-|\!|\!|_\varepsilon}{2} - \delta_2.$$

Given $\eta > 0$, we choose decompositions $x_\pm = \sum\limits_{i\in\mathbb{F}_\pm} x_i^\pm$ such that

$$\sum_{i\in\mathbb{F}_\pm}\|x_i^\pm\|_\varepsilon \le |\!|\!|x_\pm|\!|\!|_\varepsilon + \eta \le 1 + \eta \quad \text{and} \quad x_i^\pm \ne o,$$

where $\mathbb{F}_\pm := \{1, \ldots, N_\pm\}$. By symmetry, it can be assumed that $\mathbb{F}_+ \subseteq \mathbb{F}_-$. We may further arrange that

$$\|x_i^+\| = \|x_i^-\| \quad \text{whenever } i \in \mathbb{F}_+.$$

To this end, let

$$i_0 := \min\{\, i \in \mathbb{F}_+ : \|x_i^+\| \ne \|x_i^-\|\,\}.$$

If $\|x_{i_0}^+\| < \|x_{i_0}^-\|$, then we split $x_{i_0}^-$ into the parts $\lambda x_{i_0}^-$ and $(1-\lambda)x_{i_0}^-$, where $0<\lambda<1$ is defined by $\|x_{i_0}^+\| = \lambda\|x_{i_0}^-\|$. The new decomposition

$$x_- = \sum_{i<i_0} x_i^- + \lambda x_{i_0}^- + (1-\lambda)x_{i_0}^- + \sum_{i>i_0} x_i^-$$

has the advantage that we get another pair of elements with equal norms, while the number of summands is increased by 1 only. The opposite case $\|x_{i_0}^-\| < \|x_{i_0}^+\|$ can be treated analogously. This process terminates after at most $N_+ + N_-$ steps.

Let

$$\mathbb{F} := \left\{ i \in \mathbb{F}_+ : \left\| \frac{Tx_i^+}{\|x_i^+\|} - \frac{Tx_i^-}{\|x_i^-\|} \right\| \geq 2\varepsilon \right\}.$$

We have

$$\begin{aligned} \sum_{i\in\mathbb{F}_+\setminus\mathbb{F}} \|Tx_i^+ - Tx_i^-\| &\leq 2\varepsilon \sum_{i\in\mathbb{F}_+\setminus\mathbb{F}} \|x_i^+\| \\ &\leq 2\varepsilon w_\varepsilon^{-1} \sum_{i\in\mathbb{F}_+} \|x_i^+\|_\varepsilon \leq 2\varepsilon w_\varepsilon^{-1}(1+\eta) \qquad (1) \end{aligned}$$

and

$$\begin{aligned} 8\varepsilon &\leq \|Tx_+ - Tx_-\| \\ &\leq \sum_{i\in\mathbb{F}} \|Tx_i^+ - Tx_i^-\| + \sum_{i\in\mathbb{F}_+\setminus\mathbb{F}} \|Tx_i^+ - Tx_i^-\| + \sum_{i\in\mathbb{F}_-\setminus\mathbb{F}_+} \|Tx_i^-\|. \qquad (2) \end{aligned}$$

Next, we conclude from $\|x_i^+\| = \|x_i^-\|$ for $i \in \mathbb{F}_+$ that

$$\begin{aligned} \sum_{i\in\mathbb{F}_-\setminus\mathbb{F}_+} \|x_i^-\| &= \sum_{i\in\mathbb{F}_-} \|x_i^-\| - \sum_{i\in\mathbb{F}_+} \|x_i^+\| \\ &\leq w_\varepsilon^{-1} \sum_{i\in\mathbb{F}_-} \|x_i^-\|_\varepsilon - \|x_+\| \leq w_\varepsilon^{-1}(1+\eta) - 1. \qquad (3) \end{aligned}$$

Combining (1), (2) and (3) yields

$$8\varepsilon \leq 2\sum_{i\in\mathbb{F}} \|x_i^+\| + 2\varepsilon w_\varepsilon^{-1}(1+\eta) + w_\varepsilon^{-1}(1+\eta) - 1.$$

We now choose w_ε and η such that

$$(1+2\varepsilon)w_\varepsilon^{-1} \leq 1+4\varepsilon \quad \text{and} \quad (1+4\varepsilon)(1+\eta) \leq 1+6\varepsilon.$$

Then

$$\varepsilon \leq \sum_{i\in\mathbb{F}} \|x_i^+\|. \qquad (4)$$

For $i \in \mathbb{F}$, we know from the first step that

$$\left\| \frac{x_i^+}{\|x_i^+\|} + \frac{x_i^-}{\|x_i^-\|} \right\|_\varepsilon \leq \left\| \frac{x_i^+}{\|x_i^+\|} \right\|_\varepsilon + \left\| \frac{x_i^-}{\|x_i^-\|} \right\|_\varepsilon - 2\delta_1.$$

In view of $\|x_i^+\| = \|x_i^-\|$,

$$\|x_i^+ + x_i^-\|_\varepsilon \le \|x_i^+\| + \|x_i^-\| - 2\delta_1 \|x_i^+\|$$

and, by (4), we get

$$\sum_{i\in\mathbb{F}} \|x_i^+ + x_i^-\|_\varepsilon \le \sum_{i\in\mathbb{F}} \|x_i^+\| + \sum_{i\in\mathbb{F}} \|x_i^-\| - 2\delta_1\varepsilon. \tag{5}$$

Since

$$x_+ + x_- = \sum_{i\in\mathbb{F}} (x_i^+ + x_i^-) + \sum_{i\in\mathbb{F}_+\setminus\mathbb{F}} x_i^+ + \sum_{i\in\mathbb{F}_-\setminus\mathbb{F}} x_i^-,$$

it follows from (5) that

$$\begin{aligned}
|||x_+ + x_-|||_\varepsilon &\le \sum_{i\in\mathbb{F}} \|x_i^+ + x_i^-\|_\varepsilon + \sum_{i\in\mathbb{F}_+\setminus\mathbb{F}} \|x_i^+\|_\varepsilon + \sum_{i\in\mathbb{F}_-\setminus\mathbb{F}} \|x_i^-\|_\varepsilon \\
&\le \sum_{i\in\mathbb{F}} \|x_i^+\|_\varepsilon + \sum_{i\in\mathbb{F}} \|x_i^-\|_\varepsilon - 2\delta_1\varepsilon + \sum_{i\in\mathbb{F}_+\setminus\mathbb{F}} \|x_i^+\|_\varepsilon + \sum_{i\in\mathbb{F}_-\setminus\mathbb{F}} \|x_i^-\|_\varepsilon \\
&= \sum_{i\in\mathbb{F}_+} \|x_i^+\|_\varepsilon + \sum_{i\in\mathbb{F}_-} \|x_i^-\|_\varepsilon - 2\delta_1\varepsilon \le |||x_+|||_\varepsilon + |||x_-|||_\varepsilon + 2\eta - 2\delta_1\varepsilon.
\end{aligned}$$

Choosing η sufficiently small, we finally let $\delta_2 := \delta_1\varepsilon - \eta > 0$.

Third step: *For $\varepsilon > 0$ there exists $\delta_3 > 0$ such that*

$$\left\| \frac{Tx_+ - Tx_-}{2} \right\| \ge 5\varepsilon \quad \textit{and} \quad |||x_\pm|||_\varepsilon = 1$$

imply

$$\left|\left|\left| \frac{x_+ + x_-}{2} \right|\right|\right|_\varepsilon \le 1 - \delta_3.$$

In view of $w_\varepsilon \|x_\pm\| \le |||x_\pm|||_\varepsilon = 1 \le \|x_\pm\|$, we have

$$0 \le 1 - \frac{1}{\|x_\pm\|} \le 1 - w_\varepsilon.$$

Hence

$$\begin{aligned}
\|Tx_+ - Tx_-\| &\le \left\| Tx_+ - \frac{Tx_+}{\|x_+\|} \right\| + \left\| \frac{Tx_+}{\|x_+\|} - \frac{Tx_-}{\|x_-\|} \right\| + \left\| \frac{Tx_-}{\|x_-\|} - Tx_- \right\| \\
&\le w_\varepsilon^{-1} \left| 1 - \frac{1}{\|x_+\|} \right| + \left\| \frac{Tx_+}{\|x_+\|} - \frac{Tx_-}{\|x_-\|} \right\| + w_\varepsilon^{-1} \left| \frac{1}{\|x_-\|} - 1 \right|
\end{aligned}$$

gives

$$\left\| \frac{Tx_+}{\|x_+\|} - \frac{Tx_-}{\|x_-\|} \right\| \ge \|Tx_+ - Tx_-\| - 2(w_\varepsilon^{-1} - 1) \ge 8\varepsilon$$

provided that $w_\varepsilon^{-1} \le 1 + \varepsilon$. Consequently, by the second step,

$$\left|\left|\left| \frac{x_+}{\|x_+\|} + \frac{x_-}{\|x_-\|} \right|\right|\right|_\varepsilon \le \left|\left|\left| \frac{x_+}{\|x_+\|} \right|\right|\right|_\varepsilon + \left|\left|\left| \frac{x_-}{\|x_-\|} \right|\right|\right|_\varepsilon - 2\delta_2.$$

Finally,

$$\begin{aligned}|||x_+ + x_-|||_\varepsilon &\le \left|\left|\left| x_+ - \frac{x_+}{\|x_+\|}\right|\right|\right|_\varepsilon + \left|\left|\left| \frac{x_+}{\|x_+\|} + \frac{x_-}{\|x_-\|}\right|\right|\right|_\varepsilon + \left|\left|\left| x_- - \frac{x_-}{\|x_-\|}\right|\right|\right|_\varepsilon \\ &= \left|1 - \frac{1}{\|x_+\|}\right| + \left|\left|\left| \frac{x_+}{\|x_+\|} + \frac{x_-}{\|x_-\|}\right|\right|\right|_\varepsilon + \left|\frac{1}{\|x_-\|} - 1\right| \le 2 - 2\delta_2.\end{aligned}$$

So we may take $\delta_3 := \delta_2$.

Fourth step: *For $\varepsilon > 0$ there exists $\delta_4 > 0$ such that*

$$\left\|\frac{Tx_+ - Tx_-}{2}\right\| \ge 6\varepsilon\left(\frac{|||x_+|||_\varepsilon^2 + |||x_-|||_\varepsilon^2}{2}\right)^{1/2}$$

implies

$$\left|\left|\left|\frac{x_+ + x_-}{2}\right|\right|\right|_\varepsilon \le (1-\delta_4)\left(\frac{|||x_+|||_\varepsilon^2 + |||x_-|||_\varepsilon^2}{2}\right)^{1/2}.$$

This is an immediate consequence of Lemma 7.10.3.

Final step: *For $\varepsilon > 0$ there exists $\delta > 0$ such that*

$$\left\|\frac{Tx_+ - Tx_-}{2}\right\| \ge \varepsilon\left(\frac{|||x_+|||^2 + |||x_-|||^2}{2}\right)^{1/2}$$

implies

$$\left|\left|\left|\frac{x_+ + x_-}{2}\right|\right|\right| \le (1-\delta)\left(\frac{|||x_+|||^2 + |||x_-|||^2}{2}\right)^{1/2}.$$

This is uniform convexity.

Choose $\varepsilon_0 = \frac{1}{2^{m_0}}$ such that $\varepsilon \ge 8\varepsilon_0$. We conclude from 7.10.20 that

$$\tfrac{3}{4}|||x_\pm||| \le \tfrac{3}{4}\|x_\pm\| \le |||x_\pm|||_{1/2^{m_0}} \le \|x_\pm\| \le \tfrac{4}{3}|||x_\pm|||.$$

Hence

$$\left\|\frac{Tx_+ - Tx_-}{2}\right\| \ge \varepsilon\left(\frac{|||x_+|||^2 + |||x_-|||^2}{2}\right)^{1/2} \ge 6\varepsilon_0\left(\frac{|||x_+|||_{1/2^{m_0}}^2 + |||x_-|||_{1/2^{m_0}}^2}{2}\right)^{1/2}.$$

Thus, in view of the fourth step, there exists $\delta_4 > 0$ such that

$$\left|\left|\left|\frac{x_+ + x_-}{2}\right|\right|\right|_{1/2^{m_0}} \le (1-\delta_4)\left(\frac{|||x_+|||_{1/2^{m_0}}^2 + |||x_-|||_{1/2^{m_0}}^2}{2}\right)^{1/2}.$$

In the case $m \ne m_0$ we use the obvious estimate

$$\left|\left|\left|\frac{x_+ + x_-}{2}\right|\right|\right|_{1/2^m} \le \left(\frac{|||x_+|||_{1/2^m}^2 + |||x_-|||_{1/2^m}^2}{2}\right)^{1/2}.$$

Then

$$\left|\!\left|\!\left|\frac{x_+ + x_-}{2}\right|\!\right|\!\right|^2 = \sum_{m=1}^{\infty} \frac{1}{2^m} \left|\!\left|\!\left|\frac{x_+ + x_-}{2}\right|\!\right|\!\right|^2_{1/2^m}$$

$$\leq \left\{ \begin{array}{c} \displaystyle\sum_{m=1}^{\infty} \frac{1}{2^m} \frac{|||x_+|||^2_{1/2^m} + |||x_-|||^2_{1/2^m}}{2} \\ - \\ \displaystyle\frac{[1-(1-\delta_4)^2]}{2^{m_0}} \frac{|||x_+|||^2_{1/2^{m_0}} + |||x_-|||^2_{1/2^{m_0}}}{2} \end{array} \right\}$$

$$\leq \left(1 - \frac{9}{16}\frac{[1-(1-\delta_4)^2]}{2^{m_0}}\right) \frac{|||x_+|||^2 + |||x_-|||^2}{2}.$$

REMARK. For the convenience of the reader, we recapitulate which conditions must be satisfied by the weights w_ε. First of all, $\frac{3}{4} \leq w_\varepsilon < 1$. Next, we choose $n_\varepsilon \in \mathbb{N}_0$ according to the growing tree property and assume that $w_\varepsilon(1+\frac{1}{2^{n_\varepsilon}}) \geq 1+\frac{1}{2^{n_\varepsilon+1}}$. Finally, it is required that $w_\varepsilon^{-1} \leq 1+\varepsilon$ and $(1+2\varepsilon)w_\varepsilon^{-1} \leq 1+4\varepsilon$.

7.10.22 We now combine 7.10.13, 7.10.18 and 7.10.21.

THEOREM. $\mathfrak{UC} \circ \mathfrak{L} = \mathfrak{SR}$.

7.10.23 Luckily, with respect to uniform convexity, the transition from T to $[l_2, T]$ is relatively simple. The following result goes back to M. M. Day and the proof is due to E. J. McShane; see [day 2] and [mcs].

PROPOSITION. *Let $1 < r < \infty$. If an operator T is uniformly convex, then so is $[L_r, T]$.*

PROOF. Assume that $\|T\| = 1$. Let $\boldsymbol{f}_\pm \in [L_r(M,\mu), X]$ such that

$$\|\boldsymbol{f}_\pm|L_r\| = 1 \quad\text{and}\quad \left\|\frac{[L_r,T]\boldsymbol{f}_+ - [L_r,T]\boldsymbol{f}_-}{2}\right|L_r\right\| \geq \varepsilon > 0.$$

Decompose M into the subsets

$$A := \left\{ \xi \in M : \left\|\frac{T\boldsymbol{f}_+(\xi) - T\boldsymbol{f}_-(\xi)}{2}\right\|^r > \tfrac{1}{2}\varepsilon^r \frac{\|\boldsymbol{f}_+(\xi)\|^r + \|\boldsymbol{f}_-(\xi)\|^r}{2} \right\}$$

and

$$B := \left\{ \xi \in M : \left\|\frac{T\boldsymbol{f}_+(\xi) - T\boldsymbol{f}_-(\xi)}{2}\right\|^r \leq \tfrac{1}{2}\varepsilon^r \frac{\|\boldsymbol{f}_+(\xi)\|^r + \|\boldsymbol{f}_-(\xi)\|^r}{2} \right\}.$$

Then it follows from

$$\int\limits_B \left\|\frac{T\boldsymbol{f}_+(\xi) - T\boldsymbol{f}_-(\xi)}{2}\right\|^r d\mu(\xi) \leq \tfrac{1}{2}\varepsilon^r \quad\text{and}\quad \int\limits_M \left\|\frac{T\boldsymbol{f}_+(\xi) - T\boldsymbol{f}_-(\xi)}{2}\right\|^r d\mu(\xi) \geq \varepsilon^r$$

that

$$\int\limits_A \left\| \frac{T\boldsymbol{f}_+(\xi) - T\boldsymbol{f}_-(\xi)}{2} \right\|^r d\mu(\xi) \geq \tfrac{1}{2}\varepsilon^r.$$

Hence

$$\begin{aligned}
2^{1/r'}\varepsilon &\leq \Bigg(\int\limits_A \|T\boldsymbol{f}_+(\xi) - T\boldsymbol{f}_-(\xi)\|^r d\mu(\xi) \Bigg)^{1/r} \\
&\leq \Bigg(\int\limits_A \|\boldsymbol{f}_+(\xi)\|^r d\mu(\xi) \Bigg)^{1/r} + \Bigg(\int\limits_A \|\boldsymbol{f}_-(\xi)\|^r d\mu(\xi) \Bigg)^{1/r} \\
&\leq 2^{1/r'} \Bigg(\int\limits_A \Big(\|\boldsymbol{f}_+(\xi)\|^r + \|\boldsymbol{f}_-(\xi)\|^r \Big) d\mu(\xi) \Bigg)^{1/r}. \qquad (1)
\end{aligned}$$

By 7.10.3, there exists $\delta > 0$ such that

$$\left\| \frac{Tx_+ - Tx_-}{2} \right\| \leq 2^{-1/r}\varepsilon \left(\frac{\|x_+\|^r + \|x_-\|^r}{2} \right)^{1/r}$$

whenever

$$\left\| \frac{x_+ + x_-}{2} \right\| \geq (1-\delta) \left(\frac{\|x_+\|^r + \|x_-\|^r}{2} \right)^{1/r}.$$

Thus, for $\xi \in A$, we have

$$\left\| \frac{\boldsymbol{f}_+(\xi) + \boldsymbol{f}_-(\xi)}{2} \right\| < (1-\delta) \left(\frac{\|\boldsymbol{f}_+(\xi)\|^r + \|\boldsymbol{f}_-(\xi)\|^r}{2} \right)^{1/r}$$

or

$$\begin{aligned}
&\frac{\|\boldsymbol{f}_+(\xi)\|^r + \|\boldsymbol{f}_-(\xi)\|^r}{2} - \left\| \frac{\boldsymbol{f}_+(\xi) + \boldsymbol{f}_-(\xi)}{2} \right\|^r > \\
&\qquad\qquad > [1 - (1-\delta)^r] \frac{\|\boldsymbol{f}_+(\xi)\|^r + \|\boldsymbol{f}_-(\xi)\|^r}{2}. \qquad (2)
\end{aligned}$$

Finally, from (1) and (2) we get

$$\begin{aligned}
1 - \left\| \frac{\boldsymbol{f}_+ + \boldsymbol{f}_-}{2} \middle| L_r \right\|^r &= \\
&= \int\limits_M \left(\frac{\|\boldsymbol{f}_+(\xi)\|^r + \|\boldsymbol{f}_-(\xi)\|^r}{2} - \left\| \frac{\boldsymbol{f}_+(\xi) + \boldsymbol{f}_-(\xi)}{2} \right\|^r \right) d\mu(\xi) \\
&\geq \int\limits_A \left(\frac{\|\boldsymbol{f}_+(\xi)\|^r + \|\boldsymbol{f}_-(\xi)\|^r}{2} - \left\| \frac{\boldsymbol{f}_+(\xi) + \boldsymbol{f}_-(\xi)}{2} \right\|^r \right) d\mu(\xi) \\
&\geq [1 - (1-\delta)^r] \int\limits_A \frac{\|\boldsymbol{f}_+(\xi)\|^r + \|\boldsymbol{f}_-(\xi)\|^r}{2} d\mu(\xi) \geq [1 - (1-\delta)^r] \frac{\varepsilon^r}{2}.
\end{aligned}$$

That is,

$$\left\| \frac{f_+ + f_-}{2} \middle| L_r \right\|^r \le 1 - [1 - (1-\delta)^r] \frac{\varepsilon^r}{2}.$$

7.10.24 Since $\mathfrak{SR} = \mathfrak{UC} \circ \mathfrak{L}$, the above result implies the following corollary.

PROPOSITION. *If an operator T is super weakly compact, then so is $[l_2, T]$.*

REMARK. Unfortunately, no direct proof of the above implication seems to be known. This bad luck is due to the fact that nobody has been able to find isomorphic embeddings of $[l_2, X]^{\mathcal{U}}$ into some $[L_2(M,\mu), X^{\mathcal{V}}]$. This is quite understandable, since isometric embeddings cannot exist at all; see [HAY*, p. 123].

7.10.25 THEOREM. $\mathfrak{L} \circ \mathfrak{UT} = \mathfrak{AT} = \mathfrak{SR} = \mathfrak{AC} = \mathfrak{UC} \circ \mathfrak{L}$.

PROOF. Combining the foregoing implication with that in 7.7.20 yields

$$\mathfrak{SR} \subseteq \mathfrak{MT}.$$

Moreover, we have

$$\mathfrak{MT} = \mathfrak{AT} \quad \text{(by 7.7.7)},$$

$$\mathfrak{AT} \subseteq \mathfrak{SR} \quad \text{(by 7.6.15)}.$$

The rest follows from

$$\mathfrak{SR} = \mathfrak{UC} \circ \mathfrak{L} \quad \text{(by 7.10.22)},$$

$$\mathfrak{AT} = (\mathfrak{AT})' = (\mathfrak{AC})' = \mathfrak{AC} \quad \text{(by 7.5.21)},$$

$$(\mathfrak{UC} \circ \mathfrak{L})' = \mathfrak{L} \circ \mathfrak{UT} \text{ and } (\mathfrak{L} \circ \mathfrak{UT})' = \mathfrak{UC} \circ \mathfrak{L} \quad \text{(by 7.10.6)}.$$

7.10.26 We conclude this section by stating explicitly a special case of the preceding result.

ENFLO'S THEOREM. *For every Banach space X, the following are equivalent:*

(1) *X is superreflexive.*

(2) *X can be given an equivalent uniformly convex norm.*

(3) *X can be given an equivalent uniformly smooth norm.*

REMARK. We know from 7.8.16 and 7.7.8 that

$$\bigcup_{1<p\le 2} \mathsf{AT}_p = \mathsf{AT} = \mathsf{SR} = \mathsf{AC} = \bigcup_{2\le q<\infty} \mathsf{AC}_q.$$

Therefore, the above result is also a consequence of 7.9.22. In this way, we even get uniformly p-smooth and uniformly q-convex renormings. However, this method only works for spaces.

8

Unconditionality

We now modify the definitions of Riemann and Dirichlet ideal norms by allowing changes of signs. For Fourier and Walsh systems, unconditionality only occurs in Hilbertian Banach spaces. The most interesting case is that of Haar functions, in which we obtain the ideal norms $\boldsymbol{\varrho}_{\pm}(\mathcal{H}(\mathbb{D}_1^n), \mathcal{H}(\mathbb{D}_1^n))$. These can be used to treat operators and spaces with the unconditionality property for martingale differences (UMD).

The celebrated Bourgain–Burkholder theorem says that

$$\boldsymbol{\varrho}_{\pm}(X|\mathcal{H}(\mathbb{D}_1^n), \mathcal{H}(\mathbb{D}_1^n)) = O(1) \quad \text{if and only if} \quad \boldsymbol{\delta}(X|\mathcal{E}_n, \mathcal{E}_n) = O(1).$$

In other terms, the class UMD consists just of all Banach spaces which admit a vector-valued extension of the Hilbert transform.

We refer to the following publications:

Burkholder .... *Martingales and Fourier analysis in Banach spaces* [bur 8],

Deville/Godefroy/Zizler... *Smoothness and renorming in Banach spaces* [DEV*].

8.1 Unconditional Riemann ideal norms

8.1.1 For $\mathcal{A}_n = (a_1, \dots, a_n)$ and $e = (\varepsilon_1, \dots, \varepsilon_n) \in \mathbb{E}_2^n$, we write

$$e \circ \mathcal{A}_n = (\varepsilon_1 a_1, \dots, \varepsilon_n a_n).$$

8.1.2 Let $\mathcal{A}_n = (a_1, \dots, a_n)$ and $\mathcal{B}_n = (b_1, \dots, b_n)$ be orthonormal systems in the Hilbert spaces $L_2(M, \mu)$ and $L_2(N, \nu)$, respectively.

For $T \in \mathfrak{L}(X, Y)$, we denote by $\boldsymbol{\varrho}_{\pm}(T|\mathcal{B}_n, \mathcal{A}_n)$ the least constant $c \geq 0$ such that

$$\left(\int\limits_N \Big\| \sum_{k=1}^n T x_k b_k(t) \Big\|^2 d\nu(t) \right)^{1/2} \leq c \left(\int\limits_M \Big\| \sum_{k=1}^n \varepsilon_k x_k a_k(s) \Big\|^2 d\mu(s) \right)^{1/2} \qquad (\mathrm{R}_{\pm})$$

whenever $x_1, \dots, x_n \in X$ and $e = (\varepsilon_1, \dots, \varepsilon_n) \in \mathbb{E}_2^n$. More concisely, the preceding inequality reads as follows:

$$\|(Tx_k)|\mathcal{B}_n\| \le c\,\|(x_k)|e \circ \mathcal{A}_n\|.$$

We refer to

$$\boldsymbol{\varrho}_\pm(\mathcal{B}_n, \mathcal{A}_n) : T \longrightarrow \boldsymbol{\varrho}_\pm(T|\mathcal{B}_n, \mathcal{A}_n)$$

as an **unconditional Riemann ideal norm.**

8.1.3 Via the following formula, almost all results from Section 3.3 carry over to this new setting.

PROPOSITION.

$$\boldsymbol{\varrho}_\pm(\mathcal{B}_n, \mathcal{A}_n) = \max_{e \in \mathbb{E}_2^n} \boldsymbol{\varrho}(\mathcal{B}_n, e \circ \mathcal{A}_n) = \max_{e \in \mathbb{E}_2^n} \boldsymbol{\varrho}(e \circ \mathcal{B}_n, \mathcal{A}_n).$$

8.1.4 We proceed with a trivial observation which suggests viewing $\boldsymbol{\varrho}_\pm(\mathcal{A}_n, \mathcal{A}_n)$ as a measure of *conditionality* (=*non-unconditionality*). It will be shown in 8.4.13 that the Fourier system is highly conditional.

PROPOSITION. *An orthonormal system* $\mathcal{A}_n = (a_1, \dots, a_n)$ *is* $\mathbb{R}$*-unconditional if and only if* $\boldsymbol{\varrho}_\pm(\mathcal{A}_n, \mathcal{A}_n) = 1$.

8.2 Unconditional Dirichlet ideal norms

8.2.1 Let $\mathcal{A}_n = (a_1, \dots, a_n)$ and $\mathcal{B}_n = (b_1, \dots, b_n)$ be orthonormal systems in the Hilbert spaces $L_2(M, \mu)$ and $L_2(N, \nu)$, respectively.

For $T \in \mathfrak{L}(X, Y)$, we denote by $\boldsymbol{\delta}_\pm(T|\mathcal{B}_n, \mathcal{A}_n)$ the least constant $c \ge 0$ such that

$$\left(\int_N \Big\| \sum_{k=1}^n \varepsilon_k T \langle \boldsymbol{f}, \overline{a_k} \rangle\, b_k(t) \Big\|^2 d\nu(t) \right)^{1/2} \le c \left(\int_M \|\boldsymbol{f}(s)\|^2 d\mu(s) \right)^{1/2} \qquad (\mathrm{D}_\pm)$$

whenever $\boldsymbol{f} \in [L_2(M, \mu), X]$ and $e = (\varepsilon_1, \dots, \varepsilon_n) \in \mathbb{E}_2^n$. More concisely, the preceding inequality reads as follows:

$$\|(\varepsilon_k T \langle \boldsymbol{f}, \overline{a_k} \rangle)|\mathcal{B}_n\| \le c\,\|\boldsymbol{f}|L_2\|.$$

We refer to

$$\boldsymbol{\delta}_\pm(\mathcal{B}_n, \mathcal{A}_n) : T \longrightarrow \boldsymbol{\delta}_\pm(T|\mathcal{B}_n, \mathcal{A}_n)$$

as an **unconditional Dirichlet ideal norm.**

8.2.2 Via the following formula, which is an analogue of 8.1.3, almost all results from Section 3.4 carry over to this new setting.

PROPOSITION.

$$\boldsymbol{\delta}_\pm(\mathcal{B}_n, \mathcal{A}_n) = \max_{e \in \mathbb{E}_2^n} \boldsymbol{\delta}(\mathcal{B}_n, e \circ \mathcal{A}_n) = \max_{e \in \mathbb{E}_2^n} \boldsymbol{\delta}(e \circ \mathcal{B}_n, \mathcal{A}_n).$$

8.3 Random unconditionality

8.3.1 The ideal norms $\varrho(\mathcal{R}_n \otimes \mathcal{B}_n, \mathcal{A}_n)$ and $\varrho(\mathcal{B}_n, \mathcal{R}_n \otimes \mathcal{A}_n)$ can be viewed as *random versions* of $\varrho_\pm(\mathcal{B}_n, \mathcal{A}_n)$.

PROPOSITION.

$$\varrho(\mathcal{R}_n \otimes \mathcal{B}_n, \mathcal{A}_n) \le \varrho_\pm(\mathcal{B}_n, \mathcal{A}_n) \quad \textit{and} \quad \varrho(\mathcal{B}_n, \mathcal{R}_n \otimes \mathcal{A}_n) \le \varrho_\pm(\mathcal{B}_n, \mathcal{A}_n).$$

PROOF. The left-hand inequality follows from

$$\int\limits_N \Big\| \sum_{k=1}^n T x_k r_k(u) b_k(t) \Big\|^2 d\nu(t) \le \varrho_\pm(T|\mathcal{B}_n, \mathcal{A}_n)^2 \int\limits_M \Big\| \sum_{k=1}^n x_k a_k(s) \Big\|^2 d\mu(s)$$

by integration over $u \in [0,1)$, and the right-hand one can be shown similarly.

8.3.2 Next, we establish a counterpart of the preceding inequalities.

PROPOSITION. *Let $\mathcal{F}_n$ be any normalized system. Then*

$$\varrho_\pm(\mathcal{B}_n, \mathcal{A}_n) \le \varrho(\mathcal{B}_n, \mathcal{R}_n \otimes \mathcal{F}_n) \circ \varrho(\mathcal{R}_n \otimes \mathcal{F}_n, \mathcal{A}_n).$$

In particular,

$$\varrho_\pm(\mathcal{B}_n, \mathcal{A}_n) \le \varrho(\mathcal{B}_n, \mathcal{R}_n) \circ \varrho(\mathcal{R}_n, \mathcal{A}_n).$$

PROOF. Since $\mathcal{R}_n$ is $\mathbb{R}$-unconditional, we have

$$\begin{aligned}
\|(\varepsilon_k S T x_k)|\mathcal{B}_n\| &\le \varrho(S|\mathcal{B}_n, \mathcal{R}_n \otimes \mathcal{F}_n)\, \|(\varepsilon_k T x_k)|\mathcal{R}_n \otimes \mathcal{F}_n\| \\
&= \varrho(S|\mathcal{B}_n, \mathcal{R}_n \otimes \mathcal{F}_n)\, \|(T x_k)|\mathcal{R}_n \otimes \mathcal{F}_n\| \\
&\le \varrho(S|\mathcal{B}_n, \mathcal{R}_n \otimes \mathcal{F}_n)\, \varrho(T|\mathcal{R}_n \otimes \mathcal{F}_n, \mathcal{A}_n)\, \|(x_k)|\mathcal{A}_n\|
\end{aligned}$$

whenever $(\varepsilon_1, \ldots, \varepsilon_n) \in \mathbb{E}_2^n$. The rest follows by letting $\mathcal{F}_n = \mathcal{R}_n$.

8.3.3 PROPOSITION.

$$\begin{aligned}
\varrho(\mathcal{A}_n, \mathcal{I}_n) &\le \varrho(\mathcal{A}_n, \mathcal{R}_n \otimes \mathcal{A}_n) \circ \varrho(\mathcal{R}_n, \mathcal{I}_n), \\
\varrho(\mathcal{I}_n, \mathcal{A}_n) &\le \varrho(\mathcal{I}_n, \mathcal{R}_n) \circ \varrho(\mathcal{R}_n \otimes \mathcal{A}_n, \mathcal{A}_n).
\end{aligned}$$

PROOF. We only check the lower inequality. Indeed, 3.7.8 yields

$$\varrho(\mathcal{I}_n, \mathcal{A}_n) \le \varrho(\mathcal{I}_n, \mathcal{R}_n \otimes \mathcal{A}_n) \circ \varrho(\mathcal{R}_n \otimes \mathcal{A}_n, \mathcal{A}_n) \le \varrho(\mathcal{I}_n, \mathcal{R}_n) \circ \varrho(\mathcal{R}_n \otimes \mathcal{A}_n, \mathcal{A}_n).$$

8.3.4 PROPOSITION. *Let A_n be any system of characters on a compact Abelian group $\mathbb{G}$. Then*

$$\varrho(\mathcal{R}_n \otimes \mathcal{A}_n, \mathcal{A}_n) \le \varrho(\mathcal{R}_n, \mathcal{A}_n) \quad \textit{and} \quad \varrho(\mathcal{A}_n, \mathcal{R}_n \otimes \mathcal{A}_n) \le \varrho(\mathcal{A}_n, \mathcal{R}_n).$$

PROOF. By 3.8.10 and 3.6.3, we have

$$\varrho(\mathcal{R}_n \otimes \mathcal{A}_n, \mathcal{A}_n) = \varrho(\mathcal{R}_n \otimes \mathcal{A}_n, \mathcal{A}_n \otimes \mathcal{A}_n) \le \varrho(\mathcal{R}_n, \mathcal{A}_n).$$

The other inequality follows in the same way.

8.3.5 Finally, we state a similar result which is implied by 3.6.2.

PROPOSITION.
If the orthonormal system $\mathcal{A}_n$ is $\mathbb{R}$-unimodular, then

$$\boldsymbol{\varrho}(\mathcal{R}_n \otimes \mathcal{A}_n, \mathcal{A}_n) = \boldsymbol{\varrho}(\mathcal{R}_n, \mathcal{A}_n) \quad \text{and} \quad \boldsymbol{\varrho}(\mathcal{A}_n, \mathcal{R}_n \otimes \mathcal{A}_n) = \boldsymbol{\varrho}(\mathcal{A}_n, \mathcal{R}_n).$$

If the orthonormal system $\mathcal{A}_n$ is $\mathbb{C}$-unimodular, then

$$\tfrac{2}{\pi}\, \boldsymbol{\varrho}(\mathcal{R}_n, \mathcal{A}_n) \le \boldsymbol{\varrho}(\mathcal{R}_n \otimes \mathcal{A}_n, \mathcal{A}_n) \le \tfrac{\pi}{2}\, \boldsymbol{\varrho}(\mathcal{R}_n, \mathcal{A}_n)$$

and

$$\tfrac{2}{\pi}\, \boldsymbol{\varrho}(\mathcal{A}_n, \mathcal{R}_n) \le \boldsymbol{\varrho}(\mathcal{A}_n, \mathcal{R}_n \otimes \mathcal{A}_n) \le \tfrac{\pi}{2}\, \boldsymbol{\varrho}(\mathcal{A}_n, \mathcal{R}_n).$$

8.4 Fourier unconditionality

8.4.1 We now deal with the unconditional analogues of the ideal norms treated in Section 5.2. Most proofs can easily be adapted. For example, the following result is an immediate consequence of $\boldsymbol{\varrho}(\mathcal{E}_n^\circ, \mathcal{E}_n) \le 3$.

PROPOSITION.

$$\begin{aligned}
\boldsymbol{\varrho}_\pm(\mathcal{E}_n^\circ, \mathcal{E}_n) &\le 3\, \boldsymbol{\varrho}_\pm(\mathcal{E}_n, \mathcal{E}_n),\\
\boldsymbol{\delta}_\pm(\mathcal{E}_n^\circ, \mathcal{E}_n) &\le 3\, \boldsymbol{\delta}_\pm(\mathcal{E}_n, \mathcal{E}_n),\\
\boldsymbol{\delta}_\pm(\mathcal{E}_n^\circ, \mathcal{E}_n^\circ) &\le 3\, \boldsymbol{\delta}_\pm(\mathcal{E}_n, \mathcal{E}_n^\circ).
\end{aligned}$$

8.4.2 We now dualize the second and the third of the above inequalities with the help of 3.4.6.

PROPOSITION.

$$\boldsymbol{\delta}_\pm(\mathcal{E}_n, \mathcal{E}_n^\circ) \le 3\, \boldsymbol{\delta}_\pm(\mathcal{E}_n, \mathcal{E}_n) \quad \text{and} \quad \boldsymbol{\delta}_\pm(\mathcal{E}_n^\circ, \mathcal{E}_n^\circ) \le 3\, \boldsymbol{\delta}_\pm(\mathcal{E}_n^\circ, \mathcal{E}_n).$$

8.4.3 The next inequalities are obtained by the method used in 5.2.7.

PROPOSITION.

$$\begin{aligned}
\boldsymbol{\varrho}_\pm(\mathcal{E}_n, \mathcal{E}_n) &\le \boldsymbol{\varrho}_\pm(\mathcal{E}_n^\circ, \mathcal{E}_n),\\
\boldsymbol{\varrho}_\pm(\mathcal{E}_n, \mathcal{E}_n) &\le \boldsymbol{\varrho}_\pm(\mathcal{E}_n, \mathcal{E}_n^\circ),\\
\boldsymbol{\varrho}_\pm(\mathcal{E}_n, \mathcal{E}_n) &\le \boldsymbol{\varrho}_\pm(\mathcal{E}_n^\circ, \mathcal{E}_n^\circ).
\end{aligned}$$

8.4.4 We proceed with another elementary observation.

PROPOSITION. *The sequences $(\boldsymbol{\varrho}_\pm(\mathcal{E}_n,\mathcal{E}_n))$ and $(\boldsymbol{\delta}_\pm(\mathcal{E}_n,\mathcal{E}_n))$ are non-decreasing.*

REMARK. Remember that nothing is known about the monotonicity of $\boldsymbol{\delta}(\mathcal{E}_n, \mathcal{E}_n)$. But for $\boldsymbol{\delta}_\pm(\mathcal{E}_n, \mathcal{E}_n)$ we may play with signs.

8.4.5 We now state a corollary of 3.4.13.

LEMMA. $\delta_\pm(\mathcal{E}_{2n},\mathcal{E}_{2n}) \le 2\,\delta_\pm(\mathcal{E}_n,\mathcal{E}_n)$.

8.4.6 The proof of the next lemma is almost the same as that of 5.5.4.

LEMMA. $\delta_\pm(\mathcal{E}_{2m+1},\mathcal{E}_{2m+1}) \le 3\,\varrho_\pm(\mathcal{E}_{4m-1},\mathcal{E}_{4m-1})$.

PROOF. Letting

$$V_m^0 := V_m e_{2m} \quad \text{and} \quad \boldsymbol{f}^0 := \boldsymbol{f} e_{m-1},$$

the Fourier coefficients of the m-th de la Vallée–Poussin kernel V_m and of the function $\boldsymbol{f} \in [L_2[-\pi,+\pi),X]$ are shifted to the right by $2m$ steps and $(m-1)$ steps, respectively. Since the spectrum of $V_m^0 * \boldsymbol{f}^0$ is contained in $\{1,\ldots,4m-1\}$, we have

$$V_m^0 * \boldsymbol{f}^0 = \sum_{k=1}^{4m-1} \langle V_m^0 * \boldsymbol{f}^0, \overline{e_k}\rangle\, e_k.$$

If $h=1,\ldots,2m+1$ and $k=m,\ldots,3m$ are related by $h=k-m+1$, then

$$\langle V_m^0 * \boldsymbol{f}^0, \overline{e_k}\rangle = \langle V_m^0, \overline{e_k}\rangle \langle \boldsymbol{f}^0, \overline{e_k}\rangle = \langle V_m, \overline{e_{k-2m}}\rangle \langle \boldsymbol{f}, \overline{e_{k-m+1}}\rangle = \langle \boldsymbol{f}, \overline{e_h}\rangle .$$

It follows from Young's inequality 4.11.8 and $\|V_m^0|L_1\| \le 3$ that

$$\begin{aligned}\big\|\big(\langle V_m^0 * \boldsymbol{f}^0, \overline{e_k}\rangle\big)\big|\,\mathcal{E}_{4m-1}\big\| &= \big\|V_m^0 * \boldsymbol{f}^0\big|\,L_2\big\| \le \big\|V_m^0\big|\,L_1\big\|\,\big\|\boldsymbol{f}^0\big|\,L_2\big\| \\ &\le 3\,\|\boldsymbol{f}|\,L_2\| .\end{aligned}$$

Given $e=(\varepsilon_1,\ldots,\varepsilon_{2m+1}) \in \mathbb{E}_2^{2m+1}$, we define the $(4m-1)$-tuples

$$e^+ := (\underbrace{+1,\ldots,+1}_{m-1},\varepsilon_1,\ldots,\varepsilon_{2m+1},\underbrace{+1,\ldots,+1}_{m-1}),$$

and

$$e^- := (\overbrace{-1,\ldots,-1}^{m-1},\varepsilon_1,\ldots,\varepsilon_{2m+1},\overbrace{-1,\ldots,-1}^{m-1}).$$

Then

$$\begin{aligned}&2\,\|(\varepsilon_h T\langle \boldsymbol{f},\overline{e_h}\rangle)|\mathcal{E}_{2m+1}\| \le \\ &\quad\le \big\|\big(\varepsilon_k^+ T\langle V_m^0 * \boldsymbol{f}^0,\overline{e_k}\rangle\big)\big|\,\mathcal{E}_{4m-1}\big\| + \big\|\big(\varepsilon_k^- T\langle V_m^0 * \boldsymbol{f}^0,\overline{e_k}\rangle\big)\big|\,\mathcal{E}_{4m-1}\big\| \\ &\quad\le 2\,\varrho_\pm(T|\mathcal{E}_{4m-1},\mathcal{E}_{4m-1})\,\big\|\big(\langle V_m^0 * \boldsymbol{f}^0,\overline{e_k}\rangle\big)\big|\,\mathcal{E}_{4m-1}\big\| \\ &\quad\le 6\,\varrho_\pm(T|\mathcal{E}_{4m-1},\mathcal{E}_{4m-1})\,\|\boldsymbol{f}|L_2\|,\end{aligned}$$

which yields the desired inequality.

8.4.7 Next, we establish an analogue of 5.5.5.

PROPOSITION. $\delta_\pm(\mathcal{E}_n,\mathcal{E}_n) \le 6\,\varrho_\pm(\mathcal{E}_n,\mathcal{E}_n)$.

PROOF. We may assume that $n \ge 2$. Let $m := [\frac{n}{2}]$. Then $n \le 2m+1$ and

$4m-1 \le 2n$. Using the monotonicity of $\boldsymbol{\varrho}_\pm(\mathcal{E}_n,\mathcal{E}_n)$ and $\boldsymbol{\delta}_\pm(\mathcal{E}_n,\mathcal{E}_n)$, we conclude from Lemmas 8.4.5 and 8.4.6 that

$$\begin{aligned}\boldsymbol{\delta}_\pm(\mathcal{E}_n,\mathcal{E}_n) &\le \boldsymbol{\delta}_\pm(\mathcal{E}_{2m+1},\mathcal{E}_{2m+1}) \le 3\,\boldsymbol{\varrho}_\pm(\mathcal{E}_{4m-1},\mathcal{E}_{4m-1})\\ &\le 3\,\boldsymbol{\varrho}_\pm(\mathcal{E}_{2n},\mathcal{E}_{2n}) \le 3\cdot 2\,\boldsymbol{\varrho}_\pm(\mathcal{E}_n,\mathcal{E}_n).\end{aligned}$$

8.4.8 We are now in a position to prove the main result of this section, which is analogous to 5.2.9.

THEOREM. *The sequences of the following ideal norms are uniformly equivalent:*

$$\begin{gathered}\boldsymbol{\varrho}_\pm(\mathcal{E}_n,\mathcal{E}_n),\ \boldsymbol{\varrho}_\pm(\mathcal{E}_n,\mathcal{E}_n^\circ),\ \boldsymbol{\varrho}_\pm(\mathcal{E}_n^\circ,\mathcal{E}_n),\ \boldsymbol{\varrho}_\pm(\mathcal{E}_n^\circ,\mathcal{E}_n^\circ),\\ \boldsymbol{\delta}_\pm(\mathcal{E}_n,\mathcal{E}_n),\ \boldsymbol{\delta}_\pm(\mathcal{E}_n,\mathcal{E}_n^\circ),\ \boldsymbol{\delta}_\pm(\mathcal{E}_n^\circ,\mathcal{E}_n),\ \boldsymbol{\delta}_\pm(\mathcal{E}_n^\circ,\mathcal{E}_n^\circ).\end{gathered}$$

PROOF. By 8.4.2, 8.4.3 and 8.4.7, we have

$$\boldsymbol{\varrho}_\pm(\mathcal{E}_n,\mathcal{E}_n) \le \boldsymbol{\varrho}_\pm(\mathcal{E}_n^\circ,\mathcal{E}_n) \le 3\,\boldsymbol{\varrho}_\pm(\mathcal{E}_n,\mathcal{E}_n),$$

$$\boldsymbol{\varrho}_\pm(\mathcal{E}_n,\mathcal{E}_n) \le \boldsymbol{\varrho}_\pm(\mathcal{E}_n,\mathcal{E}_n^\circ) = \boldsymbol{\delta}_\pm(\mathcal{E}_n,\mathcal{E}_n^\circ) \le 3\,\boldsymbol{\delta}_\pm(\mathcal{E}_n,\mathcal{E}_n) \le 24\,\boldsymbol{\varrho}_\pm(\mathcal{E}_n,\mathcal{E}_n)$$

and

$$\begin{aligned}\boldsymbol{\varrho}_\pm(\mathcal{E}_n,\mathcal{E}_n) &\le \boldsymbol{\varrho}_\pm(\mathcal{E}_n^\circ,\mathcal{E}_n^\circ) = \boldsymbol{\delta}_\pm(\mathcal{E}_n^\circ,\mathcal{E}_n^\circ) \le 3\,\boldsymbol{\delta}_\pm(\mathcal{E}_n^\circ,\mathcal{E}_n) \le\\ &\le 9\,\boldsymbol{\delta}_\pm(\mathcal{E}_n,\mathcal{E}_n) \le 54\,\boldsymbol{\varrho}_\pm(\mathcal{E}_n,\mathcal{E}_n).\end{aligned}$$

8.4.9 We proceed with an obvious observation.

PROPOSITION. *The ideal norm $\boldsymbol{\delta}_\pm(\mathcal{E}_n^\circ,\mathcal{E}_n^\circ)$ is symmetric.*

8.4.10 Almost nothing is known about the operator ideals generated by $\boldsymbol{\varrho}_\pm(\mathcal{E}_n,\mathcal{E}_n)$.

THEOREM. $\mathfrak{L}_0[n^{-1/2}\boldsymbol{\varrho}_\pm(\mathcal{E}_n,\mathcal{E}_n)] = \mathfrak{RT}$.

PROOF. Example 8.4.13 says that $\boldsymbol{\varrho}_\pm(l_1^n|\mathcal{E}_n,\mathcal{E}_n) \asymp \sqrt{n}$. Hence, we have $\boldsymbol{\varrho}_\pm(T|\mathcal{E}_n,\mathcal{E}_n) \asymp \sqrt{n}$ for every uniformly l_1^n-injective operator; see 4.4.17. This implies that

$$\mathfrak{L}_0[n^{-1/2}\boldsymbol{\varrho}_\pm(\mathcal{E}_n,\mathcal{E}_n)] \subseteq \mathfrak{RT}.$$

By 3.10.10 and 4.14.8, the reverse inclusion follows from $\boldsymbol{\varrho}_\pm(\mathcal{E}_n,\mathcal{E}_n) \le \boldsymbol{\varrho}_n$.

8.4.11 The next result is due to M. Junge and M. Defant; see [def*].

THEOREM. $\mathsf{L}[\boldsymbol{\varrho}_\pm(\mathcal{E}_n,\mathcal{E}_n)] = \mathsf{H}$.

PROOF. Every Banach space

$$X \in \mathsf{L}[\boldsymbol{\varrho}_\pm(\mathcal{E}_n,\mathcal{E}_n)] \subset \mathsf{L}[n^{-1/2}\boldsymbol{\varrho}_\pm(\mathcal{E}_n,\mathcal{E}_n)] = \mathsf{B} \subset \mathsf{MP}$$

is MP-convex. Hence we conclude from 5.9.16 (Remark), 8.3.5 and 8.3.1 that

$$\varrho(X|\mathcal{R}_n,\mathcal{I}_n) \asymp \varrho(X|\mathcal{R}_n,\mathcal{E}_n) \asymp \varrho(X|\mathcal{R}_n \otimes \mathcal{E}_n,\mathcal{E}_n) \prec \varrho_\pm(X|\mathcal{E}_n,\mathcal{E}_n).$$

So $X \in \mathsf{RT}_2$. By duality, we also obtain $X \in \mathsf{RT}_2' \subset \mathsf{RC}_2$. Now Kwapień's theorem 4.10.7 yields

$$\mathsf{L}[\varrho_\pm(\mathcal{E}_n,\mathcal{E}_n)] \subseteq \mathsf{RT}_2 \cap \mathsf{RC}_2 = \mathsf{H}.$$

The reverse inclusion is trivial.

8.4.12 One may ask a question which is an analogue of 5.9.5.

PROBLEM. *Is it true that* $\mathfrak{L}[\varrho_\pm(\mathcal{E}_n,\mathcal{E}_n)] = \mathfrak{H}$?

8.4.13 The Fourier system turns out to be highly conditional.

EXAMPLE.

$$\varrho_\pm(l_r^n|\mathcal{E}_n,\mathcal{E}_n) \asymp \varrho_\pm(L_r|\mathcal{E}_n,\mathcal{E}_n) \asymp n^{|1/r-1/2|} \quad \textit{if } 1 \le r \le \infty.$$

PROOF. Following [bri*], we define a variant of the Rudin–Shapiro polynomials by letting

$$P_0(\zeta) := \zeta \quad \text{and} \quad P_{n+1}(\zeta) := P_n(\zeta^2) + \zeta^{-1} P_n(-\zeta^2) \quad \text{for } \zeta \in \mathbb{T}.$$

Then

$$P_n(\zeta) = \sum_{k=1}^{2^n} \alpha_k^{(n)} \zeta^k \quad \text{with} \quad \alpha_k^{(n)} = \pm 1,$$

and the parallelogram equality yields

$$|P_n(\zeta)|^2 + |P_n(-\zeta)|^2 = 2^{n+1}.$$

We now obtain

$$\|(\alpha_k^{(n)} e_k)|[\mathcal{E}_{2^n}, L_\infty]\| = \sup\{\, |P_n(\zeta)| : \zeta \in \mathbb{T} \,\} \le 2^{(n+1)/2}.$$

By interpolation, this fact combined with

$$\|(\alpha_k^{(n)} e_k)|[\mathcal{E}_{2^n}, L_2]\| = 2^{n/2}$$

yields

$$\|(\alpha_k^{(n)} e_k)|[\mathcal{E}_{2^n}, L_r]\| \le 2^{(n+1)/2}$$

whenever $2 \le r \le \infty$. On the other hand, we know from 5.1.7 that

$$\|(e_k)|[\mathcal{E}_{2^n}, L_r]\| = \|d_{2^n}|L_r\| \asymp 2^{n/r'}.$$

Now it follows from

$$\|(e_k)|[\mathcal{E}_{2^n}, L_r]\| \le \varrho_\pm(L_r|\mathcal{E}_{2^n},\mathcal{E}_{2^n})\, \|(\alpha_k^{(n)} e_k)|[\mathcal{E}_{2^n}, L_r]\|$$

that

$$2^{n(1/2-1/r)} \prec \varrho_\pm(L_r|\mathcal{E}_{2^n},\mathcal{E}_{2^n}).$$

Restricting the polynomials ζ^k and $P_n(\zeta)$ to the discrete subgroup $\mathbb{E}_2^{2^n}$, the same estimate is obtained for $\varrho_\pm(l_r^{2^n}|\mathcal{E}^\circ_{2^n},\mathcal{E}^\circ_{2^n})$; see 5.1.4. Taking into account Theorem 8.4.8 and the monotonicity of $\varrho_\pm(\mathcal{E}_n,\mathcal{E}_n)$ yields the lower estimate of $\varrho_\pm(l_r^n|\mathcal{E}_n,\mathcal{E}_n)$ if $2\le r\le\infty$.

The upper estimate is a consequence of 3.4.17. Thanks to 8.4.9, the case $1\le r\le 2$ can be treated by duality.

8.4.14 In view of 8.3.5, Examples 5.9.17 and 5.9.18 yield the following relations.

EXAMPLE.

$$\left.\begin{array}{l}\varrho(L_p|\mathcal{R}_n\otimes\mathcal{E}_n,\mathcal{E}_n)\asymp n^{1/p-1/2}\\ \varrho(L_p|\mathcal{E}_n,\mathcal{R}_n\otimes\mathcal{E}_n)\asymp 1\end{array}\right\}\ \textit{if}\ 1\le p\le 2,$$

$$\left.\begin{array}{l}\varrho(L_q|\mathcal{R}_n\otimes\mathcal{E}_n,\mathcal{E}_n)\asymp 1\\ \varrho(L_q|\mathcal{E}_n,\mathcal{R}_n\otimes\mathcal{E}_n)\asymp n^{1/2-1/q}\end{array}\right\}\ \textit{if}\ 2\le q<\infty.$$

8.5 Haar unconditionality/UMD

8.5.1 For $T\in\mathfrak{L}(X,Y)$, we denote by $\varrho_\pm(T|\mathcal{H}(\mathbb{D}_1^n),\mathcal{H}(\mathbb{D}_1^n))$ the least constant $c\ge 0$ such that

$$\|(Tx_k^{(j)})|\mathcal{H}(\mathbb{D}_1^n)\|\le c\,\|(\varepsilon_k^{(j)}x_k^{(j)})|\mathcal{H}(\mathbb{D}_1^n)\|$$

for every $\mathbb{D}_1^n$-tuple $(x_k^{(j)})$ in X and any choice of $\varepsilon_k^{(j)}=\pm1$. We refer to

$$\varrho_\pm(\mathcal{H}(\mathbb{D}_1^n),\mathcal{H}(\mathbb{D}_1^n)):T\longrightarrow\varrho_\pm(T|\mathcal{H}(\mathbb{D}_1^n),\mathcal{H}(\mathbb{D}_1^n))$$

as a **UMD-ideal norm**. Similarly, we get $\varrho_\pm(\mathcal{H}(\mathbb{D}_0^n),\mathcal{H}(\mathbb{D}_0^n))$.

8.5.2 The above definition can be modified by assuming that the signs are changed on every level simultaneously. In other words, $\varepsilon_k^{(j)}=\varepsilon_k=\pm1$ depends on k, but not on $j=1,\dots,2^{k-1}$. A still weaker concept is obtained by using only the signs $\varepsilon_k^{(j)}=(-1)^k$. The ideal norms just defined will be denoted by $\mathring{\varrho}_\pm(\mathcal{H}(\mathbb{D}_1^n),\mathcal{H}(\mathbb{D}_1^n))$ and $\mathring{\mathring{\varrho}}_\pm(\mathcal{H}(\mathbb{D}_1^n),\mathcal{H}(\mathbb{D}_1^n))$, respectively. Obviously,

$$\mathring{\mathring{\varrho}}_\pm(\mathcal{H}(\mathbb{D}_1^n),\mathcal{H}(\mathbb{D}_1^n))\le\mathring{\varrho}_\pm(\mathcal{H}(\mathbb{D}_1^n),\mathcal{H}(\mathbb{D}_1^n))\le\varrho_\pm(\mathcal{H}(\mathbb{D}_1^n),\mathcal{H}(\mathbb{D}_1^n)).$$

Surprisingly, we also have an estimate in the reverse direction; see 8.5.6. The proof of this fact requires some preparations.

8.5.3 LEMMA.

$$\mathring{\mathring{\varrho}}_\pm(\mathcal{H}(\mathbb{D}_{m+1}^{n+1}),\mathcal{H}(\mathbb{D}_{m+1}^{n+1}))\le\mathring{\mathring{\varrho}}_\pm(\mathcal{H}(\mathbb{D}_m^n),\mathcal{H}(\mathbb{D}_m^n))\quad\textit{for}\ n\ge m\ge 1.$$

PROOF. This can be seen as in 7.4.12.

8.5.4 LEMMA. $\overset{\circ\circ}{\varrho}_\pm(\mathcal{H}(\mathbb{D}_1^{2n}),\mathcal{H}(\mathbb{D}_1^{2n})) \le 3\,\overset{\circ\circ}{\varrho}_\pm(\mathcal{H}(\mathbb{D}_1^{n}),\mathcal{H}(\mathbb{D}_1^{n}))$.

PROOF. We conclude from

$$\|(x_k^{(j)})|\mathcal{H}(\mathbb{D}_1^n)\| \le \delta(X|\mathcal{H}(\mathbb{D}_0^n),\mathcal{H}(\mathbb{D}_0^n))\,\|(x_k^{(j)})|\mathcal{H}(\mathbb{D}_1^{2n})\| = \|(x_k^{(j)})|\mathcal{H}(\mathbb{D}_1^{2n})\|$$

that

$$\|(x_k^{(j)})|\mathcal{H}(\mathbb{D}_{n+1}^{2n})\| \le \|(x_k^{(j)})|\mathcal{H}(\mathbb{D}_1^n)\| + \|(x_k^{(j)})|\mathcal{H}(\mathbb{D}_1^{2n})\| \le 2\|(x_k^{(j)})|\mathcal{H}(\mathbb{D}_1^{2n})\|.$$

Thus, in view of the previous lemma,

$$\begin{aligned}\|((-1)^k T x_k^{(j)})|\mathcal{H}(\mathbb{D}_{n+1}^{2n})\| &\le \overset{\circ\circ}{\varrho}_\pm(T|\mathcal{H}(\mathbb{D}_{n+1}^{2n}),\mathcal{H}(\mathbb{D}_{n+1}^{2n}))\,\|(x_k^{(j)})|\mathcal{H}(\mathbb{D}_{n+1}^{2n})\| \\ &\le 2\,\overset{\circ\circ}{\varrho}_\pm(T|\mathcal{H}(\mathbb{D}_1^{n}),\mathcal{H}(\mathbb{D}_1^{n}))\,\|(x_k^{(j)})|\mathcal{H}(\mathbb{D}_1^{2n})\|.\end{aligned}$$

Moreover, by definition, we have

$$\|((-1)^k T x_k^{(j)})|\mathcal{H}(\mathbb{D}_1^{n})\| \le \overset{\circ\circ}{\varrho}_\pm(T|\mathcal{H}(\mathbb{D}_1^{n}),\mathcal{H}(\mathbb{D}_1^{n}))\,\|(x_k^{(j)})|\mathcal{H}(\mathbb{D}_1^{n})\|.$$

Combining these inequalities yields

$$\begin{aligned}&\|((-1)^k T x_k^{(j)})|\mathcal{H}(\mathbb{D}_1^{2n})\| \le \\ &\qquad \le \|((-1)^k T x_k^{(j)})|\mathcal{H}(\mathbb{D}_1^{n})\| + \|((-1)^k T x_k^{(j)})|\mathcal{H}(\mathbb{D}_{n+1}^{2n})\| \\ &\qquad \le 3\,\overset{\circ\circ}{\varrho}_\pm(T|\mathcal{H}(\mathbb{D}_1^{n}),\mathcal{H}(\mathbb{D}_1^{n}))\,\|(x_k^{(j)})|\mathcal{H}(\mathbb{D}_1^{2n})\|.\end{aligned}$$

8.5.5 The following proof is based on fork transforms. Their properties are summarized in 7.4.18.

LEMMA. $\varrho_\pm(\mathcal{H}(\mathbb{D}_1^{n}),\mathcal{H}(\mathbb{D}_1^{n})) \le \overset{\circ\circ}{\varrho}_\pm(\mathcal{H}(\mathbb{D}_1^{2n}),\mathcal{H}(\mathbb{D}_1^{2n}))$.

PROOF. Fix Haar polynomials

$$\boldsymbol{f} := \sum_{\mathbb{D}_1^n} x_k^{(j)}\chi_k^{(j)} \quad\text{and}\quad \boldsymbol{f}^e := \sum_{\mathbb{D}_1^n} \varepsilon_k^{(j)} x_k^{(j)}\chi_k^{(j)},$$

where $e = (\varepsilon_k^{(j)})$ is any $\mathbb{D}_1^n$-tuple of signs.

In a first step, we choose a composition of finitely many fork transforms which map the tree $\mathbb{D}_1^n$ into $\mathbb{D}_1^{2n}$ such that the levels $\mathbb{D}_{2k}$ with $k = 1, ..., n$ remain free. This can be achieved as follows:

$$\mathbb{D}_n \longrightarrow \mathbb{D}_{2n-1}, \ \ldots, \ \mathbb{D}_k \longrightarrow \mathbb{D}_{2k-1}, \ \ldots, \ \mathbb{D}_2 \longrightarrow \mathbb{D}_3, \ \mathbb{D}_1 \longrightarrow \mathbb{D}_1.$$

Note that the above order is necessary for admissibility.

In a second step, all coefficients furnished with $+$ are gradually moved to the next even level, $\mathbb{D}_{2k-1} \to \mathbb{D}_{2k}$. In this case, the order is irrelevant.

Our final outcome is transformed Haar polynomials

$$\boldsymbol{f}\circ\varphi := \sum_{\mathbb{D}_1^{2n}} x_{\varphi,k}^{(j)}\chi_k^{(j)} \quad\text{and}\quad \boldsymbol{f}^e\circ\varphi := \sum_{\mathbb{D}_1^{2n}} (-1)^k x_{\varphi,k}^{(j)}\chi_k^{(j)}$$

such that

$$\|\boldsymbol{f}\circ\varphi|L_2\| = \|\boldsymbol{f}|L_2\| \quad \text{and} \quad \|[L_2,T]\boldsymbol{f}^e\circ\varphi|L_2\| = \|[L_2,T]\boldsymbol{f}^e|L_2\|,$$

or, in terms of coefficients,

$$\|(x_{\varphi,k}^{(j)})|\mathcal{H}(\mathbb{D}_1^{2n})\| = \|(x_k^{(j)})|\mathcal{H}(\mathbb{D}_1^n)\|$$

and

$$\|((-1)^k Tx_{\varphi,k}^{(j)})|\mathcal{H}(\mathbb{D}_1^{2n})\| = \|(\varepsilon_k^{(j)}Tx_k^{(j)})|\mathcal{H}(\mathbb{D}_1^n)\|.$$

Thus the inequality

$$\|((-1)^k Tx_{\varphi,k}^{(j)})|\mathcal{H}(\mathbb{D}_1^{2n})\| \le \mathring{\mathring{\boldsymbol{\varrho}}}_\pm(\mathcal{H}(\mathbb{D}_1^{2n}),\mathcal{H}(\mathbb{D}_1^{2n}))\,\|(x_{\varphi,k}^{(j)})|\mathcal{H}(\mathbb{D}_1^{2n})\|$$

passes into

$$\|(\varepsilon_k^{(j)}Tx_k^{(j)})|\mathcal{H}(\mathbb{D}_1^n)\| \le \mathring{\mathring{\boldsymbol{\varrho}}}_\pm(\mathcal{H}(\mathbb{D}_1^{2n}),\mathcal{H}(\mathbb{D}_1^{2n}))\,\|(x_k^{(j)})|\mathcal{H}(\mathbb{D}_1^n)\|.$$

This completes the proof.

REMARK. For a similar reduction within the setting of arbitrary martingales, we refer to [bur 6, pp. 651–652].

8.5.6 We now combine Lemmas 8.5.4 and 8.5.5 to get the main result of this section; see [wen 5].

THEOREM. $\boldsymbol{\varrho}_\pm(\mathcal{H}(\mathbb{D}_1^n),\mathcal{H}(\mathbb{D}_1^n)) \le 3\,\mathring{\mathring{\boldsymbol{\varrho}}}_\pm(\mathcal{H}(\mathbb{D}_1^n),\mathcal{H}(\mathbb{D}_1^n))$.

8.5.7 Most of the ideal norms occurring below have not been defined yet. We leave this simple job to the imaginative reader. Although the subsequent undertaking looks like a waste of energy, there are situations in which one of these ideal norms is more suitable than the others.

THEOREM. *The sequences of the following ideal norms are uniformly equivalent:*

$$\boldsymbol{\varrho}_\pm(\mathcal{H}(\mathbb{D}_1^n),\mathcal{H}(\mathbb{D}_1^n)),\ \mathring{\boldsymbol{\varrho}}_\pm(\mathcal{H}(\mathbb{D}_1^n),\mathcal{H}(\mathbb{D}_1^n)),\ \mathring{\mathring{\boldsymbol{\varrho}}}_\pm(\mathcal{H}(\mathbb{D}_1^n),\mathcal{H}(\mathbb{D}_1^n)),$$
$$\boldsymbol{\varrho}_\pm(\mathcal{H}(\mathbb{D}_0^n),\mathcal{H}(\mathbb{D}_0^n)),\ \mathring{\boldsymbol{\varrho}}_\pm(\mathcal{H}(\mathbb{D}_0^n),\mathcal{H}(\mathbb{D}_0^n)),\ \mathring{\mathring{\boldsymbol{\varrho}}}_\pm(\mathcal{H}(\mathbb{D}_0^n),\mathcal{H}(\mathbb{D}_0^n)),$$
$$\boldsymbol{\delta}_\pm(\mathcal{H}(\mathbb{D}_1^n),\mathcal{H}(\mathbb{D}_1^n)),\ \mathring{\boldsymbol{\delta}}_\pm(\mathcal{H}(\mathbb{D}_1^n),\mathcal{H}(\mathbb{D}_1^n)),\ \mathring{\mathring{\boldsymbol{\delta}}}_\pm(\mathcal{H}(\mathbb{D}_1^n),\mathcal{H}(\mathbb{D}_1^n)),$$
$$\boldsymbol{\delta}_\pm(\mathcal{H}(\mathbb{D}_0^n),\mathcal{H}(\mathbb{D}_0^n)),\ \mathring{\boldsymbol{\delta}}_\pm(\mathcal{H}(\mathbb{D}_0^n),\mathcal{H}(\mathbb{D}_0^n)),\ \mathring{\mathring{\boldsymbol{\delta}}}_\pm(\mathcal{H}(\mathbb{D}_0^n),\mathcal{H}(\mathbb{D}_0^n)).$$

PROOF. The crucial point is the inequality established in the preceding paragraph. All other relations are elementary. We have, for example,

$$\begin{aligned}\boldsymbol{\varrho}_\pm(\mathcal{H}(\mathbb{D}_1^n),\mathcal{H}(\mathbb{D}_1^n)) &\le \boldsymbol{\delta}_\pm(\mathcal{H}(\mathbb{D}_1^n),\mathcal{H}(\mathbb{D}_1^n)) \le \boldsymbol{\delta}_\pm(\mathcal{H}(\mathbb{D}_0^n),\mathcal{H}(\mathbb{D}_0^n))\\ &= \boldsymbol{\varrho}_\pm(\mathcal{H}(\mathbb{D}_0^n),\mathcal{H}(\mathbb{D}_0^n)) \le 3\,\boldsymbol{\varrho}_\pm(\mathcal{H}(\mathbb{D}_1^n),\mathcal{H}(\mathbb{D}_1^n)).\end{aligned}$$

8.5.8 PROPOSITION. $\boldsymbol{\varrho}_{\pm}(\mathcal{H}(\mathbb{D}_0^n), \mathcal{H}(\mathbb{D}_0^n)) = \boldsymbol{\varrho}'_{\pm}(\mathcal{H}(\mathbb{D}_0^n), \mathcal{H}(\mathbb{D}_0^n))$.

PROOF. Since $\boldsymbol{\varrho}_{\pm}(\mathcal{H}(\mathbb{D}_0^n), \mathcal{H}(\mathbb{D}_0^n)) = \boldsymbol{\delta}_{\pm}(\mathcal{H}(\mathbb{D}_0^n), \mathcal{H}(\mathbb{D}_0^n))$, the assertion follows from 3.4.6.

8.5.9 The next result can easily be deduced from the fact that $\mathcal{I}(\mathbb{D}_1^n)$ is unconditional.

PROPOSITION.

$$\boldsymbol{\varrho}_{\pm}(\mathcal{H}(\mathbb{D}_1^n), \mathcal{H}(\mathbb{D}_1^n)) \le \boldsymbol{\varrho}(\mathcal{H}(\mathbb{D}_1^n), \mathcal{I}(\mathbb{D}_1^n)) \circ \boldsymbol{\varrho}(\mathcal{I}(\mathbb{D}_1^n), \mathcal{H}(\mathbb{D}_1^n)).$$

8.5.10 We proceed with the following corollary.

PROPOSITION. *Let* $1 < p < 2$ *and* $2 < q < \infty$. *Then*

$$\boldsymbol{\varrho}_{\pm}(X|\mathcal{H}(\mathbb{D}_1^n), \mathcal{H}(\mathbb{D}_1^n)) = O(n^{1/p-1/q}) \quad \textit{for} \quad X \in \mathsf{AT}_p^{weak} \cap \mathsf{AC}_q^{weak}.$$

REMARK. The above result is optimal, since there exists a Banach space $X_{pq} \in \mathsf{AT}_p \cap \mathsf{AC}_q$ (even a Banach lattice) such that

$$\boldsymbol{\varrho}_{\pm}(X_{pq}|\mathcal{H}(\mathbb{D}_1^n), \mathcal{H}(\mathbb{D}_1^n)) \asymp n^{1/p-1/q};$$

see [bou 5], [gei 1] and [gei 3].

8.5.11 PROPOSITION. $1 \le \boldsymbol{\varrho}_{\pm}(\mathcal{H}(\mathbb{D}_1^n), \mathcal{H}(\mathbb{D}_1^n)) \le n.$

PROOF. Recall from 7.4.6 that

$$\boldsymbol{\varrho}(\mathcal{H}(\mathbb{D}_1^n), \mathcal{I}(\mathbb{D}_1^n)) \le \sqrt{n} \quad \text{and} \quad \boldsymbol{\varrho}(\mathcal{I}(\mathbb{D}_1^n), \mathcal{H}(\mathbb{D}_1^n)) \le \sqrt{n}.$$

The upper estimate now follows from 8.5.9.

8.5.12 The definitions of **UMD-ideal norms** given in 8.5.1 and 8.5.2 can be generalized to exponents $1 < r < \infty$.

For $T \in \mathfrak{L}(X, Y)$, we denote by $\boldsymbol{\varrho}_{\pm}^{(r)}(T|\mathcal{H}(\mathbb{D}_1^n), \mathcal{H}(\mathbb{D}_1^n))$ the least constant $c \ge 0$ such that

$$\|(Tx_k^{(j)})|\mathcal{H}(\mathbb{D}_1^n)\|_r \le c\, \|(\varepsilon_k^{(j)} x_k^{(j)})|\mathcal{H}(\mathbb{D}_1^n)\|_r$$

for every $\mathbb{D}_1^n$-tuple $(x_k^{(j)})$ in X and any choice of $\varepsilon_k^{(j)} = \pm 1$.

The ideal norms $\overset{\circ}{\boldsymbol{\varrho}}{}_{\pm}^{(r)}(\mathcal{H}(\mathbb{D}_1^n), \mathcal{H}(\mathbb{D}_1^n))$ and $\overset{\circ\circ}{\boldsymbol{\varrho}}{}_{\pm}^{(r)}(\mathcal{H}(\mathbb{D}_1^n), \mathcal{H}(\mathbb{D}_1^n))$ are obtained analogously.

8.5.13 Next, we show that the asymptotic behaviour of the ideal norms just defined does not depend on the exponent r.

THEOREM. *The sequences of the following ideal norms are uniformly equivalent for all* $1 < r < \infty$:

$$\boldsymbol{\varrho}_{\pm}^{(r)}(\mathcal{H}(\mathbb{D}_1^n), \mathcal{H}(\mathbb{D}_1^n)), \;\; \overset{\circ}{\boldsymbol{\varrho}}{}_{\pm}^{(r)}(\mathcal{H}(\mathbb{D}_1^n), \mathcal{H}(\mathbb{D}_1^n)), \;\; \overset{\circ\circ}{\boldsymbol{\varrho}}{}_{\pm}^{(r)}(\mathcal{H}(\mathbb{D}_1^n), \mathcal{H}(\mathbb{D}_1^n)).$$

PROOF. For fixed r, the relations

$$\varrho_{\pm}^{(r)}(\mathcal{H}(\mathbb{D}_1^n),\mathcal{H}(\mathbb{D}_1^n)) \asymp \overset{\circ}{\varrho}{}_{\pm}^{(r)}(\mathcal{H}(\mathbb{D}_1^n),\mathcal{H}(\mathbb{D}_1^n)) \asymp \overset{\circ\circ}{\varrho}{}_{\pm}^{(r)}(\mathcal{H}(\mathbb{D}_1^n),\mathcal{H}(\mathbb{D}_1^n))$$

follow as in the proof of 8.5.6, and

$$\overset{\circ}{\varrho}{}_{\pm}^{(r)}(\mathcal{H}(\mathbb{D}_1^n),\mathcal{H}(\mathbb{D}_1^n)) \asymp \overset{\circ}{\varrho}{}_{\pm}^{(s)}(\mathcal{H}(\mathbb{D}_1^n),\mathcal{H}(\mathbb{D}_1^n)) \quad \text{whenever } 1<r,s<\infty$$

is an immediate consequence of 7.2.10.

8.5.14 EXAMPLE. $\overset{\circ}{\varrho}_{\pm}(L_r|\mathcal{H}(\mathbb{D}_1^n),\mathcal{H}(\mathbb{D}_1^n)) \asymp 1$ *if* $1<r<\infty$.

PROOF. Using the Burkholder–Gundy–Davis inequality 7.2.12 in both directions, we get

$$\begin{aligned}\Big\|\sum_{k=1}^n \varepsilon_k d_k \Big| L_r\Big\| &\le a_r \Big\|\Big(\sum_{k=1}^n |\varepsilon_k d_k|^2\Big)^{1/2}\Big| L_r\Big\| \\ &= a_r \Big\|\Big(\sum_{k=1}^n |d_k|^2\Big)^{1/2}\Big| L_r\Big\| \le a_r b_r \Big\|\sum_{k=1}^n d_k \Big| L_r\Big\|\end{aligned}$$

whenever $(\varepsilon_1,\ldots,\varepsilon_n)\in\mathbb{E}_2^n$; for an elementary proof see [bur 7]. Let

$$\boldsymbol{f}_n = \sum_{k=1}^n \boldsymbol{d}_k$$

be any $L_r(M,\mu)$-valued martingale. Then integration over $\xi\in M$ yields

$$\Big(\int_0^1\!\!\int_M \Big|\sum_{k=1}^n \varepsilon_k d_k(t,\xi)\Big|^r dt\, d\mu(\xi)\Big)^{1/r} \le a_r b_r \Big(\int_0^1\!\!\int_M \Big|\sum_{k=1}^n d_k(t,\xi)\Big|^r dt\, d\mu(\xi)\Big)^{1/r}.$$

This inequality can be written in the form

$$\Big(\int_0^1 \Big\|\sum_{k=1}^n \varepsilon_k \boldsymbol{d}_k(t)\Big| L_r\Big\|^r dt\Big)^{1/r} \le a_r b_r \Big(\int_0^1 \Big\|\sum_{k=1}^n \boldsymbol{d}_k(t)\Big| L_r\Big\|^r dt\Big)^{1/r}.$$

Hence

$$\overset{\circ}{\varrho}{}_{\pm}^{(r)}(L_r|\mathcal{H}(\mathbb{D}_1^n),\mathcal{H}(\mathbb{D}_1^n)) \asymp 1.$$

Finally, the previous theorem implies that

$$\overset{\circ}{\varrho}{}_{\pm}^{(2)}(L_r|\mathcal{H}(\mathbb{D}_1^n),\mathcal{H}(\mathbb{D}_1^n)) \asymp 1.$$

REMARK. We may deduce from [bur 6] or [bur 7] that

$$\overset{\circ}{\varrho}{}_{\pm}^{(r)}(L_r|\mathcal{H}(\mathbb{D}_1^n),\mathcal{H}(\mathbb{D}_1^n)) \le \max\{r-1, r'-1\}.$$

8.5.15 EXAMPLE.

$$\mathring{\varrho}_\pm(l_1^{2^n}|\mathcal{H}(\mathbb{D}_1^n),\mathcal{H}(\mathbb{D}_1^n)) \asymp \mathring{\varrho}_\pm(L_1|\mathcal{H}(\mathbb{D}_1^n),\mathcal{H}(\mathbb{D}_1^n)) \asymp n,$$
$$\mathring{\varrho}_\pm(l_\infty^{2^n}|\mathcal{H}(\mathbb{D}_1^n),\mathcal{H}(\mathbb{D}_1^n)) \asymp \mathring{\varrho}_\pm(L_\infty|\mathcal{H}(\mathbb{D}_1^n),\mathcal{H}(\mathbb{D}_1^n)) \asymp n.$$

PROOF. In what follows, we identify $l_\infty^{2^n}$, $l_\infty(\mathbb{D}_0^n)$ and $L_\infty([0,1),\mathcal{D}_n)$. Consider the $L_\infty([0,1),\mathcal{D}_n)$-valued Haar polynomial

$$\boldsymbol{f}_n = \sum_{k=1}^{n} \boldsymbol{d}_k = \sum_{\mathbb{D}_1^n} x_k^{(j)}\chi_k^{(j)} \quad \text{with} \quad x_k^{(j)} := \tfrac{1}{2^{k-1}}\chi_k^{(j)}.$$

Fix $t,\xi \in [0,1)$, and denote by k_0 the largest index k such that t and ξ belong to one and the same dyadic interval $\Delta_k^{(j)}$. Then we see from

$$d_k(t,\xi) = \frac{1}{2^{k-1}} \sum_{j=1}^{2^{k-1}} \chi_k^{(j)}(t)\chi_k^{(j)}(\xi)$$

that

$$d_k(t,\xi) = \begin{cases} +1 & \text{if } k < k_0+1, \\ -1 & \text{if } k = k_0+1, \\ 0 & \text{if } k > k_0+1. \end{cases} \tag{1}$$

In particular,

$$|d_k(t,\xi)| \le 1, \; d_k(t,t) = 1 \quad \text{and} \quad \Big|\sum_{k=1}^{n}(-1)^k d_k(t,\xi)\Big| \le 2.$$

Hence

$$\left(\int_0^1 \Big\|\sum_{k=1}^{n} \boldsymbol{d}_k(t)\Big|L_\infty\Big\|^2 dt\right)^{1/2} = n \tag{2}$$

and

$$\left(\int_0^1 \Big\|\sum_{k=1}^{n} (-1)^k \boldsymbol{d}_k(t)\Big|L_\infty\Big\|^2 dt\right)^{1/2} \le 2. \tag{3}$$

The estimate $n \prec \mathring{\mathring{\varrho}}_\pm(l_\infty^{2^n}|\mathcal{H}(\mathbb{D}_1^n),\mathcal{H}(\mathbb{D}_1^n))$ is now a consequence of

$$\left(\int_0^1 \Big\|\sum_{k=1}^{n} \boldsymbol{d}_k(t)\Big|L_\infty\Big\|^2 dt\right)^{1/2} \le$$
$$\le \mathring{\mathring{\varrho}}_\pm(l_\infty^{2^n}|\mathcal{H}(\mathbb{D}_1^n),\mathcal{H}(\mathbb{D}_1^n)) \left(\int_0^1 \Big\|\sum_{k=1}^{n} (-1)^k \boldsymbol{d}_k(t)\Big|L_\infty\Big\|^2 dt\right)^{1/2}.$$

The upper estimate was proved in 8.5.11 and the rest follows by duality.

8.5.16 We refer to T as a **UMD-operator** if

$$\|T|\mathfrak{UMD}\| := \sup_n \varrho_\pm(T|\mathcal{H}(\mathbb{D}_1^n), \mathcal{H}(\mathbb{D}_1^n))$$

is finite. These operators form the Banach ideal

$$\mathfrak{UMD} := \mathfrak{L}[\varrho_\pm(\mathcal{H}(\mathbb{D}_1^n), \mathcal{H}(\mathbb{D}_1^n))].$$

The letters UMD, which stand for **unconditionality of martingale differences**, will also be used when we speak of Banach spaces in UMD.

8.5.17 Quasi-UMD-operators are defined by the property

$$\varrho_\pm(T|\mathcal{H}(\mathbb{D}_1^n), \mathcal{H}(\mathbb{D}_1^n)) = o(n).$$

They form the closed ideal

$$\mathfrak{QUMD} := \mathfrak{L}_0[n^{-1}\varrho_\pm(\mathcal{H}(\mathbb{D}_1^n), \mathcal{H}(\mathbb{D}_1^n))].$$

8.5.18 PROPOSITION. *The operator ideals $\mathfrak{UMD}$ and $\mathfrak{QUMD}$ are injective, surjective, symmetric and l_2-stable.*

8.5.19 EXAMPLE. $\|l_1^n|\mathfrak{UMD}\| = \|l_\infty^n|\mathfrak{UMD}\| \asymp 1 + \log n.$

PROOF. We know from 8.5.14 (Remark) that

$$\|l_q^n|\mathfrak{UMD}\| \le c_0 q,$$

where $c_0 > 0$ is an absolute constant and $2 \le q < \infty$. The upper estimate now follows from 1.3.6 and the lower estimate from 8.5.15.

8.5.20 Examples 8.5.14 and 8.5.15 can be summarized as follows. Compare with 2.4.12 and 5.4.22.

EXAMPLE. $L_r \in \mathsf{UMD}$ *if* $1 < r < \infty$, *and* $L_1, L_\infty \notin \mathsf{QUMD}$.

8.5.21 Next, we generalize 8.5.15.

EXAMPLE.

$$\varrho_\pm(l_1^N|\mathcal{H}(\mathbb{D}_0^n), \mathcal{H}(\mathbb{D}_0^n)) = \varrho_\pm(l_\infty^N|\mathcal{H}(\mathbb{D}_0^n), \mathcal{H}(\mathbb{D}_0^n)) \asymp \min\{n, 1+\log N\}.$$

PROOF. By 8.5.11,

$$\varrho_\pm(l_\infty^N|\mathcal{H}(\mathbb{D}_1^n), \mathcal{H}(\mathbb{D}_1^n)) \le n.$$

On the other hand, we conclude from 8.5.19 that

$$\varrho_\pm(l_\infty^N|\mathcal{H}(\mathbb{D}_1^n), \mathcal{H}(\mathbb{D}_1^n)) \le \|l_\infty^N|\mathfrak{UMD}\| \asymp 1 + \log N.$$

This proves the upper estimate. The lower one follows from 8.5.15. The result is formulated with respect to the index set $\mathbb{D}_0^n$ in order to get

$$\varrho_\pm(l_1^N|\mathcal{H}(\mathbb{D}_0^n), \mathcal{H}(\mathbb{D}_0^n)) = \varrho_\pm(l_\infty^N|\mathcal{H}(\mathbb{D}_0^n), \mathcal{H}(\mathbb{D}_0^n)) \quad \text{(by 8.5.8).}$$

8.5.22 PROPOSITION. $\mathfrak{AT} \subseteq \mathfrak{QUMD}$.

PROOF. Writing $T \in \mathfrak{AT}(X,Y)$ in the form $T = TI_X$, we conclude from 8.5.9 and 7.4.6 that

$$\begin{aligned}\varrho_\pm(T|\mathcal{H}(\mathbb{D}_1^n),\mathcal{H}(\mathbb{D}_1^n)) &\le \varrho(T|\mathcal{H}(\mathbb{D}_1^n),\mathcal{I}(\mathbb{D}_1^n))\,\varrho(X|\mathcal{I}(\mathbb{D}_1^n),\mathcal{H}(\mathbb{D}_1^n))\\ &\le o(\sqrt{n})\,\sqrt{n} = o(n).\end{aligned}$$

8.5.23 We conjecture that $\mathfrak{AT}=\mathfrak{QUMD}$. By 7.6.11, 7.6.12 and 7.10.25, an affirmative answer to the following question would yield a proof.

PROBLEM. *Is it true that*

$$\varrho_\pm(\Sigma_{2^n} : l_1^{2^n} \to l_\infty^{2^n} |\mathcal{H}(\mathbb{D}_1^n),\mathcal{H}(\mathbb{D}_1^n)) \asymp n\ ?$$

8.5.24 Finally, we combine 7.6.13, 7.10.25 and 8.5.22.

THEOREM. $\mathfrak{QHM} \subseteq \mathfrak{GR} \subseteq \mathfrak{QUMD}$.

REMARK. It seems likely that we even have equalities.

8.6 Random Haar unconditionality

8.6.1 For every orthonormal system $\mathcal{A}_n = (a_1,\ldots,a_n)$ in a Hilbert space $L_2(M,\mu)$, we consider the orthonormal system

$$\mathcal{A}_n \otimes \mathcal{H}(\mathbb{D}_1^n) := \{\, a_k \otimes \chi_k^{(j)} : (k,j) \in \mathbb{D}_1^n\}$$

in $L_2(M\times[0,1))$. Moreover, $\mathcal{A}_n \otimes \mathcal{H}(\mathbb{D}_0^n)$ is obtained from $\mathcal{A}_n \otimes \mathcal{H}(\mathbb{D}_1^n)$ by adding the function $\chi_1^{(0)} \equiv 1$.

8.6.2 Elementary manipulations yield the following formula.

LEMMA. $\|(x_k^{(j)})|\mathcal{I}_n \otimes \mathcal{H}(\mathbb{D}_1^n)\| = \|(x_k^{(j)})|\mathcal{I}(\mathbb{D}_1^n)\|$.

8.6.3 PROPOSITION.

$$\varrho(\mathcal{B}_n \otimes \mathcal{R}_n, \mathcal{A}_n \otimes \mathcal{R}_n) \le \varrho(\mathcal{B}_n \otimes \mathcal{H}(\mathbb{D}_1^n), \mathcal{A}_n \otimes \mathcal{H}(\mathbb{D}_1^n)) \le \varrho(\mathcal{B}_n, \mathcal{A}_n).$$

If the orthonormal systems $\mathcal{A}_n$ and $\mathcal{B}_n$ are $\mathbb{R}$-unconditional, then the inequalities become equalities.

PROOF. In view of

$$r_k = 2^{-(k-1)/2} \sum_{j=1}^{2^{k-1}} \chi_k^{(j)},$$

the left-hand part follows by substituting

$$x_k^{(j)} := 2^{-(k-1)/2} x_k \quad \text{for } k = 1,\ldots,n$$

in the defining inequality of $\varrho(\mathcal{B}_n \otimes \mathcal{H}(\mathbb{D}_1^n), \mathcal{A}_n \otimes \mathcal{H}(\mathbb{D}_1^n))$. We have

$$\int\limits_N \Big\| \sum_{k=1}^{n} T\Big(\sum_{j=1}^{2^{k-1}} x_k^{(j)} \chi_k^{(j)}(u) \Big) b_k(t) \Big\|^2 d\nu(t) \leq$$

$$\leq \varrho(T|\mathcal{B}_n, \mathcal{A}_n)^2 \int\limits_M \Big\| \sum_{k=1}^{n} \Big(\sum_{j=1}^{2^{k-1}} x_k^{(j)} \chi_k^{(j)}(u) \Big) a_k(s) \Big\|^2 d\mu(s).$$

Integrating over $u \in [0, 1)$ yields

$$\int\limits_0^1 \int\limits_N \Big\| \sum_{k=1}^{n} \sum_{j=1}^{2^{k-1}} T x_k^{(j)} b_k(t) \chi_k^{(j)}(u) \Big\|^2 du\, d\nu(t) \leq$$

$$\leq \varrho(T|\mathcal{B}_n, \mathcal{A}_n)^2 \int\limits_0^1 \int\limits_M \Big\| \sum_{k=1}^{n} \sum_{j=1}^{2^{k-1}} T x_k^{(j)} a_k(s) \chi_k^{(j)}(u) \Big\|^2 du\, d\mu(s).$$

This means that

$$\|(T x_k^{(j)})|\mathcal{B}_n \otimes \mathcal{H}(\mathbb{D}_1^n)\| \leq \varrho(T|\mathcal{B}_n, \mathcal{A}_n)\, \|(x_k^{(j)})|\mathcal{A}_n \otimes \mathcal{H}(\mathbb{D}_1^n)\|,$$

which proves the right-hand part. In the case of $\mathbb{R}$-unconditionality, equality follows from 3.6.2.

8.6.4 By 8.6.2, the following special cases of the preceding result hold.

PROPOSITION.

$\varrho(\mathcal{R}_n \otimes \mathcal{H}(\mathbb{D}_1^n), \mathcal{I}(\mathbb{D}_1^n)) = \varrho(\mathcal{R}_n, \mathcal{I}_n)$ *and* $\varrho(\mathcal{I}(\mathbb{D}_1^n), \mathcal{R}_n \otimes \mathcal{H}(\mathbb{D}_1^n)) = \varrho(\mathcal{I}_n, \mathcal{R}_n)$.

8.6.5 We have just seen that

$$\varrho(\mathcal{R}_n \otimes \mathcal{H}(\mathbb{D}_1^n), \mathcal{I}(\mathbb{D}_1^n)) \quad \text{and} \quad \varrho(\mathcal{I}(\mathbb{D}_1^n), \mathcal{R}_n \otimes \mathcal{H}(\mathbb{D}_1^n))$$

do not give anything new. On the other hand, the ideal norms

$$\varrho(\mathcal{R}_n \otimes \mathcal{H}(\mathbb{D}_1^n), \mathcal{H}(\mathbb{D}_1^n)) \quad \text{and} \quad \varrho(\mathcal{H}(\mathbb{D}_1^n), \mathcal{R}_n \otimes \mathcal{H}(\mathbb{D}_1^n))$$

will turn out to be of some interest. Their properties (though not their definitions) justify our referring to them as **Rademacher–Haar type ideal norms** and **Rademacher–Haar cotype ideal norms**, respectively.

8.6.6 Let us begin the study of these quantities by stating a counterpart of 7.4.5 which follows at once from 8.6.4 and 3.3.5.

PROPOSITION.

$$\varrho(\mathcal{H}(\mathbb{D}_1^n), \mathcal{I}(\mathbb{D}_1^n)) \leq \varrho(\mathcal{H}(\mathbb{D}_1^n), \mathcal{R}_n \otimes \mathcal{H}(\mathbb{D}_1^n)) \circ \varrho(\mathcal{R}_n, \mathcal{I}_n),$$
$$\varrho(\mathcal{I}(\mathbb{D}_1^n), \mathcal{H}(\mathbb{D}_1^n)) \leq \varrho(\mathcal{I}_n, \mathcal{R}_n) \circ \varrho(\mathcal{R}_n \otimes \mathcal{H}(\mathbb{D}_1^n), \mathcal{H}(\mathbb{D}_1^n)).$$

8.6.7 Next, we establish another consequence of 8.6.4.

PROPOSITION.

$$\varrho(\mathcal{R}_n \otimes \mathcal{H}(\mathbb{D}_1^n), \mathcal{H}(\mathbb{D}_1^n)) \le \varrho(\mathcal{R}_n, \mathcal{I}_n) \circ \varrho(\mathcal{I}(\mathbb{D}_1^n), \mathcal{H}(\mathbb{D}_1^n)),$$
$$\varrho(\mathcal{H}(\mathbb{D}_1^n), \mathcal{R}_n \otimes \mathcal{H}(\mathbb{D}_1^n)) \le \varrho(\mathcal{H}(\mathbb{D}_1^n), \mathcal{I}(\mathbb{D}_1^n)) \circ \varrho(\mathcal{I}_n, \mathcal{R}_n).$$

8.6.8 The asymmetry in the following result is unpleasant. However, we have to accept that both upper estimates are sharp (up to a factor); see 8.6.13 and 8.6.14.

PROPOSITION.

$$1 \le \varrho(\mathcal{R}_n \otimes \mathcal{H}(\mathbb{D}_1^n), \mathcal{H}(\mathbb{D}_1^n)) \le n \quad \textit{and} \quad 1 \le \varrho(\mathcal{H}(\mathbb{D}_1^n), \mathcal{R}_n \otimes \mathcal{H}(\mathbb{D}_1^n)) \le \sqrt{n}.$$

PROOF (S. Geiss). In view of

$$\varrho(\mathcal{R}_n \otimes \mathcal{H}(\mathbb{D}_1^n), \mathcal{H}(\mathbb{D}_1^n)) \le \varrho(\mathcal{R}_n, \mathcal{I}_n) \circ \varrho(\mathcal{I}(\mathbb{D}_1^n), \mathcal{H}(\mathbb{D}_1^n)),$$

the left-hand inequality follows from

$$\varrho(\mathcal{R}_n, \mathcal{I}_n) \le \sqrt{n} \quad \text{and} \quad \varrho(\mathcal{I}(\mathbb{D}_1^n), \mathcal{H}(\mathbb{D}_1^n)) \le \sqrt{n};$$

see 3.3.2 and 7.4.6. On the other hand, 3.1.5 says that

$$\Big\| \sum_{k=1}^{n} x_k \Big\| \le \sqrt{n}\, \|(x_k)|\mathcal{A}_n\|$$

for any orthonormal system $\mathcal{A}_n = (a_1, \dots, a_n)$. Substituting

$$x_k := \sum_{j=1}^{2^{k-1}} x_k^{(j)} \chi_k^{(j)}(t)$$

yields

$$\Big\| \sum_{\mathbb{D}_1^n} x_k^{(j)} \chi_k^{(j)}(t) \Big\|^2 \le n \int_M \Big\| \sum_{\mathbb{D}_1^n} x_k^{(j)} a_k(s) \chi_k^{(j)}(t) \Big\|^2 d\mu(s).$$

Hence, by integration over $t \in [0,1)$, we obtain

$$\|(x_k^{(j)})|\mathcal{H}(\mathbb{D}_1^n)\| \le \sqrt{n}\, \|(x_k^{(j)})|\mathcal{A}_n \otimes \mathcal{H}(\mathbb{D}_1^n)\|,$$

which proves a general version of the right-hand inequality.

8.6.9 When dealing with duality, it is more convenient to use the orthonormal system $\mathcal{H}(\mathbb{D}_0^n)$, since $\boldsymbol{\delta}(\mathcal{H}(\mathbb{D}_0^n), \mathcal{H}(\mathbb{D}_0^n)) = 1$.

PROPOSITION.

$$\varrho'(\mathcal{H}(\mathbb{D}_0^n), \mathcal{R}_n \otimes \mathcal{H}(\mathbb{D}_0^n)) \le \varrho(\mathcal{R}_n \otimes \mathcal{H}(\mathbb{D}_0^n), \mathcal{H}(\mathbb{D}_0^n)),$$

$$\varrho'(\mathcal{R}_n \otimes \mathcal{H}(\mathbb{D}_0^n), \mathcal{H}(\mathbb{D}_0^n)) \le \varrho(\mathcal{H}(\mathbb{D}_0^n), \mathcal{R}_n \otimes \mathcal{H}(\mathbb{D}_0^n)) \circ \boldsymbol{\delta}(\mathcal{R}_n \otimes \mathcal{H}(\mathbb{D}_0^n), \mathcal{R}_n \otimes \mathcal{H}(\mathbb{D}_0^n)).$$

PROOF. By 3.4.6, we obtain

$$\begin{aligned}\varrho'(\mathcal{H}(\mathbb{D}_0^n),\mathcal{R}_n\otimes\mathcal{H}(\mathbb{D}_0^n)) &\le \delta'(\mathcal{H}(\mathbb{D}_0^n),\mathcal{R}_n\otimes\mathcal{H}(\mathbb{D}_0^n))\\ &= \delta(\mathcal{R}_n\otimes\mathcal{H}(\mathbb{D}_0^n),\mathcal{H}(\mathbb{D}_0^n)) = \varrho(\mathcal{R}_n\otimes\mathcal{H}(\mathbb{D}_0^n),\mathcal{H}(\mathbb{D}_0^n))\end{aligned}$$

and

$$\begin{aligned}\varrho'(\mathcal{R}_n\otimes\mathcal{H}(\mathbb{D}_0^n),\mathcal{H}(\mathbb{D}_0^n)) &= \delta'(\mathcal{R}_n\otimes\mathcal{H}(\mathbb{D}_0^n),\mathcal{H}(\mathbb{D}_0^n))\\ &= \delta(\mathcal{H}(\mathbb{D}_0^n),\mathcal{R}_n\otimes\mathcal{H}(\mathbb{D}_0^n))\\ &\le \varrho(\mathcal{H}(\mathbb{D}_0^n),\mathcal{R}_n\otimes\mathcal{H}(\mathbb{D}_0^n))\circ\delta(\mathcal{R}_n\otimes\mathcal{H}(\mathbb{D}_0^n),\mathcal{R}_n\otimes\mathcal{H}(\mathbb{D}_0^n)).\end{aligned}$$

8.6.10 The next inequalities are obvious.

PROPOSITION.

$$\begin{aligned}\mathring{\varrho}_\pm(\mathcal{H}(\mathbb{D}_1^n),\mathcal{H}(\mathbb{D}_1^n)) \le \varrho(\mathcal{H}(\mathbb{D}_1^n),\mathcal{R}_n\otimes\mathcal{H}(\mathbb{D}_1^n))\circ\varrho(\mathcal{R}_n\otimes\mathcal{H}(\mathbb{D}_1^n),\mathcal{H}(\mathbb{D}_1^n)),\\ \varrho(\mathcal{R}_n\otimes\mathcal{H}(\mathbb{D}_1^n),\mathcal{H}(\mathbb{D}_1^n)) \le \mathring{\varrho}_\pm(\mathcal{H}(\mathbb{D}_1^n),\mathcal{H}(\mathbb{D}_1^n)),\\ \varrho(\mathcal{H}(\mathbb{D}_1^n),\mathcal{R}_n\otimes\mathcal{H}(\mathbb{D}_1^n)) \le \mathring{\varrho}_\pm(\mathcal{H}(\mathbb{D}_1^n),\mathcal{H}(\mathbb{D}_1^n)).\end{aligned}$$

8.6.11 We proceed with an immediate consequence of 8.5.14 and 8.6.10.

EXAMPLE. $\varrho(L_r|\mathcal{R}_n\otimes\mathcal{H}(\mathbb{D}_1^n),\mathcal{H}(\mathbb{D}_1^n))\asymp 1$ *if* $1<r<\infty$.

8.6.12 In the case $1<r<\infty$, the next example could also be obtained from 8.5.14 and 8.6.10, or by duality. However, it turns out that we have the same estimate in the limiting case $r=1$ as well. Thus a separate proof is needed; see [gar, p. 109].

EXAMPLE. $\varrho(L_r|\mathcal{H}(\mathbb{D}_1^n),\mathcal{R}_n\otimes\mathcal{H}(\mathbb{D}_1^n)\asymp 1$ *if* $1\le r<\infty$.

PROOF. Let

$$\boldsymbol{f}_n=\sum_{k=1}^n \boldsymbol{d}_k$$

be any $L_1(M,\mu)$-valued martingale. Then the Burkholder–Gundy–Davis inequality 7.2.12 tells us that

$$\int_0^1 \Big|\sum_{k=1}^n d_k(t,\xi)\Big|\,dt \le a_1\int_0^1 \Big[\sum_{k=1}^n |d_k(t,\xi)|^2\Big]^{1/2} dt.$$

On the other hand, by Khintchine's inequality,

$$\Big[\sum_{k=1}^n |d_k(t,\xi)|^2\Big]^{1/2} \le K_1'(\mathcal{R})\int_0^1 \Big|\sum_{k=1}^n r_k(s)d_k(t,\xi)\Big|\,ds.$$

Integration over $t \in [0,1)$ and $\xi \in M$ yields

$$\int_0^1\!\!\int_M \Big|\sum_{k=1}^n d_k(t,\xi)\Big|\,dt\,d\mu(\xi) \le a_1 \int_0^1\!\!\int_M \Big[\sum_{k=1}^n |d_k(t,\xi)|^2\Big]^{1/2} dt\,d\mu(\xi)$$

$$\le a_1 K_1'(\mathcal{R}) \int_0^1\!\!\int_0^1\!\!\int_M \Big|\sum_{k=1}^n r_k(s) d_k(t,\xi)\Big|\,ds\,dt\,d\mu(\xi),$$

which means that

$$\int_0^1 \Big\|\sum_{k=1}^n \boldsymbol{d}_k(t)\Big|L_1\Big\|\,dt \le a_1 K_1'(\mathcal{R}) \int_0^1\!\!\int_0^1 \Big\|\sum_{k=1}^n r_k(s)\boldsymbol{d}_k(t)\Big|L_1\Big\|\,ds\,dt.$$

Applying 7.2.10 to the martingale transforms

$$\sum_{k=1}^n \boldsymbol{d}_k \longrightarrow \sum_{k=1}^n \boldsymbol{d}_k \quad\text{and}\quad \sum_{k=1}^n \boldsymbol{d}_k \longrightarrow \sum_{k=1}^n r_k \otimes \boldsymbol{d}_k$$

gives

$$\Big(\int_0^1 \Big\|\sum_{k=1}^n \boldsymbol{d}_k(t)\Big|L_1\Big\|^2 dt\Big)^{1/2} \le c_2 a_1 K_1'(\mathcal{R}) \Big(\int_0^1\!\!\int_0^1 \Big\|\sum_{k=1}^n r_k(s)\boldsymbol{d}_k(t)\Big|L_1\Big\|^2 ds\,dt\Big)^{1/2}.$$

This completes the proof.

8.6.13 The following example goes back to D. J. H. Garling [gar, p. 105].

EXAMPLE.

$$\boldsymbol{\varrho}(l_\infty^{2^n}|\mathcal{H}(\mathbb{D}_1^n), \mathcal{R}_n \otimes \mathcal{H}(\mathbb{D}_1^n)) \asymp \boldsymbol{\varrho}(L_\infty|\mathcal{H}(\mathbb{D}_1^n), \mathcal{R}_n \otimes \mathcal{H}(\mathbb{D}_1^n)) \asymp \sqrt{n}.$$

PROOF. We use the $L_\infty([0,1),\mathcal{D}_n)$-valued Haar polynomial

$$\boldsymbol{f}_n = \sum_{k=1}^n \boldsymbol{d}_k = \sum_{\mathbb{D}_1^n} x_k^{(j)} \chi_k^{(j)} \quad\text{with}\quad x_k^{(j)} := \tfrac{1}{2^{k-1}} \chi_k^{(j)}.$$

Then it follows from formula (1) in 8.5.15 that

$$\sup_{0 \le \xi < 1} \Big|\sum_{k=1}^n r_k(s) d_k(t,\xi)\Big| \le 3 \sup_{1 \le m \le n} \Big|\sum_{k=1}^m r_k(s)\Big|,$$

where the right-hand expression is a maximal function; see 7.1.8. Thus, by Doob's inequality 7.1.9, integration over $s, t \in [0,1)$ and $\xi \in M$ yields

$$\Big(\int_0^1\!\!\int_0^1 \Big\|\sum_{k=1}^n r_k(s)\boldsymbol{d}_k(t)\Big|L_\infty\Big\|^2 ds\,dt\Big)^{1/2} \le 3 \Big(\int_0^1 \sup_{1 \le m \le n} \Big|\sum_{k=1}^m r_k(s)\Big|^2 ds\Big)^{1/2} \prec \sqrt{n},$$

which means that $\|(x_k^{(j)})|\mathcal{R}_n \otimes \mathcal{H}(\mathbb{D}_1^n)\| \prec \sqrt{n}$. On the other hand, we know from (2) in 8.5.15 that $\|(x_k^{(j)})|\mathcal{H}(\mathbb{D}_1^n)\| = n$. Combining these results with

$$\|(x_k^{(j)})|\mathcal{H}(\mathbb{D}_1^n)\| \le \boldsymbol{\varrho}(l_\infty^{2^n}|\mathcal{H}(\mathbb{D}_1^n), \mathcal{R}_n \otimes \mathcal{H}(\mathbb{D}_1^n))\, \|(x_k^{(j)})|\mathcal{R}_n \otimes \mathcal{H}(\mathbb{D}_1^n)\|,$$

we obtain

$$\sqrt{n} \prec \boldsymbol{\varrho}(l_\infty^{2^n}|\mathcal{H}(\mathbb{D}_1^n), \mathcal{R}_n \otimes \mathcal{H}(\mathbb{D}_1^n)) \le \boldsymbol{\varrho}(L_\infty|\mathcal{H}(\mathbb{D}_1^n), \mathcal{R}_n \otimes \mathcal{H}(\mathbb{D}_1^n)).$$

The upper estimate $\boldsymbol{\varrho}(L_\infty|\mathcal{H}(\mathbb{D}_1^n), \mathcal{R}_n \otimes \mathcal{H}(\mathbb{D}_1^n)) \le \sqrt{n}$ is a consequence of 8.6.8.

8.6.14 Next, we prove a counterpart of 8.6.13.

EXAMPLE.

$$\boldsymbol{\varrho}(l_1^{2^n}|\mathcal{R}_n \otimes \mathcal{H}(\mathbb{D}_0^n), \mathcal{H}(\mathbb{D}_0^n)) \asymp \boldsymbol{\varrho}(L_1|\mathcal{R}_n \otimes \mathcal{H}(\mathbb{D}_0^n), \mathcal{H}(\mathbb{D}_0^n)) \asymp n.$$

PROOF (S. Geiss). In order to check the lower estimates, we define the $l_1^{2^n}$-valued Haar polynomial

$$\boldsymbol{f}_n = \sum_{\mathbb{D}_0^n} x_k^{(j)} \chi_k^{(j)} \quad \text{by} \quad \boldsymbol{f}_n(t) := u_i^{(2^n)} \quad \text{if } t \in \Delta_n^{(i)}.$$

Then it follows from $\|\boldsymbol{f}_n(t)|l_1^{2^n}\| \equiv 1$ that

$$\|(x_k^{(j)})|\mathcal{H}(\mathbb{D}_0^n)\| = \|\boldsymbol{f}_n|L_2\| = 1. \tag{1}$$

Write

$$\mathbb{N}_k^{(j)} := \{\, i : \Delta_n^{(i)} \subseteq \Delta_k^{(j)} \,\},$$

and note that

$$\mathbb{N}_{k-1}^{(j)} = \mathbb{N}_k^{(2j-1)} \cup \mathbb{N}_k^{(2j)} = \mathbb{N}_{k+1}^{(4j-3)} \cup \mathbb{N}_{k+1}^{(4j-2)} \cup \mathbb{N}_{k+1}^{(4j-1)} \cup \mathbb{N}_{k+1}^{(4j)}.$$

Since

$$\begin{aligned}
\boldsymbol{f}_{k-1}(0) &= \tfrac{1}{2^{n-k+1}}(1,\dots,1,1,\dots,1,1,\dots,1,1,\dots,1,0,\dots,0),\\
\boldsymbol{f}_k(0) &= \tfrac{1}{2^{n-k+1}}(2,\dots,2,2,\dots,2,0,\dots,0,0,\dots,0,0,\dots,0),\\
\boldsymbol{f}_{k+1}(0) &= \tfrac{1}{2^{n-k+1}}(\underbrace{4,\dots,4}_{\mathbb{N}_{k+1}^{(1)}},\underbrace{0,\dots,0}_{\mathbb{N}_{k+1}^{(2)}},\underbrace{0,\dots,0}_{\mathbb{N}_{k+1}^{(3)}},\underbrace{0,\dots,0}_{\mathbb{N}_{k+1}^{(4)}},0,\dots,0),
\end{aligned}$$

the 2^n-tuple

$$2\boldsymbol{d}_k(0) - \boldsymbol{d}_{k+1}(0) = 3\boldsymbol{f}_k(0) - 2\boldsymbol{f}_{k-1}(0) - \boldsymbol{f}_{k+1}(0)$$

has the form

$$\tfrac{1}{2^{n-k+1}}(\underbrace{0,\dots,0}_{\mathbb{N}_{k+1}^{(1)}},\underbrace{+4,\dots,+4}_{\mathbb{N}_{k+1}^{(2)}},\underbrace{-2,\dots,-2}_{\mathbb{N}_{k+1}^{(3)}},\underbrace{-2,\dots,-2}_{\mathbb{N}_{k+1}^{(4)}},0,\dots,0).$$

So, in view of $|\mathbb{N}_{k+1}^{(j)}| = 2^{n-k-1}$, we obtain $\|2\boldsymbol{d}_k(0) - \boldsymbol{d}_{k+1}(0)|l_1^{2^n}\| = 2$. To simplify matters, we now assume that $n = 2m$ is even. Then the unconditionality of the Rademacher system implies that

$$\|(\boldsymbol{d}_{2h-1}(0))|\mathcal{R}_m\| \le \|(\boldsymbol{d}_k(0))|\mathcal{R}_n\| \quad \text{and} \quad \|(\boldsymbol{d}_{2h}(0))|\mathcal{R}_m\| \le \|(\boldsymbol{d}_k(0))|\mathcal{R}_n\|.$$

Since the 2^n-tuples

$$2\boldsymbol{d}_{2h-1}(0) - \boldsymbol{d}_{2h}(0) \in l_1^{2^n} \quad \text{with } h = 1, \dots, m$$

have disjoint supports, we get

$$n = 2m = \|(2\boldsymbol{d}_{2h-1}(0) - \boldsymbol{d}_{2h}(0))|\mathcal{R}_m\| \le 3\,\|(\boldsymbol{d}_k(0))|\mathcal{R}_n\|.$$

By symmetry, this estimate is true not only for $t{=}0$, but for all $t{\in}[0,1)$,

$$n = 2m = \|(2\boldsymbol{d}_{2h-1}(t) - \boldsymbol{d}_{2h}(t))|\mathcal{R}_m\| \le 3\,\|(\boldsymbol{d}_k(t))|\mathcal{R}_n\|.$$

Integration over $t \in [0,1)$ now yields

$$n \le 3\,\|(x_k^{(j)})|\mathcal{R}_n \otimes \mathcal{H}(\mathbb{D}_1^n)\| \le 3\,\|(x_k^{(j)})|\mathcal{R}_n \otimes \mathcal{H}(\mathbb{D}_0^n)\|. \tag{2}$$

Combining (1) and (2), we finally arrive at

$$n \le 3\,\boldsymbol{\varrho}(l_1^{2^n}|\mathcal{R}_n \otimes \mathcal{H}(\mathbb{D}_0^n), \mathcal{H}(\mathbb{D}_0^n)).$$

By monotonicity, the same asymptotic behaviour occurs when n is odd. The upper estimates are consequences of 8.6.8.

8.6.15 The next example follows from the fact that L_∞ contains L_1 as a subspace.

EXAMPLE. $\boldsymbol{\varrho}(L_\infty|\mathcal{R}_n \otimes \mathcal{H}(\mathbb{D}_0^n), \mathcal{H}(\mathbb{D}_0^n)) \asymp n.$

8.6.16 It seems likely that the above relation also holds in the finite dimensional case.

PROBLEM. *Is it true that*

$$\boldsymbol{\varrho}(l_\infty^{2^n}|\mathcal{R}_n \otimes \mathcal{H}(\mathbb{D}_0^n), \mathcal{H}(\mathbb{D}_0^n)) \asymp n\ ?$$

8.6.17 We now treat the ideal norm $\boldsymbol{\delta}(\mathcal{R}_n \otimes \mathcal{H}(\mathbb{D}_0^n), \mathcal{R}_n \otimes \mathcal{H}(\mathbb{D}_0^n))$. First of all, a rough analogue of 4.13.1 will be proved.

PROPOSITION.

$$\boldsymbol{\delta}(\mathcal{R}_n \otimes \mathcal{H}(\mathbb{D}_0^n), \mathcal{R}_n \otimes \mathcal{H}(\mathbb{D}_0^n)) \le \boldsymbol{\varrho}(\mathcal{R}_n \otimes \mathcal{H}(\mathbb{D}_0^n), \mathcal{H}(\mathbb{D}_0^n)).$$

PROOF. Recall that

$$\boldsymbol{\varrho}(\mathcal{R}_n \otimes \mathcal{H}(\mathbb{D}_0^n), \mathcal{H}(\mathbb{D}_0^n)) = \boldsymbol{\delta}(\mathcal{R}_n \otimes \mathcal{H}(\mathbb{D}_0^n), \mathcal{H}(\mathbb{D}_0^n)),$$

and let $r_0 \equiv 1$. Using the unconditionality of the Rademacher system, we now get

$$\int_0^1\int_0^1 \Big\| \sum_{\mathbb{D}_0^n} \int_0^1 Tf(s,t) r_k(s) \chi_k^{(j)}(t)\, dt\, r_k(u) \chi_k^{(j)}(v) \Big\|^2 du\, dv =$$

$$= \int_0^1\int_0^1 \Big\| \sum_{\mathbb{D}_0^n} \int_0^1 Tf(s,t) \chi_k^{(j)}(t)\, dt\, r_k(u) \chi_k^{(j)}(v) \Big\|^2 du\, dv$$

$$\leq \varrho(T|\mathcal{R}_n \otimes \mathcal{H}(\mathbb{D}_0^n), \mathcal{H}(\mathbb{D}_0^n))^2 \int_0^1 \|f(s,t)\|^2 dt$$

for all $f \in [L_2([0,1) \times [0,1)), X]$ and $s \in [0,1)$. Moreover,

$$\Big\| \int_0^1 \Big[\sum_{\mathbb{D}_0^n} \int_0^1 Tf(s,t) r_k(s) \chi_k^{(j)}(t)\, dt\, r_k(u) \chi_k^{(j)}(v) \Big] ds \Big\|^2 \leq$$

$$\leq \int_0^1 \Big\| \sum_{\mathbb{D}_0^n} \int_0^1 Tf(s,t) r_k(s) \chi_k^{(j)}(t)\, dt\, r_k(u) \chi_k^{(j)}(v) \Big\|^2 ds.$$

Integration over $s \in [0,1)$ and $u, v \in [0,1)$, respectively, yields

$$\int_0^1\int_0^1 \Big\| \sum_{\mathbb{D}_0^n} \int_0^1\int_0^1 Tf(s,t) r_k(s) \chi_k^{(j)}(t)\, ds\, dt\, r_k(u) \chi_k^{(j)}(v) \Big\|^2 du\, dv \leq$$

$$\leq \int_0^1\int_0^1\int_0^1 \Big\| \sum_{\mathbb{D}_0^n} \int_0^1 Tf(s,t) r_k(s) \chi_k^{(j)}(t)\, dt\, r_k(u) \chi_k^{(j)}(v) \Big\|^2 ds\, du\, dv$$

$$\leq \varrho(T|\mathcal{R}_n \otimes \mathcal{H}(\mathbb{D}_0^n), \mathcal{H}(\mathbb{D}_0^n))^2 \int_0^1\int_0^1 \|f(s,t)\|^2 ds\, dt.$$

8.6.18 PROPOSITION.

$$\delta(\mathcal{R}_n \otimes \mathcal{H}(\mathbb{D}_1^n), \mathcal{R}_n \otimes \mathcal{H}(\mathbb{D}_1^n)) \leq \varrho(\mathcal{R}_n, \mathcal{I}_n) \circ \varrho'(\mathcal{R}_n, \mathcal{I}_n).$$

PROOF. In view of 8.6.4,

$$\begin{aligned}
&\delta(\mathcal{R}_n \otimes \mathcal{H}(\mathbb{D}_1^n), \mathcal{R}_n \otimes \mathcal{H}(\mathbb{D}_1^n)) \leq \\
&\qquad \leq \varrho(\mathcal{R}_n \otimes \mathcal{H}(\mathbb{D}_1^n), \mathcal{I}(\mathbb{D}_1^n)) \circ \delta(\mathcal{I}(\mathbb{D}_1^n), \mathcal{R}_n \otimes \mathcal{H}(\mathbb{D}_1^n)) \\
&\qquad = \varrho(\mathcal{R}_n \otimes \mathcal{H}(\mathbb{D}_1^n), \mathcal{I}(\mathbb{D}_1^n)) \circ \delta'(\mathcal{R}_n \otimes \mathcal{H}(\mathbb{D}_1^n), \mathcal{I}(\mathbb{D}_1^n)) \\
&\qquad = \varrho(\mathcal{R}_n, \mathcal{I}_n) \circ \varrho'(\mathcal{R}_n, \mathcal{I}_n).
\end{aligned}$$

8.6.19 We proceed with a corollary of 8.6.11 and 8.6.17.

EXAMPLE. $\boldsymbol{\delta}(L_r|\mathcal{R}_n\otimes\mathcal{H}(\mathbb{D}_1^n),\mathcal{R}_n\otimes\mathcal{H}(\mathbb{D}_1^n))\asymp 1$ *if* $1<r<\infty$.

8.6.20 Next, the limiting cases are treated.

EXAMPLE.

$$\boldsymbol{\delta}(l_1^{2^n}|\mathcal{R}_n\otimes\mathcal{H}(\mathbb{D}_1^n),\mathcal{R}_n\otimes\mathcal{H}(\mathbb{D}_1^n))\asymp\boldsymbol{\delta}(L_1|\mathcal{R}_n\otimes\mathcal{H}(\mathbb{D}_1^n),\mathcal{R}_n\otimes\mathcal{H}(\mathbb{D}_1^n))\asymp n,$$
$$\boldsymbol{\delta}(l_\infty^{2^n}|\mathcal{R}_n\otimes\mathcal{H}(\mathbb{D}_1^n),\mathcal{R}_n\otimes\mathcal{H}(\mathbb{D}_1^n))\asymp\boldsymbol{\delta}(L_\infty|\mathcal{R}_n\otimes\mathcal{H}(\mathbb{D}_1^n),\mathcal{R}_n\otimes\mathcal{H}(\mathbb{D}_1^n))\asymp n.$$

PROOF. Recall from 8.6.9, 8.6.15 and 8.6.12 that

$$\boldsymbol{\varrho}(L_\infty|\mathcal{R}_n\otimes\mathcal{H}(\mathbb{D}_0^n),\mathcal{H}(\mathbb{D}_0^n))\le$$
$$\le\boldsymbol{\varrho}(L_1|\mathcal{H}(\mathbb{D}_0^n),\mathcal{R}_n\otimes\mathcal{H}(\mathbb{D}_0^n))\boldsymbol{\delta}(L_1|\mathcal{R}_n\otimes\mathcal{H}(\mathbb{D}_0^n),\mathcal{R}_n\otimes\mathcal{H}(\mathbb{D}_0^n)),$$

$$\boldsymbol{\varrho}(L_\infty|\mathcal{R}_n\otimes\mathcal{H}(\mathbb{D}_0^n),\mathcal{H}(\mathbb{D}_0^n))\asymp n \quad\text{and}\quad \boldsymbol{\varrho}(L_1|\mathcal{H}(\mathbb{D}_0^n),\mathcal{R}_n\otimes\mathcal{H}(\mathbb{D}_0^n))\asymp 1.$$

These facts imply the lower estimate for L_1. We omit the proof in the case of $l_1^{2^n}$, which is more complicated. The upper estimate is a consequence of 8.6.17 and 8.6.8, while the rest follows by duality.

8.6.21 We now ask a question suggested by 4.8.12 and 4.12.12.

PROBLEM. *Is it true that, for every Banach space X, we have either*

$$\boldsymbol{\delta}(X|\mathcal{R}_n\otimes\mathcal{H}(\mathbb{D}_1^n),\mathcal{R}_n\otimes\mathcal{H}(\mathbb{D}_1^n))\asymp 1$$

or

$$\boldsymbol{\delta}(X|\mathcal{R}_n\otimes\mathcal{H}(\mathbb{D}_1^n),\mathcal{R}_n\otimes\mathcal{H}(\mathbb{D}_1^n))\asymp n\,?$$

8.6.22 All examples related to the Haar system are collected in the following table. The entries indicate asymptotic equivalences.

	$r=1$	$1<r\le 2$	$2\le r<\infty$	$r=\infty$
$\boldsymbol{\varrho}(L_r\|\mathcal{H}(\mathbb{D}_1^n),\mathcal{I}(\mathbb{D}_1^n))$	$\sqrt{n}$ 7.4.7	$n^{1/r-1/2}$ 7.4.8	1 7.4.9	$\sqrt{n}$ 7.4.10
$\boldsymbol{\varrho}(L_r\|\mathcal{I}(\mathbb{D}_1^n),\mathcal{H}(\mathbb{D}_1^n))$	$\sqrt{n}$ 7.4.4	1 7.4.4	$n^{1/2-1/r}$ 7.4.4	$\sqrt{n}$ 7.4.4
$\boldsymbol{\varrho}_\pm(L_r\|\mathcal{H}(\mathbb{D}_1^n),\mathcal{H}(\mathbb{D}_1^n))$	n 8.5.15	1 8.5.14	1 8.5.14	n 8.5.15
$\boldsymbol{\varrho}(L_r\|\mathcal{R}_n\otimes\mathcal{H}(\mathbb{D}_1^n),\mathcal{H}(\mathbb{D}_1^n))$	n 8.6.14	1 8.6.11	1 8.6.11	n 8.6.15
$\boldsymbol{\varrho}(L_r\|\mathcal{H}(\mathbb{D}_1^n),\mathcal{R}_n\otimes\mathcal{H}(\mathbb{D}_1^n))$	1 8.6.12	1 8.6.12	1 8.6.12	$\sqrt{n}$ 8.6.13
$\boldsymbol{\delta}(L_r\|\mathcal{R}_n\otimes\mathcal{H}(\mathbb{D}_1^n),\mathcal{R}_n\otimes\mathcal{H}(\mathbb{D}_1^n))$	n 8.6.20	1 8.6.19	1 8.6.19	n 8.6.20

Of course, the same relations hold when $\mathbb{D}_1^n$ is replaced by $\mathbb{D}_0^n$.

8.6.23 EXAMPLE.

$$\boldsymbol{\varrho}(\Sigma_{2^n}: l_1^{2^n} \to l_\infty^{2^n} | \mathcal{R}_n \otimes \mathcal{H}(\mathbb{D}_0^n), \mathcal{H}(\mathbb{D}_0^n)) \asymp$$
$$\asymp \boldsymbol{\varrho}(\Sigma : l_1 \to l_\infty | \mathcal{R}_n \otimes \mathcal{H}(\mathbb{D}_0^n), \mathcal{H}(\mathbb{D}_0^n)) \asymp \sqrt{n}.$$

PROOF (S. Geiss). By 4.3.18 and 7.4.6, we have

$$\boldsymbol{\varrho}(\Sigma : l_1 \to l_\infty | \mathcal{R}_n, \mathcal{I}_n) \le 2 \quad \text{and} \quad \boldsymbol{\varrho}(l_1 | \mathcal{I}(\mathbb{D}_1^n), \mathcal{H}(\mathbb{D}_1^n)) \le \sqrt{n}.$$

Thus the upper estimate follows from 8.6.7.

In order to check the lower estimate, we define the $l_1^{2^n}$-valued Haar polynomial

$$\boldsymbol{f}_n = \sum_{\mathbb{D}_0^n} x_k^{(j)} \chi_k^{(j)} \quad \text{by} \quad \boldsymbol{f}_n(t) := u_i^{(2^n)} \quad \text{if } t \in \Delta_n^{(i)}$$

as in the proofs of 7.6.14 and 8.6.14. Write

$$\mathbb{N}_k^{(j)} := \{ i : \Delta_n^{(i)} \subseteq \Delta_k^{(j)} \}.$$

If $t \in \Delta_k^{(2j-1)}$, then

$$\begin{aligned} \boldsymbol{f}_k(t) &= \tfrac{1}{2^{n-k+1}}(0, \ldots, 0, 2, \ldots, 2, 0, \ldots, 0, 0, \ldots, 0), \\ \boldsymbol{f}_{k-1}(t) &= \tfrac{1}{2^{n-k+1}}(0, \ldots, 0, \underbrace{1, \ldots, 1}_{\mathbb{N}_k^{(2j-1)}}, \underbrace{1, \ldots, 1}_{\mathbb{N}_k^{(2j)}}, 0, \ldots, 0). \end{aligned}$$

Hence

$$\boldsymbol{d}_k(t) = \tfrac{1}{2^{n-k+1}}(0, \ldots, 0, \underbrace{1, \ldots, 1}_{\mathbb{N}_k^{(2j-1)}}, \underbrace{-1, \ldots, -1}_{\mathbb{N}_k^{(2j)}}, 0, \ldots, 0).$$

If $t \in \Delta_k^{(2j)}$, we have

$$\boldsymbol{d}_k(t) = \tfrac{1}{2^{n-k+1}}(0, \ldots, 0, \underbrace{-1, \ldots, -1}_{\mathbb{N}_k^{(2j-1)}}, \underbrace{1, \ldots, 1}_{\mathbb{N}_k^{(2j)}}, 0, \ldots, 0).$$

For $t \in \Delta_{k-1}^{(j)}$, both formulas are summarized by

$$r_k(t)\boldsymbol{d}_k(t) = \tfrac{1}{2^{n-k+1}}(0, \ldots, 0, \underbrace{1, \ldots, 1}_{\mathbb{N}_k^{(2j-1)}}, \underbrace{-1, \ldots, -1}_{\mathbb{N}_k^{(2j)}}, 0, \ldots, 0).$$

Consequently,

$$r_k(t)\Sigma_{2^n}\boldsymbol{d}_k(t) = \tfrac{1}{2^{n-k+1}}(0, \ldots, 0, \underbrace{1, \ldots, 2^{n-k}}_{\mathbb{N}_k^{(2j-1)}}, \underbrace{2^{n-k}-1, \ldots, 1, 0}_{\mathbb{N}_k^{(2j)}}, 0, \ldots, 0).$$

We now obtain

$$r_k(t)\langle \Sigma_{2^n}\boldsymbol{d}_k(t), \boldsymbol{f}_{k-1}(t)\rangle = \frac{1}{4^{n-k+1}}\left(2^{n-k} + 2\sum_{l=1}^{2^{n-k}-1} l\right) = \frac{1}{4}.$$

Moreover, the $\mathcal{D}_{k-1}$-measurability of $r_k \Sigma_{2^n} \boldsymbol{d}_k$ implies that

$$\int_0^1 r_k(t)\langle \Sigma_{2^n} \boldsymbol{d}_k(t), \boldsymbol{f}_n(t)\rangle\, dt = \int_0^1 r_k(t)\langle \Sigma_{2^n} \boldsymbol{d}_k(t), \boldsymbol{f}_{k-1}(t)\rangle\, dt.$$

Next, it follows that

$$\begin{aligned}
\|(\Sigma_{2^n} x_k^{(j)})|\mathcal{R}_n \otimes \mathcal{H}(\mathbb{D}_1^n)\| &= \left(\int_0^1\!\!\int_0^1 \left\| \sum_{k=1}^n r_k(s)\Sigma_{2^n}\boldsymbol{d}_k(t) \Big| l_\infty^{2^n} \right\|^2 ds\, dt\right)^{1/2} \\
&\geq \left(\int_0^1\!\!\int_0^1 \left| \left\langle \sum_{k=1}^n r_k(s)\Sigma_{2^n}\boldsymbol{d}_k(t), \boldsymbol{f}_n(t)\right\rangle \right|^2 ds\, dt\right)^{1/2} \\
&= \left(\int_0^1 \sum_{k=1}^n \left|\langle \Sigma_{2^n}\boldsymbol{d}_k(t), \boldsymbol{f}_n(t)\rangle\right|^2 dt\right)^{1/2} \\
&\geq \frac{1}{\sqrt{n}} \int_0^1 \sum_{k=1}^n \left|\langle \Sigma_{2^n}\boldsymbol{d}_k(t), \boldsymbol{f}_n(t)\rangle\right| dt \\
&\geq \frac{1}{\sqrt{n}} \int_0^1 \sum_{k=1}^n r_k(t)\langle \Sigma_{2^n}\boldsymbol{d}_k(t), \boldsymbol{f}_n(t)\rangle\, dt \\
&= \frac{1}{\sqrt{n}} \int_0^1 \sum_{k=1}^n r_k(t)\langle \Sigma_{2^n}\boldsymbol{d}_k(t), \boldsymbol{f}_{k-1}(t)\rangle\, dt = \tfrac{1}{4}\sqrt{n}.
\end{aligned}$$

Taking into account that $\|(x_k^{(j)})|\mathcal{H}(\mathbb{D}_0^n)\| = \|\boldsymbol{f}_n|L_2\| = 1$, we finally get

$$\begin{aligned}
\tfrac{1}{4}\sqrt{n} &\leq \|(\Sigma_{2^n} x_k^{(j)})|\mathcal{R}_n \otimes \mathcal{H}(\mathbb{D}_1^n)\| \leq \|(\Sigma_{2^n} x_k^{(j)})|\mathcal{R}_n \otimes \mathcal{H}(\mathbb{D}_0^n)\| \\
&\leq \boldsymbol{\varrho}(\Sigma_{2^n} : l_1^{2^n} \to l_\infty^{2^n} | \mathcal{R}_n \otimes \mathcal{H}(\mathbb{D}_0^n), \mathcal{H}(\mathbb{D}_0^n)).
\end{aligned}$$

8.6.24 PROBLEM. *Is it true that*

$$\begin{aligned}
&\boldsymbol{\delta}(\Sigma_{2^n} : l_1^{2^n} \to l_\infty^{2^n} | \mathcal{R}_n \otimes \mathcal{H}(\mathbb{D}_0^n), \mathcal{R}_n \otimes \mathcal{H}(\mathbb{D}_0^n)) \asymp \\
&\qquad \asymp \boldsymbol{\delta}(\Sigma : l_1 \to l_\infty | \mathcal{R}_n \otimes \mathcal{H}(\mathbb{D}_0^n), \mathcal{R}_n \otimes \mathcal{H}(\mathbb{D}_0^n)) \asymp \sqrt{n}\ ?
\end{aligned}$$

REMARK. The upper estimate follows from 8.6.17 and 8.6.23.

8.6.25 Very little is known about the operator ideals associated with the ideal norms

$$\boldsymbol{\delta}(T|\mathcal{R}_n \otimes \mathcal{H}(\mathbb{D}_1^n), \mathcal{R}_n \otimes \mathcal{H}(\mathbb{D}_1^n)),$$

$$\boldsymbol{\varrho}(\mathcal{R}_n \otimes \mathcal{H}(\mathbb{D}_1^n), \mathcal{H}(\mathbb{D}_1^n)) \quad \text{and} \quad \boldsymbol{\varrho}(\mathcal{H}(\mathbb{D}_1^n), \mathcal{R}_n \otimes \mathcal{H}(\mathbb{D}_1^n)).$$

Thus, at least now, it makes no sense to introduce colourful names, like *operators of Rademacher–Haar type*, *Rademacher–Haar cotype*, etc.

8.6.26 Letting

$$\mathfrak{RAT}_2 := \Big\{ T \in \mathfrak{L} : \varrho(T|\mathcal{R}_n \otimes \mathcal{H}(\mathbb{D}_1^n), \mathcal{H}(\mathbb{D}_1^n)) = O(1) \Big\}$$

and

$$\mathfrak{RAC}_2 := \Big\{ T \in \mathfrak{L} : \varrho(T|\mathcal{H}(\mathbb{D}_1^n), \mathcal{R}_n \otimes \mathcal{H}(\mathbb{D}_1^n)) = O(1) \Big\},$$

the left-hand formulas, established in 7.5.16 and 4.10.7, have the right-hand counterparts:

$$\begin{array}{rcl}
\mathfrak{AT}_p \subset \mathfrak{RT}_p & \text{and} & \mathfrak{RAC}_2 \circ \mathfrak{RT}_p \subseteq \mathfrak{AT}_p, \\
\mathfrak{AT}_p^{weak} \subset \mathfrak{RT}_p^{weak} & \text{and} & \mathfrak{RAC}_2 \circ \mathfrak{RT}_p^{weak} \subseteq \mathfrak{AT}_p^{weak}, \\
\mathfrak{AC}_q \subset \mathfrak{RC}_q & \text{and} & \mathfrak{RC}_q \circ \mathfrak{RAT}_2 \subseteq \mathfrak{AC}_q, \\
\mathfrak{AC}_q^{weak} \subset \mathfrak{RC}_q^{weak} & \text{and} & \mathfrak{RC}_q^{weak} \circ \mathfrak{RAC}_2 \subseteq \mathfrak{AC}_q^{weak}, \\
\mathfrak{RC}_2 \circ \mathfrak{RT}_2 = \mathfrak{H} & \text{and} & \mathfrak{RAC}_2 \circ \mathfrak{RAT}_2 \subseteq \mathfrak{UMD}.
\end{array}$$

REMARKS. Recall from 7.6.14 that

$$\mathfrak{RT}_2 \not\subseteq \mathfrak{AT}_p \quad \text{if } 1 < p \leq 2 \quad \text{and} \quad \mathfrak{RC}_2 \not\subseteq \mathfrak{AC}_q \quad \text{if } 2 \leq q < \infty.$$

The classes RAT_2 and RAC_2 were, for the first time, considered by D. J. H. Garling; see [gar].

8.6.27 S. Geiss found the following elegant proof of a classical result, which goes back to [kad*b, p. 176] and [ros, p. 366].

PROPOSITION. *B-convex subspaces of L_1 are superreflexive.*

PROOF. In view of

$$\mathsf{B} = \bigcup_{1<p\leq 2} \mathsf{RT}_p \quad \text{and} \quad \bigcup_{1<p\leq 2} \mathsf{AT}_p = \mathsf{SR},$$

we infer from $\mathsf{RAC}_2 \cap \mathsf{RT}_p \subseteq \mathsf{AT}_p$ that $\mathsf{RAC}_2 \cap \mathsf{B} \subseteq \mathsf{SR}$. Since, by 8.6.12, subspaces of L_1 belong to RAC_2, the assertion follows.

8.6.28 The operator ideal

$$\mathfrak{RAT} := \Big\{ T \in \mathfrak{L} : \varrho(T|\mathcal{R}_n \otimes \mathcal{H}(\mathbb{D}_1^n), \mathcal{H}(\mathbb{D}_1^n)) = o(n) \Big\}$$

turns out to be a good old friend, while

$$\Big\{ T \in \mathfrak{L} : \varrho(T|\mathcal{H}(\mathbb{D}_1^n), \mathcal{R}_n \otimes \mathcal{H}(\mathbb{D}_1^n)) = o(n) \Big\}$$

coincides with $\mathfrak{L}$, because of 8.6.8.

PROPOSITION. $\mathfrak{RAT} = \mathfrak{RT}$.

PROOF. Recall that

$$\varrho(\mathcal{R}_n \otimes \mathcal{H}(\mathbb{D}_1^n), \mathcal{H}(\mathbb{D}_1^n)) \leq \varrho(\mathcal{R}_n, \mathcal{I}_n) \circ \varrho(\mathcal{I}(\mathbb{D}_1^n), \mathcal{H}(\mathbb{D}_1^n)) \qquad \text{(by 8.6.7)}$$

and

$$\varrho(\mathcal{I}(\mathbb{D}_1^n), \mathcal{H}(\mathbb{D}_1^n)) \leq \sqrt{n} \qquad \text{(by 7.4.6).}$$

We now obtain

$$\varrho(\mathcal{R}_n \otimes \mathcal{H}(\mathbb{D}_1^n), \mathcal{H}(\mathbb{D}_1^n)) \leq \sqrt{n}\, \varrho(\mathcal{R}_n, \mathcal{I}_n).$$

Hence $\mathfrak{RT} \subseteq \mathfrak{RAT}$.

If $T \in \mathfrak{L}(X, Y)$ fails to have Rademacher subtype, then it follows from 4.4.19 that there exists a diagram as described in 4.4.17. Hence, taking into account the injectivity of $\mathfrak{RAT}$ and

$$\varrho(l_1^{2^n} | \mathcal{R}_n \otimes \mathcal{H}(\mathbb{D}_0^n), \mathcal{H}(\mathbb{D}_0^n)) \asymp n,$$

we obtain $T \notin \mathfrak{RAT}(X, Y)$, which proves the reverse inclusion.

8.6.29 For $0 < \lambda < 1$, we let

$$\mathfrak{RAT}^\lambda := \Big\{ T \in \mathfrak{L} : \varrho(T|\mathcal{R}_n \otimes \mathcal{H}(\mathbb{D}_1^n), \mathcal{H}(\mathbb{D}_1^n)) = O(n^\lambda) \Big\}$$

and

$$\mathfrak{RAT}_0^\lambda := \Big\{ T \in \mathfrak{L} : \varrho(T|\mathcal{R}_n \otimes \mathcal{H}(\mathbb{D}_1^n), \mathcal{H}(\mathbb{D}_1^n)) = o(n^\lambda) \Big\}.$$

By 7.6.11 and 7.6.12, we conclude from 8.6.23 that $\mathfrak{RAT}_0^{1/2} \subseteq \mathfrak{SR}$ and $\mathfrak{RAT}^{1/2} \not\subseteq \mathfrak{SR}$. Thus these intermediate ideals can be used to classify operators $T \notin \mathfrak{SR}$.

8.6.30 Probably, we have an analogue of 8.6.28 for the operator ideal

$$\mathfrak{RAC} := \Big\{ T \in \mathfrak{L} : \varrho(T|\mathcal{H}(\mathbb{D}_1^n), \mathcal{R}_n \otimes \mathcal{H}(\mathbb{D}_1^n)) = o(\sqrt{n}) \Big\}.$$

PROBLEM. *Is it true that* $\mathfrak{RAC} = \mathfrak{RC}$ *?*

8.6.31 Almost nothing is known about

$$\mathfrak{RAP} := \Big\{ T \in \mathfrak{L} : \delta(T|\mathcal{R}_n \otimes \mathcal{H}(\mathbb{D}_1^n), \mathcal{R}_n \otimes \mathcal{H}(\mathbb{D}_1^n)) = O(1) \Big\}$$

and

$$\mathfrak{QRAP} := \Big\{ T \in \mathfrak{L} : \delta(\mathcal{R}_n \otimes \mathcal{H}(\mathbb{D}_1^n), \mathcal{R}_n \otimes \mathcal{H}(\mathbb{D}_1^n)) = o(n) \Big\}.$$

REMARK. In a rough sense, SR plays the role of B when we pass from the Rademacher to the Haar functions. Bearing in mind that RP = QRP = B, we therefore may ask whether RAP = QRAP = SR. However, this is just speculation.

8.7 The Dirichlet ideal norms $\boldsymbol{\delta}(\mathcal{W}_n, \mathcal{W}_n)$

8.7.1 First of all, we observe that $\boldsymbol{\delta}(\mathcal{W}_{2^n}, \mathcal{W}_{2^n}) = 1$. However, the full sequence $(\boldsymbol{\delta}(\mathcal{W}_m, \mathcal{W}_m))$ oscillates strongly. More precisely, by 3.8.12, it follows from the asymptotic behaviour of the Lebesgue constants $L(\mathcal{W}_m)$ described in [GOL*, pp. 40–42] and [SCHI*, pp. 34–35] that

$$0 < \limsup_{m\to\infty}(1+\log m)^{-1}\boldsymbol{\delta}(L_1|\mathcal{W}_m, \mathcal{W}_m) < \infty.$$

Thus, in order to get monotonicity, we will consider the sequence of the regularized ideal norms $\max\limits_{1\le m\le 2^n} \boldsymbol{\delta}(\mathcal{W}_m, \mathcal{W}_m)$. The main result is to be found in 8.7.7.

REMARK. Note that the classical Lebesgue constants $L(\mathcal{E}_m)$ form a strictly increasing sequence; see [sze] and [ZYG, I, p. 73].

8.7.2 Recall from 6.1.3 that

$$\varphi(\mathbb{M}) := \sum_{h\in\mathbb{M}} 2^{h-1} \quad \text{for } \mathbb{M}\in\mathcal{F}(\mathbb{N}),$$

and let $\mathbb{M}(>k) := \{\, h\in\mathbb{M} : h>k \,\}$.

LEMMA. *For $\mathbb{A},\mathbb{B}\in\mathcal{F}(\mathbb{N})$, we have $\varphi(\mathbb{A})<\varphi(\mathbb{B})$ if and only if there exists a (unique) $k\in\mathbb{B}\setminus\mathbb{A}$ such that $\mathbb{A}(>k)=\mathbb{B}(>k)$.*

PROOF. Obviously, the inequalities $\varphi(\mathbb{A})<\varphi(\mathbb{B})$ and $\varphi(\mathbb{A}\setminus\mathbb{B})<\varphi(\mathbb{B}\setminus\mathbb{A})$ are equivalent. However, the latter relation means that

$$\max\mathbb{A}\setminus\mathbb{B} =: h < k := \max\mathbb{B}\setminus\mathbb{A}.$$

8.7.3 For $m=1,2,\ldots$, the m-th **Walsh–Dirichlet projection** on $L_2[0,1)$ is defined by

$$S_m f := \sum_{h=0}^{m-1}\langle f, w_h\rangle\, w_h = \sum_{\varphi(\mathbb{A})<m}\langle f, w_{\mathbb{A}}\rangle\, w_{\mathbb{A}}.$$

Sometimes it is useful to agree that $S_0 := O$.

8.7.4 Let H_k be the 2^k-dimensional linear space of all functions that are constant on the dyadic intervals $\Delta_k^{(j)}$ with $j=1,\ldots,2^k$. Then

$$\{\, w_{\mathbb{A}\,\triangle\{k\}} : \mathbb{A}<k \,\} \quad \text{and} \quad \{\, \chi_k^{(j)} : j=1,\ldots,2^{k-1} \,\}$$

are orthonormal bases of the 2^{k-1}-dimensional subspace $H_k\cap H_{k-1}^{\perp}$. We denote the orthogonal projection onto this subspace by D_k, with the understanding that $D_0 := S_1$. Obviously,

$$S_{2^m} = \sum_{k=0}^{m} D_k.$$

REMARK. With the notation used in 7.3.7, we have $H_k = L_2([0,1), \mathcal{D}_k)$,

$$D_0 = E(\mathcal{D}_0) \quad \text{and} \quad D_k = E(\mathcal{D}_k) - E(\mathcal{D}_{k-1}) \quad \text{for } k = 1, 2, \ldots.$$

8.7.5 For $m = 1, 2, \ldots$, the m-th **Walsh–Dirichlet kernel** is given by

$$S_m := \sum_{h=0}^{m-1} w_h = \sum_{\varphi(\mathbb{A})<m} w_{\mathbb{A}}.$$

In particular, we have

$$S_{2^n} = \sum_{k=0}^{n} D_k \quad \text{with } D_0 \equiv 1 \text{ and } D_k := \sum_{\mathbb{B}<k} w_{\mathbb{B}\triangle\{k\}} \quad \text{for } k = 1, 2, \ldots.$$

If $\mathbb{A} \le n$, then we may interpret $w_{\mathbb{A}}$ as a character on the n-th Cantor group $\mathbb{E}_2^n$. This point of view has the advantage that the projections S_m and D_k become convolution operators,

$$S_m f = S_m * f \quad \text{and} \quad D_k f = D_k * f \quad \text{for } f \in L_2(\mathbb{E}_2^n).$$

REMARK. It will cause no confusion that the symbols S_m and D_k are used to denote both the kernels and the associated operators.

8.7.6 The following formula is taken from [SCHI*, p. 28].

LEMMA. *Let $\mathbb{M} \in \mathcal{F}(\mathbb{N})$ and $m = \varphi(\mathbb{M})$. Then*

$$S_m f = w_{\mathbb{M}} \sum_{k \in \mathbb{M}} D_k(f w_{\mathbb{M}}) \quad \textit{for } f \in L_2[0,1).$$

PROOF. If $\mathbb{B} < k$ and $k \in \mathbb{M}$, then the set $\mathbb{A} := \mathbb{B} \triangle \mathbb{M} \triangle \{k\}$ satisfies the condition $\varphi(\mathbb{A}) < m = \varphi(\mathbb{M})$. Conversely, every $\mathbb{A}$ with $\varphi(\mathbb{A}) < m = \varphi(\mathbb{M})$ can be obtained in this way. In fact, by 8.7.2, we choose $k \in \mathbb{M} \setminus \mathbb{A}$ such that $\mathbb{A}(>k) = \mathbb{M}(>k)$. Then $\mathbb{B} := \mathbb{A} \triangle \mathbb{M} \triangle \{k\}$ has the required property. Using this correspondence, we obtain

$$S_m = \sum_{\varphi(\mathbb{A})<m} w_{\mathbb{A}} = \sum_{k \in \mathbb{M}} \sum_{\mathbb{B}<k} w_{\mathbb{B}\triangle\mathbb{M}\triangle\{k\}} = w_{\mathbb{M}} \sum_{k \in \mathbb{M}} \sum_{\mathbb{B}<k} w_{\mathbb{B}\triangle\{k\}} = w_{\mathbb{M}} \sum_{k \in \mathbb{M}} D_k.$$

Fix $f \in L_2[0,1)$, and consider $f_n := S_{2^n} f$ as a function of $e \in \mathbb{E}_2^n$ with $2^n \ge m$. Then

$$\begin{aligned} S_m f_n(e) &= S_m * f_n(e) = \frac{1}{2^n} \sum_{d \in \mathbb{E}_2^n} \Big(w_{\mathbb{M}}(ed^{-1}) \sum_{k \in \mathbb{M}} D_k(ed^{-1}) \Big) f_n(d) \\ &= w_{\mathbb{M}}(e) \left[\frac{1}{2^n} \sum_{d \in \mathbb{E}_2^n} \Big(\sum_{k \in \mathbb{M}} D_k(ed^{-1}) \Big) f_n(d) w_{\mathbb{M}}(d^{-1}) \right] \\ &= w_{\mathbb{M}}(e) \sum_{k \in \mathbb{M}} D_k(f_n w_{\mathbb{M}})(e). \end{aligned}$$

Since $S_m f = S_m f_n$ and $D_k(f w_{\mathbb{M}}) = D_k(f_n w_{\mathbb{M}})$, this proves the required formula.

8.7.7 We now establish the main result of this section; see [wen 1].

THEOREM.

$$\max_{1\le m\le 2^n} \boldsymbol{\delta}(\mathcal{W}_m, \mathcal{W}_m) \le \mathring{\boldsymbol{\delta}}_{\pm}(\mathcal{H}(\mathbb{D}_1^n), \mathcal{H}(\mathbb{D}_1^n)) \le 2 \max_{1\le m\le 2^n} \boldsymbol{\delta}(\mathcal{W}_m, \mathcal{W}_m).$$

PROOF. Choose $\mathbb{M}$ such that $m=\varphi(\mathbb{M})$, define $d=(\delta_1,\dots,\delta_n)\in\mathbb{E}_2^n$ by

$$\delta_k := \begin{cases} +1 & \text{if } k\in\mathbb{M}, \\ -1 & \text{if } k\notin\mathbb{M}, \end{cases}$$

and let $e=(\varepsilon_1,\dots,\varepsilon_n)=(+1,\dots,+1)$. Then it follows from the previous lemma that

$$S_m f = \tfrac{1}{2} w_{\mathbb{M}} \sum_{k=1}^{n} \delta_k D_k(f w_{\mathbb{M}}) + \tfrac{1}{2} w_{\mathbb{M}} \sum_{k=1}^{n} \varepsilon_k D_k(f w_{\mathbb{M}}) \quad \text{for } f\in L_2[0,1).$$

Applying this equality in the vector-valued case yields

$$\|[S_m,T]\boldsymbol{f}|L_2\| \le \mathring{\boldsymbol{\delta}}_{\pm}(T|\mathcal{H}(\mathbb{D}_1^n),\mathcal{H}(\mathbb{D}_1^n))\, \|\boldsymbol{f} w_{\mathbb{M}}|L_2\| \quad \text{for } \boldsymbol{f}\in[L_2[0,1),X],$$

which implies the left-hand inequality.

On the other hand, given $e=(\varepsilon_1,\dots,\varepsilon_n)\in\mathbb{E}_2^n$, we let

$$\mathbb{A} := \{\, k : \varepsilon_k = +1 \,\} \quad\text{and}\quad \mathbb{B} := \{\, k : \varepsilon_k = -1 \,\}.$$

Then $a=\varphi(\mathbb{A})\le 2^n$ and $b=\varphi(\mathbb{B})\le 2^n$. So, for $\boldsymbol{f}\in[L_2[0,1),X]$ and $\boldsymbol{d}_k := [D_k,X]\boldsymbol{f}$, we obtain

$$\begin{aligned}
\Big\|\sum_{k=1}^{n} \varepsilon_k [L_2,T]\boldsymbol{d}_k \Big| L_2\Big\| &= \Big\|\sum_{k\in\mathbb{A}} [L_2,T]\boldsymbol{d}_k - \sum_{k\in\mathbb{B}} [L_2,T]\boldsymbol{d}_k \Big| L_2\Big\| \\
&= \| w_{\mathbb{A}}[S_a,T](\boldsymbol{f} w_{\mathbb{A}}) - w_{\mathbb{B}}[S_b,T](\boldsymbol{f} w_{\mathbb{B}}) | L_2\| \\
&\le \|[S_a,T](\boldsymbol{f} w_{\mathbb{A}})|L_2\| + \|[S_b,T](\boldsymbol{f} w_{\mathbb{B}})|L_2\| \\
&\le 2 \max_{1\le m\le 2^n} \boldsymbol{\delta}(\mathcal{W}_m,\mathcal{W}_m)\, \|\boldsymbol{f}|L_2\|.
\end{aligned}$$

Hence

$$\mathring{\boldsymbol{\delta}}_{\pm}(\mathcal{H}(\mathbb{D}_1^n),\mathcal{H}(\mathbb{D}_1^n)) \le 2 \max_{1\le m\le 2^n} \boldsymbol{\delta}(\mathcal{W}_m,\mathcal{W}_m).$$

8.8 The Burkholder–Bourgain theorem

8.8.1 We now compare the ideal norms

$$\boldsymbol{\varrho}_{\pm}(\mathcal{H}(\mathbb{D}_1^n),\mathcal{H}(\mathbb{D}_1^n)) \quad \text{and} \quad \boldsymbol{\kappa}(H_n).$$

In shorthand notation, the main result reads as follows.

BURKHOLDER–BOURGAIN THEOREM. UMD = HM.

This means that

$$\boldsymbol{\varrho}_{\pm}(X|\mathcal{H}(\mathbb{D}_1^n),\mathcal{H}(\mathbb{D}_1^n)) = O(1) \iff \boldsymbol{\kappa}(X|H_n) = O(1)$$

for every Banach space X. The implication $\Rightarrow$ was obtained in the pioneering work of D. L. Burkholder, and J. Bourgain contributed the reverse. The proofs are quite complicated and would need a lot of preparation including basic concepts from the theory of Brownian motion. Thus the reader is referred to the original papers and to a course given by D. L. Burkholder in Varenna; see [bou 3], [bur 5] and [bur 8].

REMARK. Usually, the Burkholder–Bourgain theorem is stated in the form UMD = HT. We have replaced HT by HM; see 5.4.19. This point of view has the advantage that both classes, UMD and HM, are generated by sequences of ideal norms.

8.8.2 Unfortunately, the Burkholder–Bourgain approach does not provide any information about the relationship between $\boldsymbol{\varrho}_{\pm}(\mathcal{H}(\mathbb{D}_1^n),\mathcal{H}(\mathbb{D}_1^n))$ and $\boldsymbol{\kappa}(H_n)$. We only get that

$$\|X|\mathfrak{UMD}\| \le c\,\|X|\mathfrak{HM}\|^2 \quad \text{and} \quad \|X|\mathfrak{HM}\| \le c\,\|X|\mathfrak{UMD}\|^2$$

for $X \in$ UMD = HM, where $c \ge 1$ is a universal constant.

PROBLEM. *Is it true that* $\boldsymbol{\varrho}_{\pm}(\mathcal{H}(\mathbb{D}_1^n),\mathcal{H}(\mathbb{D}_1^n)) \asymp \boldsymbol{\kappa}(H_{2^n})$ *?*

Only optimists will expect an affirmative answer; see also Problem 6.5.3. So far, we do not know how to produce a counterexample. Note that

$$\boldsymbol{\varrho}_{\pm}(l_\infty^N|\mathcal{H}(\mathbb{D}_1^n),\mathcal{H}(\mathbb{D}_1^n)) \asymp \min\{n, 1+\log N\}$$

and

$$\boldsymbol{\kappa}(l_\infty^N|H_{2^n}) \asymp \min\{n, 1+\log N\};$$

see 8.5.21 and 5.4.23. Moreover,

$$\boldsymbol{\varrho}_{\pm}(\Sigma_{2^n} : l_1^{2^n} \to l_\infty^{2^n}|\mathcal{H}(\mathbb{D}_1^n),\mathcal{H}(\mathbb{D}_1^n)) \overset{?}{\asymp} n$$

and

$$\boldsymbol{\kappa}(\Sigma_{2^n} : l_1^{2^n} \to l_\infty^{2^n}|H_{2^n}\| \asymp n;$$

see 8.5.23 and 2.4.4. It would be extremely important to remove the question mark in the upper relation.

Finally, we sketch the situation in the limiting cases:

$$\boldsymbol{\varrho}_\pm(X|\mathcal{H}(\mathbb{D}_1^n),\mathcal{H}(\mathbb{D}_1^n)) = O(1) \overset{\textit{Burkholder-Bourgain}}{\iff} \boldsymbol{\kappa}(X|H_n) = O(1),$$

$$\boldsymbol{\varrho}_\pm(X|\mathcal{H}(\mathbb{D}_1^n),\mathcal{H}(\mathbb{D}_1^n)) = o(n) \underset{\Longleftarrow}{\overset{?}{\Longrightarrow}} \boldsymbol{\kappa}(X|H_n) = o(1+\log n)$$

for all Banach spaces X, and

$$\boldsymbol{\varrho}_\pm(T|\mathcal{H}(\mathbb{D}_1^n),\mathcal{H}(\mathbb{D}_1^n)) = O(1) \overset{?}{\iff} \boldsymbol{\kappa}(T|H_n) = O(1),$$

$$\boldsymbol{\varrho}_\pm(T|\mathcal{H}(\mathbb{D}_1^n),\mathcal{H}(\mathbb{D}_1^n)) = o(n) \underset{\Longleftarrow}{\overset{?}{\Longrightarrow}} \boldsymbol{\kappa}(T|H_n) = o(1+\log n)$$

for all operators T. This means that just half of all possible implications are unknown. The lower direction in $\underset{\Longleftarrow}{\overset{?}{\Longrightarrow}}$ follows from 8.5.24.

REMARK. As shown in 5.3.8 and 8.7.7, we have

$$\boldsymbol{\delta}(\mathcal{E}_n,\mathcal{E}_n) \asymp \boldsymbol{\kappa}(H_n) \quad \text{and} \quad \max_{1\le m\le 2^n} \boldsymbol{\delta}(\mathcal{W}_m,\mathcal{W}_m) \asymp \mathring{\boldsymbol{\delta}}_\pm(\mathcal{H}(\mathbb{D}_1^n),\mathcal{H}(\mathbb{D}_1^n)).$$

Thus the above problems are equivalent to those raised in 6.5.3.

9

Miscellaneous

This final chapter is devoted to some overlapping subjects: interpolation, Lorentz spaces, Schatten–von Neumann spaces and ideal norms of finite rank operators. Moreover, we discuss some results and open problems concerning orthogonal polynomials. There should also be a section about 'wavelets'. Unfortunately, nobody has yet checked what happens when the Haar system is generalized in this way.

We have added some historical and philosophical remarks which are intended to put the subject of this treatise in a global perspective.

9.1 Interpolation

Throughout, $T \in \mathfrak{L}(X_0 + X_1, Y_0 + Y_1)$ is assumed to transform X_0 into Y_0 and X_1 into Y_1; see Section 0.6. We investigate how properties of

$$T_0 : X_0 \xrightarrow{T} Y_0 \quad \text{and} \quad T_1 : X_1 \xrightarrow{T} Y_1$$

carry over to

$$\underset{\text{(complex method)}}{T_\theta : X_\theta \xrightarrow{T} Y_\theta} \quad \text{and} \quad \underset{\text{(real method)}}{T_{\theta,w} : X_{\theta,w} \xrightarrow{T} Y_{\theta,w}}.$$

The parameters satisfy the conditions

$$1/p = (1-\theta)/p_0 + \theta/p_1, \; 1 < p_0 < p_1 < 2, \quad \text{and} \quad 0 < \theta < 1.$$

9.1.1 Recall from 3.4.9 that Dirichlet ideal norms and, in particular, type ideal norms behave very well under complex interpolation.

9.1.2 We proceed with an immediate consequence of this fact.

PROPOSITION.

$$T_0 \in \mathfrak{RT}_{p_0}^{weak} \quad \textit{and} \quad T_1 \in \mathfrak{RT}_{p_1}^{weak} \quad \textit{imply} \quad T_\theta \in \mathfrak{RT}_p^{weak}.$$

The same holds for $\mathfrak{ET}_p^{weak}$, $\mathfrak{WT}_p^{weak}$, $\mathfrak{KT}_p^{weak}$ *and* $\mathfrak{AT}_p^{weak}$.

9.1.3 The preceding statements are based on the interpolation formula

$$\Big[L_2,[X_0,X_1]_\theta\Big]=\Big[[L_2,X_0],[L_2,X_1]\Big]_\theta,$$

which is only a special case of the following result; see [BER*, p. 107] and [TRIE, p. 128].

LEMMA. *Let* $1\le r<\infty$. *Then*

$$\Big[L_r,[X_0,X_1]_\theta\Big]=\Big[[L_r,X_0],[L_2,X_1]\Big]_\theta,$$

and the norms on both sides coincide.

9.1.4 Referring to the diagrams in the following paragraphs, we state the next proposition without proof.

PROPOSITION.

$$T_0\in\mathfrak{RT}_{p_0}\quad\textit{and}\quad T_1\in\mathfrak{RT}_{p_1}\quad\textit{imply}\quad T_\theta\in\mathfrak{RT}_p.$$

The same holds for $\mathfrak{CT}_p$, $\mathfrak{WT}_p$ *and* $\mathfrak{AT}_p$.

REMARK. The above result remains true in the limiting cases $p_0=1$ and/or $p_1=2$.

9.1.5 For the real method, the situation is more involved. For example, Z. Altshuler/P. G. Casazza/B. L. Lin and M. Levy showed that every Lorentz space $l_{r,w}$ with $1<r<\infty$ and $1\le w\le\infty$ contains isomorphic copies of l_w; see [alt*], [LIN*1, p. 177] and [lev]. Moreover, it follows from 9.1.14 that the Rademacher type and cotype indices, defined in 4.16.4, are given by

$$p(l_{r,w})=\min\{2,r,w\}\quad\text{and}\quad q(l_{r,w})=\max\{2,r,w\}.$$

Thus, roughly speaking, we can say that the geometric properties of $l_{r,w}$ are determined by the worst of both indices r and w.

9.1.6 In what follows, we need an auxiliary result from real interpolation theory which goes back to [lio*, p. 47].

LEMMA. *Let*

$$0<\theta<1,\ \ 1\le r_0<r_1\le\infty\quad\textit{and}\quad 1/r=(1-\theta)/r_0+\theta/r_1.$$

Then, for any interpolation couple (X_0,X_1), *we have*

$$([L_{r_0},X_0],[L_{r_1},X_1])_{\theta,w}\subseteq[L_r,(X_0,X_1)_{\theta,w}]\quad\textit{if}\ 1\le w\le r\le\infty,$$
$$[L_r,(X_0,X_1)_{\theta,w}]\subseteq([L_{r_0},X_0],[L_{r_1},X_1])_{\theta,w}\quad\textit{if}\ 1\le r\le w\le\infty.$$

These embeddings are continuous.

REMARK. For $r = w$ the above inclusions turn into an equality,

$$([L_{r_0}, X_0], [L_{r_1}, X_1])_{\theta,r} = [L_r, (X_0, X_1)_{\theta,r}],$$

and this is the case that most writers have treated to date. However, the proofs given in [pee 2, p. 30], [BER*, p. 123] and [TRIE, p. 128] can easily be adapted to the more general setting. Instead of Fubini's theorem we have to use Jessen's inequality; [cob 1, p. 61]. For $r = r_0 = r_1$; see [KOE, p. 197].

9.1.7 The following result is due to F. Cobos; see [cob 1].

PROPOSITION. *If* $1 < p \le w < \infty$, *then*

$$T_0 \in \mathfrak{RT}_{p_0} \quad \textit{and} \quad T_1 \in \mathfrak{RT}_{p_1} \quad \textit{imply} \quad T_{\theta,w} \in \mathfrak{RT}_p.$$

PROOF. By definition, the Rademacher type ideal norm $\boldsymbol{\varrho}_p^{(w)}(T|\mathcal{R}_n, \mathcal{J}_n)$ is the least constant $c \ge 0$ such that

$$\left(\int_0^1 \Big\| \sum_{k=1}^n T x_k r_k(t) \Big\|^w dt \right)^{1/w} \le c \left(\sum_{k=1}^n \|x_k\|^p \right)^{1/p}$$

whenever $x_1, \ldots, x_n \in X$. This means that $\boldsymbol{\varrho}_p^{(w)}(T|\mathcal{R}_n, \mathcal{J}_n)$ coincides with the norm of the map

$$[R_n, T] : (x_k) \longrightarrow \sum_{k=1}^n T x_k r_k,$$

viewed as an operator from $[l_p^n, X]$ into $[L_w[0,1), Y]$. We now conclude from the diagram

$$\begin{array}{ccc}
[l^n_{p_0}, X_0] & \xrightarrow{[R_n,T_0]} & [L_w, Y_0] \\
[l^n_{p_1}, X_1] & \xrightarrow{[R_n,T_1]} & [L_w, Y_1] \\
\hline
([l^n_{p_0}, X_0], [l^n_{p_1}, X_1])_{\theta,w} & \xrightarrow{[R_n,T_{\theta,w}]} & ([L_w, Y_0], [L_w, Y_1])_{\theta,w} \\
{\scriptstyle p \le w}\uparrow & & \uparrow\downarrow \\
[l^n_p, (X_0, X_1)_{\theta,w}] & \xrightarrow{[R_n,T_{\theta,w}]} & [L_w, (Y_0, Y_1)_{\theta,w}]
\end{array}$$

that

$$\boldsymbol{\varrho}_p^{(w)}(T_{\theta,w}|\mathcal{R}_n, \mathcal{J}_n) \le c\, \boldsymbol{\varrho}_{p_0}^{(w)}(T_0|\mathcal{R}_n, \mathcal{J}_n)^{1-\theta}\, \boldsymbol{\varrho}_{p_1}^{(w)}(T_1|\mathcal{R}_n, \mathcal{J}_n)^{\theta},$$

where $c > 0$ does not depend on $n = 1, 2, \ldots$. In view of 4.3.3, this proves the assertion.

9.1.8 PROPOSITION. *If* $1 < p < w < \infty$, *then*

$$T_0 \in \mathfrak{RT}_{p_0}^{weak} \quad \textit{and} \quad T_1 \in \mathfrak{RT}_{p_1}^{weak} \quad \textit{imply} \quad T_{\theta,w} \in \mathfrak{RT}_p^{weak}.$$

PROOF. For any sufficiently small $\varepsilon > 0$, we define u_0, u_1 and u by

$$1/p_0 - 1/u_0 = 1/p_1 - 1/u_1 = 1/p - 1/u = \varepsilon$$

such that $1 < u_0 < u < u_1 \leq 2$. Then

$$1/u = (1-\theta)/u_0 + \theta/u_1.$$

The last inequality in the preceding proof tells us that

$$\boldsymbol{\varrho}_{u_0}^{(w)}(T_0|\mathcal{R}_n, \mathcal{I}_n) \leq c_0'\, n^{1/p_0 - 1/u_0} \quad \text{and} \quad \boldsymbol{\varrho}_{u_1}^{(w)}(T_1|\mathcal{R}_n, \mathcal{I}_n) \leq c_1'\, n^{1/p_1 - 1/u_1}$$

imply

$$\boldsymbol{\varrho}_{u}^{(w)}(T_{\theta,w}|\mathcal{R}_n, \mathcal{I}_n) \leq c'\, n^{1/p - 1/u}$$

whenever $u \leq w < \infty$. Of course, if $p < w$, then $\varepsilon > 0$ can be chosen such that $p < u < w$. This completes the proof, since we know from 4.3.9 that

$$\mathfrak{RT}_p^{weak} = \mathfrak{L}\Big[n^{1/u-1/p}\boldsymbol{\varrho}_u^{(w)}(\mathcal{R}_n, \mathcal{I}_n)\Big].$$

9.1.9 PROPOSITION. *If* $p \leq w \leq p'$, *then*

$$T_0 \in \mathfrak{ET}_{p_0} \quad \textit{and} \quad T_1 \in \mathfrak{ET}_{p_1} \quad \textit{imply} \quad T_{\theta,w} \in \mathfrak{ET}_p.$$

The same holds for Walsh type.

PROOF. By Definition 5.6.1, the Fourier type ideal norm $\boldsymbol{\varrho}_p^{(p')}(T|\mathcal{E}_n, \mathcal{I}_n)$ is the least constant $c \geq 0$ such that

$$\left(\frac{1}{2\pi}\int_{-\pi}^{+\pi}\Big\|\sum_{k=1}^{n} Tx_k \exp(ikt)\Big\|^{p'} dt\right)^{1/p'} \leq c\left(\sum_{k=1}^{n}\|x_k\|^p\right)^{1/p}$$

whenever $x_1, \ldots, x_n \in X$. This means that $\boldsymbol{\varrho}_p^{(p')}(T|\mathcal{E}_n, \mathcal{I}_n)$ coincides with the norm of the map

$$[E_n, T] : (x_k) \longrightarrow \sum_{k=1}^{n} Tx_k e_k,$$

viewed as an operator from $[l_p^n, X]$ into $[L_{p'}[-\pi, +\pi), Y]$. We now conclude from the diagram

$$\begin{array}{ccc} [l^n_{p_0}, X_0] & \xrightarrow{[E_n,T_0]} & [L_{p_0'}, Y_0] \\ [l^n_{p_1}, X_1] & \xrightarrow{[E_n,T_1]} & [L_{p_1'}, Y_1] \end{array}$$

$$\begin{array}{ccc} ([l^n_{p_0}, X_0], [l^n_{p_1}, X_1])_{\theta,w} & \xrightarrow{[E_n,T_{\theta,w}]} & ([L_{p_0'}, Y_0], [L_{p_1'}, Y_1])_{\theta,w} \\ {\scriptstyle p\le w}\uparrow & & \downarrow{\scriptstyle w\le p'} \\ [l^n_p, (X_0, X_1)_{\theta,w}] & \xrightarrow{[E_n,T_{\theta,w}]} & [L_{p'}, (Y_0, Y_1)_{\theta,w}] \end{array}$$

that

$$\boldsymbol{\varrho}_p^{(p')}(T_{\theta,w}|\mathcal{E}_n, \mathcal{I}_n) \le c\, \boldsymbol{\varrho}_{p_0}^{(p_0')}(T_0|\mathcal{E}_n, \mathcal{I}_n)^{1-\theta}\, \boldsymbol{\varrho}_{p_1}^{(p_1')}(T_1|\mathcal{E}_n, \mathcal{I}_n)^{\theta},$$

where $c > 0$ does not depend on $n = 1, 2, \ldots$. This proves the assertion.

9.1.10 PROPOSITION. *If $p < w < p'$, then*

$$T_0 \in \mathfrak{ET}^{weak}_{p_0} \quad \textit{and} \quad T_1 \in \mathfrak{ET}^{weak}_{p_1} \quad \textit{imply} \quad T_{\theta,w} \in \mathfrak{ET}^{weak}_{p}.$$

The same holds for weak Walsh type.

PROOF. For any sufficiently small $\varepsilon > 0$, we define u_0, u_1 and u by

$$1/p_0 - 1/u_0 = 1/p_1 - 1/u_1 = 1/p - 1/u = \varepsilon$$

such that $1 < u_0 < u < u_1 \le 2$. Then

$$1/u = (1-\theta)/u_0 + \theta/u_1.$$

The last inequality in the preceding proof tells us that

$$\boldsymbol{\varrho}_{u_0}^{(u_0')}(T_0|\mathcal{E}_n, \mathcal{I}_n) \le c_0\, n^{1/p_0 - 1/u_0} \quad \text{and} \quad \boldsymbol{\varrho}_{u_1}^{(u_1')}(T_1|\mathcal{E}_n, \mathcal{I}_n) \le c_1\, n^{1/p_1 - 1/u_1}$$

imply

$$\boldsymbol{\varrho}_{u}^{(u')}(T_{\theta,w}|\mathcal{E}_n, \mathcal{I}_n) \le c\, n^{1/p - 1/u}$$

whenever $u \le w \le u'$. Of course, if $p < w < p'$, then $\varepsilon > 0$ can be chosen such that $p < u < w < u' < p'$. This completes the proof, since we know from 5.6.18 that

$$\mathfrak{ET}^{weak}_p = \mathfrak{L}\Big[n^{1/u - 1/p}\, \boldsymbol{\varrho}_u^{(u')}(\mathcal{E}_n, \mathcal{I}_n)\Big].$$

9.1.11 PROPOSITION. *If $1 < p \le w < \infty$, then*

$$T_0 \in \mathfrak{AT}_{p_0} \quad \textit{and} \quad T_1 \in \mathfrak{AT}_{p_1} \quad \textit{imply} \quad T_{\theta,w} \in \mathfrak{AT}_{p}.$$

PROOF. A short look at Definition 7.5.1 shows that the Haar type ideal norm $\boldsymbol{\varrho}_p^{(w)}(T|\mathcal{H}(\mathbb{D}^n_1), \mathcal{I}(\mathbb{D}^n_1))$ coincides with the norm of the martingale transform

$$\boldsymbol{T} : \sum_{k=1}^{n} \boldsymbol{d}_k(t) \longrightarrow \sum_{k=1}^{n} T\boldsymbol{d}_k(t)$$

viewed as an operator from $\big[L_w([0,1),\mathcal{D}_n),[l_p^n,X]\big]$ into $[L_w[0,1),Y]$. In the following diagram, we simply write L_w instead of $L_w([0,1),\mathcal{D}_n)$ and $L_w[0,1)$:

$$\begin{array}{ccc} [L_w,[l_{p_0}^n,X_0]] & \xrightarrow{\ T_0\ } & [L_w,Y_0] \\ [L_w,[l_{p_1}^n,X_1]] & \xrightarrow{\ T_1\ } & [L_w,Y_1] \\ \hline \Big([L_w,[l_{p_0}^n,X_0]],[L_w,[l_{p_1}^n,X_1]]\Big)_{\theta,w} & \xrightarrow{\ T_{\theta,w}\ } & ([L_w,Y_0],[L_w,Y_1])_{\theta,w} \\ \uparrow\downarrow & & \uparrow\downarrow \\ \Big[L_w,\big([l_{p_0}^n,X_0],[l_{p_1}^n,X_1]\big)_{\theta,w}\Big] & & \\ p\le w \uparrow & & \\ [L_w,[l_p^n,(X_0,X_1)_{\theta,w}]] & \xrightarrow{\ T_{\theta,w}\ } & [L_w,(Y_0,Y_1)_{\theta,w}]. \end{array}$$

Hence

$$\begin{aligned} &\boldsymbol{\varrho}_p^{(w)}(T_{\theta,w}|\mathcal{H}(\mathbb{D}_1^n),\mathcal{I}(\mathbb{D}_1^n)) \le \\ &\qquad \le c\,\boldsymbol{\varrho}_{p_0}^{(w)}(T_0|\mathcal{H}(\mathbb{D}_1^n),\mathcal{I}(\mathbb{D}_1^n))^{1-\theta}\,\boldsymbol{\varrho}_{p_1}^{(w)}(T_1|\mathcal{H}(\mathbb{D}_1^n),\mathcal{I}(\mathbb{D}_1^n))^{\theta}, \end{aligned}$$

where $c>0$ does not depend on $n=1,2,\ldots$. In view of 7.5.5, this proves the assertion.

9.1.12 The following result is an analogue of 9.1.8. However, since we have no substitute for 4.3.9, the proof only works for $w=2$.

PROPOSITION.

$$T_0\in\mathfrak{AT}_{p_0}^{weak} \quad \textit{and} \quad T_1\in\mathfrak{AT}_{p_1}^{weak} \quad \textit{imply} \quad T_{\theta,2}\in\mathfrak{AT}_p^{weak}.$$

9.1.13 The next observation is due to J. L. Rubio de Francia; see [rub 2] and [cob 2].

PROPOSITION. *If $0<\theta<1$ and $1<w<\infty$, then*

$$T_0\in\mathfrak{UMD} \quad \textit{and} \quad T_1\in\mathfrak{UMD} \quad \textit{imply} \quad T_{\theta,w}\in\mathfrak{UMD}.$$

An analogous result holds for the complex method.

PROOF. By Definition 8.5.12, the ideal norm $\mathring{\boldsymbol{\varrho}}_{\pm}^{(w)}(T|\mathcal{H}(\mathbb{D}_1^n),\mathcal{H}(\mathbb{D}_1^n))$ is the least constant $c\ge 0$ such that

$$\|(Tx_k^{(j)})|\mathcal{H}(\mathbb{D}_1^n)\|_w \le c\,\|(\varepsilon_k x_k^{(j)})|\mathcal{H}(\mathbb{D}_1^n)\|_w$$

for every $\mathbb{D}_1^n$-tuple $(x_k^{(j)})$ in X and any choice of $\varepsilon_k=\pm1$. This means that $\boldsymbol{\varrho}_p^{(p')}(T|\mathcal{E}_n,\mathcal{I}_n)$ coincides with the norm of the map

$$[e\circ H_n,T]:\boldsymbol{f}\longrightarrow\sum_{\mathbb{D}_1^n}\varepsilon_k T\langle\boldsymbol{f},\chi_j^{(k)}\rangle\chi_j^{(k)},$$

viewed as an operator from $[L_w([0,1),\mathcal{D}_n),X]$ into $[L_w[0,1),Y]$. In the following diagram, we simply write L_w instead of $L_w([0,1),\mathcal{D}_n)$ and $L_w[0,1)$:

$$\begin{array}{ccc} [L_w,X_0] & \xrightarrow{[\mathrm{eo}H_n,T_0]} & [L_w,Y_0] \\ [L_w,X_1] & \xrightarrow{[\mathrm{eo}H_n,T_1]} & [L_w,Y_1] \\ \hline ([L_w,X_0],[L_w,X_1])_{\theta,w} & \xrightarrow{[\mathrm{eo}H_n,T_{\theta,w}]} & ([L_w,Y_0],[L_w,Y_1])_{\theta,w} \\ \uparrow\downarrow & & \uparrow\downarrow \\ [L_w,(X_0,X_1)_{\theta,w}] & \xrightarrow{[\mathrm{eo}H_n,T_{\theta,w}]} & [L_w,(Y_0,Y_1)_{\theta,w}]. \end{array}$$

Hence

$$\mathring{\boldsymbol{\varrho}}_{\pm}^{(w)}(T_{\theta,w}|\mathcal{H}(\mathbb{D}_1^n),\mathcal{H}(\mathbb{D}_1^n)) \le$$
$$\le c\,\mathring{\boldsymbol{\varrho}}_{\pm}^{(w)}(T_0|\mathcal{H}(\mathbb{D}_1^n),\mathcal{H}(\mathbb{D}_1^n))^{1-\theta}\,\mathring{\boldsymbol{\varrho}}_{\pm}^{(w)}(T_1|\mathcal{H}(\mathbb{D}_1^n),\mathcal{H}(\mathbb{D}_1^n))^{\theta},$$

where $c>0$ does not depend on $n=1,2,\ldots$. This proves the assertion, in view of 8.5.13.

REMARK. The following result can be found in [kal, p. 527]:

Let (X_0,X_1) be an interpolation couple of Köthe function spaces. Then the set

$$\Big\{\,\theta\in(0,1) : (X_0,X_1)_\theta\in\mathsf{UMD}\,\Big\}$$

is either open or a single point.

9.1.14 We now apply the previous results to Lorentz spaces.

EXAMPLE. *Let* $1<r,w<\infty$. *Then* $L_{r,w}\in\mathsf{UMD}$ *and*

$$L_{r,w}\in\mathsf{AT}_p\subset\mathsf{RT}_p \quad \textit{with} \quad p:=\min\{2,r,w\} \quad \textit{and} \quad r\neq 2,$$
$$L_{r,w}\in\mathsf{AC}_q\subset\mathsf{RC}_q \quad \textit{with} \quad q:=\max\{2,r,w\} \quad \textit{and} \quad r\neq 2.$$

In the exceptional case $r=2$, *we have*

$$L_{2,p}\in\mathsf{RT}_p \quad \textit{if } 1<p\le 2 \quad \textit{and} \quad L_{2,q}\in\bigcap_{\varepsilon>0}\mathsf{RT}_{2-\varepsilon} \quad \textit{if } 2<q<\infty,$$

$$L_{2,q}\in\mathsf{RC}_q \quad \textit{if } 2<q<\infty \quad \textit{and} \quad L_{2,p}\in\bigcap_{\varepsilon>0}\mathsf{RC}_{2+\varepsilon} \quad \textit{if } 1<p<2.$$

PROOF. Clearly, we can find $1<r_0<r_1<\infty$ and $0<\theta<1$ such that $1/r=(1-\theta)/r_0+\theta/r_1$. Then $L_{r,w}=(L_{r_0},L_{r_1})_{\theta,w}\in\mathsf{UMD}$ is implied by $L_{r_0},L_{r_1}\in\mathsf{UMD}$; see 8.5.20. The other statements are obtained similarly.

REMARK. It seems likely that

$$\varrho(L_{2,q}|\mathcal{R}_n, \mathcal{I}_n) \asymp (1+\log n)^{1/2-1/q} \qquad \text{if } 2<q<\infty,$$
$$\varrho(L_{2,p}|\mathcal{I}_n, \mathcal{R}_n) \asymp (1+\log n)^{1/p-1/2} \qquad \text{if } 1<p<2.$$

For the finite dimensional spaces $l_{2,q}^n$ and $l_{2,p}^n$, the lower estimates are obtained by substituting $x_k = k^{-1/2}u_k^{(n)}$ in the defining inequalities of the ideal norms.

9.1.15 The following table shows what is known about the complex and real interpolation spaces $X_\theta = [X_0, X_1]_\theta$ and $X_{\theta,w} = (X_0, X_1)_{\theta,w}$ if X_0 and X_1 satisfy the conditions listed in the first and last columns. As usual, for $1 < p_0 < p_1 < 2$ and $0 < \theta < 1$, we let $1/p = (1-\theta)/p_0 + \theta/p_1$.

X_0	X_θ	$X_{\theta,w}$		X_1
$\mathsf{RT}_{p_0}^{weak}$	RT_p^{weak}	RT_p^{weak},	$p<w<\infty$	$\mathsf{RT}_{p_1}^{weak}$
RT_{p_0}	RT_p	RT_p,	$p\le w<\infty$	RT_{p_1}
$\mathsf{ET}_{p_0}^{weak}$	ET_p^{weak}	ET_p^{weak},	$p<w<p'$	$\mathsf{ET}_{p_1}^{weak}$
ET_{p_0}	ET_p	ET_p,	$p\le w\le p'$	ET_{p_1}
$\mathsf{WT}_{p_0}^{weak}$	WT_p^{weak}	WT_p^{weak},	$p<w<p'$	$\mathsf{WT}_{p_1}^{weak}$
WT_{p_0}	WT_p	WT_p,	$p\le w\le p'$	WT_{p_1}
$\mathsf{AT}_{p_0}^{weak}$	AT_p^{weak}	AT_p^{weak},	$w=2$	$\mathsf{AT}_{p_1}^{weak}$
AT_{p_0}	AT_p	AT_p,	$p\le w<\infty$	AT_{p_1}
L	B	B		B
L	L	L		RC_2
L	SR	SR		SR
UMD	UMD	UMD		UMD
L	L	L		UMD

REMARKS. The reader is asked to check what happens in the limiting cases $p_0 = 1$ and/or $p_1 = 2$.

We conjecture that, with respect to real interpolation, the weak ideals behave in the same way as RT_p, ET_p, WT_p and AT_p. In other words, one should try to improve the boundary conditions on w.

G. Pisier constructed an example which shows that Rademacher cotype behaves very badly under interpolation; see [dil, p. 504].

9.2 Schatten–von Neumann spaces

This section contains a survey of Schatten–von Neumann ideals S_r, which can be viewed as the non-commutative analogues of the classical spaces L_r. For the basic theory, we refer to [SCHA 2], [DUN*2], [GOH*] and [RIN].

9.2.1 Let H be a Hilbert space, and let $1 \le r < \infty$. We denote by $\mathfrak{S}_r(H)$ the collection of all operators $T \in \mathfrak{L}(H)$ which can be written in the form

$$Tx = \sum_{i \in \mathbb{I}} \tau_i (x, u_i) v_i \quad \text{for } x \in H,$$

where (u_i) and (v_i) are orthonormal systems over a finite or countably infinite index set $\mathbb{I}$ and $(\tau_i) \in l_r(\mathbb{I})$. Taking $\mathbb{I} = \{1, 2, \ldots\}$ or $\mathbb{I} = \{1, \ldots, n\}$, we may arrange that (τ_i) is strictly positive and non-increasing. Then τ_i equals the i-th approximation number of T. Note that $\mathfrak{S}_r(H)$ is an ideal in $\mathfrak{L}(H)$ which becomes a Banach space under the norm

$$\|T|\mathfrak{S}_r\| := \left(\sum_{i \in \mathbb{I}} |\tau_i|^r \right)^{1/r}.$$

For $r = \infty$, we let $\mathfrak{S}_\infty(H) := \mathfrak{L}(H)$.

In what follows, we are exclusively interested in geometric properties like Rademacher type, Haar type, convexity, etc. Thus, forgetting the ideal structure, we look at $\mathfrak{S}_r(H)$ as a Banach space and use the shorthand notation $S_r := \mathfrak{S}_r(l_2)$ and $S_r^n := \mathfrak{S}_r(l_2^n)$.

9.2.2 The interpolation formulas 0.6.2 and 0.6.3 have the following counterparts:

$$[S_{r_0}, S_{r_1}]_\theta = S_r \quad \text{and} \quad (S_{r_0}, S_{r_1})_{\theta, r} = S_r$$

for $1/r = (1-\theta)/r_0 + \theta/r_1$, $1 \le r_0 < r_1 \le \infty$ and $0 < \theta < 1$. The left-hand equality is even an isometry.

Remark. Proofs are to be found in [pie*, p. 100], [KOE, p. 121], and [PIE 3, p. 87].

9.2.3 Assigning to every sequence $t = (\tau_k) \in l_r$ the associated diagonal operator D_t on l_2 yields an isometric embedding from l_r into S_r. Thus the subspace-invariant properties of S_r cannot be better than those of l_r.

9.2.4 On the basis of earlier work of C. McCarthy, convexity and smoothness properties of S_r were determined by N. Tomczak-Jaegermann; see [mcca] and [tom 1, p. 168].

EXAMPLE. $S_p \in \mathsf{UT}_p \subset \mathsf{AT}_p \subset \mathsf{RT}_p$ *if* $1<p\le 2$,
$S_q \in \mathsf{UT}_2 \subset \mathsf{AT}_2 \subset \mathsf{RT}_2$ *if* $2\le q<\infty$,
$S_p \in \mathsf{UC}_2 \subset \mathsf{AC}_2 \subset \mathsf{RC}_2$ *if* $1<p\le 2$,
$S_q \in \mathsf{UC}_q \subset \mathsf{AC}_q \subset \mathsf{RC}_q$ *if* $2\le q<\infty$.

Moreover, $S_1 \in \mathsf{RC}_2 \setminus \mathsf{AC}_2$.

9.2.5 The next result can be obtained by complex interpolation 2.3.9. Compare with 2.3.26.

EXAMPLE. $S_p,\, S_{p'} \in \mathsf{KT}_p^{weak}$ *if* $1<p<2$.

REMARK. Therefore S_p and $S_{p'}$ have properties **(1)**, **(2)** and **(3)** stated in 3.10.11; see [tom 3].

9.2.6 The following analogue of 8.5.20 was independently proved in three papers; see [gut, p. 150], [bou 7, p. 16] and [ber*2].

EXAMPLE. $S_r \in \mathsf{UMD}$ *for* $1<r<\infty$ *and* $S_1, S_\infty \notin \mathsf{QUMD}$.

9.2.7 All properties of S_r considered until now are shared with L_r. Thus the question arises whether S_r and L_r can be distinguished with the help of orthonormal systems. In the rest of this section, we will give an affirmative answer, which is based on results of C. McCarthy and G. Pisier; see [mcca, p. 268–269] and [pis 7, p. 7–11].

9.2.8 Recall from 3.6.4 that $\mathcal{A}_m \overset{\text{full}}{\otimes} \mathcal{B}_n$ denotes the full tensor product of the orthonormal systems $\mathcal{A}_m = (a_1, \dots, a_m)$ and $\mathcal{B}_n = (b_1, \dots, b_n)$.

LEMMA.

$$K'_p(\mathcal{A}_m \overset{\text{full}}{\otimes} \mathcal{B}_n) \le K'_p(\mathcal{A}_m)\, K'_p(\mathcal{B}_n) \qquad \textit{if } 1\le p\le 2,$$

$$K_q(\mathcal{A}_m \overset{\text{full}}{\otimes} \mathcal{B}_n) \le K_q(\mathcal{A}_m)\, K_q(\mathcal{B}_n) \qquad \textit{if } 2\le q\le \infty.$$

PROOF. For $\xi_{hk} \in \mathbb{K}$, we have

$$\Big(\sum_{h=1}^m \Big|\sum_{k=1}^n \xi_{hk} b_k(t)\Big|^2\Big)^{1/2} \le K'_p(\mathcal{A}_m) \Big(\int_M \Big|\sum_{h=1}^m \Big[\sum_{k=1}^n \xi_{hk} b_k(t)\Big] a_h(s)\Big|^p d\mu(s)\Big)^{1/p}$$

and

$$\Big(\sum_{k=1}^n |\xi_{hk}|^2\Big)^{1/2} \le K'_p(\mathcal{B}_n) \Big(\int_N \Big|\sum_{k=1}^n \xi_{hk} b_k(t)\Big|^p d\nu(t)\Big)^{1/p}.$$

Hence, in view of Jessen's inequality,

$$\begin{aligned}
\Big(\sum_{h=1}^{m}\sum_{k=1}^{n}|\xi_{hk}|^2\Big)^{1/2} &\le \\
&\le K_p'(\mathcal{B}_n)\Bigg(\sum_{h=1}^{m}\Big[\int\limits_N\Big|\sum_{k=1}^{n}\xi_{hk}b_k(t)\Big|^p d\nu(t)\Big]^{2/p}\Bigg)^{1/2}\\
&\le K_p'(\mathcal{B}_n)\Bigg(\int\limits_N\Big[\sum_{h=1}^{m}\Big|\sum_{k=1}^{n}\xi_{hk}b_k(t)\Big|^2\Big]^{p/2}d\nu(t)\Bigg)^{1/p}\\
&\le K_p'(\mathcal{A}_m)\,K_p'(\mathcal{B}_n)\Bigg(\int\limits_M\int\limits_N\Big|\sum_{h=1}^{m}\sum_{k=1}^{n}\xi_{hk}a_h(s)b_k(t)\Big|^p d\mu(s)\,d\nu(t)\Bigg)^{1/p}.
\end{aligned}$$

This proves the first inequality. The second one can be checked similarly.

9.2.9 EXAMPLE. $\varrho_\pm(L_r|\mathcal{R}_n\overset{\text{full}}{\otimes}\mathcal{R}_n,\mathcal{R}_n\overset{\text{full}}{\otimes}\mathcal{R}_n)=O(1)$ *if* $1\le r<\infty$.

PROOF. Let $\varepsilon_{hk}=\pm1$, $f_{hk}\in L_p(\Omega,\omega)$ and $1\le p\le 2$. Then the previous lemma implies that

$$\Big(\sum_{h=1}^{n}\sum_{k=1}^{n}|f_{hk}(\xi)|^2\Big)^{1/2}\le K_p'(\mathcal{R}_n)^2\Bigg(\int_0^1\int_0^1\Big|\sum_{h=1}^{n}\sum_{k=1}^{n}f_{hk}(\xi)r_h(s)r_k(t)\Big|^p ds\,dt\Bigg)^{1/p}.$$

Moreover,

$$\Bigg(\int_0^1\int_0^1\Big|\sum_{h=1}^{n}\sum_{k=1}^{n}\varepsilon_{hk}f_{hk}(\xi)r_h(s)r_k(t)\Big|^p ds\,dt\Bigg)^{1/p}\le\Big(\sum_{h=1}^{n}\sum_{k=1}^{n}|\varepsilon_{hk}f_{hk}(\xi)|^2\Big)^{1/2}.$$

Integration over $\xi\in\Omega$ yields

$$\begin{aligned}
\Bigg(\int_0^1\int_0^1\Big\|\sum_{h=1}^{n}\sum_{k=1}^{n}\varepsilon_{hk}f_{hk}r_h(s)r_k(t)\Big|L_p\Big\|^p ds\,dt\Bigg)^{1/p} &\le\\
&\le K_p'(\mathcal{R}_n)^2\Bigg(\int_0^1\int_0^1\Big\|\sum_{h=1}^{n}\sum_{k=1}^{n}f_{hk}r_h(s)r_k(t)\Big|L_p\Big\|^p ds\,dt\Bigg)^{1/p}\\
&\le K_p'(\mathcal{R}_n)^2\Bigg(\int_0^1\int_0^1\Big\|\sum_{h=1}^{n}\sum_{k=1}^{n}f_{hk}r_h(s)r_k(t)\Big|L_p\Big\|^2 ds\,dt\Bigg)^{1/2}.
\end{aligned}$$

In the last step, we twice apply the Khintchine–Kahane inequality and once Jessen's inequality to get

$$\Bigg(\int_0^1\Bigg[\int_0^1\Bigg\|\sum_{h=1}^n\Big(\sum_{k=1}^n\varepsilon_{hk}f_{hk}r_k(t)\Big)r_h(s)\Big|L_p\Bigg\|^2ds\Bigg]^{2/2}dt\Bigg)^{1/2}\leq$$

$$\leq K(p,2)\Bigg(\int_0^1\Bigg[\int_0^1\Bigg\|\sum_{h=1}^n\Big(\sum_{k=1}^n\varepsilon_{hk}f_{hk}r_k(t)\Big)r_h(s)\Big|L_p\Bigg\|^p ds\Bigg]^{2/p}dt\Bigg)^{1/2}$$

$$\leq K(p,2)\Bigg(\int_0^1\Bigg[\int_0^1\Bigg\|\sum_{k=1}^n\Big(\sum_{h=1}^n\varepsilon_{hk}f_{hk}r_h(s)\Big)r_k(t)\Big|L_p\Bigg\|^2 dt\Bigg]^{p/2}ds\Bigg)^{1/p}$$

$$\leq K(p,2)^2\Bigg(\int_0^1\Bigg[\int_0^1\Bigg\|\sum_{k=1}^n\Big(\sum_{h=1}^n\varepsilon_{hk}f_{hk}r_h(s)\Big)r_k(t)\Big|L_p\Bigg\|^p dt\Bigg]^{p/p}ds\Bigg)^{1/p}.$$

This proves the assertion for $1<p\leq 2$. The case $2\leq q<\infty$ can be treated in the same way.

9.2.10 Next, we show that the ideal norms $\boldsymbol{\varrho}_\pm(\mathcal{R}_n\overset{\text{full}}{\otimes}\mathcal{R}_n,\mathcal{R}_n\overset{\text{full}}{\otimes}\mathcal{R}_n)$ behave differently on S_r and L_r for $1\leq r<\infty$ and $r\neq 2$.

EXAMPLE. $\boldsymbol{\varrho}_\pm(S_r|\mathcal{R}_n\overset{\text{full}}{\otimes}\mathcal{R}_n,\mathcal{R}_n\overset{\text{full}}{\otimes}\mathcal{R}_n)\succ n^{|1/r-1/2|}$.

PROOF. By monotonicity, it is enough to check the assertion for powers of 2. Let $U_{2^n}^{(hk)}$ denote the $(2^n,2^n)$-matrix that has the entry 1 for the index (h,k) and 0 otherwise. Moreover, we define $A_{2^n}=\big(\alpha_{hk}^{(2^n)}\big)$ by induction:

$$A_{2^{n+1}}:=\begin{pmatrix}+A_{2^n} & +A_{2^n}\\ +A_{2^n} & -A_{2^n}\end{pmatrix}\quad\text{and}\quad A_1:=(1).$$

Then

$$\operatorname{rank}\Bigg(\sum_{h=1}^{2^n}\sum_{k=1}^{2^n}r_h(s)\,U_{2^n}^{(hk)}\,r_k(t)\Bigg)=1,$$

and

$$2^{-n/2}\sum_{h=1}^{2^n}\sum_{k=1}^{2^n}r_h(s)\,\alpha_{hk}^{(2^n)}U_{2^n}^{(hk)}\,r_k(t)$$

is unitary. Hence

$$\Bigg\|\sum_{h=1}^{2^n}\sum_{k=1}^{2^n}r_h(s)\,U_{2^n}^{(hk)}\,r_k(t)\Big|S_r\Bigg\|=\|(r_h(s)|l_2^{2^n}\|\,\|(r_k(t)|l_2^{2^n}\|=2^n$$

and

$$\Bigg\|\sum_{h=1}^{2^n}\sum_{k=1}^{2^n}r_h(s)\,\alpha_{hk}^{(2^n)}U_{2^n}^{(hk)}\,r_k(t)\Big|S_r\Bigg\|=2^{n/r+n/2}$$

for all $s,t \in [0,1)$. Since $\alpha_{hk}^{(2^n)} = \pm 1$, substituting the matrix $U_{2^n}^{(hk)}$ into the defining inequality of $\varrho_\pm(\mathcal{R}_{2^n} \overset{\text{full}}{\otimes} \mathcal{R}_{2^n}, \mathcal{R}_{2^n} \overset{\text{full}}{\otimes} \mathcal{R}_{2^n})$ yields

$$2^{n/r+n/2} \le \varrho_\pm(S_r | \mathcal{R}_{2^n} \overset{\text{full}}{\otimes} \mathcal{R}_{2^n}, \mathcal{R}_{2^n} \overset{\text{full}}{\otimes} \mathcal{R}_{2^n})\, 2^n.$$

This proves the lower estimate in the case $1 \le r \le 2$. If $2 \le r \le \infty$, then the roles of $U_{2^n}^{(hk)}$ and $\alpha_{hk}^{(2^n)} U_{2^n}^{(hk)}$ must be interchanged.

REMARKS. It seems likely that $\varrho_\pm(S_r | \mathcal{R}_n \overset{\text{full}}{\otimes} \mathcal{R}_n, \mathcal{R}_n \overset{\text{full}}{\otimes} \mathcal{R}_n) \asymp n^{|1/r-1/2|}$. Example 9.2.9 is a special case of a proposition which states that

$$\varrho_\pm(X | \mathcal{R}_n \overset{\text{full}}{\otimes} \mathcal{R}_n, \mathcal{R}_n \overset{\text{full}}{\otimes} \mathcal{R}_n) = O(1)$$

whenever X is isomorphic to a subspace of a Banach lattice and does not contain the spaces l_∞^n uniformly; [pis 7, p. 8]. Therefore S_r with $r \ne 2$ cannot have local unconditional structure; see [lew 1].

9.2.11 For completeness, we note that

$$\varrho_\pm(S_1 | \mathcal{R}_n \overset{\text{full}}{\otimes} \mathcal{R}_n, \mathcal{R}_n \overset{\text{full}}{\otimes} \mathcal{R}_n) \asymp \varrho_\pm(S_\infty | \mathcal{R}_n \overset{\text{full}}{\otimes} \mathcal{R}_n, \mathcal{R}_n \overset{\text{full}}{\otimes} \mathcal{R}_n) \asymp \sqrt{n}$$

is the worst that can happen.

PROPOSITION. $\varrho_\pm(\mathcal{R}_n \overset{\text{full}}{\otimes} \mathcal{R}_n, \mathcal{R}_n \overset{\text{full}}{\otimes} \mathcal{R}_n) \le \sqrt{2n}.$

PROOF. In the following, we use the nuclear norm $\| \cdot | \mathfrak{N} \|$ and the 1-summing norm $\| \cdot | \mathfrak{P}_1 \|$ as defined in [PIE 3, p. 64 and 46]. Note that

$$\| I_n : l_1^n \to l_2^n | \mathfrak{P}_1 \| \le K_1'(\mathcal{R}) = \sqrt{2}$$

and

$$\| A_n : X \to l_\infty^n | \mathfrak{N} \| = \| A_n : X \to l_\infty^n | \mathfrak{P}_1 \|;$$

see [PIE 2, pp. 99, 233, 264, 269] and [DIE*a, pp. 99, 115]. Given any (n,n)-matrix $A_n = (\alpha_{hk})$ with $\alpha_{hk} = \pm 1$, we consider the factorization

$$A_n : l_1^n \xrightarrow{I_n} l_2^n \xrightarrow{I_n} l_1^n \xrightarrow{A_n} l_\infty^n.$$

Then

$$\begin{aligned} \| A_n : l_1^n \to l_\infty^n | \mathfrak{N} \| &= \| A_n : l_1^n \to l_\infty^n | \mathfrak{P}_1 \| \\ &\le \| I_n : l_1^n \to l_2^n | \mathfrak{P}_1 \| \, \| I_n : l_2^n \to l_1^n \| \, \| A_n : l_1^n \to l_\infty^n \| \le \sqrt{2n}\,. \end{aligned}$$

Consequently, since the closed unit ball of l_∞^n coincides with the convex hull of $\mathbb{E}_2^n$, we find real coefficients λ_{de} such that

$$\alpha_{hk} = \sum_{d,e} \lambda_{de}\, \delta_h\, \varepsilon_k \quad \text{and} \quad \sum_{d,e} |\lambda_{de}| \le \sqrt{2n},$$

where $d = (\delta_1, \ldots, \delta_n)$ and $e = (\varepsilon_1, \ldots, \varepsilon_n)$ range over $\mathbb{E}_2^n$. The required estimate now easily follows.

REMARK. Let $\varepsilon_1=\varepsilon_2=\varepsilon_3=+1$ and $\varepsilon_4=-1$. Substituting

$$\begin{aligned}\varepsilon_1x_1&=(+1,+1,+1,+1) \quad\text{and}\quad x_1=(+1,+1,+1,+1),\\ \varepsilon_2x_2&=(+1,-1,+1,-1) \quad\text{and}\quad x_2=(+1,-1,+1,-1),\\ \varepsilon_3x_3&=(+1,+1,-1,-1) \quad\text{and}\quad x_3=(+1,+1,-1,-1),\\ \varepsilon_4x_4&=(+1,-1,-1,+1) \quad\text{and}\quad x_4=(-1,+1,+1,-1)\end{aligned}$$

into

$$\left\{\begin{matrix}\|\varepsilon_1x_1+\varepsilon_2x_2+\varepsilon_3x_3+\varepsilon_4x_4\|^2+\|\varepsilon_1x_1-\varepsilon_2x_2+\varepsilon_3x_3-\varepsilon_4x_4\|^2\\ +\\ \|\varepsilon_1x_1+\varepsilon_2x_2-\varepsilon_3x_3-\varepsilon_4x_4\|^2+\|\varepsilon_1x_1-\varepsilon_2x_2-\varepsilon_3x_3+\varepsilon_4x_4\|^2\end{matrix}\right\}\le$$

$$\le\boldsymbol{\varrho}_\pm(l_\infty^4|\mathcal{R}_n\overset{\text{full}}{\otimes}\mathcal{R}_n,\mathcal{R}_n\overset{\text{full}}{\otimes}\mathcal{R}_n)^2\left\{\begin{matrix}\|x_1+x_2+x_3+x_4\|^2+\|x_1-x_2+x_3-x_4\|^2\\ +\\ \|x_1+x_2-x_3-x_4\|^2+\|x_1-x_2-x_3+x_4\|^2\end{matrix}\right\},$$

we obtain $2\le\boldsymbol{\varrho}_\pm(l_\infty^4\,|\,\mathcal{R}_n\overset{\text{full}}{\otimes}\mathcal{R}_n,\mathcal{R}_n\overset{\text{full}}{\otimes}\mathcal{R}_n)$. Thus the upper estimate $\boldsymbol{\varrho}_\pm(\mathcal{R}_n\overset{\text{full}}{\otimes}\mathcal{R}_n,\mathcal{R}_n\overset{\text{full}}{\otimes}\mathcal{R}_n)\le\sqrt{2n}$ is sharp for $n=2$.

9.2.12 It would be interesting to characterize all Banach spaces for which

$$\boldsymbol{\varrho}_\pm(X|\mathcal{R}_n\overset{\text{full}}{\otimes}\mathcal{R}_n,\mathcal{R}_n\overset{\text{full}}{\otimes}\mathcal{R}_n)=O(1) \quad\text{or}\quad \boldsymbol{\varrho}_\pm(X|\mathcal{R}_n\overset{\text{full}}{\otimes}\mathcal{R}_n,\mathcal{R}_n\overset{\text{full}}{\otimes}\mathcal{R}_n)=o(\sqrt{n}).$$

9.2.13 The Schatten–von Neumann spaces show that Banach operator ideals can be used to produce a huge number of concrete Banach spaces. To give another example, we consider the components $\mathfrak{P}_2(l_u,l_v)$ formed by all 2-summing operators from l_u into l_v with $1\le u,v\le\infty$. Note that $\mathfrak{P}_2(l_2,l_2)=\mathfrak{S}_2(l_2,l_2)$ is a Hilbert space. Obviously, the local properties of $\mathfrak{P}_2(l_u,l_v)$ cannot be better than those of $l_{u'}$ and l_v. However, the situation is even worse, as illustrated by the following results of P.-K. Lin and G. Pisier.

EXAMPLE. *Let $1<u,v\le2$ and $p:=\min\{u,v\}$. Then*

$$\mathfrak{P}_2(l_u,l_v)\in\mathsf{AT}_p\cap\mathsf{AC}_{p'} \quad\textit{and}\quad \mathfrak{P}_2(l_u,l_v)\in\mathsf{KT}_p^{weak}.$$

In the remaining case when $\max\{u,v\}>2$, the space $\mathfrak{P}_2(l_u,l_v)$ fails to be B-convex.

REMARK. Given $1<u,v\le2$, we let

$$1/u=(1-\theta)/u_0+\theta/2,\quad 1/v=(1-\theta)/v_0+\theta/2 \quad\text{and}\quad 0<\theta<1$$

such that $\min\{u_0,v_0\}=1$. As shown in [pis 19],

$$l_u=[l_{u_0},l_2]_\theta \quad\text{and}\quad l_v=[l_{v_0},l_2]_\theta$$

imply

$$\mathfrak{P}_2(l_u, l_v) = \big[\mathfrak{P}_2(l_{u_0}, l_{v_0}), \mathfrak{P}_2(l_2, l_2)\big]_\theta.$$

Hence it follows from 9.1.4 that $\mathfrak{P}_2(l_u, l_v)$ has Haar type $p = \min\{u, v\}$. The cotype case can be treated by duality, $\mathfrak{P}_2(l_u^n, l_v^n) = \mathfrak{P}_2(l_v^n, l_u^n)'$. The negative results concerning B-convexity are to be found in [lin].

REMARK. We conjecture that $\mathfrak{P}_2(l_u, l_v)$ with $\max\{u, v\} > 2$ even fails to be MP-convex.

9.2.14 PROBLEM. *Is $\mathfrak{P}_2(l_u, l_v)$ with $1 < u, v \leq 2$ a UMD-space?*

REMARK. We know from [schu, p. 180] that the canonical basis in $\mathfrak{P}_2(l_u, l_v)$ with $1 < u, v \leq 2$ is unconditional.

9.3 Ideal norms of finite rank operators

9.3.1 Taking the logarithm of the Banach–Mazur distance

$$d(X, Y) := \inf\Big\{ \|T\|\,\|T^{-1}\| \,:\, T \in \mathfrak{L}(X, Y), \text{ bijection}\Big\}$$

we get a metric on L_N, the collection of all N-dimensional Banach spaces. The compact topological space obtained in this way is said to be the N-th **Minkowski compactum.**

CONVENTION. Isometric spaces will always be identified.

9.3.2 For every ideal norm $\boldsymbol{\alpha}$ and $N = 1, 2, \ldots$, we let

$$\boldsymbol{\alpha}(N) := \sup\Big\{ \boldsymbol{\alpha}(X) \,:\, \dim(X) = N \Big\}.$$

Obviously, the sequence $(\boldsymbol{\alpha}(N))$ is non-decreasing. Moreover, it follows from

$$\boldsymbol{\alpha}(X) \leq d(X, Y)\, \boldsymbol{\alpha}(Y) \quad \text{for } X, Y \in \mathsf{L}_N$$

that $X \to \boldsymbol{\alpha}(X)$ is a continuous function on L_N. So the above supremum is indeed a maximum.

REMARKS. The quantities $\boldsymbol{\alpha}(1)$ and $\boldsymbol{\alpha}(\infty)$ have already been defined in 1.1.4. Note that

$$\sup_N \boldsymbol{\alpha}(N) \leq \boldsymbol{\alpha}(\infty) := \sup\Big\{ \boldsymbol{\alpha}(X) \,:\, \dim(X) = \infty \Big\}.$$

For all concrete ideal norms considered in this book, we even get equality. However, this is not so in general. A counterexample is given by

$$\boldsymbol{\alpha}(T) := \|T\| + \inf\Big\{ \|T - A\| \,:\, A \in \mathfrak{F}(X, Y) \Big\}.$$

9.3.3 Next, we state an elementary inequality which can be checked by factoring finite rank operators through their range.

PROPOSITION. $\boldsymbol{\alpha}(T) \le \boldsymbol{\alpha}(N)\,\|T\|$ *whenever* $\operatorname{rank}(T)=N$.

9.3.4 In what follows, we present several examples.

EXAMPLE. *For every orthonormal system* $\mathcal{A}_n$,
$$\boldsymbol{\pi}(N|\mathcal{A}_n) = \boldsymbol{\pi}_n(N) = \min\{\sqrt{n}, \sqrt{N}\}.$$

PROOF. We have
$$\boldsymbol{\pi}(X|\mathcal{A}_n) \le \boldsymbol{\pi}_n(X) \le \sqrt{n} \quad \text{for all } X \in \mathsf{L}$$
and
$$\boldsymbol{\pi}(X|\mathcal{A}_n) \le \boldsymbol{\pi}_n(X) \le \|X|\mathfrak{P}_2\| = \sqrt{N} \quad \text{for all } X \in \mathsf{L}_N;$$
see [DIE*a, p. 87] or [PIE 2, p. 385]. Clearly, $\boldsymbol{\pi}(l_2^N|\mathcal{A}_n) = \min\{\sqrt{n}, \sqrt{N}\}$.

9.3.5 EXAMPLE. $\boldsymbol{\varrho}(N|\mathcal{R}_n, \mathcal{I}_n) = \boldsymbol{\varrho}(N|\mathcal{I}_n, \mathcal{R}_n) = \min\{\sqrt{n}, \sqrt{N}\}$.

PROOF. We have
$$\boldsymbol{\varrho}(X|\mathcal{R}_n, \mathcal{I}_n) \le \sqrt{n} \quad \text{for all } X \in \mathsf{L}.$$

Moreover, John's theorem 0.2.4 says that $d(X, l_2^N) \le \sqrt{N}$ if $X \in \mathsf{L}_N$. Since $\boldsymbol{\varrho}(l_2^N|\mathcal{R}_n, \mathcal{I}_n) = 1$, we obtain $\boldsymbol{\varrho}(X|\mathcal{R}_n, \mathcal{I}_n) \le \sqrt{N}$. By 4.2.10,
$$\boldsymbol{\varrho}(l_1^N|\mathcal{R}_n, \mathcal{I}_n) = \boldsymbol{\varrho}(l_\infty^N|\mathcal{I}_n, \mathcal{R}_n) = \min\{\sqrt{n}, \sqrt{N}\}.$$

9.3.6 PROPOSITION. $\boldsymbol{\varrho}(N|\mathcal{G}_n, \mathcal{R}_n) \asymp \min\{\sqrt{1+\log n}, \sqrt{1+\log N}\}$.

PROOF. In view of 4.8.7, we have
$$\boldsymbol{\varrho}(X|\mathcal{G}_n, \mathcal{R}_n) \prec \sqrt{1+\log n} \quad \text{for all } X \in \mathsf{L}.$$

If $X \in \mathsf{L}_N$ and $N < n$, it follows from 4.8.16 that
$$\boldsymbol{\varrho}(X|\mathcal{G}_n, \mathcal{R}_n) \le c \sum_{k=1}^{n} k^{-3/2}(1+\log k)^{-1/2} \boldsymbol{\varrho}(X|\mathcal{I}_k, \mathcal{R}_k)$$
$$\le \left\{ \begin{array}{c} c \displaystyle\sum_{k=1}^{N} k^{-3/2}(1+\log k)^{-1/2} k^{1/2} \\ + \\ c \displaystyle\sum_{k=N+1}^{n} k^{-3/2}(1+\log k)^{-1/2} N^{1/2} \end{array} \right\} \prec \sqrt{1+\log N}.$$

This proves the upper estimate. On the other hand, by 4.8.13,
$$\boldsymbol{\varrho}(l_\infty^N|\mathcal{G}_n, \mathcal{R}_n) \asymp \min\{\sqrt{1+\log n}, \sqrt{1+\log N}\}.$$

9.3.7 PROPOSITION.

$$\boldsymbol{\delta}(N|\mathcal{R}_n, \mathcal{R}_n) \asymp \max\{\sqrt{n}, 1+\log N\}.$$

PROOF. By 3.4.4, we have

$$\boldsymbol{\delta}(X|\mathcal{R}_n, \mathcal{R}_n) \le \sqrt{n} \quad \text{for all } X \in \mathsf{L}.$$

Moreover, G. Pisier proved that

$$\boldsymbol{\delta}(X|\mathcal{R}_n, \mathcal{R}_n) \prec 1+\log N \quad \text{for all } X \in \mathsf{L}_N;$$

see [pis 11, pp. 6–8] and [TOM, p. 87]. Examples due to J. Bourgain yield the lower estimate; [bou 5, p. 51].

REMARK. We know from 4.12.13 that

$$\boldsymbol{\delta}(l_1^N|\mathcal{R}_n, \mathcal{R}_n) \asymp \boldsymbol{\delta}(l_\infty^N|\mathcal{R}_n, \mathcal{R}_n) \asymp \min\{\sqrt{n}, \sqrt{1+\log N}\}.$$

As shown in [pis 11, pp. 8–10], the relation

$$\boldsymbol{\delta}(X|\mathcal{R}_n, \mathcal{R}_n) \prec \min\{\sqrt{n}, \sqrt{1+\log N}\}$$

even holds for all N-dimensional Banach lattices X. Consequently, the spaces for which the maxima are attained must be more complicated. J. Bourgain used subspaces of $L_\infty[0,1)$ spanned by randomly chosen subsets of Walsh functions.

9.3.8 PROPOSITION.

$$\min\{1+\log n, N^{1/3}\} \prec \boldsymbol{\kappa}(N|H_n) \prec \min\{1+\log n, N^{1/2}\}.$$

PROOF. The upper estimate is an immediate consequence of 2.4.3 and John's theorem 0.2.4. Examples constructed by J. Bourgain yield the lower estimate; [bou 4].

REMARK. We know from 5.4.23 that

$$\boldsymbol{\kappa}(l_1^N|H_n) = \boldsymbol{\kappa}(l_\infty^N|H_n) \asymp \min\{1+\log n, 1+\log N\}.$$

As in the foregoing paragraph, more complicated spaces are needed to get better lower estimates. J. Bourgain used subspaces of $C[-\pi, +\pi)$ spanned by randomly chosen subsets of sine functions. However, in this case there remains a gap.

9.3.9 PROBLEM. *What about* $\boldsymbol{\varrho}_\pm(N|\mathcal{H}(\mathbb{D}_1^n), \mathcal{H}(\mathbb{D}_1^n))$ *?*

REMARK. By 8.5.11 and John's theorem 0.2.4, we have

$$\boldsymbol{\varrho}_\pm(N|\mathcal{H}(\mathbb{D}_1^n), \mathcal{H}(\mathbb{D}_1^n)) \prec \min\{n, \sqrt{N}\}.$$

Unfortunately, Bourgain's technique does not yield lower estimates in this case.

9.3.10 Let $(\boldsymbol{\alpha}_n)$ be any sequence of ideal norms. A function $\varphi : \mathbb{N} \to \mathbb{N}$ is said to **stabilize** $(\boldsymbol{\alpha}_n)$ if there exists a constant $c > 0$ such that

$$\|T|\mathfrak{L}[\boldsymbol{\alpha}_n]\| := \sup_n \boldsymbol{\alpha}_n(T) \le c\, \boldsymbol{\alpha}_{\varphi(N)}(T) \quad \text{whenever} \quad \operatorname{rank}(T) = N.$$

REMARK. As shown in 4.15.10, 2.2.8 and 2.2.12, the sequences $(\boldsymbol{\pi}(\mathcal{G}_n))$, $(\boldsymbol{\pi}(I_n))$ and $(\boldsymbol{\pi}_n)$ stabilize at the index N.

9.3.11 We now state an immediate consequence of 2.1.9 and 3.10.4.

PROPOSITION. *Let $T \in \mathfrak{F}(X,Y)$. Then*

$$\sup_n \boldsymbol{\delta}_n(T) = \boldsymbol{\delta}_N(T) \quad \textit{whenever} \quad \operatorname{rank}(T) = N.$$

9.3.12 In view of 3.10.7, 3.10.8 and 3.10.9, we have the following supplement to the preceding result.

PROPOSITION. *Let $T \in \mathfrak{F}(X,Y)$. Then*

$$\sup_n \boldsymbol{\kappa}_n(T) \le 2\, \boldsymbol{\kappa}_N(T) \quad \textit{whenever} \quad \operatorname{rank}(T) = N.$$

The same statement holds for $\boldsymbol{\tau}_n$, $\boldsymbol{\omega}_n$ and $\boldsymbol{\varrho}_n$.

9.3.13 PROPOSITION. *Let $T \in \mathfrak{F}(X,Y)$. Then*

$$\sup_n \boldsymbol{\varrho}(T|\mathcal{G}_n, \mathcal{I}_n) \le \sqrt{2}\, \boldsymbol{\varrho}(T|\mathcal{G}_N, \mathcal{I}_N) \quad \textit{whenever} \quad \operatorname{rank}(T) = N.$$

PROOF. Obviously, it is enough to treat the case when $n \ge N$. Given $(x_k) \in [U_2^n, X]$, there exists $A_{nN} = (\alpha_{hi}) \in \mathbb{U}^{nN}$ such that

$$Tx_h = \sum_{i=1}^{N} \alpha_{hi} T\Big(\sum_{k=1}^{n} \alpha_{ik}^* x_k\Big) \quad \text{for } h = 1, \dots, n \qquad \text{(by 2.1.9)}.$$

In view of 2.2.16,

$$\Big(\sum_{k=1}^{n} \alpha_{ik}^* x_k\Big) \in \mathbb{L}^{Nn} \circ [U_2^n, X] \subseteq \sqrt{2}\ \operatorname{conv}(\mathbb{U}^N \circ [U_2^N, X]).$$

Thus, to get $\|(Tx_h)|\mathcal{G}_n\| \le \sqrt{2}\, \boldsymbol{\varrho}(T|\mathcal{G}_N, \mathcal{I}_N)$, we must show that

$$\Big\|\Big(\sum_{i=1}^{N} \alpha_{hi} T x_i^\circ\Big)\Big|\mathcal{G}_n\Big\| \le \boldsymbol{\varrho}(T|\mathcal{G}_N, \mathcal{I}_N)$$

whenever $(x_i^\circ) \in \operatorname{conv}(\mathbb{U}^N \circ [U_2^N, X])$. It is even sufficient to verify this property for $(u_i^\circ) \in \mathbb{U}^N \circ [U_2^N, X]$ only. To this end, pick a representation

$$u_i^\circ := \sum_{j=1}^{N} \beta_{ij} u_j \quad \text{with} \quad (u_j) \in [U_2^N, X] \quad \text{and} \quad B_N = (\beta_{ij}) \in \mathbb{U}^N.$$

It follows from $A_{nN}B_N \in \mathbb{U}^{nN}$ that $\gamma^N = B_N^t A_{nN}^t \gamma^n$. So we have indeed

$$\Big\|\Big(\sum_{i=1}^N \alpha_{hi} T u_i^\circ\Big)\Big|\mathcal{G}_n\Big\| = \Bigg(\int_{\mathbb{K}^n} \Big\|\sum_{h=1}^n \sum_{i=1}^N \alpha_{hi} T \sum_{j=1}^N \beta_{ij} u_j g_h^{(n)}(s)\Big\|^2 d\gamma^n(s)\Bigg)^{1/2}$$

$$= \Bigg(\int_{\mathbb{K}^n} \Big\|\sum_{j=1}^N T u_j \sum_{h=1}^n \sum_{i=1}^N \alpha_{hi}\beta_{ij} g_h^{(n)}(s)\Big\|^2 d\gamma^n(s)\Bigg)^{1/2}$$

$$= \Bigg(\int_{\mathbb{K}^N} \Big\|\sum_{j=1}^N T u_j g_j^{(N)}(t)\Big\|^2 d\gamma^N(t)\Bigg)^{1/2} = \|Tu_j|\mathcal{G}_N\| \le \boldsymbol{\varrho}(T|\mathcal{G}_N, \mathcal{I}_N).$$

REMARK. In view of 4.9.4, a similar result holds for Rademacher type ideal norms.

9.3.14 PROPOSITION. *Let $T \in \mathfrak{F}(X,Y)$. Then*

$$\sup_n \boldsymbol{\varrho}(T|\mathcal{I}_n, \mathcal{G}_n) \le \sqrt{2}\, \boldsymbol{\varrho}(T|\mathcal{I}_N, \mathcal{G}_N) \quad \textit{whenever} \quad \operatorname{rank}(T) = N.$$

PROOF. Obviously, it is enough to treat the case when $n \ge N$. Given $x_1, \dots, x_n \in X$, we define $A_n \in \mathfrak{L}(l_2^n, X)$ by

$$A_n(\xi_k) := \sum_{k=1}^n \xi_k x_k \quad \text{for} \quad (\xi_k) \in l_2^n.$$

Then 4.15.8 yields

$$\boldsymbol{\pi}(A_n|\mathcal{G}_N) \le \boldsymbol{\pi}(A_n|\mathcal{G}_n) = \|(x_k)|\mathcal{G}_n\|.$$

Note that $\boldsymbol{\pi}(I_N) = \boldsymbol{\pi}(\mathcal{I}_N)$. Now it follows from 2.2.14 and 3.11.5 that

$$\begin{aligned} \|(Tx_k)|\mathcal{I}_n\| &= \|(TA_n u_k^{(n)})|l_2^n\| \le \boldsymbol{\pi}(TA_n|I_n) \le \sqrt{2}\,\boldsymbol{\pi}(TA_n|I_N) \\ &= \sqrt{2}\,\boldsymbol{\pi}(TA_n|\mathcal{I}_N) \le \sqrt{2}\,\boldsymbol{\varrho}(T|\mathcal{I}_N, \mathcal{G}_N)\,\boldsymbol{\pi}(A_n|\mathcal{G}_N) \\ &\le \sqrt{2}\,\boldsymbol{\varrho}(T|\mathcal{I}_N, \mathcal{G}_N)\,\|(x_k)|\mathcal{G}_n\|, \end{aligned}$$

which completes the proof.

REMARK. It is not clear whether we have a counterpart of the preceding result for the Rademacher system.

9.3.15 Nothing seems to be known about stabilization of all other sequences of concrete ideal norms. Thus many open questions could be formulated. However, we restrict ourselves to one prominent example.

PROBLEM. *Do the sequences $(\boldsymbol{\delta}(\mathcal{R}_n, \mathcal{R}_n))$ and $(\boldsymbol{\delta}(\mathcal{G}_n, \mathcal{G}_n))$ stabilize at $(1 + [\log N])^2$?*

REMARK. By 9.3.7, we have $\boldsymbol{\delta}(N|\mathcal{R}_n, \mathcal{R}_n) \asymp \max\{\sqrt{n}, 1+\log N\}$. Thus, roughly speaking, $(1 + [\log N])^2$ cannot be underbidden.

9.4 Orthogonal polynomials

For the theory of orthogonal polynomials we refer to [SZE] and [TRIC].

9.4.1 Let w be a positive and continuous weight function on a bounded or unbounded interval Δ such that the absolute moments

$$\int\limits_{\Delta} |s|^k \, w(s) \, ds$$

are finite for $k = 0, 1, 2, \ldots$. In particular, we assume that

$$\int\limits_{\Delta} w(s) \, ds = 1.$$

Denote by $L_2(\Delta, w)$ the Hilbert space of all square-integrable functions $f : \Delta \to \mathbb{K}$ equipped with the inner product

$$(f, g)_w := \int\limits_{\Delta} f(s)\overline{g(s)}w(s) \, ds.$$

Applying the Gram–Schmidt process to $1, s, s^2, \ldots$ yields the orthonormal system

$$\mathfrak{L}^{(w)} := \{L_0^{(w)}, L_1^{(w)}, L_2^{(w)}, \ldots\},$$

which is uniquely determined by the conditions

$L_k^{(w)}(s) = \lambda_{kk}^{(w)} s^k + \ldots + \lambda_{k1}^{(w)} s + \lambda_{k0}^{(w)}$ with $\lambda_{k0}^{(w)}, \ldots, \lambda_{kk}^{(w)} \in \mathbb{R}$ and $\lambda_{kk}^{(w)} > 0$.

9.4.2 These are the most important examples.

Jacobi polynomials on $(-1, +1)$:

$$\mathfrak{L}^{(\alpha,\beta)} := \{L_0^{(\alpha,\beta)}, L_1^{(\alpha,\beta)}, L_2^{(\alpha,\beta)}, \ldots\}$$

with $w_{\alpha,\beta}(s) := c_{\alpha,\beta}^{-1}(1-s)^\alpha(1+s)^\beta$ and $\alpha, \beta > -1$,

Laguerre polynomials on $(0, \infty)$:

$$\mathfrak{L}^{(\gamma)} := \{L_0^{(\gamma)}, L_1^{(\gamma)}, L_2^{(\gamma)}, \ldots\}$$

with $w_\gamma(s) := c_\gamma^{-1} s^\gamma e^{-s}$ and $\gamma > -1$,

Hermite polynomials on $(-\infty, +\infty)$:

$$\mathfrak{L}^{(H)} := \{L_0^{(H)}, L_1^{(H)}, L_2^{(H)}, \ldots\}$$

with $w_H(s) := c_H^{-1} e^{-s^2}$.

The constants

$$c_{\alpha,\beta} := \int_{-1}^{+1} (1-s)^\alpha (1+s)^\beta \, ds = 2^{\alpha+\beta+1} \frac{\Gamma(\alpha+1)\Gamma(\beta+1)}{\Gamma(\alpha+\beta+2)},$$

$$c_\gamma := \int_0^\infty s^\gamma e^{-s} \, ds = \Gamma(\gamma+1) \quad \text{and} \quad c_H := \int_{-\infty}^{+\infty} e^{-s^2} \, ds = \Gamma(\tfrac{1}{2}) = \sqrt{\pi}$$

are chosen such that the measure of Δ equals 1; see [SZE, p. 68] and [TRIC, p. 163].

WARNING. Usually, classical polynomials are normalized differently.

9.4.3 We now treat four special cases where the quadratic averages $\|(x_k)|\mathfrak{L}_n^{(\alpha,\beta)}\|$ can be written in terms of trigonometric functions.

LEMMA.

$$\int_{-1}^{+1} \Big\| \sum_{k=1}^{n} x_k L_k^{(-\frac{1}{2},-\frac{1}{2})}(s) \Big\|^2 w_{-\frac{1}{2},-\frac{1}{2}}(s)\, ds = \frac{2}{\pi} \int_0^\pi \Big\| \sum_{k=1}^{n} x_k \cos kt \Big\|^2 dt,$$

$$\int_{-1}^{+1} \Big\| \sum_{k=0}^{n-1} x_k L_k^{(+\frac{1}{2},+\frac{1}{2})}(s) \Big\|^2 w_{+\frac{1}{2},+\frac{1}{2}}(s)\, ds = \frac{2}{\pi} \int_0^\pi \Big\| \sum_{k=1}^{n} x_k \sin kt \Big\|^2 dt,$$

$$\int_{-1}^{+1} \Big\| \sum_{k=0}^{n-1} x_k L_k^{(-\frac{1}{2},+\frac{1}{2})}(s) \Big\|^2 w_{-\frac{1}{2},+\frac{1}{2}}(s)\, ds = \frac{2}{\pi} \int_0^\pi \Big\| \sum_{k=0}^{n-1} x_k \cos \tfrac{2k+1}{2} t \Big\|^2 dt,$$

$$\int_{-1}^{+1} \Big\| \sum_{k=0}^{n-1} x_k L_k^{(+\frac{1}{2},-\frac{1}{2})}(s) \Big\|^2 w_{+\frac{1}{2},-\frac{1}{2}}(s)\, ds = \frac{2}{\pi} \int_0^\pi \Big\| \sum_{k=0}^{n-1} x_k \sin \tfrac{2k+1}{2} t \Big\|^2 dt.$$

PROOF. First of all, note that

$$c_{-\frac{1}{2},-\frac{1}{2}} = c_{+\frac{1}{2},-\frac{1}{2}} = c_{-\frac{1}{2},+\frac{1}{2}} = \pi \quad \text{and} \quad c_{+\frac{1}{2},+\frac{1}{2}} = \frac{\pi}{2}.$$

(1) Let $\alpha = -\frac{1}{2}$ and $\beta = -\frac{1}{2}$. Then

$$\int_{-1}^{+1} f(s)\,\overline{g(s)} w_{-\frac{1}{2},-\frac{1}{2}}(s)\, ds = \frac{1}{\pi} \int_{-1}^{+1} f(s)\,\overline{g(s)} \frac{1}{\sqrt{1-s^2}}\, ds$$

$$= \frac{1}{\pi} \int_0^\pi f(\cos t)\, \overline{g(\cos t)}\, dt.$$

We conclude from

$$2\cos kt \cos t = \cos(k+1)t + \cos(k-1)t$$

that

$$f(s) \longleftrightarrow f(\cos t)$$

yields a one-to-one correspondence between the linear spans of

$$\{1, s, \dots, s^n\} \quad \text{and} \quad \{1, \cos t, \dots, \cos nt\}.$$

Consequently,

$$L_k^{(-\frac{1}{2},-\frac{1}{2})}(\cos t) = \sqrt{2} \cos kt \quad \text{for } k = 1, 2, \dots$$

and

$$\int_{-1}^{+1} \Big\| \sum_{k=1}^{n} x_k L_k^{(-\frac{1}{2},-\frac{1}{2})}(s) \Big\|^2 w_{-\frac{1}{2},-\frac{1}{2}}(s)\, ds = \frac{2}{\pi} \int_0^{\pi} \Big\| \sum_{k=1}^{n} x_k \cos kt \Big\|^2 dt.$$

(2) Let $\alpha = +\frac{1}{2}$ and $\beta = +\frac{1}{2}$. Then

$$\begin{aligned} \int_{-1}^{+1} f(s)\, \overline{g(s)} w_{+\frac{1}{2},+\frac{1}{2}}(s)\, ds &= \frac{2}{\pi} \int_{-1}^{+1} f(s)\, \overline{g(s)} \sqrt{1 - s^2}\, ds \\ &= \frac{2}{\pi} \int_0^{\pi} f(\cos t)\, \overline{g(\cos t)} \sin^2 t\, dt. \end{aligned}$$

We conclude from

$$2 \cos kt \sin t = \sin(k+1)t - \sin(k-1)t$$

that

$$f(s) \longleftrightarrow f(\cos t) \sin t$$

yields a one-to-one correspondence between the linear spans of

$$\{1, s, \dots, s^{n-1}\} \quad \text{and} \quad \{\sin t, \sin 2t, \dots, \sin nt\}.$$

Consequently,

$$L_k^{(+\frac{1}{2},+\frac{1}{2})}(\cos t) \sin t = \sin(k+1)t \quad \text{for } k = 0, 1, \dots$$

and

$$\int_{-1}^{+1} \Big\| \sum_{k=0}^{n-1} x_k L_k^{(+\frac{1}{2},+\frac{1}{2})}(s) \Big\|^2 w_{+\frac{1}{2},+\frac{1}{2}}(s)\, ds = \frac{2}{\pi} \int_0^{\pi} \Big\| \sum_{k=1}^{n} x_k \sin kt \Big\|^2 dt.$$

(3) Let $\alpha = -\frac{1}{2}$ and $\beta = +\frac{1}{2}$. Then

$$\begin{aligned} \int_{-1}^{+1} f(s)\, \overline{g(s)} w_{-\frac{1}{2},+\frac{1}{2}}(s)\, ds &= \frac{1}{\pi} \int_{-1}^{+1} f(s)\, \overline{g(s)} \sqrt{\frac{1+s}{1-s}}\, ds \\ &= \frac{2}{\pi} \int_0^{\pi} f(\cos t)\, \overline{g(\cos t)} \cos^2 \tfrac{1}{2} t\, dt. \end{aligned}$$

We conclude from

$$2\cos kt\cos\tfrac{1}{2}t=\cos\tfrac{2k+1}{2}t+\cos\tfrac{2k-1}{2}t$$

that

$$f(s)\longleftrightarrow f(\cos t)\cos\tfrac{1}{2}t$$

yields a one-to-one correspondence between the linear spans of

$$\{1,s,\ldots,s^{n-1}\}\quad\text{and}\quad\{\cos\tfrac{1}{2}t,\cos\tfrac{3}{2}t,\ldots,\cos\tfrac{2n-1}{2}t\}.$$

Consequently,

$$L_k^{(-\frac{1}{2},+\frac{1}{2})}(\cos t)\cos\tfrac{1}{2}t=\cos\tfrac{2k+1}{2}t\quad\text{for } k=0,1,\ldots$$

and

$$\int_{-1}^{+1}\Big\|\sum_{k=0}^{n-1}x_kL_k^{(-\frac{1}{2},+\frac{1}{2})}(s)\Big\|^2w_{-\frac{1}{2},+\frac{1}{2}}(s)\,ds=\frac{2}{\pi}\int_0^{\pi}\Big\|\sum_{k=0}^{n-1}x_k\cos\tfrac{2k+1}{2}t\Big\|^2dt.$$

(4) Let $\alpha=+\frac{1}{2}$ and $\beta=-\frac{1}{2}$. Then

$$\int_{-1}^{+1}f(s)\,\overline{g(s)}w_{+\frac{1}{2},-\frac{1}{2}}(s)\,ds=\frac{1}{\pi}\int_{-1}^{+1}f(s)\,\overline{g(s)}\sqrt{\frac{1-s}{1+s}}\,ds$$

$$=\frac{2}{\pi}\int_0^{\pi}f(\cos t)\,\overline{g(\cos t)}\sin^2\tfrac{1}{2}t\,dt.$$

We conclude from

$$2\cos kt\sin\tfrac{1}{2}t=\sin\tfrac{2k+1}{2}t-\sin\tfrac{2k-1}{2}t$$

that

$$f(s)\longleftrightarrow f(\cos t)\sin\tfrac{1}{2}t$$

yields a one-to-one correspondence between the linear spans of

$$\{1,s,\ldots,s^{n-1}\}\quad\text{and}\quad\{\sin\tfrac{1}{2}t,\sin\tfrac{3}{2}t,\ldots,\sin\tfrac{2n-1}{2}t\}.$$

Consequently,

$$L_k^{(+\frac{1}{2},-\frac{1}{2})}(\cos t)\sin\tfrac{1}{2}t=\sin\tfrac{2k+1}{2}t\quad\text{for } k=0,1,\ldots$$

and

$$\int_{-1}^{+1}\Big\|\sum_{k=0}^{n-1}x_kL_k^{(+\frac{1}{2},-\frac{1}{2})}(s)\Big\|^2w_{+\frac{1}{2},-\frac{1}{2}}(s)\,ds=\frac{2}{\pi}\int_0^{\pi}\Big\|\sum_{k=0}^{n-1}x_k\sin\tfrac{2k+1}{2}t\Big\|^2dt.$$

9.4.4 Let

$$e_k^{\times}(t):=\exp(i\tfrac{2k-1}{2}t)\quad\text{for } t\in\mathbb{R}\text{ and } k\in\mathbb{Z}.$$

The modified **Fourier system**

$$\mathcal{E}_n^{\times}:=(e_1^{\times},\ldots,e_n^{\times})$$

is orthonormal in $L_2[-\pi,+\pi)$ equipped with the inner product

$$(f,g)=\frac{1}{2\pi}\int_{-\pi}^{+\pi} f(t)\overline{g(t)}\,dt.$$

9.4.5 We write

$$c_k^\times(t):=\sqrt{2}\cos\tfrac{2k-1}{2}t \text{ and } s_k^\times(t):=\sqrt{2}\sin\tfrac{2k-1}{2}t \quad \text{for } t\in\mathbb{R} \text{ and } k\in\mathbb{N}.$$

The modified **cosine system** and the modified **sine system,**

$$\mathcal{C}_n^\times:=(c_1^\times,\ldots,c_n^\times) \quad\text{and}\quad \mathcal{S}_n^\times:=(s_1^\times,\ldots,s_n^\times),$$

are orthonormal in $L_2[0,\pi)$ equipped with the inner product

$$(f,g):=\frac{1}{\pi}\int_0^{\pi} f(t)\overline{g(t)}\,dt.$$

9.4.6 We are now able to establish a supplement to 5.5.16.

THEOREM. *The sequences of the following ideal norms are uniformly equivalent:*

$$\begin{gathered}\varrho(\mathcal{E}_n,\mathcal{I}_n),\ \varrho(\mathcal{E}_n^\times,\mathcal{I}_n),\ \varrho(\mathcal{C}_n^\times,\mathcal{I}_n),\ \varrho(\mathcal{S}_n^\times,\mathcal{I}_n),\\ \varrho(\mathcal{I}_n,\mathcal{E}_n),\ \varrho(\mathcal{I}_n,\mathcal{E}_n^\times),\ \varrho(\mathcal{I}_n,\mathcal{C}_n^\times),\ \varrho(\mathcal{I}_n,\mathcal{S}_n^\times).\end{gathered}$$

PROOF. We conclude from

$$\exp(i\tfrac{2k-1}{2}t)=\exp(ikt)\exp(-\tfrac{1}{2}it) \text{ and } \cos\tfrac{2k-1}{2}(\pi-t)=(-1)^k\sin\tfrac{2k-1}{2}t$$

that

$$\|(x_k)|\mathcal{E}_n^\times\|=\|(x_k)|\overline{\mathcal{E}}_n^\times\|=\|(x_k)|\mathcal{E}_n\| \text{ and } \|(x_k)|\mathcal{C}_n^\times\|=\|((-1)^k x_k)|\mathcal{S}_n^\times\|.$$

This implies that

$$\begin{aligned}\varrho(\mathcal{E}_n,\mathcal{I}_n)&=\varrho(\mathcal{E}_n^\times,\mathcal{I}_n), & \varrho(\mathcal{C}_n^\times,\mathcal{I}_n)&=\varrho(\mathcal{S}_n^\times,\mathcal{I}_n),\\ \varrho(\mathcal{I}_n,\mathcal{E}_n)&=\varrho(\mathcal{I}_n,\mathcal{E}_n^\times), & \varrho(\mathcal{I}_n,\mathcal{C}_n^\times)&=\varrho(\mathcal{I}_n,\mathcal{S}_n^\times).\end{aligned}$$

Moreover, we have

$$\|(x_k)|\mathcal{C}_n^\times\|\le\tfrac{1}{\sqrt{2}}\Big(\|(x_k)|\mathcal{E}_n^\times\|+\|(x_k)|\overline{\mathcal{E}}_n^\times\|\Big)$$

and

$$\|(x_k)|\mathcal{E}_n^\times\|\le\tfrac{1}{\sqrt{2}}\Big(\|(x_k)|\mathcal{C}_n^\times\|+\|(x_k)|\mathcal{S}_n^\times\|\Big).$$

Hence

$$\varrho(\mathcal{C}_n^\times,\mathcal{I}_n)\le\sqrt{2}\,\varrho(\mathcal{E}_n^\times,\mathcal{I}_n)\le 2\,\varrho(\mathcal{C}_n^\times,\mathcal{I}_n)$$

and

$$\varrho(\mathcal{I}_n,\mathcal{C}_n^\times)\le\sqrt{2}\,\varrho(\mathcal{I}_n,\mathcal{E}_n^\times)\le 2\,\varrho(\mathcal{I}_n,\mathcal{C}_n^\times).$$

9.4.7 In view of 9.4.3, we get an immediate consequence of 5.5.16 and the previous theorem.

THEOREM. *The sequences of the following ideal norms are uniformly equivalent:*

$$\boldsymbol{\varrho}(\mathcal{L}_n^{(+\frac{1}{2},+\frac{1}{2})},\mathcal{I}_n),\ \boldsymbol{\varrho}(\mathcal{L}_n^{(-\frac{1}{2},-\frac{1}{2})},\mathcal{I}_n),\ \boldsymbol{\varrho}(\mathcal{L}_n^{(+\frac{1}{2},-\frac{1}{2})},\mathcal{I}_n),\ \boldsymbol{\varrho}(\mathcal{L}_n^{(-\frac{1}{2},+\frac{1}{2})},\mathcal{I}_n),$$
$$\boldsymbol{\varrho}(\mathcal{I}_n,\mathcal{L}_n^{(+\frac{1}{2},+\frac{1}{2})}),\ \boldsymbol{\varrho}(\mathcal{I}_n,\mathcal{L}_n^{(-\frac{1}{2},-\frac{1}{2})}),\ \boldsymbol{\varrho}(\mathcal{I}_n,\mathcal{L}_n^{(+\frac{1}{2},-\frac{1}{2})}),\ \boldsymbol{\varrho}(\mathcal{I}_n,\mathcal{L}_n^{(-\frac{1}{2},+\frac{1}{2})}).$$

9.4.8 Next, we provide an analogue of 5.2.23.

LEMMA. *Let $x_1,\dots,x_n\in X$ and $x_0=x_{n+1}=o$. Then, for $t\in\mathbb{R}$,*

$$2\cos\tfrac{1}{2}t\sum_{k=1}^{n}x_k\cos\tfrac{2k-1}{2}t=\sum_{k=0}^{n}(x_{k+1}+x_k)\cos kt, \tag{1}$$

$$2\sin\tfrac{1}{2}t\sum_{k=1}^{n}x_k\sin\tfrac{2k-1}{2}t=\sum_{k=0}^{n}(x_{k+1}-x_k)\cos kt, \tag{2}$$

$$2\cos\tfrac{1}{2}t\sum_{k=1}^{n}x_k\sin\tfrac{2k-1}{2}t=\sum_{k=1}^{n}(x_k+x_{k+1})\sin kt, \tag{3}$$

$$2\sin\tfrac{1}{2}t\sum_{k=1}^{n}x_k\cos\tfrac{2k-1}{2}t=\sum_{k=1}^{n}(x_k-x_{k+1})\sin kt, \tag{4}$$

$$2\cos\tfrac{1}{2}t\sum_{k=1}^{n}x_k\cos kt=\sum_{k=1}^{n+1}(x_k+x_{k-1})\cos\tfrac{2k-1}{2}t, \tag{$1^\times$}$$

$$2\sin\tfrac{1}{2}t\sum_{k=1}^{n}x_k\sin kt=\sum_{k=1}^{n+1}(x_k-x_{k-1})\cos\tfrac{2k-1}{2}t, \tag{$2^\times$}$$

$$2\cos\tfrac{1}{2}t\sum_{k=1}^{n}x_k\sin kt=\sum_{k=1}^{n+1}(x_{k-1}+x_k)\sin\tfrac{2k-1}{2}t, \tag{$3^\times$}$$

$$2\sin\tfrac{1}{2}t\sum_{k=1}^{n}x_k\cos kt=\sum_{k=1}^{n+1}(x_{k-1}-x_k)\sin\tfrac{2k-1}{2}t. \tag{$4^\times$}$$

9.4.9 We now obtain a supplement to 5.2.32.

THEOREM. *The sequences of the following ideal norms are uniformly equivalent:*

$$\boldsymbol{\delta}(\mathcal{E}_n,\mathcal{E}_n),\ \boldsymbol{\delta}(\mathcal{E}_n^\times,\mathcal{E}_n^\times),\ \boldsymbol{\delta}(\mathcal{C}_n^\times,\mathcal{C}_n^\times),\ \boldsymbol{\delta}(\mathcal{S}_n^\times,\mathcal{S}_n^\times),\ \boldsymbol{\varrho}(\mathcal{C}_n^\times,\mathcal{S}_n^\times),\ \boldsymbol{\varrho}(\mathcal{S}_n^\times,\mathcal{C}_n^\times).$$

PROOF. It follows from

$$\|(x_k)|\mathcal{E}_n\|=\|(x_k)|\mathcal{E}_n^\times\|\quad\text{and}\quad\|(x_k)|\mathcal{S}_n^\times\|=\|((-1)^k x_k)|\mathcal{C}_n^\times\|$$

that $\boldsymbol{\delta}(\mathcal{E}_n,\mathcal{E}_n)=\boldsymbol{\delta}(\mathcal{E}_n^\times,\mathcal{E}_n^\times)$ and $\boldsymbol{\varrho}(\mathcal{C}_n^\times,\mathcal{S}_n^\times)=\boldsymbol{\varrho}(\mathcal{S}_n^\times,\mathcal{C}_n^\times)$, respectively.

If $\boldsymbol{g}(s):=\boldsymbol{f}(\pi-s)$ for $\boldsymbol{f}\in[L_2[0,\pi),X]$, then

$$\langle \boldsymbol{f}, s_k^\times\rangle=(-1)^{k+1}\langle \boldsymbol{g}, c_k^\times\rangle .$$

Consequently, $\boldsymbol{\delta}(\mathcal{S}_n^\times,\mathcal{S}_n^\times)=\boldsymbol{\delta}(\mathcal{C}_n^\times,\mathcal{C}_n^\times)$.

As in 5.2.11 we get $\boldsymbol{\delta}(\mathcal{C}_n^\times,\mathcal{C}_n^\times)\le 2\,\boldsymbol{\delta}(\mathcal{E}_n^\times,\mathcal{E}_n^\times)$, and

$$\sum_{k=-n+1}^{+n}\exp(i\tfrac{2k-1}{2}(s-t))=2\sum_{k=1}^{n}\cos\tfrac{2k-1}{2}s\cos\tfrac{2k-1}{2}t+2\sum_{k=1}^{n}\sin\tfrac{2k-1}{2}s\sin\tfrac{2k-1}{2}t$$

yields $\boldsymbol{\delta}(\mathcal{E}_{2n}^\times,\mathcal{E}_{2n}^\times)\le\boldsymbol{\delta}(\mathcal{C}_n^\times,\mathcal{C}_n^\times)+\boldsymbol{\delta}(\mathcal{S}_n^\times,\mathcal{S}_n^\times)$.

We now come to the only non-trivial parts of the proof. First of all, formulas (1) and (4) of the previous lemma imply

$$\|(x_{k+1}+x_k)|\mathcal{C}_n\|\le 2\,\|(x_k)|\mathcal{C}_n^\times\|+\sqrt{2}\,\|x_1\|$$

and

$$\|(x_{k+1}-x_k)|\mathcal{S}_n\|\le 2\,\|(x_k)|\mathcal{C}_n^\times\|.$$

Moreover, combining formulas (2) and (3) gives

$$2\exp(i\tfrac{1}{2}t)\sum_{k=1}^{n}x_k\sin\tfrac{2k-1}{2}t=\sum_{k=1}^{n}(x_{k+1}+x_k)\sin kt+i\sum_{k=0}^{n}(x_{k+1}-x_k)\cos kt.$$

Hence, in view of $\|x_1\|\le\|(x_k)|\mathcal{C}_n^\times\|$ and $\|Tx_1\|\le\boldsymbol{\varrho}(T|\mathcal{C}_n,\mathcal{S}_n)\|(x_k)|\mathcal{C}_n^\times\|$,

$$\begin{aligned}2\,\|(Tx_k)|\mathcal{S}_n^\times\| &\le \left\{\begin{matrix}\|(Tx_{k+1}+Tx_k)|\mathcal{S}_n\|\\ +\\ \|(Tx_{k+1}-Tx_k)|\mathcal{C}_n\|\end{matrix}\right\}+\sqrt{2}\,\|Tx_1\|\\ &\le \left\{\begin{matrix}\boldsymbol{\varrho}(T|\mathcal{S}_n,\mathcal{C}_n)\|(x_{k+1}+x_k)|\mathcal{C}_n\|\\ +\\ \boldsymbol{\varrho}(T|\mathcal{C}_n,\mathcal{S}_n)\|(x_{k+1}-x_k)|\mathcal{S}_n\|\end{matrix}\right\}+\sqrt{2}\,\|Tx_1\|\\ &\le (2+\sqrt{2})\big(\boldsymbol{\varrho}(T|\mathcal{S}_n,\mathcal{C}_n)+\boldsymbol{\varrho}(T|\mathcal{C}_n,\mathcal{S}_n)\big)\|(x_k)|\mathcal{C}_n^\times\|.\end{aligned}$$

Therefore, since $2+\sqrt{2}=3.4142\ldots<4$, we have

$$\|(Tx_k)|\mathcal{S}_n^\times\|\le 4\,\max\{\boldsymbol{\varrho}(T|\mathcal{S}_n,\mathcal{C}_n),\boldsymbol{\varrho}(T|\mathcal{C}_n,\mathcal{S}_n)\}\,\|(x_k)|\mathcal{C}_n^\times\|,$$

which proves that

$$\boldsymbol{\varrho}(\mathcal{C}_n^\times,\mathcal{S}_n^\times)=\boldsymbol{\varrho}(\mathcal{S}_n^\times,\mathcal{C}_n^\times)\le 4\,\max\{\boldsymbol{\varrho}(\mathcal{S}_n,\mathcal{C}_n),\boldsymbol{\varrho}(\mathcal{C}_n,\mathcal{S}_n)\}.$$

Similarly, formulas $(1^\times)$ and $(4^\times)$ of the previous lemma imply

$$\|(x_k+x_{k-1})|\mathcal{C}_{n+1}^\times\|\le 2\,\|(x_k)|\mathcal{C}_n\|$$

and

$$\|(x_k-x_{k-1})|\mathcal{S}_{n+1}^\times\|\le 2\,\|(x_k)|\mathcal{C}_n\|.$$

Moreover, combining formulas ($2^\times$) and ($3^\times$) gives

$$2\exp(i\tfrac{1}{2}t)\sum_{k=1}^{n} x_k \sin kt = \sum_{k=1}^{n+1}(x_k - x_{k-1}) \sin \tfrac{2k-1}{2}t + i\sum_{k=1}^{n+1}(x_k + x_{k-1}) \cos \tfrac{2k-1}{2}t.$$

Hence, in view of 3.3.6,

$$\begin{aligned} 2\,\|(Tx_k)|\mathcal{S}_n\| &\le \left\{ \begin{matrix} \|(Tx_k - Tx_{k-1})|\mathcal{S}_{n+1}^\times\| \\ + \\ \|(Tx_k + Tx_{k-1})|\mathcal{C}_{n+1}^\times\| \end{matrix} \right\} \\ &\le \left\{ \begin{matrix} \varrho(T|\mathcal{S}_{n+1}^\times, \mathcal{C}_{n+1}^\times)\,\|(x_k - x_{k-1})|\mathcal{C}_{n+1}^\times\| \\ + \\ \varrho(T|\mathcal{C}_{n+1}^\times, \mathcal{S}_{n+1}^\times)\,\|(x_k + x_{k-1})|\mathcal{S}_{n+1}^\times\| \end{matrix} \right\} \\ &\le 6\left(\varrho(T|\mathcal{S}_n^\times, \mathcal{C}_n^\times) + \varrho(T|\mathcal{C}_n^\times, \mathcal{S}_n^\times)\right)\|(x_k)|\mathcal{C}_n\|. \end{aligned}$$

Therefore we have

$$\|(Tx_k)|\mathcal{S}_n\| \le 6\,\max\left\{\varrho(T|\mathcal{S}_n^\times, \mathcal{C}_n^\times), \varrho(T|\mathcal{C}_n^\times, \mathcal{S}_n^\times)\right\}\|(x_k)|\mathcal{C}_n\|,$$

which proves that

$$\varrho(\mathcal{S}_n, \mathcal{C}_n) \le 6\,\varrho(\mathcal{S}_n^\times, \mathcal{C}_n^\times) = 6\,\varrho(\mathcal{C}_n^\times, \mathcal{S}_n^\times).$$

The inequality

$$\varrho(\mathcal{C}_n, \mathcal{S}_n) \le 6\,\varrho(\mathcal{C}_n^\times, \mathcal{S}_n^\times) = 6\,\varrho(\mathcal{S}_n^\times, \mathcal{C}_n^\times)$$

can be obtained in the same way. This completes the proof, by 5.2.31.

REMARK. We see from

$$\cos \tfrac{2k-1}{2}(\pi - t) = (-1)^k \sin \tfrac{2k-1}{2}t$$

that $\varrho(T|\mathcal{C}_n^\times, \mathcal{S}_n^\times)$ is the least constant $c \ge 0$ such that

$$\left(\frac{2}{\pi}\int_0^\pi \left\|\sum_{k=1}^n (-1)^k Tx_k \sin \tfrac{2k-1}{2}t\right\|^2 dt\right)^{1/2} \le c\left(\frac{2}{\pi}\int_0^\pi \left\|\sum_{k=1}^n x_k \sin \tfrac{2k-1}{2}t\right\|^2 dt\right)^{1/2}$$

whenever $x_1, \ldots, x_n \in X$. Consequently, there is a certain analogy with the ideal norm $\mathring{\mathring{\varrho}}_\pm(\mathcal{H}(\mathbb{D}_1^n), \mathcal{H}(\mathbb{D}_1^n))$, defined in 8.5.2. However, in the present situation we have $\varrho(\mathcal{C}_n^\times, \mathcal{S}_n^\times) \not\asymp \varrho_\pm(\mathcal{S}_n^\times, \mathcal{S}_n^\times)$.

9.4.10 By 9.4.3, the next result can be deduced from 5.2.18 and the previous theorem.

THEOREM. *The sequences of the following ideal norms are uniformly equivalent:*

$$\boldsymbol{\delta}(\mathcal{L}_n^{(+\frac{1}{2},+\frac{1}{2})}, \mathcal{L}_n^{(+\frac{1}{2},+\frac{1}{2})}),\ \boldsymbol{\delta}(\mathcal{L}_n^{(-\frac{1}{2},+\frac{1}{2})}, \mathcal{L}_n^{(-\frac{1}{2},+\frac{1}{2})}),$$
$$\boldsymbol{\delta}(\mathcal{L}_n^{(+\frac{1}{2},-\frac{1}{2})}, \mathcal{L}_n^{(+\frac{1}{2},-\frac{1}{2})}),\ \boldsymbol{\delta}(\mathcal{L}_n^{(-\frac{1}{2},-\frac{1}{2})}, \mathcal{L}_n^{(-\frac{1}{2},-\frac{1}{2})}).$$

9.4.11 Apart from the preceding results, almost nothing is known about ideal norms associated with orthogonal polynomials. Thus there exists a huge number of unsolved problems. In what follows, we only list a small selection.

PROBLEMS.

(1) *For which exponents* $\alpha, \beta > -1$ *do we have*
$$\boldsymbol{\varrho}(\mathcal{L}_n^{(\alpha,\beta)}, \mathcal{J}_n) \asymp \boldsymbol{\varrho}(\mathcal{J}_n, \mathcal{L}_n^{(\alpha,\beta)}) \ ?$$

(2) *For which pairs* $\alpha_0, \beta_0 > -1$ *and* $\alpha_1, \beta_1 > -1$ *do we have*
$$\boldsymbol{\varrho}(\mathcal{L}_n^{(\alpha_0,\beta_0)}, \mathcal{J}_n) \asymp \boldsymbol{\varrho}(\mathcal{L}_n^{(\alpha_1,\beta_1)}, \mathcal{J}_n) \ ?$$

(3) *For which pairs* $\alpha_0, \beta_0 > -1$ *and* $\alpha_1, \beta_1 > -1$ *do we have*
$$\boldsymbol{\delta}(\mathcal{L}_n^{(\alpha_0,\beta_0)}, \mathcal{L}_n^{(\alpha_0,\beta_0)}) \asymp \boldsymbol{\delta}(\mathcal{L}_n^{(\alpha_1,\beta_1)}, \mathcal{L}_n^{(\alpha_1,\beta_1)}) \ ?$$

(4) *Is it true that*
$$\boldsymbol{\varrho}(\mathcal{L}_n^H, \mathcal{J}_n) \asymp \boldsymbol{\varrho}(\mathcal{J}_n, \mathcal{L}_n^H) \ ?$$

(5) *Is it true that*
$$\boldsymbol{\varrho}(\mathcal{L}_n^H, \mathcal{J}_n) \asymp \boldsymbol{\varrho}(\mathcal{E}_n, \mathcal{J}_n) \ ?$$

(6) *Is it true that*
$$\boldsymbol{\delta}(\mathcal{L}_n^H, \mathcal{L}_n^H) \asymp \boldsymbol{\delta}(\mathcal{E}_n, \mathcal{E}_n) \ ?$$

REMARKS. It follows from [koe*a, p. 186] that
$$\boldsymbol{\varrho}(X|\mathcal{L}_n^{(\alpha_0,\beta_0)}, \mathcal{L}_n^{(\alpha_1,\beta_1)}) = O(1)$$
for any choice of $\alpha_0, \beta_0 > -1$ and $\alpha_1, \beta_1 > -1$ if and only if $X \in \mathsf{UMD}$. The condition
$$\boldsymbol{\delta}(X|\mathcal{L}_n^H, \mathcal{L}_n^H) = O(1)$$
also characterizes UMD-spaces; see [koe 4]. Consequently, counterexamples to the relations considered in **(1)**, **(2)**, **(3)** and **(6)** can only be found outside UMD.

9.4.12 The situation becomes much more delicate when we pass to the L_r-theory with $r \neq 2$. Then the r-th average should be defined by
$$\|(x_k)|\mathcal{L}_n^{(w)}\|_r := \left(\int_0^1 \Big\| \sum_{k=0}^{n-1} x_k L_k^{(w)}(s) w(s)^{1/2} \Big\|^r ds \right)^{1/r},$$
and not by
$$\|(x_k)|\mathcal{L}_n^{(w)}\|_r := \left(\int_0^1 \Big\| \sum_{k=0}^{n-1} x_k L_k^{(w)}(s) \Big\|^r w(s)\, ds \right)^{1/r}.$$

This means, for example, that it is recommendable to deal with Hermite functions instead of Hermite polynomials. Moreover, natural boundaries occur for the exponent r. Already in the scalar-valued case, the Jacobi polynomials form an L_r-basis if and only if

$$|1/r - 1/2| < \min\{1/4, (\alpha+1)/2, (\beta+1)/2\};$$

see [muc]. For $\alpha, \beta \geq -1/2$, we obtain $4/3 < r < 4$. The same condition guarantees that the Hermite functions yield a basis of $L_r(-\infty, +\infty)$; see [ask*].

9.5 History

References are listed in chronological order. Sometimes we have added the numbers of pages on which specific definitions and results are to be found.

9.5.1 Let us begin with a quotation from the Habilitationsschrift of B. Riemann:

FOURIER *bemerkte, dass in der trigonometrischen Reihe*

$$f(x) = \begin{cases} \quad a_1 \sin x + a_2 \sin 2x + \dots \\ +\frac{1}{2}b_0 + b_1 \cos x + b_2 \cos 2x + \dots \end{cases}$$

die Coefficienten sich durch die Formeln

$$a_n = \frac{1}{\pi}\int_{-\pi}^{\pi} f(x) \sin nx \, dx, \quad b_n = \frac{1}{\pi}\int_{-\pi}^{\pi} f(x) \cos nx \, dx$$

bestimmen lassen.

In other words, J. B. J. Fourier knew the **orthogonality relations** satisfied by sine and cosine functions. He presented his results to the Académie des Sciénces (Paris) on December 21st, 1807. Further important contributions to the theory of Fourier expansions are due to P. G. Lejeune Dirichlet (1829) and B. Riemann (1854). In the middle of the eighteenth century, **trigonometric series** were already used by D. Bernoulli (1700–1782) and L. Euler to treat the problem of vibrating strings. However, most mathematicians of their time believed that only 'analytic' functions could be represented in this form.

By purely formal manipulations, M. A. Parseval (1799) obtained a formula that became known as **Parseval's equality**, and **Bessel's inequality** was proved in 1828.

Ch. Sturm (1836) and J. Liouville (1837) discovered that eigenfunctions of the differential operator $u(x) \to u''(x) + q(x)u(x)$ associated with different eigenvalues are orthogonal. Thus the door was opened to the consideration of general **orthogonal systems**. After the introduction of Lebesgue's integral (1904), this development culminated in the Fischer–Riesz theorem (1907).

REFERENCES: [par], [FOU], [bes], [dir], [riem], [fis] and [rie a].

9.5.2 **Haar functions** were defined, for the first time, in the Thesis of A. Haar (1910), supervised by D. Hilbert.

REFERENCE: [haar].

9.5.3 The introduction of **Walsh functions** in harmonic analysis is attributed to J. L. Walsh (1923), although they appeared much earlier in electrical engineering. The most common enumeration was given by R. E. A. C. Paley (1932). He recognized that Walsh functions are products of Rademacher functions. Finally, N. Y. Vilenkin (1947) and N. J. Fine (1949) identified Walsh functions as characters on the Cantor group. This fact explains why there are so many parallel results for the trigonometric system and the Walsh system.

REFERENCES: [wal] and [pal].

9.5.4 **Rademacher functions** were introduced in 1922. They play an important role not only in the theory of orthogonal systems but also in stochastics, since $r_1, r_2, \ldots$ are independent identically distributed **Bernoulli random variables**, which are the basic tool for modelling the process of coin tossing (J. Bernoulli, 1654–1705). In this context, it is common to use the Cantor group as the underlying probability space. The complex analogue of the Rademacher system was first investigated by H. Steinhaus (1930).

A. Khintchine (1922) established the estimate

$$\text{measure}\left\{ t \in [0,1) : \left| \sum_{k=1}^{n} \xi_k r_k(t) \right| > u \left(\sum_{k=1}^{n} |\xi_k|^2 \right)^{1/2} \right\} \le c \exp(-u^2/2)$$

for $\xi_1, \ldots, \xi_n \in \mathbb{R}$ and $u > 0$, where c is an absolute constant. This was the origin of the famous (upper) **Khintchine inequality**. Without reference to A. Khintchine, the two-sided inequality

$$A_r \left(\sum_{k=1}^{n} |\xi_k|^2 \right)^{1/2} \le \left(\frac{1}{2^n} \sum_{\mathbb{E}_2^n} \left| \sum_{k=1}^{n} \varepsilon_k \xi_k \right|^r \right)^{1/r} \le B_r \left(\sum_{k=1}^{n} |\xi_k|^2 \right)^{1/2}$$

for $\xi_1, \ldots, \xi_n \in \mathbb{R}$ and $0 < r < \infty$ was obtained by J. E. Littlewood (1930). At the same time, R. E. A. C. Paley/A. Zygmund gave an independent proof of the upper Khintchine inequality. The best possible constants were computed by U. Haagerup (1982).

In view of Khintchine's inequality, all averages

$$\left(\frac{1}{2^n} \sum_{\mathbb{E}_2^n} \left| \sum_{k=1}^{n} \varepsilon_k \xi_k \right|^r \right)^{1/r}$$

with $0 < r < \infty$ are equivalent. G. Pisier (1977) observed that, by a result of J.-P. Kahane (1964), this fact remains true in the vector-valued case as well; **Khintchine–Kahane inequality** 4.1.10.

REFERENCES: [rad], [khin], [lit, p. 170], [pal*, p. 340], [stei], [kah], [pis 6], [haag], [KAH, pp. 18 and 282] and [lat*].

9.5.5 The function

$$\varphi(t) = \frac{1}{\sigma\sqrt{2\pi}} \mathrm{e}^{-t^2/2\sigma^2}$$

first appeared in the work of P. S. Laplace and C. F. Gauss, at the beginning of the nineteenth century. Due to the **central limit theorem** (de Moivre 1733, Laplace 1820, Tschebyscheff 1867/87, Lindberg 1922), the normal distribution is now one of the most important objects in probability theory. In the course of time, the notion of independent standard **Gaussian random variables** came into existence.

REFERENCE: [LED*].

9.5.6 The concept of a **complete normed linear space** was introduced by S. Banach, H. Hahn and N. Wiener at the beginning of the 1920s. The name **Banach space** is due to M. Fréchet. The basic reference was and is Banach's monograph

Théorie des opérations linéaires, Warszawa, 1932.

Further information about the history of Banach space theory is to be found in [DUN*1, p. 85], [PIE 3, pp. 279–311] and [pie 6].

9.5.7 The **Banach–Mazur distance** $d(X,Y)$ of isomorphic Banach spaces X and Y was already defined in [BAN, Remarques, p. 242]. Later, it turned out that this notion is of particular significance in the finite dimensional case. The famous **John theorem** (1948) asserts that $d(X, l_2^n) \leq \sqrt{n}$ for all n-dimensional spaces X, and V. I. Gurarii/ M. I. Kadets/V. I. Macaev (1966) computed the asymptotic behaviour of $d(l_u^n, l_v^n)$ as $n \to \infty$. Another important result is due to D. R. Lewis

(1978), who proved that $d(X, l_r^n) \le n^{|1/r-1/2|}$ for all n-dimensional subspaces of any L_r.

REFERENCES: [joh], [gur*], [lew 2] and [TOM].

9.5.8 It follows from the Hahn–Banach theorem and Auerbach's lemma [BAN, p. 238] that every n-dimensional subspace of a Banach space X is the range of a projection $P \in \mathfrak{L}(X)$ with $\|P\| \le n$. Significant progress is due to M. I. Kadets/M. G. Snobar (1971), who got the bound $\sqrt{n}$. Furthermore, D. R. Lewis (1976) showed that in L_r there exist projections with $\|P\| \le n^{|1/r-1/2|}$. Banach spaces characterized by this property were investigated by H. König/J. R. Retherford/N. Tomczak-Jaegermann (1980) and A. Pietsch (1991) in connection with eigenvalue distributions of nuclear operators. We refer to [pis 17] for an excellent survey of **finite rank projections**.

REFERENCES: [kad*c], [lew 2], [fig*c], [koe*b] and [pie 4].

9.5.9 The theory of **Banach operator ideals** emerged from the theory of **tensor products** which was created by A. F. Ruston, R. Schatten and A. Grothendieck around 1950. The concept of an **ideal norm** corresponds to that of a $\otimes$-norm (tensor norm). Most ideal norms are not defined for all operators. However, A. Pełczyński (1980) observed that, quite often, they can be generated as limits of their gradations. A. Pietsch (1990) used this idea as a general principle for constructing Banach operator ideals; see 1.2.8.

Sequences of ideal norms (defined for all operators) were considered, for the first time, by B. Maurey/G. Pisier (1976) in their theory of Rademacher type and cotype. In particular, they discovered that **submultiplicativity** is a very useful property.

REFERENCES: [SCHA 1], [gro 1], [mau*, pp. 55, 71], [PIE 2], [pel 1, p. 165], [TOM, p. 175], [pie 2] and [DEF*].

9.5.10 A classical theorem of P. Jordan/J. von Neumann (1935) says that the parallelogram equality

$$\|x+y\|^2 + \|x-y\|^2 = 2\|x\|^2 + 2\|y\|^2$$

characterizes Hilbert spaces isometrically. It was much harder to find isomorphic criteria. The first result along these lines, implicitly contained in a paper of J. Lindenstrauss (1963), was obtained by J. I. Joichi (1966); see 3.10.11 (Remarks).

The Banach ideal of operators factoring through a Hilbert space (**Hilbertian operators**) was investigated by A. Grothendieck in 1956.

His studies were carried on in a famous paper of J. Lindenstrauss/ A. Pełczyński (1968). In particular, we refer to their Theorem 7.3. Based on this result, S. Kwapień (1972) found another important characterization of Hilbertian Banach spaces which was later extended to the setting of operators; see 2.3.23.

REFERENCES: [gro 1, p. 40], [lin 1, p. 246], [joi], [lin*b, pp. 292–294 and 313], [kwa 2, p. 586], [mau 3, p. 329], [PIS 1, pp. 21–30] and [pis 20, pp. 254–256].

9.5.11 Operators $T \in \mathfrak{L}(X, Y)$ factoring in the form $T : X \to C \to H \to Y$ were already considered in Grothendieck's Résumé (1956, applications prohilbertiennes gauches, propriété de prolongement hilbertien). In 1967, it turned out that these are just the **2-summing operators**. Of particular interest for the purpose of this monograph is a result due to N. Tomczak-Jaegermann (1979) which says that, whenever $\text{rank}(T) = N$, the 2-summing norm $\|T|\mathfrak{P}_2\|$ can be computed with N vectors, up to the factor $\sqrt{2}$; see 2.2.8.

REFERENCES: [gro 1, pp. 36, 52], [pie 1] and [tom 2].

9.5.12 S. Kwapień (1970) characterized Hilbertian Banach spaces X by the property that the dual of every 2-summing operator from X into l_2 is 2-summing as well. This result was modified by G. Pisier (1979) in various directions. The final outcome is Theorem 3.10.4.

REFERENCES: [kwa 1], [pis 8] and [pis 10, p. 278].

9.5.13 Based on work of B. Maurey, the concept of a **γ-summing operator** was introduced by W. Linde/A. Pietsch (1974). Later, T. Figiel/ N. Tomczak-Jaegermann (1979) found out that the γ-summing norm of operators $T \in \mathfrak{L}(l_2^n, X)$, usually denoted by $\ell(T)$, plays an important role in Banach space geometry. In particular, there is a close connection with Dvoretzky's theorem; see 4.15.19.

REFERENCES: [lin*a], [fig*c], [MIL*, pp. 54 and 106–113] and [TOM, pp. 79–94].

9.5.14 Dvoretzky's theorem (1959) is the most important result in the local theory of Banach spaces. It grew out of the Dvoretzky–Rogers lemma (1950) which was the main tool in proving that absolute and unconditional convergence of series coincide in finite dimensional Banach spaces only. The existence of almost spherical n-dimensional sections in Banach spaces with sufficiently high dimension was conjectured by

A. Grothendieck (1956). For the history of this subject, we refer to [lin 3] and [mil c].

REFERENCES: [dvo*], [gro 2, p. 108] and [dvo].

9.5.15 The concept of **B-convexity** was introduced by A. Beck in 1962 when he characterized Banach spaces X with the property that, mutatis mutandis, the strong law of large numbers holds for sequences of X-valued random variables. His Ph. D. student D. P. Giesy (1966) developed the basic theory. It turned out that B is invariant under the formation of subspaces, quotients and duals. Moreover, Banach spaces $X \in \mathsf{B}$ were characterized by the property of **not containing the spaces l_1^n uniformly**. Giesy also knew that the reflexive space $[l_2, l_1^n]$ fails to be B-convex. Finally, G. Pisier (1973) proved that a Banach space is B-convex if and only if it has **non-trivial Rademacher type**. B. Beauzamy (1975) extended most of these considerations to the setting of operators. A characterization of B-convex Banach spaces in terms of eigenvalue distributions of nuclear operators was given by A. Pietsch (1990).

REFERENCES: [bec], [gie, pp. 123, 125, 127 and 130], [pis 2], [pis 3, p. 12], [bea 2], [PAJ] and [pie 3].

9.5.16 Banach spaces X **not containing the spaces l_∞^n uniformly** were first considered by B. Maurey and G. Pisier. In particular, G. Pisier (1973) showed that this geometric property is equivalent to $\varrho(X|\mathcal{G}_n, \mathcal{R}_n) = o(\sqrt{1+\log n})$. Moreover, a limiting case of the celebrated Maurey–Pisier theorem 4.16.5 asserts that these are just those spaces having **non-trivial Rademacher cotype**. A simple proof of the last fact was given by A. Hinrichs. Further improvements of his reasoning have been achieved in our treatise by using ultraproduct techniques.

REFERENCES: [mau 1], [pis 1] and [hin 2].

9.5.17 W. Orlicz (1933) was the first who used Khintchine's inequality in Banach space theory. He considered spaces in which $\sum_{k=1}^\infty \|x_k\|^2 < \infty$ whenever $\sum_{k=1}^\infty x_k$ converges unconditionally, a property that is very close to being of Rademacher cotype 2.

In 1972, S. Kwapień gave an isomorphic characterization of inner product spaces via two-sided Khintchine inequalities for Rademacher and/or Gaussian random variables,

$$\mathsf{RT}_2 \cap \mathsf{RC}_2 = \mathsf{GT}_2 \cap \mathsf{GC}_2 = \mathsf{H}.$$

At the same time, E. Dubinsky, A. Pełczyński and H. P. Rosenthal showed that all operators from l_∞ into a Banach space X are 2-summing if the dual space has **subquadratic Gaussian average**, which means that X' has **Gaussian type** 2. This result was improved by B. Maurey (1973), who only required that X has **Gaussian cotype** 2.

When dealing with sums of independent vector-valued random variables, J. Hoffmann-Jørgensen (1972) defined the notions of **Rademacher type** p and **Rademacher cotype** q with $1 \le p \le 2$ and $2 \le q \le \infty$. In this context, the **Rademacher type index** $p(X)$ and the **Rademacher cotype index** $q(X)$ were introduced. Based on previous results of G. Nordlander (1961), he computed $p(L_r)$ and $q(L_r)$ for $1 \le r \le \infty$. We also owe to him the observation that, in the case of type, the Rademacher system can be replaced by the Gauss system. The true significance of these new parameters only became apparent through the work of B. Maurey/G. Pisier (1976); see 4.16.5. They also introduced the quantities $\boldsymbol{\varrho}(X|\mathcal{R}_n, \mathcal{I}_n)$ and $\boldsymbol{\varrho}(X|\mathcal{I}_n, \mathcal{R}_n)$ (our notation).

REFERENCES: [orl], [nor 2], [kwa 2], [dub*], [HOF, pp. 55, 59 and 74], [mau 4, p. 116] and [mau*].

9.5.18 The concepts of Rademacher type and cotype admit various modifications. R. C. James (1978) introduced Banach spaces of **equal-norm Rademacher type** p and presented a result of G. Pisier which says that, for $p = 2$, we get nothing new. On the contrary, if $1 < p < 2$, then a construction of L. Tzafriri (1979) yields Tsirelson spaces that have equal-norm Rademacher type p, but fail to have the corresponding type in the ordinary sense. The situation for cotype is similar.

Another direction was opened by V. D. Milman/G. Pisier (1986) when they considered Banach spaces of weak cotype 2. As shown in [PIS 2, p. 163], this class is not stable under the formation of l_2-multiples. So, unfortunately for us, such spaces cannot be characterized in terms of Riemann or Dirichlet ideal norms. However, excluding the limiting cases $p = 2$ and $q = 2$, an 'idealistic' theory of **weak Rademacher type** p and **weak Rademacher cotype** q was developed by V. Mascioni (1988). He proved that the modifications 'weak' and 'equal-norm' are equivalent. These results were completed by J. Bourgain/N. Kalton/L. Tzafriri (1989) when they showed that $\boldsymbol{\varrho}(\mathcal{R}_n, \mathcal{I}_n) \asymp \boldsymbol{\varrho}^{eq}(\mathcal{R}_n, \mathcal{I}_n)$. Surprisingly, the analogous relation fails in the case of Rademacher cotype; see [hin 6].

Similar generalizations are possible for the Gauss system. We also refer to a paper of G. Geiss/M. Junge, which deals with ideal norms of the form $\boldsymbol{\varrho}(\mathcal{A}_n, \mathcal{G}_n)$ and $\boldsymbol{\varrho}(\mathcal{G}_n, \mathcal{A}_n)$, where $\mathcal{A}_n$ is any orthonormal system.

REFERENCES: [jam 7], [tza], [mil*a], [mas, pp. 85, 98], [bou*, p. 160] and [gei*].

9.5.19 Letting $1/p + 1/q = 1$, the concepts of Rademacher type p and Rademacher cotype q are *almost* dual to each other. The typical exception is given by the space l_1 which has cotype 2, while c_0 and l_∞ fail to have any non-trivial type p. This phenomenon caused B. Maurey and G. Pisier (1976) to introduce the concept of **K-convexity** via Rademacher projections. The quantities $\boldsymbol{\delta}(X|\mathcal{R}_n, \mathcal{R}_n)$ and $\boldsymbol{\delta}(X|\mathcal{G}_n, \mathcal{G}_n)$ first appeared in a paper of T. Figiel/N. Tomczak-Jaegermann (1979). However, the most important success was achieved when G. Pisier (1980) showed the equivalence with **B-convexity**. In the setting of operators, the connection between these properties was illuminated by A. Hinrichs (1996).

REFERENCES: [mau*, p. 88], [pis 12], [pis 14], [pis 15], [pis 16] and [hin 1].

9.5.20 Rademacher, Steinhaus and lacunary trigonometric systems have many properties in common. The typical phenomena were observed by S. Sidon in a series of papers devoted to Fourier series with gaps, which appeared between 1926 and 1935. For a historical survey of **lacunarity** and **Sidonicity**, we refer the reader to [HEW*2, pp. 445–449]. In the vector-valued case, the most important result is due to G. Pisier and A. Pełczyński; see 4.11.9.

REFERENCES: [pis 6], [pis 13], [pel 2] and [sei].

9.5.21 The notion of **uniform convexity** was defined by J. A. Clarkson (1936) when he looked for Banach spaces in which the Radon–Nikodým theorem holds. Eight years later, M. M. Day (1944) introduced the dual concept of **uniform smoothness** (flattening). His main tools were the **modulus of convexity** and the **modulus of smoothness**. The precise connection between these quantities was discovered by J. Lindenstrauss (1963). The moduli of L_r were computed by O. Hanner (1955), and G. Nordlander (1960) showed that the behaviour of Hilbert spaces is, in a sense, optimal.

REFERENCES: [cla], [day 3], [han], [nor 1] and [lin 1].

9.5.22 Independently of each other, D. P. Milman and B. J. Pettis proved in 1938/39 that uniformly convex Banach spaces are reflexive. Whereas reflexivity is preserved when passing to an equivalent norm, uniform convexity is not. Moreover, M. M. Day (1941) observed that, for $1 < r_n < \infty$ and $\lim\limits_{n\to\infty} r_n = 1$, the reflexive space $[l_2, l_{r_n}]$ is not isomorphic to a uniformly convex space. This phenomenon raised the problem of characterizing all Banach spaces that admit a uniformly convex **renorming**. The solution took 30 years.

REFERENCES: [mil a], [pet] and [day 1].

9.5.23 The Milman–Pettis theorem was generalized by R. C. James (1964) when he showed that not only uniformly convex, but also uniformly non-square spaces are reflexive. This result suggested the conjecture that all B-convex spaces could be reflexive. For Banach spaces with unconditional bases the answer is indeed affirmative. However, R. C. James (1974) found a uniformly non-octahedral and non-reflexive space. The above facts can be summarized as follows: $\mathsf{UC} \subset \mathsf{B}_2 \subset \mathsf{R}$ and $\mathsf{B}_3 \not\subset \mathsf{R}$. Luckily enough, J. J. Schäffer and K. Sundaresan observed that James's proof given in the case of $\mathsf{J}_2 = \mathsf{B}_2^{\mathbb{R}}$ and J_3 also worked for all classes J_n. Hence $\mathsf{J} := \bigcup_{n=2}^{\infty} \mathsf{J}_n \subset \mathsf{R}$, and, for obvious reasons, the underlying concept is now called **J-convexity**. Another proof of the inclusion $\mathsf{J} \subset \mathsf{R}$ was given by A. Brunel/L. Sucheston based on spreading models. In this book, we present an approach via Ramsey's theorem, which is due to J. Wenzel; see [wen 4].

REFERENCES: [jam 1], [scha*], [jam 6], [bru*1], and [bru*2].

9.5.24 In modern mathematics, a theory is very seldom developed as a one-man show. Respectfully, we note that this happened in the case of James's work on **superreflexivity**. Beginning in about 1964, he has built a theory stone by stone. His main ideas were presented at a symposium held in Baton Rouge (Spring 1967). The final stone was put in place by P. Enflo (1972) by showing that every superreflexive space can be given an equivalent uniformly convex norm.

REFERENCES: [jam 1], [jam 3], [jam 4], [jam 5] and [enf].

9.5.25 There are various possibilities for defining **integrals of vector-valued functions**. The one which is relevant for our purpose was developed by S. Bochner (1933), who also introduced the spaces $[L_r, X]$. In the course of his research, Bochner observed that not all classical theorems remain true in the vector-valued setting. A prominent example is Bessel's inequality. Moreover, in a joint paper with A. E. Taylor, he showed that the Fourier series of an L_∞-valued function need not converge. That is, $L_\infty \notin \mathsf{UMD}$. This fact raised the question: *Under what conditions do the classical transformations admit vector-valued extensions?* Some general results were obtained by C. Herz (1971), M. Defant (1986) and J. L. Rubio de Francia/J. L. Torrea (1987).

REFERENCES: [boc 1], [boc 2], [boc*, p. 934], [her], [def 1, pp. 31–45] and [rub*].

9.5.26 Classical results about the **Fourier transform** state that

$$F : L_1(\mathbb{R}) \longrightarrow C_0(\mathbb{R}) \qquad \text{(B. Riemann 1854, H. Lebesgue 1903)},$$

$$F : L_2(\mathbb{R}) \longrightarrow L_2(\mathbb{R}) \quad \text{(M. A. Parseval 1799, M. Plancherel 1910)},$$

$$F : L_p(\mathbb{R}) \longrightarrow L_{p'}(\mathbb{R}) \qquad \text{(W. H. Young 1912 , F. Hausdorff 1923)},$$

where $1 < p < 2$.

J. Peetre (1969) was the first who looked at Banach spaces of **Fourier type** p with $1 \le p \le 2$. In particular, he proved interpolation theorems. M. Milman (1984) continued these investigations. However, neither author was aware of the local nature of this notion. More important is the work of S. Vági (1969). In the case when there is an unconditional basis, he showed that Banach spaces of Fourier type 2 are Hilbertian. S. Kwapień (1972) was able to remove the additional assumption.

In this treatise, we have proved that all trigonometric systems $\{e^{ikt}\}$, $\{\sqrt{2}\cos kt\}$ and $\{\sqrt{2}\sin kt\}$ as well as their discrete counterparts yield the same notions of Fourier type; see 5.5.16.

REFERENCES: [riem], [leb], [pla], [you], [hau], [pee 1], [vág, p. 307], [kwa 2], [kwa 3], [mil b], [koe 3] and [wen 2].

9.5.27 The theory of **vector-valued martingales** was initiated by S. D. Chatterji (1960) and F. S. Scalaro (1961). In this context, the Radon–Nikodým property turned out to be quite important. An excellent history of this subject is to be found in [DIE*b, pp. 141–146]. For the local theory of Banach spaces, a slightly different aspect became relevant when B. Maurey (1974) observed that the **trees** in James's theory of superreflexivity can be viewed as X-valued martingales. On the basis of this observation, G. Pisier (1975) refined Enflo's renorming theorem of superreflexive spaces. The qualitative concepts of uniform convexity and uniform smoothness were replaced by the quantitative concepts of uniform q-convexity and uniform p-smoothness, respectively. It turned out that the uniformly 2-smoothable and uniformly 2-convexifiable Banach spaces are just the Banach spaces of Haar type 2 and Haar cotype 2, respectively. In a joint paper, T. Figiel/G. Pisier (1974) showed that

$$\mathsf{UT}_2 \cap \mathsf{UC}_2 \subset \mathsf{AT}_2 \cap \mathsf{AC}_2 = \mathsf{H}.$$

This solved a problem raised in [lin 1, p. 245]. A nice survey of the interplay between geometry and martingales in Banach spaces was given by W. A. Woyczyński.

Based on Enflo's approach, B. Beauzamy (1976) studied **uniformly convexifiable operators** (uniformly convexifying operators in his terminology), and S. Heinrich studied super weakly compact operators.

The state of affairs can be summarized by the formula

$$\mathfrak{L} \circ \mathfrak{UT} = \mathfrak{AT} = \mathfrak{SR} = \mathfrak{AC} = \mathfrak{UC} \circ \mathfrak{L}.$$

REFERENCES: [cha], [sca], [mau 5, Exp. 1, pp. 4–6], [ass 1], [ass 2], [fig*b], [pis 5], [woy], [bea 3], [hei 1] and [pin].

9.5.28 Banach spaces having the **unconditionality property for martingale differences** were considered for the first time by B. Maurey (1974). His exposition contains a fundamental observation of G. Pisier which says, inter alia, that $L_r \in \mathsf{UMD}$ if $1 < r < \infty$. Moreover, B. Maurey stated that the class UMD is stable when passing to subspaces, quotients and duals. He already knew that UMD-spaces are superreflexive. On the other hand, G. Pisier (1975) constructed a superreflexive space $X \notin \mathsf{UMD}$. Thus the theory of UMD-spaces seemed to be more or less completed in 1974/75. However, the real breakthrough only came some years later. In a series of papers D. L. Burkholder (1981/83) characterized UMD-spaces by a geometric property, called **ζ-convexity**. As a consequence he proved that $\mathsf{UMD} \subseteq \mathsf{HT}$. Finally, the reverse inclusion was obtained by J. Bourgain (1983).

REFERENCES: [mau 5, Exposé 2, pp. 11–12], [pis 4, p. 10], [ald], [bur 3], [bur 4], [bur 5] and [bou 3].

9.5.29 The vector-valued versions of the two-sided Khintchine and the two-sided Burkholder–Gundy–Davis inequality can be separated into an upper and a lower part. This yields the formulas

$$\mathsf{RT}_2 \cap \mathsf{RC}_2 = \mathsf{H} \quad \text{and} \quad \mathsf{AT}_2 \cap \mathsf{AC}_2 = \mathsf{H}, \qquad (*)$$

which characterize Hilbertian Banach spaces by joining a type 2 and a cotype 2 condition. D. J. H. Garling (1990) found another useful splitting of the Burkholder–Gundy–Davis inequality. Using his idea, S. Geiss (1995) defined a new kind of ideal norm, namely $\boldsymbol{\varrho}(\mathcal{R}_n \otimes \mathcal{H}(\mathbb{D}_1^n), \mathcal{H}(\mathbb{D}_1^n))$ and $\boldsymbol{\varrho}(\mathcal{H}(\mathbb{D}_1^n), \mathcal{R}_n \otimes \mathcal{H}(\mathbb{D}_1^n))$. As a counterpart of $(*)$, he got

$$\mathsf{RAT}_2 \cap \mathsf{RAC}_2 = \mathsf{UMD}.$$

REFERENCES: [gar] and [gei 1].

9.5.30 One of the most famous results in harmonic analysis is the M. Riesz theorem (1926) which asserts that the periodic **Hilbert transform** yields an operator on $L_r[-\pi, +\pi)$ whenever $1 < r < \infty$. Analogous statements hold for the Hilbert transform on the real line and for the discrete Hilbert transform.

Banach spaces which allow vector-valued extensions of Hilbert transforms were first considered by D. L. Burkholder and B. Virot (1981). We also refer to the Thesis of J. A. Gutierrez (1982). The main success was achieved by Burkholder, who discovered the connection with the UMD-property. Systematic studies of the operator ideal associated with the Hilbert transform were made by M. Defant.

The Hilbert transform is closely connected with the **Riesz projection**. In this monograph, we have studied the related ideal norms in order to get quantitative results. As in the setting of type and cotype, the trigonometric systems $\{e^{ikt}\}$, $\{\sqrt{2}\cos kt\}$ and $\{\sqrt{2}\sin kt\}$ as well as their discrete counterparts can, in most cases, be replaced by each other; see 5.2.32. Of special interest is the ideal norm $\boldsymbol{\varrho}(\mathcal{E}_n, \mathcal{E}_n^\circ)$ which measures the difference between

$$\left(\frac{1}{2\pi}\int_{-\pi}^{+\pi}\Big\|\sum_{k=1}^{n} x_k \exp(ikt)\Big\|^2 dt\right)^{1/2} \text{ and } \left(\frac{1}{n}\sum_{h=1}^{n}\Big\|\sum_{k=1}^{n} x_k \exp\left(\tfrac{2\pi i}{n}hk\right)\Big\|^2\right)^{1/2}.$$

REFERENCES: [rie b], [tit], [LIN*2, p. 166], [vir 1], [vir 2], [gut], [bur 5], [bou 3], [bur 8], [def 1], [def 2], [cle*] and [wen 2].

9.5.31 G. Pisier observed that his criterion of B-convexity remains true when non-trivial Rademacher type is replaced by **non-trivial Walsh type**; see [bou 1, p. 174]. Finally, J. Bourgain (1988) got the same characterization in terms of **non-trivial Fourier type**. For Rademacher and Walsh functions, the proofs were based on the submultiplicativity of the underlying sequences of type ideal norms; see 4.2.6 and 6.2.7. In this book, we have succeeded in extending this approach to the trigonometric case as well; see 5.5.11.

REFERENCES: [pis 2], [pis 3], [bou 1], [bou 2] and [bou 8].

9.5.32 The notions of **Haar type** and **cotype** go back to G. Pisier (1975) who, however, formulated his results in terms of Walsh–Paley martingales. Further progress was achieved by J. Wenzel (1997). The main observation is that **non-trivial Haar type** characterizes J-convexity (=superreflexivity) in the same way as non-trivial Rademacher, Fourier and Walsh type characterize B-convexity.

REFERENCES: [pis 5], [bea 3] and [wen 3].

9.5.33 In **interpolation theory** we conclude from properties at the endpoints of an interval to properties valid for all interior points. The classical pattern is given by the scale $\{L_r(M,\mu),\ 1 \leq r \leq \infty\}$. Abstract intermediate spaces can be constructed by the real and complex methods, which are due to J.-L. Lions/J. Peetre (1961) and A. P. Calderón (1963), respectively. In particular, we refer to the pioneering work of J. Peetre.

The most important fact, for the purpose of our book, is the interpolation property of the Dirichlet ideal norms; see 3.4.9. A survey of results can be found in 9.1.15.

REFERENCES: [PEE], [cal], [lio*], [pee 1], [bea 1, p. 13], [BEA 1, pp. 75–81], [cre], [cwi*] and [pis*1].

9.5.34 In contrast to interpolation, **extrapolation theory** extends conclusions from properties at one point to properties at all points of a whole interval. Of course, this can only work under very strong assumptions.

In our treatise, we have used two results of this nature. One is the Benedek–Calderón–Panzone–Schwartz theorem (1961/62), which deals with vector-valued convolution operators. Secondly, we refer to extrapolation of martingale inequalities. This theory was founded by D. L. Burkholder in 1966. Further important contributions are due to Burkholder in collaboration with R. F. Gundy, as well as to B. Davis, P. Hitczenko and S. Geiss.

REFERENCES: [schw], [ben*], [bur 1], [bur*], [dav], [hit 1], [hit 2], [gei 1] and [gei 2].

9.5.35 The concept of **finite representability** was already known to A. Grothendieck (1956). Its first application in the theory of super-reflexivity and its naming go back to R. C. James (1972). The use of **ultraproducts** in Banach space theory was initiated by D. Dacunha-Castelle/J. L. Krivine (1972). A crucial observation is due to B. Maurey (1974), who showed that Y is finitely representable in X if and only if there exists an isometric embedding from Y into some ultrapower $X^{\mathcal{U}}$; see 0.8.3. Also quite useful is the fact that non-trivial ultrapowers of operators attain their norm. Thus one may work with equalities instead of inequalities, as done in the proofs of 3.7.12, 3.7.13 and 7.8.4.

REFERENCES: [gro 2, p. 109], [dac], [dac*], [jam 5, p. 896], [mau 2, p. 6], [ster] and [hei 2, p. 85].

9.6 Epilogue

The INTRODUCTION was written for readers with a basic knowledge of Banach space theory. Having no precise definitions at our disposal, we explained the objective of this treatise quite vaguely. Now, we welcome all those who have held out to the very end. Hopefully, there remain at least some enthusiasts. It is the intention of this EPILOGUE to discuss the *philosophical background* of our studies. Of course, most of the following statements are trivial for experts.

9.6.1 There are many reasons to extend classical theorems to the vector-valued setting:

- With every scalar-valued function u depending on the variables x and t we may associate the function $\boldsymbol{u} : t \to u(\cdot, t)$ whose values are functions of x. In this way, the heat equation $\frac{\partial u}{\partial t} = \frac{\partial^2 u}{\partial x^2}$ can be viewed as an ordinary differential equation $\frac{d\boldsymbol{u}}{dt} = T\boldsymbol{u}$, where T is an abstract operator.
- As a substitute for kernels, X-valued functions quite naturally occur when we look for representations of operators, say from $L_1(M, \mu)$ into X; see [DUN*1, pp. 498–511].
- The resolvent $(\zeta I - T)^{-1}$ of any operator $T \in \mathfrak{L}(X)$ is a holomorphic $\mathfrak{L}(X)$-valued function. The historical starting point of this observation was Fredholm's determinant theory.
- The non-linear map which assigns to $f \in L_2[-\pi, +\pi)$ the maximal function

$$S_n^*(f,t) := \max_{0 \le m \le n} |S_m(f,t)| \quad \text{with} \quad S_m(f,t) := \sum_{|k| \le m} (f, e_k) e_k(t)$$

can be 'linearized' by taking pointwise the l_∞^{n+1}-norm of the vector-valued function

$$\boldsymbol{S}_n(f,t) := \Big(S_0(f,t), \ldots, S_n(f,t) \Big).$$

That is,

$$S_n^*(f,t) = \| \boldsymbol{S}_n(f,t) | l_\infty^{n+1} \|.$$

- Most deterministic problems have a stochastic counterpart. Then scalar variables turn into random variables. The result is 'Probability in Banach spaces'.

All these facts motivate the study of vector-valued orthogonal expansions.

9.6.2 Local theory deals with those properties of Banach spaces and operators that are stable when passing to ultrapowers. Since this definition is quite abstract, we try to explain the underlying philosophy. Roughly speaking, localization reduces the study of infinite dimensional objects to that of finite dimensional objects. The most usual way is to consider **quantitative** relations that are assumed to hold uniformly for any choice of n elements and/or functionals. The involved constants may or may not depend on n. Typical examples are the parallelogram equality $\|x+y\|^2+\|x-y\|^2=2\|x\|^2+2\|y\|^2$ and the inequalities defining B–convexity; see 4.4.4.

Let us point out a significant difference. The parallelogram equality holds with respect to a given norm (metric), while B-convexity is preserved under equivalent renormings. Thus we must carefully distinguish between **isometric** and **isomorphic** statements. Using these concepts, we may say that our book is concerned with the *local and isomorphic theory of Banach spaces and their operators.* At first sight, it is very surprising that local conditions influence the global structure of an infinite dimensional Banach space or an operator of infinite rank. But this is indeed so, and James's theory yields an impressive example: *superreflexivity is a local property, while reflexivity is not.*

9.6.3 Our basic tools are **ideal norms** defined with the help of orthonormal systems. These are parameters which describe local properties of operators and spaces. Apart from extreme point arguments, there is no general method to evaluate these quantities in concrete cases. The main tools are experience and imagination. Luckily enough, for most purposes estimates suffice. Indeed, knowing the exact value of only one ideal norm provides almost no information about the object under consideration. We must consider the asymptotic behaviour of a whole sequence of ideal norms. Therefore, the main problem treated throughout the book is the following: *Decide whether two sequences of ideal norms are uniformly equivalent.* Here are some typical examples:

$$\max_{1\le m\le 2^n} \boldsymbol{\delta}(\mathcal{W}_m,\mathcal{W}_m)\asymp \boldsymbol{\varrho}_{\pm}(\mathcal{H}(\mathbb{D}_1^n),\mathcal{H}(\mathbb{D}_1^n)),$$

$$\boldsymbol{\varrho}(\mathcal{R}_n,\mathcal{I}_n)\asymp \boldsymbol{\varrho}(\mathcal{G}_n,\mathcal{I}_n) \quad\text{and}\quad \boldsymbol{\delta}(\mathcal{E}_n,\mathcal{E}_n)\asymp \boldsymbol{\kappa}(H_n).$$

Sometimes we only have weaker statements like

$$\boldsymbol{\varrho}_{\pm}(X|\mathcal{H}(\mathbb{D}_1^n),\mathcal{H}(\mathbb{D}_1^n))=O(1) \iff \boldsymbol{\kappa}(X|H_n)=O(1)$$

and

$$\boldsymbol{\varrho}(T|\mathcal{R}_n,\mathcal{I}_n)=o(\sqrt{n}) \iff \boldsymbol{\varrho}(T|\mathcal{E}_n,\mathcal{I}_n)=o(\sqrt{n}).$$

9.6.4 Suppose we have found that a given Banach space is isomorphic to some L_r, but the exact value of the index r remains unknown. Is it possible to determine r from the asymptotic behaviour of sequences of suitable ideal norms? The following are true:

$$\boldsymbol{\varrho}(L_r|\mathcal{G}_n,\mathcal{R}_n)\asymp 1 \quad \text{if } 1\le r<\infty \quad \text{and} \quad \boldsymbol{\varrho}(L_\infty|\mathcal{G}_n,\mathcal{R}_n)\asymp\sqrt{1+\log n},$$

$$\boldsymbol{\varrho}(L_r|\mathcal{S}_n,\mathcal{C}_n)\asymp 1 \quad \text{if } 1<r<\infty \quad \text{and} \quad \left.\begin{cases}\boldsymbol{\varrho}(L_1|\mathcal{S}_n,\mathcal{C}_n)\\ \boldsymbol{\varrho}(L_\infty|\mathcal{S}_n,\mathcal{C}_n)\end{cases}\right\}\asymp 1+\log n,$$

$$\boldsymbol{\varrho}(L_r|\mathcal{E}_n,\mathcal{I}_n)\asymp n^{|1/r-1/2|} \quad \text{if } 1\le r\le\infty,$$

$$\boldsymbol{\varrho}(L_r|\mathcal{R}_n,\mathcal{I}_n)\asymp n^{1/r-1/2} \quad \text{and} \quad \boldsymbol{\varrho}(L_r|\mathcal{I}_n,\mathcal{R}_n)\asymp 1 \quad \text{if } 1<r\le 2,$$

$$\boldsymbol{\varrho}(L_r|\mathcal{R}_n,\mathcal{I}_n)\asymp 1 \quad \text{and} \quad \boldsymbol{\varrho}(L_r|\mathcal{I}_n,\mathcal{R}_n)\asymp n^{1/2-1/r} \quad \text{if } 2\le r<\infty,$$

$$\boldsymbol{\varrho}(L_1|\mathcal{R}_n,\mathcal{I}_n)\asymp\sqrt{n} \quad \text{and} \quad \boldsymbol{\varrho}(L_1|\mathcal{I}_n,\mathcal{R}_n)\asymp 1,$$

$$\boldsymbol{\varrho}(L_\infty|\mathcal{R}_n,\mathcal{I}_n)\asymp\sqrt{n} \quad \text{and} \quad \boldsymbol{\varrho}(L_\infty|\mathcal{I}_n,\mathcal{R}_n)\asymp\sqrt{n}.$$

Therefore the ideal norms $\boldsymbol{\varrho}(\mathcal{G}_n,\mathcal{R}_n)$ and $\boldsymbol{\varrho}(\mathcal{S}_n,\mathcal{C}_n)$ are of no use for our purpose, whereas $\boldsymbol{\varrho}(\mathcal{E}_n,\mathcal{I}_n)$ is unable to distinguish between L_r and $L_{r'}$. Using $\boldsymbol{\varrho}(\mathcal{R}_n,\mathcal{I}_n)$ and $\boldsymbol{\varrho}(\mathcal{I}_n,\mathcal{R}_n)$ in tandem, however, we succeed.

Next, let us assume that we have either L_r or S_r. In this case, the above method yields the missing exponent r, and $\boldsymbol{\varrho}_\pm(\mathcal{R}_n\overset{\text{full}}{\otimes}\mathcal{R}_n,\mathcal{R}_n\overset{\text{full}}{\otimes}\mathcal{R}_n)$ can distinguish between L_r and S_r.

On the other hand, if the Banach spaces X and Y are finitely representable in each other, then we have $\boldsymbol{\varrho}(X|\mathcal{B}_n,\mathcal{A}_n)=\boldsymbol{\varrho}(Y|\mathcal{B}_n,\mathcal{A}_n)$ for all orthonormal systems $\mathcal{A}_n$ and $\mathcal{B}_n$. In particular, there is no chance of distinguishing l_r from $L_r(0,1)$. This shows the limits of the local theory. Nevertheless: *Many Banach spaces can be classified with the help of orthonormal systems.*

9.6.5 We now consider the reverse situation. The problem consists in realizing the difference between orthonormal systems $\mathcal{A}=(a_1,a_2,\dots)$ and $\mathcal{B}=(b_1,b_2,\dots)$ by the growth of $\boldsymbol{\varrho}(X|\mathcal{B}_n,\mathcal{A}_n)$ or $\boldsymbol{\varrho}(X|\mathcal{A}_n,\mathcal{B}_n)$, where X is a suitable Banach space. Of course, l_∞, c_0 and $C[0,1]$ are the best 'test spaces'. For example, $\boldsymbol{\varrho}(l_\infty|\mathcal{E}_n,\mathcal{W}_n)\succ\sqrt{1+\log n}$ shows that $\mathcal{E}_n$ and $\mathcal{W}_n$ behave differently.

Recall from 3.1.7 that $\boldsymbol{\varrho}(\mathcal{B}_n,\mathcal{A}_n)=1$ whenever $\mathcal{A}_n$ and $\mathcal{B}_n$ have identical distributions. Thus, we cannot discover all differences. Another example is given by the Rademacher system $\mathcal{R}_n$ and the lacunary system $\mathcal{S}(\mathbb{L}_n)=(\sqrt{2}\sin 2t,\dots,\sqrt{2}\sin 2^n t)$, which behave similarly with respect to Banach spaces; see 5.9.20. Moreover, we do not know whether the exponents α and β of any Jacobi system can be reconstructed from suitable ideal norms associated with $\mathcal{L}_n^{(\alpha,\beta)}$. Nevertheless: *Many orthonormal systems can be classified with the help of Banach spaces.*

9.6.6 There are numerous parallel results for concrete orthonormal systems. So, for $1 < p < p_0 < 2$, we have

$$\mathfrak{RT}_{p_0} \subset \mathfrak{RT}_{p_0}^{weak} \subset \mathfrak{RT}_p \quad \text{and} \quad \mathfrak{AT}_{p_0} \subset \mathfrak{AT}_{p_0}^{weak} \subset \mathfrak{AT}_p,$$

$$\mathfrak{ET}_{p_0} \subset \mathfrak{ET}_{p_0}^{weak} \subset \mathfrak{ET}_p \quad \text{and} \quad \mathfrak{WT}_{p_0} \subset \mathfrak{WT}_{p_0}^{weak} \subset \mathfrak{WT}_p.$$

Another example is the following chain of equalities:

$$\bigcup_{1<p\leq 2} \mathsf{RT}_p = \bigcup_{1<p\leq 2} \mathsf{ET}_p = \bigcup_{1<p\leq 2} \mathsf{WT}_p = \mathsf{B}. \qquad (*)$$

However, we stress that these relations were not obtained within a general theory, but needed specific proofs. Thus the main use of this parallelism is to provide information about the common structure of different theories. Having proved a statement for Rademacher type, we may try to get the corresponding statement for Haar type. In this way, $\mathsf{RT}_2 \cap \mathsf{RC}_2 = \mathsf{H}$ suggests $\mathsf{AT}_2 \cap \mathsf{AC}_2 = \mathsf{H}$, which is true. On the other hand, the analogue of $(*)$ fails:

$$\xcancel{\bigcup_{1<p\leq 2} \mathsf{AT}_p = \mathsf{B}}\,.$$

Nevertheless, we learn from this parallelism that Banach spaces of Haar subtype should have interesting properties, and this is indeed so:

$$\bigcup_{1<p\leq 2} \mathsf{AT}_p = \mathsf{SR}.$$

9.6.7 Our major aim is to convince the reader that instead of asking: Which is more important, *spaces or operators?* it is much better to deal with *spaces and operators!*

9.6.8 H. Minkowski (1896) introduced the concept of a finite dimensional normed linear space when he developed his geometry of numbers; [MIN, p. 102]. This was the birth of Banach space geometry. However, the decisive step to the infinite dimensional setting became necessary when Banach spaces were needed as the background on which operators live. Clearly, large parts of the theory of convex bodies can be developed without any use of operators. On the other hand, we stress that the titles of Banach's classical works *Sur les opérations dans les ensembles abstraits et leur application aux équations intégrales* and *Théorie des opérations linéaires* refer to operators and not to spaces.

Of course, operators cannot exist without the underlying spaces. Given linear equations in infinitely many unknowns

$$\sum_{k=1}^{\infty} \sigma_{hk} \xi_k = \eta_h,$$

integral equations

$$\int_M K(s,t)f(t)\,dt = g(s) \quad \text{and} \quad f(s) + \int_M K(s,t)f(t)\,dt = g(s)$$

or differential equations

$$\operatorname{div}(p \operatorname{grad} u) + qu = f,$$

in each case we must specify which sequences or functions are under consideration. In a sense, spaces are the base, the static elements. The dynamics comes from operators.

9.6.9 The subject of this book is well suited to illustrate the fruitful interplay which we have in mind. Every operator ideal $\mathfrak{A}$ determines a class of Banach spaces:

$$\mathsf{A} := \{\, X \in \mathsf{L} : I_X \in \mathfrak{A} \,\}.$$

Thereby it happens quite often that different ideals yield one and the same kind of spaces. So we have

$$\begin{array}{ccc} \mathfrak{RT}_2 \cap \mathfrak{RC}_2 & & \mathfrak{AT}_2 \cap \mathfrak{AC}_2 \\ & \searrow \quad \swarrow & \\ & \mathsf{H} & \end{array}$$

Indeed, there even exist infinitely many ideals $\mathfrak{A}$ with $\mathsf{A} = \mathsf{H}$ that can be obtained by interpolation. As another example, we mention that the operator ideals $\mathfrak{L}[n^{-\theta}\boldsymbol{\delta}(\mathcal{R}_n, \mathcal{R}_n)]$ increase with $0 \le \theta < 1/2$, but all of them yield the class of B-convex Banach spaces. Of particular interest are the equalities

$$\bigcup_{1<p\le 2} \mathsf{RT}_p = \mathsf{RT}, \qquad \bigcup_{1<p\le 2} \mathsf{ET}_p = \mathsf{ET}, \qquad \bigcup_{1<p\le 2} \mathsf{AT}_p = \mathsf{AT}$$

which correspond to the strict inclusions

$$\bigcup_{1<p\le 2} \mathfrak{RT}_p \subset \mathfrak{RT}, \qquad \bigcup_{1<p\le 2} \mathfrak{ET}_p \subset \mathfrak{ET}, \qquad \bigcup_{1<p\le 2} \mathfrak{AT}_p \subset \mathfrak{AT}.$$

Thus, as in bifurcation theory, one point splits into various branches when we pass from spaces to operators. This yields a refinement of the whole picture.

9.6.10 All classes of Banach spaces treated in this book were obtained via operator ideals. For example,

super weakly compact operators	$\longrightarrow$	J-convexity,
operators of Rademacher subtype	$\longrightarrow$	B-convexity,
operators of Rademacher subcotype	$\longrightarrow$	MP-convexity.

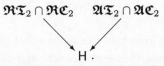

We also refer to the concepts of uniform p-smoothness/smoothability and uniform q-convexity/convexifiability, which have been extended from spaces to operators.

9.6.11 Most ideals considered in this book are very large, since they contain, among others, all operators acting on Hilbert spaces. Thus it makes no sense to search for good spectral properties as in the case of 2-summing operators; see [PIE 3]. However, as described above, these large ideals can be used to define various classes of Banach spaces. In this way, we are able to systematize, at least to some extent, the huge collection of all Banach spaces.

For specialists, classification of Banach spaces is a mathematical problem in its own right. Since this undertaking may be considered a waste of energy, it is more convincing to point out some consequences. What is the use of knowing that a space is superreflexive? Generally speaking, we are looking for theorems that hold only if the underlying spaces enjoy additional properties.

9.6.12 In fact, spectral theory and Banach space geometry were developed quite separately. The sole geometric tool in the theory of compact linear operators is the Riesz lemma, which says that Banach spaces with a compact unit ball are finite dimensional. To study Fredholm operators one merely needs the observation that finite dimensional and finite codimensional subspaces are complemented. The Riesz–Dunford functional calculus even works in abstract Banach algebras, which means that the underlying spaces have disappeared. Clearly, these are the most famous parts of operator theory. However, we really hope that this is not the end of the story.

Here are some examples to illustrate what we mean:

- If a Banach space X is B-convex, then the law of large numbers generalizes to X-valued sequences of random variables.
- If a Banach space X has weak Kwapień type p, then the eigenvalues of any nuclear operator on X converge to zero as $O(n^{-2/p'})$.
- If a Banach space X has weak Kwapień type p, then the approximation numbers of any 2-summing operator from X into an arbitrary Banach space Y converge to zero as $O(n^{-1/p'})$.
- If a Banach space X has Rademacher cotype 2, then all operators from l_∞ into X are even 2-summing.

REFERENCES: [bec], [mau 4, p. 116], [LIN*2, p. 96], [koe*b], [pie 4] and [pie 5].

9.6.13 UMD is by far the most useful and flexible class of Banach spaces. Though rather small, it contains the classical spaces L_r, their non-commutative analogues S_r and the Lorentz spaces $L_{r,w}$ provided that $1 < r, w < \infty$. The most prominent spaces lacking the UMD-property are L_1, $C(K)$ and $\mathfrak{L}(l_2)$.

Many results about scalar-valued functions carry over to functions having values in a UMD-space X. Let $1 < r < \infty$.

- X-valued martingale difference sequences are unconditional.
- Not only Hilbert transforms, but many other singular integral operators admit X-valued extensions.
- The classical multiplier theorems of Mikhlin, Littlewood–Paley and Marcinkiewicz generalize to the X-valued case.
- Certain orthonormal bases are equivalent in $[L_r, X]$. This is true, among others, for the Haar system and the Franklin system. Moreover, all Jacobi systems are mutually equivalent.
- Various characterizations of scalar-valued functions belonging to the Hardy space H_1 also hold for $[H_1, X]$, and Fefferman's duality theorem extends, $[H_1, X]' = [BMO, X']$.
- Under the additional assumption that X has an unconditional basis, Carleson's theorem about almost everywhere convergence of Fourier series in L_r is true in $[L_r, X]$ as well.
- There is a spectral decomposition for power-bounded, invertible operators on X.
- UMD-spaces can successfully be applied in the theory of some specific semigroups of operators and for solving the abstract Cauchy problem $\boldsymbol{u}'(t) + T\boldsymbol{u}(t) = \boldsymbol{f}(t)$ under the initial condition $\boldsymbol{u}(0) = o$ in $[L_r, X]$.

REFERENCES: [mcco], [rub 1], [ber*1], [bou 7], [dor*], [bla], [fig 1], [zim], [fig 2] and [koe*a].

9.6.14 Another facet of the fruitful relation *spaces* ↔ *operators* is the fact that components $\mathfrak{A}(X, Y)$ of a Banach operator ideal provide a rich source of concrete finite and infinite dimensional Banach spaces, which has not yet been really exploited. The most important examples are the Schatten–von Neumann spaces S_r^n and S_r; see Section 9.3.

9.6.15 We have just presented some paradigms illustrating the interplay between spaces and operators. But this is not enough! Thus it would be a worthwhile aim to remedy the situation by using more intensively the deep knowledge about Banach space geometry:

Operators are needed to understand spaces,
spaces are needed to understand operators.

Summaries

For the convenience of the reader, we finally collect some important results and examples.

Classes of Banach spaces

As described in Chapter 1, the following process yields various classes of Banach spaces. Given any sequence of ideal norms $\boldsymbol{\alpha}_1, \boldsymbol{\alpha}_2, \ldots$, we let

$$\boldsymbol{\alpha}_n(1) := \inf\Big\{\, \boldsymbol{\alpha}_n(X) \,:\, X \in \mathsf{L} \,\Big\} = \boldsymbol{\alpha}_n(I_{\mathbb{K}})$$

and

$$\boldsymbol{\alpha}_n(\infty) := \sup\Big\{\, \boldsymbol{\alpha}_n(X) \,:\, X \in \mathsf{L} \,\Big\}.$$

For simplicity, we assume that $\boldsymbol{\alpha}_n(I_{\mathbb{K}}) = 1$. Then there is a smallest (non-trivial) class

$$\mathsf{L}[\boldsymbol{\alpha}_n] := \Big\{\, X \in \mathsf{L} \,:\, \boldsymbol{\alpha}_n(X) = O(1) \,\Big\}$$

and a largest counterpart

$$\mathsf{L}_0[\boldsymbol{\alpha}_n(\infty)^{-1}\boldsymbol{\alpha}_n] := \Big\{\, X \in \mathsf{L} \,:\, \boldsymbol{\alpha}_n(X) = o(\boldsymbol{\alpha}_n(\infty)) \,\Big\},$$

where O and o are the Landau symbols. Moreover, we can define many classes in between. For example,

$$\mathsf{L}^{\theta}[\boldsymbol{\alpha}_n] := \Big\{\, X \in \mathsf{L} \,:\, \boldsymbol{\alpha}_n(X) = O(\boldsymbol{\alpha}_n(\infty)^{\theta}) \,\Big\}$$

with $0 < \theta < 1$.

The most important facts and problems are listed in a table.

$\mathcal{A}_n$	$\mathcal{G}_n$	$\mathcal{R}_n$	$\mathcal{H}(\mathbb{D}_1^n)$	$\mathcal{W}_n$	$\mathcal{E}_n$
$\boldsymbol{\varrho}(\infty\|\mathcal{A}_n, \mathcal{I}_n)$	$\sqrt{n}$	$\sqrt{n}$	$\sqrt{n}$	$\sqrt{n}$	$\sqrt{n}$
subtype	B	B	SR	B	B
type $1<p<2$	**(1)**	**(1)**	**(5)**	**(7)**	**(7)**
type 2	**(1)**	**(1)**	**(3)**	H	H
type 2+cotype 2	H	H	H	H	H
cotype 2	**(2)**	**(2)**	**(4)**	H	H
cotype $2<q<\infty$	**(2)**	**(2)**	**(6)**	**(7)**	**(7)**
subcotype	MP	MP	SR	B	B
$\boldsymbol{\varrho}(\infty\|\mathcal{I}_n, \mathcal{A}_n)$	$\sqrt{\frac{n}{1+\log n}}$	$\sqrt{n}$	$\sqrt{n}$	$\sqrt{n}$	$\sqrt{n}$
$\delta_n := \boldsymbol{\delta}(\infty\|\mathcal{A}_n, \mathcal{A}_n)$	$\sqrt{n}$	$\sqrt{n}$	1	$1+\log n$ (*)	$1+\log n$
$\boldsymbol{\delta}(X\|\mathcal{A}_n, \mathcal{A}_n) = O(1)$	B	B	L	UMD	UMD
intermediate	B	B	L	**(8)**	**(8)**
$\boldsymbol{\delta}(X\|\mathcal{A}_n, \mathcal{A}_n) = o(\delta_n)$	B	B	L	**(9)**	**(9)**
$\varrho_n := \boldsymbol{\varrho}_\pm(\infty\|\mathcal{A}_n, \mathcal{A}_n)$	$\sqrt{n}$	$\sqrt{n}$	n	$1+\log n$	$1+\log n$
$\boldsymbol{\varrho}_\pm(X\|\mathcal{A}_n, \mathcal{A}_n) = O(1)$	L	L	UMD	H	H
intermediate	L	L	**(8)**	**(8)**	**(8)**
$\boldsymbol{\varrho}_\pm(X\|\mathcal{A}_n, \mathcal{A}_n) = o(\varrho_n)$	L	L	**(9)**	B	B

REMARKS.

(*) Replace $\boldsymbol{\delta}(\mathcal{W}_n, \mathcal{W}_n)$ by $\max_{m\leq n} \boldsymbol{\delta}(\mathcal{W}_m, \mathcal{W}_m)$.

(1) Recall from 4.16.5 that $p(X) = \min \operatorname{spec}(X)$.

(2) Recall from 4.16.5 that $q(X) = \max \operatorname{spec}(X)$.

(3) These are the uniformly 2-smoothable Banach spaces; 7.9.21.

(4) These are the uniformly 2-convexifiable Banach spaces; 7.9.20.

(5) Uniform p-smoothable Banach spaces have weak Haar type p.

(6) Uniform q-convexifiable Banach spaces have weak Haar cotype q.

(7) No geometric characterization is known.

(8) Almost nothing is known.

(9) We conjecture that this is the class of superreflexive Banach spaces.

Relations between classes of Banach spaces

$$\mathsf{B}=\mathsf{KT}=\mathsf{ET}=\mathsf{WT}=\mathsf{GT}=\mathsf{RT}=\mathsf{QGP}=\mathsf{QRP}=\mathsf{GP}=\mathsf{RP}$$ 4.14.9, 5.6.30, 6.3.13

$$\mathsf{MP}=\mathsf{GR}=\mathsf{QGR}=\mathsf{RC}=\mathsf{GC}$$ 4.8.22

$$\mathsf{RT}_2\cap\mathsf{RC}_2=\mathsf{H}$$ 4.10.7

$$\bigcup_{1<p\le 2}\mathsf{RT}_p=\bigcup_{1<p<2}\mathsf{RT}_p^{weak}=\mathsf{RT}$$ 4.4.1

$$\bigcup_{2\le q<\infty}\mathsf{RC}_q=\bigcup_{2<q<\infty}\mathsf{RC}_q^{weak}=\mathsf{RC}$$ 4.8.21

$$\mathsf{ET}_2=\mathsf{EC}_2=\mathsf{H}$$ 5.9.4

$$\bigcup_{1<p\le 2}\mathsf{ET}_p=\bigcup_{1<p<2}\mathsf{ET}_p^{weak}=\mathsf{ET}$$ 5.6.27

$$\mathsf{WT}_2=\mathsf{WC}_2=\mathsf{H}$$ 6.4.3

$$\bigcup_{1<p\le 2}\mathsf{WT}_p=\bigcup_{1<p<2}\mathsf{WT}_p^{weak}=\mathsf{WT}$$ 6.4.17

$$\mathsf{AT}_2\cap\mathsf{AC}_2=\mathsf{H}$$ 7.5.17

$$\bigcup_{1<p\le 2}\mathsf{AT}_p=\bigcup_{1<p<2}\mathsf{AT}_p^{weak}=\mathsf{AT}$$ 7.7.8

$$\mathsf{AT}=\mathsf{AC}=\mathsf{SR}$$ 7.8.16

$$\mathsf{RAT}_2\cap\mathsf{RAC}_2=\mathsf{UMD}$$ 8.6.27

Relations between operator ideals

$$\mathfrak{RT}_2\subset\mathfrak{RT}_{p_0}\subset\mathfrak{RT}_{p_0}^{weak}\subset\mathfrak{RT}_p \text{ if } 1<p<p_0<2$$ 4.3.10

$$\mathfrak{ET}_2\subset\mathfrak{ET}_{p_0}\subset\mathfrak{ET}_{p_0}^{weak}\subset\mathfrak{ET}_p \text{ if } 1<p<p_0<2$$ 5.6.23

$$\mathfrak{WT}_2\subset\mathfrak{WT}_{p_0}\subset\mathfrak{WT}_{p_0}^{weak}\subset\mathfrak{WT}_p \text{ if } 1<p<p_0<2$$ 6.3.7

$$\mathfrak{AT}_2\subset\mathfrak{AT}_{p_0}\subset\mathfrak{AT}_{p_0}^{weak}\subset\mathfrak{AT}_p \text{ if } 1<p<p_0<2$$ 7.5.12

$$\mathfrak{RC}_2\subset\mathfrak{RC}_{q_0}\subset\mathfrak{RC}_{q_0}^{weak}\subset\mathfrak{RC}_q \text{ if } 2<q_0<q<\infty$$ 4.5.10

$$\bigcup_{2\le q<\infty}\mathfrak{RC}_q=\bigcup_{2<q<\infty}\mathfrak{RC}_q^{weak}=\bigcup_{2\le q<\infty}\mathfrak{GC}_q=\bigcup_{2<q<\infty}\mathfrak{GC}_q^{weak}$$ 4.10.2

$$\mathfrak{GR}\subset\mathfrak{QGR}=\mathfrak{RC}=\mathfrak{GC}$$ 4.10.11

$$\bigcup_{1<p\le 2}\mathfrak{WT}_p = \bigcup_{1<p<2}\mathfrak{WT}_p^{weak} \subset \mathfrak{RP} \subset \mathfrak{QRP} = \mathfrak{WT} \qquad 6.4.16$$

$$\mathfrak{KT} = \mathfrak{ET} = \mathfrak{WT} = \mathfrak{GT} = \mathfrak{RT} = \mathfrak{QGP} = \mathfrak{QRP}$$

4.14.8, 5.6.29, 6.3.12

$$\mathfrak{L} \circ \mathfrak{UT} = \mathfrak{AT} = \mathfrak{SR} = \mathfrak{AC} = \mathfrak{UC} \circ \mathfrak{L} \qquad 7.10.25$$

Inequalities between ideal norms

$$\boldsymbol{\kappa}_n \le \boldsymbol{\varrho}(\mathcal{I}_n, \mathcal{R}_n) \circ \boldsymbol{\varrho}(\mathcal{R}_n, \mathcal{I}_n) \qquad 4.10.5$$

$$\boldsymbol{\varrho}(\mathcal{I}_n, \mathcal{R}_n) \le \boldsymbol{\varrho}'(\mathcal{R}_n, \mathcal{I}_n) \qquad 4.2.4$$

$$\boldsymbol{\varrho}(\mathcal{R}_n, \mathcal{I}_n) \le c\,\boldsymbol{\varrho}(\mathcal{E}_n, \mathcal{I}_n) \qquad 5.9.2$$

$$\boldsymbol{\varrho}(\mathcal{R}_n, \mathcal{I}_n) \le \boldsymbol{\varrho}(\mathcal{W}_n, \mathcal{I}_n) \qquad 6.4.1$$

$$2^{-n/2}\,\boldsymbol{\varrho}(\mathcal{I}_{2^n}, \mathcal{R}_{2^n}) \le n^{-1/2}\,\boldsymbol{\varrho}(\mathcal{R}_n, \mathcal{I}_n) \qquad 4.2.17$$

$$2^{-n/2}\,\boldsymbol{\varrho}(\mathcal{E}_{2^n}, \mathcal{I}_{2^n}) \le c\,n^{-1/2}\,\boldsymbol{\varrho}(\mathcal{R}_n, \mathcal{I}_n) \qquad 4.11.18$$

$$2^{-n/2}\,\boldsymbol{\varrho}(\mathcal{W}_{2^n}, \mathcal{I}_{2^n}) \le n^{-1/2}\,\boldsymbol{\varrho}(\mathcal{R}_n, \mathcal{I}_n) \qquad 4.2.16$$

$$2^{-n/2}\boldsymbol{\varrho}(\mathcal{R}_{2^n}, \mathcal{I}_{2^n}) \le c\,n^{-1/2}\boldsymbol{\delta}(\mathcal{R}_n, \mathcal{R}_n) \qquad 4.13.2$$

$$2^{-n/2}\boldsymbol{\varrho}(\mathcal{H}(\mathbb{D}_1^{2^n}), \mathcal{I}(\mathbb{D}_1^{2^n})) \le 3\,n^{-1/2}\boldsymbol{\varrho}(\mathcal{I}(\mathbb{D}_0^n), \mathcal{H}(\mathbb{D}_0^n)) \qquad 7.5.19$$

$$2^{-n/2}\boldsymbol{\delta}_{2^n} \overset{?}{\le} c\,n^{-1/2}\boldsymbol{\varrho}(\mathcal{R}_n, \mathcal{I}_n) \qquad \text{Problem 4.13.7}$$

$$\boldsymbol{\delta}(\mathcal{R}_n, \mathcal{R}_n) \le c\,\boldsymbol{\varrho}(\mathcal{R}_{2^n}, \mathcal{I}_{2^n}) \qquad 4.13.1$$

$$\boldsymbol{\delta}(\mathcal{R}_n, \mathcal{R}_n) \not\le c_p\,\boldsymbol{\varrho}_p^{(2)}(\mathcal{R}_{2^n}, \mathcal{I}_{2^n}) \ \text{ if } \ 1<p<2$$

$$\boldsymbol{\delta}(\mathcal{R}_n, \mathcal{R}_n) \overset{?}{\le} c_p\,\boldsymbol{\varrho}_p^{(p')}(\mathcal{E}_{2^n}, \mathcal{I}_{2^n}) \qquad \text{6.4.15, Remark}$$

$$\boldsymbol{\delta}(\mathcal{R}_n, \mathcal{R}_n) \le c_p\,\boldsymbol{\varrho}_p^{(p')}(\mathcal{W}_{2^n}, \mathcal{I}_{2^n}) \qquad 6.4.15$$

$$\boldsymbol{\varrho}(\mathcal{G}_n, \mathcal{R}_n) \le c_q\,\boldsymbol{\varrho}_2^{(q)}(\mathcal{I}_n, \mathcal{R}_n) \qquad 4.8.16$$

Diagonal operators

The diagonal operators

$$\begin{aligned} B_\alpha &: (x_k) \to ((1+\log k)^{-\alpha} x_k), \\ C_\alpha &: (x_k) \to (k^{-\alpha} x_k), \\ D_\alpha &: (x_k) \to (2^{-k\alpha} x_k). \end{aligned}$$

act on $[l_2, l_\infty^{2^k}]$; see 1.2.12.

$$\boldsymbol{\varrho}(B_\alpha|\mathcal{R}_n,\mathcal{I}_n) \asymp n^{1/2}(1+\log n)^{-\alpha} \qquad 4.4.2$$

$$\boldsymbol{\varrho}(C_\alpha|\mathcal{R}_n,\mathcal{I}_n) \asymp n^{1/2-\alpha} \qquad 4.6.21$$

$$\boldsymbol{\varrho}(C_\alpha|\mathcal{I}_n,\mathcal{R}_n) \asymp n^{1/2}(1+\log n)^{-\alpha} \qquad 4.6.2$$

$$\boldsymbol{\varrho}(C_\alpha|\mathcal{E}_n,\mathcal{I}_n) \asymp n^{1/2}(1+\log n)^{-\alpha} \qquad 5.6.28$$

$$\boldsymbol{\varrho}(C_\alpha|\mathcal{W}_n,\mathcal{I}_n) \asymp n^{1/2}(1+\log n)^{-\alpha} \qquad 6.3.11$$

$$\boldsymbol{\varrho}(C_\alpha|\mathcal{G}_n,\mathcal{R}_n) \asymp (1+\log n)^{1/2-\alpha} \qquad 4.8.14$$

$$\boldsymbol{\varrho}(D_\alpha|\mathcal{I}_n,\mathcal{R}_n) \asymp n^{1/2-\alpha} \qquad 4.10.4$$

$$\boldsymbol{\varrho}(D_\alpha|\mathcal{I}_n,\mathcal{G}_n) \asymp \frac{n^{1/2-\alpha}}{(1+\log n)^{1/2}} \qquad 4.10.4$$

$$\boldsymbol{\delta}(C_\alpha|\mathcal{R}_n,\mathcal{R}_n) \asymp n^{1/2-\alpha} \qquad 4.12.14$$

$$\boldsymbol{\delta}(C_\alpha|\mathcal{G}_n,\mathcal{G}_n) \asymp n^{1/2-\alpha} \qquad 4.12.14$$

Summation operators

$$\|\Sigma_n : l_1^n \to l_\infty^n|\mathfrak{H}\| \asymp 1+\log n \qquad 2.4.14$$

$$\boldsymbol{\varrho}(\Sigma_n : l_1^n \to l_\infty^n|\mathcal{R}_n,\mathcal{I}_n) \asymp 1 \qquad 4.3.17$$

$$\boldsymbol{\delta}(\Sigma_n : l_1^n \to l_\infty^n|\mathcal{R}_n,\mathcal{R}_n) \asymp 1 \qquad 4.13.1$$

$$\boldsymbol{\varrho}(\Sigma_n : l_1^n \to l_\infty^n|\mathcal{E}_n,\mathcal{I}_n) \asymp \text{???} \qquad \text{open}$$

$$\boldsymbol{\kappa}(\Sigma_n : l_1^n \to l_\infty^n|H_n) \asymp 1+\log n \qquad 2.4.4$$

$$\boldsymbol{\delta}(\Sigma_n : l_1^n \to l_\infty^n|\mathcal{E}_n,\mathcal{E}_n) \asymp 1+\log n \qquad 5.3.8$$

$$\boldsymbol{\varrho}(\Sigma_n : l_1^n \to l_\infty^n|\mathcal{W}_n,\mathcal{I}_n) \asymp \text{???} \qquad \text{open}$$

$$\boldsymbol{\varrho}(\Sigma_{2^n} : l_1^{2^n} \to l_\infty^{2^n}|\mathcal{I}(\mathbb{D}_0^n),\mathcal{H}(\mathbb{D}_0^n)) \asymp \sqrt{n} \qquad 7.6.14$$

$$\boldsymbol{\varrho}_\pm(\Sigma_{2^n} : l_1^{2^n} \to l_\infty^{2^n}|\mathcal{H}(\mathbb{D}_0^n),\mathcal{H}(\mathbb{D}_0^n)) \overset{?}{\asymp} n \qquad 8.5.23$$

$$\boldsymbol{\varrho}(\Sigma_{2^n} : l_1^{2^n} \to l_\infty^{2^n}|\mathcal{R}_n \otimes \mathcal{H}(\mathbb{D}_0^n),\mathcal{H}(\mathbb{D}_0^n)) \asymp \sqrt{n} \qquad 8.6.23$$

$$\boldsymbol{\varrho}(\Sigma_{2^n} : l_1^{2^n} \to l_\infty^{2^n}|\mathcal{H}(\mathbb{D}_0^n),\mathcal{R}_n \otimes \mathcal{H}(\mathbb{D}_0^n)) \asymp 1 \qquad 8.6.12$$

$$\boldsymbol{\delta}(\Sigma_{2^n} : l_1^{2^n} \to l_\infty^{2^n}|\mathcal{R}_n \otimes \mathcal{H}(\mathbb{D}_0^n),\mathcal{R}_n \otimes \mathcal{H}(\mathbb{D}_0^n)) \overset{?}{\asymp} \sqrt{n} \qquad 8.6.24$$

List of symbols

Conventions

If not stated otherwise, we use the following conventions:

(1) The lowercase letter o denotes the zero element in a Banach space, while the uppercase letter O stands for the zero operator.

(2) We also use o and O to denote the Landau symbols, defined in 0.10.1. However, there is no risk of confusion.

(3) Letters r, u and v denote exponents and indices such that $1 \leq r, u, v \leq \infty$, while p and q are assumed to satisfy the extra conditions $1 \leq p \leq 2$ and $2 \leq q \leq \infty$. The limiting cases are sometimes excluded and sometimes not.

(4) To distinguish between scalar-valued and vector-valued functions, we use the symbols $f, g, \ldots$ and $\boldsymbol{f}, \boldsymbol{g}, \ldots$, respectively.

(5) When Banach function spaces $L_r(M, \mu)$ occur as examples, we simply write L_r and assume that they are infinite dimensional.

(6) When we deal with Khintchine constants $K'_p(\mathcal{A}_n)$ and $K_q(\mathcal{A}_n)$ for some orthonormal system $\mathcal{A}_n = (a_1, \ldots, a_n)$ in a Hilbert space $L_2(M, \mu)$, then μ is assumed to be a probability measure.

(7) The symbol $\mathcal{P}_n$ is used as a synonym of $\mathcal{P}_n(\pm 1)$ and $\mathcal{P}_n(e^{i\tau})$. Similarly, $\mathbb{E}^n$ stands for $\mathbb{E}_2^n$ and $\mathbb{T}^n$.

(8) If $\mathcal{A}_n = (a_1, \ldots, a_n)$ is a system of characters, then we assume that its elements are pairwise different, $a_h \neq a_k$ for $h \neq k$.

(9) Bold Gothic capitals (possibly decorated with super- or subscripts) stand for operator ideals, and the associated classes of Banach spaces are denoted by the same bold sans serif capitals; see 1.3.1.

Miscellaneous

$\mathbb{A} \leq n$	subset of $\{1, \dots, n\}$	6.1.1
$\mathbb{C}$	complex field	
$\mathbb{D}_m^n$	dyadic tree	7.3.1
$\mathbb{D}_m^n(t)$	branch of dyadic tree	7.3.2
$D(u)$	distribution function of γ	4.8.15
$\mathbb{E}_n$	cyclic group	3.8.3
f^*	non-increasing rearrangement	0.5.1
$\lvert\mathbb{F}\rvert$	cardinality of a finite set $\mathbb{F}$	0.3.1
$\mathcal{F}(\mathbb{N})$	collection of finite subsets of $\mathbb{N}$	6.1.1
$\mathbb{K}$	synonym of $\mathbb{R}$ and $\mathbb{C}$	
$\mathbb{L}^{mn}$	set of contracting matrices	2.1.5
$\mathbb{L}_n$	lacunary set $\{2^1, 2^2, \dots, 2^n\}$	5.9.20
$\mathbb{N}$	set of natural numbers without 0	
$\mathbb{N}_0$	set of natural numbers with 0	
O, o	Landau symbols	0.10.1
$\mathbb{R}$	real field	
$\mathbb{R}_+$	set of positive real numbers	
$\mathbb{S}^n$	$(n-1)$-dimensional unit sphere	4.7.9
$\mathbb{T}$	circle group	3.8.3
$\mathbb{U}^{mn}$	set of isometric matrices	2.1.6
$\mathbb{Z}$	set of integers	
γ, γ^n, $\gamma_{\mathbb{R}}^n$, $\gamma_{\mathbb{C}}^n$	Gaussian measures	4.7.1
$\Delta_k^{(j)}$	dyadic interval	7.2.2
χ^n	Haar measure on $\mathbb{U}^n$	4.12.9
ω^n	invariant measure on $\mathbb{S}^n$	4.7.9
$\asymp$, $\prec$	asymptotic relations	0.10.1

Linear spaces

$S_0(M, \mu)$	space of simple functions	0.4.3
$[WP_n^0, X]$	collection of Walsh–Paley martingales	7.2.3

Banach spaces

$C(\mathbb{G})$	space of continuous functions	3.8.2
$C_0(\mathbb{G})$	space of functions vanishing at infinity	5.8.2
$C_{exp}(\mathbb{R}^n)$	space of slowly growing functions	4.7.15
$\mathrm{COD}(X)$	set of finite codimensional subspaces	0.2.2
$\mathrm{DIM}(X)$	set of finite dimensional subspaces	0.2.2
$D(X,\varepsilon)$	Dvoretzky dimension	4.15.15
$l_r(\mathbb{I})$, l_r, l_r^n	classical sequence space	0.3.2
$[l_r(\mathbb{I}), X_i]$	l_r-sum	0.3.2
$L_r(M,\mu)$	classical function space	0.4.5
$L_r(M,\mu)\otimes X$	tensor product	0.4.5
$[L_r(M,\mu), X]$	vector-valued case	0.4.5
$l_{r,w}(\mathbb{I})$, $l_{r,w}$, $l_{r,w}^n$	Lorentz sequence space	0.5.3
$L_{r,w}(M,\mu)$	Lorentz function space	0.5.2
$p(X)$	Rademacher type index	4.16.4
$q(X)$	Rademacher cotype index	4.16.4
$\mathrm{spec}(X)$	Minkowski spectrum	4.16.3
U_X	closed unit ball in X	0.1.1
U_X°	closed unit ball in X'	0.1.3
$[U_2^n, X]$	closed unit ball in $[l_2^n, X]$	2.2.16
$u_k^{(n)}$	unit vector $(0,\ldots,0,1,0,\ldots,0)$	3.1.9
u_k	unit sequence $(0,\ldots,0,1,0,\ldots)$	3.1.9
$\|x\|$, $\|x\|X\|$	norm	0.1.1
X'	dual Banach space	0.1.3
$X^{\mathcal{U}}$	ultrapower of a Banach space	0.8.2
$X_\theta := [X_0, X_1]_\theta$	complex interpolation space	0.6.2
$X_{\theta,w} := [X_0, X_1]_{\theta,w}$	real interpolation space	0.6.3

Matrices

C_n°	cosine matrix	5.1.5
E_n°	Fourier matrix	5.1.2
H_n, K_n	Hilbert matrices	2.4.1
I_n	unit matrix	2.1.1

S_n°	sine matrix	5.1.5
S_{mn}	arbitrary (m,n)-matrix, $S_n = S_{nn}$	2.1.1
S_{mn}^t	transpose of S_{mn}	2.1.1
$\|S_{mn}\|$	operator norm of $S_{mn} = (\sigma_{hk})$	2.1.2
$\|\|S_{mn}\|\|$	operator norm of $\|S_{mn}\| := (\|\sigma_{hk}\|)$	2.1.2
$\|S_{mn}\|\mathfrak{S}\|$	Hilbert–Schmidt norm of $S_{mn} = (\sigma_{hk})$	2.1.2
W_n	Walsh matrix	6.1.5

Operators

$a_n(T)$	n-th approximation number	4.13.4
B_α, C_α, D_α	diagonal operators	1.2.12
$E(\mathcal{F})$	conditional expectation	7.1.3
$F^{\mathbb{G}}$, $F^{\mathbb{R}}$, $F^{\mathbb{T}}$, $F^{\mathbb{Z}}$	Fourier transforms	5.8.3
$H^{\mathbb{R}}$, $H^{\mathbb{T}}$, $H^{\mathbb{Z}}$, $H^{\mathbb{Z}}_{\Delta}$	Hilbert transforms	5.4.1–5.4.3
I_X	identity map of X	0.1.4
$Id : X \to Y$	identity map from X into Y	2.2.15, 3.11.8
J_n	James operator	7.8.1
J_M^X	injection from M into X	0.1.9
K_X	embedding of X into X''	0.1.7
$[l_2(\mathbb{I}), T_i]$	diagonal operator	0.3.4
P_ε, Q_ε	Poisson transforms	5.4.8
P_E	Riesz projection	5.4.1
P_R	Rademacher projection	4.14.3
Q_N^X	surjection from X onto X/N	0.1.9
Σ	infinite summation operator	0.7.1
Σ_n	finite summation operator	0.7.3
S_a, J, Q, $R_{2\pi}$, $E_{2\pi}$, M	auxiliary operators	5.4.11, 5.8.9
$\|T\|$, $\|T : X \to Y\|$	operator norm	0.1.4
T'	dual operator	0.1.6
$T^{\mathcal{U}}$	ultrapower of an operator	0.8.4
$W \otimes T$	tensor product	0.4.9
$[W,T]$, $[W,X]$, $[L_r,T]$	vector-valued extensions	0.4.9

Orthonormal systems

$\mathcal{A}_n$	arbitrary orthonormal system $(a_1, \ldots, a_n)$	3.1.1
$\mathcal{N}\mathcal{A}_n$	modification $(a_1/a, \ldots, a_n/a)$	3.1.8
$\mathcal{A}_m \overset{\text{full}}{\otimes} \mathcal{B}_n$	full tensor product	3.6.4
$\mathcal{A}_n \otimes \mathcal{F}_n$	tensor product	3.6.1
$e \circ \mathcal{A}_n$	$(\varepsilon_1 a_1, \ldots, \varepsilon_n a_n)$	8.1.1
$\pi \circ \mathcal{A}_n$	$(a_{\pi(1)}, \ldots, a_{\pi(n)})$	3.8.17
$A_n \circ \mathcal{A}_n$	$(\sum_{k=1}^n \alpha_{1k} a_k, \ldots, \sum_{k=1}^n \alpha_{nk} a_k)$	4.15.2
$\mathcal{C}_n$	cosine system $(c_1, \ldots, c_n)$	5.1.3
$\mathcal{C}_n^\circ$	discrete cosine system $(c_1^{(n)}, \ldots, c_n^{(n)})$	5.1.5
$\mathcal{D}_n$	Dirichlet system $(d_1^{(n)}, \ldots, d_n^{(n)})$	5.1.8
$\mathcal{E}_n$	Fourier system $(e_1, \ldots, e_n)$	5.1.1
$\mathcal{E}_n^\circ$	discrete Fourier system $(e_1^{(n)}, \ldots, e_n^{(n)})$	5.1.2
$\mathcal{G}_n$	Gauss system $(g_1^{(n)}, \ldots, g_n^{(n)})$	4.7.3
$\mathcal{H}(\mathbb{D}_0^n)$	Haar system $(\chi_k^{(j)} : (k,j) \in \mathbb{D}_0^n)$	7.3.6
$\mathcal{I}_n$	unit vector system $(u_1^{(n)}, \ldots, u_n^{(n)})$	3.1.9
$\mathcal{P}_n(\pm 1)$	Bernoulli system $(p_1^{(n)}, \ldots, p_n^{(n)})$	4.1.5
$\mathcal{P}_n(e^{i\tau})$	Steinhaus system $(e^{i\tau_1}, \ldots, e^{i\tau_n})$	4.1.6
$\mathcal{P}_n(a)$	$(a \circ p_1^{(n)}, \ldots, a \circ p_n^{(n)})$	4.1.3
$\mathcal{Q}_n$	spherical system $(q_1^{(n)}, \ldots, q_n^{(n)})$	4.7.10
$\mathcal{R}_n$	Rademacher system $(r_1, \ldots, r_n)$	4.1.1
$\mathcal{S}_n$	sine system $(s_1, \ldots, s_n)$	5.1.3
$\mathcal{S}_n^\circ$	discrete sine system $(s_1^{(n)}, \ldots, s_n^{(n)})$	5.1.5
$\mathcal{W}_n$	Walsh system $(w_1, \ldots, w_n)$	6.1.2
$\boldsymbol{\Delta}_n$	collection of discrete orthonormal systems	3.9.4
$\boldsymbol{\Omega}_n$	collection of orthonormal systems	3.9.4

Norms of n-tuples

$\|(x_k)\|\mathcal{A}_n\|$, $\|(x_k)\|[\mathcal{A}_n, X]\|$	average with respect to $\mathcal{A}_n$	3.1.3
$\|(x_k)\|S_{mn}\|$, $\|(x_k)\|[S_{mn}, X]\|$	average with respect to S_{mn}	2.1.7
$\|(x_k)\|l_r^n\|$, $\|(x_k)\|[l_r^n, X]\|$	l_r^n-norm	0.3.2
$\|(x_k)\|w_2^n\|$, $\|(x_k)\|[w_2^n, X]\|$	weak l_2^n-norm	2.2.1

Parameters associated with orthonormal systems

$K_q(\mathcal{A}_n)$	upper Khintchine constant	3.2.1
$K'_p(\mathcal{A}_n)$	lower Khintchine constant	3.2.2
$L(\mathcal{A}_n)$	Lebesgue constant	3.8.12
M_r, $M_r^{\mathbb{R}}$, $M_r^{\mathbb{C}}$	absolute Gaussian moments	4.7.2
$M_r(\mathcal{A}_n)$, $N_r(\mathcal{A}_n)$	n-dimensional moments	3.2.3
$S(\mathcal{A}_n)$	Sidon constant	4.11.1
$S(\mathcal{A}_n, \mathcal{B}_n)$	relative Sidon constant	4.11.7

Ideal norms

$\boldsymbol{\alpha}$, $\boldsymbol{\beta}$	arbitrary ideal norms	1.1.1
$\boldsymbol{\alpha}'$, $\boldsymbol{\beta}'$	dual ideal norms	1.1.9
$\boldsymbol{\alpha} \le \boldsymbol{\beta} \circ \boldsymbol{\gamma}$	inequality between ideal norms	1.1.7
$\boldsymbol{\alpha}_n \prec \boldsymbol{\beta}_n$	asymptotic inequality	1.1.8
$\boldsymbol{\alpha}_n \asymp \boldsymbol{\beta}_n$	uniform equivalence	1.1.8
$\boldsymbol{\alpha}(1)$, $\boldsymbol{\alpha}(\infty)$	$\boldsymbol{\alpha}(N) := \sup\Big\{\boldsymbol{\alpha}(X) : \dim(X)=N\Big\}$	1.1.4, 9.3.2
$\boldsymbol{\alpha}(T)$, $\boldsymbol{\alpha}(T : X \to Y)$	ideal norm	1.1.1
$\boldsymbol{\delta}(\mathcal{B}_n, \mathcal{A}_n)$	Dirichlet ideal norm	3.4.1
$\boldsymbol{\delta}_\pm(\mathcal{B}_n, \mathcal{A}_n)$	unconditional Dirichlet ideal norm	8.2.1
$\boldsymbol{\delta}_n$	universal Dirichlet ideal norm	3.10.1
$\boldsymbol{\kappa}(S_{mn})$	Kwapień ideal norm	2.3.1
$\boldsymbol{\kappa}_n$	universal Kwapień ideal norm	2.3.15
$\boldsymbol{\pi}(\mathcal{A}_n)$	Parseval ideal norm	3.11.1
$\boldsymbol{\pi}(S_{mn})$	Parseval ideal norm	2.2.2
$\boldsymbol{\pi}_n$	universal Parseval ideal norm	2.2.9
$\boldsymbol{\varrho}(\mathcal{B}_n, \mathcal{A}_n)$	Riemann ideal norm	3.3.1
$\boldsymbol{\varrho}_\pm(\mathcal{B}_n, \mathcal{A}_n)$	unconditional Riemann ideal norm	8.1.2
$\boldsymbol{\varrho}_n$	universal Riemann ideal norm	3.10.1
$\boldsymbol{\tau}_n$	universal type ideal norm	3.10.1
$\boldsymbol{\omega}_n$	universal cotype ideal norm	3.10.1

Operator ideals

$\mathfrak{A}$, $\mathfrak{B}$	arbitrary operator ideals	1.2.1
$\mathfrak{A}'$, $\mathfrak{B}'$	dual operator ideals	1.2.5
$\mathfrak{A} \circ \mathfrak{B}$	product of $\mathfrak{A}$ and $\mathfrak{B}$	1.2.4
$\mathfrak{A}(X,Y)$	component of an operator ideal, $\mathfrak{A}(X) = \mathfrak{A}(X,X)$	1.2.1
$\mathfrak{AC}$	operators of Haar subcotype	7.5.20
$\mathfrak{AC}_q$	operators of Haar cotype q	7.5.8
$\mathfrak{AC}_q^{weak}$	operators of weak Haar cotype q	7.5.10
$\mathfrak{AT}$	operators of Haar subtype	7.5.20
$\mathfrak{AT}_p$	operators of Haar type p	7.5.7
$\mathfrak{AT}_p^{weak}$	operators of weak Haar type p	7.5.9
$\mathfrak{EC}$	operators of Fourier subcotype	5.7.4
$\mathfrak{EC}_q$	operators of Fourier cotype q	5.7.4
$\mathfrak{EC}_q^{weak}$	operators of weak Fourier cotype q	5.7.4
$\mathfrak{EP}$	operators compatible with the Riesz projection	5.2.36
$\mathfrak{ET}$	operators of Fourier subtype	5.6.26
$\mathfrak{ET}_p$	operators of Fourier type p	5.6.6
$\mathfrak{ET}_p^{weak}$	operators of weak Fourier type p	5.6.7
$\mathfrak{F}$	finite rank operators	0.2.5
$\mathfrak{FT}_p$	operators compatible with the Fourier transform	5.8.13
$\mathfrak{FT}_p^{\mathbb{G}}$	operators compatible with the Fourier transform	5.8.6
$\mathfrak{FT}_p^{\mathbb{G},weak}$	operators compatible with the Fourier transform	5.8.6
$\mathfrak{GC}$	operators of Gauss subcotype	4.10.8
$\mathfrak{GC}_q$	operators of Gauss cotype q	4.10.2
$\mathfrak{GC}_q^{weak}$	operators of weak Gauss cotype q	4.10.2
$\mathfrak{GP}$	operators compatible with the Gauss projection	4.14.1
$\mathfrak{GR}$	GR-operators	4.8.17
$\mathfrak{GT}$	operators of Gauss subtype	4.10.1
$\mathfrak{GT}_p$	operators of Gauss type p	4.10.1
$\mathfrak{GT}_p^{weak}$	operators of weak Gauss type p	4.10.1

$\mathfrak{H}$	Hilbertian operators	2.3.12
$\mathfrak{HM}$	HM-operators	2.4.7
$\mathfrak{HT}_r$	operators compatible with the Hilbert transform	5.4.5
$\mathfrak{KC}_q^{weak}$	operators of weak Kwapień cotype q	3.10.10
$\mathfrak{KT}$	operators of Kwapień subtype	2.3.27
$\mathfrak{KT}_p^{weak}$	operators of weak Kwapień type p	2.3.24
$\mathfrak{L}$	all operators	0.1.5
$\mathfrak{L}[\alpha_n]$	operators such that $\alpha_n(T) = O(1)$	1.2.8
$\mathfrak{L}_0[\alpha_n]$	operators such that $\alpha_n(T) = o(1)$	1.2.8
$\mathfrak{L} \circ \mathfrak{UT}_p$	uniformly p-smoothable operators	7.9.9
$\mathfrak{MT}$	operators of martingale subtype	7.7.6
$\mathfrak{P}_2$	2-summing operators	2.2.6
$\mathfrak{P}_\gamma$	γ-summing operators	4.15.7
$\mathfrak{QEP}$	quasi-EP-operators	5.2.37
$\mathfrak{QHM}$	quasi-HM-operators	2.4.9
$\mathfrak{QGR}$	quasi-GR-operators	4.8.18
$\mathfrak{QGP}$	quasi-GP-operators	4.14.7
$\mathfrak{QRAP}$	quasi-RAP-operators	8.6.31
$\mathfrak{QRP}$	quasi-RP-operators	4.14.7
$\mathfrak{QUMD}$	quasi-UMD-operators	8.5.17
$\mathfrak{R}$	weakly compact operators	7.6.1
$\mathfrak{RAP}$	RAP-operators	8.6.31
$\mathfrak{RC}$	operators of Rademacher subcotype	4.5.19
$\mathfrak{RC}_q$	operators of Rademacher cotype q	4.5.6
$\mathfrak{RC}_q^{weak}$	operators of weak Rademacher cotype q	4.5.7
$\mathfrak{RP}$	operators compatible with the Rademacher projection	4.14.1
$\mathfrak{RT}$	operators of Rademacher subtype	4.3.19
$\mathfrak{RT}_p$	operators of Rademacher type p	4.3.6
$\mathfrak{RT}_p^{weak}$	operators of weak Rademacher type p	4.3.7
$\mathfrak{SR}$	super weakly compact operators	7.6.7

$\mathfrak{UC}_q$	uniformly q-convex operators	7.9.2
$\mathfrak{UC}_q \circ \mathfrak{L}$	uniformly q-convexifiable operators	7.9.7
$\mathfrak{UMD}$	UMD-operators	8.5.16
$\mathfrak{UT}_p$	uniformly p-smooth operators	7.9.4
$\mathfrak{WC}$	operators of Walsh subcotype	6.3.9
$\mathfrak{WC}_p$	operators of Walsh cotype p	6.3.5
$\mathfrak{WC}_p^{weak}$	operators of weak Walsh cotype p	6.3.6
$\mathfrak{WT}$	operators of Walsh subtype	6.3.9
$\mathfrak{WT}_p$	operators of Walsh type p	6.3.5
$\mathfrak{WT}_p^{weak}$	operators of weak Walsh type p	6.3.6

Classes of Banach spaces

We omit those classes of Banach spaces $\mathsf{A} := \{ X \in \mathsf{L} : I_X \in \mathfrak{A} \}$ which are obtained from operator ideals; see 1.3.1.

B	B-convex Banach spaces	4.4.4
B_n	B_n-convex Banach spaces	4.4.4
J	J-convex Banach spaces	7.8.10
J_n	J_n-convex Banach spaces	7.8.10
L_N	N-th Minkowski compactum	9.3.1
MP	MP-convex Banach spaces	4.6.4
MP_n	MP_n-convex Banach spaces	4.6.4

Bibliography

An abbreviation of the author's name is used to refer to publications. In order to distinguish between books and articles, we use uppercase and lowercase letters, respectively. A * marks publications by several authors. Classical treatises are labeled by •. Numbers in [...] on the right side refer to the pages where items are quoted.

Aldous, D. J.

[ald] *Unconditional bases and martingales in $L_p(F)$.* Math. Proc. Cambr. Phil. Soc. **85** (1979), 117–123. [p. 499]

Alon, N.; Milman, V. D.

[alo*] *Embedding of l_∞^k in finite dimensional Banach spaces.* Israel J. Math. **45** (1983), 265–289. [p. 163]

Altshuler, Z.; Casazza, P. G.; Lin, B. L.

[alt*] *On symmetric basic sequences in Lorentz sequence spaces.* Israel J. Math. **15** (1973), 140–155. [p. 462]

Amir, D.

[AMI] Characterization of inner product spaces. Birkhäuser, Basel, 1986. [p. 1]

Askey, R.; Waigner, S.

[ask*] *Mean convergence of expansions in Laguerre and Hermite series.* Amer. J. Math. **87** (1965), 695–708. [p. 489]

Assouad, P.

[ass 1] *Espaces p-lissés et q-convex, inégalités de Burkholder.* Séminaire Maurey–Schwartz 1974/75, Exposé 15, École Polytechnique, Paris, 1975. [p. 499]

[ass 2] *Espaces p-lissés, réarrangements.* Séminaire Maurey–Schwartz 1974/75, Exposé 16, École Polytechnique, Paris, 1975. [p. 499]

• **Banach, S.**

[BAN] Théorie des opérations linéaires. Warszawa, 1932. [pp. 491, 492, 505]

[ban] *Sur les opérations dans les ensembles abstraits et leur application aux équations intégrales.* Fund. Math. **3** (1922), 133–181. [p. 505]

Beauzamy, B.

[BEA 1] Espaces d'interpolation réelles: topologie et géométrie. Lecture Notes in Mathematics, vol. **666**, Springer, Berlin–Heidelberg, 1978. [p. 501]

[BEA 2] Introduction to Banach spaces and their geometry. 2nd edition, Math. Studies **68**, North Holland, Amsterdam–London–New York–Tokyo, 1985. [pp. 4, 21, 345, 401]

[bea 1] *Propriétés géométriques des espaces d'interpolation.* Séminaire Maurey–Schwartz 1974/75, Exposé 14, École Polytechnique, Paris, 1975. [p. 501]

[bea 2] *Opérateurs de type Rademacher entre espaces de Banach.* Séminaire Maurey–Schwartz 1975/76, Exposés 6 et 7, École Polytechnique, Paris, 1976. [pp. 163, 494]

[bea 3] *Opérateurs uniformément convexifiants.* Studia Math. **57** (1976), 103–139. [pp. 499, 500]

Beck, A.

[bec] *A convexity condition in Banach spaces and the strong law of large numbers.* Proc. Amer. Math. Soc. **13** (1962), 329–334. [pp. 494, 507]

Benedek, A.; Calderón, A. P.; Panzone, R.

[ben*] *Convolution operators on Banach space valued functions.* Proc. Nat. Acad. Sci. USA **48** (1962), 356–365. [pp. 271, 501]

Benke, H.; Sommer, F.

[BENK*] Theorie der analytischen Funktionen einer komplexen Veränderlichen. Springer, Berlin–Heidelberg, 1955. [p. 308]

Bennett, C.; Sharpley, R.

[BENN*] Interpolation of operators. Academic Press, Orlando (Florida), 1988. [pp. 13, 18, 101, 269]

Bergh, J.; Löfström, J.

[BER*] Interpolation spaces. Springer, Berlin–Heidelberg, 1976. [pp. 13, 18, 50, 315, 462, 463]

Berkson, E.; Gillespie, T. A.; Muhly, P. S.

[ber*1] *Abstract spectral decompositions guaranteed by the Hilbert transform.* Proc. London Math. Soc. (3) **53** (1986), 489–517. [p. 508]

[ber*2] *A generalization of Macaev's theorem to non-commutative L_p-spaces.* Integral Equations Operator Theory **10** (1987), 164–186. [p. 470]

• **Bessel, F. W.**

[bes] *Über die Bestimmung des Gesetzes einer periodischen Erscheinung.* Astron. Nachr. **6** (1828), 333–348. [p. 490]

Beurling, A.

[beu] *On analytic extensions of semi-groups of operators.* J. Funct. Anal. **6** (1970), 387–400. [p. 331]

Blasco, O.

[bla] *Hardy spaces of vector-valued functions: duality.* Trans. Amer. Math. Soc. **308** (1988), 495–507. [p. 508]

• **Bochner, S.**

[boc 1] *Integration von Funktionen, deren Werte die Elements eines Vektorraumes sind.* Fund. Math. **20** (1933), 262–276. [p. 497]

[boc 2] *Abstrakte Funktionen und die Besselsche Ungleichung.* Nachr. Ges. Wiss. Göttingen **40** (1933), 178–180. [p. 497]

• **Bochner, S.; Taylor, A. E.**

[boc*] *Linear functionals on certain spaces of abstractly-valued functions.* Ann. Math. (2) **39** (1938), 913–944. [p. 497]

Bonsall, F. F.

[bon] *The decomposition of continuous linear functionals into non-negative components.* Proc. Univ. Durham Phil. Soc. Ser. A **13:2** (1957), 6–11. [p. 52]

Bourgain, J.

[bou 1] *On trigonometric series in super reflexive spaces.* J. London Math. Soc. (2) **24** (1981), 165–174. [pp. 303, 500]

[bou 2] *A Hausdorff–Young inequality for B-convex Banach spaces.* Pac. J. Math. **101** (1982), 255–262. [pp. 303, 304, 500]

[bou 3] *Some remarks on Banach spaces in which martingale difference sequences are unconditional.* Ark. Mat. **22** (1983), 163–168. [pp. 459, 499, 500]

[bou 4] *Sur les sommes de sinus.* Publ. Math. Orsay, **83**-01, Exposé 3 (1983). [p. 477]

[bou 5] *On martingale transforms in finite dimensional lattices with an appendix on the K-convexity constant.* Math. Nachr. **119** (1984), 41–53. [pp. 439, 477]

[bou 6] *Subspaces of L_N^∞, arithmetical diameter and Sidon sets.* Probability in Banach spaces (Medford, 1984). Lecture Notes in Mathematics, vol. **1153**, Springer, Berlin–Heidelberg, 1985, pp. 96–127. [pp. 201, 314]

[bou 7] *Vector-valued singular integrals and the H^1-BMO duality.* Probability theory and harmonic analyis (Chao, J.-A.; Woyczyński, W. A., editors), pp. 1–19, Dekker, New York–Basel 1986. [pp. 470, 508]

[bou 8] *Vector-valued Hausdorff–Young inequalities and applications.* GAFA 1986/87, Lecture Notes in Mathematics, vol. **1317**, Springer, Berlin–Heidelberg, 1988, pp. 239–249. [pp. 293, 295, 303, 326, 330, 500]

Bourgain, J.; Kalton, N. J.; Tzafriri, L.

[bou*] *Geometry of finite dimensional subspaces and quotients of L_p.* GAFA 1987/88, Lecture Notes in Mathematics, vol. **1376**, Springer, Berlin–Heidelberg, 1989, pp. 138–175. [pp. 149, 495]

Brillhart, J.; Carlitz, L.

[bri*] *Note on the Shapiro polynomials.* Proc. Amer. Math. Soc. **25** (1970), 114–118. [p. 435]

Bromwich, T. J. I'A.

[BRO] An introduction to the theory of infinite series. 2nd edition. Macmillan, London, 1926. [p. 267]

Brunel, A.; Sucheston, L.

[bru*1] *On B-convex Banach spaces.* Math. Systems Theory **7** (1974), 294–299. [p. 497]

[bru*2] *On J-convexity and some ergodic super-properties of Banach spaces.* Trans. Amer. Math. Soc. **204** (1975), 79–90. [p. 497]

Burkholder, D. L.

[bur 1] *Martingale transforms.* Ann. Math. Statist. **37** (1966), 1494–1504. [p. 501]

[bur 2] *Distribution function inequalities for martingales.* Ann. Probab. **1** (1973), 19–42. [p. 350]

[bur 3] *A geometric characterization of Banach spaces in which martingale difference sequences are unconditional.* Ann. Probab. **9** (1981), 997–1011. [p. 499]

[bur 4] *Martingale transforms and the geometry of Banach spaces.* Probability in Banach spaces (Medford, 1980). Lecture Notes in Mathematics, vol. **860**, Springer, Berlin–Heidelberg, 1981, pp. 35–50. [p. 499]

[bur 5] *A geometric condition that implies the existence of certain singular integrals of Banach-space-valued functions.* Conference on Harmonic Analysis in Honor of Antoni Zygmund (Chicago, 1981), vol. I, pp. 270–286, Wadsworth, Belmont, 1983. [pp. 459, 499, 500]

[bur 6] *Boundary value problems and sharp inequalites for martingale transforms.* Ann. Probab. **12** (1984), 647–702. [pp. 438, 440]

[bur 7] *An elementary proof of an inequality of R. E. A. C. Paley.* Bull. London Math. Soc. **17** (1985), 474–478. [p. 440]

[bur 8] *Martingales and Fourier analysis in Banach spaces.* Probability and Analysis (Varenna, Italy, 1985). Lecture Notes in Mathematics, vol. **1206**, Springer, Berlin–Heidelberg, 1986, pp. 61–108. [pp. 429, 459, 500]

Burkholder, D. L.; Gundy, R. F.

[bur*] *Extrapolation and interpolation of quasilinear operators on martingales.* Acta Math. **124** (1970), 249–304. [p. 501]

Butzer, P. L.; Berens, H.

[BUT*a] Semi-groups of operators and approximation. Springer, Berlin–Heidelberg, 1967. [p. 18]

Butzer, P. L.; Nessel, R. J.

[BUT*b] Fourier analysis and approximation. Birkhäuser, Basel–Stuttgart, 1971. [pp. 269, 272]

Calderón, A. P.

[cal] *Intermediate spaces and interpolation, the complex method.* Studia Math. **24** (1964), 113–190. [p. 501]

Casazza, P. G.; Shura, T. J.

[CAS*] Tsirelson's space. Lecture Notes in Mathematics, vol. **1363**, Springer, Berlin–Heidelberg, 1989. [pp. 142, 158]

Chatterji, S. D.

[cha] *Martingales of Banach-valued random variables.* Bull. Amer. Math. Soc. **66** (1960), 395–398. [p. 499]

Clarkson, J. A.

[cla] *Uniformly convex spaces.* Trans. Amer. Math. Soc. **40** (1936), 396–414. [p. 496]

Clement, P.; de Pagter, B.

[cle*] *Some remaks on the Banach space valued Hilbert transform.* Indag. Math., N.S. **2** (1991), 453–460. [p. 500]

Cobos, F.

[cob 1] *On the type of interpolation spaces and $S_{p,q}$.* Math. Nachr. **113** (1983), 59–64. [p. 463]

[cob 2] *Some spaces in which martingale difference sequences are unconditional.* Bull. Pol. Acad. Sci. Math. **34** (1986), 695–703. [p. 466]

Creekmore, J.

[cre] *Type and cotype in Lorentz L_{pq} spaces.* Indag. Math. **43** (1981), 145–153. [p. 501]

Cwikel, M.; Reisner, S.

[cwi*] *Interpolation of uniformly convex Banach spaces.* Proc. Amer. Math. Soc. **84** (1982), 555–559. [p. 501]

Dacunha-Castelle, D.

[dac] *Ultraproduits d'espaces de Banach.* Séminaire Goulaouic–Schwartz 1971/72, Exposé 9, École Polytechnique, Paris, 1972. [p. 501]

Dacunha-Castelle, D.; Krivine, J. L.

[dac*] *Applications des ultraproduits à l'étude des espaces et des algèbres de Banach.* Studia Math. **41** (1972), 315–344. [p. 501]

Davis, B.

[dav] *On the integrability of the martingale square function.* Israel J. Math. **8** (1970), 187–190. [p. 501]

Davis, W. J.; Figiel, T.; Johnson, W. B.; Pełczyński, A.

[dav*] *Factoring weakly compact operators.* J. Funct. Anal. **17** (1974), 311–327. [p. 382]

Day, M. M.

[DAY] Normed linear spaces. 3rd edition. Springer, Berlin–Heidelberg, 1973. [p. 4]

[day 1] *Reflexive spaces not isomorphic to uniformly convex spaces.* Bull. Amer. Math. Soc. **47** (1941), 313–317. [p. 497]

[day 2] *Some more uniformly convex spaces.* Bull. Amer. Math. Soc. **47** (1941), 504–507. [p. 426]

[day 3] *Uniform convexity in factor and conjugate spaces.* Ann. Math. (2) **45** (1944), 375–385. [p. 496]

Defant, A.; Floret, K.

[DEF*] Tensor norms and operator ideals. Math. Studies **176**, North Holland, Amsterdam–London–New York–Tokyo, 1993. [pp. 6, 25, 492]

Defant, M.

[def 1] Zur vektorwertigen Hilberttransformation. Thesis, Univ. Kiel, 1986. [pp. 497, 500]

[def 2] *On the vector-valued Hilbert transform.* Math. Nachr. **141** (1989), 251–265. [pp. 273, 500]

Defant, M.; Junge, M.

[def*] *Unconditional orthonormal systems.* Math. Nachr. **158** (1992), 233–240. [p. 434]

Deville, R.; Godefroy, G.; Zizler, V.

[DEV*] Smoothness and renorming in Banach spaces. Longman, Harlow, 1993. [pp. 345, 429]

Diestel, J.

[DIE] Geometry of Banach spaces. Selected topics. Lecture Notes in Mathematics, vol. **485**, Springer, Berlin–Heidelberg, 1975. [p. 417]

Diestel, J.; Jarchow, H.; Tonge, A.

[DIE*a] Absolutely summing operators. Cambridge Univ. Press, 1995. [pp. 3, 21, 35, 40, 119, 127, 178, 220, 230, 233, 473, 476]

Diestel, J.; Uhl, J. J.

[DIE*b] Vector measures. Math. Surveys, No. 15, Amer. Math. Soc., Providence, 1977. [pp. 11, 25, 417, 498]

Dilworth, S. J.

[dil] *Complex convexity and the geometry of Banach spaces.* Math. Proc. Cambr. Phil. Soc. **99** (1986), 495–506. [p. 468]

Dinculeanu, N.

[DIN] Vector measures. Deutscher Verlag der Wissenschaften, Berlin, 1966. [pp. 11, 79]

• **Dirichlet, P. G. Lejeune**
[dir] *Sur la convergence des séries trigonométriques qui servent à représenter une fonction arbitraire entre des limites données.* J. Reine Angew. Math. **4** (1829), 157–169. [p. 490]

Doob, J. L.
[DOO] Stochastic processes. Wiley, 1953. [p. 347]

Dore, G.; Venni, A.
[dor*] *On the closedness of the sum of two closed operators.* Math. Z. **196** (1987), 189–201. [p. 508]

Dubinsky, E.; Pełczyński, A.; Rosenthal, H. P.
[dub*] *On Banach spaces X for which* $\Pi_2(\mathcal{L}_\infty, X) = B(\mathcal{L}_\infty, X)$. Studia Math. **44** (1972), 617–648. [p. 495]

van Dulst, D.
[DUL] *Reflexive and superreflexive Banach spaces.* Math. Centre Tracts 102, Amsterdam, 1978. [p. 345]

Dunford, N.; Schwartz, J. T.
[DUN*1] Linear operators, vol. I. Interscience, New York, 1958. [pp. 4, 10, 172, 373, 415, 417, 418, 491, 502]
[DUN*2] Linear operators, vol. II. Interscience, New York, 1963. [pp. 79, 271, 469]

Dvoretzky, A.
[dvo] *Some results on convex bodies and Banach spaces.* Proc. Symp. on Linear Spaces (Jerusalem, 1961), pp. 123–160. [pp. 232, 494]

Dvoretzky, A.; Rogers, C. A.
[dvo*] *Absolute and unconditional convergence in normed linear spaces.* Proc. Nat. Acad. Sci. USA **36** (1950), 192–197. [p. 494]

Elton, J.
[elt] *Sign-embeddings of* l_1^n. Trans. Amer. Math. Soc. **279** (1983), 113–124. [p. 151]

Enflo, P.
[enf] *Banach spaces which can be given an equivalent uniformly convex norm.* Israel J. Math. **13** (1972), 281–288. [p. 497]

Figiel, T.
[fig 1] *On equivalence of some bases to the Haar system in spaces of vector-valued functions.* Bull. Pol. Acad. Sci. Math. **36** (1988), 119–131. [p. 508]
[fig 2] *Singular integral operators: a martingale approach.* Geometry of Banach spaces (Strobl, 1989). London Math. Soc. Lect. Note Ser. **158** (1990), 95–110. [p. 508]

Figiel, T.; Lindenstrauss, J.; Milman, V. D.
[fig*a] *The dimension of almost spherical sections of convex bodies.* Acta Math. **139** (1977), 53–94. [pp. 230, 232]

Figiel, T.; Pisier, G.

[fig*b] *Séries aléatoires dans les espaces uniformément convexes ou uniformément lisses.* C. R. Acad. Sci. Paris **279 A** (1974), 611–614. [pp. 370, 499]

Figiel, T.; Tomczak-Jaegermann, N.

[fig*c] *Projections onto Hilbertian subspaces of Banach spaces.* Israel J. Math. **33** (1979), 155–171. [pp. 492, 493]

• **Fischer, E.**

[fis] *Sur la convergence en moyenne.* C. R. Acad. Sci. Paris **144** (1907), 1022–1024. [p. 490]

• **Fourier, J. B. J.**

[FOU] Théorie analytique de la chaleur. Didot, Paris, 1822. [p. 490]

Garling, D. J. H.

[gar] *Random martingale transform inequalities.* Probability in Banach spaces **6**, pp. 101–119, Birkhäuser, Basel, 1990. [pp. 446, 447, 454, 499]

Garcia-Cuerva, J.; Rubio de Francia, J. L.

[GARC*] Weighted norm inequalities and related topics. Math. Studies **104**, North Holland, Amsterdam–London–New York–Tokyo, 1985. [p. 272]

Garcia-Cuerva, J.; Kazarian, K. S.; Kolyada, V. I.

[garc*] *Paley type inequalities for the orthogonal series with vector-valued coefficients* (Russian). To be published in Mat. Sbornik.

Garsia, A.

[GARS] Martingale inequalites. Seminar Notes on Recent Progress, Benjamin, Reading (Mass.), 1973. [pp. 345, 347, 352]

Geiss, S.

[gei 1] Martingale transforms and applications to the Banach space theory. Habilitationsschrift (Thesis), Univ. Jena, 1995. [pp. 351, 439, 499, 501]

[gei 2] *BMO_ψ-spaces and applications to the extrapolation theory.* Studia Math. **122** (1997), 235–274. [p. 501]

[gei 3] *A counterexample concerning the relation between decoupling constants and UMD-constants.* Trans. Amer. Math. Soc. [p. 439]

Geiss, S.; Junge, M.

[gei*] *Type and cotype with respect to arbitrary orthonormal systems.* J. Approx. Theory **82** (1995), 399–433. [pp. 318, 495]

Giesy, D. P.

[gie] *On a convexity condition in normed linear spaces.* Trans. Amer. Math. Soc. **125** (1966), 114–146. [pp. 146, 494]

Gohberg, I. C.; Krein, M. G.

[GOH*] Introduction to the theory of linear non-self adjoint operators (Russian). Nauka, Moscow, 1965. Engl. transl.: Amer. Math. Soc., Providence, 1969. [p. 469]

Golubov, B. I.; Efimov, A. V.; Skvortsov, V. A.

[GOL*] Walsh series and transforms (Russian). Nauka, Moscow, 1987. Engl. transl.: Kluwer, Dordrecht–Boston–London, 1991. [pp. 321, 456]

Graham, R. L.; Rothschild, B. L.; Spencer, J. H.

[GRA*] Ramsey theory. Wiley, New York–Chichester–Brisbane–Toronto, 1980. [p. 392]

Grothendieck, A.

[gro 1] *Résumé de la théorie métrique des produits tensoriels topologiques.* Bol. Soc. Mat. São Paulo **8** (1956), 1–79. [pp. 492, 493]

[gro 2] *Sur certaines classes de suites dans les espaces de Banach, et le théorème de Dvoretzky–Rogers.* Bol. Soc. Mat. São Paulo **8** (1956), 81–110. [pp. 494, 501]

Guerre-Delabrière, S.

[GUE] Classical sequences in Banach spaces. Dekker, New York–Basel–Hong Kong 1992. [pp. 21, 233]

Gurarii, V. I.; Kadets, M. I.; Macaev, V. I.

[gur*] *Distances between finite dimensional analogues of L_p-spaces* (Russian). Mat. Sbornik **70** (1966), 481–489. [p. 492]

Gutierrez, J. A.

[gut] On the boundedness of the Banach space-valued Hilbert transfrom. Thesis, Univ. of Texas, Austin, 1982. [pp. 470, 500]

Haagerup, U.

[haag] *The best constants in the Khintchine inequality.* Studia Math. **70** (1982), 231–283. [pp. 128, 491]

• **Haar, A.**

[haar] *Zur Theorie der orthogonalen Funktionensysteme.* Math. Ann. **69** (1910), 331–371. [p. 490]

Halász, G.

[hal] *On a result of Salem and Zygmund concerning random polynomials.* Studia Sci. Math. Hungar. **8** (1973), 369–377. [p. 319]

Halmos, P. R.

[HAL] A Hilbert space problem book. Van Nostrand, Princeton, 1967. [p. 37]

Hanner, O.

[han] *On the uniform convexity of L^p and l^p.* Ark. Mat. **3** (1955), 239–244. [p. 496]

• **Hardy, G. H.; Littlewood, J. E.; Pólya, G.**

[HAR*] Inequalities. Cambridge Univ. Press, 1934. [pp. 59, 246]

• **Hausdorff, F.**

[hau] *Eine Ausdehnung des Parsevalschen Satzes über Fourierreihen.* Math. Z. **16** (1923), 163–169. [pp. 306, 498]

Haydon, R.; Levy, M.; Raynaud, Y.

[HAY*] Randomly normed spaces. Travaux en cours **41**, Hermann, Paris, 1991. [p. 428]

Heinrich, S.

[hei 1] *Finite representability and super-ideals of operators.* Dissertationes Math. **172** (1980). [p. 499]

[hei 2] *Ultraproducts in Banach space theory.* J. Reine Angew. Math. **313** (1980), 72–104. [p. 501]

Herz, C.

[her] *The theory of p-spaces with an application to convolution operators.* Trans. Amer. Math. Soc. **154** (1971), 69–82. [p. 497]

Hewitt, E.; Ross, K. R.

[HEW*1] Abstract harmonic analysis I. Springer, Berlin–Heidelberg, 1963. [pp. 65, 99, 127]

[HEW*2] Abstract harmonic analysis II. Springer, Berlin–Heidelberg, 1970. [pp. 65, 127, 201, 496]

Hinrichs, A.

[hin 1] *On the type constants with respect to systems of characters of a compact abelian group.* Studia Math. **118** (1996), 231–243. [pp. 135, 496]

[hin 2] Inequalities between ideal norms formed with orthonormal systems and applications to the theory of operator ideals. Thesis, Univ. Jena, 1996. [p. 494]

[hin 3] *Type norms with respect to characters of compact abelian groups and algebraic relations.* Archiv Math. (Basel) **68** (1997), 265–273. [p. 105]

[hin 4] *K-convex operators and Walsh type norms.* Math. Nachr. [p. 331]

[hin 5] *Rademacher and Gaussian averages and Rademacher cotype of operators between Banach spaces.* To be published in Proc. Amer. Math. Soc.. [p. 178]

[hin 6] *Operators of Rademacher and Gaussian subcotype.* To be published. [pp. 162, 163, 495]

Hitczenko, P.

[hit 1] *Upper bound for the L_p-norms of martingales.* Probab. Theory Rel. Fields **86** (1990), 225–238. [p. 501]

[hit 2] *Domination inequality for martingale transforms of a Rademacher sequence.* Israel J. Math. **84** (1993), 161–178. [p. 501]

Hoffmann-Jørgensen, J.

[HOF] Sums of independent Banach space valued random variables. Aarhus Universitet, Matematisk Institut, Preprint Series 1972/73, No. **15**. [p. 495]

Huppert, B.

[HUP] Endliche Gruppen, Band I. Springer, Berlin–Heidelberg, 1983. [pp. 99, 110]

James, R. C.

[jam 1] *Uniformly non-square Banach spaces.* Ann. Math. (2) **80** (1964), 542–550. [pp. 144, 497]

[jam 2] *Weak compactness and reflexivity.* Israel J. Math. **2** (1964), 101–119. [p. 375]

[jam 3] *Some self-dual properties of linear normed spaces.* Symposium on infinite dimensional topology (Baton Rouge, 1967), pp. 159–175. Princeton Univ. Press, New Jersey 1972. [pp. 397, 497]

[jam 4] *Super-reflexive spaces with bases.* Pac. J. Math. **41** (1972), 409–419. [pp. 397, 497]

[jam 5] *Super-reflexive Banach spaces.* Can. J. Math. **24** (1972), 896–904. [pp. 397, 497, 501]

[jam 6] *A nonreflexive Banach space that is uniformly nonoctahedral.* Israel J. Math. **18** (1974), 145–155. [p. 497]

[jam 7] *Nonreflexive spaces of type 2.* Israel J. Math. **30** (1978), 1–13. [pp. 312, 370, 495]

Jameson, G. J. O.

[JAM] Summing and nuclear norms in Banach space theory. London Math. Soc. Student Texts **8**, Cambridge Univ. Press, 1987. [pp. 35, 52]

• **Jessen, B.**

[jes] *Om Uligheder imellem Potensmiddelværdier.* Mat. Tidsskrift **B** (1933), 1–19. [p. 13]

John, F.

[joh] *Extremum problems with inequalities as subsidiary conditions.* Courant Anniversary Volume, pp. 187–204. Interscience, New York, 1948. [pp. 7, 492]

Joichi, J. I.

[joi] *Normed linear spaces equivalent to inner-product spaces.* Proc. Amer. Math. Soc. **17** (1966), 423–426. [pp. 122, 493]

Junge, M.

[jun] *Cotype and summing properties in Banach spaces.* [p. 140]

• **Kaczmarz, S.; Steinhaus, H.**

[KAC*] Theorie der Orthogonalreihen. Warszawa–Lwów, 1935. [pp. 65, 95, 128]

Kadets (Kadec), V. M.; Kadets, M. I.

[KAD*] Rearrangement of series in Banach spaces (Russian). Tartu, 1988. Engl. transl.: Amer. Math. Soc., Providence, 1991. [p. 127]

Kadets, M. I.; Mityagin, B. S.

[kad*a] *Complemented subspaces in Banach spaces* (Russian). Uspehi Mat. Nauk **28:6** (1973), 77–94. [p. 122]

Kadets , M. I.; Pełczyński, A.

[kad*b] *Bases, lacunary sequences and complemented subspaces in the spaces L_p.* Studia Math. **21** (1962), 161–176. [p. 454]

Kadets, M. I.; Snobar, M. G.

[kad*c] *Certain functionals on the Minkowski compactum* (Russian). Mat. Zametki **10** (1971), 453–457. [p. 492]

Kadison, R. V.

[kadi] *Isometries of operator algebras.* Ann. Math. (2) **54** (1951), 325–338. [p. 37]

Kahane, J.–P.

[KAH] Some random series of functions. Heath Math. Monographs, 1968, 2nd edition, Cambridge Univ. Press, 1985. [pp. 85, 491]

[kah] *Sur les sommes vectorielles $\sum \pm u_n$.* C. R. Acad. Sci. Paris **259** (1964), 2577–2580. [p. 491]

Kak, S. C.

[kak] *The discrete Hilbert transform.* Proc. IEEE **78** (1970), 585–586. [p. 270]

Kalton, N. J.

[kal] *Differentials of complex interpolation processes for Köthe function spaces.* Trans. Amer. Math. Soc. **333** (1992), 479–529. [p. 467]

Kashin, B. S.; Saakyan, A. A.

[KAS*] Orthogonal series (Russian). Nauka, Moscow 1984. Engl. transl.: Amer. Math. Soc., Providence, 1989. [pp. 65, 95]

Kato, T.

[kat] *A characterization of holomorphic semi-groups.* Proc. Amer. Math. Soc. **25** (1970), 495–498. [p. 331]

Kelly, J. L.

[KEL] General topology. Van Nostrand, New York, 1955. [p. 108]

• **Khintchine, A.**

[khin] *Über dyadische Brüche.* Math. Z. **18** (1923), 109–116. [p. 491]

König, H.

[KOE] Eigenvalue distribution of compact operators. Birkhäuser, Basel–Stuttgart, 1986. [pp. 25, 463, 469]

[koe 1] *Weyl-type inequalities for operators in Banach spaces.* Functional analysis: surveys and recent results II (Paderborn, 1979). Math. Studies **38**, pp. 297–317. North Holland, Amsterdam–London–New York–Tokyo, 1980. [p. 186]

[koe 2] *On the tensor stability of s-number ideals.* Math. Ann. **269** (1984), 77–93. [p. 312]

[koe 3] *On the Fourier-coefficients of vector-valued functions.* Math. Nachr. **152** (1991), 215–227. [pp. 308, 498]

[koe 4] *Vector-valued Lagrange interpolation and mean convergence of Hermite series.* pp. 227–247, Lecture Notes in Pure and Appl. Math. **150**, Dekker, New York–Basel–Hong Kong, 1993. [p. 488]

König, H.; Nielsen, N. J.

[koe*a] *Vector-valued L_p-convergence of orthogonal series and Lagrange interpolation.* Forum Math. **6** (1994), 183–207. [pp. 488, 508]

König, H.; Retherford, J. R.; Tomczak-Jaegermann, N.

[koe*b] *On the eigenvalues of $(p, 2)$-summing operators and constants associated with normed spaces.* J. Funct. Anal. **37** (1980), 88–126. [pp. 122, 193, 492, 507]

König, H.; Tzafiri, L.

[koe*c] *Some estimates for type and cotype constants.* Math. Ann. **256** (1981), 85–94. [p. 164]

Kreĭn, S. G.; Petunin, J. I.; Semenov, E. M.

[KRE*] Interpolation of linear operators (Russian). Nauka, Moscow 1978. Engl. transl.: Amer. Math. Soc., Providence, 1982. [pp. 18, 315]

Krivine, J. L.

[kri] *Sous-espaces de dimension finie des espaces de Banach réticulés.* Ann. Math. (2) **104** (1976), 1–29. [p. 233]

Kwapień, S.

[kwa 1] *A linear topological characterization of inner product spaces.* Studia Math. **38** (1970), 277–278. [p. 493]

[kwa 2] *Isomorphic characterizations of inner product spaces by orthogonal series with vector valued coefficients.* Studia Math. **44** (1972), 583–595. [pp. 57, 171, 192, 308, 313, 493, 495, 498]

[kwa 3] *Isomorphic characterizations of Hilbert spaces by orthogonal series with vector valued coefficients.* Séminaire Maurey–Schwartz 1972/73, Exposé 8, École Polytechnique, Paris, 1973. [p. 498]

Lang, S.

[LAN] Algebra. 2nd edition. Addison-Wesley, Menlo Park, Cal., 1984. [p. 99]

Latała, R.; Oleszkiewicz, K.

[lat*] *On the best constant in the Khinchin–Kahane inequality.* Studia Math. **109** (1994), 101–104. [pp. 128, 130, 491]

• **Lebesgue, H.**

[leb] *Sur les séries trigonométriques.* Ann. Sci. École Norm. Sup. (3) **20** (1903), 453–485. [p. 498]

Ledoux, M.; Talagrand, M.

[LED*] Probability in Banach spaces. Springer, Berlin–Heidelberg, 1991. [pp. 127, 130, 167, 178, 491]

Levy, M.

[lev] *L'espace d'interpolation réel* $(A_0, A_1)_{\theta,p}$ *contient* l^p. C. R. Acad. Sci. Paris **289 A** (1979), 675–677. [p. 462]

Lewis, D. R.

[lew 1] *An isomorphic characterization of the Schmidt class.* Compositio Math. **30** (1975), 293–297. [p. 473]

[lew 2] *Finite dimensional subspaces of* L_p. Studia Math. **63** (1978), 207–212. [p. 492]

Lin, P.-K.

[lin] *B-convexity of the space of 2-summing operators.* Israel J. Math. **37** (1980), 139–150. [p. 475]

Linde, W.; Pietsch, A.

[lin*a] *Mappings of Gaussian measures of cylindrical sets in Banach spaces* (Russian). Teor. Verojatnost. i Primenen. **19** (1974), 472–487. [pp. 230, 493]

Lindenstrauss, J.

[lin 1] *On the modulus of smoothness and divergent series in Banach spaces.* Michigan Math. J. **10** (1963), 241–252. [pp. 493, 496, 498]

[lin 2] *On operators which attain their norm.* Israel J. Math. **1** (1963), 139–148. [p. 417]

[lin 3] *Almost spherical sections; their existence and their applications.* Jahresber. Deutsch. Math.-Verein., Jubiläumstagung, 100 Jahre DMV (1992), 39–61. [p. 494]

Lindenstrauss, J.; Pełczyński, A.

[lin*b] *Absolutely summing operators in* $\mathcal{L}_p$*-spaces and applications.* Studia Math. **29** (1968), 275–326. [pp. 375, 493]

Lindenstrauss, J.; Tzafriri, L.

[LIN*1] Classical Banach spaces I. Springer, Berlin–Heidelberg, 1977. [pp. 4, 462]

[LIN*2] Classical Banach spaces II. Springer, Berlin–Heidelberg, 1979. [pp. 4, 127, 128, 130, 269, 347, 405, 415, 500, 507]

[lin*c] *On the complemented subspace problem.* Israel J. Math. **9** (1971), 263–269. [p. 122]

Lions, J.-L.; Peetre, J.

[lio*] *Sur une classe d'espaces d'interpolation.* Publ. Math. Inst. Hautes Études Sci. **19** (1964), 5–68. [pp. 312, 462, 501]

Littlewood, J. E.

[lit] *On bounded bilinear forms in an infinite number of variables.* Quart. J. Math. Oxford **1** (1930), 164–174. [p. 491]

López, J. M.; Ross, K. A.

[LOP*] Sidon sets. Lecture Notes in Pure and Appl. Math. **13**, Dekker, New York, 1975. [p. 127]

Mascioni, V.

[mas] *On weak cotype and weak type in Banach spaces.* Note di Mat. (Lecce) **8** (1988), 67–110. [p. 495]

Maurey, B.

[mau 1] *Une nouvelle démonstration d'un théorème de Grothendieck.* Séminaire Maurey–Schwartz 1972/73, Exposé 22, École Polytechnique, Paris, 1973. [pp. 160, 494]

[mau 2] *Type et cotype dans les espaces munis de structures locales inconditionnelles.* Séminaire Maurey–Schwartz 1973/74, Exposés 24 et 25, École Polytechnique, Paris, 1974. [p. 501]

[mau 3] *Un théorème de prolongement.* C. R. Acad. Sci. Paris **279 A** (1974), 329–332. [p. 493]

[mau 4] Théorèmes de factorisation pour les opérateurs linéaires à valeur dans les espaces L_p. Astérisque **11**, 1974. [pp. 495, 507]

[mau 5] *Système de Haar.* Séminaire Maurey–Schwartz 1974/75, Exposés 1 et 2, École Polytechnique, Paris, 1975. [p. 499]

Maurey, B.; Pisier, G.

[mau*] *Séries de variables aléatoires vectorielles indépendantes et propriétés géométriques des espaces de Banach.* Studia Math. **58** (1976), 45–90. [pp. 160, 178, 233, 492, 495, 496]

McCarthy, C. A.

[mcca] c_p. Israel J. Math. **5** (1967), 249–271. [pp. 469, 470]

McConnell, T. R.

[mcco] *On Fourier multiplier transformations of Banach-valued functions.* Trans. Amer. Math. Soc. **285** (1984), 739–757. [p. 508]

McShane, E. J.

[mcs] *Linear functionals on certain Banach spaces.* Proc. Amer. Math. Soc. **1** (1950), 402–408. [p. 426]

Meir, A.

[mei] *On the uniform convexity of L^p spaces, $1 < p \leq 2$.* Illinois J. Math. **28** (1984), 420–424. [p. 405]

• **Milman, D. P.**
[mil a] *On some criteria for the regularity of spaces of type (B)* (Russian). Dokl. Akad. Nauk SSSR **20** (1938), 243–246. [p. 497]

Milman, M.
[mil b] *Complex interpolation and geometry of Banach spaces.* Ann. Mat. Pura Appl. (4) **136** (1984), 317–328. [p. 498]

Milman, V. D.
[mil c] *Dvoretzky's theorem – thirty years later.* Geometric and Functional Analysis **2** (1992), 455–479. [p. 494]

Milman, V. D.; Pisier, G.
[mil*a] *Banach spaces with a weak cotype 2 property.* Israel J. Math. **54** (1986), 139–158. [p. 495]

Milman, V. D.; Schechtman, G.
[MIL*] Asymptotic theory of finite dimensional normed spaces. Lecture Notes in Mathematics, vol. **1200**, Springer, Berlin–Heidelberg, 1986. [pp. 3, 127, 229, 230, 232, 233, 493]

• **Minkowski, H.**
[MIN] Geometrie der Zahlen. Teubner, Leipzig 1896. [p. 505]

Muckenhoupt, B.
[muc] *Mean convergence of Jacobi series.* Proc. Amer. Math. Soc. **23** (1969), 306–310. [p. 489]

Namioka, I.; Asplund, E.
[nam*] *A geometric proof of Ryll-Nardzewski's fixed point theorem.* Bull. Amer. Math. Soc. **73** (1967), 443–445. [p. 417]

Neveu, J.
[NEV] Martingales à temps discret. Masson, Paris, 1973. [pp. 345, 347]

Nordlander, G.
[nor 1] *The modulus of convexity in normed linear spaces.* Ark. Mat. **4** (1960), 15–17. [p. 496]
[nor 2] *On sign-independent and almost sign-independent convergence in normed linear spaces.* Ark. Mat. **4** (1962), 287–296. [p. 495]

Novikov, I.; Semenov, E.
[NOV*] Haar series and linear operators. Kluwer. Dordrecht, 1997 [p. 345]

Olevskii, M. A.
[OLE] Fourier series with respect to general orthonormal systems. Springer, Berlin–Heidelberg, 1975. [pp. 65, 95]

• **Orlicz, W.**
[orl] *Über unbedingte Konvergenz in Funktionenräumen.* Studia Math. **4** (1933), 33–37, 41–47. [p. 495]

Pajor, A.

[PAJ] Sous-espaces l_1^n des espaces de Banach. Travaux en cours **16**, Hermann, Paris, 1985. [pp. 151, 494]

• **Paley, R. E. A. C.**

[pal] *A remarkable system of orthogonal functions.* Proc. London Math. Soc. **34** (1932), 241–279. [p. 490]

• **Paley, R. E. A. C.; Zygmund, A.**

[pal*] *On some series of functions, (1).* Math. Proc. Cambr. Phil. Soc. **26** (1930), 337–357. [p. 491]

• **Parseval, M. A.**

[par] *Mémoire sur les séries et sur l'intégration complète d'une équation aux différences partielles linéaires du second ordre, à coefficients constants.* Mémoires présentés à l'Institut des Sciences, Lettres et Arts (1799), 638–648. [p. 490]

Peetre, J.

[PEE] A theory of interpolation of normed spaces. Lecture Notes, Brasilia, 1963; IMPA, Notas de Matemática **39**, Rio de Janeiro, 1968. [p. 501]

[pee 1] *Sur la transformation de Fourier des fonctions à valeurs vectorielles.* Rend. Sem. Mat. Univ. Padova **42** (1969), 15–26. [pp. 498, 501]

[pee 2] *A new approach in interpolation spaces.* Studia Math. **34** (1970), 23–42. [p. 463]

Pełczyński, A.

[pel 1] *Geometry of finite dimensional Banach spaces and operator ideals.* Notes in Banach spaces. University of Texas Press, Austin–London, 1980. [p. 492]

[pel 2] *Commensurate sequences of characters.* Proc. Amer. Math. Soc. **104** (1988), 525–531. [pp. 199, 496]

• **Pettis, B. J.**

[pet] *A proof that every uniformly convex space is reflexive.* Duke Math. J. **5** (1939), 249–253. [p. 497]

Pichorides, S. K.

[pic] *On the best values of the constant in the theorem of M. Riesz, Zygmund and Kolmogorov.* Studia Math. **46** (1972), 164–179. [p. 270]

Pietsch, A.

[PIE 1] Nuclear locally convex spaces. Springer, Berlin–Heidelberg, 1972. [p. 128]

[PIE 2] Operator ideals. Deutscher Verlag der Wissenschaften, Berlin, 1978. [pp. 6, 7, 25, 30, 35, 40, 119, 197, 219, 230, 382, 473, 476, 492]

[PIE 3] Eigenvalues and s-numbers. Cambridge Univ. Press, 1987. [pp. 25, 40, 41, 118, 219, 220, 222, 469, 473, 491, 507]

[pie 1] *Absolut p-summierende Abbildungen in normierten Räumen.* Studia Math. **28** (1967), 333–353. [p. 493]

[pie 2] *Sequences of ideal norms.* Note di Mat. (Lecce) **10** (1990), 411–441. [p. 492]

[pie 3] *Eigenvalue criteria for the* B-*convexity of Banach spaces.* J. London Math. Soc. (2) **41** (1990), 310–322. [p. 494]

[pie 4] *Eigenvalue distribution and geometry of Banach spaces.* Math. Nachr. **150** (1991), 41–81. [pp. 122, 492, 507]

[pie 5] *Approximation numbers of nuclear operators and geometry of Banach spaces.* Archiv Math. (Basel) **57** (1991), 155–168. [p. 507]

[pie 6] *Banach-Räume.* Jahrbuch Überblicke Mathematik 1992, pp. 1–26, Vieweg, Braunschweig–Wiesbaden, 1992. [p. 491]

Pietsch, A.; Triebel, H.

[pie*] *Interpolationstheorie für Banachideale von beschränkten linearen Operatoren.* Studia Math. **31** (1968), 95–109. [p. 469]

Pinelis, I.

[pin] *Optimum bounds for the distributions of martingales in Banach spaces.* Ann. Probab. **22** (1994), 1679–1706. [p. 499]

Pisier, G.

[PIS 1] Factorization of linear operators and geometry of Banach spaces. Regional Conf. Series in Math. **60**, Amer. Math. Soc., Providence, 1986. [pp. 25, 35, 493]

[PIS 2] The volume of convex bodies and Banach space geometry. Cambridge Univ. Press, 1989. [pp. 231, 495]

[pis 1] *Sur les espaces qui ne contiennent pas de l_∞^n uniformément.* Séminaire Maurey–Schwartz 1972/73, Annexe, École Polytechnique, Paris, 1973. [pp. 160, 494]

[pis 2] *Sur les espaces de Banach qui ne contiennent pas uniformément de l_1^n.* C. R. Acad. Sci. Paris **277:A** (1973), 991–994. [pp. 146, 494, 500]

[pis 3] *Sur les espaces qui ne contiennent pas de l_1^n uniformément.* Séminaire Maurey–Schwartz 1973/74, Exposé 7, École Polytechnique, Paris, 1974. [pp. 146, 494, 500]

[pis 4] *Un exemple concernant la super-reflexivité.* Séminaire Maurey–Schwartz 1974/75, Annexe 2, École Polytechnique, Paris, 1975. [p. 499]

[pis 5] *Martingales with values in uniformly convex spaces.* Israel J. Math. **20** (1975), 326–350. [pp. 408, 499, 500]

[pis 6] *Les inégalités de Khintchine–Kahane d'après C. Borell.* Séminaire sur la Géométrie des Espaces de Banach 1977/78, Exposé 16, École Polytechnique, Palaiseau, 1978. [pp. 130, 199, 320, 491, 496]

[pis 7] *Some results on Banach spaces without local unconditional structure.* Compositio Math. **37** (1978), 3–19. [pp. 470, 473]

[pis 8] *Sur les espaces de Banach de dimension finie à distance extrémale d'un espace euclidien.* d'après V.D. Milman et H. Wolfson,

Séminaire d'Analyse Fonctionelle 1978/79, Exposé 16, École Polytechnique, Palaiseau, 1979. [pp. 58, 117, 122, 493]

[pis 9] *La méthode d'interpolation complexe: applications aux treillis de Banach.* Séminaire d'Analyse Fonctionelle 1978/79, Exposé 17, École Polytechnique, Palaiseau, 1979. [p. 193]

[pis 10] *Some applications of the complex interpolation method to Banach lattices.* J. Analyse Math. **35** (1979), 264–281. [pp. 117, 122, 193, 493]

[pis 11] *Remarques sur un résultat non publié de B. Maurey.* Séminaire d'Analyse Fonctionelle. Exposé 5, École Polytechnique, Palaiseau, 1980/81. [p. 477]

[pis 12] *Semi-groupes holomorphes et K-convexité.* Séminaire d'Analyse Fonctionelle, Exposés 2 et 7, École Polytechnique, Palaiseau, 1980/81. [pp. 331, 496]

[pis 13] *Des nouvelles caractérisations des ensembles de Sidon.* Mathematical analysis and applications, Part B, Advances in Math., Suppl. Studies **7B** (1981), 685–726. [p. 496]

[pis 14] *On the duality between type and cotype.* Martingale Theory in Harmonic Analysis and Banach spaces (Cleveland, 1981). Lecture Notes in Mathematics, vol. **939**, Springer, Berlin–Heidelberg, 1982, pp. 131–144. [p. 496]

[pis 15] *Holomorphic semi-groups and the geometry of Banach spaces.* Ann. of Math. **115** (1982), 375–392. [pp. 331, 496]

[pis 16] *K-convexity.* Proc. Res. Workshop on Banach Space Theory, (1981). Univ. of Iowa Press, 1982, pp. 139–151. [pp. 225, 331, 496]

[pis 17] *Finite rank projections on Banach spaces and a conjecture of Grothendieck.* Proc. Int. Congr. Math. Warszawa **2** (1983), 1027–1039. [p. 492]

[pis 18] *Probabilistic methods in the geometry of Banach spaces.* Probability and Analysis (Varenna, Italy, 1985). Lecture Notes in Mathematics, vol. **1206**, Springer, Berlin–Heidelberg, 1986, pp. 167–241. [pp. 178, 231]

[pis 19] *A remark on $\Pi_2(l_p, l_p)$.* Math. Nachr. **148** (1990), 243–245. [p. 474]

[pis 20] *Completely bounded maps between sets of Banach space operators.* Indiana Univ. Math. J. **39** (1990), 249–277. [pp. 52, 493]

Pisier, G.; Xu, Q.

[pis*1] *Random series in the real interpolation spaces between the spaces v_p.* GAFA 1985/86. Lecture Notes in Mathematics, vol. **1267**, Springer, Berlin–Heidelberg, 1987, pp. 185–209. [pp. 64, 370, 501]

[pis*2] *The strong p-variation of martingales and orthonormal series.* Probab. Theory Related Fields **77** (1988), 497–514. [p. 57]

• **Plancherel, M.**
[pla] *Contribution à l'étude de la représentation d'une fonction arbitraire par les intégrales définies.* Rend. Circ. Math. Palermo **30** (1910), 289–335. [pp. 306, 498]

• **Rademacher, H.**
[rad] *Einige Sätze über Reihen von allgemeinen Orthogonalfunktionen.* Math. Ann. **87** (1922), 112–138. [p. 491]

• **Riemann, B.**
[riem] Über die Darstellbarkeit einer Funktion durch trigonometrische Reihen. Habilitationsschrift (Thesis), Univ. Göttingen. Abhandlungen der Königl. Gesellschaft der Wiss. in Göttingen **13** (1854), 259–287. [pp. 490, 498]

• **Riesz, F.**
[rie a] *Sur les systèmes orthogonaux des fonctions.* C. R. Acad. Sci. Paris **144** (1907), 615–619. [p. 490]

• **Riesz, M.**
[rie b] *Sur les maxima des formes bilinéaires et sur les fonctionelles linéaires.* Acta Math. **49** (1926), 465–497. [p. 500]

Ringrose, J. R.
[RIN] Compact non-self-adjoint operators. Van Nostrand, London 1971. [p. 469]

Rockafellar, R. T.
[ROC] Convex Analysis. Princeton Univ. Press, Princeton, 1970. [pp. 21, 22]

Rosenthal, H. P.
[ros] *On subspaces of L^p.* Ann. Math. (2) **97** (1973), 344–373. [p. 454]

Rubio de Francia, J. L.
[rub 1] *Fourier series and Hilbert transforms with values in UMD Banach spaces.* Studia Math. **81** (1985), 95–105. [p. 508]
[rub 2] *Martingale and integral transfroms of Banach spaces valued functions.* Probability and Banach spaces, Zaragoza 1985. Lecture Notes in Mathematics, vol. **1221**, Springer, Berlin–Heidelberg, 1986, pp. 195–202. [p. 466]

Rubio de Francia, J. L.; Torrea, J. L.
[rub*] *Some Banach techniques in vector-valued Fourier analysis.* Coll. Mat. **54** (1987), 273–284. [p. 497]

Rudin, W.
[RUD] Fourier analysis on groups. Wiley, New York–London–Sydney, 1961. [pp. 65, 98, 306]

Sawa, J.

[saw] *The best constant in the Khintchine inequality for complex Steinhaus variables, the case $p = 1$.* Studia Math. **81** (1986), 107–126. [p. 130]

Scalaro, F. S.

[sca] *Abstract martingale convergence theorems.* Pac. J. Math. **11** (1961), 347–374. [p. 499]

Schäffer, J. J.; Sundaresan, K.

[scha*] *Reflexivity and the girth of spheres.* Math. Ann. **184** (1970), 163–168. [p. 497]

Schatten, R.

[SCHA 1] A theory of cross-spaces. Annals of Math. Studies **26**, Princeton Univ. Press, 1950 [p. 492]

[SCHA 2] Norm ideals of completely continuous operators. Springer, Berlin–Heidelberg, 1970. [p. 469]

Schipp, F.; Wade, W. R.; Simon, P.

[SCHI*] Walsh series. Adam Hilger, Bristol–New York, 1990. [pp. 321, 323, 456, 457]

Schütt, I.

[schu] *Unconditional bases in Banach spaces of absolutely p-summing operators.* Math. Nachr. **146** (1990), 175–194. [p. 475]

Schwartz, J.

[schw] *A remark on inequalities of Calderón–Zygmund type for vector-valued functions.* Comm. Pure Appl. Math. **14** (1961), 785–799. [pp. 271, 501]

Schwartz, L.

[SCHW] Geometry and probability in Banach spaces. Lecture Notes in Mathematics, vol. **852**, Springer, Berlin–Heidelberg, 1981. [p. 233]

Seigner, J.

[sei] Über eine Klasse von Idealnormen, die durch Orthonormalsysteme gebildet sind. Thesis, Univ. Jena, 1995. [p. 496]

Stein, E. M.; Weiss, G.

[STEI*] Fourier analysis in euclidian spaces. Princeton Univ. Press, New Jersey, 1971. [pp. 13, 269]

• **Steinhaus, H.**

[stei] *Über die Wahrscheinlichkeit dafür, daß der Konvergenzkreis einer Potenzreihe ihre natürliche Grenze ist.* Math. Z. **31** (1930), 408–416. [p. 491]

Stern, J.

[ster] *Propriétés locales et ultrapuissances d'espaces de Banach.* Séminaire Maurey–Schwartz 1974/75, Exposés 7 et 8, École Polytechnique, Paris, 1975. [p. 501]

• **Szegö, G.**

[SZE] Orthogonal polynomials, revised edition. Colloquium Publ. **23**, Amer. Math. Soc., Providence, 1959. [pp. 480, 481]

[sze] *Über die Lebesgueschen Konstanten bei den Fourierschen Reihen.* Math. Z. **9** (1921), 163–166. [p. 456]

• **Titchmarsh, E. C.**

[TIT] Introduction to the theory of Fourier integrals. Oxford Univ. Press, 1937. [p. 269]

[tit] *Reciprocal formulae involving series and integrals.* Math. Z. **25** (1926), 321–347. / Correction. Math. Z. **26** (1927), 496. [pp. 270, 500]

Tomczak-Jaegermann, N.

[TOM] Banach–Mazur distances and finite dimensional operator ideals. Longman, Harlow, 1989. [pp. 3, 7, 25, 35, 127, 178, 219, 229, 314, 477, 492, 493]

[tom 1] *The moduli of smoothness and convexity and the Rademacher averages of trace classes S_p $(1 < p < \infty)$.* Studia Math. **50** (1974), 163–182. [p. 469]

[tom 2] *Computing 2-summing norms with few vectors.* Ark. Mat. **17** (1979), 273–277. [pp. 40, 493]

[tom 3] *Finite-dimensional subspaces of uniformly convex and uniformly smooth Banach lattices and trace classes S_p.* Studia Math. **66** (1980), 261–281. [p. 470]

Tricomi, F. G.

[TRIC] Vorlesungen über Orthogonalreihen. Springer, Berlin–Heidelberg, 1970 [pp. 480, 481]

Triebel, H.

[TRIE] Interpolation theory, function spaces, differential operators, North Holland, Amsterdam–London–New York–Tokyo, 1978. [pp. 13, 18, 462, 463]

Tzafriri, L.

[tza] *On the type and cotype of Banach spaces.* Israel J. Math. **32** (1979), 32–38. [pp. 142, 158, 495]

Vági, S.

[vág] *A remark on Plancherel's theorem for Banach space valued functions.* Ann. Scuola Norm. Sup. Pisa Cl. Sci. (3) **23** (1969), 305–315. [p. 498]

Virot, B.

[vir 1] *Extensions vectorielles d'opérateurs linéaires bornés sur L^p.* C. R. Acad. Sci. Paris **293** (1981), 413–415. [p. 500]

[vir 2] *Quelques inégalités concernant les transformées de Hilbert des fonctions à valeurs vectorielles.* C. R. Acad. Sci. Paris **293** (1981), 459–463. [p. 500]

Wallis, W. D.; Street, A. P.; Wallis, J. S.

[WAL*] Combinatorics: room squares, sum-free sets, Hadamard matrices. Lecture Notes in Mathematics, vol. **292**, Springer, Berlin–Heidelberg, 1972. [p. 56]

• **Walsh, J. L.**

[wal] *A closed set of normal orthogonal functions.* Amer. J. Math. **55** (1923), 5–24. [p. 490]

Wenzel, J.

[wen 1] *Mean convergence of vector-valued Walsh series.* Math. Nachr. **162** (1993), 117–124. [p. 458]

[wen 2] *Ideal norms and trigonometric orthonormal systems.* Studia Math. **112** (1994), 59–74. [pp. 251, 498, 500]

[wen 3] *Vector-valued Walsh–Paley martingales and geometry of Banach spaces.* Israel J. Math. **97** (1997), 29–49. [p. 500]

[wen 4] *Superreflexivity and J-convexity of Banach spaces.* Acta Math. Univ. Com. **66** (1997), 135–147. [p. 497]

[wen 5] *Ideal norms associated with the UMD-property.* Arch. Math. **69** (1997), 327–332. [p. 438]

Williams, D.

[WIL] Probability with martingales. Cambridge Univ. Press, 1991. [pp. 345, 347]

Woyczyński, W. A.

[woy] *Geometry and martingales in Banach spaces.* Winter School in Probalility (Karpacz). Lecture Notes in Mathematics, vol. **1200**, Springer, Berlin–Heidelberg, 1975, pp. 229–275. [p. 499]

• **Young, W. H.**

[you] *Sur la généralisation du théorème de Parseval.* C. R. Acad. Sci. Paris **155** (1912), 30–33. [pp. 306, 498]

Zimmermann, F.

[zim] *On vector-valued Fourier multiplier theorems.* Stud. Math. **43** (1989), 201–222. [p. 508]

• **Zygmund, A.**

[ZYG] Trigonometric series. Cambridge Univ. Press, 1959. [pp. 10, 236, 240, 241, 456]

Index